SYSTEMS BIOLOGY AND SYNTHETIC BIOLOGY

SYSTEMS BIOLOGY AND SYNTHETIC BIOLOGY

Edited by

Pengcheng Fu

China University of Petroleum Beijing, China

Sven Panke

ETH Zurich, Institute of Process Engineering Zurich, Switzerland

WILEY

A JOHN WILEY & SONS, INC., PUBLICATION

Library of Congress Cataloging-in-Publication Data:

Systems biology and synthetic biology / [edited by] Pengcheng Fu, Sven Panke.
    p. ; cm.
  Includes bibliographical references.
  ISBN 978-0-471-76778-7 (cloth)
  1. Biotechnology. 2. Genetic engineering. 3. Biological systems. I.
Fu, Pengcheng. II. Panke, Sven.
  [DNLM: 1. Genetic Engineering–methods. 2. Molecular Biology–methods.
3. Genes, Synthetic. 4. Genomics–methods. 5. Models, Genetic. 6.
Systems Biology–methods.   QU 450 S995 2009]
    TP248.2.S97 2009
    660.6 –dc22

                                                        2008027981

Printed in the United States of America
10 9 8 7 6 5 4 3 2 1

To my wife Juan Huang and sons, Eugene and Edgar

CONTENTS

FOREWORD

The popular use of the term "systems biology" arose following the appearance of the first full genome sequences. These genome sequences suggested that we would have a full delineation of the molecular components of an organism. Expression profiling and proteomic data then could tell us when these components were actually used in a context-specific manner.

The need to track the interrelationship of all such components created the need to develop networks of the interactions of such components. Protein–protein interaction maps are one manifestation of this need, stoichiometric models are another; they are, however, amenable to rigorous mathematical analysis and prospective uses. Network reconstruction took center stage in systems biology, as networks describe the interactions between the gene products and the chemical compounds they make, provide context for high-throughput data mapping, and give the basis for mechanistic models that can compute phenotypic functions.

Having molecular manipulation tools and mathematical models in turn provides tools that allow the synthesis of biological components and biological functions. We thus witnessed the emergence of "synthetic biology." It is practiced on multiple scales, from component design, that is akin to classical molecular biology, to design of whole cell functions, such as metabolic engineering.

Thus, in retrospect we can state that genomics gave rise to systems, and systems biology in turn gave rise to synthetic biology. This of course is a simplified view, but provides a first-order approximation to the historical origin and appearance of these popularly used terms. This volume contains a series of chapters that highlight the development and status of the various aspects of systems and synthetic biology.

BERNHARD Ø. PALSSON

University of California, San Diego, CA

CONTRIBUTORS

Wenlong Cheng, Department of Biological and Environmental Engineering, Cornell University, Ithaca, New York 14853

Hidde de Jong, INRIA, Grenoble-Rhône-Alpes, 655 avenue de l'Europe, Montbonnot, 38334 Saint Ismier Cedex, France

Liang Ding, Department of Biological and Environmental Engineering, Cornell University, Ithaca, New York 14853

Pengcheng Fu, Faculty of Chemical Science and Engineering, China University of Petroleum, 18 Fuxue Road, Changping District, Beijing 102249, People's Republic of China

Hisakage Funabashi, Department of Biological and Environmental Engineering, Cornell University, Ithaca, New York 14853

Martin Fussenegger, Department of Biosystems Science and Engineering, ETH Zurich, Mattenstrasse 26, CH-4058 Basel, Switzerland

Johannes Geiselmann, Laboratoire Adaptation et Pathogénie des Micro-organismes, Université Joseph Fourier, UMR CNRS 5163, Bâtiment Jean Roget, Faculté de Médecine-Pharmacie, Domaine de la Mercie, La Tronche 38700, France

Marcus Graf, GENEART AG, Josef-Engert-Str. 11, Regensburg 93053 Germany

David Greber, Department of Biosystems Science and Engineering, ETH Zurich, Mattenstrasse 26, CH-4058 Basel, Switzerland

Claes Gustafsson, DNA2.0, Inc., 1430 O'Brien Drive, Menlo Park, California 94025

Matthias Heinemann, ETH Zurich, Institute of Molecular Systems Biology, Wolfgang-Pauli-Str. 16, 8093 Zurich, Switzerland

Philipp Holliger, MRC Laboratory of Molecular Biology, Cambridge CB2 2QH, UK

Cliff Hooker, Faculty of Education and Arts, School of Humanities and Social Science, The University of Newcastle, NSW, Australia

Mitsuhiro Itaya, Laboratory of Genome Designing Biology, Institute for Advanced Biosciences, Keio University, 403-1 Nipponkoku, Daihoji, Tsuruoka, Yamagata 997-0017, Japan

Andrew R. Joyce, Bioinformatics Program, University of California, San Diego, 9500 Gilman Drive, La Jolla, California 92093

Hyun Uk Kim, Department of Chemical and Biomolecular Engineering (BK21 Program), Center for Systems and Synthetic Biotechnology, Institute for the BioCentury, KAIST, Daejeon 305-701, Korea

Jin Sik Kim, Department of Chemical and Biomolecular Engineering (BK21 Program), Center for Systems and Synthetic Biotechnology, Institute for the BioCentury, KAIST, Daejeon 305-701, Korea

Tae Yong Kim, Department of Chemical and Biomolecular Engineering (BK21 Program), Center for Systems and Synthetic Biotechnology, Institute for the BioCentury, KAIST, Daejeon 305-701, Korea

Hiroaki Kitano, Sony Computer Science Laboratories, Inc., 3-14-13 Higashi-Gotanda, Shinagawa, Tokyo 141-0022, Japan; The Systems Biology Institute, 6-31-15 Jingumae, Shibuya, Tokyo 150-0001, Japan; Department of Cancer Systems Biology, Cancer Institute of Japanese Foundation for Cancer Research, Tokyo, Japan

Sang Yup Lee, Department of Chemical and Biomolecular Engineering (BK21 Program), Center for Systems and Synthetic Biotechnology, Institute for the BioCentury, KAIST, Daejeon 305-701, Korea; Department of Bio and Brain Engineering, BioProcess Engineering Research Center and Bioinformatics Research Center, KAIST, Daejeon 305-701, Korea

James C. Liao, Chemical and Biomolecular Engineering Department, University of California, Box 951592, Los Angeles, California 90095

David Loakes, MRC Laboratory of Molecular Biology, Cambridge CB2 2QH, UK

Dan Luo, Department of Biological and Environmental Engineering, Cornell University, Ithaca, New York 14853

Juanita Mathews, Department of Molecular Biosciences and Bioengineering, University of Hawaii at Manoa, Honolulu, Hawaii 96822

Jens Nielsen, Department of Chemical and Biological Engineering, Chalmers University of Technology, Kemivägen 10, SE-412 96, Gothenburg, Sweden

Gordon S. Okimoto, Bioinformatics Shared Resource, Cancer Research Center of Hawaii, University of Hawaii at Manoa, 1236 Lauhala Street, Honolulu, Hawaii 96813

Bernhard Ø. Palsson, Department of Bioengineering, University of California, San Diego, 9500 Gilman Drive, La Jolla, California 92093

Sven Panke, ETH Zurich, Institute of Process Engineering, Universitätsstr. 6, 8092 Zurich, Switzerland

Nokyoung Park, Department of Biological and Environmental Engineering, Cornell University, Ithaca, New York 14853

Delphine Ropers, INRIA Grenoble-Rhône-Alpes, 655 avenue de l'Europe, Montbonnot, 38334 Saint Ismier Cedex, France

Thomas Schoedl, GENEART AG, Josef-Engert-Str. 11, Regensburg 93053, Germany

Huidong Shi, Department of Pathology and Anatomical Sciences, Ellis Fischel Cancer Center, University of Missouri School of Medicine, M263 Medical Science Building, One Hospital Drive, Columbia, Missouri 65203; Department of Biochemistry and Molecular Biology, Medical College of Georgia, MCG Cancer Center CN4130, 1120 15th Street, Augusta, Georgia 30912

Seung Bum Sohn, Department of Chemical and Biomolecular Engineering (BK21 Program), Center for Systems and Synthetic Biotechnology, Institute for the BioCentury, KAIST, Daejeon 305-701, Korea

Dorothea K. Thompson, Department of Biological Sciences, Purdue University, West Lafayette, Indiana 47907

Soong Ho Um, Department of Biological and Environmental Engineering, Cornell University, Ithaca, New York 14853

Jordi Vallverdú, Philosophy Department, Universitat Autònoma de Barcelona, E-08193 Bellaterra (BCN), Catalonia

Goutham Vemuri, Center for Microbial Biotechnology, Denmark Technical University, Biocentrum-DTU, Lyngby DK2800, Denmark

Ralf Wagner, University of Regensburg, Molecular Microbiology and Gene Therapy, Franz-Josef-Strauss Allee 11, Regensburg 93053, Germany

Xiu-Feng Wan, Department of Microbiology, Miami University, Oxford, Ohio 45056

Guangyi Wang, Department of Oceanography, University of Hawaii at Manoa, 1000 Pope Road, Honolulu, Hawaii 96811

Michael X. Wang, Department of Pathology and Anatomical Sciences, Ellis Fischel Cancer Center, University of Missouri School of Medicine, M263 Medical Science Building, One Hospital Drive, Columbia, Missouri 65203

Wilson W. Wong, Chemical and Biomolecular Engineering Department, University of California, Box 951592, Los Angeles, California 90095

Jianfeng Xu, Department of Biological and Environmental Engineering, Cornell University, Ithaca, New York 14853

Hongseok Yun, Department of Chemical and Biomolecular Engineering (BK21 Program), Center for Systems and Synthetic Biotechnology, Institute for the BioCentury, KAIST, Daejeon 305-701, Korea

1

INTRODUCTION

Pengcheng Fu

Faculty of Chemical Science & Engineering, China University of Petroleum, Beijing, 18 Fuxue Road, Changping District, Beijing 102249, China

In the twentieth century, engineering sciences have inspired numerous successful applications in the fields of manufacturing, electronics, communications, transportation, computer and networks, and so on. Compared to the engineering systems, biological systems are more complex and their mechanisms are less known. Historically, biological questions have been approached by a reductionist paradigm that is completely different from methodologies being applied to engineering systems. This reductionist way of thinking was based on the assumption that by unraveling the function of all the different components the information gained could be used to piece together the puzzle of complex cellular networks [1]. The research paradigm has dominated mainstream biology with enormous progresses in accumulating biological information at genetic and protein levels. However, this is a slow and exhaustive process that fails to adequately approach the true complexities of living phenomena and is of limited relevance to biological systems as a whole.

The fast-growing applications of genomics and high-throughput technologies have led to recognition of the limitations of the reductionist/atomistic view of the world. It is realized that a new systems biology paradigm is needed for the next level of understanding of the functions of the genes and proteins, and the regulation of intracellular networks that cannot be obtained by studying the individual constituents on a part-by-part basis. It is also realized that there is great similarity between biology and engineering at the system level, despite their obviously different physical implementation, and that important research challenges in biology may have parallels with those in complicated engineering systems [2]. This similarity forms a basis for the introduction of synthetic biology or the engineering applications within biological systems.

Systems biology attempts to investigate the behavior and relations of all the elements in a particular biological system while it is functioning [3,4]. It aims at system-level understanding of biological processes and biochemical networks as a whole. This "system-oriented" new biology is shifting our focus from examining particular molecular details to studying the information flows at all biological levels: Genomic DNA, mRNA, proteins, informational pathways, and regulatory networks. Systems biology approaches seek to study the complexity of life to help in understanding how the cellular networks work together. To this end, the approach emphasizes the investigation of biological phenomena by considering system structures, system dynamics, control methods, and design methods [5,6]. It requires a broad interdisciplinary integration of molecular and cell biology, biochemistry, informatics, mathematics, computing, and engineering.

Synthetic biology is a recently emerging field that applies engineering formalisms to design and construct new biological parts, devices, and systems for novel functions or life forms that do not exist in nature. This "engineering" biology relies on and shares tools from genetic engineering, bioengineering, systems biology, and many other engineering disciplines. Synthetic biology is also different from these subjects, in both insights and approach. The synthetic biology study will not only investigate the effects of genetic and pathway modification or the cellular responses on genetic variation/environmental perturbation, but also design and build biological systems with novel cellular functions, combining *in silico* and *in vivo* experimental approaches.

Recently, synthetic biology has been redefined as (1) the design and construction of new biological parts, devices, and systems that do not already exist in the nature and (2) the redesign of existing, natural biological systems for useful purposes (http://syntheticbiology.org/). More specifically, synthetic biology aims to design and build engineered biological systems that process information, manipulate chemicals, fabricate materials, produce energy, provide food, and maintain and enhance human health and our environment (see the Wikipedia: http://en.wikipedia.org/wiki/Synthetic_biology).

Synthetic biology is a "bottom-up" approach, in which basic functional elements of replication, self-assembly, growth, metabolism, repair, signaling, and regulation are defined and assembled into life forms or biomaterials with new properties and behavior. Synthetic biology makes use of and is complementary to systems biology, which focuses "from the top down" on the fully integrated networks of function and control in living cells, and their responses to various perturbations. While preceded by some pioneering work, systems biology and synthetic biology are the new subdisciplines that were invisible 20 years ago. A Google search under these categories now yields more than a million web pages and systems and synthetic biology research departments have grown from scratch to 100 + staff in a decade or less. The aspiration of systems biology is no less than the full mastery of cellular and multicellular dynamics, both epistemic and technical. This has made enormous implications for health, through the development of new treatments and (as importantly) new preventive measures, and for agriculture and ecology more generally through the development of new strains, disease controls, and management processes. It also promises/threatens a host of new security and military applications. The complementary mission of synthetic biology is no less than the systematic extension of engineering, in all its aspects, so as to encompass the biological realm and, conversely, the systematic integration of biological elements into engineered systems. This encompasses practical applications from biofuel to beer production, from sensors to cyborgs.

There are several foundational and illustrative examples regarding applications of systems biology and synthetic biology available in the literature.

A gene deletion perturbation experimental paradigm has emerged in systems biology beginning with research conducted by Trey Ideker, Timothy Galitski, and Leroy Hood [3]. The authors used a systems approach to explore, expand, and refine the understanding of the yeast galactose utilization (GAL) system. A regulatory network model was used to predict changes in gene expression. The authors perturbed the galactose pathway by deleting each of the nine galactose genes of interest. They carried out replicate hybridizations in four different DNA microarrays for each perturbed condition to obtain robust estimates of how the gene expression profile of each knockout strain differed from that of the wild type. Expression data for each perturbation were visually superimposed on the metabolic network. The predicted and observed cellular responses to the perturbation were found to be consistent for most cases. The authors then used the discrepancies between the predicted and observed expression responses to suggest possible refinement to the model [3].

As another example, Pamela Silver, Professor in the Department of Systems Biology at Harvard Medical School, and her group have successfully created a "memory" loop in yeast cells, using two synthesized, transcription factor coding genes. The first gene, which was designed to switch on when exposed to galactose, created a transcription factor that grabbed on to, and thus activated, the second gene. Cellular memory may enable the yeast cells to remember their functional network states. In addition to the reconstruction of the dynamics of cell memory, the researchers have also established a mathematical model for the prediction on the functional outputs of the transcription factor to control how much of a particular protein the gene should make (see BiomimicryNews.com: Scientists synthesize memory in yeast cells, 9/15/2007, http://www.biomimicrynews.com/research/Scientists_synthesize_memory_in_yeast_cells.as). This application exemplifies the core concept that systems biology and synthetic biology should be integrated for understanding life phenomena and creating novel biological modules to modify an existing biological system.

More recently, Craig Venter and his synthetic biology research team have constructed a largest man-made DNA structure on the Earth out of laboratory chemicals. They used the bacterium *Mycoplasma genetalium* as the template to build the synthetic chromosome that contains 381 genes, that is, 582,970 bp long. This wholly artificial genetic circuit was then transplanted into a living *M. genetalium* cell. The resultant "*Mycoplasma laboratorium*" is expected to be able to replicate itself with its synthetically reconstructed DNA, making it the most fully synthetic organism to date, although the molecular machinery and chemical environment that would allow it to replicate would not be synthetic (http://www.guardian.co.uk/science/2007/oct/06/

genetics.climatechange). The building blocks of DNA—adenine (A), guanine (G), cytosine (C), and thiamine (T)—are not easy chemicals to artificially synthesize into chromosomes. A key point for the Craig Venter Institute to achieve their goal of synthetic biology is that they could use homologous recombination (a process that cells use to repair damage to their chromosomes) in the yeast *Saccharomyces cerevisiae* to rapidly build the entire bacterial chromosome from large subassemblies (http://www.syntheticgenomics.com/press/2008-01-24.htm). The team attempts to continue to work for the creation of a living bacterial cell based entirely on the synthetically made genome. It is obvious that using the same approach, existing life forms can be modified by adding components to them or by taking components away from them.

These examples illustrate that we are able to not only "read" the genetic code to understand living systems but also "write" the message for the creation of new life forms. All this fuels the need to frame these latest developments that promise to revolutionize our understanding of biology, blur the boundaries between the living and the engineered in a vital new bioengineering, and transform our daily relationship to the living world.

Systems biology and synthetic biology are emerging as two complementary approaches that embody the breakthrough in biology and invite application of engineering principles. Although systems biology and synthetic biology approach biological problems with different emphasis, they are indeed the two sides of the same coin. Borrowing from ancient Chinese intellectual thinking, we can see systems biology and synthetic biology as Yin and Yang of a research and development framework in the new biological paradigm. For systems biology, all of the "omics" experiments are used to discover life phenomena and to empower scientists to increase fundamental understanding of complex living systems. For synthetic biology, existing life forms can be modified based on the existing body of knowledge. Since systems biology and synthetic biology are like Yin and Yang in nature, we will keep the mutuality of the two approaches in mind when writing the book. We will demonstrate that systems biology relies on synthetic biology technologies to perturb and monitor responses of the biological systems, while synthetic biology depends on the knowledge obtained from systems biology approaches for design and implementation. The development of systems biology and synthetic biology is thus cyclic; understanding (by systems biology) and creation (by synthetic biology) will continuously enhance and transform each other and thus converge in an iterative fashion.

We have assembled a group of investigators/educators who are at the cutting edge of biological research in bioinformatics, functional genomics, genome-scale modeling, and systems biology and synthetic biology. This book will benefit undergraduate and graduate students, scientific researchers, university lectures, and those with larger managerial responsibilities. It should be of especial interest to business development professionals working in biotechnological and pharmaceutical companies since novel products/functions are the goals of systems biology and synthetic biology. The first of its kind, this book aims to become an important reference book for researchers in various engineering fields, such as chemical engineering, mechanical engineering, and civil engineering, who consider systems analysis and engineering applications in biology as their next frontier.

ORGANIZATION OF THE BOOK

This book is organized into modular, stand-alone topics related to systems biology and synthetic biology. Chapter 2, by Michael Wang and Huidong Shi, provides an overview of the modern molecular biology to the readers from backgrounds other than biology. Chapter 3, by Xiu-Feng Wan and Dorothea Thompson, introduces recent advances in high-throughput technologies and functional genomics, with application examples to illustrate how the "omics" data can be used in aid of the establishment of the linkages between transcriptome profiling and biological functions. In Chapter 4, Gordon Okimoto explores and discusses mathematical modeling for "omics" data fusions to predict the global behavior of complex biological systems as networks of interacting genes, proteins, and metabolites. They use algorithms for cancer diagnosis and identification of subcategories. Chapter 5, by Mitsuhiro Itaya, provides a novel approach where a whole genome is cloned into another species. In Chapter 6, Andrew Joyce and Bernhard Palsson present and discuss *in silico* genome-scale metabolic models for microorganisms, such as *Escherichia coli* and the yeast *S. cerevisiae*, using the constraints-based approach. The constraints-based modeling approach they describe, using flux balance analysis and linear programming, is currently the only methodology capable of delivering a high, indeed a surprisingly high, degree of correlation between the predictions of genome-scale models and independently obtained experimental data.

Chapter 7, by Delphine Ropers, Hidde de Jong, and Johannes Geiselmann, discusses the application of mathematical modeling to study the genetic regulatory networks that control gene expression in an organism and the adaptation of its cells to the environment. Illustrative examples using *E. coli* show that many aspects of their structure and dynamics can be fruitfully compared with the principles governing man-made systems. In Chapter 8, Vallverdú and Gustafsson bring up bioethic issues caused by our systems biology and synthetic biology efforts. It is obvious that the benefits from these emerging fields will far outweigh the negative impacts. However, as they point out, the technologies for systems biology and synthetic biology which were not thinkable a couple of decades ago have the potential for misuse. Systems biology and synthetic biology must be operated within a framework of safety, ethics and public acceptance. Chapter 9 is contributed by Goutham Vemuri and Jens Nielsen. In this chapter, the authors review the use of the yeast *S. cerevisiae* as a prototype for systems biology and synthetic biology research. In Chapter 10, Sang Yup Lee et al. examine the construction of genome-scale metabolic models in global understanding of metabolism and physiological characteristics, and also in designing metabolic engineering strategies for the enhanced production of various bioproducts. This chapter can be related in part as application examples to the genome-scale modeling framework outlined in Chapter 6.

In Chapter 11, Matthias Heinemann and Sven Panke sketch a draft picture of synthetic biology as a new bio-based discipline. In their essay, synthetic biology is put into perspective with its scientific counterpart, the field of systems biology, and also the fundamental differences to other "bioengineering" areas such as metabolic engineering or protein engineering. In Chapter 12, Marcus Graf, Thomas Schoedl, and Ralf Wagner present their work for rational gene design and *de novo* gene construction. Since such genes do not exist in nature, they have to be constructed and cloned *de novo* from synthetic oligonucleotides. Advanced synthetic biology approaches are thus needed for generation of proteins with novel functions, new metabolic pathways, or even artificial organisms.

Self-replication is defined as the ability of a system to direct the synthesis of accurate copies of itself from dispersed building blocks. Chapter 13, written by Philipp Holliger and David Loakes, describes the opportunities and challenges for the engineering and bottom-up assembly of artificial self-replicating systems with quasibiotic properties and their potential applications in molecular sensing, computing, and the manufacture of nanodevices. In Chapter 14, Wilson Wong and James Liao present and discuss synthetic approaches similar to the design of engineering machinery for construction of biological circuits based on physical concepts, guided by mathematical models, and constrained by biological and chemical realities. In Chapter 15, David Greber and Martin Fussenegger discuss the state of the art in the field of synthetic genetic networks with particular emphasis on relating network architecture and design to network characteristics. They examine engineered devices such as toggle switches, oscillating networks, and molecular sensors that possess increasingly sophisticated functionality.

Chapter 16 by Hiroaki Kitano focuses on biological robustness as a fundamental feature of living systems, whereby its relationship with evolution, trade-offs among robustness, fragility, resource demands, and performance provides a possible framework for how biological systems have evolved and become organized. In this way, the understanding of robustness and its intrinsic properties provides us with a deeper understanding of biological systems, their anomalies, and countermeasures to reduce these. In Chapter 17, Wenlong Cheng et al. address the properties and functions of oligonucleic acids used as generic, instead of genetic, materials via nucleic acid engineering to utilize a myriad of molecular tools, mostly enzymes, to design and build novel biological systems with desired functions. They also discuss various approaches to the characterization and manipulation of oligonucleic acids and applications of nucleic acid engineering. Chapter 18, by Guangyi Wang and Juanita Mathews, reviews the use of genetic material from marine microbes to engineer conventional hosts for biotechnological and ecological benefits. The major goal is to illustrate the application of synthetic biology in oceanography and marine biotechnology research. In Chapter 19, in the light of these discussions, Cliff Hooker reflects at the revolution in biology that the rise of systems and synthetic biology represents, setting it against the still larger revolution across sciences as a whole that is the rapid expansion of complex systems ideas, principles, methods, and models. Finally, in Chapter 20, Fu and Hooker discuss future directions for systems biology and synthetic biology and the challenges posed to our conception of scientific understanding and bioengineering practices.

REFERENCES

1. Hofmeyr JH, Westerhoff HV. Building the cellular puzzle: control in multi-level reaction networks. *J Theor Biol* 2001;208:261–285.
2. Doyle JC. Robustness and dynamics in biological networks. In: *The First International Conference on Systems Biology, Japan Science and Technology Corporation*. New York: MIT Press, 2000.
3. Ideker T, Galitski T, Hood L. A new approach to decoding life: systems biology. *Annu Rev Genomics Hum Genet* 2001;2:343.
4. Palsson BO. The challenge of *in silico* biology. *Nat Biotechnol* 2000;18:1147–1150.
5. Kitano H. Systems biology: a brief overview. *Science* 2002;295:1662–1664.
6. Wolkenhauer O. Systems biology: the reincarnation of systems theory applied in biology? *Brief Bioinform* 2001;2(3):258–270.

BASICS OF MOLECULAR BIOLOGY, GENETIC ENGINEERING, AND METABOLIC ENGINEERING

Michael X. Wang[1] and Huidong Shi[1,2]

[1]Department of Pathology and Anatomical Sciences, Ellis Fischel Cancer Center,
University of Missouri School of Medicine, M263 Medical Science Building,
One Hospital Drive, Columbia, Missouri 65203
[2]Department of Biochemistry and Molecular Biology, Medical College of Georgia,
MCG Cancer Center CN4130, 1120 15th Street, Augusta, Georgia 30912

2.1 BIOMOLECULES IN LIVING CELLS

Carbon, hydrogen, oxygen, and nitrogen are the most abundant elements in living organisms. Carbon can covalently bond to hydrogen, oxygen, and nitrogen to form biomolecules. Small biomolecules can combine to form more complex macromolecules such as nucleic acids, proteins, and carbohydrates. All living cells are built with these biomolecules.

2.1.1 Nucleic Acids

Nucleic acids carry the genetic information in the cell. The major types of nucleic acids are deoxyribonucleic acid (DNA) and ribonucleic acid (RNA). Both of them are polymers of nucleotides.

Systems Biology and Synthetic Biology   Edited by Pengcheng Fu and Sven Panke
Copyright © 2009 John Wiley & Sons, Inc.

2.1.1.1 Nucleotides Nucleotides are building blocks of nucleic acids. Nucleotides have three characteristic structural components: (1) base, (2) pentose, and (3) phosphate. The bases are derivatives of two parental compounds, purine and pyrimidine. The two major purine bases are adenine (A) and guanine (G), and three major pyrimidines are cytosine (C), thymine (T), and uracil (U). Two types of pentose are 2'-deoxy-D-ribose and D-ribose. Deoxyribonucleotides (deoxyribonucleoside 5'-monophosphate), the structural units of DNA, contain 2'-deoxy-D-ribose. Ribonucleotides (ribonucleoside 5'-monophosphates), the structural units of RNA, contain D-ribose. The phosphate group gives the nucleic acid a negative charge property.

2.1.1.2 DNA In 1953, Watson and Crick postulated a three-dimensional model of the DNA molecule based on the available data at the time. It consists of two helical polynucleotide strands twisted around the same axis to form a right-handed double helix structure. The hydrophilic backbones of deoxyribose and phosphate groups are outside the double helix, whereas purine and pyrimidine bases are stacked inside the double helix. Each purine base of one strand is paired in the same plane with a pyrimidine base of the other strand by hydrogen bonds. There are three hydrogen bonds between G and C and only two between A and T. As a result, the two antiparallel strands are not identical but complementary to each other (Fig. 2-1a and b).

The double helix strands of DNA can be separated from each other (denatured or melting) by heating or at extremes of pH *in vitro*. The temperature at which 50 percent of the double-stranded DNA molecules separate into single strand is the melting temperature (T_m). DNAs rich in G/C pairs have higher melting points than DNAs rich in A/T pairs. The T_m can be calculated according to G/C content of a given DNA fragment. On the contrary, denatured single-stranded DNAs can anneal to form a double helix (renaturation or hybridization). High G/C content, decreasing temperature, increasing the ion concentration, or neutralizing the pH are favorable to DNA renaturation. The nucleotide sequences of DNA can be determined. The human genome and many other genomes of organisms have been successfully sequenced. These sequences are available in the public database (http://www.ncbi.nlm.nih.gov/). DNA can also be synthesized with simple, automated protocols involving chemical and enzymatic methods such as polymerase chain reaction (PCR).

The biological significances of the double-stranded helical structure of DNA are threefold. It stores genetic information in a form of linear nucleotide sequence with chemically stable features; it allows the genetic information to be passed on to the next generation of cells by semiconservative DNA replication with very high fidelity during cell division; and it acts as the template to transfer genetic information into messenger RNAs (mRNAs) and then amino acid sequences of proteins.

2.1.1.3 RNA RNA is usually single-stranded polynucleotides. The chemical compositions of RNA differ from DNA in two ways: (1) sugar-phosphate backbone contains D-ribose rather than 2'-deoxyribose in DNA, and (2) nucleotide base thymine (T) in DNA is replaced in RNA with uracil (U), which is paired with

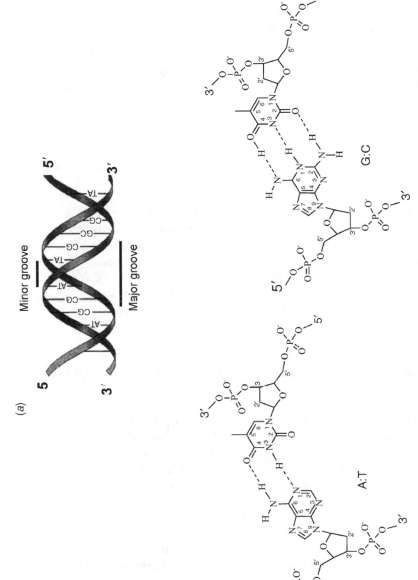

Figure 2-1 (a) Chemical structure of DNA double helix and base pairs defined by Watson and Crick. Major and minor grooves on the surface of double helix are indicated. (b) Formation of hydrogen bonds between A and T and C and G bases. The antiparallel complementary strands of double helix structure of DNA are held by hydrogen bonds.

11

Table 2-1 Types and functions of RNAs

Type of RNA	Function
mRNA (messenger RNA)	RNA that functions as the intermediary (transcript) between DNA in the nucleus and protein production in the cytoplasm
tRNA (transfer RNA)	RNA that transfers an amino acid to a growing polypeptide chain during translation
rRNA (ribosomal RNA)	RNA that is a component of ribosomes for mRNA processing
Ribozyme RNA	Functions as an enzyme by catalyzing chemical reactions in the cell
Small RNAs	Include miRNA and siRNA. Functions as translational repression and RNA degradation (RNAi)

adenine (A). RNA can form complex three-dimensional structures by intramolecular base pairing.

There are five types of RNAs with distinct biological functions (Table 2-1). They are mRNA, transfer RNA (tRNA), ribosomal RNA (rRNA), ribozyme RNA, and small RNAs (micro RNA (miRNA) and short interfering RNA (siRNA)). RNAs are cell specific. All cells have identical DNA content in a given organism; however, RNA levels and types differ in different cell types. RNA can be synthesized from a DNA template (transcription). In contrast, RNA can also function as a template to synthesize DNA (complementary or cDNA) by a reverse transcriptase. RNA is chemically unstable compared with DNA and is degraded easily *in vitro*.

2.1.2 Proteins

There are thousands of different proteins that perform the bulk of cellular activities. Proteins are the polymers of amino acids.

2.1.2.1 Amino Acids   Amino acids are the building blocks of proteins. There are a total of 20 amino acids in cells. Amino acids share a common chemical structure. They have a carboxyl group and an amino group bonded to the same carbon atom (α-carbon). But they differ from each other in their side chains or R-groups, which vary in structure, size, and electric charge (Table 2-2). Each amino acid has its own chemical features determined by the R-group. For example, cysteine ($R-CH_2-SH$) is readily oxidized to form a covalently linked dimeric amino acid called cystine by forming a disulfide bond. Disulfide bonds play a special role in the structures of many proteins either in the same polypeptide (intra) or between two different polypeptide chains (inter). Both tryptophan and tyrosine have similar light absorption spectra with the maximal wavelength of 280 nm, which gives the spectroscopic properties of proteins. Amino acids have been assigned three-letter and one-letter abbreviations (Table 2-2).

2.1.2.2 Protein Structure   Protein is a polypeptide in which amino acids are linked by a peptide bond. The linkage is formed by removing a water molecule (dehydration) from the α-carboxyl group of one amino acid and the α-amino group of

Table 2-2 Amino acids, abbreviations, and the R-groups

Amino Acids	Three Letter	Single Letter	R-Group, Basic Structure:
Alanine	Ala	A	$CH_3-CH(NH_2)-COOH$
Arginine	Arg	R	$HN=C(NH_2)-NH-(CH_2)_3-CH(NH_2)-COOH$
Asparagine	Asn	N	$H_2N-CO-CH_2-CH(NH_2)-COOH$
Aspartic acid	Asp	D	$HOOC-CH_2-CH(NH_2)-COOH$
Cysteine	Cys	C	$HS-CH_2-CH(NH_2)-COOH$
Glutamine	Gln	Q	$H_2N-CO-(CH_2)_2-CH(NH_2)-COOH$
Glutamic acid	Glu	E	$HOOC-(CH_2)_2-CH(NH_2)-COOH$
Glycine	Gly	G	NH_2-CH_2-COOH
Histidine	His	H	$NH-CH=N-CH=C-CH_2-CH(NH_2)-COOH$
Isoleucine	Ile	I	$CH_3-CH_2-CH(CH_3)-CH(NH_2)-COOH$
Leucine	Leu	L	$(CH_3)_2-CH-CH_2-CH(NH_2)-COOH$
Lysine	Lys	K	$H_2N-(CH_2)_4-CH(NH_2)-COOH$
Methionine	Met	M	$CH_3-S-(CH_2)_2-CH(NH_2)-COOH$
Phenylalanine	Phe	F	$Ph-CH_2-CH(NH_2)-COOH$
Proline	Pro	P	$NH-(CH_2)_3-CH-COOH$
Serine	Ser	S	$HO-CH_2-CH(NH_2)-COOH$
Threonine	Thr	T	$CH_3-CH(OH)-CH(NH_2)-COOH$
Tryptophan	Trp	W	$Ph-NH-CH=C-CH_2-CH(NH_2)-COOH$
Tyrosine	Tyr	Y	$HO-p-Ph-CH_2-CH(NH_2)-COOH$
Valine	Val	V	$(CH_3)_2-CH-CH(NH_2)-COOH$

Note: All 20 amino acids share a common chemical structure. They have a carboxyl group and an amino group bonded to the same carbon atom (α-carbon), but differ from each other in their side chains or R-groups. The R-groups vary in structure, size, and electric charge. The chemical features of each amino acid are determined by its R-group.

another amino acid. The polypeptide terminates in an amino group at one end (N-terminal) and a carboxyl group at the other end (C-terminal). The length of polypeptide chain varies considerably. For example, human cytochrome c has 104 amino acid residues with the molecular weight of 13 kDa, whereas human titin protein has 26,926 amino acid residues with molecular weight of 2993 kDa. The average size of human proteins is about 50 kDa.

The structure of a protein is categorized into four levels: primary, secondary, tertiary, and quaternary structures (Fig. 2-2). *Primary structure* is the linear order of amino acid sequence in the polypeptide chain. *Secondary structure* is the local steric interaction resulting from the hydrogen bonding between O and N of the C=O and the N=H of peptide backbone. The most prevalent elements of the secondary structure are α-*helix*, β-*sheet, and turn.* Particular combinations of these secondary structure elements form *motifs* such as helix–loop–helix. *The tertiary structure* is a folded three-dimensional shape of a polypeptide resulting from long-distance interactions between different regions of the protein molecule. The three-dimensional globular region is known as a *domain*, the functional unit of proteins. The large protein contains multiple domains. The final folded protein structure or *conformation* is largely stabilized by weak interactions such as hydrogen bonds and ionic interactions by

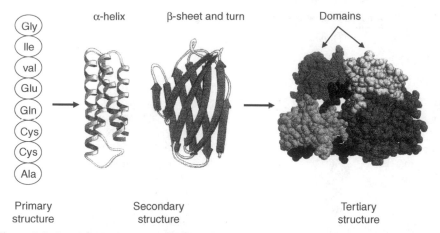

Figure 2-2 Levels of the hierarchical structure in protein. The *primary structure* is a sequence of amino acids linked by peptide bonds; the resulting peptide is coiled into *secondary structure*, α-helix and β-sheet; they are folded to form the *tertiary structure* that contains several functional domains (illustrated are α-helix, β-sheet, and spacing–filling model of Src protein kinase (Copyright 2002 from Molecular Biology of the Cell by Albert et al, reproduced by permission of Garland Science).

which free energy is minimal. Interruption of the hydrogen bonds results in loss of the secondary and tertiary structure or denaturation. Heat and extremes of pH or high concentration of salt induce the protein denaturation. The two general classes of proteins are fibrous and globular, based on their tertiary structures. *The quaternary structure* is the arrangement of two or more polypeptide *subunits* that fit together in space to form a single functional complex. For example, hemoglobin contains four subunits, two α- and two β-subunits, with a symmetrical arrangement to form a functional multisubunit protein to carry oxygen. Furthermore, multiple proteins can assemble spontaneously into complex structures as functional cellular machinery, such as replisome, ribosome, and proteasome.

The protein can be purified from the tissues by various techniques such as chromatography and electrophoresis. The amino acid composition and sequence of proteins can be determined chemically or by mass spectrometry. Three-dimensional structures of proteins are determined by X-ray crystallography or nuclear magnetic resonance (NMR) spectroscopy. The protein can also be synthesized with recombinant DNA technology in significant quantity. Many useful protein databases are also available. SwissProt, PDB, and SCOP are examples of such databases.

2.1.2.3 Protein Function   Proteins are the most abundant biomolecules in mammalian cells and consist of 18 percent of total cell weight compared with 0.25 percent of DNA and 1 percent of RNA. Proteins are the basic structural and functional molecules of cells, whereas DNA and RNA simply serve as vehicles to store and express genetic information. The proteins carry out almost all biological activities in the cells. There are about 200,000 different proteins in human body. According to its particular structure, each protein has its specific function. However, it usually takes more than one protein to accomplish a biologic task. All biological activities are archived by interaction of multiple proteins and other biomolecules such as RNA or small molecules. Often these interactions are reversible binding a *ligand* through the binding site of protein; simple examples include enzyme binding to its substrate and the receptor binding to a hormone ligand.

The largest group of proteins with a related function is the *enzymes*. These proteins specialize in catalyzing chemical reactions within cellular compartments. Enzymes increase the rate at which a chemical reaction reaches equilibrium, but they do not alter the end point of the chemical equilibrium. Enzymes can enhance reaction rates by a factor of 10^5–10^{17} in a very heterogeneous biochemical mixture. Their highly effective and specific catalytic properties largely determine the metabolic capacity of any given cell type. The catalytic properties and specificity of an enzyme are determined by the specific chemical configuration on the protein surface, *active site*. These sites are associated with a pocket, a cleft, or a pit on the surface of the enzyme, which binds the reactants or substrates facilitating chemical change by reducing activation energy. Enzymatically catalyzed reactions control metabolic activities in all cells.

There are many proteins other than enzymes that are basic structural components of cells or critical functional molecules in the organisms. These include such diverse examples as collagen, the connective tissue molecule; actin and myosin, the contractile proteins; insulin, the pancreatic hormone for glucose metabolism; and immunoglobulins (Igs), the antibody molecules of the immune system; histones, the proteins integral to chromosome structure in eukaryotes, and so on.

The potential for such diverse functions rests with the enormous variation of three-dimensional conformation that may be achieved by proteins. The final conformation of a protein is the direct result of the unique linear sequence of amino acids. To come full circle, the amino acid sequence of protein is determined by DNA sequence.

Any given biological function is the sum of work involving hundreds of different related proteins. Life depends on thousands of proteins with specific properties and functions. The activity of each protein component as well as the whole network is highly regulated to meet physiological needs in any given time and condition. These dynamic biochemical processes define the forms of the life. For example, the budding yeast *Saccharomyces cerevisiae* contain 6000 genes. By using the two-hybrid screening or the double mutation scoring method, a large scale of protein–protein interactions has been mapped. Elucidating the whole set of proteins and the interaction of the proteins in living cells is an important task of systems biology.

2.1.3 Polysaccharides

Polysaccharides (glycans) are carbohydrate polymers made up of many monosaccharides joined together by glycosidic linkages. The most abundant monosaccharide is D-glucose (dextrose), a six-carbon sugar containing an aldose group and five hydroxyl groups. Glucose and other hexose derivatives usually form a ring structure in aqueous solution with either α- or β-anomer. Thousands of monomers of the same type, such as glucose, link together to form homopolysaccharides. Examples include storage polysaccharides such as starch and glycogen. Heteropolysaccharides contains two or more different monomers. Polysaccharides are very large, often branched, molecules. They tend to be amorphous, insoluble in water, and have no sweet taste. Polysaccharides have a general formula of $(CH_2O)_n$; therefore, they are sometimes called carbohydrate.

Carbohydrates are not only the primary source of fuel and structural components of cells but also important informational molecules. Monosaccharides can be assembled into an almost unlimited variety of oligosaccharides, which differ in the stereochemistry and position of glycosidic bonds, the type and orientation of substituent groups, and the number and type of branches. These oligosaccharides are covalently linked with proteins or lipids to form glycoproteins or glycolipids on cell surface. The specific configuration of these oligosaccharides provides recognition sides for cell–cell interaction, bacterial toxin, or viral adhesion onto the cells. For example, blood types are determined by different oligosaccharides on the red blood cells.

2.1.4 Lipids

Biological lipids comprise a diverse group of molecules that are relatively water insoluble or nonpolar. Lipids commonly found in animals or plants include fatty acids and fatty acid-derived phospholipids, sphingolipids, glycolipids, sterols, and waxes. Some lipids are linear aliphatic molecules, whereas others have ring structures. Some are flexible, whereas others are rigid. The biological functions are as diverse as their chemistry. Fats and oils are the principal storage forms of fuel. Too much storage of fats results in obesity, which is becoming a severe problem in public health in the modern society. Phospholipids and sterols are major structural components of biomembrane. Steroid hormones, eicosanoids, and phosphorylated derivatives are important molecules in cell signaling.

In addition to being largely nonpolar molecules, most lipids have some degree of polar property. Generally, the bulk of their structure is nonpolar or hydrophobic acyl chains consisting of an even number of 10–22 hydrocarbon units (CH_2). They are either saturated or unsaturated. Another part of their structure is polar or hydrophilic containing carboxyl, hydroxyl, or phosphorated group. The bipolar feature of lipids (polar head and nonpolar tail) makes them amphophilic molecules. In the case of cholesterol, the polar group is a mere hydroxyl group. In the case of phospholipids, the polar groups are considerably larger.

Phospholipids, or, more precisely, glycerophospholipids or phosphoglycerides, have a glycerol core where two fatty acid-derived "tails" are linked to the first two carbons by ester bonds and one "head" group is linked to the third carbon by a phosphodiester bond. The phospholipids found in biological membranes are phosphatidylcholine (lecithin), phosphatidylethanolamine, phosphatidylserine, and phosphatidylinositol.

2.2 MICROSTRUCTURE AND FUNCTION OF CELLS

Biomolecules can be defined as any molecules found in living organisms. They can be either macromolecules or small molecules. Macromolecules include proteins, nucleic acids, polysaccharides, and lipids as already described. Small molecules are water, inorganic ions, and hundreds of organic metabolites. A simple mixture of these molecules cannot make life. All living organisms are highly organized structural and functional systems that are characterized by their ability to metabolize and self-replicate.

Since Leeuwenhoek first observed cells with his simple microscope in 1674, it has been confirmed that all living organisms are composed of cells. Many animals have trillions of cells, whereas bacteria are single-celled organisms. Cells are basic structural and functional units of life. The defined function of a living cell is determined by its particular structure.

Cells are small and complex. Under light microscope, cells can be divided into three parts: cell membrane, cytoplasm, and nucleus. There are many distinct functional structures or organelles in cells observed through electron microscope.

2.2.1 Prokaryotic Versus Eukaryotic Cells

There are 10–100 million living species in the biological universe on Earth. They consist of two basic cell types, prokaryotic and eukaryotic. Prokaryotic cells are less complex with a single compartment surrounded by cell membrane and with no defined nucleus. In contrast, eukaryotic cells have defined membrane-bounded nucleus and extensive internal membrane compartments or organelles. Bacteria are single-celled prokaryotes, whereas numerous animals, plants, and fungi are eukaryotes. Since the genetic code in DNA is same in all living organisms, prokaryotes and eukaryotes probably evolved from a common single-celled progenitor. Prokaryotic cells may represent the primitive cell type on Earth and eukaryotic cell types evolved from them. Table 2-3 shows the comparison between the two cell types. Figure 2-3 shows the microstructure of a typical animal cell.

2.2.2 Cell Membrane

A eukaryotic cell is classically divided into three compartments: cell membrane, cytoplasm, and nucleus. The basic architecture of cell membranes consists of a lipid bilayer associated with peripheral extrinsic proteins and intrinsic integral proteins (fluid mosaic model). Peripheral proteins are loosely associated with membrane through electrostatic and hydrogen bonds or by covalently attached lipid anchors. Integral proteins associate firmly with the membrane by hydrophobic interactions between the interior of the lipid bilayer and nonpolar amino acid side chains. The transmembrane sequences consist of about 20 or more amino acid residues in either α-helix or β-barrel structure. The composition of both lipids and proteins in the inner and outer lefts of the membrane is asymmetric. Many extrinsic proteins on the outside surface of the cell membrane are attached by oligosaccharides that are molecules for cell–cell interaction. The plasma membrane defines the external boundaries of cells and regulates the molecular traffic across the boundary. In eukaryotic cells, the biomembrane also divides cytoplasmic space into many distinct functional compartments (organelles). Biomembranes are crucial for life;

Table 2-3   Comparison of prokaryotic and eukaryotic cells

Characteristics	Prokaryotic Cell	Eukaryotic Cell
Prototype	Bacteria	Animal and plant cells
Structure	Simple	Complex
Cell size	1–10 μm	1–500 μm
Genome makeup	DNA with nonhistone proteins	DNA with histone and nonhistone proteins forming chromosomes
DNA size	$1–4 \times 10^6$ bp	$1–3 \times 10^9$ bp
Gene number	~4300 (*E. coli*)	~25,000 (human)
Nucleus envelope	Absent	Present
Cell division	Fission or budding	Mitosis
Membrane-bounded organelles	Absent	Present such as mitochondria or chloroplasts (plants), endoplasmic reticulum, Golgi apparatus, and lysosomes
Energy metabolism	Variable metabolic patterns, no mitochondria	More unified oxidative metabolism in mitochondria
Cytoskeleton	None	Complex with microtubules, intermediate filaments, and actins

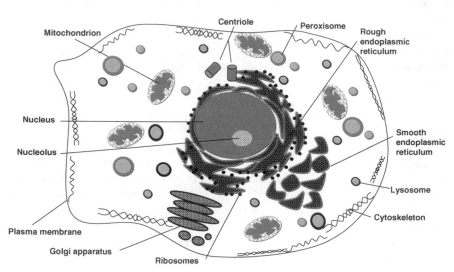

Figure 2-3   Substructures of a typical animal cell.(a) Plasma membrane with bilayer fluid mosaic structure. (b) Nucleus is filled with chromatin composed of DNA and nuclear proteins. (c) Nucleolus is a nuclear subcompartment area where rRNA is synthesized. (d) Mitochondria are surrounded by a double membrane, where ATP is generated. (e) Centriole for mitosis. (f) Peroxisomes. (g) Rough endoplasmic reticulum, attached with many ribosomes. (h) Smooth endoplasmic reticulum. (i) Lysosomes, biomembrane structure with an acidic lumen. (j) Cytoskeletal fibers form networks and bundles. (k) Ribosome. (l) Golgi apparatus. (m) Secretory vesicles store secreted proteins. (n) Nuclear envelope, a double membrane, outer membrane, is continuous with the rough ER .

there are two major reasons why the cell must separate itself from the outside environment. First, it must keep its biomolecules (DNA, RNA, proteins, and metabolites) inside the cell and keep foreign material outside the cell. Second, it must communicate with the environment to continuously monitor the external conditions and adapt to them and exchange the materials and energy with its surroundings. For example, when an *Escherichia coli* bacterium detects a high concentration of lactose in medium, it begins synthesizing proteins for metabolism of lactose. It needs to pump lactose in through a lactose transporter and release toxic metabolic products (see Fig. 2-8a). Thus, the cell membrane uses the lipid bilayer to function as a physical barrier and the integral proteins to function as selective biochemical transporters. Some types of transporters have ATPase activity and pump ions (channels) or small molecules (transporters) against electrochemical or concentration gradient.

In addition, some membrane proteins on the cell surface have specific functions in signal reception or receptors (see Fig. 2-12). The ability of cells to receive and act on signals in their surrounding environment is important for survival and cell–cell interaction in multicellular organisms. Each of the cells in all tissues communicates with dozens, if not hundreds, of other types of cells about a variety of important issues, such as when it should grow or differentiate or die, when it should release certain protein products such as growth factors or hormones needed by other cells at distant sites in the body, and what other cells it should associate with to build complex tissue architectures. These crucial decisions are made at tissue or whole body level to maintain a dynamic living system by cell receptors and signaling.

2.2.3 Cytoplasm

The homogeneous region of the cell between the plasma membrane and the nucleus is defined as the cytoplasm. In fact, the cytoplasm is not "homogeneous," but when viewed under an electron microscope, it is a highly compartmentalized structure. After centrifugation at high speed, the cytoplasm is separated into two fractions, supernatant aqueous phase (cytosol) and pellet phase (organelles). The cytosol is composed of water, ions, nutrients, and soluble macromolecules such as enzymes, carbohydrates, RNA, and a vast variety of metabolites. The cytosol makes up some 50 percent of the cell volume and functions as a perfect biochemical matrix in which hundreds of metabolic reactions occur in any given moment. All protein synthesis and glycolysis are carried out within the cytosol. On the contrary, pellet phase is rich in organelles with diverse functions (see subsequent sections).

2.2.4 Nucleus

The nucleus is the central compartment formed with two layers of concentric continuous biomembrane that is punctured with *nuclear pores*. This is where DNA is stored and RNA is synthesized. DNA is the inherited genetic material containing all information for the cell to live and to function. Synthesized RNAs are transported out of the nucleus through the nuclear pores. Proteins needed inside the nucleus are imported through the nuclear pores. The *nucleolus* is usually visible as a dark or red region in the nucleus where much of rRNA is synthesized and the ribosome is assembled.

2.2.5 Organelles

Organelles are membrane-bound small structures within eukaryotic cells that perform dedicated functions. Structural compartmentalization creates a stable environment and increases local concentration of reactive molecules, thus improving the biochemical efficiency. There are a dozen different types of organelles commonly found in eukaryotic cells, including mitochondria, chloroplast (in plants), lysosomes, peroxisomes, ribosomes, endoplasmic reticulum (ER), Golgi apparatus, and some vacuoles. Here, we will focus on only a handful of organelles and their roles at a molecular level in the cell with a brief description of the structure.

2.2.5.1 Cytoskeleton Network   The cytoplasm contains numerous filaments that form an interlocking three-dimensional network or the cytoskeleton. There are three types of cytoplasmic filaments: microfilaments (actins), intermediate filaments, and microtubules. They differ in diameter (from about 6 to 22 nm), protein subunits, and specific function. The cytoskeleton provides a framework for the traffic of intracellular organelles, the organization of enzyme pathways, mitosis, cell shape, and cell movement.

2.2.5.2 Mitochondria   Mitochondria are oval-shaped organelles formed with two layers of biomembrane, an inner and an outer layer resulting in two internal compartments, central matrix and intermembrane space. Aerobic respiration occurs in the mitochondria. Many enzymes for oxidative reactions including oxidation of pyruvate, fatty acid, and citric acid cycle are enriched in the central matrix. The inner membrane contains 80 percent of proteins where adenosine triphosphate (ATP) is generated by linking oxidative phosphorylation. The outer membrane contains 50 percent of proteins similar to cytoplasm membrane. The mitochondria contain its own circular DNA (16,569 bp) that encodes some proteins used in oxidative phosphorylation. The existence of the double membrane and complete genetic system have led many biologists believe that mitochondria are the descendants of some bacteria that has been endocytosed by a larger cell a billion years ago and coexist in an endosymbiotic relationship.

2.2.5.3 Chloroplasts   These organelles are the site of photosynthesis in plants and other photosynthesizing organisms. Similar to mitochondria, they also have a double membrane.

2.2.5.4 Endoplasmic Reticulum   The ER is an extensive network of biomembrane-bounded sacs, the site for many biosynthesis. There are two types of ER, rough and smooth. Rough ER is attached with many *ribosomes* where RNA is translated into protein. The smooth ER is the site for fatty acid, steroid, and phospholipid synthesis.

2.2.5.5 *Golgi Apparatus (Complex)*   Like ER, Golgi apparatus is a stack of flattened membrane vesicles. This organelle modifies (such as glycosylation) and packages newly synthesized proteins from rough ER into small membrane-bound vesicles. These vesicles can be targeted to various locations in the cell and secreted out of the cell.

2.2.5.6 *Lysosome*   This organelle contains a group of enzymes to digest large biomolecules into small monomeric subunits. All lysosomal enzymes have high activity at acid pH in lumen and collectively termed acid hydrolases. When the enzymes are released into cytoplasm, their activity is diminished.

2.2.5.7 *Peroxisomes*   Unlike mitochondria, peroxisomes contain several oxidases for fatty acid oxidation, but do not produce ATP. Instead, the energy is released into heat. In addition, H_2O_2 can be formed and degraded by catalase, important reactions for detoxification, as in the cases of ethanol and other toxic molecules.

2.3 SYNTHESIS OF BIOMOLECULES IN LIVING CELLS

2.3.1 Bioenergetics: The Law of Order

Each cell can be viewed as a tiny chemical factory. Cells require an ongoing supply of energy to carry out various kinds of work, including synthesis, movement, concentration, charge separation, the generation of heat, and bioluminescence. The energy required for these processes comes either from the sun or from the organic molecules such as carbohydrates, fats, and proteins. Organisms including plants, algae, and certain groups of bacteria are capable of capturing light energy by means of photosynthetic reaction. This group of organisms is called phototrophs. Another group of organisms is called chemotrophs because they require the intake of chemical compounds such as carbohydrates, fats, and proteins. All animals, fungi, and most bacteria are chemotrophs.

The flow of energy through cells is followed by the laws of thermodynamics. *Bioenergetics* is the application of thermodynamic principle in the biological system. The first law of thermodynamics (conservation) states that energy is always conserved, it cannot be created or destroyed, but energy can be converted from one form into another. The second law of thermodynamics states that "in all energy exchanges, if no energy enters or leaves the system, the potential energy of the state will always be less than that of the initial state." The second law provides a measure of thermodynamic spontaneity, although this only means that a reaction can go and says nothing about whether it will actually go or at what rate. Free energy change, ΔG, is a measure of thermodynamic spontaneity and is defined so that negative values correspond to favorable reaction and positive values represent unfavorable reaction. A negative ΔG is a necessary prerequisite for a reaction to proceed, but it does not guarantee that the reaction will actually occur at a reasonable rate. The presence or absence of an appropriate catalyst such as an enzyme will determine the rate at which a reaction can occur.

2.3.1.1 Energy Carrier Adenosine triphosphate is a universal "energy currency" in the cell and it can store and release the energy efficiently. Adenosine may occur in the cell in the unphosphorylated form or with one, two, or three phosphates attached, forming adenosine monophosphate (AMP), adenosine diphosphate (ADP), and adenosine triphosphate (ATP), respectively. ATP is extremely rich in chemical energy, in particular between the second and third phosphate groups. The net change in energy of the decomposition of ATP into ADP and an inorganic phosphate is -12 kcal/mol *in vivo* and -7.3 kcal/mol *in vitro*. This massive release in energy makes the decomposition of ATP extremely exergonic, and hence useful as a means for chemically storing energy. Many biochemical reactions that occur inside a cell are coupled with the formation or decomposition of ATP.

2.3.1.2 Electron Carriers Most of the energy of eukaryotic cells is generated from oxidizing fuel molecules, which involves the transfer of electrons from fuel molecules to oxygen. The fuel molecules are oxidized, while the oxygen is reduced. The electron carrier molecules of choice are nicotinamide adenine dinucleotide (NAD) and its relative nicotinamide adenine dinucleotide phosphate (NADP), two of the most important coenzymes in the cell. The oxidized forms, NAD^+ and $NADP^+$, serve as electron acceptors by acquiring two electrons, thereby generating the reduced form NADH and NADPH. However, only one proton accompanies the reduction. The other proton, produced as two hydrogen atoms are removed from the molecule being oxidized, is liberated into the surrounding medium.

2.3.2 Enzymes as Catalysts of Life

Enzymes allow many chemical reactions to occur within the homeostasis constraints of a living system. Enzymes function as biological catalysts. The use of enzyme can decrease the free energy of activation of chemical reactions. The first step in catalysis is the formation of an enzyme–substrate complex. By bringing the reactants closer together, chemical bonds may be weakened and reactions will proceed faster than without the enzyme. An enzyme-catalyzed reaction proceeds via an enzyme–substrate intermediate and follows Michaelis–Menten kinetics, which is characterized by a hyperbolic relationship between the initial reaction rate (velocity, v) and the substrate concentration [s].

Enzymes are regulated by many ways to adjust their intracellular concentrations and activity levels to meet the cellular needs. First, all protein enzymes are sensitive to temperature and pH. Changes in temperature or pH may denature the enzyme and most enzymes are adapted to operate at a specific pH or pH range. Second, enzyme activity is influenced not only by substrate availability but also by products, alternative substrates, substrate analogues, cofactors, and coenzymes. The binding of the substrate to the active site of an enzyme alters the structure of the enzyme, placing some strain on the substrate and further facilitating the reaction. Cofactors are nonproteins essential for enzyme activity. Ions such as Zn^{2+} and Cu^{2+} are cofactors. Coenzymes are nonprotein organic molecules such as NAD^+ or $NADP^+$ bound to enzymes near the active site. Additional control mechanisms include allosteric regulation. Most allosterically regulated enzymes catalyze the first step in a reaction sequence and are multisubunit proteins with multiple catalytic subunits and multiple regulatory subunits. Each of the catalytic subunits has an active site that recognizes substrates and products, whereas each regulatory subunit has one or more allosteric sites that recognize specific effector molecules. A given effector may either inhibit or activate the enzyme, depending on which form of the enzymes is favored by effector binding. Such a mechanism is commonly employed in feedback inhibition. Often one of the products, either an end or near-end product, acts as an allosteric effector, blocking or shunting the pathway. The biosynthesis of enzymes in living cells is , also subjected to various regulations at transcriptional and translational levels described subsequently.

2.3.3 Metabolism and Metabolic Pathways

Metabolism is all of the biochemical reactions that occur within a cell. This includes the biosynthesis of complex organic molecules such as nucleic acids, proteins, lipids, and carbohydrates (anabolism) and the degradation of these large molecules into smaller, simpler ones with the release of chemical energy (catabolism) in the form of ATP. Catabolism can be carried out either in the presence of oxygen (aerobic conditions) or in the absence of oxygen (anaerobic conditions). The energy yield is much greater in the presence of oxygen, which probably explains the preponderance of aerobic organisms in the world. However, anaerobic catabolism is also important, both for organisms in environments that are always devoid of oxygen and for organisms and cells that are temporarily deprived of oxygen. Photosynthesis, a phototrophic energy metabolism, is an important biochemical process in which plants, algae, and some bacteria acquire the energy of sunlight to produce food. Ultimately, nearly all living things depend on energy produced from photosynthesis for their nourishment, making it vital to life on Earth.

Metabolism usually consists of sequences of enzymatic steps, also called metabolic pathways. Metabolic pathways are of two general types: anabolic pathways are connected with the synthesis of cellular components and are usually involved in a substantial increase in molecular order and require energy, whereas catabolic pathways are involved in the breakdown of cellular constituents and release energy. Catabolic pathways play two roles in cells: they give rise to the small organic molecules or metabolites that are the building blocks for biosynthesis and the production of energy that is used to synthesize the macromolecules and other cellular function.

A very large number of metabolic pathways including both catabolic and anabolic pathways have been discovered and they can be found in various databases on the Internet. The KEGG collection (http://www.genome.jp/kegg/) of metabolic and regulatory databases currently has a record of 54,622 metabolic pathways. As an example, Figure 2-4 shows the most common and well-known carbohydrate catabolic pathway, including *glycolysis*, *pentose–phosphate* (*PP*) *pathway*, *fermentation*, and *aerobic respiration* in recombinant *E. coli*. Using glucose as a prototype substrate, catabolism under both anaerobic and aerobic conditions begins with the glycolytic

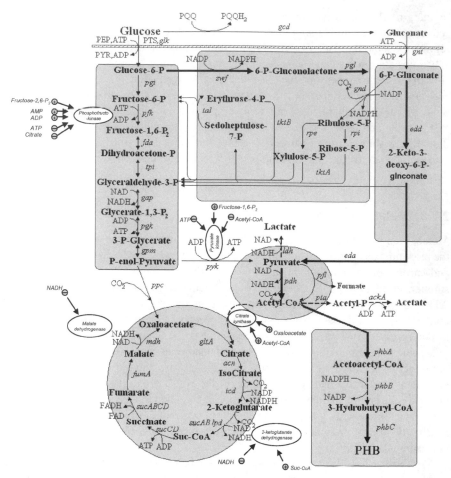

Figure 2-4 Central metabolic pathways in a recombinant *E. coli* expressing three genes for poly-3-hydroxybutyrate (PHB) synthesis. Glycolysis, fermentation, TCA cycle, and PHB synthesis pathways are shown. Genes encoding the important metabolic enzymes are also indicated (italic letters). Major regulatory effects are indicated as either activation ($+$) or inhibition ($-$). Kinetic regulators are highlighted: fructose-2, 6-P2, citrate, and acetyl-CoA for glycolysis; acetyl-CoA, oxaloacetate, and succinyl-CoA for TCA cycles. Key allosteric effectors are NAD$^+$, ADP, and AMP as activator and acetyl-CoA, NADH, and ATP as inhibitors.

pathway, a 10-step sequence of reactions in which glucose is converted to pyruvate with the net production of two molecules of ATP per molecules of glucose. In the absence of oxygen, reduced coenzyme generated during glycolysis must be reoxidized at the expense of pyruvate, leading to fermentation end products such as acetate, lactate or ethanol, and carbon dioxide. The most common and well-known type of glycolysis is the *Embden–Meyerhof* (*EMP*) pathway. Compared to fermentative process, *aerobic respiration* gives the cell access to much more of the free energy that is released by the oxidation of organic substrates. The completion of catabolism of carbohydrates begins with the glycolytic pathways mentioned above, but the pyruvate is converted to acetyl CoA. The acetyl CoA is then oxidized fully by enzymes of the *tricarboxylic acid* (*TCA*) cycle. The reduced coenzymes (NADH, $FADH_2$) are further reoxidized by the *electron transport system* and generated additional ATP. A total of 38 molecules of ATP are generated per molecule of glucose in most prokaryotic and some eukaryotic cells. This is a factor of 19 times more energy per sugar molecule than what the typical anaerobic reaction generates. The small molecules such as acetyl-CoA and energy such as ATP, NADH, and NADPH generated by these catabolic reactions can be used for biosynthesis of other cellular products and cell mass. ATP, NADH, and NADPH are continually generated and consumed. NADPH, which carries two electrons at a high potential, provides reducing power in the biosynthesis of cell components from more oxidized precursors. Figure 2-4 also illustrates a biopolymer PHB synthetic pathway that is introduced into *E. coli* by recombinant DNA techniques. In recombinant *E. coli*, PHB can be synthesized from acetyl-CoA by a sequence of three enzymatic reactions catalyzed by β-ketothiolase, acetoacetyl-CoA reductase, and PHB synthase. Moreover, key reaction types are used repeatedly in metabolic pathways.

2.3.4 Regulation of Metabolism

As already discussed, metabolic pathways form as a result of the common occurrence of a series of dependent chemical reactions. These reactions are carefully regulated to ensure that the rate of product formation is tuned to actual cellular need. The end product of the pathway depends on the successful completion of all sequential reactions, each mediated by a specific enzyme. Also, intermediate products tend not to accumulate, making the process more efficient. The metabolism is regulated at two levels to achieve the overall cellular "fitness" and balance. First, catalytic activities of many enzymes are directly regulated by allosteric interactions (as in feedback inhibition) and by covalent modification. These processes regulate the activity of preexisting enzymes in both catabolic and anabolic pathways. Most of the constitutive enzymes such as the enzymes operating in glycolysis and TCA cycles are regulated at this level. Several key regulatory steps in glycolysis and TCA cycles are highlighted in Figure 2-4. Second, the amounts of many enzymes are controlled by regulation of the rate of protein synthesis and degradation. The processes of end product repression, enzyme induction, and catabolite repression are involved in the control of synthesis of enzymes. End product repression and enzyme induction are mechanisms of negative control that lead to a decrease in the transcription of proteins. Catabolite repression is considered a form of positive control because it leads to an increase in transcription of proteins. Many inducible or repressible enzymes are regulated at this level. Such examples of enzyme induction and catabolite repression (i. e., repression and induction of *lac* operon, see Fig. 2-8a) will be discussed subsequently. In addition, the movement of many substrates into cells and subcellular compartments is also controlled. Distinct pathways for biosynthesis and degradation contribute to metabolic regulation. The energy charge, which depends on the relative amounts of ATP, ADP, and AMP, plays an important role in metabolic regulation. A high-energy charge inhibits ATP-generating (catabolic) pathways and stimulates ATP-utilizing (anabolic) pathways.

2.4 THE INFORMATION FLOW IN LIVING CELLS

As we discussed, the cell is the basic structural and functional unit for all living species. In cells the genetic information is carried by DNA molecules and its specificity is determined by the sequence of nucleotides. The proteins are functional forms of life and perform most cellular activities. The largest group of proteins with a related function is enzymes. In this section, we will discuss how genetic information flow transmits from one generation to the next; how genetic information is expressed as functional proteins; and how expression is controlled in a living cell by cell signaling. On the other hand, genetic information can also be inherited epigenetically. In addition, genetic, epigenetic, and biochemical information flow are integrated for final cellular controls.

2.4.1 Genetic Information Flow

The so-called central dogma of molecular biology comprises the three major biological processes: replication, transcription, and translation. Replication is the copying of parental DNA to form daughter DNA molecules with identical nucleotide sequences as well as identical epigenetic modifications. Transcription is the process by which the genetic information stored in DNA sequence is copied precisely into messenger RNA. Translation is the genetic information encoded in messenger RNA being translated into a polypeptide. With the completion of human genome project and with genome sequences of many other species available, these processes can be studied at genome levels (Fig. 2-5).

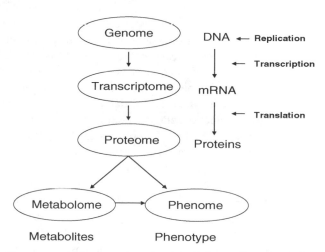

Figure 2-5 New central dogma of molecular biology. The classical central dogma (DNA → RNA → protein in single gene) has been redefined at whole genome level. Ultimately, the total phenotypic characteristic of an organism is determined by the proteome and the whole set of functional proteins and their interactions.

2.4.1.1 DNA Replication Genome is defined as the complete set of genetic information carried by a cell in the organism. DNA is the primary material that carries genetic information. For many viruses and prokaryotes, the genome consists of one linear or circular DNA molecule. In the eukaryotic cell, DNA is packed in multiple chromosomes and confined by nuclear envelope in the nucleus. The size of genome is quite different among species. The human genome contains 3.2×10^9 nucleotide pairs, divided into 22 different autosomal chromosomes and 2 sex chromosomes, while prokaryote *E. coli* contains 4.6×10^6 nucleotide pairs in a single circular DNA. Eukaryotic chromosomes consist of numerous highly coiled DNA/histone complex or nucleosome connected by linker DNA. Each nucleosome contains the protein core of eight histones (two copies each of H2A, H2B, H3, and H4) and a 200 bp segment of DNA. The histone core is encircled by DNA fiber (Fig. 2-11).

With these basic concepts in mind, we will discuss the chemical mechanism of heredity in which the genetic information is passed from a cell to its daughter cells at cell division and from one generation to the next through the reproduction of organisms.

DNA replication is semiconservative; each strand of double helix of DNA can be used as the template to synthesize a new complementary strand of DNA (Fig. 2-6a). It

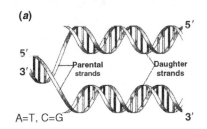

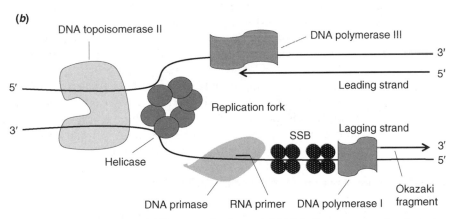

Figure 2-6 DNA replication. (a) Two strands of parental DNA helix are unwound and used as templates to produce new daughter strands. The outcome is two copies of identical DNA, each containing one of the original strands and one new complementary strand (semiconservative replication). (b) The main biochemical steps (see text) and the proteins involved in DNA replication are illustrated.

is carried out in three identifiable phases: initiation, elongation, and termination. The replication starts at the origin and usually precedes bidirectionally. Parental double helix DNA must be separated from one another (denaturation) to be the templates. This process is accomplished by many enzymes such as helicases and topoisomerases under normal physiological condition. DNA is then synthesized in the $5' \rightarrow 3'$ direction by DNA polymerases. Since the two strands of a DNA are antiparallel, this $5' \rightarrow 3'$ DNA synthesis can take place continuously on only one of the strands at a replication fork (the leading strand). On the other strand (lagging strand), the short DNA fragments are synthesized discontinuously as Okazaki fragments, which are subsequently ligated and the gaps are filled by ligase. The complex of many proteins and enzymes at the replication fork is called replisome. The key component in replisome is DNA polymerase. Most cells have several DNA polymerases. In *E. coli*, DNA polymerase III is the primary replication enzyme (Fig. 2-6b). Eukaryotic chromosomes have many replication origins and proceed at multiple sites by utilizing DNA polymerase α.

The fidelity of DNA replication is maintained by several mechanisms: (1) base selection by DNA polymerase according to template nucleotide following the role of Watson–Crick base pairing (A/T; C/G); (2) $3' \rightarrow 5'$ proofreading exonuclease activity that is part of most DNA polymerases, and (3) specific DNA repair systems for mismatch correction.

Epigenetic information such as DNA methylation is also inherited during DNA replication. As a result, the two replicated DNA molecules in parental cell are exactly the same sequences and methylcytosine content are equally divided into two daughter cells. Once the two daughter cells receive the same genetic material and epigenetic information, heredity is pursued.

2.4.1.2 *From DNA to RNA—Transcription*   As discussed above, genomic DNA contains all information to build a cell or organism. Although the genome in all somatic cells (except lymphocytes) in a given multicellular organism is same, the structure and function of the different types of cells are totally different. Hepatocytes are different from neurons. Myocytes are different from epithelial cells. The difference is not due to DNA, but mRNAs and proteins.

The central dogma is an early attempt to understand how the amino acid sequence of the protein is determined by nucleic acid sequence of DNA based on following observations. First, the DNA is confined in the nucleus, while protein synthesis occurs in association with ribosomes in the cytoplasm. Second, RNA is synthesized in the nucleus and then transported to the cytoplasm. Third, RNA is chemically similar to DNA. Collectively, these observations suggest that genetic information, stored in DNA, is transferred to an RNA intermediate (mRNA), which directs the synthesis of proteins.

The process by which RNA molecules are synthesized on a DNA template is called transcription. It results in an mRNA molecule complementary to the DNA sequence of one strand (template strand) of the double helix DNA.

Like DNA replication, transcription is also a complicated biochemical process and many protein factors and enzymes are involved. However, transcription occurs only in particular DNA regions. The process of transcription can be divided into four phases: initiation, elongation, termination, and processing. During initiation, RNA polymerase binds to a specific site in DNA (the promoter), locally melts the double-stranded DNA to reveal the unpaired template strand, and polymerizes the first two nucleotides. There are as many as 30 polypeptides (general transcription factors) assembled as an initiation complex facilitating the initiation. During strand elongation, RNA polymerase (II in eukaryotic DNA) moves along the DNA, melting sequential segments of the DNA and adding nucleotides to the growing RNA strand. When RNA polymerase reaches a termination sequence in the DNA (terminator), the enzyme stops transcription, leading to the release of the completed RNA and dissociation of the enzyme from the template DNA. RNA polymerase and transcription factors can be reused for the next round of transcription.

In eukaryotic DNA, the initial primary transcript (pre-mRNA) very often contains noncoding regions (introns) interspersed along coding regions (exons). Transcripts from genes containing introns undergo splicing, the removal of the introns and joining of the exons, which is catalyzed by small nuclear RNAs in the spliceosome. During processing, the ends of nearly all primary transcripts are also modified by addition of a 5′-cap and 3′-poly (A). Then the mature mRNAs are exported through the nuclear pores to cytoplasm for protein synthesis (Fig. 2-7).

Not all DNA sequences in genome are used for coding RNAs or proteins. In fact, only a small fraction of genome is coding sequence in eukaryotes. For instance, only

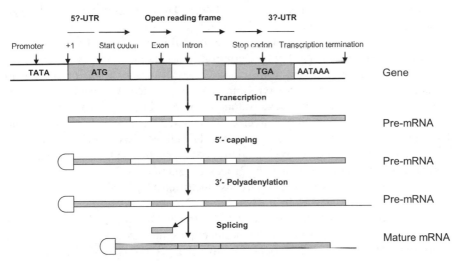

Figure 2-7 Gene transcription. The regulatory region and the coding region of a gene are illustrated in upper panel. The steps of transcription and mRNA processing are illustrated. In both prokaryotes and eukaryotes, a promoter such as TATA box, located upstream of transcription start site (+ 1), is required for RNA polymerase binding and transcription initiation. The mRNA processing occurs in eukaryotes only. The newly synthesized pre-mRNA needs to be processed, which includes capping 5′-end with 7-methylguanylate, adding poly (A) tail at 3′-end, and removal of introns and connection of exons.

1.2 percent of human genome DNA codes RNAs then proteins. In genome, the region that directs the synthesis of a single polypeptide or functional RNA (such as tRNA) is called a gene. A gene is the physical and functional genetic unit composed of a segment of DNA. A gene consists of a regulatory region and a coding region. A human has proximally 25,000 genes, whereas *E. coli* contains about 4288 genes. However, these genes are not always expressed in the cells. Hepatocytes express only liver function-related genes, whereas neurons express only brain function-required genes. However, some housekeeping genes needed for basic cellar activities express in all cells.

Gene expression is tightly controlled in a given cell type at any given moment in the organisms. Transcription is the most important control point, which can be activated or repressed. In prokaryotes, the classical example of transcriptional control is *lac* operon in *E. coli*. The *lac* operon encodes three enzymes for the metabolism of lactose. For transcription of the *lac* operon to begin, the σ^{70} subunit of the RNA polymerase must bind to the *lac* promoter, which lies just upstream of the transcription start site. When no lactose is present, binding of the *lac* repressor protein to a sequence called the *lac* operator, which overlaps the transcription start site, blocks transcription initiation by the polymerase. When lactose is present, lactose molecules bind to specific binding sites in each subunit of the tetrameric *lac* repressor, causing a conformational change in the protein that makes it dissociate from the *lac* operator. As a result, the polymerase binds to promoter to initiate transcription of the *lac* operon (Fig. 2-8a).

Transcriptional control in eukaryotes is much more complicated than that in prokaryotes. There are at least three types of promoter proximal elements: silencer element, upstream activator sequence (UAS), and core promoter (TATA box, initiator (INR), and downstream promoter elements (DPE)). These proximal elements, located within 200 bp of transcription start sites, direct the basal level of transcription. In contrast, enhancers and suppressors may be located up to 10 kb either upstream or downstream from a promoter. The transcription activator or repressor protein binds to enhancers or suppressors and interacts with basal transcriptional machinery to enhance or repress the transcription. In most cases, the action is mediated by a set of other proteins, the so-called coactivators or corepressor (Fig. 2-8b). Furthermore, transcription activity can also be affected by changing the chromatin configuration. Histone acetylase and deacetylase complex, as well as DNA methylation in CpG islands, can regulate transcription initiation. The epigenetic regulation is important in normal development and pathological processes (Section 2.4.2).

The discovery of miRNA and siRNA has elucidated additional mechanisms of the posttranscriptional control. Both miRNAs and siRNAs contain 21–23 nucleotides that are generated from longer hairpin-like double-stranded precursor RNA ($\sim$70 bp) by DICER ribonuclease. One strand of the shorter duplex intermediate RNA is assembled into a multiprotein, RNA-induced silencing complex (RISC). The miRNA in complex forms imperfect hybrids with sequences in 3'-untranslated region (3'-UTR) of specific target mRNAs and represses the translation initiation. The siRNA in complex forms a perfect hybrid with target mRNA and leads to degradation of target mRNA. Hundred types of miRNA and siRNA have been identified in higher eukaryotes. The double-stranded RNA (dsRNA) containing these small RNA sequences can be constructed or synthesized and then introduced into a cell *in vitro* or *in vivo*. It provides a huge

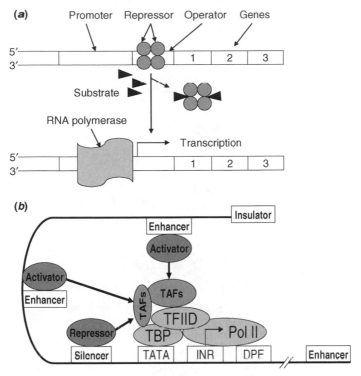

Figure 2-8 Transcription regulation in prokaryotes and eukaryotes. (a) A prokaryotic regulatory unit, operon. Transcription is induced by the substrate that binds to and releases the repressor from the operator, and then RNA polymerase binds to the promoter and initiates the transcription. (b) A eukaryotic regulatory unit. The core promoter contains TATA box (TATA), initiator sequences (INR), and downstream promoter elements. A complex arrangement of multiple enhancers interspersed with silencer and insulator elements that can be located 10–50 kb either upstream or downstream of the core promoter. Transcription is initiated at core promoter regions by interactions of the RNA polymerase II core complex, general transcription factors (TFIID), and multiple subunits of cofactors including TBP-associated factors (TAFs). Transcription factors binding enhancer or silencer regulate the transcription by interaction with TAFs or other cofactors.

potential to specifically knockdown or silence a gene at posttranscriptional level. The target mRNA can be viral (such as HIV) or oncogenic gene transcripts (Fig. 2-9).

2.4.1.3 *From RNA to Protein—Translation*   There are three basic types of RNAs in cells, namely, mRNA, tRNA, and rRNA. Other RNA species include miRNA, siRNA, and snRNA (Table 2-1). All these RNAs somehow play a specific role in mRNA-directed protein synthesis or translation.

mRNA   Genetic information stored in DNA in the form of a nucleotide sequence is transcribed into mRNA. The nucleotide sequence in mature mRNA contains

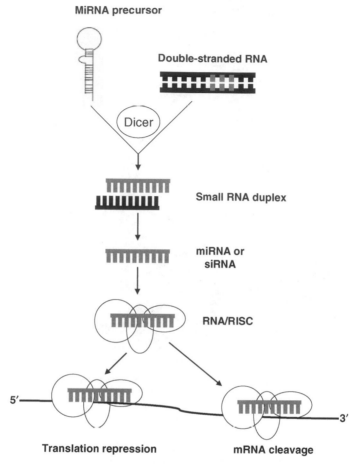

Figure 2-9   RNA interference. Both miRNA and siRNA are cleaved from their precursors, the large double-stranded RNAs, by the Dicer ribonuclease. After binding to multiple proteins and forming RISC, one strand of 21–23 nucleotides hybridizes the target mRNA and degrades mRNA (by siRNA) or inhibits the translation (by miRNA). The gene expression is inhibited by this posttranscriptional regulation.

continuous degenerated triplet nucleotides that determine specific amino acids called codons. Each triplet nucleotide (or codon) in the mRNA is, in turn, complementary to a triplet nucleotide (anticodon) corresponding tRNA. Each tRNA carries a specific amino acid that is correctly inserted into the polypeptide chain during translation. The complete genetic codes have now been elucidated (Table 2-4). Many amino acids are encoded by more than one codon. The AUG codon for methionine is the most common start codon, while three codons (UAA, UGA, UAG) function as stop codons specifying no amino acids. The region of mRNA from the start codon to a stop codon is called the reading frame. The 5′-cap (7-methylguanylate) and poly (A) tail define the 5′- and 3′-ends of mRNA. The region between the 5′-cap and the start codon AUG is known as the 5′-untranslated region (5′-UTR). The region from the stop codon to the start point

Table 2-4 The genetic code in mRNA

First Position	Second Position				Third Position
	U	C	A	G	
U	Phe	Ser	Tyr	Cys	U
	Phe	Ser	Tyr	Cys	C
	Leu	Ser	Stop	Stop	A
	Leu	Ser	Stop	Trp	G
C	Leu	Pro	His	Arg	U
	Leu	Pro	His	Arg	C
	Leu	Pro	Gln	Arg	A
	Leu	Pro	Gln	Arg	G
A	Ile	Thr	Asn	Ser	U
	Ile	Thr	Asn	Ser	C
	Ile	Thr	Lys	Arg	A
	Met	Thr	Lys	Arg	G
G	Val	Ala	Asp	Gly	U
	Val	Ala	Asp	Gly	C
	Val	Ala	Glu	Gly	A
	Val	Ala	Glu	Gly	G

Note: The first base of the codon (5'-end) is shown in the left column and the second base is shown in the third row. The third base in the right column plays lesser specific role; AUG (for Met) is most common initiator codon. Sometimes GUG and UUG are also used. The three termination codons (UAA, UGA, and UAG) match no aminoacyl-tRNA and are recognized by termination factors for translation termination.

of poly (A) tail is known as 3'-untranslation region (3'-UTR). The pre-mRNAs are matured (processing) in nucleus, transported to cytoplasm, and used as templates for protein synthesis during translation. The processes of pre-mRNA maturation include adding 5'-cap, poly (A) tail, and splicing (Fig. 2-7).

tRNA   There are 64 types of tRNAs found in eukaryotes. All tRNAs have a similar three-dimensional structure including an acceptor arm for attachment of a specific amino acid and a stem-loop with a three-base anticodon. Each type of amino acid has its own set of tRNAs, which bind the specific amino acid and carry it to the growing end of a polypeptide. The anticodons in each tRNA can base-pair with its complementary codon in the mRNA, by which the nucleotide sequence in the mRNA is translated into amino acid sequence in the peptides.

rRNA   The rRNA associates with a set of proteins to form ribosomes. These large ribonucleoprotein complexes move along an mRNA and catalyze the assembly of each amino acid into peptide chain. Both prokaryotic (70S) and eukaryotic (80S) ribosomes consist of a small and a large subunits. Each subunit contains numerous different proteins and one major rRNA molecule. The topological structure of the ribosome has been elucidated in detail. The genes encoding rRNA are in the nucleolus region.

Protein Synthesis   Similar to DNA replication and RNA transcription, translation is a highly organized and regulated biochemical process in which many proteins factors are involved. It can be divided into three phases: initiation, elongation, and termination. Before translation initiation, each amino acid needs to be activated by one of the 20 specific aminoacyl-tRNA synthetases. As the result, the amino acid is linked to the acceptor arm of tRNA by a high-energy bond, aminoacyl-tRNA.

During initiation, the small ribosomal subunit binds to mRNA near translation start site with the initiator tRNA carrying the amino-terminal methionine (Met-tRNAi^{Met}). Then the large subunit and multiple initiation factors (eIF2, 3, 4, 5) also bind to form initiation complex. The complex precedes the scan along mRNA ($5' \rightarrow 3'$) until it encounters the start codon AUG. The anticodon ($5'CAU3'$) of Met-tRNAi^{Met} is base-paired with the start codon AUG in mRNA at P-site of ribosome.

During chain elongation, each incoming aminoacyl-tRNA moves through three ribosome sites, A, P, and E. First, the new aminoacyl-tRNA binds to the A-site that makes Met-tRNAi^{Met} move to the P-site, the large rRNA subunit catalyzed peptide bond formation between Met and incoming amino acid. At the same time, the ribosome undergoes conformational change and moves one codon down along the mRNA, the unacylated tRNAi^{Met} is shifted to the E-site from the P-site and the peptidyl-tRNA from the A-site to the P-site. In the next step, the incoming aminoacyl-tRNA binds to the A-site and unacylated tRNAi^{Met} is ejected from the E-site. The cycle repeats and the chain is elongated. This process is fast and accurate. A peptide of 100 amino acid residues needs only 5 s to synthesize (Fig. 2-10).

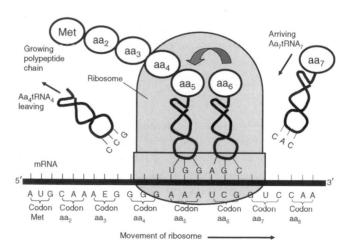

Figure 2-10   Translation. Translation is initiated from formation of initiation complex with fMet-tRNA^{met} binding to start codon AUG in mRNA. The second aminoacyl-tRNA enters the A-site of the ribosome. The peptidyl transferase of rRNA ribozyme in large subunit catalyzes the peptide bond formation between fMet-tRNA^{met} and the second aminoacyl-tRNA, The ribosome moves one codon toward the 3'-end of mRNA. The third incoming aminoacyl-tRNA binds the A-site of ribosome. The cycle is repeated and newly synthesized peptide is elongated until it hits the stop codons. The translation is then terminated. Many protein factors are involved in translation.

At the termination phase, when the peptide chain-bearing ribosome reaches a stop codon (UAA, UGA, UAG), the elongation stops since no tRNA molecules with anticodons matches the stop codons. The release factor eRF1 then enters the ribosomal complex and cleaves the peptide chain from tRNA at P-site. The peptides spontaneously fold into the active three-dimensional forms. The tRNA and two ribosomal subunits are dissociated and released. The ribosome is recycled for the next round of translation. In most reactions, GTP binding proteins hydrolyzing GTP to GDP provide the energy.

2.4.1.4 *Posttranslational Modification of Proteins*   Although proteins are the end products of the genetic information flow, they need additional modifications to become the functional molecules. Posttranslational modification means the chemical modification of a protein after being translated. It is the last step in protein biosynthesis for many functional proteins. Posttranslational modification may involve the formation of disulfide bridges and attachment of any number of biochemical functional groups, such as carbohydrates, acetate, phosphate, and various lipids. Enzymes may also remove one or more amino acids from the ends of the polypeptide chain or cut the polypeptide in the middle of the chain. For instance, proinsulin is cut twice after disulfide bond formation to form the active form of insulin. In other cases, two or more polypeptide chains that are synthesized separately may associate to form the quaternary large protein, such as immunoglobulin and hemoglobulin. The most common posttranslational modification is glycosylation by which the carbohydrate chains are added to the side chains of the peptides to form glycoprotein. The carbohydrate chains are important for cell–cell recognition (sugar code). Notably, glycosylation is absent in bacteria and is somewhat different in each type of eukaryotic cells. Protein phosphorylation is part of common mechanisms for controlling the function of a protein, for instance, activating or inactivating an enzyme by protein kinase in cell signaling pathways (see Fig. 2-12). All proteins are eventually degraded by ubiquitin-dependant proteolysis in a large protein complex, proteasome.

2.4.2 Epigenetic Inheritance

In classic genetic inheritance, traits are passed from one generation to the next via DNA sequences in the genome. Differences in a DNA sequence specify differences in a trait. Epigenetic inheritance involves passing a trait from one generation to the next without the difference in DNA sequence, but DNA structure. Known mechanisms of epigenetic inheritance include changes in molecular structures in the DNA (such as DNA methylation) or histones (such as histone methylation and acetylation), chromosome remolding, and RNA interference (RNAi) so that while the gene (DNA sequence) is the same, the gene expression is different (Fig. 2-11). For example, genes switch on and off in response to hormonal signals. Changes in molecular conformation around the gene can influence the gene transcription. This can change developmental processes and can alter the course of diseases or result in genomic imprinting. The study of epigenetic inheritance is known as epigenetics.

Epigenetic inheritance systems allow cells of different phenotype but of identical genotype to transmit their phenotype to their offspring. Proteins or chemical groups

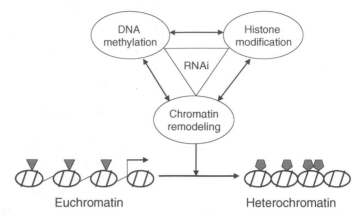

Figure 2-11 Cell epigenetics and gene silencing mechanisms. Cell epigenetic system consists of DNA methylation and histone methylations (pentagon), histone acetylation (triangle), chromatin (nucleosome) remodeling, and RNAi. Interaction between these components results in transition of euchromatin to heterochromatin and the transcription is inactivated (silenced) permanently.

that are attached to DNA and modify its activity are called chromatin marks. These marks are copied when DNA replicates. For example, cytosines in eukaryotic DNA can be methylated (5-methylcytosine). The number and pattern of such methylated cytosines influence the functional state of the gene: low levels of methylation correspond to high level of gene expression, whereas high levels of methylation correspond to low levels of gene expression. Although there are random changes in the DNA methylation pattern, specific changes induced by environmental factors do occur. After DNA replication, maintenance DNA methyltransferases (DNMT 1) at replication fork make sure that the methylation pattern of the parental DNA is copied to the daughter strand of DNA precisely. In such a way, the pattern of DNA methylation is maintained from parental cells to the daughter cells. If the DNA methylation occurs in the germ cells, the pattern can be inherited from one generation to the next. In a few cases, expression of a gene solely depends on whether it is inherited from the mother or father. This phenomenon is called genomic imprinting. The molecular mechanism of genomic imprinting is DNA methylation. The insulin-like growth factor-2 (Igf-2) gene is one example of an imprinting gene. In this case, only the copy of Igf-2 gene from paternal side is transcribed, whereas the maternal copy of Igf-2 gene is silenced by DNA methylation. The loss of Igf-2 imprint is associated with carcinogenesis, especially in colorectal cancer.

During the course of evolution, accidental deamination of unmethylated C gives rise to U, which is recognized by DNA repair system, uracil DNA glycosylase. The U in DNA sequence is excised, then replaced with C, and again restored in the original DNA sequence. However, deamination of a methylated C in the genome tends to be eliminated and replaced by a T by DNA repair system. As a result, most dinucleotide C–G sequences have been lost because of the elimination of methylcytosine. The remaining residual C–G sequences are distributed very unevenly in the genome.

In some regions, dinucleotide C–G sequences are present at 10–20 times more than their average density, called CpG islands. The CpG dinucleotide differs from C–G base pairing. The CpG islands are often located in the promoters of the housekeeping genes. The housekeeping genes encode many proteins that are essential for cell basic metabolism and viability and are therefore expressed in most cells all of the time. In most cases, the CpG islands seem to remain in an unmethylated state in all cell types. However, some tissue-specific genes, which code for proteins needed only in selected types of cells, are also associated with CpG islands. For instance, DNA methylation in the dividing fibroblasts gives rise to only new fibroblasts rather than some other cell types, even though the genome is identical in all cells.

If the CpG islands in promoter regions are abnormally methylated, the transcription will be blocked by the binding of methyl binding protein complex and the gene will be silenced. In the case of cancer cells, many tumor suppressor genes are inactivated by DNA methylation in CpG islands surrounding the promoters. Abnormal epigenetics including DNA methylation pattern is a hallmark of cancer cells (Fig. 2-11).

2.4.3 Cell Signaling and Integrated Controls

As we have discussed so far, there are four basic components in a living cell: functional structure, metabolism, energetic transfer, and information flow. The information flow can be divided into genetic flow and biochemical signaling. The integration of these two pathways results in the final control of the system. The genetic flow has been discussed in detail. This section will focus on the biochemical signaling.

For single-celled organisms to survive, cells must sense the changes in their environment and make adaptive responses constantly. These responses include a movement toward the nutrients or away from the toxin, changes of the metabolic patterns, and induction of certain protein expression. The *lac* operon in bacterial genome is an example of this type of control (Fig. 2-8a).

For a multicellular organism, cells not only need to adapt to their surrounding environment but also need to communicate with their neighboring cells and to adjust their behaviors and function to fit the whole system needs. The human body consists of about 50 trillion cells with 200 different cell types; the whole body control is carried out by a nerve system and an endocrine system conducted by cell signal transduction or cell signaling. In general, a cell signaling pathway consists of seven components: ligand, receptor, transducer, effector, second message, amplifier, and target (Fig. 2-12a). The ligands can be a hormone, neurotransmitter, growth factor, or even a gas (such as nitric oxide or NO). The concentration of the ligands is usually extremely low and it requires very specific binding to its receptor. There are various types of cellular receptors, including nuclear receptors and cell surface receptors. The nuclear receptors are proteins specific for steroid hormone binding. The cell surface receptors usually are integrated cell membrane proteins consisting of extracellular domain, transmembrane domain, and cytoplasmic domain. There are at least seven super-families of surface receptors binding different types of ligands to conduct different signaling pathways (Fig. 2-12b). The specificity of binding to the ligand is determined by the specific three dimensions of extracellular domain, while the cytoplasmic domain usually has kinase activity. The transducer transmits the signal from the membrane into the cells. The various G-proteins functioning as the transducer trigger the effecter enzymes to produce the second message. The second message is usually small molecules, such as cAMP, cGMP, Ca^{2+}, inositol 1,4,5-triphosphate (IP3), or 1,2-diacylglycerol (DAG). An amplifier is a series of protein kinases, such as protein kinases A, B, C, or tyrosine protein kinase. The signal is amplified by a cascade of kinase reactions. The target is either an enzyme, ion channel, or a gene. Signal transduction research is concerned with the mechanisms by which cells receive, interpret, integrate, and act upon information received. There are extensive interactions or cross talks between the pathways. The final output of cell signaling is the result of changes of cell metabolism, function, or gene expression.

There has been an explosion of information in recent years related to the signal transduction mechanisms whereby cell surface receptors transmit external stimuli, delivered in the form of hormonal or other environmental cues, to the intracellular response machinery in the cytoplasm and nucleus. Particularly, it is now widely recognized that signal transduction abnormalities involving the changes of gene

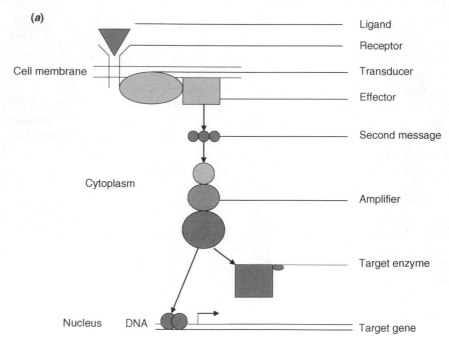

Figure 2-12 Cell signal transduction pathways. (a) A simplified model of a cell signal transduction pathway includes multiple components: ligand, receptor, transducer, effector, second message, amplifier, and targets. A ligand molecule binds to a receptor protein, thereby activating an intracellular signaling pathway that is mediated by a series of signaling proteins. Finally, one or more of these intracellular signaling proteins interacts with the target proteins, either in cytoplasm or nucleus, altering cell metabolism or gene expression. (b) Examples of signal transduction pathways in eukaryote. These pathways are important for cell metabolism and gene regulation. A detailed description can be found in the references.

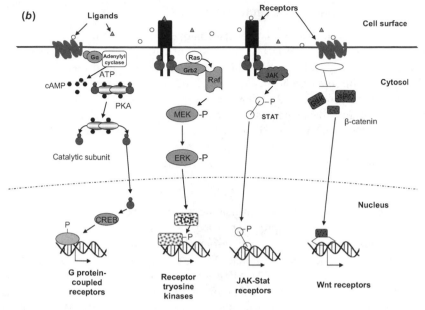

Figure 2-12 (*Continued*).

expression play important roles in several major human diseases. These findings have fueled a massive scientific effort aimed toward the identification and functional dissection of various signaling pathways. Furthermore, a substantial proportion of the current effort in modern drug discovery is founded on the premise that pharmacologic manipulation of signaling proteins will prove beneficial in the prevention and treatment of major human diseases.

2.5 GENETIC ENGINEERING OF LIVING CELLS

The term genetic engineering refers to the process of manipulating genes, usually outside the organism's normal reproductive process. It often involves the use of recombinant DNA technologies for the isolation, manipulation, and reintroduction of DNA into cells or model organisms and ultimately to express a protein. The goal is to introduce new characteristics such as increasing the yield of a crop species, introducing a novel trait, or producing a new protein or enzyme. The completions of the sequencing of the human genome, as well as the genomes of many agriculturally and scientifically important plants and animals, have significantly increased the opportunities for genetic engineering research. Expedient and inexpensive access to comprehensive genetic data has become a reality, with billions of sequenced nucleotides already online and annotated. Genetic engineering has become the gold standard in biotechnology research, and major research progress has been made using a wide variety of techniques.

2.5.1 Recombinant DNA Technology

Recombinant DNA technology is a set of techniques used for cutting apart and splicing together different pieces of DNA. The pieces of foreign DNA are introduced into another cell or organism and continue to produce their own coded proteins or substances within the new host cell. The cell becomes a factory for the production of these foreign proteins. The techniques for the introduction of foreign genes into bacteria were first developed in the early 1970s. In 1978, Herbert Boyer used recombinant DNA technology to produce recombinant human insulin, the first product of biotechnology, as we know it today. Although many methods are available, the basic procedure of making recombinant DNA involves the following steps (see Fig. 2-13 for an overview):

Isolating DNA. The first step in making recombinant DNA is to isolate donor and vector DNA. The procedure used for obtaining donor and vector DNA depends on the nature of the resource. The bulk of DNA extracted from the donor will be nuclear genomic DNA in eukaryotic cells or the main genomic DNA in prokaryotic cells; these types of DNA are generally the ones required for analysis. Bacterial plasmids are commonly used vectors, and these plasmids must be purified from the bacterial genomic DNA. DNA isolation is now simple with various commercial kits allowing a fast and chemically safe technique.

Cutting and Joining DNA. The cornerstone of recombinant DNA technology is a class of bacteria enzymes called restriction endonucleases. Restriction enzymes recognize a specific nucleotide sequence and cut both strands of the DNA within that sequence. To date, over 3000 restriction crosses have been identified and they can be found in the restriction enzyme database (REBASE, http://rebase. neb.com/rebase/). The first restriction enzyme named *EcoRI* was identified in *E. coli*. The DNA fragments produced by *EcoRI* digestion have overhanging single-stranded tails (called sticky ends) that reanneal with complementary single-stranded tails on other DNA fragments. If two pieces of DNA digested with same restriction enzyme are mixed under the proper conditions, DNA fragments from two sources form recombinant molecules by hydrogen bonding of their sticky ends. The enzyme DNA ligase covalently links these fragments to form recombinant DNA molecules.

Vectors or Plasmids. "Vector" is a carrier DNA molecule that can bring a foreign DNA fragment into a host cell. "Cloning vector" is used for reproducing the DNA fragment and "expression vector" is used for expressing certain genes in the DNA fragment. Commonly used vectors include plasmid, Lambda phage, cosmid, and yeast artificial chromosome (YAC). Vector DNA molecules are able to independently replicate themselves and the DNA segment they carry. They also contain a number of restriction enzyme cleavage sites that are present only once in the vector. One site is cleaved with a restriction enzyme and is used to insert a DNA segment cut with the same enzyme. Vectors usually carry a selectable marker such as antibiotic resistance genes or genes for

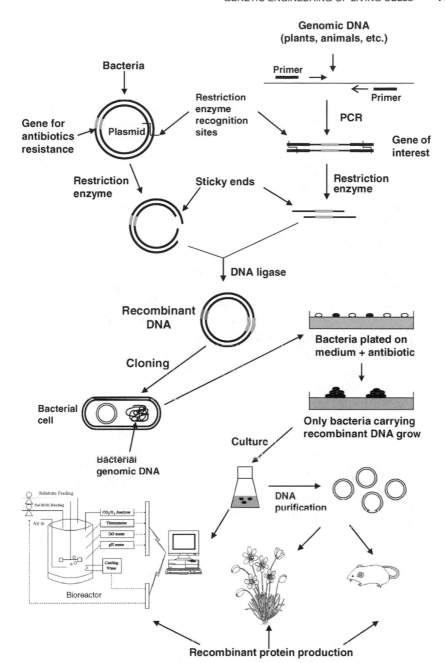

Figure 2-13 The recombinant DNA technology enables individual fragments of DNA from any genome to be inserted into vector DNA molecules such as plasmids and transformed into bacteria. Each of such recombinant DNA molecules can then be used for the production of foreign proteins in microorganisms and for creating transgenic plants and animals.

enzymes missing in the host cell. These markers can distinguish host cells that carry vectors from host cells that do not carry vectors. Many genetically engineered plasmid vectors are now available, and certain features make it easier to identify host cells carrying a plasmid with an inserted DNA fragment. Although only a single plasmid may enter a host cell, many plasmids can replicate themselves in the host cell so that several hundred copies are present. When used as vectors, such plasmids allow more copies of cloned DNA to be produced.

DNA Library. DNA library is a collection of many cloned DNA fragments. There are two types of DNA library. The genomic library is made of DNA fragments representing the entire genome of an organism. The cDNA library is generated from complementary DNA molecules synthesized from mRNA molecules in a cell. Therefore, the cDNA library contains only the coding region of a genome. To prepare a cDNA library, the first step is to isolate the total mRNA from the cell type of interest. Because eukaryotic mRNAs consist of a poly-A tail, they can easily be separated. A DNA strand complementary to each mRNA molecule is then synthesized by the enzyme called reverse transcriptase. After the single-stranded DNA molecules are converted into double-stranded DNA molecules by DNA polymerase, they are inserted into vectors and cloned.

Polymerase Chain Reaction. PCR is an enzyme reaction that targets a segment of DNA and then produces multiple amounts of the same segment; it is based on the ability of a DNA polymerase enzyme that can synthesize a complementary strand to a targeted segment of DNA. The PCR reaction mixture contains appropriate amounts of four deoxyribonucleotides and two short DNA oligo-nucleotides (each about 20 bases long), called primers, which have sequences complementary to areas adjacent to each side of the target sequence. If chosen well, the primer sequences will be unique in the entire genome and match only the place specifically chosen, thus limiting and defining the area to be copied. Figure 2-14 illustrates step-by-step the process of PCR reaction. Repeated heating and cooling cycles multiply the target DNA exponentially, since each new double strand separates to become two templates for further synthesis. Theoretically, the number of target sequences produced equals 2^n; that is, 20 PCR cycles can amplify the target by a million fold.

DNA Sequencing. The ability to sequence cloned recombinant DNA has greatly enhanced our understanding of gene structure, gene function, and the mechanisms of regulation. The most common method of DNA sequencing (Sanger sequencing) is based on dideoxy chain termination. In this procedure, a single-stranded DNA molecule whose sequence is to be determined is extended by DNA polymerase, similar to elongation during DNA replication. In addition, each tube contains a small amount of one of the four base-specific analogues called dideoxynucleotides (e.g., ddCTP). When ddCTP is incorporated into the extension, termination takes place at each of the C nucleotides in the newly synthesized DNA. A similar stop occurs with the other ddNTPs (Fig. 2-15). As the reaction proceeds, the tubes accumulate a series of DNA molecules that

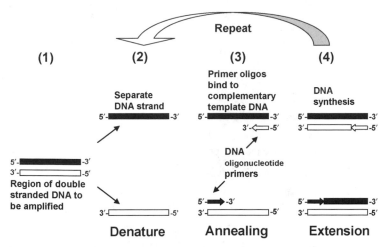

Figure 2-14 The polymerase chain reaction. (1) Double-stranded DNA containing the target sequence. (2) The strands are separated by heating at 94°C. (3) Two primers have sequence complementing primer binding sites at the 3′-ends of the target gene on the two strands. The DNA is allowed to cool to about 55°C, which allows the primers to stick to the single-stranded DNA at either end. (4) Taq polymerase then synthesizes the first set of complementary strands in the reaction. The final product is double-stranded DNA, which comes from the region defined by the primers. Steps 2–4 can be repeated to produce (in theory, if the process is 100 percent efficient) 2^n times the amount of template DNA (where $n =$ number of cycles).

differ in length by one nucleotide at their 3′-ends. The fragments from each reaction tube are separated in four adjacent lanes (one for each tube) by gel electrophoresis (Fig. 2-15). Electrophoresis separates DNA fragments in each lane that differ in size by a single nucleotide. The nucleotide sequence of the DNA can be read directly from bottom to top, corresponding to the 5′–3′ sequence of the DNA strand complementary to the template. The newly developed florescence labeling and multiple channel capillary electrophoresis technology have greatly enhanced the speed and capacity of sequencing. Most recently, massively parallel sequencing technologies developed by several companies such as 454 Life Science, Solxa, and Applied Biosystems have dramatically increased the sequencing throughput. Now it is possible to sequence a bacterial genome in several hours to several days with the automated sequencing machine.

2.5.2 Genetically Modified Microorganisms

Organisms containing introduced foreign DNA in their genome are referred to as being *transgenic*. The introduced foreign gene is called a *transgene*. Hence, bacteria containing eukaryotic gene are transgenic bacteria. A major use for many of these transgenic microorganisms is to produce proteins that have immense commercial value. Numerous studies have focused on finding ways to produce them efficiently and in a functional form. As illustrated in Figure 2-13, the gene encoding a foreign protein

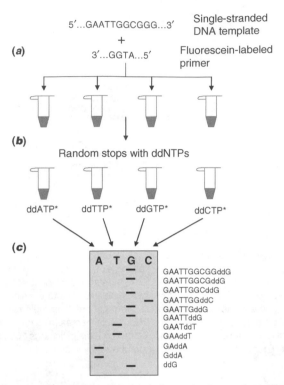

Figure 2-15 DNA sequencing with the dideoxy chain termination method. Single-stranded DNA template and single-stranded primer labeled with fluorescein are mixed and aliquoted into four tubes.(a) In the presence of DNA polymerase and a mixture of the four deoxynucleotides (dGTP, dATP, dTTP, and dCTP), primer extension occurs from the primer/template annealing site. (b) Random stops in extension are then generated by adding to each tube one of the dideoxynucleotides (ddNTP). The reaction results in a mixture containing variable lengths of extended DNA segments. (c) Finally each reaction mixture is separated electrophoretically in the gel track (or a capillary electrophoresis system) corresponding to the dideoxynucleotides added. As illustrated in the first lane, ddATP will produce two random stops where there is an adenine nucleotide and so two double-stranded DNA products are formed at corresponding sites. The remaining three dideoxynucleotides will do likewise in their individual reactions. The DNA sequence is read from bottom to top as illustrated in the figure.

can be cloned into an expression vector and transformed *E. coli* for protein synthesis under the control of a specific promoter, that is, *lac* promoter. Now various expression systems are available for overproducing foreign proteins in *E. coli*. The first human gene product manufactured using recombinant DNA and licensed for therapeutic use was human insulin, which is produced in *E. coli*. There are hundreds of therapeutic recombinant protein products in the market since then.

The use of genetically engineered microorganisms is a cost-effective, scalable technology for the production of recombinant proteins. However, the production of a functional protein is intimately related to the cellular machinery of the organism producing the protein. Posttranslational modifications usually performed in higher eukaryotes, for example, correct folding, disulphide bond formation, O- and N-linked glycosylation, and processing of signal sequences are absent in the bacterial expression system. Therefore, eukaryotic expression systems have been developed, such as yeast (*S. cerevisiae* and *Pichia pastoris*), the insect cell/baculovirus system, and the plant cell and mammalian cell culture systems.

2.5.3 Transgenic Plants

Because of their economic significance, plants have long been the subject of genetic engineering aimed at developing improved varieties. Progress is being made on several fronts to introduce new traits into plants using recombinant DNA technology. Since it is able to grow a whole plant from a single cell, researchers can engage in the genetic manipulation of the cell, let the cell develop into a completely mature plant, and examine the whole spectrum of physiological and growth effects of the genetic manipulation within a relatively short period of time. The most commonly used cloning vector for making transgenic plants is the "Ti" plasmid. This plasmid is carried by the bacterium known as *Agrobacterium tumefaciens*, which has the ability to infect plants. When these bacteria infect a plant cell, a 30,000 base pair segment of the Ti plasmid, called T DNA, separates from the plasmid and incorporates into the host cell genome (Fig. 2-16). Therefore, the Ti plasmid can be used to shuttle exogenous genes into host plant cells. Foreign genes such as bacterial, plant, or mammalian DNA engineered with plant regulatory elements can be inserted into the Ti plasmid and then be placed back into the *A. tumefaciens* cell. That cell can be put into plant cells either by the process of infection or by direct insertion. The foreign DNA (T DNA and the inserted gene) can be incorporated into the host plant genome and passed on to future generations of the plant. For the plant cell types that are not susceptible to *A. tumefaciens* transfection, naked DNA molecules can be delivered into the target cells by using other gene delivery methods such as microinjection, electroporation, and particle bombardment, which will be discussed subsequently. These developments, important in the commercial application of plant genetic engineering, render the valuable food crops of corn, rice, and wheat susceptible to a variety of manipulations by the techniques of recombinant DNA and biotechnology. In recent years, progress has been made to improve nutritional quality, increase insect, disease, and herbicide resistance, and salt tolerance.

Moreover, the production of foreign proteins in transgenic plants has become a viable alternative to conventional production systems such as microbial fermentation or mammalian cell culture. Transgenic plants, acting as bioreactors, can efficiently produce recombinant proteins in larger quantities than those produced using mammalian cell systems. Plant-derived proteins are particularly attractive, since they are free of human diseases and mammalian viral vectors. Large quantities of biomass can be easily grown in the field and may permit storage of material prior to processing. Thus, plants offer the potential for efficient, large-scale production of recombinant proteins with increased freedom from contaminating human pathogens. A wide variety of other therapeutic agents have been derived from plants, including antibodies,

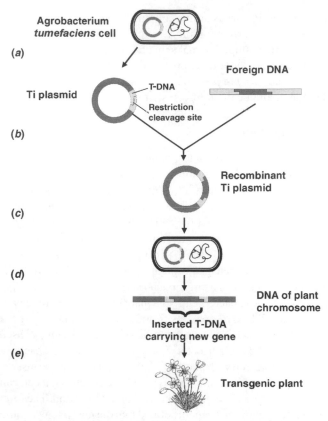

Figure 2-16 Generation of a transgenic plant through the growth of a cell transformed by T-DNA. (a) The Ti plasmid is isolated from *Agrobacterium* cells, (b) subjected to standard recombinant DNA procedures to insert the desired DNA into the T-DNA region of the plasmid, and then (c) put back into *Agrobacterium*. (d) Cultured plant cells are infected with bacteria containing the recombinant plasmid, and (e) these plant cells are then used to regenerate whole plant. The resulting transgenic plants contain the recombinant T-DNA region stably integrated into the genome of every cell. (modified with permission from Becker WM, Reece JB, and Poenie MF. *The World of the Cell*, 3rd ed. Copyright 1996 The Benjamin/Cummings Publishing Company).

vaccines, hormones, enzymes, interleukins, interferons (IFN), and human serum albumin (HSA).

2.5.4 Transgenic Animals

A transgenic animal is one that carries a foreign gene that has been inserted into its genome. The foreign gene is constructed using recombinant DNA methodology. Transgenic sheep and goats that express foreign proteins in their milk have been produced. Transgenic chickens are now able to synthesize human proteins in the egg whites. These animals should eventually prove to be valuable sources of proteins for therapeutic purpose.

Mice are typically the most important models for mammals. Much of the general technology developed in mice can be applied to humans and other animals. Transgenic mice are produced by microinjection of recombinant DNA into the pronucleus of a fertilized oocytes. Figure 2-17 illustrates step-by-step the generation of a transgenic mouse. First, the DNA molecule containing the gene of interest (e.g., the insulin gene) is constructed by using recombinant DNA methods. The vector DNA also contains promoter and enhancer sequences to enable the gene to be expressed by host cells. The recombinant DNA molecule is then transfected into the cultured embryonic stem (ES) cells and successfully transfected ES cells will be selected. Second, the transformed

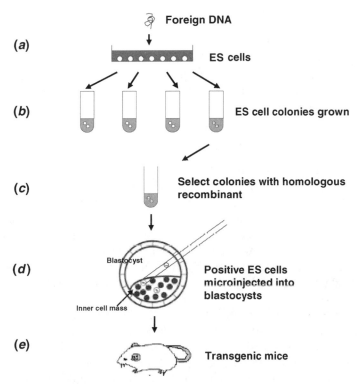

Figure 2-17 Generation of transgenic mouse carrying recombinant DNA.(a) Embryonic stem cells are transfected with foreign DNA. Many ES cells will take up the DNA, but this will involve different sites in the mouse genome because of random integration. In a very rare case, the integration will involve the correct part of the genome by a process of homologous recombination. (b) Colonies of ES cells are grown. (c) DNA is isolated from pools of colonies. The colony that has DNA integrated into the corrected position in the genome by homologous recombination can be identified by PCR. (d) ES cells that have the homologous recombined DNA are injected into mouse blastocysts. (e) If the transgene has also integrated into the germline, then some transgenic eggs or sperm will be produced and the next generation of mice will be fully transgenic—where every cell contains a copy of the foreign DNA (modified with permission from Trent RJ. *Molecular Medicine*, 3rd ed. Copyright 2005 Elsevier Academic Press).

ES cells are microinjected into an early embryo. The resulting progenies are chimeric, having tissue derived from either recipient or transplanted ES cells. The early vectors used for gene insertion could place the gene (from 1 to 200 copies of it) anywhere in the genome. However, if you know some of the DNA sequence flanking a particular gene, it is possible to design vectors that replace the gene. The replacement gene can be one that restores function in a mutant animal or knocks out the function of a particular locus.

All genes are associated with specialized DNA sequences such as promoters and enhancers that support transcription. Once in the nucleus, exogenous DNA requires a strong promoter and enhancer upstream of the transgene for its expression. The human cytomegalovirus (CMV) promoter has been used extensively to drive the expression of transgenes in mammalian cells. However, these viral promoters result in uncontrolled expression of transgenes. The uncontrolled expression of the transgene during embryonic development could also be lethal if it is not normally expressed at this time. Attempts controlling gene expression in transgenics involve the use of inducers, for example, chemical or hormonal signaling molecules. This was possible because the inserted gene had a promoter element that was inducible when exposed to the signal. There are also techniques with which transgenic mice can be made where a particular gene gets knocked out or introduced in only one type of cell.

2.5.5 New Tools for Gene Knockdown

The sequence-specific interaction of short nucleic acids with target RNAs or DNAs can be exploited as a tool for targeted inhibition of gene expression. These methods have been used in a wide variety of applications ranging from understanding the function of a given gene to molecular therapeutics. Although the mechanisms of these two methods differ, the problems for effective application are quite similar since both antisense oligonucleotides (AOs) and RNA interference are subject to artifacts such as off-target effects and CpG motif immune stimulation. It may be necessary to validate findings made using one method through use of a second approach.

2.5.5.1 Antisense Oligonucleotides   The evidence that small AOs could specifically inhibit gene expression was discovered in 1978 by Zamecnik and Stevenson, who demonstrated that viral replication could be blocked by treating infected cells with an AO that was complementary to a portion of the viral mRNA. Since then, there have been thousands of published reports describing the application of AOs both as research tools and as medicines. Synthetic 15–25mer nucleotides could enter living cells and might be tailored to target and elicit RNAse H-mediated cleavage of mRNA molecules, and subsequently lead to reduced production of a target protein. Using this technology, researchers could quickly determine specific gene functions in organisms at any stage of their life cycle. Although AOs have became a great research tool, some difficulties had been anticipated, such as finding ways to make AOs resistant to extracellular nucleases that would degrade them before they entered cells and devising efficient means of gene delivery. Other challenges were less anticipated, including finding a systematic way to predict which portion of a selected mRNA would make the best target. Still other hurdles came in the form of startling surprises. Foremost among them was the discovery that the DNA dinucleotide sequence C–G has immune stimulatory action that has nothing to do with the intended antisense effects.

2.5.5.2 RNA Interference   As we discussed earlier, a newly discovered group of small RNA (miRNA and siRNA) regulate gene expression at posttranscriptional level (Fig. 2-9). RNAi refers to the inhibition of gene expression in a sequence-specific fashion by double-stranded RNA. Although this method has only been in routine use in mammalian systems for 7–8 years, it is currently being successfully used in thousands of labs and ambitious whole genome screening projects. The most straightforward approach to RNAi is use of duplex RNAs made by chemical synthesis. Like AOs, RNAi occurs when oligonucleotides form base pairs with target sequences within specific mRNAs to silence them. Although the RNAi processes are complex, the end results are conceptually similar to what occurs when using AOs: The presence of an oligonucleotide inhibits a specific mRNA function, leading to a decrease in the amount of protein translated by that mRNA. In contrast to the antisense process using AOs, however, RNAi uses synthetic RNAs to mimic naturally occurring processes. These natural processes involve, for example, genomically encoded microRNAs for fine-tuning the regulation of gene expression and siRNAs for potentiating certain host–microbe interactions. Researchers have uncovered examples of siRNAs or microRNAs acting at transcriptional and posttranscriptional levels. As molecules that interfere with posttranscriptional (i.e., translational) events, the microRNAs may interact with their mRNA targets either destructively or reversibly, depending upon the biological circumstance. Each of these new RNAi functions inspires novel applications for research and treatment tools. In addition, the fact that applied RNAi mimics real cellular processes has led some researchers to argue that harnessing RNAi for therapeutic uses has a much greater likelihood of succeeding than an unnatural process such as AO-mediated mRNA silencing. The level of enthusiasm for RNAi has been extraordinary. In 2002, *Science* proclaimed RNAi to be the "breakthrough of the year"—and that was just an early milestone.

2.5.6 Gene Delivery

Both viral and nonviral vectors can be used for delivering a gene into the cell nucleus. These vectors carry exogenous DNA or RNA (including siRNA and AOs) across the cell membrane into the nucleus to allow transcription. Since the nucleases in the endolysosomes and cytoplasm actively degrade free nucleic acid, the vectors have to escape from the degradation process before unpackaging their genetic materials and inserting into the host cell genome. The vectors also need to cross the plasma membrane into the nucleus of target cells, either by passive diffusion through the nuclear pore complex or by an energy-dependent translocation that requires a group of proteins called importins. Therefore, for a successful gene delivery, a number of cellular barriers must be overcome.

Viruses use multiple mechanisms to infect their host cells efficiently, either by fusion with the cell membrane or by receptor-mediated endocytosis, followed by nuclear localization of the viral genome. As a result, viral vectors are able to mediate gene transfer with higher efficiency and possibility of long-term gene expression. Most modern viral vectors are unable to replicate freely, owing to the deletion of essential genes, and carry little risk of proliferation or reversion to wild type. Their main drawback is that they tend to be immunogenetic and cause the acute immune response, which limits their *in vivo* potential. In addition, viral vectors can alter cellular function after transduction. The limitation in the size of the transgene that recombinant viruses can carry and issues related to the production of viral vectors present additional practical challenges.

Approaches of nonviral gene delivery have also been explored using physical and chemical approaches. Physical approaches, including needle injection, electroporation, gene gun, ultrasound, and hydrodynamic delivery, employ a physical force that permeates the cell membrane and facilitates intracellular gene transfer. For instance, electroporation uses brief pulses of high-voltage electricity to induce the formation of transient pores in the membrane of the host cell. Such pores appear to act as passage ways through which the naked DNA can enter the host cell; microinjection involves the direct injection of material into a host cell using a finely drawn micropipette needle. In addition, particle bombardment actually shoots DNA-coated microscopic pellets through a plant cell wall and delivers the foreign DNA into the host cell. The chemical approaches use synthetic or naturally occurring compounds as carriers to deliver the transgene into cells. For instance, the cell and nuclear membranes can be made more permeable to DNA following coprecipitation of DNA with calcium phosphate and DNA can also be packaged into cationic liposomes that are able to cross the cell membrane. Nonviral vectors are less able than their viral counterparts to overcome the problems of binding, escape from endosomes, uncoating, and transport into the nucleus. Thus, they possess lower transfection efficiency. As a result, new strategies to improve internalization and endosomal escape of nonviral vectors are being developed.

2.6 METABOLIC ENGINEERING OF LIVING CELLS

Metabolic engineering is a powerful approach to the understanding and utilization of metabolic processes and has become a new paradigm for the improvement of cellular properties or metabolite production. As the name implies, metabolic engineering emphasizes the targeted and purposeful modification of metabolic pathways found in an organism. Built largely on the theoretical and computational analysis of a biosystem, the field has embraced a growing number of genome-scale experimental tools. The rapid expansion of genomics information across various species and the integration of system biology approach have transformed our ability to carry out metabolic engineering approaches. As such, we are on the cusp of a new age of metabolic engineering involving many applications that address the new challenges in the 21th century including energy, pollution, global warming, food and human health.

2.6.1   Principle of Metabolic Engineering

Metabolic engineering can be defined as directed modification of cellular metabolism and properties through the introduction, deletion, and modification of metabolic pathways by using recombinant DNA technologies. Much of this effort has focused on microbial organisms, but important work has been done in plants, insects, and animals. In a global sense, this is not different from what genetic engineers have been doing for years with phenotypic improvements resulting from the manipulation of genes directly involved in creating the product of interest. However, with metabolic engineering, the focus is placed on understanding the larger metabolic network inside the cell in a systematic fashion. Thousands of chemical and biological reactions occur in a typical cell, which serve a multitude of purpose critical for maintaining cellular physiology and fitness within its environment; this reinforces the need for a systematic approach to understand the cellular activities as a whole. As we discussed earlier, various known and unknown regulatory (i.e., transcriptional, translational, enzymatic, signal transduction) mechanisms exist in a cell to manage and direct the resource to process that optimize cellular fitness. Thus, changing pathways that do not improve fitness or even detract from the fitness within a population often lead to relatively small improvements in product formation despite large increase in specific enzymatic activities. Metabolic engineering approaches embrace techniques that fill the gaps between genetic engineering and classical strain improvement. Metabolic engineers place the emphasis on understanding the mechanistic features that genetic modifications confer, thereby adding knowledge that can be used for relational approaches while searching the metabolic landscape.

Metabolic engineering also employs concepts from reaction engineering and thermodynamics for the analysis of biochemical reaction pathways. Although it shares common fundamentals with traditional biochemical engineering, the focus has shifted away from equipment to analysis of cells as integral units. Another novel aspect of metabolic engineering is the emphasis it places on integrated metabolic pathways as opposed to individual reactions. As such, metabolic engineering is concerned with complete bioreaction networks and issues of pathway synthesis, thermodynamic feasibility, and pathway flux and flux control. An enhanced perspective of metabolism and cellular function can be obtained by considering reactions in their entirety rather than in isolation from one another; this is of central importance in understanding the metabolic network. The main issue is amplification and/or redistribution of pathway flux. This is very different from chemical plant scale-up issues. Instead of increasing the capacity of a processing plant by increasing the capacity of its units, one now attempts at increasing the capacity of a single cell by amplifying the activity of some key enzymes.

Various metabolic engineering strategies have been widely applied for the more efficient production of desired metabolites and biomolecules. Metabolic engineering is an iterative process: cycle of genetic modification–analysis of metabolic consequences of change (identifying limitations)–choice of next genetic modification (see Fig. 2-18 for a hypothetical example). Measurement requires the ability to assay a large part of the network and extract as much information about the effect of an imposed network perturbation as possible. Another important part of metabolic

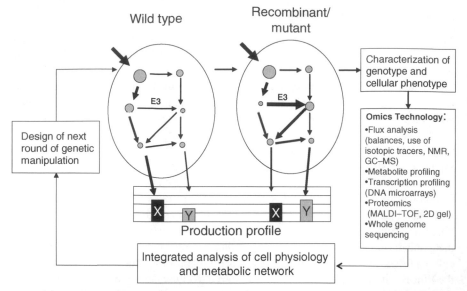

Figure 2-18 The iterative approach of metabolic engineering. Iterative perturbations and systematic phenotype and genotype characterizations yield system insight into the integration of high-throughput data sets. In this schematic, wild-type cells are engineered to overexpress enzyme E3 with the goal of increasing the low yield of product Y. However, because of network interactions, overexpression of E3 has a minimal effect on the accumulation rates of either products Y or X. To improve the yield of product Y, multiple steps in the network will have to be targeted and genetically modified. To identify these targets, various omics technologies will be used to generate an integrated profile of cellular networks in the recombinant or mutant strain. And then, comparative analyses of omics profiles are conducted to identify the target pathways for the overproduction of Y. Gene manipulation is carried out within the suggested candidate genes and characterizations of the new strain will be repeated. The gray circles indicate the pool size of metabolites in the network. Arrow thickness depicts relative flux magnitude of the corresponding reactions (modified with permission from Stephanopoulos G, et al. *Nature Biotechnology* 2004; 32:1262 Copyright 2004 Nature Publishing Group).

engineering is being able to perform the desired genetic perturbations efficiently. A variety of molecular engineering techniques are currently available to create gene deletion or overexpress genes of interest. However, it is essential for metabolic engineer to be able to precisely change the activities of certain enzymes in a desired pathway, as the desired change in activity may not be deletion (no activity) or overexpression driven by a strong promoter (order of magnitude change in activity). In some cases, a deletion is not possible as the enzyme is required for cell survival. Likewise, strong overexpression can result in deleterious outcomes such as the accumulation of toxic intermediates in a pathway.

2.6.2 Molecular and Computational Tools for Metabolic Engineering

Metabolic engineering relies upon methods that perturb the genome, measure fluxes, and analyze the state of the cell, such that the cell's network architecture can be elucidated and effective targets for genetic manipulation can be identified. Both experimental and computational tools have been developed for identification of targets of metabolic engineering.

2.6.2.1 Mathematical Tools for Analysis

Metabolic flux analysis (MFA) is based on a known biochemistry framework. A linearly independent metabolic matrix is constructed based on the law of mass conservation and on the pseudo-steady-state hypothesis on the intracellular metabolites. The accumulation rate of metabolites in the metabolic network may be expressed as

$$x_i = \sum a_j r_j \qquad (i = 1, 2, \ldots, m; \quad j = 1, 2, \ldots, n) \qquad (2\text{-}1)$$

where x_i is the accumulation rate of metabolite i, a_j is a stoichiometric coefficient, and r_j is the flux through reaction j. Equation 2-1 can be expressed in matrix form as

$$Ar = x \qquad (2\text{-}2)$$

where A is an $m \times n$ matrix of stoichiometric coefficients, r is an n-dimensional flux vector, and x is an m-dimensional metabolite accumulation rate vector. Typically, the system that results is an underdetermined system where $m > n$. However, under certain conditions, some pathways are inoperative and can be neglected. The system may become completely determined or overdetermined and can be solved along with the measurements of external or internal fluxes. Determining the fluxes often requires the measurements made by incorporating ^{13}C-labeling; as the labeled substrate proceeds through the metabolic network, the pools of metabolites that are downstream from the substrate become labeled. At the steady state, the fraction of labeled substrate in a given pool can be used to calculate the flux through the pathway. Noninvasive methods of analysis such as nuclear magnetic resonance can also provide information on the structure of the biochemical network as well as flux measurements.

Metabolic control theory (MCT) was independently developed by Kacser and Burns and Heinrich et al. to identify the kinetic constraints in a biochemical network. With MCT, the control structures are quantified as mathematical formulation based on the so-called elasticity coefficients and control coefficients. Especially useful are the flux control coefficients (FCCs), which quantify the influence of the individual reaction rates (or enzyme activities) on the overall flux through the pathway. This approach is used to determine the rate-limiting reaction in a network. However, a single rate-limiting step may not exist and several steps may share the control of the metabolic network. The FCC of an enzyme is defined as the relative effect of modulating the amount of an enzyme on the flux through the desired pathway. Equation 2-3 shows the flux control coefficient C_i^J of an enzyme E_i on the flux J.

$$C_i^J = \frac{dJ}{dE_i} \left(\frac{E_i}{J} \right) \qquad (2\text{-}3)$$

The FCC is essentially a sensitivity coefficient of the flux with respect to various enzymes. An important property of the FCC is that summation of all the FCCs affecting a particular flux must equal unity (Eq. 2-4).

$$\sum_i C_i^J = 1 \tag{2-4}$$

An FCC that approached unity would imply a rate-limiting step. MCT is a powerful tool for qualitative studies of metabolic pathways. A serious drawback of the method is, however, the requirement that either the elasticity coefficients or the control coefficients have to be measured. This is not an easy task because it requires independent variation of the activity *in vivo* of all the enzymes within the pathway.

2.6.2.2 Metabolic Profiling and Metabolomics
Metabolomics is the "systematic study of the unique chemical fingerprints that specific cellular processes leave behind"—specifically, the study of their small-molecule metabolite profiles. Small-molecule metabolites are critical in regulating transcriptional and translational processes and measuring the abundance of small metabolites provides a broad glimpse of the metabolic cellular state. Small molecules possess a wider range of chemical characteristics and are more difficult to measure. Complex profiles obtained using techniques such as nuclear magnetic resonance, liquid chromatography–mass spectrometry (LC–MS) and electrochemical array (EC) are typically analyzed for differences and changes in patterns of small molecules. Statistical analyses, using pattern recognition software even prior to identification of specific metabolites, allow a rapid means of finding specific markers for disease, toxicity, or some other process. Metabolic profiling typically involves the generation of patterns of analysts, containing both known and unknown compounds, to differentiate one sample group from another using these statistical analysis and pattern recognition software tools. However, unlike previously mentioned isotopic-labeling methods, metabolic profiling does not attempt to establish the intracellular flux, making these experiments more convenient. Nevertheless, it may be that the metabolite profiles provide enough similar information such that, when combined with protein and transcript profiles, a fairly complete picture of the cell is obtained that can be used to solve some more complex systemic problems.

2.6.2.3 High-Throughput Methods of Gene Manipulation
The recent development in high-throughput methods of gene manipulation provides a way of rapidly screening for new targets of metabolic engineering. In the case of bacteria, the large libraries of knockout mutants can be generated quickly by using transposable elements and subsequently screened for improved physiological performance. High-efficiency transformations can generate libraries of as many as 10^9 genetic variants. Transposon mutagenesis also enables a high-throughput form of mutagenesis where there is only one mutation introduced per cell. In a similar manner for mammalian cells, large-scale screening techniques using genome-wide RNAi provide an

opportunity for metabolic engineering in human and animal cells. This technique can be used to ablate all specific genes in the genome of an organism. Bacteria library containing vectors expressing siRNA against all genes of several model organisms are now commercially available. This technique was found to be very efficient for the study of genes involved in cellular metabolism, confirming not only the several genes already known to be involved in this process but also finding several others not previously linked to the control of cellular metabolism in *Caenorhabditis elegans*.

2.6.3 Application of Metabolic Engineering

Metabolic engineering has many areas of applications. Introduction of new pathways enables us to use nature's diversity to meet human needs in a sustainable way. Strain improvement of microorganisms through metabolic engineering will improve productivity, lower costs, reduce environmental pollution, and generally improve results in a wide variety of industries and areas, including pharmaceuticals, chemical bioprocess, food technology, agriculture, environmental bioremediation, and biomass conversion. The prospective categories of chemicals to be produced by metabolically engineered microorganisms include nutriceuticals, fine chemicals, vitamins, preservatives, sweeteners, minerals, nutrisupplements, pharma intermediates, pharmaceuticals, amino acids, acidulents, cosmetics, and food ingredients.

Using metabolic engineering to redesign plants could potentially improve nutritional value of crops (e.g., essential amino acid supply for storage proteins, vitamin content, modifying lignin to enhance forage digestibility), create new industrial crops (e.g., modified fatty acid composition of seed triglycerides, pharmaceuticals, polyhydroxybutyrate synthesis), alter photosynthate partitioning to increase economic yield, enhance resistance to biotic and abiotic stresses such as infectious disease, and reduce undesired (toxic or unpalatable) metabolites. Furthermore, the potential exists to produce therapeutic proteins (i.e., antibody) in plants, which could eliminate the need for large-scale fermentation or cell culture facilities and could only require purification and formulation processes. There are many opportunities and challenges for metabolic engineers in this area, including increasing protein production, controlling glycosylation, and altering desirable metabolic pathway.

In addition to its application in industrial and agricultural biotechnology, metabolic engineering principles can also be applied to medical research and practice. For instance, flux measurements and metabolite profiling can be conducted on primary cells and/or body fluids isolated from patients' sample for investigating disease initiation, progression and treatment effect. These types of work may lead to the identification of surrogate markers for diagnosis or prognosis of certain diseases as well as molecular targets for new drug development. Metabolic profiling of urine or blood plasma samples can also be used to detect the physiological changes caused by toxic insult of a chemical or drug. Advances in this area promise to contribute to personalized medicine by incorporating patient-specific genetic and metabolite profiles.

2.7 SUMMARY

In this chapter, we first discussed the major types of biomolecules including nucleic acids, carbohydrate, proteins, and lipids; assembly of the biomolecules into subcellular level (organelles) and cell; and the information flow including genetic and epigenetic inheritance and biochemical signaling. We further discussed the major metabolic pathways of biomolecules, linked energy production (bioenergetics) and metabolism, and integrated controls of metabolism and gene expression. Finally, we discussed the application of molecular technology in industry, agriculture, and medicine. We highlighted the important new discoveries and developments in each field. We hope that it will provide a primer and a framework to understand systems biology. The postgenomic era presents new opportunities as new challenges to all fields related to biology. A variety of new technologies allow studying the basic biological processes in the central dogma at whole genome level. Comparative genomics, transcriptomics, proteomics, structural genomics, and metabolomics become active research subjects. However, how to integrate transcriptomics, proteomics, and metabolomics information to draw a complete picture of a cell or a given organism is one of the big challenges. Thus, systems biology emerged. It is likely that the use of postgenomic tools will allow identification of far more complicated functional interactions between protein–protein, protein–DNA, and even protein–metabolite. The incorporation of spectroscopic approaches for metabolic profiling and flux analysis combined with mathematical modeling will contribute to the development of rational metabolic engineering strategies and will lead to the development of new tools to assess temporal and subcellular changes in metabolite pools. New technologies for pathway engineering, including use of heterologous systems, directed enzyme evolution, engineering of transcription factors, and application of molecular/genetic techniques for controlling biosynthetic pathways will move the metabolic engineering approach to the next level. The accumulated new data will provide the basis for systems biology to interpret the function of a biological system at molecular level in the quantitative manner; this in turn will allow modification or redesigning of the biosystems for future applications.

ACKNOWLEDGMENTS

We thank Miss Angel Surdin for her expert editorial assistance and contributions to accomplishing this valuable book. We also thank Miss Elise C. Welsh for her help in preparation of the figures.

SUGGESTED READING

General

1. Alberts B, Johnson A, Lewis J, Raff M, Roberts K, Walter P. *Molecular Biology of the Cell*, 4th ed. Garland Science, Taylor & Francis Group, 2002.

2. Allis CD, Jenuwein T, Reinberg D, Caparros M. *Epigenetics.* Cold Spring Harbor Laboratory Press, 2007.

3. Cooper GM, Hausman RE. *The Cell: A Molecular Approach,* 4th ed. ASM Press, Sinauer Associates, Inc, 2007.

4. Lodish H, Berk A, Matsudaira P, Kaiser CA, Krieger M, Darnell J. *Molecular Cell Biology,* 5th ed. W. H. Freeman and Company, 2004.

5. Nelson DL, Cox MM. *Lehninger Principles of Biochemistry,* 4th ed. W. H. Freeman and Company, 2005.

Biomolecules in Living Cells

Nucleic Acids

1. Arnhein N, Erlich H. Polymerase chain reaction strategy. *Annu Rev Biochem* 1992;61:131–156.

2. Brenner S, Jacob F, Meselson M. An unstable intermediate carrying information from genes to ribosomes for protein synthesis. *Nature* 1961;190:576–581.

3. Brenner S. The end of the beginning. *Science* 2000;24:2173–2174.

4. Collins FS, Green ED, Guttmacher AE, Guyer MS. A vision for the future of genomics research. *Nature* 2003;422:835–847.

5. Dickerson RE. The DNA helix and how it is read. *Sci Am* 1983;249:94–111.

6. Douand JA, Cech TR. The chemical repertoire of natural ribozymes. *Nature* 2002;418:222–228.

7. International Human Genome Sequencing Consortium, Finishing the euchromatic sequence of the human genome. *Nature* 2004;431:931–945.

8. Lander ES, Linton LM, Birren B, Nusbaum C, Zody MC, Baldwin J, Devon K, Dewar K, Doyle M, FitzHugh W, et al. Initial sequencing and analysis of the human genome. *Nature* 2001;409:860–921.

9. Lehman IR. Discovery of DNA polymerase. *J Biol Chem* 2003;278:34733–34738.

10. Lehman IR. Wanderings of a DNA enzymologist: from DNA polymerase to viral latency. *Annu Rev Biochem* 2006;75:1–17.

11. Mello CC, Conte D Jr. Revealing the world of RNA interference. *Nature* 2004;431:338–342.

12. Moore PB. Structural motifs in RNA. *Annu Rev Biochem* 1999;68:287–300.

13. Novina CD, Sharp PA. The RNAi revolution. *Nature* 2004;430:161–164.

14. Wang JC, Superhelical DNA. *Trends Biochem Sci* 1980;5:219–221.

15. Watson JD, Crick FHC. Molecular structure of nucleic acids: a structure for deoxyribose nucleic acid. *Nature* 1953;171:737–738.

Proteins

1. Ahn NG, Shabb JB, Old WM, Resing KA. Achieving in-depth proteomics profiling by mass spectrometry. *ACS Chem Biol* 2007;2:39–52.

2. Alberts B. The cell as a collection of protein machines: preparing the next generation of molecular biologists. *Cell* 1998;92:291–294.

3. Berman HM. The past and future of structure databases. *Curr Opin Biotechnol* 1999;10:76–80.

4. Branden C, Tooze J. *Introduction to Protein Structure*, 2nd ed. New York: Garland Publishing, 1999.

5. Doolittle RF. The multiplicity of domains in proteins. *Annu Rev Biochem* 1995;64:287–314.

6. Dougherty DA. Unusual amino acids as probes of protein structure and function. *Curr Opin Chem Biol* 2000;4:645–652.

7. Gygi SP, Aebersold R. Mass spectrometry and proteomics. *Curr Opin Chem Biol* 2000;4:489–494.

8. Hazbun TR, Fields S. Networking proteins in yeast. *Proc Natl Acad Sci USA* 2001;98:4277–4278.

9. Jones S, Thornton JM. Principles of protein–protein interaction. *Proc Natl Acad Sci USA* 1996;93:13–20.

10. Khosla C, Harbury PB. Modular enzymes. *Nature* 2001;409:247–252.

11. Luque L, Leavitt SA, Freire E. The linkage between protein folding and functional cooperativity: two sides of the same coin? *Annu Rev Biophys Biomol Struct* 2002;31:235–256.

12. Marcotte EM, Pellegrini M,et al. Detecting protein function and protein–protein interactions from genome sequences. *Science* 1999;285:751–753.

13. Ponting CP, Russel RR. The natural history of protein domains. *Annu Rev Biophys Biomol Struct* 2002;31:45–71.

14. Resing KA, Ahn NG. Proteomics strategies for protein identification. *FEBS Lett* 2005;579:885–889.

15. Sali A, Kuriyan J. Challenges at the frontiers of structural biology. *Trends Cell Biol* 1999;9: M20–M24.

16. Sanger F. Sequences, sequences, and sequences. *Annu Rev Biochem* 1988;57:1–28.

17. Teichmann SA, Murzin AG, Chothia C. Determination of protein function, evolution and interactions by structural genomics. *Curr Opin Struct Biol* 2001;11:354–363.

18. Toyoshima C, Nakasako M, Nomura H, Ogawa H. Crystal structure of the calcium pump of sarcoplasmic reticulum at 2.6 Å resolution. *Nature* 2000;405:647–655.

19. Vale RD. The molecular motor toolbox for intracellular transport. *Cell* 2003;112:467–480.

20. Walsh C. Enabling the chemistry of life. *Nature* 2001;409:226–231.

Polysaccharides

1. Bertozzi CR, Kiessling LL. Chemical glycobiology. *Science* 2001;291:2357–2363.

2. Esko JD, Selleck SB. Order out of chaos: assembly of ligand binding sites in heparan sulfate. *Annu Rev Biochem* 2002;71:435–471.

3. Gabius H-J, Andre S, Kaltner H, Siebert HC. The sugar code: functional lectinomics. *Biochim Biophys Acta* 2002;1572:165–177.

4. Gahmberg CG, Tolvanen M. Why mammalian cell surface proteins are glycoproteins. *Trends Biochem Sci* 1996;21:308–311.

5. Helenius A, Aebi M. Intracellular functions of N-linked glycans. *Science* 2001;291:2364–2369.

6. Loris R. Principles of structures of animal and plant lectins. *Biochim Biophys Acta* 2002;1572:198–208.

7. Selleck S. Proteoglycans and pattern formation: sugar biochemistry meets developmental genetics. *Trends Genet* 2000;16:206–212.

8. Varki A, Cummings R, Esko J, Freeze H, Hart G, Marth J. *Essentials of Glycobiology.* Cold Spring Harbor, NY: Cold Spring Harbor Laboratory Press, 1999.

Lipids

1. Christie WW. *Lipid Analysis*, 3rd ed. Bridgwater, UK: The Oily Press, 2003.

2. Clouse SD. Brassinosteroid signal transduction: clarifying the pathway from ligand perception to gene expression. *Mol Cell* 2002;10:973–982.

3. Dowhan W. Molecular basis for membrane phospholipid: why are there so many lipids? *Annu Rev Biochem* 1997;66:199–232.

4. Lemmon MA, Ferguson KM. Signal-dependent membrane targeting by pleckstrin homology (PH) domains. *Biochem J* 2000;350:1–18.

5. Vance DE, Vance JE,editors. *Biochemistry of Lipids, Lipoproteins and Membranes*, New Comprehensive Biochemistry, Vol. 36. New York: Elsevier Science Publishing Co, Inc, 2002.

6. Weber H. Fatty acid-derived signals in plants. *Trends Plant Sci* 2002;7:217–224.

Microstructure and Function of Cells and Organelles

1. Altan-Bonnet N, Sougrat R, Liu W, Snapp EL, Ward T, Lippincott-Schwartz J. Golgi inheritance in mammalian cells is mediated through endoplasmic reticulum export activities. *Mol Biol Cell* 2006;17:990–1005.

2. Battye FL, Shortman K. Flow cytometry and cell-separation procedures. *Curr Opin Immunol* 1991;3:238–241.

3. Baumeister W, Steven AC. Macromolecular electron microscopy in the era of structural genomics. *Trends Biochem Sci* 2000;25:624–630.

4. Bray D. *Cell Movements: From Molecules to Motility.* Garland, 2001.

5. Chen XJ, Butow RA. The organization and inheritance of the mitochondrial genome. *Nat Rev Genet* 2005;6:815–825.

6. Cuervo AM, Dice JF. Lysosomes: a meeting point of proteins, chaperones, and proteases. *J Mol Med* 1998;76:6–12.

7. de Duve C. Exploring cells with a centrifuge. *Science* 1975;189:186–194.

8. Engelman DM. Membranes are more mosaic than fluid. *Nature* 2005;438:578–580.

9. Higgins CF, Linton KJ. Structural biology. The xyz of ABC transporters. *Science* 2001;293:1782–1784.

10. Howell KE, Devaney E, Gruenberg J. Subcellular fractionation of tissue culture cells. *Trends Biochem Sci* 1989;14:44–48.

11. Hruz PW, Mueckler MM. Structural analysis of the GLUT1 facilitative glucose transport. *Mol Membr Biol* 2001;18:183–193.

12. Lamond A, Earnshaw W. Structure and function in the nucleus. *Science* 1998;280:547–553.

13. Kaushik S, Kiffin R, Cuervo AM. Chaperone-mediated autophagy and aging: a novel regulatory role of lipids revealed. *Autophagy* 2007;3:387–389.

14. MacKenzie KR, Prestegard JH, Engelman DM. A transmembrane helix dimer: structure and implications. *Science* 1997;276:131–133.

15. Mackinnon R. Structural biology. Membrane protein insertion and stability. *Science* 2005;307:1425–1426.

16. Masters C, Crane D. Recent developments in peroxisome biology. *Endeavour* 1996;20:68–73.

17. Minor DL. Potassium channels: life in the post-structural world. *Curr Opin Struct Biol* 2001;11:408–414.

18. Mason WT. *Fluorescent and Luminescent Probes: Biological Activity*, 2nd ed. Academic Press, 1999.

19. Matsumoto B,editor. *Methods in Cell Biology: Biological Applications of Confocal Microscopy*, Vol. 70 G. Academic Press, 2002.

20. Sachs JN, Engelman DM. Introduction to the membrane protein reviews: the interplay of structure, dynamics, and environment in membrane protein function. *Annu Rev Biochem* 2006;75:707–712.

21. Schekman R, Peroxisomes: another branch of the secretory pathway? *Cell* 2005;122:1–2.

22. Schliwa M, Woehlke G. Molecular motors. *Nature* 2003;422:759–765.

23. Sluder G, Wolf D,editors. *Methods in Cell Biology: Video Microscopy*, Vol. 56. Academic Press, 1998.

24. Sprong H, van der Sluijs P, van Meer G. How proteins move lipids and lipids move proteins. *Nature Rev Mol Cell Biol* 2001;2:504–513.

25. Subramani S. Components involved in peroxisome import, biogenesis, proliferation, turnover, and movement. *Physiol Rev* 1998;78:171–188.

26. Taddei A. Active genes at the nuclear pore complex. *Curr Opin Cell Biol* 2007;19:305–310.

27. Tamm LK, Kiessling VK, Wagner ML. *Membrane Dynamics: Encyclopedia of Life Sciences*. Nature Publishing Group, 2001.

28. Vance DE, Vance JE, *Biochemistry of Lipids, Lipoproteins and Membranes*, 4th ed. Elsevier, 2002.

29. Wickner W, Schekman R. Protein translocation across biological membranes. *Science* 2005;310:1452–1456.

Metabolism and Regulation

1. Fell D. *Understanding the Control of Metabolism*. Portland Press, 1997.

2. van de Werve G, Lange A, Newgard C, Mechin MC, Li Y, Berteloot A. New lessons in the regulation of glucose metabolism taught by the glucose 6-phosphatase system. *Eur J Biochem* 2000;267:1533–1549.

3. Harris DA. *Bioenergetics at a Glance*. Blackwell Scientific, 1995.

4. Hurley JH, Dean AM, Sohl JL, Koshland DJ, Stroud RM. Regulation of an enzyme by phosphorylation at the active site. *Science* 1990;249:1012–1016.

5. Lu HP, Xun L, Xie XS. Single-molecule enzymatic dynamics. *Science* 1998;282:1877–1882.

6. Martin BR. *Metabolic Regulation: A Molecular Approach.* Blackwell Scientific, 1987.

7. Miziorko HM. Phosphoribulokinase: current perspectives on the structure/function basis for regulation and catalysis. *Adv Enzymol Relat Areas Mol Biol* 2000;74:95–127.

8. Moser CC, Keske JM, Warncke K, Farid RS, Dutton PL, Nature of biological electron transfer. *Nature* 1992;355:796–802.

9. Page MI, Williams A,editors. *Enzyme Mechanisms.* Royal Society of Chemistry, 1987.

10. Schilling CH, Letscher D, Palsson BO. Theory for the systemic definition of metabolic pathways and their use in interpreting metabolic function from a pathway-oriented perspective. *J Theor Biol* 2000;203:229–248.

Genetic Inheritance

DNA Replication

1. Bullock PA. The initiation of simian virus 40 DNA replication *in vitro. Crit Rev Biochem Mol Biol* 1997;32:503–568.

2. Cvetic C, Walter JC. Eukaryotic origins of DNA replication: could you please be more specific? *Semin Cell Dev Biol* 2005;16:343–353.

3. Kornberg A, Baker TA. *DNA Replication*, 3rd ed. University Science Books, 2005.

4. Kunkel TA. DNA replication fidelity. *J Biol Chem* 2004;279:16895–16898.

5. Pursell ZF, Isoz I, Lundstrom EB, Johansson E, Kunkel TA. Yeast DNA polymerase epsilon participates in leading-strand DNA replication. *Science* 2007;317:127–130.

6. Waga S, Stillman B. The DNA replication fork in eukaryotic cells. *Annu Rev Biochem* 1998;67:721–751.

Transcriptional Control

1. Bell CE, Lewis M. The *lac* repressor: a second generation of structural and functional studies. *Curr Opin Struct Biol* 2001;11:19–25.

2. Blackwood EM, Kadonaga JT. Going the distance: a current view of enhancer action. *Science* 1998;281:60 63.

3. Boeger H, Bushnell DA, Davis R, Griesenbeck J, Lorch Y, Strattan JS, Westover KD, Kornberg RD. Structural basis of eukaryotic gene transcription. *FEBS Lett* 2005;579:899–903.

4. Borukhov S, Lee J, Laptenko O. Bacterial transcription elongation factors: new insights into molecular mechanism of action. *Mol Microbiol* 2005;55:1315–1324.

5. Brasset E, Vaury C. Insulators are fundamental components of the eukaryotic genomes. *Heredity* 2005;94:571–576.

6. Bushnell DA, Westover KD, Davis RE, Kornberg RD. Structural basis of transcription: an RNA polymerase II-TFIIB cocrystal at 4.5 Angstroms. *Science* 2004;303:983–988.

7. Butler JE, Kadonaga JT. The RNA polymerase core promoter: a key component in the regulation of gene expression. *Genes Dev* 2002;16:2583–2592.

8. Cramer P. Multisubunit RNA polymerases. *Curr Opin Struct Biol* 2002;12:89–97.

9. Courey AJ, Jia S. Transcriptional repression: the long and the short of it. *Genes Dev* 2001;15:2786–2796.

10. Darst SA. Bacterial RNA polymerase. *Curr Opin Struct Biol* 2001;11:155–162.

11. Green MR. Eukaryotic transcription activation: right on target. *Mol Cell* 2005;18:399–402.

12. Harbison CT, et al. Transcriptional regulatory code of a eukaryotic genome. *Nature* 2004;431:99–104.

13. Isogai Y, Takada S, Tjian R, Keles S. Novel TRF1/BRF target genes revealed by genome-wide analysis of *Drosophila* Pol III transcription. *EMBO J* 2007;26:79–89.

14. Kim DK,ct al. The regulation of elongation by eukaryotic RNA polymerase II: a recent view. *Mol Cell* 2001;11:267–274.

15. Levine M, Tjian R. Transcription regulation and animal diversity. *Nature* 2003;424:147–151.

16. Luscombe NM,et al. An overview of the structures of protein–DNA complexes. *Genome Biol* 2000;1:1–37.

17. Marr MT, Isogai Y, Wright KJ, Tjian R. Coactivator cross-talk specifies transcriptional output. *Genes Dev* 2006;20:1458–1469.

18. Nirenberg M. Historical review: deciphering the genetic code—a personal account. *Trends Biochem Sci* 2004;29:46–54.

19. Roeder RG. Transcriptional regulation and the role of diverse coactivators in animal cells. *FEBS Lett* 2005;579:909–915.

20. Tupler R, Perini G, Green MR. Expressing the human genome. *Nature* 2001;409:832–833.

21. Woychik NA, Hampsey M. The RNA polymerase II machinery: structure illuminates function. *Cell* 2002;108:453–463.

22. Young BA, Gruber TM, Gross CA. Views of transcription initiation. *Cell* 2002;109:417–420.

Translational and Posttranslational Controls

1. Alexander RW, Schimmel P. Domain–domain communication in aminoacyl-tRNA synthetases. *Prog Nucleic Acid Res Mol Biol* 2001;69:317–349.

2. Garrett RA, et al., editors. *The Ribosome: Structure, Function, Antibiotics, and Cellular Interactions.* ASM Press, 2000.

3. Green R. Ribosomal translocation: EF–G turns the crank. *Curr Biol* 2000;10: R369–R373.

4. Holcik M, Pestova TV. Translation mechanism and regulation: old players, new concepts. Meeting on Translational Control and Non-Coding RNA. *EMBO J* 2007;8:639–643.

5. Horan LH, Noller HF. Intersubunit movement is required for ribosomal translocation. *Proc Natl Acad Sci USA* 2007;104:4881–4885.

6. Maguire BA, Zimmermann RA. The ribosome in focus. *Cell* 2001;104:813–816.

7. Nirenberg M, et al. The RNA code in protein synthesis. *Cold Spring Harbor Symp Quant Biol* 1966;31:11–24.

8. Pestova TV, et al. Molecular mechanisms of translation initiation in eukaryotes. *Proc Natl Acad Sci USA* 2001;98:7029–7036.

9. Pickart CM, Cohen RE, Proteasomes and their kin: proteases in the machine age. *Nat Rev Mol Cell Biol* 2004;5:177–187.

10. Ramakrishnan V. Ribosome structure and the mechanism of translation. *Cell* 2002;108:557–572.

11. Sonenberg N, Hershey JWB, Mathews MB,editors. *Translational Control of Gene Expression*. Cold Spring Harbor Laboratory Press, 2000.

12. Welchman RL, Gordon C, Mayer RJ. Ubiquitin and ubiquitin-like proteins as multifunctional signals. *Nat Rev Mol Cell Biol* 2005;6:599–609.

Epigenetic Inheritance

1. Berger SL. Histone modifications in transcriptional regulation. *Curr Opin Genet Dev* 2002;8:142–148.

2. Bestor TH. The DNA methyltransferases of mammals. *Hum Mol Genet* 2000;9: 2395–2402.

3. Bernstein BE, Meissner A, Lander ES. The mammalian epigenome. *Cell* 2007;128:669–681.

4. Bird A. The essentials of DNA methylation. *Cell* 1992;70:5–8.

5. Craig JM, Bickmore WA. The distribution of CpG islands in mammalian chromosomes. *Nat Genet* 1994;7:376–382.

6. Goldberg AD, Allis CD, Bernstein E. Epigenetics: a landscape takes shape. *Cell* 2007;128:635–638.

7. Jenuwein T, Allis CD. Translating the histone code. *Science* 2001;293:1074–1080.

8. Jones PA, Baylin SB. The fundamental role of epigenetic events in cancer. *Nat Rev Genet* 2002;3:415–428.

9. Murrell A, Rakyan VK, Beck S. From genome to epigenome. *Hum Mol Genet* 2005;15: R3–R10

10. Richards EJ, Elgin SC. Epigenetic codes for heterochromatin formation and silencing: rounding up the usual suspects. *Cell* 2002;108:489–500.

11. Robertson KD, Jones PA. DNA methylation: past, present and future directions. *Carcinogenesis* 2000;27:461–467.

Cell Signaling and Integrated Controls

1. Andres A, et al. Protein tyrosine phosphatases in the human genome. *Cell* 2004;117:699–711.

2. Bell J. Predicting disease using genomics. *Nature* 2004;429:453–456.

3. Cantley LC, Transcription: translocating tubby. *Science* 2001;292:2019–2021.

4. Cantley L. The phosphoinositide 3-kinase pathway. *Science* 2002;296:1655–1657.

5. Chang L, Karin M. Mammalian MAP kinase signaling cascades. *Nature* 2001;410:37–40.

6. Delmas P, Brown D. Junctional signaling microdomains: bridging the gap between neuronal cell surface and Ca_2^+ stores. *Neuron* 2002;36:787–790.

7. Dhillon AS, Hagan S, Rath O, Kolch W. MAP kinase signaling pathways in cancer. *Oncogene* 2007;26:3279–3290.

8. Kirschner MW. The meaning of systems biology. *Cell* 2005;121:503–504.

9. Leonard W. Role of Jak kinases and STATs in cytokine signal transduction. *Int J Hematol* 2001;73:271–277.

10. Lum L, Beachy PA. The Hedgehog response network: sensors, switches, and routers. *Science* 2004;304:1755–1759.

11. Luttrell LM, Daaka Y, Lefkowitz RJ. Regulation of tyrosine kinase cascades by G-protein-coupled receptors. *Curr Opin Cell Biol* 1999;11:177–183.

12. Ma'ayan et al. Formation of regulatory patterns during signal propagation in a mammalian cellular network. *Science* 2005;309:1078–1083.

13. Michel JJ, Scott JD. AKAP mediated signal transduction. *Annu Rev Pharmacol Toxicol* 2002;42:235–257.

14. Papin JA, Hunter T, Palsson BO, Subramaniam S. Reconstruction of cellular signaling networks and analysis of their properties. *Nat Rev Mol Cell Biol* 2005;6:99–111.

15. Pierce KL, Premont RT, Lefkowitz RJ. Seven-transmembrane receptors. *Nat Rev Mol Cell Biol* 2002;3:639–652.

16. Rual JF, et al. Towards a proteome-scale map of the human protein–protein interaction network. *Nature* 2005;437:1173–1178.

17. Simon M. Receptor tyrosine kinases: specific outcomes from general signals. *Cell* 2000;103:13–15.

18. Vo N, Goodman R. CREB-binding protein and p300 in transcriptional regulation. *J Biol Chem* 2001;276:13505–21351.

19. Vogelstein B, Kinzler KW. Cancer genes and the pathways they control. *Nat Med* 2004;10:789–799.

Genentic Engineering and Recombinant DNA

1. Achenbach TV, Brunner B, Heermeier K. Oligonucleotide-based knockdown technologies, antisense versus RNA interference. *ChemBioChem* 2003;4:928–935.

2. Arnheim N, Erlich H. Polymerase chain reaction strategy. *Annu Rev Biochem* 1992;61:131–156.

3. Capecchi MR. Generating mice with targeted mutations. *Nat Med* 2001;7:1086–1090.

4. Cohen S, Chang A, Boyer H, Helling R. Construction of biologically functional bacterial plasmids *in vitro*. *Proc Natl Acad Sci USA* 1973;70:3240–3244.

5. Felgner PL, Gadek TR, Holm M, Roman R, Chan HW, Wenz M, et al. Lipofection: a highly efficient, lipid-mediated DNA-transfection procedure. *Proc Natl Acad Sci USA* 1987;84:7413–7417.

6. Freshney RI. *Culture of Animal Cells: A Manual of Basic Technique*, 4th ed. New York: Wiley, 2000.

7. Golgstein DA, Thomas JA. Biopharmaceuticals derived from genetically modified plants. *Q J Med* 2004;97:705–716.

8. Hansen G, Wright MS. Recent advances in the transformation of plants. *Trends Plant Sci* 1999;4:226–231.

9. Hunkapiller T, Kaiser RJ, Koop BK, Hood L. Large-scale and automated DNA sequence determination. *Science* 254:59–67.

10. Lander ES, Linton LM, Birren B, Nusbaum C, Zody MC, Baldwin J, et al. Initial sequencing and analysis of the human genome. *Nature* 2001;409:860–921.

11. Maniatis T, et al. The isolation of structural genes from libraries of eucaryotic DNA. *Cell* 1978;15:687–701.

12. Sanger F, Nickle SN, Coulson AR. DNA sequencing with chain-terminating inhibitors. *Proc Natl Acad Sci USA* 1977;74:5463–5467.

13. Scherer LJ, Rossi JJ. Approaches for the sequence-specific knockdown of mRNA. *Nat Biotech* 2003;21:1457–1465.

14. Venter JC, Adams MA, Myers EW, et al. The sequence of the human genome. *Science* 2000;291:1304–1351.

15. Weigel D, Glazebrook J. *Arabidopsis: A Laboratory Manual.* Cold Spring Harbor, NY: Cold Spring Harbor Laboratory Press, 2001.

16. Yang NS, Sun WH. Gene gun and other non-viral approaches for cancer gene therapy. *Nat Med* 1995;1:481–483.

Metabolic Engineering

1. Askenazi M, et al. Integrating transcriptional and metabolite profiles to direct the engineering of lovastatin-producing fungal strains. *Nat Biotechnol* 2003;21:150–156.

2. Bailey JE. Toward a science of metabolic engineering. *Science* 1991;252:1668–1675.

3. Bailey JE. Lessons from metabolic engineering for functional genomics and drug discovery. *Nat Biotechnol* 1999;17:616–618.

4. Bailey JE. Complex biology with no parameters. *Nat Biotechnol* 2001;19:503–504.

5. Bailey JE. Reflections on the scope and the future of metabolic engineering and its connections to functional genomics and drug discovery. *Metab Eng* 2001;3:111–114.

6. Cullen LM, Arndt GM. Genome-wide screening for gene function using RNAi in mammalian cells. *Immunol Cell Biol* 2005;83:217–223.

7. Edwards JS, Palsson B. The *Escherichia coli* MG1655 *in silico* metabolic genotype: its definition, characteristics, and capabilities. *Proc Natl Acad Sci USA* 2000;97:5528–5533.

8. Fussenegger M, Betenbaugh MJ. Metabolic engineering II. Eukaryotic systems. *Biotechnol Bioeng* 2002;79:509–531.

9. Ideker T, et al. Integrated genomic and proteomic analyses of a systematically perturbed metabolic network. *Science* 2001;292:929–934.

10. Kacser H, Burns JA. The control of flux. *Symp Soc Exp Biol* 1973;27:65–104.

11. Klapa MI, Aon JC, Stephanopoulos G. Systematic quantification of complex metabolic flux networks using stable isotopes and mass spectrometry. *Eur J Biochem* 2003;270:3525–3542.

12. Pandey A, Mann M. Proteomics to study genes and genomes. *Nature* 2000;405:837–846.

13. Park SM, Shaw-Reid CA, Sinskey AJ, Stephanopoulos G. Elucidation of anaplerotic pathways in *Corynebacterium glutamicum* via ^{13}C-NMR spectroscopy and GC–MS. *Appl Microbiol Biotechnol* 1997;47:430–440.

14. Raab RM, Tyo K, Stephanopoulos G. Metabolic engineering. *Adv Biochem Eng Biotechnol* 2005;100:1–17.

15. Segre D, Vitkup D, Church GM. Analysis of optimality in natural and perturbed metabolic networks. *Proc Natl Acad Sci USA* 2002;99:15112–15117.

16. Stephanopoulos G, Vallino JJ. Network rigidity and metabolic engineering in metabolite overproduction. *Science* 1991;252:1675–1681.

17. Stephanopoulos G, Aristidou A, Nielsen J. *Metabolic Engineering: Principles and Methodologies.* San Diego, CA: Academic Press, 1998.

18. Stephanopoulos G, Alper H, Moxley J. Exploiting biological complexity for strain improvement through systems biology. *Nat Biotechnol* 2004;22:1261–1267.

19. Wiechert W, Mollney M, Isermann N, Wurzel M, De Graaf AA. Bidirectional reaction steps in metabolic networks: III. Explicit solution and analysis of isotopomer labeling systems. *Biotechnol Bioeng* 1999;66:69–85.

3

HIGH-THROUGHPUT TECHNOLOGIES AND FUNCTIONAL GENOMICS

Xiu-Feng Wan[1] and Dorothea K. Thompson[2]

[1]*Department of Microbiology, Miami University, Oxford, Ohio 45056*

[2]*Department of Biological Sciences, Purdue University, West Lafayette, Indiana 47907*

3.1 INTRODUCTION

Traditional biological studies generally target the structure and function of a specific gene or protein. Generally, a specific hypothesis is generated for a specific biological problem and then tested by an experimental design (Fig. 3-1). In the 1990s, the first high-throughput technologies were invented for biological studies and included genome sequencing, proteomics, DNA chips, and protein chips. These technological advancements have created a new field of bioinformatics and computational biology. The combination of bioinformatics and high-throughput technologies has re-shaped traditional biological studies; through these technologies, biologists will be able to generate better biological hypotheses, and also streamline the traditional methods, which has proven to be much more efficient than traditional biological study (Fig. 3-1). As these technologies become increasingly more mature and economically feasible, more and more laboratories are using these methods. In this chapter, we briefly introduce four high-throughput technologies and then focus on the details of three technologies: genomic sequencing, proteomics, and DNA and protein chip technologies. We also illustrate chip technologies using two applications.

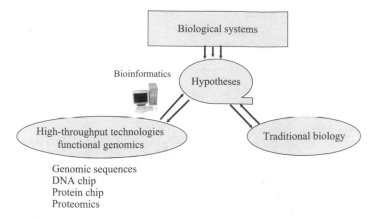

Figure 3-1 New scientific technologies for high-throughput measurements and functional genomics.

3.2 HIGH-THROUGHPUT TECHNOLOGIES

3.2.1 Genomic Sequencing

Genomic sequencing, which ultimately revolutionized the field of biology, was invented by Nobel Laureate Frederick Sanger in 1981 [1]. This technique involves the separation of fluorescently labeled DNA fragments according to length on polyacrylamide gels via electrophoresis (PAGE). Through automation, each sequencing run can yield 500 bp to 1 kb of sequence data with a modern sequence machine. DNA-sequencing technology is another milestone in understanding the evolution, structure, and function of biological systems since the discovery of the DNA structure by Watson and Crick in 1953. However, additional complementary technologies are required for sequencing complete genomes since genomic sequences may be as long as billions of bases (Table 3-1). For instance, the human genome is about 3.3 billion bases. A typical bacterial genome ranges from several hundred kilobases to more than 10 million bases. The model bacterium *Escherichia coli* K12, for example, has a genome comprising about 4.67 million bases.

In 1983, Frederick Sanger invented the shotgun-sequencing strategy and sequenced the first complete genome, bacteriophage λ, which has 48,502 bases [2]. Without parallel advances in super computational techniques, the applications of shotgun

Table 3-1 Wide ranges in genome size

Species	Genome Size (Mb)	Species	Genome Size (Mb)
HIV	0.0097	*S. cerevisiae*	11.72
SARS-CoV	0.030	*C. elegans*	~100
Mycoplasma genitalium	0.59	*A. thaliana*	~125
Escherichia coli	4.67	*Homo sapien*	~3,300

genome sequencing were not fully realized until 12 years later. In 1995, through the use of super computational facilities, the *Haemophilus influenzae* genomic sequence was published by J. Craig Venter from the Institute for Genomic Research (TIGR) and Nobel Laureate Hamilton Smith of Johns Hopkins University [3]. The human genome sequence project was launched in 1990 and was eventually completed in the year 2003 with a large collaboration of international effort by the International Sequencing Consortium (http://www.intlgenome.org/). It should be noted that the contribution of computational biology was crucial to the successful completion of the human genome sequence project.

Figure 3-2 shows a simplified procedure for the shotgun genome-sequencing strategy. To sequence a large sequence, shotgun genomic sequencing first breaks the sequence randomly several times into small fragments of about 1,500 bases by enzymes or physical shearing and then sequences these individual fragments. The computer is able to connect the sequences based on the overlapping ends between these sequences. The size of this large sequence will be less than 150 kb. This is because, for a large genome, the sequence needs to be separated into smaller fragments of about 150 kb, each of which will be cloned into bacterial artificial chromosome (BAC) vectors. Each of these large fragments is called a contig and can be sequenced using the shotgun-sequencing method. By using BACs, these contigs can be mapped, as BAC records the positions where the contigs come from the genomic sequence.

3.2.2 DNA Microarray

The DNA microarray, also known as a DNA chip, contains thousands of arrayed probes, each composed of a short oligonucleotide or cDNA fragment. The invention of DNA chip technology has made it possible to study the functions of thousands of genes at the same time, allowing for biological study in a more systematic way. The

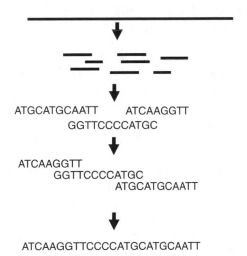

Figure 3-2 Simplified shotgun genome sequencing strategy.

fundamental mechanism underlying the DNA chip methodology is nucleotide hybridization, which had been previously deduced. However, the most important concept for DNA chips (as well as protein chips) is that these technologies facilitate the automation of evenly spotted DNA molecules onto a surface, which will allow for quantification of the hybridization signal. Thus, the first DNA arrays originated from the development in the late 1980s of robotic devices (gridding robots) that make it possible to array bacterial colonies in compact and regular patterns [4]. The original DNA chip had approximately 10,000 spots on a $22 \times 22 \, cm^2$ surface. This array allowed for rapid genomic library scanning. The functional genomics for expression analysis with quantitative acquisition of hybridization signals was first reported in 1992 [5]. This technology was based in part on integrated mapping and sequencing analysis of genomes.

The massive DNA sequences generated by shotgun sequencing have given us an opportunity as well as a challenge to study the evolution, structure, and function of these genes. Most notably, the complete genomic sequences allow us to evaluate expression patterns on a genomic scale. DNA chip technologies and functional genomics have been applied widely in many different fields, such as pathogenesis, drug discovery, cancer research, cell development, cell structure, agricultural seed selection, and even in the environmental community study [6–18]. For instance, in drug discovery, functional genomics can be applied in basic research and target discovery, biomarker determination, pharmacology, toxicogenomics, target selectivity, development of prognostic tests, and disease subclass determination [6]. Further details regarding DNA chips and functional genomics are discussed in Section 3.3.

3.2.3 Protein Microarrays

DNA microarrays are used to monitor global gene expression levels based on intracellular RNA concentration. However, the corresponding protein expression may be different from RNA abundance due to gene regulation at the translational level and alternative gene splicing. Protein chip technology was invented for this purpose. Different from DNA chips, protein chips have been used to detect the quantity of specific proteins by measuring signals from the interactions between protein versus protein and protein versus antibody. The target molecules can not only be traditional protein molecules but also be other types of molecules, such as artificial proteins [19], RNA or DNA aptamers [20], allosteric ribozymes [21], peptides, and other small molecules [22,23]. With these extensions, protein chips can be applied to monitor the interactions between protein versus ligand, protein versus drug, enzyme versus substrate, and so on.

Haab et al. [24] printed a set of 115 antibody–antigen pairs to evaluate the use of protein microarrays for specific detection and quantification of multiple proteins in complex mixtures. About 50 percent of the arrayed antigens and 20 percent of the arrayed antibodies provided specific and accurate measurements of their cognate ligands at or below concentrations of 0.34 and 1.6 µg/mL, respectively. Their studies suggest that protein microarrays can provide a practical means to characterize patterns of variation in hundreds of thousands of different proteins in clinical or research applications. Some companies have developed antibody arrays for both investigational usage as well as clinical use for monitoring allergies and small therapeutic drug monitoring. Similar to DNA chips, protein chips are able to perform thousands of reactions in parallel. Thus, by using a specific antibody, we will be able to screen for the presence of a specific protein from a specific reaction. Protein chips have become an important proteomics technology in addition to mass spectrometry and two-dimensional gel electrophoresis, both of which are however less sensitive than protein chip technology.

The original idea for protein chips followed from miniaturized immunoassay technology. In the 1980s, the development of ELISA introduced the concept of ambient analysis, which is able to quantify the antigen–antibody reaction through a specific enzyme-labeling assay. Similar to the DNA chip, development of this technology was accelerated by the genome project and improved technologies in recombinant proteins. Since most proteins used for protein arrays are made by recombination, the protein array would be able to be connected with DNA sequence and protein structural analysis. The functional analysis of the DNA-coding genes can reflect their functions.

Similar to the DNA chip, the protein chip uses covalent interactions to immobilize protein molecules onto solid surfaces by randomly conjugating the lysine residues on proteins to amine-reactive surfaces. In many cases, the recombination proteins are preferred since amino- or carboxy-terminal tags can be introduced so that the protein's functional sites can be away from the immobilization surface, which can increase the sensitivity of protein chips via the reduction of steric hindrance. The printing technologies for protein chips are similar to those for DNA chips, described in Section 3.3. However, the challenge for printing processes is how to prevent dehydration of the protein spots. Improvement in this area seems to be needed for further development [22].

Protein chips have become an important tool for biological study. Protein chips are mainly applied in micro-immunoassay, in which arrays of different capture antibodies are immobilized and subsequently exposed to a biological sample. These types of protein chips can be used for diagnostics as well as protein-profiling analysis. Specific antibodies can be immobilized on the chip to monitor the protein expression levels in a tissue or a cell. The parallel analyses would be able to monitor the protein-profiling changes for a patient and to determine the disease status or monitor the treatment or therapy through a minimum of biopsy material. In reverse immunoassays, the purified small antigens can be immobilized on the chip so that the specific antibody responses in the blood or local tissues can be evaluated. The reverse immunoassay can be used for diagnosis of various autoimmune diseases [25] or allergies [26]. These types of analyses can be used for examining binding receptor properties as well as antibody cross-reactivity and specificity.

The protein array has a very promising application for drug scanning since it directly monitors the interaction between drug and a target protein. Protein chips may be used in binding/screening assays for other small molecules, such as ligands, RNA–DNA molecules, and some artificial proteins. They can also be used for isolation of individual candidate molecules from a large pool. For instance, protein chips may be used for studying protein–DNA interactions, especially for promoter analysis, for investigating enzyme activity with different substrates, and for epitope mapping.

3.3 DNA CHIPS AND FUNCTIONAL GENOMICS

In this section, we first discuss the details about DNA chip manufacturing technologies, focusing primarily on how the probe is printed on the slides. Then we discuss the probe design, sample labeling and hybridization, scanning and image analysis, data analysis, experimental design and data interpretation, challenges of DNA chips, and applications of DNA chips.

3.3.1 Microarray Manufacturing Technologies

Current fabrication technologies for DNA microarrays can be grouped into photolithography, mechanical microspotting, or ink-jet ejection [27]. For a photolithography array, the oligonucleotide probe is synthesized directly onto a solid surface (e.g., the Affymetrix and NimbleGene arrays) based on a combination of chemistry and photolithographic methods [28]. To produce the array, the reactive amine groups from a silane reagent are attached to a glass or fused silica surface, and then the amine groups are modified via methylnitropoperonyloxycarbonyl (MeNPOC) photoprotection. A single base can then be added to the hydroxyl groups of these MeNPOC using a standard phosphoramidite DNA synthesis method after exposure to light. The photoprotection and nucleotide insertion are repeated to obtain a desired probe [27,29]. The lengths of these probes are generally 20–25 bases. The photolithographic array can have a much higher probe density than other types of arrays. Affymetrix chips can contain about 250,000 oligonucleotides in an area of 1 cm^2 while the spotted cDNA array generally only has about 1,000 oligonucleotides in the same area. This feature offers an important advantage for the Affymetrix array over spotted arrays, which have much lower probe densities. In addition, the Affymetrix system is more stable and reproducible, since it lacks the problems associated with printing spotted arrays. However, the current price of Affymetrix arrays is still too high to be widely accepted as a biological tool.

A mechanically microspotted array is called a spotted array. This type of array utilizes pins, tweezers, or capillaries to print the molecules onto glass or other solid surfaces. The molecules can be oligonucleotides, genomic DNA, or polymerase chain reaction (PCR) fragments (DNA or cDNA). For protein chips, we can even print antibodies, small drug molecules, and other small molecules. The printing process is generally achieved by a robot monitored by a computer. Compared with the array constructed on the basis of photolithography, a spotted array is more economical as well as easily implemented. In addition, the spotted array can have many more applications than the photolithography array since the array for the latter is limited to short oligonucleotides. However, preparation of the printing material and the printing process require considerable control, as the printing quality will directly affect the analysis.

Similar to a spotted array, the ink-jet ejection array prints the molecules to the solid surface by ejecting the sample from the print head. Different from the spotted array, the print head during printing does not contact the slides, which can reduce the probability of contamination. Currently, two types of noncontact ink-jet print technologies, piezoelectric pumps and syringe-solenoid, are used for printing microarrays. Similar to the spotted array, the ink-jet ejection array can print various molecules on a slide. On the other hand, the ink-jet ejection array prints at even lower densities than the spotted array.

3.3.2 Probe Design and DNA/cDNA Synthesis

Based on the DNA molecules on the slides, DNA chips can be categorized as either DNA/cDNA microarrays or oligonucleotide microarrays. Generally, the DNA fragment on DNA/cDNA microarrays is synthesized by the polymerase chain reaction while oligonucleotides are synthesized directly by machine. To synthesize the DNA fragment or cDNA primers, we need to design unique primers, which are generally 20–28 bases long. For genes shorter than 1,000 bases, the PCR-amplified fragments should be as long as possible. For genes longer than 1,000-bases, the optimal amplified fragments should be within the range of 500–1,200 bases. Xu et al. [30] developed PRIMEGENS for primer design for cDNA amplification. PRIMEGENS finds the unique fragments from a group of gene fragments or genes in a complete genome, and then applies the Primer3 algorithm [31] to design the left and right primers for each unique fragment. The user can change the primer specification based on their PCR requirement. The biggest challenge for production of the DNA or cDNA array is that occasionally PCR amplification may not be able to generate an expected yield for a given gene. Since we generally perform the reaction in 96- or 384-well plates, one may have to amplify individual genes separately. In addition, for complete genomic analysis, it is difficult to ensure complete coverage, due to the cross-hybridization between PCR fragments on the DNA/cDNA array. Furthermore, sample contamination or mishandling during amplification may generate other problems.

Due to the laborious processes for producing DNA/cDNA arrays, many laboratories are utilizing the oligo array. It should be emphasized that oligonucleotide design is not a trivial process. A program for designing optimal probes will need: (1) to minimize hybridization free energy for the target gene and maximize hybridization energy for all other genes, yet the hybridization energy depends on the concentration of the genes, which is unknown; (2) to avoid secondary structure; (3) to consider both strands of the genome as well as the cross-hybridization of the coding region and noncoding region. Many oligo design algorithms and software packages have been developed during the last several years: ProbeSelect [32], PROBEmer [33], CommonOligo [34], Oligo Design [35], Picky [36], OligoPicker [37], OligoArray [38], ROSO [39], and GoArrays [40]. Most of these methods are for a complete genome array. ProbeSelect [32] is one of the most popular methods used for oligo design for complete genome arrays. ProbeSelect first makes a suffix array of the coding sequences from a whole genome and then builds a sequence landscape for every gene based on the sequence suffix array. Based on sequence features and the sequence word rank values, ProbeSelect chooses probe candidates and then searches for matching sequences in the whole genome, allowing for a certain number of mismatches. After locating match sequence positions in all genes, ProbeSelect calculates the free energy and melting temperature for each valid target sequence. Finally, ProbeSelect matches sequences that have stable hybridization structures with a probe based on free-energy data and maintains high discrimination against other targets in the genome.

For environmental functional genomics, it will be even more challenging to design specific probes due to the high similarity between genes. Some algorithms have been designed for environmental community study [41,42]. The hierarchical probe design (HPD) program is an oligo design program especially suited for long oligo design, allowing for analyses of functional gene diversity in environmental samples [41]. HPD designs both sequence-specific probes and hierarchical cluster-specific probes from sequences of a conserved functional gene based on the clustering tree of the genes.

In general, DNA arrays have two advantages over oligo arrays: (1) DNA arrays have a higher sensitivity; and (2) DNA arrays do not need detailed sequence information; thus, DNA arrays are especially useful for environmental community study for which we generally do not know the exact sequence information. However, oligo arrays have two distinct advantages over DNA arrays: (1) oligo arrays have reduced cross-hybridization, thus providing higher specificity; and (2) unlike DNA arrays, oligo arrays do not require the intensive labor involved in PCR amplification and DNA purification. Oligo arrays are especially popular as the cost of custom array fabrication is steadily declining.

3.3.3 Sample Labeling and Hybridization

Sample labeling can be categorized as either direct or indirect labeling. The direct-labeling approach directly incorporates the fluorescent tags into the nucleic acid when preparing the hybridization samples. The fluorescent tags may be present in labeled nucleotides (e.g., Cy3- or Cy5-dCTP) or PCR primers. PCR and reverse transcription (RT)-PCR are common approaches to synthesize the labeled samples for hybridization. To detect the mRNA concentration, we can use RT-PCR to incorporate fluorescently labeled nucleotides into the transcribed cDNA during first-strand cDNA synthesis. Alternatively, mRNA can be amplified by 1,000–10,000-fold using T7 polymerase to obtain antisense mRNA (aRNA). The aRNA is then reverse-transcribed to obtain labeled cDNA [43]. One of the advantages of the T7 polymerase-based amplification method over other methods is that because amplification is a linear process, all mRNAs are amplified almost equally. Another advantage is that mRNA can be easily labeled with reverse transcriptase, which incorporates fluorescent tags much more readily than DNA polymerase [27].

The indirect labeling approach labels the sample with fluorescence after hybridization. To label the samples, indirect labeling requires epitope insertion into the target samples during cDNA synthesis. After hybridization, the epitopes can be bound by specific proteins to produce the signal. Biotin is one of the commonly used epitopes, which can be stained by a fluorescent streptavidin–phycoerythrin conjugate and detected via laser [44]. Some other types of indirect hybridizations are discussed by Zhou and Thompson [27].

After labeling, the sample will be hybridized with the probes on the slide. Before hybridization, the slide requires postprocessing, which will use ultraviolet (UV) radiation or heat to cross-link probes to the slide. For example, postprocessing can be done by exposing the slides to 120 mJ/cm^2 using a UV cross-linker or by baking the printed slides for 80 min at 80°C in a drying oven. Similar to traditional membrane-based hybridization, the microarray will also need prehybridization to reduce non-specific binding. The unbound DNA on the slides can be washed away during prehybridization to reduce the competition of unbound DNA for the labeled samples.

After postprocessing and prehybridization, the microarray is hybridized with labeled samples at a certain temperature, generally 42°C to 50°C, for a period of time (overnight to several days). A key to successful hybridization is that the hybridization solution needs to evenly cover the slide. After hybridization, the slides need to be washed to eliminate unbound samples.

3.3.4 Scanning and Image Analysis

The next step after hybridization is quantification of the hybridization signal from the slides. The scanning devices are generally categorized into two types: the confocal scanning microscope and CCD camera. In general, a confocal scanner uses laser excitation of a small region of the glass slide ($\sim$100 µm^2), and the entire array image is acquired by moving the glass slide, the confocal lens, or both across the slide in two directions [45]. The fluorescence emitted from the hybridized target molecule is gathered with an objective lens and converted to an electrical signal with a photo-multiplier tube (PMT) or an equivalent detector. The confocal scanning microscope is the most common one used to scan microarray slides. The main drawback of this type of technique is that this type of device may be very expensive since each excitation wavelength must have its own laser. In addition, the confocal scanning microscope is also very sensitive to any nonuniformity of the glass slide surface [27]. The CCD camera typically utilizes broadband xenon bulb technology and spectral filtration. The CCD system allows simultaneous acquisition of relatively large images of a slide (1 cm^2), thus, it does not require moving stages and optics. On the other hand, several images need to be captured from different areas and then combined to be representative of the complete information on the slide. Since most commonly used dyes have similar excitation and emission maxima, spectral filtration processes may have difficulty separating excitation and emission wavelengths, resulting in a possible source of error.

During the scanning process, the power of the excitation light is critical since the emitted fluorescence is generally correlated with the power of the excitation light. If the power of the excitation light is too low, the scanning sensitivity will be too long and many empty spots may be generated. However, if the power of the excitation light is too high, the incoming photons can damage the dyes and reduce the fluorescent signals during successive scans. More powerful light sources and/or longer laser exposure time can lead to significant photobleaching. Generally, photobleaching should be less than 1 percent per scan.

Since different dyes have different quantum yields and photostabilities, the PMT needs to be justified for each different channel prior to scanning. The order of channel scanning may be an additional variable to gain a better image. For example, Cy5 is more sensitive to photobleaching than Cy3. To minimize photobleaching, the Cy5 channel is always scanned first, followed by the Cy3 channel [27].

After the scanning process, we need to transform the image into quantitative signals. Many software packages, such as *Imagene, GPC VisualGrid, TIGR SpotFinder, GenePix*, have been developed to automatically quantify the images. Most of these software packages are effective. The common challenges for image quantification include (1) irregular or non-uniform spot geometries (e.g., not round, donut shape); (2) uneven hybridization (e.g., only a portion of the scanned image is quantifiable); (3) hybridization with high background; and/or (4) weak or saturated hybridization signals. Thus, for better quantification, one generally needs to use the following parameters: (1) signal/noise ratio should be more than $\sigma + 1.96\,\mu$; (2) background area selection should be local instead of global; and (3) bad spots should be removed.

3.3.5 Data Analysis

After obtaining the quantification hybridization signals from different biological replicates, we need to perform data normalization and statistical analysis.

3.3.5.1 Data Normalization The data normalization before statistical analysis is important to obtain reliable results. Data normalization can control many of the experimental sources of variability (systematic, not random or gene specific) and bring each image to the same average intensity. Data normalization is necessary to correct for the following variabilities: (1) the use of unequal quantities of starting RNA; (2) differences in dye incorporation; (3) differences in detection efficiencies of the fluorescent dyes; (4) variations in the image saturation extent for different channels; and (5) systematic biases in the measured expression levels.

Generally, there are several assumptions underlying data normalization. (1) The average mass of each molecule is approximately the same, thus the molecule number in each sample will be the same. (2) The arrayed elements represent a random sampling of the genes in the organism; and (3) the number of molecules from each sample available for hybridization is similar, thus, the total intensity for each sample will be the same.

Data normalization includes two steps: normalization within slides and normalization between slides. Normalization within slides is generally achieved by different options [46]. First, the signals can be scaled (scale normalization) by total intensity, mean, median, or the intensity of a group of genes. Second, normalization can be achieved by linear regression normalization. The most popular method for normalization with slides is the locally weighted linear regression (Lowess) normalization. Most normalization methods correct for differences in intensities between channels and do not take into account systematic bias that may appear within the data. For instance, the $\log_2$(ratio) values can have a systematic dependence on intensity. Lowess may remove the intensity-dependent effects in the

$\log_2$(ratio) values since Lowess normalizes the value point by point and generally requires a defined percent for the local area (e.g., 20 percent). Lowess normalization requires the ratio (two dyes). While normalization adjusts the mean of the $\log_2$(ratio) measurements, stochastic processes can cause the variance of the measured $\log_2$ (ratio) values to differ from one region of an array to another or between arrays. One approach to dealing with this problem is to adjust the $\log_2$(ratio) measures so that the variance is the same. This method is called variance regularization. Interested readers are encouraged to read an excellent review on data normalization by Quackenbush [46] for more details.

Since the hybridization may vary between slides (replicates) as well as channels, normalization is extremely important. Generally, normalization between slides uses scale normalization (e.g., medium).

3.3.5.2 *Statistical Analysis*   After normalization, we will be able to perform statistical analysis to rank results by confidence with significance metrics (e.g., p-value). The statistical analysis will estimate the false positive (Type I errors) and false negatives (Type II errors), achieve the desired balance of sensitivity and specificity, and result in a certain amount of flexibility (and arbitrariness) for interpreting significance metrics generated by a test.

The methods for statistical analysis depend on the experimental design. For example, for two sample statistical tests, we can utilize parametric statistical methods (t-test for paired and unpaired t-test) or nonparametric methods (Mann–Whitney test for independent samples or Wilcoxon signed-rank test for paired data). We generally assume the variations between biological replicates and technical replicates are the same to apply the two-sample statistical test. Otherwise, we can use multivariate statistics, such as one-way versus two-way analysis of variance (ANOVA) or the Kruskal–Wallis method. For multiple comparison corrections, we can use Bonferroni Correction or False Discovery Rate [47]. More details about these methods can be obtained from the book *Statistical Analysis of Gene Expression Microarray Data* by Terry Speed [48].

Many software packages, such as *GeneSpring* (Silicon Genetics), *SAM* (Stanford), and *ArrayStat* (Imaging Research), have been developed for microarray data analysis. GeneSpring is one of the most widely used microarray data analysis software tools since it has an easy-to-use interface as well as powerful normalization and statistical analysis capabilities (t-test, two-way ANOVA tests, one-way posthoc tests for reliably identifying differentially expressed genes, and so on). Different computational analysis tools for clustering, visual filtering, and pathway viewing have also been included. Also, the user can incorporate their own scripts/programming into GeneSpring to complete their analysis.

3.3.6 Experimental Design and Data Interpretation

Correct experimental design is the key to generation of meaningful biological results. A good experimental design will be more economic since it may save resources as well as slides. However, a corresponding statistical analysis should be proposed as well to

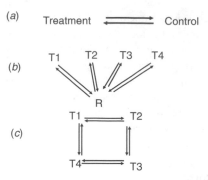

Figure 3-3 Experimental design for microarray. (a) Direct comparison. (b) Reference design. (c) Loop design.

analyze and interpret the data scientifically. Speed [48] provides a very good illustration for experimental design.

The simplest experimental design is pairwise direct comparison between treatment/experiment and control (Fig. 3-3a). For instance, we can compare the gene expression profiles for a wild type and a mutant under a certain condition to study the function of the mutated gene [49]; we can test the treatment effectiveness of a drug by comparing the gene expression profiles of the treatment group to the control group.

However, in most cases, we have to compare multiple experimental conditions. In this case, pairwise direct comparison will not meet the requirement. For example, we need to compare the gene expression profiles at different time points during bacterial growth [50]. In the drug experiments, we need to compare the effectiveness between different drugs. Obviously, it will not be wise to design all pairwise comparisons since it will be too expensive. For instance, to compare 10 conditions, one would have to design 45 pairwise experiments. In this case, we can apply common reference design (Fig. 3-3b).

More complicated designs include loop design (Fig. 3-3c) and pool design. The pool design should be very carefully used since it involves the mixture of all of the treatments and control samples as a reference sample to compare. The statistics with different experimental designs are described in the review by Yang and Terry [51].

After the statistical analysis, reconciliation between statistical results and biological functions is not a trivial matter since thousands of genes are involved in the data analysis. Generally, one overlays functional information and allows biological context to help decide what is of interest and what is not. We can use computational methods (classification, clustering, promoter prediction, and so on) to assist this analysis (Section 3.3.7). Microarray data are required to link to various public identifiers, such as Genbank, Swiss-Prot, and Gene Ontology (GO) database. GO is the most commonly used public domain sources of gene classification, and it provides controlled vocabulary hierarchies for molecular functions, biological processes, and cellular components. Other common databases include LocusLink, HomologGene, RefSeq, and UniGene.

3.3.7 Bioinformatics and Functional Genomics

The massive information that microarray profiling generates provides a great challenge for how to extract biologically meaningful information from the raw data. Thus, the discipline of bioinformatics plays an important role in microarray data analysis.

The most common approach is to deduce the coregulated genes (regulons) that have similar expression patterns. Further, the regulatory motifs are expected to be a predictor for each regulon. Many different algorithms have been used for the clustering process as well as for regulatory motif prediction. Within a single experimental condition, MotifRegressor [52] can be used to find a sequence motif. MotifRegressor first predicts all of the possible motifs and then performs regression analysis between microarray data and motif strength. MotifRegressor has an advantage in that it does not require the selection of a group of genes to predict the motif, which may generate a bias for motif prediction since some highly expressed genes are also indirectly regulated genes. For multiple experimental conditions, we can apply clustering methods, such as k-means, hierarchical clustering, self-organizing maps, and minimum spanning tree (EXCAVATOR) [53], to identify a group of potential genes with the same trend in expression pattern. EXCAVATOR [53] is based on a new framework for representing gene expression data, that is, the minimum spanning tree in graph theory. Through this data representation, an expression data-clustering problem is reduced to a tree-partitioning problem without losing information essential for the purpose of clustering. EXCAVATOR then applies an algorithm that mathematically guarantees to find globally optimal clustering efficiently, for a general objective function. After identifying the coexpressed genes in a cluster, we can apply motif prediction programs to predict the DNA binding motifs. The most commonly used *cis* regulatory motif and transcription factor DNA binding site prediction algorithms include such programs as Gibbs sampler [54], AlignACE [55,56], and BioProspector [57].

During the past several years, transcriptional regulatory networks have attracted substantial interest from both the computational and biological science communities. A number of statistical and computational methods have been applied in the modeling of gene regulation networks [58–64]. A few regulatory networks have been defined [60,65–68]. Despite this, regulatory network construction remains a great challenge due to the requirement of large experimental data sets.

The storage and management of microarray data is critical for efficient analysis. This, however, is a very challenging undertaking, since the many details of microarray analysis will affect the final results. The information about the samples hybridized, the hybridization images and their extracted data matrices, information about the physical array, and the features and reporter molecules all need to be included in the database. BioArray software environment (BASE) is a Web-based customizable bioinformatics solution for the management and analysis of all areas of microarray experimentation [69]. BASE manages biomaterial information, raw data and images, and provides integrated and "plug-in"-able normalization, data viewing and analysis tools. The organization and interface of BASE was designed to closely follow the natural workflow of the microarray biologist, and is compatible with most types of array platforms and data types (e.g., cDNA/oligos spotted on any substrate, Affymetrix, CGH on arrays, and so on).

3.3.8   Challenges of DNA Chips

Although DNA chips have advantages of high-throughput features, these technologies have several other disadvantages and challenges:

(1) Cost of diagnostic microarrays. Currently, the cheapest chips still cost the users at least $100 per experiment even for a noncustomed array. An Affymetrix array costs more than $400 per experiment. Currently, it is cost-prohibitive to apply microarrays as a routine diagnostic tool.

(2) The robustness of the microarray technologies must be improved. For SNP screening in particular, the sensitivity and specificity will need to be improved.

(3) The chip technologies need to be performed in a simplified and sturdy format without errors. A standard package includes the experimental protocol. These packages should tell the user how to justify the array quality in addition to giving the intensity of chip array data. A highly efficient quality control needs to be set up for microarray data analysis as well.

3.3.9   Development and Applications of DNA Chips

DNA chips have been widely used in many different fields. Most DNA chips focus on the protein-coding region to study the gene expression values. In addition, other types of arrays are designed to study the function of other elements in the genomes, such as small gene prediction, antisense gene study, gene alternative splicing, and so on.

The first significant application of DNA chips were serial analysis of gene expression (SAGE) for expression profiles [70]. SAGE was designed based on two principles: (1) a short nucleotide sequence tag can uniquely identify the transcript from an individual gene provided it is from a defined position within the transcript. For example, although the total number of human genes is expected to be of the order of 30,000, a sequence tag of only 9 nucleotides can, in principle, distinguish $4^9 = 262,144$ different transcripts. (2) Concatenation of short sequence tags allows the efficient analysis of transcripts in a serial manner. The tags from different transcripts can be covalently linked together within a single clone, and the clone can then be sequenced to identify the different tags in that clone. SAGE has been applied successfully in malarial parasite, yeast, plant, and animal systems [71].

ChIP-on-chip is a DNA array technique for isolation and identification of specific protein binding sites in genomic DNA [72]. ChIP-on-chip is useful for regulatory binding site identification, and thus, for regulatory network construction [73,74]. These regulatory binding sites can help identify the functions of the transcriptional regulatory protein during cell development and disease progression. The identified binding sites may also be used as a basis for annotating functional elements in genomes. The types of functional elements that one can identify using ChIP-on-chip include promoters, enhancers, repressor and silencing elements, insulators, boundary elements, and sequences that control DNA replication (http://www.chiponchip.org/).

Tiling array is a DNA array covering whole genome sequences using overlapped fragments, and it can be applied to examine not only upstream sequences of genes but

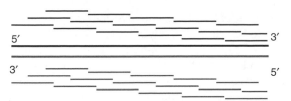

Figure 3-4 Probe design for tiling array. The oligonucleotide probes are tiled across the whole genomic sequence.

also intragenic and intergenic regions [75]. Tiling arrays use millions of DNA probes evenly spaced, or "tiled" across the genome, including coding and noncoding regions (Fig. 3-4). Tiling array has been a very useful tool for genome-wide analysis of many important biological functions, including transcription [76], antisense gene expression [77], protein binding sites [78], sites of chromatin modification [79], sites of DNA methylation [80,81], experimental genome annotation, and regulatory pathway discovery [82].

3.4 TRANSCRIPTOME PROFILING OF AN *ArcA* Mutant of *Shewanella oneidensis*

In this section, discussion will focus on cDNA microarray technology applied for the purpose of characterizing the ArcA regulon in the bacterium *S. oneidensis* MR-1. This section first introduces the background of this study (Section 3.4.1) and then describes the experimental design for this study (Section 3.4.2). The cDNA microarray and microarray hybridization procedure are followed next in Section 3.4.3. Section 3.4.4 describes the roles of bioinformatics in this study. Section 3.4.5 describes the transcriptome profiling of an *arcA* mutant. Finally, the conclusions are presented in Section 3.4.6.

3.4.1 Background

In *E. coli* and other bacteria, the Arc (anoxic redox control) two-component signal transduction system, which consists of the ArcB transmembrane sensor kinase and the cytosolic ArcA response regulator, modulates gene expression in response to changing redox conditions [83]. Under anaerobic or microaerobic respiratory conditions, ArcB autophosphorylates and then transphosphorylates the global transcriptional regulator ArcA, thereby enhancing the affinity of the latter protein for its target promoters [84–87]. ArcA is a transcriptional regulator that can act as an activator or repressor in regulating different genes in redox metabolism such as several dehydrogenases of the flavoprotein class, terminal oxidases, tricarboxylic acid cycle enzymes, enzymes of the glyoxylate shunt and enzymes in fatty acid degradation pathways [83,88]. Recently, ArcA was predicted to directly regulate 55 new genes involved in many different functional categories in *E. coli* [89].

S. hewanella oneidensis MR-1, a facultative gram-negative bacterium, is remarkable for its ability to utilize a diverse array of terminal electron acceptors during anaerobic respiration (e.g., fumarate, nitrate, nitrite, thiosulfate, elemental sulfur, trimethylamine *N*-oxide (TMAO), dimethyl sulfoxide (DMSO), Fe(III), Mn(III) and (IV), Cr(VI), and U(VI)). Because of this exceptional metabolic versatility and the potential use of this organism for bioremediation of metal/radionuclide contaminants in the environment, the approximately 5 Mb chromosome and the 0.16 Mb megaplasmid sequences comprising the *S. oneidensis* MR-1 genome were deciphered by TIGR [60]. Sequence annotation of the MR-1 genome revealed the presence of an *arcA* homologue (SO3988) but not an *arcB* homologue. In this study, whole-genome DNA microarrays for *S. oneidensis* MR-1 were used to define the *arcA* regulon under both aerobic and anaerobic batch growth conditions. Transcriptome analysis of an *arcA* null mutant and the occurrence of a predicted sequence motif for promoter recognition by ArcA suggested that ArcA functions as a global regulator in *S. oneidensis*.

3.4.2 Microarray Construction and Hybridization

3.4.2.1 *Microarray Construction*   The *S. oneidensis* microarray contained a total of 4,761 distinct elements, representing about 99 percent of the total protein-coding capacity of the MR-1 genome [49,90] (Fig. 3-5). Of the array elements that were spotted, 4,310 constituted PCR-amplified DNA fragments corresponding to unique segments of individual MR-1 ORFs, whereas gene-specific oligonucleotide probes (50-mers) were designed and synthesized for 451 predicted genes (9 percent of the total DNA probes arrayed) that did not yield either single products or any products in PCR amplifications. PCR primers and oligonucleotide probes were designed using the program PRIMEGENS [30]. PCR products and oligonucleotides were printed in duplicate onto SuperAmine glass slides (TeleChem International, Inc.). The microarray also consisted of 32 elements corresponding to *S. oneidensis* genomic DNA (positive controls) and 42 spots representing nine genes (amplicons) from *Arabidopsis thaliana* (negative controls).

3.4.2.2 *RNA Isolation, cDNA Labeling, Microarray Hybridization, and Scanning*   Cultures of *S. oneidensis* wild-type and *arcA* mutant strains were harvested at the mid-exponential point under both aerobic and anaerobic conditions, and total cellular RNA was isolated using the TRIzol reagent (Invitrogen, Carlsbad, CA) according to the manufacturer's instructions. RNA samples were treated with RNase-free DNase I (Ambion, Inc., Austin, TX) to digest residual chromosomal DNA and then purified with the QIAGEN RNeasy Mini kit prior to spectrophotometric quantitation at 260 and 280 nm.

Fluorescein-labeled cDNA copies of total cellular RNA extracted from wild-type and mutant cells were prepared, with the exception that Cy3/Cy5-dUTP (Perkin–Elmer/NEN Life Science Products, Boston, MA) was used in the first-strand reverse transcription (RT) reaction. Two sets of duplicate reactions were carried out in

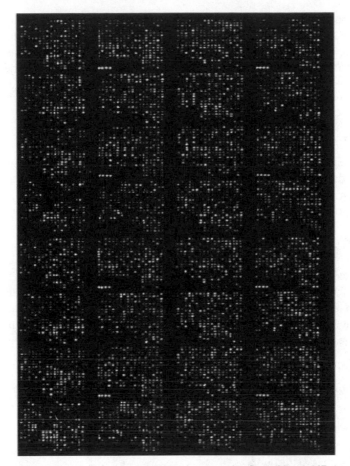

Figure 3-5 Whole genome cDNA microarray for *S. oneidensis* MR-1.

which the fluorescent dyes were reversed during cDNA synthesis to minimize gene-specific dye effects. The labeled cDNA probe was purified and concentrated by following the manufacturer's protocols.

The two labeled cDNA pools (wild type and mutant) to be compared were mixed and hybridized simultaneously to the array in a solution containing $3 \times SSC$ ($1 \times SSC$ is 0.15 M NaCl plus 0.015 M sodium citrate), 0.3 percent sodium dodecyl sulfate, 1 µM dithiothreitol (DTT), 40 percent (v/v) formamide, 0.8 µg of unlabeled herring sperm DNA (Gibco BRL)/µL, and 8.6 percent distilled H_2O. Hybridization was carried out in a 50°C water bath for 12–15 h.

To determine the fluorescence intensity (pixel density) and background intensity, 16-bit TIFF scanned images were analyzed using the software ImaGene version 5.5 (Biodiscovery, Inc., Los Angeles, CA). Microarray outputs were first filtered to remove spots with poor signal quality by excluding those data points with a mean intensity of <2 standard deviations above the overall background for both channels.

Empty spots and spots flagged as poor were removed from subsequent analyses by using ImaGene. Data transformation and normalization were carried out using GeneSite Light (Biodiscovery, Inc.). Normalized expression ratios were imported into ArrayStat (Imaging Research, Inc., Ontario, Canada) to determine the common error and to remove outliers. Only those genes with an expression ratio of ≥2 were included in further analyses.

3.4.3 Experimental Design and Data Analysis

3.4.3.1 Experimental Design Figure 3-6 illustrates the experimental design for this study. To study the *arcA* gene in *S. oneidensis*, we first constructed an in-frame deletion *arcA* mutant (designated ARCA) based on the method described earlier [49] using the primers 3988-5I (5′-<u>TGTTTAAACTTAGTGGATGGG</u>CCTCAGTTACCA CATACCC-3′), 3988-3I (5′-<u>CCCATCCACTAAGTTTAAACA</u>CCAGATACGCCAG AAATCATCG-3′), 3988-5O (5′-GCTTCTGTCGATAAACACGGC-3′), and 3988-3O (5′-TTACCCAATACTTAGTTCAGCAAGG-3′). To monitor global changes in gene expression in response to the *arcA* deletion, we compared the ARCA strain with the DSP10 parental strain grown under aerobic and anaerobic conditions using batch cultures. *S. oneidensis* parental and mutant strains were grown in Luria-Bertani (LB) medium at 30°C under aerobic or anaerobic (with 20 mM fumarate as the electron acceptor) respiratory conditions. For the aerobic condition, cells were grown (60 mL in 250-mL flasks) with agitation (200 rpm). For the anaerobic condition, the media (80 mL in 100-mL bottle) was purged with nitrogen gas while boiling for at least 30 min prior to inoculation. To minimize differences in gene expression caused by

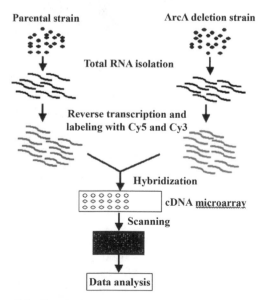

Figure 3-6 Experimental design for arcA regulon characterization.

growth-related effects, samples for transcriptome measurements were taken from exponentially growing cultures at mid-log phase.

For each growth condition tested, gene expression analysis was performed using six independent microarray experiments, including dye swapping, which yielded a total of 12 expression measurements per gene (three biological replicates, with each different mRNA preparation having four technical replicates).

3.4.3.2 *Phenotype Characterization of the ARCA Mutant Strain*   To

determine whether inactivation of the *S. oneidensis arcA* affects anaerobic metabolism, the ability of the ARCA mutant strain to grow on and/or reduce a variety of electron acceptors under anaerobic respiratory conditions was compared to that of the parental DSP10 strain [49]. The ARCA and DSP10 strains were cultured anaerobically in Luria-Bertani media with various electron acceptors, including fumarate (20 mM), colloidal Mn (5 mM), MnO_2 (2 mM), nitrite (20 mM), $MgCl_2$ (10 mM), CrO_4 (150 µM), cobalt (50 µM), FeO_2 (5 mM), ferric citrate (5 mM), $FeCl_3$ (5 mM), or Fe-NTA (10 mM). The culture turbidity was monitored spectrophotometrically at 600 nm. A growth curve was measured for the culture containing fumarate. For other electron acceptors, the growth of the culture was evaluated using end-point culture turbidity measurements.

The results indicated that the growth of ARCA in LB is slightly slower than the parent DSP10 strain (Fig. 3-7) under anaerobic conditions with fumarate (20 mM) as the electron acceptor. Based on the end-point culture turbidity, the *arcA* deletion mutant exhibits slower growth than the DSP10 parental strain under anaerobic respiratory conditions with the following electron acceptors: colloidal Mn (5 mM), MnO_2 (2 mM), nitrite (20 mM), $MgCl_2$ (10 mM), CrO_4 (150 µM), cobalt (50 µM),

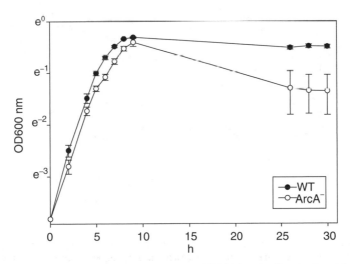

Figure 3-7 Comparison between the growth curves of ARCA (*arcA* null mutant) and the wild-type *S. oneidensis* DSP10 strain grown in Luria-Bertani medium at 30°C under anaerobic (with 20 mM fumarate as the electron acceptor) respiratory conditions.

FeO_2 (5 mM), ferric citrate (5 mM), $FeCl_3$ (5 mM). The growth of the ARCA and DSP10 strains was also evaluated in M4 minimum medium with ferric citrate (10 mM, 20 mM, and 50 mM, respectively) based on the culture turbidity at 24, 48, and 72 h, and the results demonstrated that ARCA grew slower than the parent DSP10 strain (data not shown).

The effect of hydrogen peroxide (H_2O_2) treatment at concentrations of 500 and 2500 µM on mid-exponential growth of the parental and mutant strains under aerobic conditions was also assessed. As shown in Figure 3-8, ARCA was shown to be more sensitive to H_2O_2-induced oxidative stress at different concentrations compared to the parental DSP10 strain, suggesting that ArcA might play a regulatory role in oxidative stress resistance in *S. oneidensis*. This observation agrees with the finding that *arc*A increases resistance of *Salmonella enterica* serovar Enteritidis to H_2O_2 [91].

3.4.4 Data Interpretation

3.4.4.1 Overview of Transcriptome Profiling of the ARCA Mutant under Different Respiratory Conditions
A total of 654 (294 downregulated; 360 upregulated) and 504 (135; 369) genes were identified as being differentially expressed in response to the *arc*A deletion mutation under aerobic and anaerobic respiratory conditions, respectively. Comparison of the two microarray data sets indicated that the expression levels for 248 of these genes were affected under both aerobic and anaerobic growth conditions. The differentially expressed genes encode a broad variety of functions, with the majority (44–52 percent) encoding hypothetical or conserved hypothetical proteins (Fig. 3-9a and b). Genes showing changes in

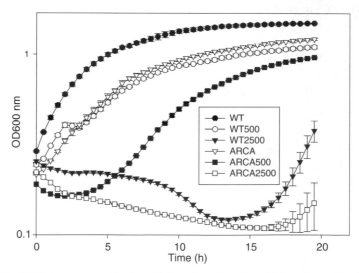

Figure 3-8 ARCA is more sensitive to oxidative stress than the wild-type *S. oneidensis* DSP10 strain. Growth was measured kinetically with a Microbiology Reader Bioscreen C (Growth Curves USA, Piscataway, NJ) [34]. WT and ArcA mutant were grown aerobically up to the mid-log phase and then treated immediately with 500 and 2500 µM H_2O_2, respectively. The cells were grown at 30°C with continuously extensive shaking. The OD_{600} nm units were read with an interval of 30 min.

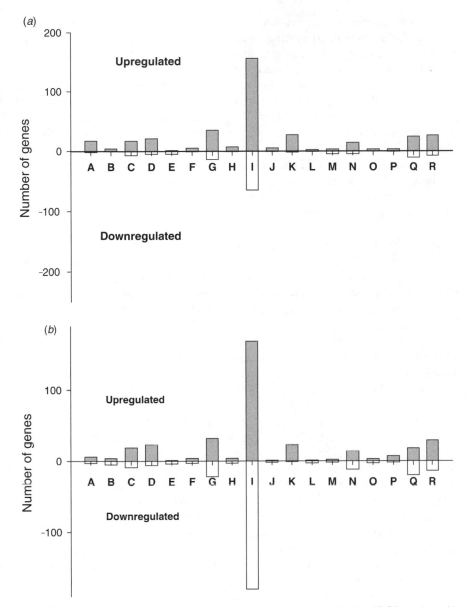

Figure 3-9 Functional distribution of differentially expressed genes in the ARCA strain under anaerobic (a) and aerobic conditions (b). Genes are grouped in their corresponding functional/homology classes (number of genes/percentage of genes) according to TIGR's annotation (http://www.tigr.org/tigr-scripts/CMR2/gene_attribute_results_org_or_role.dbi): (A) amino acid biosynthesis; (B) biosynthesis of cofactors, prosthetic groups, and carriers; (C) cell envelope; (D) cellular processes; (E) central intermediary metabolism; (F) DNA metabolism; (G) energy metabolism; (H) fatty acid and phospholipid metabolism; (I) hypothetical proteins; (J) other categories; (K) protein fate; (L) protein synthesis; (M) purines, pyrimidines, nucleosides, and nucleotides; (N) regulatory functions; (O) signal transduction; (P) transcription; (Q) transport and binding proteins; and (R) unknown function.

transcript abundance in the *arcA* deletion mutant under both growth conditions that have annotated functions are involved in a number of cellular processes including cell envelope, energy metabolism, protein fate, regulatory functions, and transport/binding proteins. Under anaerobic growth conditions, a number of genes belonging to the functional categories of protein synthesis and purines/pyrimidines/nucleosides/nucleotides were also upregulated in the *arcA* mutant (Fig. 3-9a). There were 27 and 19 predicted regulatory genes that showed significant differences in expression in a Δ*arcA* genetic background under aerobic and anaerobic conditions, respectively, and 13 genes with annotated functions in regulation were differentially expressed under both respiratory conditions. These results suggest that ArcA functions as a global regulator in *S. oneidensis*, exerting a pleiotropic effect on a number of cellular functions, and that the transcriptional effect of an *arcA* deletion was most profound under anaerobic growth conditions.

3.4.4.2 Genes with Functions in Energy Metabolism

A total of 66 of 87 (~76 percent) genes with annotated functions in energy metabolism showed altered expression profiles in the *arcA* deletion mutant (49 under anaerobic conditions and 54 under aerobic conditions). More than half of these genes (34 genes) are involved in electron transport function. Except for the *napAGHB* operon, which was upregulated under anaerobic conditions but downregulated under aerobic conditions, all other genes showed similar expression trends under anaerobic and aerobic conditions. For 19 cytochrome *b* or *c* genes, 13 were upregulated under either or both anaerobic and aerobic conditions including SO4483 (cytochrome *b*, putative), cytochrome *c* family proteins (SO1782, SO1659, SO4079–SO4078 operon, SO4142, SO4144, SO4484), cytochrome *c* oxidase *ccoPONQ*, and diheme cytochrome *c* (SO4485). However, the other six genes, cytochrome *c* (*scyA*, SO3300, SO4572, SO2727, SO0845) and decaheme cytochrome *c* (SO1427) were downregulated at either or both of these two experimental conditions. Among iron–sulfur clustering binding proteins, SO1364, *napG*, and *napH* were downregulated 2.98-, 4.44-, and 7.50-fold under aerobic conditions, respectively. The other three iron–sulfur cluster proteins (SO1519, SO1521, SO4404) were instead upregulated under aerobic conditions. Genes involved in anaerobic metabolism such as *torC* (tetraheme cytochrome *c*), *cat2* (4-hydroxybutyrate coenzyme A transferase), and *fdhB* (formate dehydrogenase) were upregulated under anaerobic conditions, but *dmaAB* (anaerobic dimethyl sulfoxide reductase), *ifcA* (fumarate reductase flavoprotein), and SO4513 (formarate dehydrogenase) were downregulated under anaerobic conditions.

Among operons/genes-encoding enzymes involved in the TCA cycle, malate synthase (*aceBA*), and aconitate hydratase 1 (*acnA*) were upregulated in ARCA under anaerobic fumarate-reducing conditions, which is similar to the situation in *E. coli* [89,92]. Up- or downregulation of the genes associated with the TCA cycle will affect redox generation, which reflects the role of ArcA in redox metabolism. However, seven other operons, including citrate synthase (*gltA*) [93], succinate dehydrogenase operon (*sdhCAB*) [94], 2-oxoglutarate dehydrogenase–succinyl–CoA synthase operon (*sucABDC*) [95,96], malate dehydrogenase (*mdh*) [94], aconitate hydratase 2 (*acnB*) [97], isocitrate dehydrogenase (*icd*) [98], and SO2222 (fumarate hydratase) were not affected in the *S. oneidensis* ARCA mutant under anaerobic fumarate-reducing conditions. Another fermentation gene, D-lactate dehydrogenase (*ldhA*), was not affected significantly under anaerobic conditions in *S. oneidensis*, but was upregulated 1.85-fold in an *E. coli arcA* mutant in MOPS-buffered LB with 20 mM D-xylose [89,99].

3.4.4.3 Expression of Genes from Other Functional Categories   Here we discuss genes known to be regulated directly or indirectly by *arcA* in other microbes. About 30 operons (including the reported gene/operons discussed above), most of which are involved in respiratory metabolism, are presently known to be regulated by phosphorylated *arc*A in other organisms [99].

The glutamate synthase operon (*gltDB*) was upregulated under anaerobic conditions, which is similar to their up-regulation in *E. coli arcA* mutants under anaerobic conditions [83,89,99]. In contrast, nine other operons related to redox metabolism, including formate acetyltransferase (*pflB*), cytochrome *d* ubiquinol oxidase operon (*cydAB*) [100–102], aldehyde dehydrogenase (*aldA*), fatty acid oxidation complex (*fadBA*), NADH dehydrogenase (*nuo*) operon, the ATP binding protein operon (*cydDC*) [100,101], glycerol kinase (*glpK*), anaerobic C4-dicarboxylate membrane transporter (*dcuB*), and lipoamide dehydrogenase (*lpdA*) [83,92,99,103,104], were not affected under the growth conditions tested in this study. These nine genes were shown to be regulated by ArcA in other organisms in previous studies [83,92,99–102,104]. The transport and binding protein, C4-dicarboxylate binding periplasmic protein (*dctP*) [92], was upregulated about 2.3-fold under aerobic conditions, which is similar to the reported trend (*dctA*, up-regulated 1.58) from *E. coli arcA* mutant microarray data [89].

3.4.4.4 Resistance of S. oneidensis arcA Null Mutant to H₂O₂ Oxidative Stress   As described earlier, the ARCA mutant strain is hypersensitive to H_2O_2 relative to the DSP10 parental strain under aerobic conditions. The *oxyR* gene encodes a transcriptional binding protein that regulates oxidative stress resistance in *S. enterica* scrovar Typhimurium and *E. coli* [105]. In this study, the expression of the *S. oneidensis oxyR* homologue, gene SO1328, was not affected by the *arcA* deletion. Also of interest was the observation that the MR-1 counterparts for such known OxyR-controlled genes as *katG* (hydroperoxidase I), *ahpF* (alkyl hydroperoxide reductase), *gor* (glutathione reductase), *grx* (glutaredoxin, SO2745), *fur* (Fur repressor of ferric ion uptake), *dps* family protein (SO1158), and *hemH* (SO2018 and SO3348) [105,106], were not affected in the ARCA strain, even though the deletion mutant exhibited H_2O_2 hypersensitivity.

Nystrom et al. [107] also showed that an *E. coli arcA* deletion mutant was not able to decrease the synthesis of the TCA enzymes malate dehydrogenase (*mdh*), isocitrate dehydrogenase (*aceB*), lipoamide dehydrogenase E3 (*lpdA*), and succinate dehydrogenase (*sdh*). Similarly, our transcriptome profiling of ARCA under aerobic conditions demonstrated that the transcription of enzymes within the TCA cycle was not affected significantly (Table 3-2). These enzymes are encoded by *gltA*, *lpdA*, *sdhCAB*, *sucABCD*, *mdh*, *acnB*, *aceB*, *icd*, *acnA*, and *acnB*. The microarray data for the *arcA*

Table 3-2 Microarray expression ratio and DNA motif prediction for the genes verified to be regulated directly by ArcA in E. coli

Functional Category	ORF	Gene Product	Anaerobic[a] (arcA/WT)	Aerobic (arcA/WT)	Start[b]	Strand	Z[c]	Predicted Motif
Amino acid biosynthesis	SO1325	Glutamate synthase, large subunit (gltB)	8.31	3.21	−227	+	2.06	GTTCTTTATTTTTTA
Energy metabolism	SO0432	Aconitate hydratase 2 (acnB)	1	1.36	−266	−	2.66	GTTCATCAATTTAAA
	SO1021	NADH dehydrogenase I, A subunit (nuoA)	1.18	0.63	−1452	−	2.02	TTTAATTGAAATTTA
					−616	+	2.35	GTTCAGGAAGTGTTT
					−140	+	2.14	GTTAATCCACTTAGT
					−240	−	2.3	GTTAAGCAATACATT
	SO1483	Malate synthase A (aceB)	3.72	1.57	−712	−	2.37	GTTAACCGTTTTCCA
	SO2629	Isocitrate dehydrogenase, NADP dependent (icd)	0.79	0.75	−493	−	2.69	GTTAATTCTAATAGA
	SO3286	Cytochrome d ubiquinol oxidase, subunit I (cydA)	1.5	1.82	−331	+	2.16	GTTCATGCTTTGGCT
					−1035	+	2.67	GTTCACACAATTGAA
	SO4230	Glycerol kinase (glpK)	1.34	1.03	−590	−	2.9	GTTAATCAAAATAAA
					−255	−	2.66	GTTAACAATATCCAT
					−203	+	2.22	GTTAATTAGATTTTT
	SO0426	Pyruvate dehydrogenase complex, E3 component, lipoamide dehydrogenase (lpdA)	1.08	1.04	NF	NF	NF	NF
	SO1926	Citrate synthase (gltA)	0.86	1.33	NF	NF	NF	NF
	SO1927[e]	Operon sdhCAB	0.68	0.55	NF	NF	NF	NF

	SO1930[e]	Operon *sucABCD*	1.03	1.5	NF	NF	NF	NF
	SO2912[e]	Operon *pflBA*	1.18	1.05	NF	NF	NF	NF
	SO4480	Aldehyde dehydrogenase	1.34	0.95	NF	NF	NF	NF
Fatty acid and phospholipid metabolism	SO0021	Fatty oxidation complex, alpha subunit (fadB)	1.12	0.78	−384	+	2.05	GTTAATAATAAATAT
Protein fate	SO3637	Survival protein surA (surA)	1.17	1.8	−151	+	2.19	GTTAATGAAAGCCGT
Regulatory functions	SO3988	Aerobic respiration control protein ArcA (arcA)	0.11	0.07	−392	+	2.57	GTTAACAAATGCCTA
Transport and binding proteins	SO0827	L-lactate permease (lldF)	0.99	1.06	−246	−	2.57	GTTAATCAAGGTATA
	SO3134	C4-dicarboxylate binding periplasmic protein (dctP)	1.62	2.31	−151	−	2.03	GTTAATAAAGTGTAG
	SO3780	ABC transporter, ATP binding protein CydD (cydD)	1.1	1.17	−35	+	2.32	GTTAAGCCTATTTCT

[a] Relative gene expression is presented as the mean ratio of the fluorescence intensity of the *arcA* deletion mutant (ARCA) to that of the parental strain (WT), and NA means that there is not any expression value available.

[b] The start site of the predicted ArcA motif away from the gene translation start site.

[c] The statistical Z-score of the predicted motif from the genomic-wide scan.

[d] No ArcA-P binding motif found in *S. oneidensis* MR-1.

[e] Only the gene expression value for the first gene in the operon was shown.

deletion mutant under H_2O_2 stress also indicated that the expression of these genes was not changed significantly (T. Li and J. Zhou, unpublished data, personal communication). Therefore, these TCA cycle enzymes may still produce the reactive oxygen species (ROS) after exposure to H_2O_2. Nystrom et al. [107] demonstrated that *E. coli* was able to overproduce superoxide dismutase to scavenge superoxide radicals generated from aerobic respiration to defend against oxidative stress. The deletion of *sodB* in *Helicobacter pylori* results in hypersensitivity of the mutant to oxidative stress and a defect in host colonization [108]. Here we found that superoxide dismutase (*sodB*) was downregulated about 1.7-fold under aerobic respiratory conditions. Our experiments also demonstrated that the gene encoding periplasmic nitrate reductase (*nap*A), which has been reported to be associated with oxidative stress resistance in *H. pylori* [108,109], was downregulated 11.3-fold under aerobic respiratory conditions. This might explain the hypersensitivity of the ARCA mutant to H_2O_2 oxidative stress. In addition, we found that two heavy metal efflux pump operons (SO4597–SO4598 and SOA0154–SOA0153) were downregulated 6.7- and 25-fold, respectively. It is unknown whether the low levels of expression of these two operons will affect the ARCA strain's H_2O_2 stress resistance capability.

3.4.5 Bioinformatics Analysis

3.4.5.1 *Sequence Analysis and Structural Modeling of S. oneidensis arcA* The putative *arcA* gene of *S. oneidensis* MR-1 encodes a 238-amino acid protein with a predicted molecular mass of 27,220 Da and a pI of 5.43. Comparison of the deduced amino acid sequence showed that *S. oneidensis* MR-1 ArcA shares a high degree of identity to its homologues in *E. coli* (81 percent), *S. enterica* (81 percent), *Yersinia pestis* (81 percent), *V. cholerae* (81 percent), and a lower degree of sequence identity to ArcA in *Pasteurella. multocida* (75 percent) and *H. influenzae* (72 percent) (Fig. 3-10). This high level of homology at the primary sequence level strongly suggests that these proteins share similar biological functions. Moreover, analysis of the deduced amino acid sequence of *S. oneidensis* MR-1 ArcA also revealed the conservation of the Asp^{54} residue in the N-terminal receiver domain and the helix–turn–helix (HTH) DNA binding motif in the carboxy-terminal effector domain (Fig. 3-10) [85,88]. Based on structure predictions using PROSPECT-PSPP [110], ArcA shows high homology to the response regulator Drrb present in *Thermotoga maritima* (PDB id 1p2f, 30). Drrb is a multidomain response regulator of the OmpR/PhoB subfamily that may regulate gene transcription by binding as a dimer to σ^{70} promoter elements [111]. In contrast to *E. coli*, *S. oneidensis arcA* is predicted to be monocistronic, and there is no obvious cognate *arcB* encoded in the MR-1 genome based on the sequence annotation [112]. This suggests that a less conserved sensor histidine kinase might be employed by the Arc two-component signal transduction system.

3.4.5.2 *Scanning the S. oneidensis MR-1 Genome with the ArcA-P Positional Weight Matrix* Structure modeling of the deduced protein encoded by the MR-1 *arcA* gene indicated a strong degree of conservation between the DNA

```
S_oneidensis   MQNPHILIVEDEAVIRNTLRSIFEAEGYVVTEANDGAEMHKAMQENKINLVVMDINLPGK  60
E_coliO157_H7  MQTPHILIVEDELVTRNTLKSIFEAEGYDVFEATDGAEMHQILSEYDINLVIMDINLPGK  60
V_cholerae     MQTPQILIVEDEQVTRNTLKSIFEAEGYAVFEASNGEEMHQVLSDYPINLVIMDINLPGK  60
Salmonella     MQTPHILIVEDELVTRNTLKSIFEAEGYDVFEATDGAEMHQILSEYDINLVIMDINLPGK  60
Y_pestis       MQTPHILIVEDEIVTRNTLKSIFEAEGYVVEANDGAEMHHILSENDINLVIMDINLPGK   60
P_multocida    MGTPQILIVEDEAITRNTLKSIFEAEGYEVFEAADGAQMHRILSNKVINLVIMDINLPGK  60
H_influenzae   MTTPKILVVEDEIVTRNTLKGIFEAEGYDVFEAENGVEMHHILANHNINLVVMDINLPGK  60
               *  * *:.*:****..********..* *** .*  **: :  : :***  ::*:********* *


S_oneidensis   NGLLLARELRELNNIGLIFLTGRDNEVDKILGLEIGADDYITKPFNPRELTIRARNLLTR 120
V_cholerae     NGLLLARELREQADVALMFLTGRDNEVDKILGLEIGADDYITKPFNPRELTIRARNLLSR 120
E_coliO157_H7  NGLLLARELREQANVALMFLTGRDNEVDKILGLEIGADDYITKPFNPRELTIRARNLLSR 120
Salmonella     NGLLLARELREQANVALMFLTGRDNEVDKILGLEIGADDYITKPFNPRELTIRARNLLSR 120
Y_pestis       NGLLLARELREQASVALMFLTGRDNEVDKILGLEIGADDYITKPFNPRELTIRARNLLSR 120
P_multocida    NGLMLARELRETTNTALMFLTGRDNEVDKILGLEIGADDYITKPFNPRELTIRARNLLQR 120
H_influenzae   NGLLLARELREELSLPLIFLTGRDNEVDKILGLEIGADDYLTKPFNPRELTIRARNLLHR 120
               ***:*******   . : *:***************************:***********


S_oneidensis   VNSAGNEVEEKSSVEYYRFNDWSLEINSRSLVSPQGESYKLPRSEFRAMLHFVENPGKIL 180
E_coliO157_H7  MNLGTVSEERRSVESYKFNGWELDINSRSLIGPDGEQYKLPRSEFRAMLHFCENPGKIQ  180
V_cholerae     SMHAGTTQEEKRSVEKYVFNGWELDINSRSLVSPDGDSYKLPRSEFRALLHFCENPGKIQ 180
Salmonella     TMNLGTVSEERRSVEKYKFNGWELDINSRSLIGPDGEQYKLPRSEFRAMLHFCENPGKIQ 180
Y_pestis       TMNLSSVGEERRLVESYKFNGWTLDLNSRTLINPEGEYKLPRSEFRAMLHFCENPGKIQ  180
P_multocida    TMQE-NSKDSHPIEQYRFNGWTLDLNSRTLINPEGEYKLPRSEFRAMLHFCENPGKIQ   179
H_influenzae   AMPH-QEKENTFGREFYRFNGWKIDLNSHSLITPEGQEFKLPRSEFRAMLHFCENPGKLQ 179
               * * *.       *: :*:*...:*  * ...: .*::*:* .*********** ******:


S_oneidensis   TRADLLMKMTGRELKPHDRTVDVIIRRIRKHFESLPDTPEIIATIHGEGYRFCGNLED 238
E_coliO157_H7  SRAELLKMTGRELKPHDRTVDVIIRRIRKHFESTPDTPEIIATIHGEGYRFCGDLED  238
V_cholerae     TRADLIFKMTGRELKPHDRTVDVIIRRIRKHFESVSGTPEIIATIHGEGYRFCGDLED 238
Salmonella     SRAELLKKMTGRELKPHDRTVDVIIRRIRKHFESTPDTPEIIATIHGEGYRFCGDLQD 238
Y_pestis       SRGELLKKMTGRELKPHDRTVDVTIRRIRKHFESTPDTPEIIATIHGEGYRFCGDLEE 238
P_multocida    TREELLKKMTGRELKPQDRTVDVTIRRIRKHFEDHPETPEIIATIHGEGYRFCGELE- 236
H_influenzae   TREELLKKMTGRELKPQDRTVDVTIRRIRKHFEDHPNTPNIIMTIHGEGYRFCGDIE- 236
               :.* :*:***********.*****.********:    .**.**  ***********..: :
```

Figure 3-10 Multiple sequence alignments for ArcA from *E. coli* (GeneBank ID: 15804972), *S. enterica* (GeneBank ID: 16763386), *Y. pestis* (GeneBank ID: 16120787), *V. cholerae* (GeneBank ID: 16272824), *P. multocida* (GeneBank ID: 15602084), and *H. influenzae* (GeneBank ID: 15642365). The multiple alignments were performed using ClustalW [119]. The shading areas are the predicted helix by PROSPECT-PSPP [110] indicating the DNA binding domain, and the bold letters are the phosphorate sites.

binding domains of the *E. coli* and *S. oneidensis* ArcA proteins (Fig. 3-10). Thus, we utilized the experimentally verified ArcA-P binding sites from 10 *arc*A-regulated proteins to construct the ArcA-P positional weight matrix [89]. The score function of positional weight matrices was adapted from the log transformation method described by Berg and Von Hippel [113]. Both strands of the *S. oneidensis* MR-1 genome sequence were scanned using a sliding window size of 15 nucleotides. The motif with the highest matrix score was selected among all of the overlapping motifs from both plus and minus strands. Scores of all potential ArcA-P recognition sites were statistically analyzed using the Z test, and only those sites with 95 percent or greater significance are presented as potential ArcA-P binding sites in *S. oneidensis*. For each gene, only the promoters located within the upstream sequence from the gene start codons are counted in this paper.

By scanning the *S. oneidensis* genome with the ArcA-P recognition weight matrix, 13 tRNA and 668 protein-encoding genes were predicted to contain potential ArcA binding sites in their upstream regions. The predicted ArcA regulon in *S. oneidensis* includes 12 ORFs shown to be controlled by ArcA in *E. coli*: *fadB, acnB, nuoA, gltB, aceB, icd, dctP, cydA, cydD, arcA, glpK*, and *lldP* [83,89,92] (Table 3-2). The *surA* gene, which is predicted to contain an ArcA-P binding site in *E. coli* [89], also has a strong *arc*A motif in *S. oneidensis*. However, the other six operons, *lpdA, gltA-sdhCAB, sucABDC, pflBA*, and *aldA*, which are regulated by *arc*A in *E. coli*, do not possess strong ArcA binding motifs based on this search. Among the 668 protein-coding genes in the *S. oneidensis* ArcA regulon, 148 genes (about 3 percent of all the predicted genes in *S. oneidensis*) exhibited significant differences in transcript levels in ARCA relative to DSP10 under aerobic and/or anaerobic conditions. Table 3-2 shows a subset of genes in *S. oneidensis* with ArcA-P binding sites, which are similar to ArcA-P binding sites in *E. coli* [83,89,92]. A sequence logo representation of the predicted conserved ArcA-P binding motif for these 148 genes is shown in Figure 3-11a. Compared with the ArcA motif in *E. coli*, the predicted motif has a weaker consensus. For example, the first, third, and fifth positions in the motif have smaller bit scores, which reflect the conservation status for each consensus position. Another genomic scanning in *S. oneidensis* using the positional weight matrix constructed from the predicted 190 binding sites for 148 genes resulted in a similar consensus (data not shown). Most (81 percent) of the motifs are located within 300 nucleotides upstream of the translation start codon (Fig. 3-11b).

Similar to the ArcA regulon in *E. coli* (20), the ArcA regulon in *S. oneidensis* is associated with 17 functional categories. Among the 148 genes with differences in expression in the *arcA* deletion mutant (Table 3-3), 46 genes were predicted to be positively regulated and 102 negatively regulated. Our results also showed that the genes controlled by ArcA are involved in functions beyond redox metabolism. These genes belong to broad functional categories and most of these genes have not been reported previously to be members of ArcA regulons from other bacterial species, such as *E. coli*. Eight of these genes with expression changes with more than 3-fold (up or down) have strong predicted ArcA binding motifs ($Z > 3.0$) and encode HoxK (SO2099), Pal/histidase family protein (SO3299), decaheme cytochrome *c* (SO1427), putative long-chain fatty acid transport protein (SO3099), TonB-dependent receptor domain

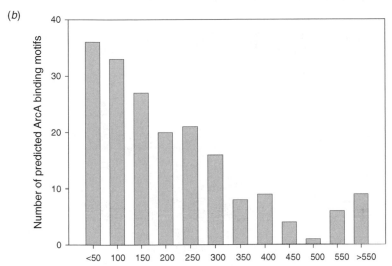

Figure 3-11 Identification of a predicted consensus ArcA binding motif in *S. oneidensis* MR-1 using computational methods. (a) Sequence logo representation of the predicted ArcA binding motif in *S. oneidensis* MR-1. (b). Position distribution of the predicted ArcA motifs.

protein (SO2907), MaoC domain protein (SO0599), PspF (SO1806), and a hypothetical protein (SO2930). Among these eight genes, *hoxK* is the first gene in the quinone-reactive Ni/Fe hydrogenase operon (*hoxK–hydB–hydC*), which catalyzes the reversible oxidation of H_2 [114]. Deletion of *hoxK* was shown to inactivate the membrane-bound hydrogenase in *Alcaligenes eutrophus* [115]. These three genes (*hoxK–hydB–hydC*) are upregulated more than 3.9-fold under aerobic conditions. Under anaerobic conditions, *hoxK* is also upregulated more than 3.5-fold. In addition, a putative undecaprenol kinase (SO4274) was also predicted to have a strong ArcA-P binding site with a Z-score larger than 3. The undecaprenol kinase (*so4274*) is a cell wall synthesis gene and has been associated with biofilm formation in *Mycobacterium smegmatis* [116]. Recently, *arcA* was found to be related to biofilm formation in *S. oneidensis* MR-1 [112] in which the undecaprenol kinase might be the functional gene target. These wide-ranging functions of ArcA are also supported by its requirement for virulence in *Haemophilus influenzae* [117] and *Vibrio cholerae* [118] as well as a recent genome-wide study of the ArcA regulon in *E. coli* [89].

Table 3-3 The genes in *S. oneidensis* MR-1 with expression differences that also have predicted binding sites

ORF	Gene Product	Ratio (ARCA/WT)		Start[b]	Strand	Z^c	Motif
		Anaerobic[a]	Aerobic				
Amino acid biosynthesis							
SO2483	Aspartate aminotransferase, putative	0.18	NA	−127	−	2.15	TTTAACTAAGTGTTA
SO1325	Glutamate synthase, large subunit (gltB)	8.31	3.21	−227	+	2.06	GTTCTTTATTTTTA
SO4245	Amino acid acetyltransferase (argA)	2.01	1.16	−162	+	2.2	GTTAAAAAAATGTGA
				−62	+	2.07	GTGAATTAAAATGCA
SO2071	Imidazoleglycerol phosphate dehydratase/histidinol-phosphatase (hisB)	2.63	1.2	−60	−	2.08	GTTAATACTTGCACT
SO4349	Ketol acid reductoisomerase (ilvC)	1.21	2.41	−44	+	2.21	GTTAACAATAAGTTG
Biosynthesis of cofactors, prosthetic groups, and carriers							
SO1198	Dihydropteroate synthase (folP)	1.17	2.25	−138	+	2.75	GTTAATTGAAAGAGA
SO2445	Thiamin biosynthesis protein ThiC (thiC)	0.66	19.66	−371	−	2.13	GTTATTAAATTTTAA
				−118	−	2.18	GTTAATAGACGGCGA
Cell envelope							
SO4179	Glycosyl transferase, group 2 family protein	0.41	0.47	−55	−	2.15	GTTCAGCCAATTGAT
SO1199	Phosphoglucosamine mutase (glmM)	1.33	2.31	−47	+	2.57	GTTAAGTATTTCATT
SO2757	Membrane protein, putative	NA	0.29	−794	−	2.56	GTTAACGATCTGCCA
SO1102	TonB-dependent receptor C-terminal region domain lipoprotein	6.34	11.8	−110	−	2.22	GTTAAGCCTGATATA
				−11	+	2.02	GCTAACAAAAAGTTT
SO1673	Outer membrane protein OmpW, putative	5.42	11.04	−272	+	2.17	GTAAATGAAATGTAA

Cellular processes

Locus	Description						Sequence
SO1278	Methyl-accepting chemotaxis protein	2.79	2	−46	−	2.01	GTTCATTTTATTTTT
				−324	+	2.01	GTTAATGTTAATGTA
SO1434	Methyl-accepting chemotaxis protein	0.47	0.37	−1012	+	2.36	GTTCACACAATCCCA
SO4557	Methyl-accepting chemotaxis protein	0.11	0.35	−155	−	2.05	ATTAATAAATATTTA
SO1961	Maltose O-acetyltransferase (maa)	1.56	2.77	−98	+	2.29	GTTAGCTAAATGGTA
SO0866	Minor curlin subunit CsgB, putative	51.27	7.13	−150	−	2.72	GTTAATCGTATGAAA
SO4317	RTX toxin, putative	1.72	4.18	−291	+	2.07	ATTCATAAAATTTTA
SO2389	Multidrug resistance protein D (emrD)	0.07	0.9	−217	+	2.58	GTTAACAAACAGCTT
				−65	−	2.54	GTTAATCATCTTGAT
SO4146	Toxin secretion ABC transporter protein, HlyB family	3.1	3.73	−53	+	2.76	GTTAATTAAGATGTT
				−160	−	2.13	GTTAACATTATGTTT
SO4274	Undecaprenol kinase, putative	2.39	1.11	−81	−	3.28	GTTAATTATATTTAA
SO1917	Multidrug resistance protein, putative	2.07	0.92	−10	+	2.37	GTTAATTGAGGCAAA

Central intermediary metabolism

Locus	Description						Sequence
SO3705	5-Methylthioadenosine nucleosidase/ S-adenosylhomocysteine nucleosidase, putative	0.23	0.22	−300	−	2.04	GTTAAGCCTTTGAGT
SO2185	Exopolyphosphatase (ppx)	0.92	2.51	−623	−	2.37	GTTCACGATTTCATT
SO0314	Ornithine decarboxylase, inducible (speF)	0.1	0.02	−682	−	2.76	GTTAATTCATTTTGA
				−506	+	2.37	GTTAATTAAAAATAA
SO1870	Biosynthetic arginine decarboxylase (speA)	0.43	0.93	−815	+	2.39	GTTTATTATATTTTA
				−912	−	2.12	GTTCAATATTTTTTA
SO3872	Arylsulfate sulfotransferase	0.38	0.45	−167	+	2.05	GTTAATTTATATATAA

(continued)

Table 3.3 *(Continued)*

ORF	Gene Product	Ratio (ARCA/WT) Anaerobic[a]	Aerobic	Start[b]	Strand	Z[c]	Motif
DNA metabolism							
SO1066	Extracellular nuclease	2.42	8.07	−368	−	2.05	TTTAATTATTTTGAA
				−29	+	2.16	GTTAATAAAAATTAG
SO1844	Extracellular nuclease, putative	2.2	1.23	−48	−	2.22	GTTAAGACTTTTCGA
Energy metabolism							
SO1812	Methionine gamma-lyase (mdeA)	11.36	21.69	−68	+	2.25	GTTATTTAAAAGATA
SO4674	2-Amino-3-ketobutyrate coenzyme A ligase (kbl)	0.4	0.63	−139	+	2.53	GTTAAGCATTAGTTT
SO3299	Pal/histidase family protein	0.25	0.22	−50	+	3.02	GTTAATTAATTTTGA
SO1421	Fumarate reductase flavoprotein subunit (ifcA-1)	0.46	0.72	−89	+	2.69	GTTAAGTGAATTTTT
SO2099	Quinone-reactive Ni/Fe hydrogenase, small subunit precursor (hoxK)	3.55	1.56	−307	−	2.86	GTTAATAAATTCAAA
				−131	+	3.26	GTTAATTAAAATGTCA
SO2727	Cytochrome c3	0.4	0.47	−204	+	2.17	GTTATTCAAATGTAA
SO1427	Decaheme cytochrome c	0.07	0.17	−319	−	2.23	ATTAATTAAATGAAA
				−230	−	3.1	GTTAATAAAATGTTT
				−71	−	2.72	GTTAACGAATGTTAA
				−5	+	2.84	GTTAACCATAAGGCA
SO0344	Methylcitrate synthase (prpC)	2.07	1.07	−13	−	2.59	GTTAACGCTATGGTT
SO3496	Aldehyde dehydrogenase	2.48	1.46	−219	+	2.03	GTTAAGGGTGGCTAA
				−188	−	2.85	GTTCATTATTTCTAA
SO1006	Dienelactone hydrolase family protein	2.26	1.87	−227	+	2.17	GTTATTGAAATGTAA
SO1483	Malate synthase A (aceB)	3.72	1.57	−712	−	2.37	GTTAACCGTTTTCCA

Fatty acid and phospholipid metabolism						
SO0572	Enoyl–CoA hydratase/isomerase family protein	7.98	13.05	−113	−	2.55	GTTAAGTATTAGTGT
SO2395	Acyl–CoA dehydrogenase family protein	1.07	7.11	−198	+	2.37	GTTAATCGTGAGCTA

Hypothetical protein							
SO0563	Hypothetical protein	NA	0.21	−349	+	2.99	GTTAACAGAATTTTA
SO0912	Hypothetical protein	NA	0.23	−507	+	2.63	GTTCATGGTAAGTTA
SO1546	Hypothetical protein	NA	0.26	−261	−	2.01	GTTAACATTGTTTAA
SO0787	Hypothetical protein	NA	0.46	−225	+	2.32	GTTAACTTAAAGTAA
				−125	+	2.02	ATTAAGTAAAATTAA
SO3511	Hypothetical protein	NA	0.47	−122	+	2.07	TTTAATAATATTAAA
SO2460	Hypothetical protein	28.07	0.32	−23	+	2.16	GTAAACAAATTGTAA
SO1479	Hypothetical protein	19.47	5.62	−13	−	2.8	GTTAATAATGATTCA
SO2930	Hypothetical protein	15.88	1.57	−19	+	3.01	GTTAACGAAATTATA
SO0306	Hypothetical protein	8.12	5.45	−176	+	2.01	GTTATTAAATTGTGA
SO3395	Hypothetical protein	7.11	15.52	−534	−	2.32	GTTAATGCTTGGCTA
SO2446	Hypothetical protein	6.04	0.23	−33	+	2.29	GTTCAGAGTTGTGA
SO1970	Hypothetical protein	5.45	5.19	−116	−	2.58	GTTAACTGTGAGGCA
SO4592	Hypothetical protein	3.68	1.29	−296	+	2.17	ATTAATTAAAACTTA
SO2199	Hypothetical protein	3.33	1.76	−85	−	2.1	GTGAATAAAATGTTT
SO2002	Hypothetical protein	2.95	1.87	−358	+	2.96	GTTAATAAAATGCCA
				−242	+	2.77	GTTAACTAACGGAAA
				−13	+	3.01	GTTAACTATATCTTT
SO1517	Hypothetical protein	2.66	0.28	−204	−	2.55	GTTAATCCATAGAAA
SO1944	Hypothetical protein	2.1	3.57	−49	−	3	GTTAAGTAATTGTAA
SO2076	Hypothetical protein	0.9	4.59	−42	−	2.88	GTTAATTAAAGCGAA
SO4018	Hypothetical protein	1.91	2.07	−99	+	2.26	GTTAAGTGTTTGAGT

(continued)

Table 3.3 *(Continued)*

ORF	Gene Product	Ratio (ARCA/WT) Anaerobic[a]	Aerobic	Start[b]	Strand	Z[c]	Motif
SO0181	Hypothetical protein	0.47	0.61	-61	−	2.49	GTTCATGAATATCAA
SO0076	Hypothetical protein	0.4	0.51	1	+	2.37	GTTAAGGGTAGTTAA
SO0712	Hypothetical protein	0.4	0.52	-240	−	2.06	ATTAATCAAAATTTA
SO0091	Hypothetical protein	0.37	0.19	-117	−	2.44	GTTAAGACTTATTCA
SO4180	Hypothetical protein	0.3	0.23	-256	+	2.04	GTTACTTATTTGTTT
SO1004	Hypothetical protein	0.21	0.53	-183	−	2.17	GTTATTGAAATGTAA
SO0120	Hypothetical protein	0.12	0.12	-233	+	2.61	GTTAACCATGTCGAA
SO0403	Hypothetical protein	0.08	0.03	-724	+	2.2	GTTATTTATTTTTAA
				-466	+	2.06	GTTAATTAAATTAGC
SO1188	Conserved hypothetical protein	NA	0.44	-302	−	2.77	GTTCATTAATTCAAA
SO0946	Conserved hypothetical protein	30.91	4.29	-150	−	2.1	GTAAATAAAATGTAT
SO3480	Hypothetical protein phosphatase	11.35	4.02	-52	−	2.95	GTTAATAATTAGATA
				-294	+	2.28	TTTAATTAAAATTTA
SO3278	Conserved hypothetical protein	9.63	3.91	-25	−	2.38	GTTAATTTTATGTAA
SO4145	Conserved hypothetical protein	7.41	5.16	-193	−	2.76	GTTAATTAAGATGTT
				-86	+	2.13	GTTAACATTATGTTT
SO3846	Conserved hypothetical protein	6.05	5.61	-180	−	2.56	GTTAACTATTGTCTT
				-163	−	2.57	GTTAACTATTGCGTT
SO3514	Hypothetical TonB-dependent receptor	3.97	6.9	-87	+	2.88	GTTAATTATTGTGTA
SO3091	Conserved hypothetical protein	3.76	4.56	-17	+	2.09	GTTAAATAAATGGCT
SO1064	Hypothetical WD domain protein	3.7	1.91	-43	+	2.71	GTTAACTGTGATTAA
SO2042	Conserved hypothetical protein	3.17	0.65	-129	+	2.03	GTTGATAAAGTGTAA
SO3280	Conserved hypothetical protein	3.06	2.1	-234	+	2.56	GTTAACGCTAATCTA
SO0440	Conserved hypothetical protein	2.96	2.94	-14	+	2.07	GTTAATTATCTAAAA
SO1267	Hypothetical glutamine amidotransferase	2.93	1.97	-263	+	2.09	GTAAATAATATTTTT
SO2711	Conserved hypothetical protein	2.81	3.34	-41	−	2.23	GTTAACAAAAGTTG

ID	Description						Sequence
SO4366	Conserved hypothetical protein	2.77	2.23	−179	−	3.03	GTTAATAAAATGCAA
SO0308	Conserved hypothetical protein	2.77	3.49	−133	−	3.03	GTTAATTAAAAGGGA
SO1399	Conserved hypothetical protein	2.2	3.13	−97	+	2.17	GTTATCTCAATGTTA
SO3507	Conserved hypothetical protein	2.06	1.69	−74	−	3.12	GTTAACTCAATGTTA
SO4563	Conserved hypothetical protein	2.03	1.65	−281	−	2.55	GTTCATCCATTGTAA
SO2041	Conserved hypothetical protein	0.77	3.62	−117	−	2.03	GTTGATAAAGTGTAA
SO2144	Conserved hypothetical protein	0.74	2.31	−370	+	2.57	GTTAACCAATACACA
SO1443	Conserved hypothetical protein	0.48	0.47	−173	+	2.86	GTTAATAGTATTATA
SO4196	Conserved hypothetical protein	0.43	0.61	−196	+	2.48	GTTAACACTTATGTT
				−68		2.55	GTTAAGTAAGTTAAT
SO1873	Hypothetical short chain dehydrogenase	0.39	0.24	−44	+	2.15	GTTAATTGTCGCAAT
SO4512	Conserved hypothetical protein	0.27	0.09	−295	−	2.84	GTTAATAATGTGTTT
				−211	+	2.42	GTTAACGCTTTTGGA
				−63	−	2.07	GTTATCAAAAAGTGA
SO3085	Conserved domain protein	2.05	2.39	−19	−	2.01	GTTGATAATATTTAT
SO2064	Conserved domain protein	0.6	3.33	−157	−	2.39	GTTAATCATCTTGGT
Other categories							
SO0643	Transposase, putative	2	1.81	−332	+	2.06	GTTAAATAATTCAAA
Protein fate							
SO3106	Cold-active serine alkaline protease (aprE)	19.93	19.72	−242	−	2.59	GTTAAGAGAATTTTT
				−111	−	2.3	GTAAATTAATTGTTA
				−24	−	2.66	GTTAATCATTATTAT
SO3942	Serine protease, HtrA/DegQ/DegS family	7.04	6.66	−133	−	2.92	GTTCATTAATATTAA
SO1915	Serine protease, subtilase family	4.84	3.8	−135	+	2.84	GTTAATAATGTGTTT
SO0491	Peptidase, M13 family	2.98	1.55	−128	−	2.52	GTTAATGCAAAGTCT
SO0867	Serine protease, subtilase family	2.72	2.68	−282	+	2.72	GTTAATCGTATGAAA
SO4537	Hypothetical Zn-dependent peptidase	2.18	1.33	−402	−	2.52	GTTAATAGATTTAAT

(*continued*)

101

Table 3.3 *(Continued)*

ORF	Gene Product	Ratio (ARCA/WT)		Start[b]	Strand	Z^c	Motif
		Anaerobic[a]	Aerobic				
SO3844	Peptidase, M13 family	1.14	2.2	−524	+	2.59	GTTCAGCATAAGGTA
SO2093	Hydrogenase accessory protein HypB (hypB)	1.32	6.12	−91	−	2.32	GTTAAGTGTGATACA
SO1065	FKBP-type peptidyl–prolyl cis–trans isomerase FkpA (fkpA)	1.6	2.92	−93	−	2.71	GTTAACTGTGATTAA
SO1127	Chaperone protein DnaJ (dnaJ)	1.64	3.02	−67	−	2.06	ATTAACTATTTGAAA
SO3659	Thiol/disulfide interchange protein, putative	3.8	2.41	−103	−	2.25	GTTAACAATGGCGCT
SO1062	Polypeptide deformylase (def-2)	2.21	1	−20	−	2.16	GTTCATCGTTTGGCT
Protein synthesis							
SO2085	Phenylalanyl-tRNA synthetase, alpha subunit (pheS)	1.06	2.08	−364	−	2.99	GTTAACAATAATAAA
Purines, pyrimidines, nucleosides, and nucleotides							
SO2001	5′-Nucleotidase (ushA)	1.12	2.66	−266	+	2.82	GTTCATTATTTTTTT
SO3565	2′,3′-Cyclic-nucleotide 2′-phosphodiesterase (cpdB)	0.31	0.4	−64	−	2.22	GTTAATTCATGCGCT
				−106	−	2.42	GTTAACGAACGGGGA
				−241	−	2.27	GTTAATTAATTGTTG
				−266	−	2	GTTGATGATAATTAA
				−538	+	2.36	GTTCACAGTCATTCA

Locus	Description						
Regulatory functions							
SO1946	Transcriptional regulatory protein PhoP (phoP)	2.53	1.72	−105	−	2.19	GTTCATCAATGTCGA
SO3988	Aerobic respiration control protein ArcA (arcA)	0.11	0.07	−392	+	2.57	GTTAACAAATGCCTA
SO1661	Transcriptional regulator, LysR family	6.49	5.2	−203	+	2.36	GTTAAATAATTGTTA
SO0864	Transcriptional regulator, LuxR family	3.35	3.72	−175	+	2.1	GTTAATTTTTTGTTT
SO3516	Transcriptional regulator, LacI family	1 7	3.73	−321	−	2.88	GTTAATTATTGTGTA
SO1422	Transcriptional regulator, LysR family	0.26	0.12	−11	−	2.69	GTTAAGTGAATTTTT
SO1935	Regulator of nucleoside diphosphate kinase (rnk)	0.38	0.5	−290	+	2.24	GTTAAGTGTGAGAAT
SO1806	psp operon transcriptional activator (pspF)	5.86	3.11	−74	−	3.1	GTTAATAAAATGTTT
SO3689	Sigma-54 dependent nitrogen response regulator	2.64	3.55	−17	+	2.72	GTTAATAAATTTGCT
Signal transduction							
SOO570	Response regulator	2.59	1.47	−443	+	2.55	GTTAAGTATTAGTGT
Transcription							
SOO208	RNA binding protein	2.36	3.32	−302	−	2.45	GTTCATCAAGTGATT
				−30	+	2.76	GTTAACAGTTATTTA
SO3840	RNA polymerase sigma-70 factor, ECF subfamily	0.39	0.98	−73	+	2.04	GTGAATAAATATTTT
Transport and binding proteins							
SO3063	Sodium/alanine symporter family protein	11.32	9.27	−158	−	2.1	GTTAAATGTATGATA
				−92	+	2.18	GTTCACACAAGTCTT

(continued)

Table 3.3 *(Continued)*

ORF	Gene Product	Ratio (ARCA/WT) Anaerobic[a]	Ratio (ARCA/WT) Aerobic	Start[b]	Strand	Z^c	Motif
SO1560	Phosphate binding protein	1.18	2.11	−17	−	2.13	GCTAACTAATTTGTA
SO3134	C4-dicarboxylate binding periplasmic protein (dctP)	1.62	2.31	−151	−	2.03	GTTAATAAAGTGTAG
SO1522	L-lactate permease, putative	2.01	1.21	−619	−	2.88	GTTAATAAAAATATT
				−227	+	2.6	GTTAAGAAAATCCCA
SO3099	Long-chain fatty acid transport protein, putative	0.07	0.16	−265	−	3.24	GTTAATTAAATTATA
				−240	−	2.68	GTTAATACTTTGTGA
				−144	−	2.86	GTTAATTAAAACATT
SO1072	Chitin binding protein, putative	2.3	0.9	−237	−	2.05	GTTAACAGGATGTAA
SO1307	Aquaporin Z (aqpZ)	3.13	0.75	−535	−	2.71	GTTAATCATGTGTTT
				−417	−	2.07	TTTAATTATTTGAAA
SO1100	Extracellular solute binding protein, family 7	2.86	2.36	−84	+	2.28	GTTAACAATATGTTC
SO1750	ABC transporter, ATP binding protein	2.12	2.28	−176	+	2.71	GTTAATTATTGTCAA
SO0450	Major facilitator family protein	0.33	0.79	−142	−	2.13	GATAATGAAATTTAA
Unclassified							
SO4570	Conserved domain protein	NA	0.49	−131	−	3.1	GTTAATAAAATGTTT
Unknown function							
SO0900	Oxidoreductase, aldo/keto reductase family	4.34	1.25	−397	−	2.74	GTTAACAAACATCTA
				−233	−	2	GTTAACTCTAAATCA
SO3497	Aminotransferase, class III	1.14	2.35	−45	−	2.03	GTTAAGGGTGGCTAA
				−76	+	2.85	GTTCATTATTTCTAA

Gene	Description						Motif
SO3301	Flavocytochrome c flavin subunit	0.43	0.44	–	–367	2.75	GTTAATAGTATGCAA
SO4136	Decarboxylase, pyridoxal dependent	0.38	0.65	+	–204	2.69	GTTAACAAACATCAA
				+	–71	2.05	GTTAACTAAAAAAAT
SO4463	Prolyl 4-hydroxylase, alpha subunit domain protein	7.02	3.52	+	–61	2.15	GTAAATAAAATGTTT
SO4457	GGDEF domain protein	5.4	5.68	–	–58	2.22	GTTAACGCAATCCCT
SO3489	GGDEF domain protein	3.55	3.41	+	–148	2.8	GTTAACTATCTGTCT
SO3976	GAF domain protein	2.38	2.46	+	–366	2.14	GTGAATTAATAGTAA
SO3912	TIM-barrel protein, yjbN family	2.26	1.26	+	–19	2.08	GTTAGCAAATTTTAA
SO4324	GGDEF domain protein	0.75	2.03	+	–35	2.71	GTTAATTATAGCGTT
SO0559	MaoC domain protein	0.32	0.38	+	–39	3.22	GTTAATTAAAAGGTA
SO2907	TonB-dependent receptor domain protein	0.25	0.48	+	–441	2.03	GTTAATTCAACGAAA
				–	–267	2.1	GTCAATAAAATGTTT
				–	–163	3.1	GTTAATAAAATGTTT
SO2469	Hypothetical TonB-dependent receptor	0.1	NA	+	–86	2.69	GTTAATGATAGTTTT
				+	–3	2.54	GTTAAGGGAATGAAA

[a]Relative gene expression is presented as the mean ratio of the fluorescence intensity of the *arcA* deletion mutant (ARCA) to that of the parental strain (WT), and NA means that there is not any expression value available.

[b]The start site of the predicted ArcA motif away from the gene translation start site.

[c]The statistical Z-score of the predicted motif from the genomic-wide scan.

Further DNA binding experiments have confirmed the ArcA-P binding motifs predicted here, which include a transcriptional regulator (SO1661), decaheme cytochrome *c* (SO1427), and *hoxK* [120]. Further investigation will be required to verify the functionality of the other predicted ArcA binding motifs in *S. oneidensis* MR-1. We believe these 148 genes with altered expression in the *arcA* deletion strain are still a subset of the ArcA regulon in *S. oneidensis* since some ArcA-regulated genes may not show expression differences under the culture conditions tested. For example, a malate oxidoreductase (*sfcA*) was predicted with a strong binding motif, but the gene expression values were lower than 2-fold and higher than 0.5-fold under both anaerobic and aerobic conditions. The binding motif predicted in *sfcA* was also confirmed by DNA binding experiments [120]. It is also worth mentioning that *pflBA* has the ArcA-P binding site in *E. coli* but no predicted ArcA-P binding sites in *S. oneidensis* MR-1 [92]. The DNA binding experiment also demonstrated that *pflBA* does not have a strong ArcA-P binding site [120].

3.4.6 Conclusions

In summary, we used microarray-based gene expression profiling to examine the transcriptome for an *arcA* null mutant compared to the parental *S. oneidensis* DSP10 strain under both aerobic and anaerobic growth conditions. Transcriptome profiling revealed a total of 654 (294 down regulated; 360 upregulated) and 504 (135; 369) open reading frames (ORFs) that were differentially expressed in an *arcA* deletion mutant relative to the parental strain under aerobic and anaerobic respiratory conditions, respectively. By integrating computational motif prediction tools and microarray analyses, we predicted an *S. oneidensis* ArcA regulon consisting of as many as 148 *S. oneidensis* genes (46 as a positive regulator and 102 as a negative regulator), which included a number of genes shown to be under the direct control of ArcA in other bacteria. Our results also demonstrated that ArcA in *S. oneidensis* acts as both a positive and negative regulator for genes associated with various other functional categories. Both transcriptome data analysis and motif predictions suggest the Arc two-component signal transduction system in *S. oneidensis* regulates a large number of genes that are different from those regulated by ArcA in *E. coli*, although they do have overlapping regulatory functions for a small subset of genes. *S. oneidensis* is typically found at oxic–anoxic interfaces in nature such as sediments and bodies of water where oxygen is limited or absent [60] whereas *E. coli* primarily lives in the mammalian gut [83]. Different living environments for *S. oneidensis* and *E. coli* might result in the observed differences in ArcA regulon compositions during evolution for environmental adaptation. Finally, phenotype characterization indicated that ArcA enables *S. oneidensis* to resist oxidative stress.

3.5 FUTURE PROSPECTS OF CHIP TECHNOLOGY

The applications of high-throughput technologies and functional genomics have proven to be great successes in biological studies. Array technology can be considered a milestone since it has revolutionized biological research (Fig. 3-1). Future development of economically feasible custom chips will permit functional genomics techniques to become routine lab tools. A standardized protocol for data analysis and information mining needs to be completed in the future as well.

ACKNOWLEDGMENTS

The authors acknowledge Dr. Jizhong Zhou for the construction of *S. oneidensis* MR-1 microarrays and the Miami University CFR grant.

REFERENCES

1. Sanger F. Determination of nucleotide sequences in DNA. *Science* 1981;214:1205–1210.
2. Sanger F, Coulson AR, Hong GF, Hill DF, Petersen GB. Nucleotide sequence of bacteriophage lambda DNA. *J Mol Biol* 1982;162:729–773.
3. Smith HO, Tomb JF, Dougherty BA, Fleischmann RD, Venter JC. Frequency and distribution of DNA uptake signal sequences in the *Haemophilus influenzae* Rd genome. *Science* 1995;269:538–540.
4. Jordan B. Historical background and anticipated developments. *Ann NY Acad Sci* 2002;975:24–32.
5. Gress TM, Hoheisel JD, Lennon GG, Zehetner G, Lehrach H. Hybridization fingerprinting of high-density cDNA-library arrays with cDNA pools derived from whole tissues. *Mamm Genome* 1992;3:609–619.
6. Butte A. The use and analysis of microarray data. *Nat Rev Drug Discov* 2002;1:951–960.
7. Bard F, Casano L, Mallabiabarrena A, Wallace E, Saito K, Kitayama H, et al. Functional genomics reveals genes involved in protein secretion and Golgi organization. *Nature* 2006;439:604–607.
8. Cassell GH, Mekalanos J. Development of antimicrobial agents in the era of new and reemerging infectious diseases and increasing antibiotic resistance. *JAMA* 2001;285: 601–605.
9. Aharoni A, Vorst O. DNA microarrays for functional plant genomics. *Plant Mol Biol* 2002; 48:99–118.
10. Chin KV, Kong AN. Application of DNA microarrays in pharmacogenomics and toxicogenomics. *Pharm Res* 2002;19:1773–1778.
11. Katsuma S, Tsujimoto G. Genome medicine promised by microarray technology. *Expert Rev Mol Diagn* 2001;1:377–382.
12. Macgregor PF. Gene expression in cancer: the application of microarrays. *Expert Rev Mol Diagn* 2003;3:185–200.
13. Nocito A, Kononen J, Kallioniemi OP, Sauter G. Tissue microarrays (TMAs) for high-throughput molecular pathology research. *Int J Cancer* 2001;94:1–5.
14. Smith L, Greenfield A. DNA microarrays and development. *Hum Mol Genet* 2003;12 Spec No 1: R1–R8.
15. Stenger DA, Andreadis JD, Vora GJ, Pancrazio JJ. Potential applications of DNA microarrays in biodefense-related diagnostics. *Curr Opin Biotechnol* 2002;13: 208–212.

16. Stoughton RB. Applications of DNA microarrays in biology. *Annu Rev Biochem* 2005; 74:53–82.

17. Ye RW, Wang T, Bedzyk L, Croker KM. Applications of DNA microarrays in microbial systems. *J Microbiol Methods* 2001;47:257–272.

18. Zammatteo N, Hamels S. De Longueville F, Alexandre I, Gala JL, Brasseur F, et al. New chips for molecular biology and diagnostics. *Biotechnol Annu Rev* 2002;8:85–101.

19. Nygren PA, Uhlen M. Scaffolds for engineering novel binding sites in proteins. *Curr Opin Struct Biol* 1997;7:463–469.

20. Brody EN, Willis MC, Smith JD, Jayasena S, Zichi D, Gold L. The use of aptamers in large arrays for molecular diagnostics. *Mol Diagn* 1999;4:381–388.

21. Seetharaman S, Zivarts M, Sudarsan N, Breaker RR. Immobilized RNA switches for the analysis of complex chemical and biological mixtures. *Nat Biotechnol* 2001;19: 336–341.

22. Wilson DS, Nock S. Recent developments in protein microarray technology. *Angew Chem Int Ed Engl* 2003;42:494–500.

23. Kodadek T. Protein microarrays: prospects and problems. *Chem Biol* 2001;8:105–115.

24. Haab BB, Dunham MJ, Brown PO. Protein microarrays for highly parallel detection and quantitation of specific proteins and antibodies in complex solutions. *Genome Biol* 2001;2:RESEARCH0004.

25. Joos TO, Bachmann J. The promise of biomarkers: research and applications. *Drug Discov Today* 2005;10:615–616.

26. Wiltshire S, O'Malley S, Lambert J, Kukanskis K, Edgar D, Kingsmore SF, et al. Detection of multiple allergen-specific IgEs on microarrays by immunoassay with rolling circle amplification. *Clin Chem* 2000;46:1990–1993.

27. Zhou JT, Dorothea K. *DNA Microarray Technology*. Hoboken, NJ: John Wiley & Sons, Inc., 2004.

28. Fodor SP, Read JL, Pirrung MC, Stryer L, Lu AT, Solas D. Light-directed, spatially addressable parallel chemical synthesis. *Science* 1991;251:767–773.

29. McGall GH, Fidanza JA. Photolithographic synthesis of high-density oligonucleotide arrays. In: Rampal JB, editor. *DNA Arrays—Methods and Protocols*. Totowa, NJ: Nuts & Bolts, Humana Press, 2001.

30. Xu D, Li G, Wu L, Zhou J, Xu Y. PRIMEGENS: robust and efficient design of gene-specific probes for microarray analysis. *Bioinformatics* 2002;18:1432–1437.

31. Skaletsky SRaHJ. Primer3 on the WWW for general users and for biologist programmers. In: Krawetz SMS, editor. *Bioinformatics Methods and Protocols: Methods in Molecular Biology*. Totowa, NJ: Humana Press, 2000, pp. 365–386.

32. Li F, Stormo GD. Selection of optimal DNA oligos for gene expression arrays. *Bioinformatics* 2001;17:1067–1076.

33. Emrich SJ, Lowe M, Delcher AL. PROBEmer: a Web-based software tool for selecting optimal DNA oligos. *Nucleic Acids Res* 2003;31:3746–3750.

34. Li X, He Z, Zhou J. Selection of optimal oligonucleotide probes for microarrays using multiple criteria, global alignment and parameter estimation. *Nucleic Acids Res* 2005;33: 6114–6123.

35. Herold KE, Rasooly A. Oligo design: a computer program for development of probes for oligonucleotide microarrays. *Biotechniques* 2003;35:1216–1221.

36. Chou HH, Hsia AP, Mooney DL, Schnable PS. Picky: oligo microarray design for large genomes. *Bioinformatics* 2004;20:2893–2902.

37. Wang X, Seed B. Selection of oligonucleotide probes for protein coding sequences. *Bioinformatics* 2003;19:796–802.

38. Rouillard JM, Zuker M, Gulari E. OligoArray 2.0: design of oligonucleotide probes for DNA microarrays using a thermodynamic approach. *Nucleic Acids Res* 2003;31: 3057–3062.

39. Reymond N, Charles H, Duret L, Calevro F, Beslon G, Fayard JM. ROSO: optimizing oligonucleotide probes for microarrays. *Bioinformatics* 2004;20:271–273.

40. Rimour S, Hill D, Militon C, Peyret P. GoArrays: highly dynamic and efficient microarray probe design. *Bioinformatics* 2005;21:1094–1103.

41. Chung WH, Rhee SK, Wan XF, Bae JW, Quan ZX, Park YH. Design of long oligonucleotide probes for functional gene detection in a microbial community. *Bioinformatics* 2005;21:4092–4100.

42. Borneman J, Chrobak M, Della Vedova G, Figueroa A, Jiang T. Probe selection algorithms with applications in the analysis of microbial communities. *Bioinformatics* 2001;17(Suppl. 1): S39–S48.

43. Dent GW, O'Dell DM, Eberwine JH. Gene expression profiling in the amygdala: an approach to examine the molecular substrates of mammalian behavior. *Physiol Behav* 2001;73:841–847.

44. Warrington JA, Dee S, Trulson M. Large-scale genomic analysis using Affymetrix GeneChip® probe arrays. In: Schena M, editor. *Microarray Biochip Technology.* Natick, MA: Eaton Publishing, 2000.

45. Schermer MJ. Confocal scanning microscopy in microarray detection. In: Schena M, editor. *DNA Microarrays.* New York: Oxford University Press, 1999.

46. Quackenbush J. Microarray data normalization and transformation. *Nat Genet* 2002;32 (Suppl.): 496–501.

47. Li SS, Bigler J, Lampe JW, Potter JD, Feng Z. FDR-controlling testing procedures and sample size determination for microarrays. *Stat Med* 2005;24:2267–2280.

48. Speed T. *Statistical analysis of gene expression microarray data.* Chapman&Hall/CRC, 2003.

49. Wan XF, Verberkmoes NC, McCue LA, Stanek D, Connelly H, Hauser LJ, et al. Transcriptomic and proteomic characterization of the Fur modulon in the metal-reducing bacterium *Shewanella oneidensis. J Bacteriol* 2004;186:8385– 8400.

50. Clark ME, He Q, He Z, Huang KH, Alm EJ, Wan X-F, Hazen TC, Arkin AP, Wall JD, Zhou J, Fields MW. Temporal transcriptomic analysis of *Desulfovibrio vulgaris* Hildenborough transition into stationary phase growth during electron donor depletion. *Appl Environ Microbiol* 2006;72:5578–5588.

51. Yang YH, Speed T. Design issues for cDNA microarray experiments. *Nat Rev Genet* 2002;3:579–588.

52. Conlon EM, Liu XS, Lieb JD, Liu JS. Integrating regulatory motif discovery and genome-wide expression analysis. *Proc Natl Acad Sci USA* 2003;100:3339–3344.

53. Xu D, Olman V, Wang L, Xu Y. EXCAVATOR: a computer program for efficiently mining gene expression data. *Nucleic Acids Res* 2003;31:5582–5589.

54. Thompson W, Rouchka EC, Lawrence CE. Gibbs Recursive Sampler: finding transcription factor binding sites. *Nucleic Acids Res* 2003;31:3580–3585.

55. Hughes JD, Estep PW, Tavazoie S, Church GM. Computational identification of cis-regulatory elements associated with groups of functionally related genes in *Saccharomyces cerevisiae*. *J Mol Biol* 2000;296:1205–1214.

56. Roth FP, Hughes JD, Estep PW, Church GM. Finding DNA regulatory motifs within unaligned noncoding sequences clustered by whole-genome mRNA quantitation. *Nat Biotechnol* 1998;16:939–945.

57. Liu X, Brutlag DL, Liu JS. BioProspector: discovering conserved DNA motifs in upstream regulatory regions of co-expressed genes. *Pac Symp Biocomput* 2001:127–138.

58. Wang T, Stormo GD. Identifying the conserved network of cis-regulatory sites of a eukaryotic genome. *Proc Natl Acad Sci USA* 2005;102:17400–17405.

59. Zou M, Conzen SD. A new dynamic Bayesian network (DBN) approach for identifying gene regulatory networks from time course microarray data. *Bioinformatics* 2005;21:71–79.

60. Heidelberg JF, Paulsen IT, Nelson KE, Gaidos EJ, Nelson WC, Read TD,et al. Genome sequence of the dissimilatory metal ion-reducing bacterium *Shewanella oneidensis*. *Nat Biotechnol* 2002;20:1118–1123.

61. Xing B, van der Laan MJ. A statistical method for constructing transcriptional regulatory networks using gene expression and sequence data. *J Comput Biol* 2005;12:229–246.

62. Chang WC, Li CW, Chen BS. Quantitative inference of dynamic regulatory pathways via microarray data. *BMC Bioinformatics* 2005;6:44.

63. Zhou X, Wang X, Pal R, Ivanov I, Bittner M, Dougherty ER. A Bayesian connectivity-based approach to constructing probabilistic gene regulatory networks. *Bioinformatics* 2004;20:2918–2927.

64. Mehra S, Hu WS, Karypis G. A Boolean algorithm for reconstructing the structure of regulatory networks. *Metab Eng* 2004;6:326–339.

65. Missal K, Cross MA, Drasdo D. Gene network inference from incomplete expression data: transcriptional control of hematopoietic commitment. *Bioinformatics* 2006;22:731–738.

66. Carninci P, Kasukawa T, Katayama S, Gough J, Frith MC, Maeda N,et al. The transcriptional landscape of the mammalian genome. *Science* 2005;309:1559–1563.

67. Aracena J, Gonzalez M, Zuniga A, Mendez MA, Cambiazo V. Regulatory network for cell shape changes during Drosophila ventral furrow formation. *J Theor Biol* 2006;239:49–62.

68. Gutierrez-Rios RM, Rosenblueth DA, Loza JA, Huerta AM, Glasner JD, Blattner FR,et al. Regulatory network of *Escherichia coli*: consistency between literature knowledge and microarray profiles. *Genome Res* 2003;13:2435–2443.

69. Saal LH, Troein C, Vallon-Christersson J, Gruvberger S, Borg A, Peterson C. BioArray Software Environment (BASE): a platform for comprehensive management and analysis of microarray data. *Genome Biol* 2002;3:SOFTWARE0003.

70. Velculescu VE, Zhang L, Vogelstein B, Kinzler KW. Serial analysis of gene expression. *Science* 1995;270:484–487.

71. Tuteja R, Tuteja N. Serial analysis of gene expression: applications in malaria parasite, yeast, plant, and animal studies. *J Biomed Biotechnol* 2004;2004:106–112.

72. Horak CE, Snyder M. ChIP–chip: a genomic approach for identifying transcription factor binding sites. *Methods Enzymol* 2002;350:469–483.

73. Qian J, Lin J, Luscombe NM, Yu H, Gerstein M. Prediction of regulatory networks: genome-wide identification of transcription factor targets from gene expression data. *Bioinformatics* 2003;19:1917–1926.

74. Hanlon SE, Lieb JD. Progress and challenges in profiling the dynamics of chromatin and transcription factor binding with DNA microarrays. *Curr Opin Genet Dev* 2004;14: 697–705.

75. Mockler TC, Chan S, Sundaresan A, Chen H, Jacobsen SE, Ecker JR. Applications of DNA tiling arrays for whole-genome analysis. *Genomics* 2005;85:1–15.

76. Li L, Wang X, Stolc V, Li X, Zhang D, Su N,et al. Genome-wide transcription analyses in rice using tiling microarrays. *Nat Genet* 2006;38:124–129.

77. Stolc V, Samanta MP, Tongprasit W, Sethi H, Liang S, Nelson DC,et al. Identification of transcribed sequences in *Arabidopsis thaliana* by using high-resolution genome tiling arrays. *Proc Natl Acad Sci USA* 2005;102:4453–4458.

78. Sun LV, Chen L, Greil F, Negre N, Li TR, Cavalli G,et al. Protein-DNA interaction mapping using genomic tiling path microarrays in Drosophila. *Proc Natl Acad Sci USA* 2003;100:9428–9433.

79. Selzer RR, Richmond TA, Pofahl NJ, Green RD, Eis PS, Nair P,et al. Analysis of chromosome breakpoints in neuroblastoma at sub-kilobase resolution using fine-tiling oligonucleotide array CGH. *Genes Chromosomes Cancer* 2005;44:305–319.

80. Schumacher A, Kapranov P, Kaminsky Z, Flanagan J, Assadzadeh A, Yau P,et al. Microarray-based DNA methylation profiling: technology and applications. *Nucleic Acids Res* 2006;34:528–542.

81. Lippman Z, Gendrel AV, Colot V, Martienssen R. Profiling DNA methylation patterns using genomic tiling microarrays. *Nat Methods* 2005;2:219–224.

82. Bertone P, Gerstein M, Snyder M. Applications of DNA tiling arrays to experimental genome annotation and regulatory pathway discovery. *Chromosome Res* 2005;13: 259–274.

83. Lynch AS, Lin ECC. Responses to molecular oxygen. In: Neidhardt FC, Curtiss R III, Ingraham JL, Lin ECC, Low KB, Magasanik B, Reznikoff WS, Riley M, Schaechter M, Umbarger HE, editors. *Escherichia coli* and Salmonella: cellular and molecular biology. Washington, DC: American Society for Microbiology, 1996; pp. 1526–1538.

84. Georgellis D, Lynch AS, Lin EC. *In vitro* phosphorylation study of the arc two-component signal transduction system of *Escherichia coli*. *J Bacteriol* 1997;179:5429–5435.

85. Iuchi S, Lin EC. Mutational analysis of signal transduction by ArcB, a membrane sensor protein responsible for anaerobic repression of operons involved in the central aerobic pathways in *Escherichia coli*. *J Bacteriol* 1992;174:3972–3980.

86. Kwon O, Georgellis D, Lin EC. Phosphorelay as the sole physiological route of signal transmission by the arc two-component system of *Escherichia coli*. *J Bacteriol* 2000;182:3858–3862.

87. Kwon O, Georgellis D, Lynch AS, Boyd D, Lin EC. The ArcB sensor kinase of *Escherichia coli*: genetic exploration of the transmembrane region. *J Bacteriol* 2000; 182:2960–2966.

88. Bauer CE, Elsen S, Bird TH. Mechanisms for redox control of gene expression. *Annu Rev Microbiol* 1999;53:495–523.

89. Liu X, De Wulf P. Probing the ArcA-P modulon of *Escherichia coli* by whole genome transcriptional analysis and sequence recognition profiling. *J Biol Chem* 2004;279: 12588–12597.

90. Gao H, Wang Y, Liu X, Yan T, Wu L, Alm E, et al. Global transcriptome analysis of the heat shock response of *Shewanella oneidensis*. *J Bacteriol* 2004;186:7796–7803.

91. Lu S, Killoran PB, Fang FC, Riley LW. The global regulator ArcA controls resistance to reactive nitrogen and oxygen intermediates in *Salmonella enterica* serovar Enteritidis. *Infect Immun* 2002;70:451–461.

92. Lynch AS, Lin EC. Transcriptional control mediated by the ArcA two-component response regulator protein of *Escherichia coli*: characterization of DNA binding at target promoters. *J Bacteriol* 1996;178:6238–6249.

93. Park SJ, McCabe J, Turna J, Gunsalus RP. Regulation of the citrate synthase (gltA) gene of *Escherichia coli* in response to anaerobiosis and carbon supply: role of the *arcA* gene product. *J Bacteriol* 1994;176:5086–5092.

94. Park SJ, Tseng CP, Gunsalus RP. Regulation of succinate dehydrogenase (sdhCDAB) operon expression in *Escherichia coli* in response to carbon supply and anaerobiosis: role of ArcA and Fnr. *Mol Microbiol* 1995;15:473–482.

95. Wilde RJ, Guest JR. Transcript analysis of the citrate synthase and succinate dehydrogenase genes of *Escherichia coli* K12. *J Gen Microbiol* 1986;132: 3239–3251.

96. Wood D, Darlison MG, Wilde RJ, Guest JR. Nucleotide sequence encoding the flavo-protein and hydrophobic subunits of the succinate dehydrogenase of *Escherichia coli*. *Biochem J* 1984;222:519–534.

97. Cunningham L, Gruer MJ, Guest JR. Transcriptional regulation of the aconitase genes (acnA and acnB) of *Escherichia coli*. *Microbiology* 1997;143(Pt 12): 3795–3805.

98. Chao G, Shen J, Tseng CP, Park SJ, Gunsalus RP. Aerobic regulation of isocitrate dehydrogenase gene (icd) expression in *Escherichia coli* by the *arcA* and *fnr* gene products. *J Bacteriol* 1997;179:4299–4304.

99. Lynch AS, Lin ECC. Regulation of aerobic and anaerobic metabolism by the Arc system. In: Lynch AS, Lin ECC, editors. *Regulation of gene expression in Escherichia coli*. Austin, TX: Landes Co., 1996; pp. 361–373.

100. Cotter PA, Chepuri V, Gennis RB, Gunsalus RP. Cytochrome *o* (cyoABCDE) and *d* (cydAB) oxidase gene expression in *Escherichia coli* is regulated by oxygen, pH, and the fnr gene product. *J Bacteriol* 1990;172:6333–6338.

101. Cotter PA, Gunsalus RP. Contribution of the fnr and arcA gene products in coordinate regulation of cytochrome *o* and *d* oxidase (cyoABCDE and cydAB) genes in *Escherichia coli*. *FEMS Microbiol Lett* 1992;70:31–36.

102. Cotter PA, Melville SB, Albrecht JA, Gunsalus RP. Aerobic regulation of cytochrome d oxidase (cydAB) operon expression in *Escherichia coli*: roles of Fnr and ArcA in repression and activation. *Mol Microbiol* 1997;25:605–615.

103. Kuritzkes DR, Zhang XY, Lin EC. Use of phi(glp-lac) in studies of respiratory regulation of the *Escherichia coli* anaerobic sn-glycerol-3-phosphate dehydrogenase genes (glpAB). *J Bacteriol* 1984;157:591–598.

104. Quail MA, Guest JR. Purification, characterization and mode of action of PdhR, the transcriptional repressor of the pdhR-aceEF-lpd operon of *Escherichia coli*. *Mol Microbiol* 1995;15:519–529.

105. Zheng M, Wang X, Doan B, Lewis KA, Schneider TD, Storz G. Computation-directed identification of OxyR DNA binding sites in *Escherichia coli*. *J Bacteriol* 2001;183: 4571–4579.

106. Zheng M, Wang X, Templeton LJ, Smulski DR, LaRossa RA, Storz G. DNA microarray-mediated transcriptional profiling of the *Escherichia coli* response to hydrogen peroxide. *J Bacteriol* 2001;183:4562–4570.

107. Nystrom T, Larsson C, Gustafsson L. Bacterial defense against aging: role of the *Escherichia coli* ArcA regulator in gene expression, readjusted energy flux and survival during stasis. *EMBO J* 1996;15:3219–3228.

108. Seyler RW Jr, Olson JW, Maier RJ. Superoxide dismutase-deficient mutants of *Helicobacter pylori* are hypersensitive to oxidative stress and defective in host colonization. *Infect Immun* 2001;69:4034–4040.

109. Cooksley C, Jenks PJ, Green A, Cockayne A, Logan RP, Hardie KR. NapA protects *Helicobacter pylori* from oxidative stress damage, and its production is influenced by the ferric uptake regulator. *J Med Microbiol* 2003;52:461–469.

110. Guo JT, Ellrott K, Chung WJ, Xu D, Passovets S, Xu Y. PROSPECT-PSPP: an automatic computational pipeline for protein structure prediction. *Nucleic Acids Res* 2004;32: W522–W525.

111. Robinson VL, Wu T, Stock AM. Structural analysis of the domain interface in DrrB, a response regulator of the OmpR/PhoB subfamily. *J Bacteriol* 2003;185:4186–4194.

112. Thormann KM, Saville RM, Shukla S, Spormann AM. Induction of rapid detachment in *Shewanella oneidensis* MR-1 biofilms. *J Bacteriol* 2005;187:1014–1021.

113. Berg OG. Selection of DNA binding sites by regulatory proteins. Functional specificity and pseudosite competition. *J Biomol Struct Dyn* 1988;6:275–297.

114. Menon AL, Mortenson LE, Robson RL. Nucleotide sequences and genetic analysis of hydrogen oxidation (hox) genes in *Azotobacter vinelandii*. *J Bacteriol* 1992;174: 4549–4557.

115. Kortlukc C, Friedrich B. Maturation of membrane bound hydrogenase of *Alcaligenes eutrophus* H16. *J Bacteriol* 1992;174:6290–6293.

116. Rose L, Kaufmann SH, Daugelat S. Involvement of *Mycobacterium smegmatis* undeca-prenyl phosphokinase in biofilm and smegma formation. *Microbes Infect* 2004;6: 965–971.

117. De Souza-Hart JA, Blackstock W, Di Modugno V, Holland IB, Kok M. Two-component systems in *Haemophilus influenzae*: a regulatory role for ArcA in serum resistance. *Infect Immun* 2003;71:163–172.

118. Sengupta N, Paul K, Chowdhury R. The global regulator ArcA modulates expression of virulence factors in *Vibrio cholerae*. *Infect Immun* 2003;71:5583–5589.

119. Thompson JD, Higgins DG, Gibson TJ. CLUSTAL W: improving the sensitivity of progressive multiple sequence alignment through sequence weighting, position-specific gap penalties and weight matrix choice. *Nucleic Acids Res* 1994;22:4673–4680.

120. Gao H, Wang X, Yang ZK, Palzkill, Zhou J. Probing regulon of ArcA in Shewanella oneidensis MR-1 by integrated genomic analyses. *BMC Genomics* 2008;9:42.

4

GENOMIC SIGNAL PROCESSING OF DNA MICROARRAY DATA FOR THE ENHANCED PREDICTION OF AXILLARY LYMPH NODE STATUS OF BREAST CANCER TUMORS

Gordon S. Okimoto

Bioinformatics Shared Resource, Cancer Research Center of Hawaii, University of Hawaii, Manoa, 1236 Lauhala Street, Honolulu, Hawaii 96813

4.1 INTRODUCTION

Breast cancer is the second leading cause of cancer deaths in the women today (after lung cancer) and is the most common cancer among women, excluding nonmelanoma skin cancers. Early detection and more effective treatments have decreased the mortality rate from breast cancer in recent years [1]. Still, according to the World Health Organization, more than 1.2 million people will be diagnosed with breast cancer each year worldwide. The American Cancer Society estimates that each year

Systems Biology and Synthetic Biology   Edited by Pengcheng Fu and Sven Panke
Copyright © 2009 John Wiley & Sons, Inc.

178,000 Americans will be diagnosed with breast cancer with 44,000 deaths expected. Moreover, breast cancer is the leading cause of death among women between 40 and 55 years of age and is the second overall cause of death among American women (exceeded only by lung cancer).

The term breast cancer refers to a collection of cells of the breast that undergo uncontrolled growth, differentiation, and proliferation. Such a collection of cells is known as a malignant breast cancer tumor. Malignant tumors penetrate and destroy healthy tissues of the breast. In addition, a group of cells within a malignant tumor may also break away and spread to other parts of the body. Breast cancer tumor cells that spread from one region of the body into another are called metastases. One goal of this chapter is to characterize the metastatic potential of breast cancer tumors in terms of their global gene expression profiles.

Clinically, the presence of metastatic breast cancer in axillary lymph nodes is the most significant factor in the overall survival of breast cancer patients [2,3]. Although the determination of lymph node status is routine, the surgical procedure is invasive, and the selection of lymph nodes for examination can introduce biases that result in false negative results. Hence, the ability to assess the lymph node status of a breast cancer tumor based on quantitative measurements derived from the tumor itself may obviate the need for axillary lymph node dissection and the morbidity associated with the procedure [4].

Previous attempts to correlate characteristics of primary breast cancer tumors such as S-phase fraction, tumor grade, ploidy, hormone receptor status, and ERBB2 overexpression with lymph node status have been less than successful in terms of the sensitivity and specificity required in clinical settings [5]. Multivariable gene expression profiling appears to have the analytical resolution necessary to complement the known clinical markers currently used for tumor characterization [6]. In addition, the genes, pathways, and predictive models that result from a global analysis of gene expression in breast cancer tumors provide biological hypotheses for highly focused studies to identify new molecular targets that may contribute to improved treatment and personalized care, and a deeper understanding of the systems biology underlying breast cancer metastasis and tumor growth [7].

In this chapter, modern signal processing and pattern recognition techniques that employ the wavelet transform (WT), singular value decomposition (SVD), and neural networks (NNs) are used to analyze microarray data to predict the spread of breast cancer to the axillary lymph nodes based solely on the gene expression profiles of the primary breast cancer tumor. In Section 4.2, background knowledge on breast cancer and genomic signal processing and a description of the main clinical problem of interest are provided; that is, assessing the distant spread of breast cancer to the axillary lymph nodes based on the molecular characteristics of primary tumor. In Section 4.3, a microarray data set based on normal tissue and breast cancer tumor samples is described. In Section 4.4, results of a prior analysis on the Huang data set by Huang et al. are summarized [4]. In Section 4.5, genomic signal processing techniques such as WT and SVD are defined and discussed. The expression data matrix is discussed in Section 4.6 and its connection to Bellman's curse of dimensionality. Experimental design issues for the current study are discussed in Section 4.7. Data preprocessing and data quality issues are discussed in Section 4.8. In Section 4.9, the modeling of phenotypic variation using features extracted from the Huang breast cancer data set using genomic signal processing techniques is described. Validation of pattern recognition models derived from the Huang microarray data is described in Section 4.10. Section 4.11 summarizes the main results of the overall study. Finally, a discussion of the main results is presented in Section 4.12.

4.2 BACKGROUND ON METHODS AND APPROACH

The central dogma of molecular biology states that a gene is transcribed into messenger RNA (mRNA) that in turn is translated into protein [8]. Networks of interacting genes and proteins then give rise to emergent states and system dynamics on these states that characterize the complex biological processes in cells, tissues, organs, and organisms [7]. Although the central dogma has been modified somewhat over the years, the core idea is still valid—the flow of information from genes to mRNA to proteins—and the underlying information processing that it implies forms the basis for life, death, and disease.

A crucial step in the information processing described by the central dogma is the transcription of a gene into mRNA, a process also known as transcription or gene expression. The expression level of every known gene represents the global gene expression pattern of a biological sample. This pattern is in constant flux over space and time and, in particular, changes as a normal cell is transformed into a cancer cell [9]. In this light, it is reasonable to assume that global gene expression patterns of normal and cancerous cells are quite different.

An important goal in cancer systems biology is the proper quantification of differences in global gene expression between normal or cancer cells [4]. The genes that underlie such differences serve as explanatory variables of quantitative models of cancer that are predictive of clinical outcomes or discriminative between different cancer subtypes [10]. For the first time, biologists are now able to measure the expression of every known gene in a tissue sample using a technology called the DNA microarray or chip. In a marriage of integrated circuit manufacturing, nanotechnology, photonics, materials science, biochemistry, and molecular biology, DNA microarrays are able to measure the activity of thousands of genes simultaneously using thousands of distinct probes that are positioned randomly on the surface of a small glass slide, plastic wafer, or silicon chip [11–13]. Each probe is composed of millions of strands of DNA that are complementary to specific mRNA target strand that we wish to quantify. Fluorescent tags are attached to the mRNA strands contained in a special "cocktail" prepared from a biological sample. The chip is immersed in the cocktail for a period of time under stringent conditions to allow the different mRNA target strands to attach, or hybridize, to their complementary DNA probes. The amount of tagged mRNA that hybridizes to a specific probe is quantified based on the intensity of the light that is emitted by the fluorescent probe when illuminated by a beam of laser light. This measure of light intensity serves as a surrogate measure of expression for the gene associated with the probe.

Each probe is interrogated in this way and the individual expression measurements are assembled into a high-dimensional vector that in aggregate provides a global snapshot of gene activity in a given tissue sample [14].

When multiple chips are hybridized to different samples, we have a microarray experiment. In the context of this chapter, the microarray experiment of interest compares the global gene expression patterns of tissue samples composed of normal cells to tissue samples composed of cancerous cells. Since not all 25,000 or so genes of the human genome are associated with cancer, it is reasonable to assume that only a relatively small number of genes will be differentially expressed (DE) between the normal and tumor samples. Here, we view the collection of DE genes as a characterizing biological state of tumor cells in terms of gene expression [15].

DE genes that interact in the context of a known signaling or regulatory pathways can be used as features for pattern recognition applications that are capable of identifying cancer subtypes and predicting clinical outcomes prior to and during treatment [5,9]. Indeed, a list of DE genes most likely intersects with multiple signaling pathways that control the transformation of a normal cell into a cancer cell [16]. Deconvolution of this list into pathways of functionally related, interacting genes helps to elucidate causal mechanisms that may lead to a more personalized treatment of cancer through early diagnosis and drugs targeted to the specific genes in specific pathways [17].

As with any other sensor system, data collected by DNA microarrays are contaminated with significant amounts of systematic (low-frequency) and random (high-frequency) variation. The primary sources for such unwanted variation include experimental error introduced by the data acquisition process unique to DNA microarrays and biological variation that exists between different tissue samples. The resulting $p \times n$ data matrix of a typical microarray experiment, where p equals the number of genes and n equals the number of samples, is ill posed in that p is greater than n ($p \gg n$) by several orders of magnitude. This situation is analogous to having many more unknowns than equations in a system of linear equations whereby the system in question has no solution. Standard statistical analysis of such ill-posed data results in models that mistake noise for signal, and hence, fail to capture the underlying biological processes that give rise to the observed patterns of differential gene expression [18,19]. Finally, background noise in microarray data is multiplicative instead of additive, which can confound standard statistical analysis and modeling techniques [20]. Modern signal processing, pattern recognition, and machine learning techniques provide the means to properly analyze and model the noisy, high-dimensional data sets generated by microarray experiments [21].

4.3 THE HUANG BREAST CANCER DATA SET

The global transcriptional profiles of 37 primary breast cancer tumor samples were measured using Affymetrix U95-AV-5 GeneChip microarrays [4]. Each microarray profiled the steady-state mRNA levels of 12,625 genes simultaneously in a single tumor sample. Of the 37 samples that were profiled, 19 were labeled as "negative" or low-risk samples and 18 as "positive" or high-risk samples based on microscopic examination of lymph node samples obtained by axillary lymph node dissection. Among ER positive patients, the high-risk (or positive) clinical profile was represented by metastases involving 10 or more lymph nodes. The low-risk (or negative) profile was defined by node negative patients of age greater than 40 years with tumor size less than 2 cm. The main hypothesis for this experimental design asserts the existence of global gene expression patterns capable of discriminating between the high- and low-risk tumor samples.

Microarray data were acquired using protocols established by Affymetrix Corporation for the U95-AV-5 GeneChip. The amount of starting total RNA for each GeneChip hybridization was 20 µg. First-strand cDNA synthesis was generated using a T7-linked oligo-dT primer, followed by second-strand synthesis. An *in vitro* transcription reaction was performed to generate the cRNA containing biotinylated UTP and CTP, which was subsequently chemically fragmented at 95°C for 35 min. The fragmented, biotinylated cRNA was hybridized in MES buffer (2-[N-morpholino] ethansulfonic acid) containing 0.5 mg/mL acetylated bovine serum albumin to Affymetrix GeneChip HumanU95Av2 arrays at 45°C for 16 h, according to the Affymetrix protocol. The arrays contained probes that measured the expression of over 12,000 genes and ESTs. Arrays were washed and stained with streptavidin phycoerythrin (SAPE, Molecular Probes). Signal amplification was performed using a biotinylated antistreptavidin antibody (Vector Laboratories, Burlingame, CA) at 3 µg/mL. This was followed by a second staining with SAPE. Normal goat IgG (2 mg/mL) was used as a blocking agent.

Each hybridized GeneChip was scanned using an Affymetrix GeneChip scanner, and the expression value for each gene was calculated using the Affymetrix Microarray Analysis Suite (v5.0), computing the expression intensities in "signal" units defined by the software. Scaling factors were determined for each hybridization based on an arbitrary target intensity of 500. Scans were rejected if the scaling factor exceeded a factor of 25, resulting in only one reject. Files containing the computed signal intensity value for each probe cell on the arrays, files containing experimental and sample information, and files providing the signal intensity values for each probe set, as derived from the Affymetrix Microarray Analysis Suite (v5.0) software, were generated and posted on the Huang study Web site.

4.4 RESULTS OF THE HUANG STUDY

Using k-means clustering, SVD, and statistical tree models, Huang et al. discovered a gene expression signature based on 200 genes that was able to discriminate between high-risk and low-risk samples with 90 percent accuracy [4,5]. Moreover, they showed that many of the genes that defined the prognostic signature mapped to biological processes related to breast cancer. In particular, an interferon-mediated immune response was identified in the list of DE genes significantly changed in expression between the positive and negative sample groups of the experiment.

In brief, the data analysis employed by Huang et al. first removed genes with fold change less than two and maximum intensity less than nine on a $\log_2$ scale. This filtering step resulted in a reduction in the number of genes available for downstream processing from 12,625 to 7030. K-means clustering was then applied to the filtered genes to obtain 496 gene clusters. Singular value decomposition was used to extract the first principal component of each gene cluster. This principal component was called the "metagene" associated with the gene cluster. The 496 metagenes were presented as input to a classification tree, where the sample space is recursively partitioned into subsets that best fit the data based on a Bayesian measure of association between metagenes and a binary variable encoding the lymph node status of the samples [5]. Lists of genes were generated from the top four metagenes having the largest marginal Bayes' association. The list was extended by adding additional genes that are highly correlated with any one of the top four metagenes.

Metagenes were discovered that were highly associated with lymph node status. These discriminative metagenes were capable of predicting lymph node status in individual patients with about 90 percent accuracy using the classification tree model based on the microarray data. The metagenes also defined distinct groups of genes that participated in biological processes related to metastatic breast cancer. It was concluded that gene expression patterns can be used to accurately predict the lymph node status of a primary breast cancer tumor based solely on the gene expression patterns of the tumor itself [4].

4.5 GENOMIC SIGNAL PROCESSING

An important goal of bioinformatics and systems biology in cancer research is to improve the diagnosis, prognosis, and treatment of cancer through more accurate disease classification and patient stratification using quantitative techniques that take full advantage of the genome-wide data generated by new technologies such as DNA microarrays [10,22,23]. This comprehensive approach to understanding cancer allows for the design of therapeutic strategies that are targeted to the specific cancer subtypes that are unique to an individual patient. The hope is that a deeper understanding of the molecular heterogeneity of cancer could potentially improve the effectiveness of existing treatment regimens based on the ability to predict therapeutic response and adverse effects, as well as suggest new strategies based on the identification of new molecular targets susceptible to pharmacological intervention [9,23].

By genomic signal processing (GSP) we mean the identification, isolation, and extraction of information from high-dimensional data, such as that produced by DNA microarrays, that are useful for modeling and/or explaining observed changes in well-defined clinical or biological phenotypes. In this chapter, we describe a number of GSP techniques that in aggregate enable the minimally invasive prediction of distant changes in lymph node status based solely on the gene expression profile of the primary breast cancer tumor.

To facilitate the application of GSP techniques, we view the prediction of lymph node status as a problem in pattern recognition where the raw data are preprocessed, informative genes are identified, feature patterns are extracted from the expression profiles of these genes, and finally a pattern recognition (PR) model is formulated

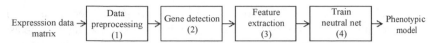

Figure 4-1 Modeling phenotypic variation using DNA microarrays. (1) Raw expression data matrix is preprocessed to remove systematic error and equalize noise. (2) Differentially expressed genes are selected from the preprocessed data matrix. (3) Feature patterns with reduced dimensionality and noise content are extracted from the data matrix of significant genes. (4) The extracted feature patterns are used to train a neural network to discriminate between phenotypic classes. The trained neural network constitutes a model of phenotypic variation defined in terms of gene expression.

based on the extracted feature patterns [24,25]. Figure 4-1 shows a high-level flowchart of the information processing chain used to formulate a NN model of breast cancer metastasis based on GSP features extracted from whole genome expression profiles. In brief, the following steps are involved: (1) the microarray data are normalized and equalized; (2) differentially expressed genes are detected; (3) feature patterns are extracted from the list of DE genes; (4) and a NN classifier is trained on the extracted feature patterns. An important step in the modeling process that is absent in Figure 4-1 is the objective assessment of predictive power of the resultant model using cross-validation techniques, which is discussed in Section 4.11.

4.6 THE EXPRESSION DATA MATRIX

Specifically, a $p \times n$ expression data matrix A_{raw} is formed where each of the n columns of A_{raw} represents the expression profile over p genes of a tumor sample. It follows that each of the p rows of A_{raw} represents the expression profile of a gene over the n samples of the microarray experiment [18]. We assume the n columns of A_{raw} are grouped so that the lymph node negative samples comprise first n_1 columns of A_{raw} for $j = 1, 2, \ldots, n_1$ and the lymph node positive samples comprise the next n_2 columns of A_{raw} where $n = n_1 + n_2$.

Typically, $p \gg n$, (p much greater than n) where, for example, $p = 12,625$ and $n = 37$ for the Huang microarray data set. This situation is known as "Bellman's curse of dimensionality," which states that the number of samples needed to adequately model phenotypic variation grows exponentially with the number of input variables [24,26]. Hence, the Huang data matrix is mathematically ill posed for analysis using standard statistical approaches since the number of variables (genes) exceeds the number of equations (microarrays) by several orders of magnitude.

NN models based on a large number of input genes (and a relatively small number of samples) admit a large number of possible solutions that vary widely in terms of prediction performance, and hence, generalize badly from a finite set of training data to the general population that were unseen during training. Methods must be used to reduce the number of variables (i.e., dimensionality) without losing information that is relevant to solving the discrimination or prediction problem at hand [27,28]. Standard statistical techniques based on optimality arguments where the number of samples grow asymptotically without bound relative to the number of variables are inadequate for the so-called "large p, small n" problems. Indeed, microarray experiments require statistical techniques based on asymptotics where the number of variables increases without bound relative to the number of samples [29]. Unfortunately, the statistical analysis of ill-posed problems is less well developed than for situations where the number of samples is plentiful and the number of variables small.

One approach is to use Bayesian statistics to constrain the space of possible solutions on a finite training set and automatically select parsimonious models that generalize to a larger population [21]. Another approach is to use signal processing techniques to select only highly informative features that reduce input space dimensionality, which in turn alleviates the negative impact of Bellman's curse on the ability of the derived model to generalize [30]. In this chapter, we describe methods more closely aligned to the latter approach where signal processing techniques are used to extract highly informative, low-dimensional features from expression data matrix. These feature patterns are then used to train NN classifiers that are capable of distinguishing benign breast cancer tumors from tumors that have spread to the axillary lymph nodes.

4.7 EXPERIMENTAL DESIGN

Global gene expression profiles are obtained using Affymetrix HU-95 GeneChip. Each hybridized GeneChip was "vectorized" into column vectors composed of 12,625 components, where each component represents the relative expression level of a single transcript on a given chip [4]. As described above, the vectorized chips were arranged to form the columns of a $12,625 \times 37$ expression data matrix A_{raw} of raw expression values where columns 1–19 represented the negative samples and columns 20–37 the positive samples. The data matrix A_{raw} was quantile normalized to obtain the preprocessed data matrix A_{nrm}.

A *sample response function* (SRF) for the Huang microarray experiment is a mapping $h:\{1,2,\ldots,n\} \rightarrow L$ defined on the columns of A_{nrm} where $L = \{-1,1\}$. Note that h reflects the phenotypic grouping of the samples such that $h(i) = -1$ for $1 \le i \le 19$ and $h(i) = -1$ for $20 \le i \le 37$. Note that h has the shape of a step function on the column indices of A_{nrm}. The ordered triple (A_{nrm}, L, h) represents the experimental design for the microarray experiment based on the Huang data set. Figure 4-2 visualizes the components of the microarray experiment (A_{nrm}, L, h). Here, Figure 4-2a is the step-like SRF defined by h for the microarray experiment, (A_{nrm}, L, h) and Figure 4-2b is a z-scored image of the data matrix $\log_2 (A_{\text{nrm}})$ [31].

The fundamental hypothesis of microarray data analysis (FHMD) for (A_{nrm}, L, h) asserts the existence of a set of genes that are highly correlated with step function h shown in Figure 4-2a. For example, a numerical measure, t_g, can be computed for each gene g defined by

$$t_g = t(x_g, h) \equiv \frac{x_g^T h}{s_g} \tag{4-1}$$

(a)

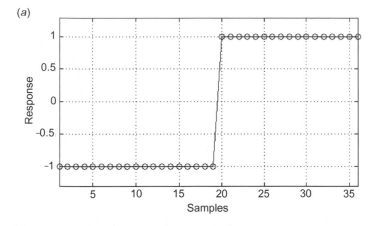

(b)

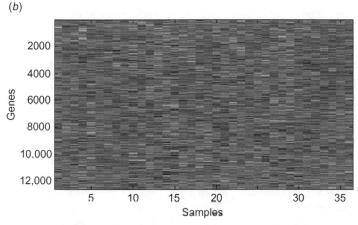

Figure 4-2 The experimental design for the Huang breast cancer microarray experiment. (a) Step-like sample response function h defined on the columns of A_{raw} that groups the columns of the data matrix into lymph node negative (columns 1–19) and positive (columns 20–36) sample groups. (b) The z-scored image of the $12,625 \times 36$ expression data matrix A_{raw} after quantile normalization and $\log_2$ transformation.

where (1) x_g is the expression profile of gene g over n samples; (2) s_g is the standard deviation of x_g; and (3) $x_g^T h$ is the correlation between x_g and h. Note that Equation 4-1 is equal to the correlation between x_g and h normalized by the standard deviation of x_g, which is also known as the t-score for g. The genes are ordered by absolute t-scores, and a subset of genes with the largest absolute t-scores is chosen based on some statistical threshold such as p-value or false discovery rate (FDR). The resulting list of genes is necessarily correlated with the unit step function h in accordance with Equation 4-1 and, in the this case, represents the genes that show the most consistent differential expression between the positive and negative samples of the Huang breast cancer data set. Such a set of genes is said to be differentially expressed between the two sample groups in accordance with the t-test.

4.8   DATA PREPROCESSING AND DATA QUALITY ASSESSMENT

The columns of A_{raw} are quantile normalized to facilitate comparison between the samples represented by the columns of A_{raw}, which results in the data matrix A_{nrm} [32]. Each entry of A_{nrm} is then $\log_2$ transformed to equalize variation over the entire range of expression values resulting in the $p \times n$ preprocessed data matrix $A_{\log 2}$ [20]. The preprocessed data matrices A_{nrm} and $A_{\log 2}$ form the basis for further downstream information processing depending on the algorithms used. For the purposes of this chapter, our focus will be on the normalized expression data matrix A_{nrm}. Note that quantile normalization essentially models and removes a low-frequency, correlated signal that corresponds to the systematic experimental error in the raw microarray data.

The primary goal of the preprocessing step is the removal of systematic nonrandom variation from the raw data to facilitate the comparison of gene expression across multiple chips. The quantile normalization procedure can be described in two steps:

- Create a mapping between ranks and expression values; that is, for rank k, find the n genes, one per array, that have rank k in terms of gene expression and compute their average expression over the n samples;
- For each gene on each array, replace the measured expression value with the rank-average expression for that gene.

Note that quantile normalization is an aggressive strategy that produces identical distributions for each array. On the contrary, quantile normalization is extremely fast, since it only requires a single sort of the data matrix, a computation of means across sorted rows, and a single pass through the data [33]. Note that other normalization schemes exist that employ nonparametric modeling techniques such as locally weighted polynomial regression (lowess) to characterize the systematic error in raw microarray data. One such method identifies genes that are invariant in terms of variation between normal and disease sample classes and models the low-frequency, correlated signal in these invariant genes using a lowess-type smoother. The resulting error model is then used to correct all the raw data for systematic error. Normalization based on lowess smoothing of invariant genes tends to be a less aggressive a procedure than quantile normalization.

Another important preprocessing step is the $\log_2$ transformation of the data matrix A_{nrm} to decouple variation in fold change from expression level. This decoupling makes the data appear more bell shaped and hence improves the performance of downstream statistical analysis algorithms designed to detect DE genes. Indeed, raw microarray data have essentially a log-normal distribution, which implies that the $\log_2$ transform of the data should be more or less normally distributed or at least unimodal [20]. Other more powerful variance stabilization methods have been proposed that view microarray data as having a normal distribution at low expression levels, a log-normal model at high intensities, and a mixture of both at intermediate intensities. The impact of such mixture models on classification and prediction performance is currently being evaluated.

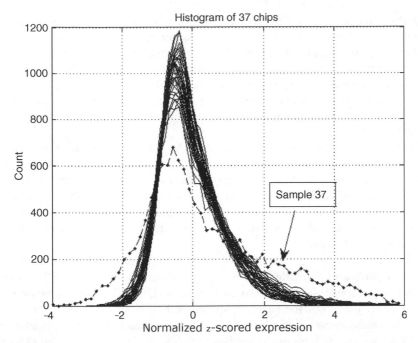

Figure 4-3 Histogram plots for quantile normalized, $\log_2$ transformed, z-scored microarray data from the Huang breast cancer data set. Note that most of the samples have similar histograms, while sample 37 is an outlier in distribution. Therefore, we removed sample 37 from the MANINI analysis since an outlier of that magnitude would distort the final performance results.

The chips that comprise the Huang microarray experiment are assessed for data quality using a number of standard statistical techniques. For example, histograms of all 37 columns of the quantile normalized data matrix A_{nrm} after z-scoring are shown in Figure 4-3. Note that sample 37 is clearly an "outlier" in distribution when compared to the other 36 chips of the experiment. Moreover, pairwise correlation analysis of the raw and normalized data indicates that sample 37 has relatively low correlation with the other 36 chips. These results suggest that the expression values for sample 37 were corrupted at some point in the data acquisition process. Hence, sample 37 was removed from this study, although we note that sample 37 was retained by Huang et al. in their study.

4.9 THE MODELING OF PHENOTYPIC VARIATION

The modeling of phenotypic variation in terms of gene expression is a pattern recognition problem that can be solved by mapping gene expression patterns directly to phenotypic states using NN classifiers [21]. Such models are known as discriminant classifiers. Note that it is not necessary to delineate the biological mechanism underlying the observed variation in disease phenotypes since the predictive model is implemented based solely on the *association* between gene expression and phenotype [15]. The proposed formulation of a discriminant pattern recognition model for the prediction of lymph node status involves a four-step information processing chain shown in Figure 4-1. Each step of the processing chain requires the use of GSP techniques. Data preprocessing was described in the previous section. Details on the remaining steps of the processing chain shown in Figure 4-1 are described below.

The proposed GSP processing chain includes the following signal processing components: (1) Microarray Analysis of Intensities and Ratios (MANINI) detection algorithm for identifying DE genes, (2) pathway compression for data reduction, (3) wavelet transformation for the separation of signal from noise, (4) singular value decomposition for further dimensionality reduction and filtering, and (5) neural networks for encoding the information contained in features derived from microarray data using GSP techniques.

The signal processing steps outlined above can be combined in different ways leading to different information processing algorithms. For example, the SVD of the wavelet transformed data is known as wavelet/SVD (WSVD) signal processing. Alternatively, the SVD of the wavelet transform of the data matrix of genes confined to a specific pathway specified by Ingenuity Pathway Analysis (IPA) is denoted by WSVD/IPA signal processing. The following sections describe the different signal processing components and how they are combined to form analysis pipelines that lead to robust predictors of lymph node status based solely on the gene expression profiles of the primary breast cancer tumor.

4.9.1 The MANINI Detection Algorithm

An alternative to the *t*-test for the supervised detection of genes highly correlated to a given SRF is the so-called MANINI detection algorithm. The MANINI detector was specially designed to handle small numbers of samples often encountered in real-world case/control microarray experiments, since in the absence of a large number of chips, one is hard-pressed to do better than use fold change to detect DE genes [34]. MANINI was also designed to detect DE genes with expression profiles that are inconsistent or highly variable over the samples of the experiment [35]. There is a growing trend in microarray data analysis toward detecting genes in heterogeneous samples (e.g., tumor samples) with expression patterns that may be too inconsistent for more conventional detector designs such as the *t*-test [35,36].

For example, assume that a single signaling pathway is modulated between two biological conditions. Due to sample heterogeneity and biological variation, it may be that different components of the pathway are DE for different samples. In this case, individual genes that participate in the pathway would be difficult to detect using the *t*-test since the associated expression profiles would be highly variable over the samples of the experiment. The MANINI detector on the other hand would be able to select such genes for downstream ontological and pathway analysis, where different ensembles of functionally related genes are assigned to the common pathway to which they belong [7,37].

Let x and y be $p \times 1$ column vectors representing the geometric averages of the control and disease chips, respectively. Let

$$M = \log 2(y) - \log 2(x) = \log 2\left(\frac{y}{x}\right)$$

and

$$A = \tfrac{1}{2}[\log 2(y) + \log 2(x)] = \log 2\left(\sqrt{xy}\right)$$

Figure 4-4 shows the Minus–Add (MA) scatter plot of M versus A where fold change is plotted versus average expression in $\log_2$–$\log_2$ space.

We note three things about the MA scatter plot: (1) the MA plot can be viewed as a visualization of differential expression; (2) genes located on the periphery of the MA data cloud are likely to be DE; and (3) the vertical variation of the MA data cloud is a function of expression level. This suggests a strategy for detecting DE genes by selecting only those genes that "live" on the edge of the MA data

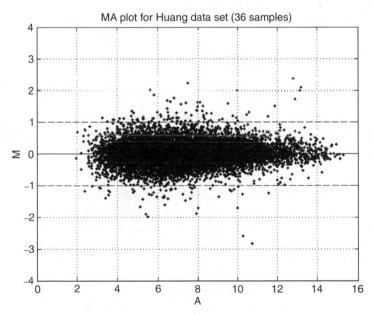

Figure 4-4 Minus–Add scatter plot for Huang breast cancer microarray data set. Each point represents the geometric average of expression (A-axis) and fold change (M-axis) for a gene in $\log_2$–$\log_2$ space. Horizontal dotted lines at $M = \pm 1$ represent constant twofold change in expression between positive and negative samples. Note the dependence of the variance of M on the intensity of A.

cloud based on a threshold that adapts to the spread of the cloud as a function of expression level [32,38]. Note that using standard twofold change as a constant threshold for differential expression over the entire range of expression values (as represented by the horizontal dashed lines in Fig. 4-4) is clearly inappropriate since too many genes are called DE at lower expression levels and too few genes are called DE at higher expression levels. A better strategy would be to adaptively threshold the genes of the MA scatter plot based on expression level.

The MANINI detector implements this idea by "binning" the horizontal axis of the MA plot into k quantiles. Each bin contains about the same number of genes that have similar expression intensities. This quantization scheme also implies that all genes in a bin have about the same degree of variation since variation is a function of intensity. For a given bin containing m genes, we assume the signal model $y_i = s_i + \eta$ for $i = 1,2, \ldots, m$. Here, s_i is the true expression level of the ith gene, and $\eta \sim N(0,\sigma)$ is normally distributed random variable with mean zero and known variance σ^2. Empirical studies show that this is a reasonable assumption for wide range of real-world data sets. It follows that differential expression within the bin can be modeled as a m-dimensional random vector, $y = [y_1,y_2, \ldots, y_m]^T$, where $E(y) = [s_1,s_2, \ldots, s_m]^T$ is sparse [29]. By sparse it is meant that most of the components of the true signal vector $E(y) = [s_1,s_2, \ldots, s_m]^T$ are zero.

In the field of wavelet denoising, Donoho and Johnstone showed that a signal contaminated by zero-mean Gaussian noise can be optimally filtered by thresholding the wavelet coefficients of the noisy signal [39]. The wavelet transform of a noisy signal localizes the information content of a signal simultaneously over time and scale. In this case, high-frequency noise is usually confined to the high-resolution scales and low-frequency coherent signal is concentrated in the low-resolution scales. Donoho and Johnstone found that by simply thresholding the higher resolution wavelet coefficients of the noisy signal and then applying the inverse wavelet transform, one can optimally estimate the true underlying signal assuming that it is sparse [29]. Thresholding in this manner to estimate a signal embedded in noise is called testimation. Note that the wavelet coefficients at a given scale of resolution form a Gaussian random vector where only a few of the coefficients are different from zero; that is, the wavelet coefficients at each scale form a sparse random vector.

Based on the properties of the MA scatter plot, the $\log_2$ ratios within an expression bin of a MA plot can be viewed as a sparse Gaussian random vector where only a small number of genes within the bin are truly DE. This observation suggests the application of the Donoho–Johnstone (DJ) universal threshold directly to the $\log_2$ ratios of a given intensity bin using

$$\hat{s}_i = \begin{cases} y_i & \text{if } |y_i| > \sigma\sqrt{2(1-\beta)\log(m)} \\ 0 & \text{otherwise} \end{cases} \tag{4-2}$$

to select those genes with "true" nonzero differential expression for $i = 1,2, \ldots, m$, where $\hat{s}_i$ is an estimate of the ith component of the true signal s and $0 < \beta < 1$ [29,39].

Note that $1 - \beta$ represents a measure of the "sparseness" of y. It can be shown that this disarmingly simple procedure is asymptotically optimal in a statistical sense (i.e., it minimizes the maximum expected risk) as m grows without bound and that its application to a noisy high-dimensional data vector amounts to a Bonferroni-type correction for multiple comparisons [29].

Note the optimality of the estimate $\hat{s}$ depends on the number of variables m (or genes) growing without bound. This is in sharp contrast to the situation in classical statistics where the number of samples is assumed to grow without bound [18]. Hence, Equation 4-2 actually becomes more accurate when the number of genes is large, provided the random vector remains sparse. This is exactly the situation for most whole genome expression profiling studies and precisely the opposite of what is required for standard statistical algorithms to work properly. Hence, the MANINI detection algorithm takes advantage of the large number of genes interrogated in a typical microarray experiment by binning the genes into subgroups containing equal numbers of genes with similar expression levels. Since the total number of genes is large, each subgroup will have enough genes for the DJ universal threshold to work (for that particular subgroup). For example, the Affymetrix U133 Plus 2.0 GeneChip uses over 54,000 probe sets to interrogate over 47,000 transcripts that represent approximately 38,000 genes and gene variants. Binning the horizontal axis of the MA plot into 51 quantiles results in 50 subgroups of genes where each subgroup contains 1094 measurements with similar expression levels. Note also that each bin of the MA plot contains only a few genes that are truly DE; that is, the $\log_2$ ratios in each bin contain a sparse signal for DE. This allows the application of the DJ threshold to most microarray experiments where the signal for DE is sparse both locally and globally.

Note that Equation 4-2 was implemented for each bin using the mean absolute deviation (MAD) statistic in place of σ to provide a robust estimate of the variation within the bin. Genes that exceeded the Donoho–Johnstone threshold for the bin were called DE. The union of all genes called DE over all bins of the MA scatter plot resulted in a list of genes that are globally DE for the microarray experiment [31].

The MANINI detection algorithm was used to analyze the $12{,}625 \times 26$ $\log_2$ transformed, quantile normalized expression data matrix denoted by A. We summarize the MANINI results in Figure 4-5. Differentially upregulated genes are marked by up-triangles, differentially downregulated genes are marked by down-triangles, and genes unchanged in expression are represented by points. The black dashed lines located at $M = \pm 1$ represent constant thresholds for a twofold change in expression in either the up ($M = 1$) or down ($M = -1$) direction. The quantiles of A-axis are represented by dark and gray vertical bands shown in the body of the MA plot. The MANINI algorithm calls a gene within a given quantile, or bin, differentially expressed if its absolute M-value exceeds the DJ noise-adjusted threshold for that bin. The union of genes called DE over all bins represents the global signal for DE detected by MANINI.

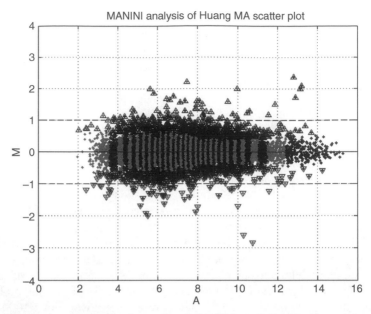

Figure 4-5 MA plot summarizing the results of a MANINI analysis of the Huang breast cancer microarray data set. Each point of the RI plot is a gene represented by average expression and fold change. Genes represented by up-triangles were called significantly upregulated by MANINI, while genes represented by the down-triangles were called significantly downregulated. The vertical bands in the body of the MA plot represent the 50 quantiles used to segment the A axis into disjoint bins containing approximately the same number of genes. Each bin represents a separate and distinct DE detection problem for genes that have comparable expression levels.

In Figure 4-6, we show scaled images of the data matrices U and D composed of genes that MANINI called significantly up- and downregulated, respectively. The rows of U shown in Figure 4-6a represents the samples of the experiment clustered by similarity of their expression profiles over genes called significantly upregulated by MANINI. A similar interpretation applies to the rows of D shown in Figure 4-6a. The results of the cluster analysis are summarized by a dendrogram shown on the left side of U and D. Moreover, the cluster structure over all samples is shown on the right side of Figure 4-6a and b, where the negative samples are labeled 1–19 and positive samples 20–36. Both data matrices U and D are quantile normalized, $\log_2$ transformed, and z-scored by rows.

Note that in Figure 4-6a and b, the samples to a large extent segregate by lymph node status with some erroneous classifications that are probably due to the inclusion of genes that are falsely called DE by the MANINI detector. This suggests that we may be able to identify a subset of genes that are able to do a better job of discriminating between positive and negative breast cancer tumor samples based on gene expression. We also note that although the MANINI detector is designed to detect genes with expression profiles that conform to a step-like response, it is also less sensitive to

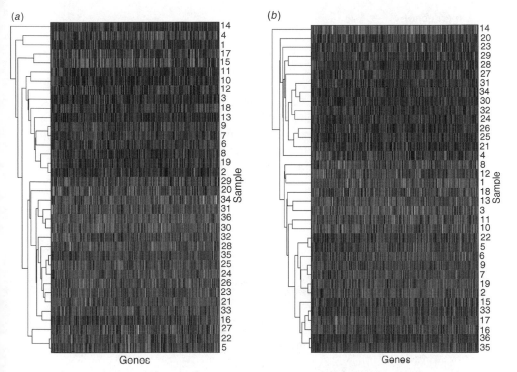

Figure 4-6 Scaled images of the expression data matrices for up- and downregulated DE genes called by the MANINI detector. The rows of each data matrix are hierarchically clustered by gene expression profile over the samples. (a) Data matrix composed of DE genes called upregulated by the MANINI algorithm. (b) Data matrix composed of DE genes called downregulated by the MANINI algorithm. Note that each set of genes approximately segregates the samples into two distinct clusters containing positive and negative breast cancer tumor samples.

deviations from the ideal response than standard statistics such as the two-sample t-score and hence will call a broader range of expression patterns as statistically significant.

4.9.2   MANINI and Signal Detection Theory

Let A_{nrm} be a $p \times n$ normalized expression data matrix and let r represent a one-to-one mapping of the column indices $\{1, 2, \ldots, n\}$ of A_{nrm} to $\{-1, 1\}$ defined by

$$r(i) = \begin{cases} -1 & \text{if the } i\text{th sample is a control} \\ 1 & \text{if the } i\text{th sample is a case sample} \end{cases}$$

Here, the function r is called a response function for the experiment. Figure 4-7 shows a response function equal to the unit step function h defined for a 64-chip microarray experiment by

$$h(i) = \begin{cases} -1 & \text{if } i = 1, 2, \ldots, n_1 \\ 1 & \text{if } i = n_1 + 1, n_1 + 2, \ldots, n_1 + n_2 \end{cases}$$

where $n_1 + n_2 = n$. Note samples 1–32 are controls and samples 33–64 are cases. Let g_i denote the row expression profile of the ith gene of A_{nrm} for $i = 1, 2, \ldots, p$. Then the ith gene is said to be differentially upregulated in the treated group if g_i is positively correlated with the step function h. Conversely, the ith gene is differentially downregulated in the treated group if g_i is negatively correlated

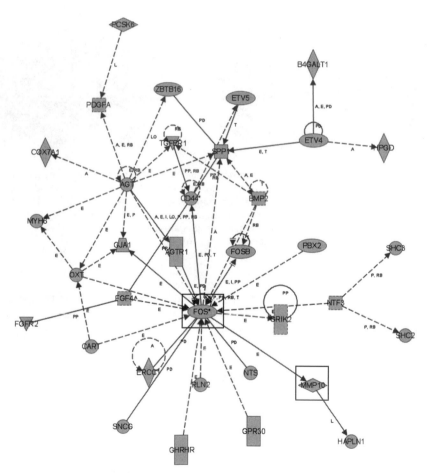

Figure 4-7 Top IPA network for downregulated genes. The network contains 35 genes with a score of 55 (p-value ~1.0E-55). An inferred function of the network is *Cancer*. The highly connected hub gene FOS has been implicated in the regulation of cell proliferation, differentiation and apoptosis, and MMP10 has also been implicated in tumor metastases.

with h. We note that transcript degradation in the control and/or case classes can also generate step-like gene expression profiles. For example, degradation of message for the ith gene in only the control samples can falsely produce a gene expression profile that suggests upregulation of the gene in question in the treated samples. We will assume that strict quality control of sample preparation and hybridization protocols will reduce message degradation and experimental variation to a minimum.

Let $y = [y_1, y_2, \ldots, y_n]^T \in \mathbb{R}^n$ be a noisy gene expression profile where each y_i represents the measured expression level of a gene in the ith sample for $i = 1, 2, \ldots, n$. In the context of statistical hypothesis testing, let

$$H_0 : y = \eta$$

and

$$H_1 : y = h + \eta$$

where $\eta \in \mathbb{R}^n$ is a Gaussian, zero-mean, independent, identically distributed random vector representing noise in the data. Then the Neyman–Pearson Lemma states that uniformly the most powerful test for H_0 versus H_1 is defined by [40]

$$\delta_h(y) = \begin{cases} 1 & \text{if} \quad f(y) > \tau \\ \gamma & \text{if} \quad f(y) > \tau \\ 0 & \text{if} \quad f(y) > \tau \end{cases} \tag{4-3}$$

where

$$f(y) - \log[L(y)] - \log\left(\frac{p_1}{p_0}\right) = \log\left[\frac{p_\eta(y-h)}{p_\eta(y)}\right] = \sum_{k=1}^{n} h_k y_k$$

By Equation 4-3, the test δ_h is called the correlation detector for the known signal h in the noisy signal y. The step response h is called a template signal for δ_h. In other words, the best strategy for detecting the presence of the known template signal h in a noisy gene expression profile y is to correlate the two together [40,41]. Here, a large absolute correlation implies that h is present in y, otherwise h is not present. The key point here is to determine the threshold τ on f that optimizes the balance between sensitivity and the false positive rate defined by $(1 - \text{specificity})$. The response function need not be confined to be the unit step function, and in fact Equation 4-3 is quite general and holds for arbitrary y and r. But a step-like response is the standard template function for detecting what is commonly known as differential expression between the case and control classes of a microarray experiment. This step-like response represents a large and consistent difference in expression between the case and control samples of the experiment.

Note that fold change can be viewed as a correlation detector for the unit step function h in the noisy gene expression profile y. Indeed, we have

$$h^T y = \sum_i h_i y_i = \sum_{controls} h_i y_i + \sum_{cases} h_i y_i = \sum_{controls} -y_i + \sum_{cases} y_i$$

$$= \left(\frac{n}{2}\right)\left[\left(\frac{2}{n}\right)\sum_{cases} \log_2(x_i) - \left(\frac{2}{n}\right)\sum_{controls}\log_2(x_i)\right]$$

$$= \left(\frac{n}{2}\right)\left[\log_2\left(\sqrt[\frac{n}{2}]{\sum_{cases} x_i}\right) - \log_2\left(\sqrt[\frac{n}{2}]{\sum_{controls} x_i}\right)\right]$$

$$= \left(\frac{n}{2}\right)\log_2\left[\frac{\left(\sum_{cases} x_i\right)^{n/2}}{\left(\sum_{controls} x_i\right)^{n/2}}\right] = \left(\frac{n}{2}\right)\log_2\left(\frac{\text{geometric average of cases}}{\text{geometric average of controls}}\right)$$

$$= \left(\frac{n}{2}\right)\log_2(\text{fold change})$$

and hence, fold change is an correlation detector of the unit step response in noisy data. Since the t-test for the difference in mean expression between two sample groups is

$$t(y) = \frac{\left(\frac{2}{n}\right)\sum_{cases}\log_2(x_i) - \left(\frac{2}{n}\right)\sum_{controls}\log_2(x_i)}{s} = \left(\frac{2}{ns}\right)\log_2(\text{fold change})$$

where s is the pooled sample standard deviation of y, it follows that the t-test is simply log fold change penalized for "within-group" variations in fold change through the estimated value for s.

Note that if the true response r deviates significantly from the step function h, then both fold change and the t-test become suboptimal tests for differential expression (as defined by h.) Note however, that the t-test is further penalized for large variations in fold change through s, that is, the t-test is biased toward step-like signals that are strong and consistent within each two-sample groups. Hence, in situations where the true response of a gene is highly variable over the samples, as in tumor samples with heterogeneous composition, the t-test will fail to detect these genes [42]. Also, genes that are up- or downregulated on only a fraction of the case samples may not be detected [36]. On the other hand, MANINI will detect genes that are modulated intermittently across the samples of the experiment since fold change by itself is not penalized for excessive variation. Finally, note that statistical validation of the resulting gene list is deferred until ontological (PANTHER) or pathway analysis (INGENUITY) can be conducted to determine the statistically significant functional categories and pathways contained in the gene lists. A gene is then called significant if it is contained in a significant functional category or pathway [31].

In summary, many genes have response profiles that deviate significantly from the step function due to intermittent up- or downregulation across the samples of the microarray experiment [43]. Hence, such genes will remain undetected by standard *t*-like tests that are designed to detect a consistent and strong step-like change in expression across the samples of the experiment. The MANINI detection algorithm attempts to circumvent this problem by selecting genes that exhibit high fold change relative to a noise-adjusted threshold that varies with expression level. This detection strategy exploits the observed relationship between variation in fold change and expression level in microarray data. Hence, a gene is not penalized for high variation so long as its average fold change exceeds a "universal" detection threshold that adapts to the noise background for the gene. Significant genes are subsequently defined as those genes that are contained in significant functional categories that signaling pathways that are contained in the gene lists as identified using IPA and Onto-Express [37].

4.9.3 Pathway Compression

We assume that the overrepresentation of known cancer-related pathways (as explained below) in the list of DE genes derived from the Huang breast cancer data set represents coherent structure that characterizes the underlying biology of lymph node positive breast cancer tumors in terms of gene expression. Conversely, we assume the absence of such coherent structure suggests that the gene list has little in common with what is known about gene function and is essentially composed of randomly selected genes representing mostly noise. We target genes contained in overrepresented pathways as means of "drilling down" to those genes that are at once the most biologically relevant and discriminative between positive and negative breast cancer tumors. This idea is called pathway compression since it serves to reduce the dimensionality of the resulting gene expression signature that will be used to train a NN model for classifying breast cancer tumors as lymph node negative or positive.

IPA was used to identify the biological networks that were perturbed in the Huang lymph node positive breast cancer samples in the context of what is currently known about mammalian biology derived from basic and clinical research [15,44]. Research findings presented in peer-reviewed scientific publications are manually encoded into a comprehensive knowledge base of gene function and gene–gene interactions. The IPA knowledge base contains over 200,000 full text scientific articles, a gene ontology of more than 9800 human, 7900 mouse, and 5000 rat genes that were manually curated and parsed from MEDLINE abstracts. A global interaction network of direct physical, transcriptional, and enzymatic interactions observed between mammalian orthologues as described in the literature—the so-called global "interactome"—was overlayed on the gene ontology. The resulting global interactome contained molecular interactions involving over 8000 orthologues with a high degree of connectivity. On average, individual genes have 11.5 interaction partners, of which 7.2 represent direct physical interactions.

Every gene interaction in an IPA network is supported by published articles. Furthermore, the original literature detailing the genetic interactions can be accessed to further examine and verify the findings. The global interactome provides a framework for structuring existing knowledge regarding mammalian biology and enables the objective validation of experimental data in the context of known genome-wide interactions to identify significant functional pathways. This method is applicable to data of high-throughput platforms such as microarray expression profiling, polymorphism analysis, and proteomics.

Significant pathways contained in MANINI-derived gene lists were identified by first overlaying the genes identified as DE onto the global interactome. Focus genes were then identified as those genes having direct interactions with other MANINI-significant genes in the database. The specificity of connections for each focus gene was calculated by the percentage of its connections to other significant genes. The initiation and growth of pathways proceeded from genes with the highest specificity of connections, where each pathway had a maximum of 35 genes. Pathways of highly interconnected genes were identified by statistical likelihood based on the following formula:

$$\text{IPA\_Score} = -\log_{10}\left[1 - \sum_{i=0}^{f-1} \frac{C(F,i)C(N-F,s-i))}{C(N,s)}\right]$$

where $C(n,k)$ is the binomial coefficient, N is the number of genes in the global interactome, F are the number of significant genes detected by MANINI, and s is the number of genes in the inferred pathway of which f are focus genes. Depending on the data set, pathways with a score greater than 5 (p-value $<1.0E-05$) are considered significant.

IPA was used to identify biologically significant pathways contained in the gene list derived by MANINI for the Huang breast cancer data set. A summary of an IPA analysis of 413 downregulated DE genes selected by MANINI is shown in Table 4-1. The list is composed of pathways that were found to be statistically overrepresented in the gene list based on gene function and gene–gene interactions contained in the IPA knowledge base. The networks were rank ordered by IPA significance score and gene descriptions, and the number of focus genes for each network was also provided. Only networks derived from the downregulated genes were targeted since they resulted in the most robust NN models. Note that the top two IPA networks have p-values on the order of 10^{-55}. Figure 4-7 shows a diagram of the top interaction network from the list (*dnet1*).

The network diagram for *dnet1* in Figure 4-7 details the internal interactions between the 35 genes that are contained in the network. Here, each node of the diagram represents a gene and each edge connecting two genes represents a documented interaction between them. We selected only the genes contained in *dnet1* for further downstream information processing. IPA pathway analysis can be viewed as a feature selection procedure where the genes in a significant IPA pathway are used as features for classifying the samples of the experiment. This gene selection process is known as pathway compression [31]. In fact, we show in Section 4.11 that genes contained in *dnet1* are able to accurately distinguish between the positive and negative samples of the Huang breast cancer data set when analyzed using WSVD signal processing and modeled using neural networks. Note that diseases and biological processes

Table 4-1   IPA analysis of genes called downregulated by MANINI

Rank	IPA Source	Focus Genes	Top Functions
1	55	35	Gene expression, cell-to-cell signaling and interaction, cancer
2	55	35	Gene expression, dermatological diseases and conditions, genetic disorder
3	16	16	Cellular development, cellular growth and proliferation, hematological system development and function
4	14	15	Cellular compromise, Dermatological diseases and conditions, gastrointestinal disease
5	14	15	Inflammatory disease, viral function, immunological disease
6	14	15	Protein synthesis, lipid metabolism, small molecule biochemistry
7	13	14	Cell signaling, cancer, cell death
8	13	14	Nervous system development and function, organ development, cancer
9	13	14	Organismal development, lipid metabolism, molecular transport
10	13	14	Carbohydrate metabolism, molecular transport, small molecule biochemistry
11	13	14	Nervous system development and function, cell-to-cell signaling and interaction, neurological disease
12	11	13	Cellular growth and proliferation, hair and skin development and function, cell signaling
13	11	13	Viral function, gene expression, cell cycle
14	10	12	Cellular movement, connective tissue development and function, cell cycle
15	10	12	Energy production, nucleic acid metabolism, small molecule biochemistry
16	9	11	Cell cycle, cellular assembly and organization, DNA replication, recombination, and repair

Each row represents a significant IPA gene network. Note the networks are ordered by p-value. The top network, *dnet*, has a p-value $\sim$1.0E-55 and contains 35 genes from the MANINI gene list. An inferred function for *dnet1* based on the function of genes contained in the network includes *Cancer*.

associated with the gene network *dnet1* include *Cancer, Cell-to-Cell Signaling* and *Gene Expression*. Table 4-2 shows a list of the 35 genes contained in *dnet1*. Although, many cancer-related genes such as FOS and MMP10 are included in the *dnet1* gene list, and the network topology of *dnet1* suggests specific biological mechanisms that may have relevance to metastatic breast cancer, we are primarily concerned with how well the *dnet1* genes are able to discriminate between positive and negative breast cancer tumors, ignoring for now the underlying biology.

4.9.4   Continuous Wavelet Transform

Wavelet signal processing analyzes a noisy signal, for example, the expression profile of a gene over a range of scales using wavelets of different locations and time durations.

Table 4-2   Gene list for IPA network *dnet1*

Affy Tag	Name	Description
684_at	AGT	Angiotensinogen (serpin petidase inhibitor, clade A, member 8)
37983_at	AGTR1	Angiotensin II receptor, type 1
40960_at	B4GALT1	UDP-Gal betaGlcNAc beta 1,4-galactosyltransferase, polypeptide 1
40367_at	BMP2	Bone morphogenetic protein 2
35457_at	CART	CART prepropeptide
2036_s_at	CD44	CD44 molecule (Indian blood group)
39031_at	COX7A1	Cytochrome *c* oxidase subunit Vlla polypeptide 1 (muscle)
1878_g_at	ERCC1	excision repair cross-complementing rodent repair deficiency, complementation group 1
2084_s_at	ETV4	ets variant gene 4 (E1A enhancer binding protein, E1AF)
34818_at	ETV5	ets variant gene 5 (ets-related molecule)
1408_at	FGF4	Fibroblast growth factor 4 (heparin secretory transforming protein 1, Kaposi sarcoma oncogen
1363_at	FGFR2	Fibroblast growth factor receptor 2 (bacteria-expressed kinase, keratinocyte growth factor receptor
2094_s_at	FOS	v-fos FBJ murine osteosarcoma viral oncogene homologue
36669_at	FOSB	FBJ murine osteosarcoma viral oncogene homologue B
32383_at	GHRHR	Growth hormone releasing hormone receptor
32531_at	GJA1	Gap junction protein, alpha 1, 43 kDa (connexin 43)
37447_at	GPR30	G-protein-coupled receptor 30
39573_at	GRIK2	Glutamate receptor, ionotropic, kainate 2
39618_at	HAPLN1	Hyaluronan and proteoglycan link protein 1
32570_at	HPGD	Hydroxyprostaglandin dehydrogenase 15-(NAD)
1006_at	MMP10	Matrix metallopeptidase 10 (stromelysin 2)
38602_at	MYH6	Myosin, heavy chain 6, cardiac muscle, alpha (cardimyopathy, hypertrophic 1)
35041_at	NTF3	Neurotrophin 3
33998_at	NTS	Neurotensin
32472_at	OXT	Oxytocin, prepro-(neurophysin I)
38295_at	PBX2	Pre-B-cell leukemia transcription factor 2
32001_s_at	PCSK6	Proprotein convertase subtilisin/kexin type 6
35703_at	PDGFA	Platelet-derived growth factor alpha polypeptide
31732_at	RLN2	Relaxin 2
35622_at	SHC2	SHC (Src homology 2 domain containing) transforming protein 2
1511_at	SHC3	SHC (Src homology 2 domain containing) transforming protein 3
36555_at	SNCG	Synuclein, gamma (breast cancer-specific protein 1)
34342_s_at	SPP1	Secreted phosphoprotein 1 (osteopontin, bone sialoprotein I, early T-lymphocyte activation 1)
32903_at	TGFBR1	Transforming growth factor, beta receptor I (activin A receptor type II-like kinase, 53 kDa)
39681_at	ZBTB16	Zinc finger and BTB domain containing 16

Shown are Affymetrix gene tags, gene name, and gene description. Cancer-related genes include FOS, MMP10, and FGF4. The interaction structure of these genes is shown in Figure 4-7.

In other words, the wavelet transform provides a *timescale* decomposition of a signal of interest [45]. Here, the term "time" is used loosely referring to some agreed-upon sequential ordering of multiple measurements (e.g., gene expression values) that may or may not reflect an actual temporal ordering. The main point is that all samples have their components ordered in the same way.

An underlying assumption of wavelet signal processing is that coherent signal and random noise "live" at different scales of resolution and hence are often well separated after wavelet transformation. Moreover, noise in real-world data sets are often better "equalized" after wavelet transformation, thus making the distinction between signal and noise even more pronounced in the wavelet transform domain. Indeed, it can be shown that the wavelet transform "diagonalizes" a scale-invariant signal in much the same way that the Fourier transform diagonalizes time-invariant signals. Since scale invariance generalizes time invariance, the wavelet transform can be viewed as a generalization of the Fourier transform. Researchers have confirmed that the improved signal/noise separation provided by the wavelet transform results in real-world pattern recognition applications with enhanced classification and predictive capabilities [27,28,46].

Wavelets at different scales and times $\psi_{s,t}$ are derived from a single "mother" wavelet ψ via scaling and translation operations. The wavelet transform of a given signal f is defined by correlating f with each wavelet $\psi_{s,t}$ and summing the correlations

$$\tilde{f}(s,t) \equiv \int_{-\infty}^{\infty} \psi\left(\frac{u-t}{s}\right) f(u)du = \int_{-\infty}^{\infty} \psi_{s,t}(u)f(u)du$$

where $\tilde{f}(s,t)$ is the CWT of f at scale s and time t with respect to the mother wavelet ψ. As a function of t for a fixed scale s, $\tilde{f}(s,t)$ represents information in the signal having frequencies that are localized to a spectral region that is centered on some frequency that depends on the fixed scale s. As a function of s for a fixed time t, $\tilde{f}(s,t)$ represents information in the signal of all frequencies that is localized in some temporal region centered on the fixed time t. Larger scale values capture coarse signal detail (e.g., global trends), while the smaller scale values capture finer detail (e.g., transient fluctuations and random noise). Hence, the CWT provides a means of characterizing both local and global variation in a single signal representation.

There are an infinite number of mother wavelets to choose from depending on the characteristic of the signal being analyzed. The mother wavelet used for the microarray data analysis in this study is known as the Daubechies mother wavelet. Computational studies using real and simulated data sets have shown that this particular wavelet results in the best classification performance on the Huang breast cancer tumor samples. This is mainly due to the shape of the Daubechies mother wavelet that simultaneously smoothes the expression profile for a given sample while capturing localized variations in the data over multiple scales of resolution.

Figure 4-8 shows the wavelet transform of quantile normalized, $\log_2$ transformed, z-scored expression profiles of a lymph node negative tumor sample (Fig. 4-9a) and a lymph node positive tumor sample (Fig. 4-9b). The expression profiles were generated by the downregulated genes in IPA network *dnet1*. In each case, the actual sample

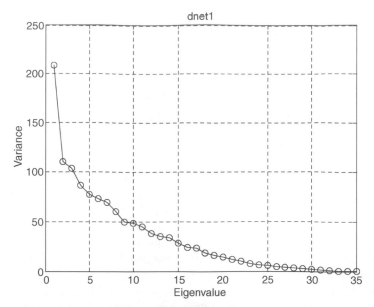

Figure 4-8 Eigenvalue plot for the data matrix of genes from IPA network *dnet1*. Each eigenvalue represents the variation in the direction of the corresponding eigencomponent. Eigenvalues 1–10 are most likely to correspond to coherent signal. Eigenvalues 11–36 probably correspond to noise.

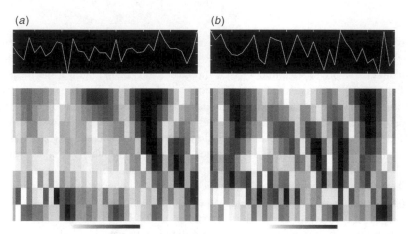

Figure 4-9 The continuous wavelet transform based on the Daubechies mother wavelet of order 4 for two sample expression profiles from the data matrix of pathway compressed genes.(a) The sample expression profile of a lymph node negative tumor over 35 genes is shown above an image of the CWT of the profile. The vertical axis of the CWT image represents scale, while the horizontal axis represents samples ordered by lymph node status with negative samples first. Note that the CWT of a one-dimensional signal is two-dimensional image. (b) The sample profile and CWT image of a lymph node positive tumor. Note how coefficients with high magnitude show different distribution over scale and samples. Coefficients contained in rows near the bottom of the image contain mostly noise, while coefficients in rows near the top of the image represent coherent signal characterizing differences between negative and positive tumor samples.

expression profile is shown above the CWT of the profile. Note that the CWT of a one-dimensional signal is a two-dimensional image. The vertical axis of each CWT image represents the scale of resolution of the CWT. Not that the scale of resolution becomes progressively coarser from the bottom up along the vertical axis. The horizontal axis represents the tumor samples ordered by lymph node status with the negative samples occupying columns 1–19 and the positive samples occupying columns 20–36. Hence, each row of the CWT image represents the information content of the profile at a particular scale of resolution, while each column represents the information content of a particular sample over eight scales of resolution.

The main idea of CWT signal processing is that random noise is concentrated at the higher scales of resolution near the bottom of the CWT image, while lower frequency coherent signal is concentrated in the coarser scales of resolution near the top. This separation of signal and noise by scale enhances subsequent compression and denoising of the data using SVD. Each CWT "image" is vectorized to generate a one-dimensional wavelet transformed expression profile that is used to form a column of a CWT data matrix A_{wave}.

In particular, let A_{dnet1} be the 35×36 data matrix of quantile normalized, $\log_2$ transformed, z-scored gene expression values defined by the genes of *dnet1*. Each 35-dimensional column of A_{dnet1} is wavelet transformed over 64 scales in increments of 8 using the continuous wavelet transform (CWT) based on the Daubechies mother wavelet of order 4 (Daub4) that results (after vectorization) in a 240×36 wavelet data matrix A_{wave}. SVD was used to compress A_{wave} down to a 12×36 data matrix $A_{\text{feats}} = [f_1, f_2, \ldots, f_{36}]$ as explained below. Each column f_j of A_{feats} represents a 12-dimensional vector of WSVD features that characterizes the jth sample of the experiment for $j = 1, 2, \ldots, 36$. Previous research has shown that WSVD features significantly enhance the performance of pattern recognition algorithms in real-world applications [27,28]. In this study, we show that pathway compression coupled with WSVD signal processing enhances the classification of high- and low-risk breast cancer samples based solely on gene expression of the primary tumor using neural network classifiers.

4.9.5 Singular Value Decomposition

SVD is a classical statistical technique for characterizing the linear correlation that exists in a data matrix [47]. It is closely related to the Karhunen–Loeve transform (random processes), principal component analysis (matrix diagonalization), and factor analysis (correlation structure of multivariate stochastic observations). SVD is used in many areas of science and engineering as a means of extracting features for pattern recognition, data compression, signal detection, and sample classification applications. Essentially, the primary goal of SVD is to find a linear transformation that maps a vector of noisy, correlated "time domain" measurements into a much smaller vector of denoised, uncorrelated feature components [27,28].

Let $A = [x_1, x_2, \ldots, x_K]^T$ be a $p \times n$ expression data matrix with p-dimensional columns x_i composed of noisy, correlated expression measurements (superscript T is the matrix transpose operator). We desire a linear transformation $L : \mathbb{R}^p \to \mathbb{R}^k$ such that $y_i = Lx_i$ is a compressed eigenarray of dimensionality k ($k \ll n$) composed of uncorrelated, denoised *eigenexpression* values. The *fundamental theorem of linear algebra* states that under very general conditions, there exists orthogonal matrices U and V and a diagonal matrix Σ such that $A = U\Sigma V^T$ [48]. Here, U is $p \times p$, U is $n \times n$, and S is $p \times n$. The columns of U are the *eigenarrays* of A, and they provide an orthonormal basis for $\mathbb{R}^p$. Similarly, the columns of V are the eigengenes of A, and they form an orthonormal basis for $\mathbb{R}^n$. The square of the diagonal entries of S are the eigenvalues, λ_i, of A and are ordered so that $\lambda_j > \lambda_{j+1}$ for $j = 1, 2, \ldots, n-1$. We choose the first k eigenarrays of U that correspond to the k largest eigenvalues (where usually $k \ll n$) and form the matrix U_{trunc} with columns equal to the selected eigenarrays. It follows that $L : \mathbb{R}^p \to \mathbb{R}^k$ defined by $L = U_{trunc}^T$ is the linear transformation we seek since it maps a p-dimensional vector into a k-dimensional vector where $k \ll n < p$. The k components of $y_i = Lx_i$ are known as the principal components of x_i [47].

We note that the resulting feature vector y_i is denoised due to the truncation of those eigenarrays of U that are associated with the remaining $(n - k)$ eigenvalues. It is assumed that the truncated eigenarrays span a $(n - k)$-dimensional subspace containing the random noise component of the data. We note that the subspace spanned by the truncated eigenarrays may contain information that is useful for classification, and one needs to be careful that this information is not lost in the dimensionality reduction process. Usually, though, a visual analysis of a plot of the eigenvalues makes it clear where the threshold should be set using, say, Kaiser's rule [49].

Figure 4-10 shows a plot of the 36 eigenvalues obtained for the 240×36 wavelet data matrix A_{wave}. Note that the eigenvalue plot becomes linear starting at about the 12th eigenvalue, so that $\sum_{i=1}^{12} \lambda_i$ represents the variation associated with coherent signal, which accounts for 79 percent of the total variation in A_{wave} based on the top 12 eigenarrays. The sum of the remaining eigenvalues $\sum_{i=13}^{36} \lambda_i$ represents the variation associated with the noise, which accounts for the remaining 21 percent of the energy in the data. The above analysis of the eigenvalues suggests that we retain the first 12 eigenarrays (i.e., columns) of U to form a 240×12 transformation matrix U_{trunc}, which is used to "compress" the 240×36 data matrix A_{dnet1} down to a 12×36 data matrix y of WSVD/IPA feature vectors using

$$y = (U_{trunc})^T x$$

where x is a 240 component column vector of A_{wave}. Note that each of the 36 tumor samples is now characterized by 12 numbers instead of the original 12,625 expression measurements. This is a huge reduction in dimensionality with a theoretically minimal loss of information accompanied by a theoretically maximal reduction in noise [15].

4.9.6 Combining Wavelets, SVD, and IPA (WSVD/IPA)

WSVD/IPA signal processing combines wavelet signal processing SVD and IPA pathway compression to extract signal features from the Huang microarray data set. The WSVD/IPA feature patterns are then used to train a NN classifier to robustly

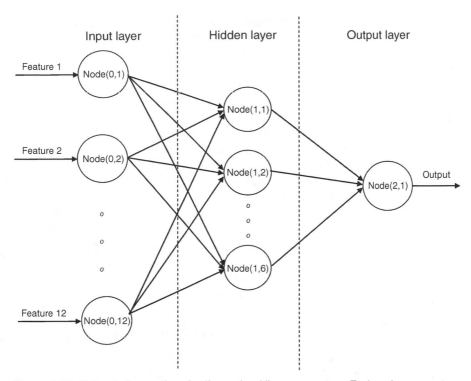

Figure 4-10 Network diagram for a feedforward multilayer perceptron. Each node represents a processing unit (artificial neuron) that computes an output according to Equation 4-4. Each arrow represents the passing of information from one node to another. Note that the output of a given node is passed as an input to the every node in the next layer. Nodes in the same layer do not pass information to each other. The diagram shown has one input layer of 12 nodes, one hidden layer of 6 nodes, and a single note in the output layer. A 12-dimensional feature vector (e.g., a WSVD/IPA expression feature pattern) is presented to the input layer of the FFMLP. The input cascades forward through the network layer by layer and eventually results in a single output value at the output layer of the FFMLP. This output vector is thresholded to determine the decision made by the FFMLP regarding the lymph node status of the tumor that is represented by the 12-dimensional input feature pattern.

predict lymph node status based on gene expression patterns from primary breast cancer tumors. The NN classifier is validated using leave-one-out cross-validation starting with the $12,625 \times 36$ raw data matrix A_{raw} with an outlier sample removed.

We basically follow the flowchart given in Figure 4-1. The columns of A_{raw} are first quantile normalization to obtain the normalized data matrix A_{nrm}. This normalization step facilitates comparisons between the microarrays of the experiment. The MANINI detection algorithm is then applied to A_{nrm} to obtain lists of up- and downregulated genes that are DE between the positive and negative samples of the Huang microarray data set. This completes the gene detection phase of the information processing chain.

Ingenuity Pathway Analysis is then used to extract statistically significant pathways from the MANINI gene lists. The genes in the most statistically significant IPA networks are then intersected with the MANINI gene lists to generate individual data matrices of normalized, $\log_2$ transformed, z-scored gene expression profiles. This step is called IPA pathway compression, since a gene selected by MANINI is passed on for downstream processing only if it is also contained in a significant IPA network. Let A_{net} be the data matrix associated with a significant IPA network. The CWT is then used to transform the columns of A_{net} to obtain the 240×36 wavelet data matrix A_{wave}. The Daubechies mother wavelet over 64 scales in increments of 8 was used since computational experiments suggested that these parameters resulted in the best NN performance. Note the CWT better separates signal and noise in the wavelet domain and enhances data compression and denoising using SVD [39,50,51]. Application of the SVD to A_{wave} resulted in a 12×36 matrix of WSVD/IPA feature patterns $A_{feats} = [f_1, f_2, \ldots, f_{36}]$. Note the columns of A_{feats} represent 12-dimensional feature patterns (derived from the original 12,625-dimensional gene expression profiles) that characterize each of the 36 samples of the Huang data set. Note the greatly reduced dimensionality of the WSVD/IPA feature vectors f_j enhances NN modeling of lymph node status by alleviating the adverse impact of Bellman's curse of dimensionality. The WSVD/IPA feature vectors f_j for $j = 1, 2, \ldots, 36$ were used to train a $12 \times 6 \times 1$ NN model with 12 input nodes, 6 hidden nodes, and 1 output node to discriminate between the positive and negative lymph node samples.

4.9.7 Neural Network Modeling of Lymph Node Status

NN models are useful for situations where there is much data, but a theory is lacking that explains the data [24]. The focus then shifts to patterns within the data that are associated with a quantifiable attribute measured for each sample. Neural networks are also known as data-driven models or machine learning models. In microarray data analysis, for example, we are usually given a finite number of ordered pairs (x, y) that associate a k-dimensional gene expression pattern $x \in \mathbb{R}^k$ to a unique phenotypic state $y \in \{0, 1\}$. In this case, we wish to "discover" a mapping from expression patterns to phenotypic states $\{0, 1\}$ that "generalizes" to new expression patterns not contained in the original data set [21]. This mapping allows the prediction of phenotypic states of new patterns that were "unseen" during training of the NN model. The discovered mapping can be implemented as a NN, which represents a massively parallel, highly distributed computational model of observed phenotypic variation based on gene expression patterns.

A NN is composed of elementary computational units that can be "connected" to each other to form a network that is capable of global information processing arising from the local interactions between the computational units. The strength of the interaction between two computational units is encoded in a "synaptic" weight that quantities the "strength" of the interaction between the two units [52]. The collection of all synaptic weights can be adjusted in parallel to realize almost any continuous mapping between two sets of variables. This emergent computational behavior can be highly nonlinear in nature, and as a result, a NN model is capable of solving very difficult classification and prediction problems that require complex boundaries between the different sample classes. The form of the resulting connection diagram is called the architecture of the network, and the computations performed by the network are highly dependent on this architecture. That is, the modeler has control over how the connections are evolved so that the architecture is plastic and trainable. Indeed, the synaptic weights of the MLP can be adjusted using a machine learning algorithm such as backpropagation to realize an arbitrarily good approximation to any mapping between expression patterns and phenotype suggested by the data.

A neural network architecture known as the *feedforward multilayer perceptron* (FFMLP) is used to model phenotypic variation over the samples of the microarray experiment in terms of gene expression. The FFMLP architecture arranges the computational nodes of the network into a "hidden" layer and an output layer. There is an additional input layer of nodes, but these nodes merely pass the input values on to the hidden layer without computation. The hidden layer is described as such because it is shielded from direct contact with the outside world by the input and output layers of nodes. The FFMLP is fully connected in that every node of a given layer is connected to every node of the next layer by an arrow representing the direction of information flow. Finally, to every arrow is assigned an adaptive weight parameter that mediates the flow of information between the two nodes. Training data consisting of empirical input–output pairs are used to adjust the weights using a nonlinear optimization algorithm (e.g., error backpropagation) to approximate the input–output mapping that is "implied" by the data. In this way, a FFMLP can approximate any mapping between any two sets of quantities to an arbitrary degree of accuracy [49].

Figure 4-10 shows a network diagram for a two-layer FFMLP, where each node represents a computational unit and each arrow represents the flow of information from one node to another. In this case, the FFMLP has an input layer consisting of 12 nodes, a hidden layer consisting of 6 nodes, and an output layer consisting of a single node. Usually, the input layer is excluded from a count of layers that compose a given FFMLP. If the dimension of the input layer is k, then the training data set consists of ordered pairs $\{(x, y) \in \mathbb{R}^k \times \mathbb{R}^2\}$. For a given training data pair (x,y), the k-dimensional feature vector x is presented to the input layer of the NN and the resulting output vector $y_{NN}(x)$ is compared with the target vector y where the dimensionality of y and y_{NN} are equal.

For this study, x is the WSVD/IPA feature vector for a chip hybridized to a sample from the Huang breast cancer data set and y is either 0.05 or 0.95, depending on whether the sample is negative or positive for lymph node involvement, respectively. The error between y and y_{NN} is propagated back through the NN to adjust the adaptive weights of the NN to reduce the error in the output layer using the *error back-propagation* algorithm. The learning process is iterated over all training pairs and repeated until the aggregate error for the training data is reduced to an acceptable level. Note that training is terminated in such a way as to balance accuracy on the training data with the ability of the FFMLP to generalize classification performance to data not seen during training. To enhance generalization to new data, the output of the FFMLP is smoothed using Bayesian regularization during training.

The input to the kth node of a given layer is a linear weighted sum of all inputs to the node from all nodes in the previous layer, given by

$$\sum_j w_{kj} x_j$$

where the sum runs over all nodes in the previous layer that sends a signal x_j to node k (we assume that a bias parameter is included in the summation) and w_{kj} is the adaptive weight for the connection between the two nodes [53]. The output of the kth node is obtained by transforming the weighted linear sum with a nonlinear activation function g using

$$z_k = g\left(\sum_j w_{kj}x_j\right) \tag{4-4}$$

For a vector of values presented to the input layer of the FFMLP, the output of each computational node in a given layer can be computed in a *feedforward* manner in terms of the outputs of all the nodes in the previous layer.

From a theoretical perspective, the FFMLP is known as a universal approximator; that is, it can uniformly approximate any continuous function on a compact domain to an arbitrary degree of accuracy provided the network has a sufficiently large number of hidden units and enough data. The key problem remains how to find suitable parameter values given a set of training data. There exist effective solutions to this machine learning problem based on both maximum likelihood and Bayesian approaches. The error backpropagation algorithm is probably the most widely used method to train a FFMLP.

In particular, the WSVD/IPA feature patterns x_j extracted for each sample were used to train a FFMLP model to discriminate between the positive and negative lymph node samples. The FFMLP had an architecture shown in Figure 4-10 compose of 12 input nodes, 6 hidden nodes, and a single output node. The 12-dimensional WSVD/IPA input feature vector is passed without any processing to the nodes of the hidden layer, where each hidden node computes an output value in accordance with Equation 4-4. The output of each hidden node is then fed into the single output node, which computes a weighted linear combination of inputs and transforms the result using the sigmoidal logistic function g. A fixed threshold (equal to 0.5) is applied to the output value to determine whether the input feature pattern was associated with a positive or negative breast cancer tumor.

The hidden nodes employed hyperbolic tangent activation functions with range confined to the interval $[-1,1]$. The single output node employed a sigmoidal logistic activation with range confined to the interval $[0,1]$. The Levenberg–Marquardt training algorithm with Bayesian regularization was used to train the FFMLP to output a value of 0.95 for lymph node positive samples and 0.05 for the lymph node negative samples. A sample was classified as lymph node positive if its associated FFMLP output exceeded a threshold of 0.5, otherwise it was classified as lymph node negative. The FFMLP was trained for 20 epochs with a targeted error goal of 0.005. Training was usually completed in less than 30 s, which facilitated validation of the GSP algorithms.

4.10   MODEL VALIDATION

The robustness and accuracy of FFMLP models trained on the Huang microarray data set was evaluated using leave-one-out cross-validation (LOOCV) analysis [4,19].

LOOCV analysis begins by removing, say, the kth column from the raw data matrix A_{raw}. This results in the column-reduced $12,625 \times 35$ data matrix A^k_{raw} from which WSVD/IPA features are extracted as described above to train a NN model $y_k(x)$, where x represents a 12-dimensional WSVD/IPA feature vector. Let x_k denote the WSVD/IPA feature vector associated with the kth column of A_{raw} that was left out. Recall that x_k was unseen during training of y_k and we want to see if y_k can correctly classify this sample. By design, we say that x_k is lymph node negative if $y_k(x_k) < 0.5$ and lymph node positive otherwise. The classification result is duly recorded and compared with the known lymph node status for the left-out sample. The entire process is repeated 36 times for each column of A_{raw}. The correct classification rate (CCR) is defined as the percentage of left-out samples that were correctly classified. Note that for each sample left out during the LOOCV analysis, a different set of downregulated genes is selected by the MANINI algorithm. This variation in the gene lists reflects the variation in the population of all breast tumor expression profiles. We utilize this variation to assess the robustness of NN models trained on such data.

Figure 4-11 shows a flowchart of the information processing used to validate feature patterns for classifying breast cancer samples into high- and low-risk groups using FFMLP classifiers. Major signal processing occurs in the orange boxes labeled "preprocess data," "select genes," "extract features," and "train classifier" as shown in Figure 4-12. In particular, gene selection based on MANINI detection and IPA pathway compression occurs in the "select genes" box of the flowchart, and WSVD features are extracted in the "extract features" box. Note also the parallel chain (in green) that processes the "left-out" sample for eventual classification by the NN

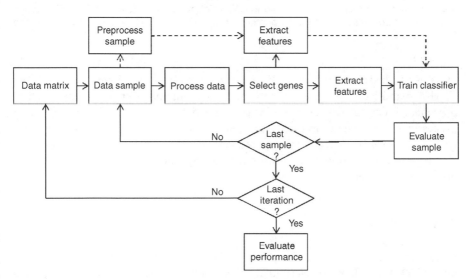

Figure 4-11 Flowchart for leave-one-out cross-validation of NN models of lymph node status based on pathway compression and WSVD signal processing. LOOCV analysis estimates the impact of sampling variation on the prediction performance of the neural network classifier. The robustness and accuracy of the proposed prediction model depends to a large extent on the quality of the feature patterns extracted from the raw data.

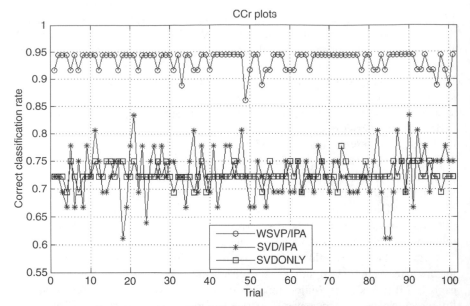

Figure 4-12   A comparison of WSVD/IPA, SVD/IPA, and SVDOnly features for predicting lymph node status using neural networks. LOOCV analysis of FFMLPs trained on each feature set was performed 100 times and plotted to assess the variation in CCR values. WSVD/IPA and SVD/IPA features were extracted from the 35 genes that were contained in the top IPA network for all genes called downregulated genes by the MANINI detector. SVDOnly features were extracted from all 413 downregulated genes detected by the MANINI detector. Note that the CCR plot for WSVD/IPA features (open circles) lies uniformly above the CCR plots for SVD/IPA (stars) and SVDOnly (open squares) features. Median CCR is 94 percent for WSVD/IPA and 72 percent for SVD/IPA and SVDOnly.

classifier trained on the remaining 35 samples. Because the algorithm used to train the FFMLP classifier is subject to entrapment in local minima, we repeated the LOOCV procedure 100 times to obtain some idea of the variability introduced into the training process due to sampling variation in the data. The maximum, median, and minimum CCR values over 100 LOOCV trails were used to evaluate the effectiveness of different signal processing algorithms in discriminating between positive and negative lymph node samples from the Huang data set.

4.11   RESULTS

In this section, we present results of a LOOCV evaluation of FFMLP models trained on WSVD/IPA features to discriminate between breast cancer tumor samples based on lymph node status. We also assessed and compared the performance of FFMLP models trained on two additional feature types derived using (1) SVD combined with pathway compression (SVD/IPA) and (2) SVD compression applied to all down-regulated genes (SVDOnly).

Figure 4-12 shows plots of 100 CCR values for FFMLP classifiers trained using WSVD/IPA (open circles), SVD/IPA (stars), and SVDOnly (open squares) feature patterns. Note that FFMLP models trained on WSVD/IPA feature patterns uniformly outperform models trained on SVD/IPA and SVDOnly features. Indeed, the median CCR value associated with WSVD/IPA features was 94 percent, while both SVD/IPA and SVDOnly features had median CCRs of 72 percent. Hence, wavelet signal processing improved FFMLP classification performance by 31 percent when used in conjunction with SVD analysis and IPA pathway compression. SVD/IPA and SVDOnly had the same median performance, but SVD/IPA had greater variation and attained a maximum CRR of 83 versus 78 percent for SVDOnly. This suggests that IPA pathway compression does not provide any improvement in classification performance over SVD alone. Table 4-3 summarizes the maximum, median, and minimum CCR values over 100 LOOCV trials for each feature type.

Overall, wavelet signal processing and pathway compression combined to significantly enhance the prediction of lymph node status when compared to more conventional signal processing based on SVD alone. Note the option of combining SVD and wavelet signaling processing (without the benefit of pathway compression) was not considered since this would have required wavelet processing of data vectors of length 413 instead of 35, which is computationally intensive. Hence, the success of wavelet signal processing in the context of this study was a direct result of the data reduction provided by pathway compression; that is, wavelets, pathway compression, and SVD worked hand in hand to reduce dimensionality without loss of information related to lymph node status, and thus significantly enhancing the classification of the Huang breast cancer samples.

Note wavelet signal processing together with pathway compression achieved a 94 percent CCR based on only 35 genes. This result compares favorably with the 90 percent CCR achieved by Huang et al. on the same data set using 200 genes and different statistical methodology. These results suggest that the lymph node status of breast cancer samples can be predicted in an accurate and robust manner using a relatively small number of genes when the appropriate signal processing and data

Table 4-3   Comparison of the classification performance of the different feature types over 100 LOOCV trials

Feature Type	Max CCR	Median CCR	Min CCR
WSVD/IPA	94%	94%	86%
SVD/IPA	83	72	61
SVDOnly	78	72	69

The maximum, median, and minimum CCR values are shown for each feature set. Note that FFMLP models trained on WSVD/IPA features attained a median CCR of 94 versus 72 percent for SVD/IPA and SVDOnly features. This result represents a 31 percent improvement in prediction performance, suggesting that wavelet signal processing significantly enhances the classification of breast cancer tumors that have spread to the lymph nodes. Note the median CCR for both the SVD/IPA and SVDOnly feature sets were equivalent, although the CCR variation for the SVD/IPA features is greater than the CCR variation for the SVDOnly features. This suggests that pathway compression without wavelet processing does not result in any appreciable improvement in classification performance.

compression algorithms are utilized. In this particular case, "appropriate" means wavelet signal processing, SVD and pathway compression, and neural networks.

4.12 DISCUSSION

We have shown that combining different signal processing techniques, namely, MANINI detection, pathway compression, wavelets, and SVD results in feature patterns that are able to accurately and robustly discriminate between high- and low-risk breast cancer tumors using as few as 35 genes. Specifically, FFMLP classifiers trained on WSVD/IPA feature patterns showed a 31 percent increase in CCR in comparison to FFMLP classifiers trained on SVD/IPA or SVDOnly features.

A key step in the robust modeling of lymph node status was the selection of genes for downstream feature extraction and machine learning using the MANINI detection algorithm. Recall that the MANINI detector assigns genes having similar expression levels into "bins" where each bin defines a separate signal detection problem based on data that is approximately normally distributed. A "test-and-estimate" or testimation procedure based on the DJ universal threshold (from wavelet denoising theory) was applied to the fold change values of the genes in each bin. Genes that exceeded the threshold were called differentially expressed. The union of genes over all intensity bins was called differentially expressed over bins represents a global gene expression signature for metastatic breast cancer tumors.

Recent research has shown a deep connection between statistical testimation, sparse signal estimation, and multiple hypothesis testing with adjustments for multiple comparisons [29]. Hence, the MANINI detector can be viewed as an optimal estimator of a sparse signal that automatically accounts for the approximately 600 statistical tests performed simultaneously within each of the 50 bins of the MANINI detector. Note also that MANINI detection throws a wide net and detects genes based on similarity rather than absolute magnitude or consistency of expression. That is, MANINI may call a gene DE even though it is altered in expression on only a fraction of the positive samples of the Huang data set, so long as the average fold change over all samples exceeds the DJ threshold for the bin to which it belongs. The ability to detect genes that are "intermittently" DE is important when dealing with heterogeneous data sets often encountered in cancer research that contain weak signals for DE, or samples that are temporally and/or developmentally out of phase.

The observed increase in FFMLP classification performance using WSVD/IPA features is due in large part to the fact that pathway compression drastically reduces the number of genes that must be processed while at the same time preserving important information related to lymph node status [16]. Since significant IPA pathways extracted from the MANINI gene list are biologically relevant in terms of known gene function and gene–gene interactions as embodied in the IPA knowledge base, pathway compression provides a biologically driven method for selecting a small number of highly informative genes from which feature patterns can be extracted for diagnostic, prognostic, and predictive applications in the clinic. Because the resultant features are biologically based, the NN models trained on such features are likely to be more robust over a larger population of samples. The results shown in Table 4-3 suggests that NN models trained on gene expression signatures derived from pathway compression generalize well to a larger population of samples.

Table 4-3 also indicates that wavelet signal processing significantly enhances the prediction of lymph node status using NN models. Indeed, wavelet signal processing raised the median CCR from 72 to 94 percent for an overall improvement of 31 percent. The main reason for this result is that wavelet analysis of a sample's expression profile exhibits better separation between coherent signal and random noise in the CWT domain [45,52]. The subsequent analysis of the wavelet transformed data matrix using SVD results in a better estimate of the intrinsic dimensionality of the wavelet denoised data matrix. SVD compression in the wavelet domain has been used to solve a number of difficult pattern recognition problems, including the automated classification of underwater buried mines and the detection of cervical pre-cancer in three-dimensional hyperspectral images of the cervix [27,28].

Note that wavelet signal processing of microarray data was made feasible to a large extent by pathway compression. Indeed, without pathway compression, sample expression profiles of 413 genes would have to be wavelet transformed, which is computationally onerous. In contrast, after pathway compression, the resulting sample expression profiles are no more than 35 genes long, thereby enabling the efficient use of the CWT for data preconditioning and denoising. Moreover, the information processing described in this chapter applies wavelet signal processing to the columns of the data matrix instead of the rows. Hence, the resolution of the wavelet transform is limited not by the number of samples in the experiment, but by the number of genes included in the pathway-compressed data matrix. Since the number of genes is intrinsically large, this approach circumvents the problem of having too few samples for the wavelet transform to work properly.

NN classification performance can probably be improved by including genes contained in different IPA networks. Future work involves the use of genetic algorithms to globally search for the best combination of genes, significant IPA networks, wavelets, and model parameters that maximizes the CCR score for predicting lymph node status in the Huang breast cancer data set. Once the optimal genes and pathways for distinguishing positive and negative breast cancer tumors are identified, close examination of individual genes and their interactions with other genes contained in the selected networks could very well lead to new insights into the molecular mechanisms underlying metastatic breast cancer. Here, the overarching assumption is that predictive performance is equivalent to biological significance, thus enabling the use of machine learning models such as FFMLPs to identify the genes and pathways that are most biologically relevant to the spread of breast cancer to the axillary lymph nodes. Are there networks or combinations of networks that result in even more robust and accurate predictions of lymph node status than those produced by the IPA network *dnet1*? It is expected that modern signal processing, FFMLP/NN models, and genetic search algorithms will provide an answer to this question.

Note that IPA assigned the biological function of cancer to *dnet1*, and moreover, simulations suggest that the topology of *dnet1* with the FOS gene as a highly connected "hub" gene is invariant to minus-one perturbations of the data. The FOS gene family has been implicated in the regulation of cell proliferation, differentiation, and transformation. Other cellular roles include transformation, apoptosis, growth, activation, motility, and cell cycle progression. FOS has also been associated with cardiovascular disease. Note that the ability to accurately classify breast cancer tumors according to lymph node status is quite different from attaining a deep understanding of the biological mechanisms underlying the spread of breast cancer. Be that as it may, a close examination of predictive networks and the genes they contain could well lead to a better understanding of the molecular mechanisms underlying metastatic breast cancer. Such mechanistic models could lead to new diagnostics and therapeutics that significantly improve the way breast cancer is treated and managed in the clinic.

ACKNOWLEDGMENTS

I would like to acknowledge my colleagues at the Cardiovascular Research Center and the Cancer Research Center of Hawaii and, in particular, Drs Charles Boyd, Richard Girton, Loic Le Marchand, and Patrick Fu for their support, encouragement, and many discussions during the research and writing of this chapter.

REFERENCES

1. Jemal A, et al. *Cancer Facts & Figures 2007*. Atlanta: American Cancer Society, 2007.

2. Veronesi U, Paganelli G, Galimberti V, Viale G, Zurrida S, Bedone M, Costa A, Decicco C, Geraghty JG, Luini A, Sacchini V, Veronesi P. Sentinel-node biopsy to avoid axillary dissection in breast cancer with clinically negative lymph-nodes. *Lancet* 1997; 349:1864–1867.

3. Jatoi I, Hilsenbeck SG, Clark GM, Osborne CK. Significance of axillary lymph node metastasis in primary breast cancer. *J Clin Oncol* 1997;17:2334–2340.

4. Huang E, et al. Gene expression predictors of breast cancer outcomes. *Lancet* 2003; 361:1590–1596.

5. West M, et al. Predicting the clinical status of human breast cancer by using gene expression profiles. *Proc Natl Acad Sci USA* 2001;98(20):11462–11467.

6. Ma XJ, Salunga R, Tuggle JT, Gaudet J, Enright E, Mcquary P, Payette T, Pistone M, Stecker K, Zhang BM, Zhou YX, Varnholt H, Smith B, Gadd M, Chatfield E, Kessler J, Baer TM, Erlander MG, Sgroi DC. Gene expression profiles of human breast cancer progression. *Proc Natl Acad Sci USA* 2003;100:5974–5979.

7. Mircean C, et al. Pathway analysis of informative genes from microarray data reveals that metabolism and signal transduction genes distinguish different subtypes of lymphomas. *Int J Oncol* 2004;24:497–504.

8. Alberts B, Johnson A, Lewis J, Raff M, Roberts K, Walter P. *Molecular Biology of the Cell*, 4th ed. New York: Garland Science, 2002.

9. van't Veer L, et al. Gene expression profiling predicts clinical outcome of breast cancer. *Nature* 2002;415:530–536.

10. Simon R. DNA microarrays for diagnostic and prognostic prediction. *Expert Rev Mol Diagn* 2003;3(5):587–595.

11. Schena M. Genome analysis with gene expression microarrays. *Bioessays* 1996;18(5): 427–431.

12. Kohane IS, Kho AT, Butte AJ. Microarrays for an Integrative Genomics. In: Istrail S, Pevzner P, Waterman M, editors. *Computational Molecular Biology.* Cambridge, MA: The MIT Press, 2003.

13. Fodor SP, Read JL, Pirrung MC, Stryer L, Lu AT, Solas D. Light-directed, spatially addressable parallel chemical synthesis. *Science* 1991;251(4995):767–773.

14. Zweiger G. *Transducing the Genome.* New York: McGraw-Hill, 2001.

15. Mertens B. Microarrays, pattern recognition and exploratory data analysis. *Stat Med* 2006;22:1879–1899.

16. Slonim D. From patterns to pathways: gene expression data analysis comes of age. *Nat Genet* 2002;32:502–508.

17. Wong DJ, Chang HY. Learning more from microarrays: insights from modules and networks. *J Invest Dermatol* 2005;125:175–182.

18. Donoho D. High-dimensional data analysis: the curses and blessings of dimensionality. In: *Mathematical Challenges of the 21st Century.* University of California, Los Angeles: American Mathematical Society, 2000.

19. Simon R, et al. Pitfalls in the use of DNA microarray data for diagnostic and prognostic classification. *J Natl Cancer Inst* 2003;95:14–18.

20. Rocke DM, Durbin B. A model for measurement error for gene expression arrays. *J Comput Biol* 2001;8(6):557–569.

21. Bishop C, editor. *Pattern Recognition and Machine Learning,* Information Science and Statistics, Jordan M, Kleinberg J, Scholkopf B, editors. New York: Springer Science + Media, 2006.

22. Vanchieri C. National Cancer Act: a look back and forward. *J Natl Cancer Inst* 2007;99(5): 342–344.

23. Yang SX, et al. Gene expression patterns and profile changes pre- and post-erlotinib treatment in patients with metastatic breast cancer. *Clin Cancer Res* 2005;11(17): 6226–6232.

24. Bishop C. *Neural Networks for Pattern Recognition.* Oxford, NY: Oxford University Press, 1998.

25. Valafar F. Pattern recognition techniques in microarray data analysis: a survey. *Ann N Y Acad Sci* 2002;980:41–64.

26. Bellman RE. *Adaptive Control Processes.* Princeton, NJ: Princeton University Press, 1961.

27. Okimoto GS, Lemonds D. Principal components analysis in the wavelet domain: new features for underwater object recognition. In: *Proceedings of SPIE on Aerosense: Detection and Remediation Technologies for Mines and Mine-like Targets IV, Orlando, FL,* 1998.

28. Okimoto GS, et al. New features for detecting cervical pre-cancer using hyperspectral diagnostic imaging. In: *Proceedings of SPIE on Clinical Diagnostic Systems, San Jose, CA,* 2001.

29. Sabatti C, Karsten SL, Geshwind DH. Thresholding rules for recovering a sparse signal from microarray experiments. *Math Biosci* 2002;176:17–34.

30. Schalkoff R. *Pattern Recognition: Statistical, Structural and Neural Approaches.* New York: John Wiley and Sons, Inc., 1992.

31. Okimoto GS. On the analysis of microarray data with applications to cancer and cardio-vascular disease. Doctoral Dissertation for the Department Molecular Biosciences and Bioengineering (in Bioinformatics), 2006.

32. Quackenbush J. Microarray data normalization and transformation. *Nat Genet* 2002;32:496–501.

33. Ballman KV, et al. Faster cyclic loess: normalizing RNA arrays via linear models. *Bioinformatics* 2004;20(16):2778–2786.

34. Quackenbush J, Computational analysis of microarray data. *Nat Rev Genet* 2001;2:418–427.

35. Lyons-Weiler J, et al. Test for finding complex patterns of differential expression in cancers: towards individualized medicine. *BMC Bioinformatics* 2004;5110.

36. Tibshirani R, Hastie T. Outlier sums for differential gene expression analysis. Stanford Technical Report, 2006.

37. Draghici S, et al. Global functional profiling of gene expression. *Genomics* 2003;81:98–104.

38. Mariani TJ, et al. A variable fold change threshold determines significance for expression microarrays. *FASEB J* 2003;17:321–323.

39. Donoho D, Johnstone I. Ideal spatial adaptation by wavelet shrinkage. *Biometrika* 1994;81:425.

40. Poor HV. *An Introduction to Signal Detection and Estimation*, Springer Texts in Electrical Engineering, Thomas JB, editor. New York: Springer Verlag, 1994.

41. Kay SM. *Fundamentals of Statistical Signal Processing, Volume 2: Detection Theory.* New York: Prentice Hall, 1998.

42. Dudoit S, et al. Statistical methods for identifying differentially expressed genes in replicated cDNA microarray experiments. Technical Report #578, University of California at Berkeley, 2000.

43. Subramanian A, et al. Gene set enrichment analysis: a knowledge-based approach for interpreting genome-wide expression profiles. *Proc Natl Acad Sci USA* 2005;102(43): 15545–15550.

44. Zimmerman Z, Golden JB III. The return on investment for Ingenuity Pathways Analysis within the pharmaceutical value chain. Unpublished white paper, 2004.

45. Kaiser G. *A Friendly Guide to Wavelets*. Boston, MA: Birkhauser, 1995.

46. Parker M, et al. Initial neural net construction for the detection of cervical intraepithelial neoplasia by fluorescence imaging. *Am J Obstet Gynecol* 2002;187:398–402.

47. Kalman D. A singularly valuable decomposition: the SVD of a matrix. *College Math Journal* 1996;27(1):2–23.

48. Strang G. *Linear Algebra and Its Applications*. Orlando, FL: Harcourt Brace Javanovich, Inc., 1988.

49. Masters T. *Advanced Algorithms for Neural Networks: A C++ Sourcebook*. New York: John Wiley & Sons, Inc., 1995.

50. Mallat S. *A Wavelet Tour of Signal Processing*. New York: Academic Press, 1999.

51. Lio P. Wavelets in bioinformatics and computational biology: state of art and perspectives. *Bioinformatics* 2003;19(1):2–9.

52. Masters T. *Signal and Image Processing with Neural Networks*. New York, NY: John Wiley and Sons, Inc., 1994.

53. Ripley BD. *Pattern Recognition and Neural Networks*. Cambridge, UK: Cambridge University Press, 1996.

5

RECOMBINANT GENOMES: NOVEL RESOURCES FOR SYSTEMS BIOLOGY AND SYNTHETIC BIOLOGY

Mitsuhiro Itaya

Laboratory of Genome Designing Biology, Institute for Advanced Biosciences, Keio University, 403-1 Nipponkoku, Daihoji, Tsuruoka, Yamagata 997-0017, Japan

5.1 INTRODUCTION

Cellular life is elaborately managed by linear as well as branched biochemical pathways and all underlying information required for these systems are contained in the nucleotide sequence of the genome. Modern genomes possess tremendous amounts of information selected and accumulated during responses to altering natural environmental conditions. "Genome" nomenclature for proliferating species on earth is normally given to cells higher than bacteria as illustrated in Figure 5-1. The term "Genome" is also used to describe DNA possessed by bacteriophages, viruses, plasmids, mitochondria [1], and chloroplasts [2]. The last two are believed to have an ancient bacterial origin; with these two systems, additional essential informative molecules must be supplemented by the host. Given a genomic scope limited to unicellular bacteria, which are generally regarded as simple, diversity is observed in species variations manifested not only in taxonomic classifications but also in physical structure as illustrated in Figure 5-1. Due to the vast exploring technologies of cloning that emerged in the last quarter-century, our knowledge of genes and their products including RNA molecules and

Systems Biology and Synthetic Biology Edited by Pengcheng Fu and Sven Panke
Copyright © 2009 John Wiley & Sons, Inc.

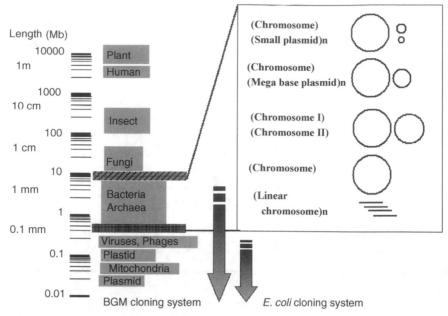

Figure 5-1 Genome variation—sizes and structure. Approximate genome size ranges are indicated for eukaryotes, bacteria, and archaea. Categorized structural types of bacterial genomes are illustrated in the insert. Sizes are in mega base pairs and the linearized lengths are scaled in the left. DNA size covered by *E. coli* plasmid vector (Section 5.8.2), and the BGM vectors (Section 5.4) are indicated.

associated metabolites has tremendously increased. Gene expressions are controlled by the complex and dynamic actions of thousands of factors as described in other chapters of this book. Perturbing small portions of gene-circuit frameworks are studied using genetically well-manipulatable organisms/cells [3].

Genome design, one of the topics highlighted in this chapter, requires at least two foundational roles: (1) Writers of novel nucleotide sequence at any length to be used as blueprint and (2) builders of high-molecular DNA who refer to the blueprint. Recombinant genome technology, not well documented elsewhere to date, has covered mostly the latter and will develop as the necessary tools for building any size DNA molecule. The former must be and is being conducted through current study of cellular life. Global attempts are described in some of other chapters where collaborative approaches among various disciplines are seen.

The DNA described in this chapter is larger in size and includes a greater number of genes than any other genetic entities handled in conventional recombinant gene technologies. Topics of this chapter will focus on emerging methodology in which complexity associated with an elevated number of genes as well as molecular constraints imposed by the increased DNA size must be treated at a time. The method utilizing *Bacillus subtilis* as a cloning host has been exploited independently from the conventional and familiar gene cloning technologies using *Escherichia coli* as a major cloning host (Fig. 5-2). Readers will be offered an overview of technical

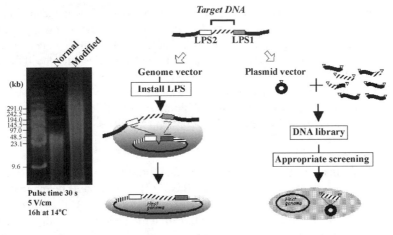

Figure 5-2 Cloning principle differs in *B. subtilis* genome vector: Essences for the *B. subtilis* genome vector (left) and the *E. coli* cloning vector (right) are in comparison. In the left path, designated/target DNA is finally guided in the host *B. subtilis* genome. Higher molecular weight DNA obtained by modified method are used for BGM cloning. *Source*: Referred to Ref. [17] and Section 5.4.2.

breakthroughs used to create recombinant genomes that would otherwise be considered impossible to obtain. The *B. subtilis* cloning system permits not only an increase in the clonable number of genes but also flexible postcloning modifications in DNA sequences, such as content, order, and orientations of genes. It will be assumed that *B. subtilis* host and its intrinsic genetic features are not familiar to most of the readers. Therefore, sections of this chapter will be attributed to basic explanation underlying this novel genome vector system. I must mention that the technological breakthroughs of several of the achievements described here are currently being employed for practical research and industrial use. Still others remain largely in nascent forms such as "Recombinant genome able to sustain life in diverse growth conditions." The intent is that the content in this chapter will not only address the genome vector protocols but also supply enough information to establish conceptual significance.

5.2 DNA (GENE) CLONING

All the primary information to sustain life is printed in the present genome DNA sequences of every organism, from bacterial to human cells, without exception [4]. According to DNA sequence determination technology with significant higher throughput [5], whole genome sequencing has extended to genomes even from nonclonal bacteria [6,7].

In parallel with the ability to determine DNA sequences with reduced cost and time, DNA cloning has been one of the most basic tools in biology to comprehend genes and gene functions. The most conventional cloning method has been developed using

E. coli as a cloning host and is summarized in Chapter 2. Tremendous accumulation of genetic and biochemical information on *E. coli* has supported versatile applications. This allows recombinant gene technologies to flourish and has made *E. coli* compatible with most research initiatives, including omics technologies introduced throughout this book. Manipulated recombinant genes that confer genetic variation, modulation, and perturbation of gene-circuit networks *in vivo* are also necessary tools for further development of systems biology. In addition, emerging *de novo* chemical synthesis technology may offer more opportunities in preparation of DNA pieces from scratch [8], yet the gene (DNA) cloning remains an inevitable step in most life science fields.

E. coli has the potential to harbor huge DNA molecules up to 350 kb in size assuming an appropriate vector choice, yet this size range remains below that of the smallest bacterial genome, 585 kb, of *Mycoplasma genitarium* [9,10]. When dealing with DNA fragments of this size, two technical skills become critically important, target cloning of the long DNA and flexibility in target sequence manipulation.

In cloning of DNAs particularly above dozens of kilo base pair, the size limit of PCR-mediated amplification method, preparation of nonsheared source DNAs is crucial. Our answer will be given in Sections 5.3–5.5 where repeated assembly of overlapping small segments leads to, via gradual elongation, final reconstruction of the target full-length DNA.

5.3 A GENOME VECTOR SUITED FOR RECOMBINANT GENOMES

Use of the 4215 kb *B. subtilis* 168 genome [11,12] as a stable cloning vector was first proposed in 1995 [13], and was supported by preceding works initiated early in 1990s [14–16]. The BGM, standing for *Bacillus GenoMe* vector and first coined in our related article [17], inherits a number of features absent in *E. coli* plasmid vector systems. After completion of this chapter one should be able to recognize many advantageous features of the BGM vector such as simplicity in daily handling, technical linkage to the conventional methods. Particularly, coverage for giant DNA segments and innovative potential for novel research in both fundamental and applied fields should be acknowledged.

The BGM cloning steps do not employ conventional enzymes such as restriction endonucleases and ligases that are vital in current DNA cloning methodology. On the contrary, homologous recombination plays a central role in this cloning system as well as in subsequent manipulations. As indicated in Figure 5-2, the conventional ligation step *in vitro* to directly connect DNA fragment to plasmid vector is replaced by inherent nature of the *B. subtilis* natural competence development/induced homologous recombination *in vivo* [18,19].

Readers may need some explanation as to why DNA goes into the genome of *B. subtilis* and not into the *E. coli* genome. The fundamental nature of this process is outlined in Figure 5-3. The key difference between the DNA uptake steps of both strains is clear: Only *B. subtilis* is able, under specific culture conditions, to develop a competent state where DNA outside the cell is actively incorporated in

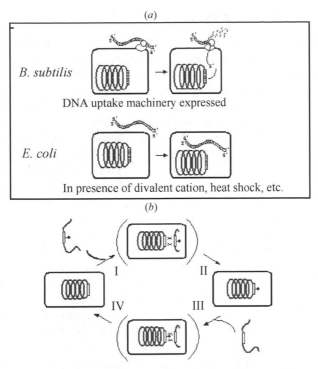

Figure 5-3 How *B. subtilis* incorporates DNA. (a) *B. subtilis* develops protein complex to actively incorporate DNA. Single-stranded DNA is the final substrate through uptake. These active mechanisms are not present in *E. coli*. (b) Insertion [I, II] and removal [III, IV] of DNA, shown by a needle with closed circle, via homologous recombination at the flunking regions of the *B. subtilis* genome.

cytoplasm [19,20]. *E. coli*, in sharp contrast, never actively transfers DNA through membranes and therefore import of DNA into cytoplasm must be induced via physicochemical treatment [21]. Another fundamental feature associated with competent *B. subtilis* is that double stranded DNA (dsDNA) taken up by the protein complex on the cell's membrane surface undergoes processing so as to deliver a single-stranded DNA (ssDNA) molecule into the cytoplasm. The cleavage site by the nuclease and the strand selection by the transformation complex are basically random [19]. The resultant highly recombinogenic single-DNA strand promptly recombines with the counter homologous sequences if present in the genome (Fig. 5-3). The mechanism for the BGM vector to integrate/clone the target DNA is simple. As illustrated in Figures 5-2 and 5-3, two DNA sequences that sandwich the target DNA, generally termed as LPS, standing for landing pad sequence, must be present/installed in the BGM genome. Homologous recombinations between the LPS sequences of incoming DNA and of the genome result in concomitant integration of the internal target DNA region. Methods for preparing the embedded LPS and how a large target DNA integrates effectively are keys to understanding the precise cloning path of the BGM vector.

5.3.1 Domino Method: A Prototype

The first successful full-length target cloning reported was the complete *E. coli* prophage lambda DNA, one of the earliest genomes fully sequenced [22]. During the cloning of the 48.5 kb lambda *c*I857sam7 genome [13], most ideas and relevant experimental technologies that lead to what we would later call the domino method were established. The lambda DNA fragment alone never integrates in the *B. subtilis* genome because no homologous sequence is present. Also, lambda DNA possesses no inherent selection marker effective in *B. subtilis*. Therefore, appropriate LPS as well as general selection markers to navigate the lambda DNA segment had to be prepared.

The lambda DNAs were segmented into several pieces of DNA fragments ranging from 2.4 to 16.8 kb by either of the two restriction enzymes *Bam*HI or *Eco*RI. They are cloned in the *E. coli* pBR322 plasmid by conventional DNA cloning method of this host (Fig. 5-4). These pBR322-based clones are collectively called domino clones. All share two common structural features possession of sequence overlap with the adjacent domino clone and two identical regions of the pBR322 sequence. The two pBR halves, *amp* and *tet* half illustrated in Figure 5-4, play essential roles and are vital in BGM cloning at any stage. The same two pBR halves were integrated earlier in particular loci of the *B. subtilis* genome [15]. This genome integrated pBR form, termed as GpBR, was proven to accommodate DNA sandwiched by the two pBR halves used as LPS. The combination of the pBR part of domino clones and the GpBR assures that DNA integrated into the *B. subtilis* genome always remains flanked by *amp* and *tet* halves of GpBR. The first domino integration is illustrated in Figure 5-5.

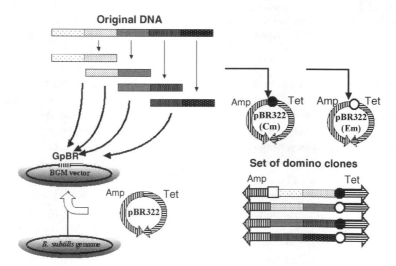

Figure 5-4 Domino method. Materials needed: Preparation of domino clones (right) and cloning locus in the BGM vector (left): Domino clones are shown in linear form. The selection markers, Cm (●) and Em (○), are set up alternately. pBR322-based domino clones all share common structural features. They are two halves of the pBR322 sequence, described as *amp* and *tet* and sequence overlap with the previous and next dominos. The two pBR halves play essential roles as illustrated in Figure 5-5.

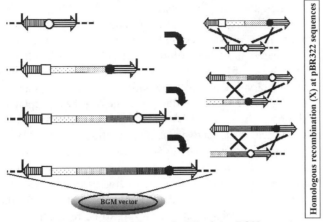

Reconstructed/assembled in the GpBR of the BGM vector

Figure 5-5 Domino method. Bottoming-up DNA cloning in the BGM vector: The first domino integrates via double homologous recombinations, indicated by X, at two pBR sequences of the domino and GpBR (shown top in the right). Integration of the second domino uses internal overlap region as one of the homologous recombination. This elongates the internal DNA and exchanges the marker for selection from Cm (●) to Em (○) (middle in the right). As elongation continues, two pBR halves always remain flunking the insert and exert reusable LPS (bottom in the right). The first domino only possesses another marker (□) to label the other end. I-*Ppol* site indicated by short vertical bar is preinstalled and used to extract cloned DNA as described in Section 5.6.1.

If an adjacent domino clone comes in, either of the pBR halves then can serve as a reusable LPS. This automatically positions the other distal end sequence, not the other pBR half, as another LPS to result in elongation of the target region as indicated in Figure 5-5.

The full-length lambda genome DNA reconstructed using four domino clones drew little interest, probably because of unfamiliarity of the host *B. subtilis* to most *E. coli* users. Given that this success occurred immediately before several whole bacterial genome sequences become available, it may have been too early to arouse interest in the manipulation of large DNA molecules. Through the lambda-cloning experiment and the subsequent trials and errors in our laboratory, related technical problems have been refined making the domino method applicable for larger sequenced genomes.

5.3.2   Domino Method: Applications to Organelle Genomes

The domino method simply requires a full set of domino clones that cover the entire target genome. Any gaps due to lack of an available domino clone should be avoided. This problem was resolved when a PCR methodology is combined with the preparation of dominos from sequenced genomes.

Innovative applications of the domino method have been clearly proven by the challenges to obtain chloroplast and mitochondria genomes whose complete sequences are known [4]. Chloroplast genomes (cpGenome) were chosen as candidates by (1) their sizes ranging between 100 and 200 kb [2], (2) circular form, and (3) no complete cloning report had been made. Due to their larger size, it is difficult to prepare intact unsheared circular chloroplast DNA suitable for one-step cloning. Instead, sheared shortened chloroplast DNA during biochemical isolation serves as template DNA sufficient for PCR reaction.

The set of domino clones that cover the entire sequence of cpGenome from rice (134.5 kb) were designed. A total of 31 domino clones for the rice cpGenome, 6 kb on average and 1 kb of which serve as LPS with the adjacent domino were prepared in *E. coli* via PCR amplification. Effectiveness of the method in the BGM vector was fully demonstrated for complete cloning of the rice cpGenome [23]. The recombinant rice cpGenome was stably maintained regardless of the two identical 21 kb-long inverted repeat (IR) sequences that are characteristic of higher plant cpGenome. The rice cpGenome example strongly indicates that the domino method permits any known DNA sequence up to and probably above 150 kb, to be reconstructed in the BGM vector.

Mitochondria genome (mtGenome), from another organelle present in nearly all eukaryotic cells, was similarly approached. The size range of mtGenomes exhibit great diversity in a species-dependent manner [1], compared with the limited size range of the cpGenome. Thus, our attempts to date have been limited to the well-studied mouse mtGenome. Mouse mtGenome (16.3 kb) is far smaller than other mitochondria species [1] and was sequenced at approximately the same time as the lambda genome. Interestingly, no report of its full-length cloning was found until recently, in spite of the small size, as small as one domino clone (16.8 kb) prepared for the lambda genome cloning [13]. The only previous success of mouse mtGenome cloning in pBR322 based plasmid appeared to be fully dependent on careful preparation of intact mtGenome from the mouse cell and a fortuitous insertion site selected by the transposon vectors used [24]. However, this pioneering work demonstrated that cloning of mouse mtGenome is not problematic. Cloning of the mouse mtGenome by our domino method, separated into four domino clones and incrementally reconstructed in the BGM vector, also had few problems [23].

These complete cpGenome and mtGenome stably cloned in the BGM vector were converted to a circular DNA form via a unique positional cloning method described and briefly mentioned in Section 5.6.3. The circular form of mtGenome was optionally propagated in an *E. coli* host using the shuttling nature of the plasmid, showing consistent results with those who had performed similar work [24].

5.3.3 Domino Method: General Application to Gene Assembly

The primary requirement for a domino clone is possession of overlapping sequence with the adjacent DNA. This requirement does not exclude the domino elongation step with an unlinked DNA segment; so long as an adjacent domino clone possesses the LPS portion as shown in Figure 5-4. In other words, target reconstructed DNA is not limited to continuous DNA segments of the present genome. In extreme cases, all the domino clones composed of two DNA blocks might be made from designed sequence.

The domino elongation as illustrated in Figure 5-5 does not exclude such designed DNA blocks, and consequently the final DNA is reconstructed according to the designed blueprint.

The designed assembly of DNA blocks was demonstrated for genes involved in *de novo* pigment synthesis. Eight cDNAs were prepared from *Arabidopsis thaliana* coding enzymes that catalyze a series of biochemical reactions from tryptophan to anthocyanidine, a violet-colored pigment made and stored by certain plants [25]. In our first assembly design, the order of these eight genes, dispersed in five different chromosomal loci, was as the same order of biochemical reactions as illustrated in Figure 5-6a. This primitive operon-like construct was built by progressive integrational elongation using eight gene blocks as domino clones. Similarly, the general domino method was applied to biosynthetic genes for another pigment, carotenoids, that is synthesized and stored in orange colored plants such as carrots as well as certain bacteria [26]. In a reassemble step of the genes included in the natural construct plasmid pACCAR25(ΔcrtX) from *Erwinia uredovora* [27] presented in Figure 5-6b, the expected intermediate substance lycopene was produced in *B. subtilis* by the assembly/insertion of the first three genes in the biochemical reaction [28]. Thus, the domino method has several examples of general applications and offers rational design in DNA assembly protocol for systems biology as well as synthetic biology. The domino method will be evaluated by comparing it with other assembly methods also exploited in our groups described in Section 5.5.4.

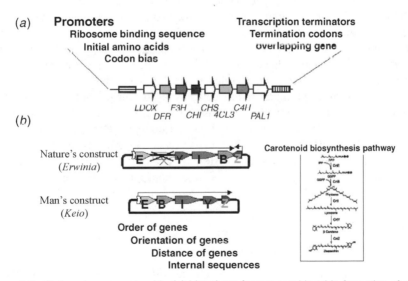

Figure 5-6 Options in gene assembly. (a) Listed are factors considered in formation of antocyanine biosynthesis cassette by the domino method in Section 5.3.3. Location of eight genes in *Arabidopsis* chromosome is indicated in Figure 5-18. (b) Carotenoid cassette (Keio form) was constructed by the domino method [61] in Section 5.5.4. Carotenoid biosynthesis by the OGAB method in Section 5.5.4.

5.4 DIRECT TARGET CLONING: PROTOTYPE

The domino method clarified two points, (1) any DNA segments can go into the *B. subtilis* genome via the intrinsic homologous recombination system utilizing preinstalled LPS sequences, and (2) the integrated/cloned DNA segment exhibits high structural stability. The latter is accounted for by precise replication as part of the *B. subtilis* genome and subsequent accurate segregation during *B. subtilis* cell division. Sequence indiscrimination during all the integration processes of *B. subtilis* guarantees the former; which is in contrast to sequence-dependent incorporation as seen for other Gram-positive strains [20,29]. Given a quantitative evaluation for nucleotide sequence fidelity of the reconstructed mtGenome and cpGenome [23], we had come to a putative conclusion that the *B. subtilis* genome has potential to harbor significantly large DNA repertoire with great fidelity in nucleotide sequence. The primary concern on LPS-mediated integration/cloning protocol was how to select foreign DNA possessing no selection markers for *B. subtilis*. Besides the progressive mode in the domino method and an exceptional gap-sealing protocol by unmarked DNA segment practiced in the lambda DNA cloning [13], the generalized selection marker scheme was a prerequisite.

5.4.1 Counter Selection Markers for Cloning Unmarked DNA Segments

Counter selection marker makes direct and rapid isolation of the correct integrant more promising. A neomycin resistance gene has been developed as a counter selection method [30]. A protein coding sequence of the neomycin resistance gene regulated under the Pr promoter (Pr-*neo*) was constructed in *E. coli* and integrated in the BGM vector at unlinked locus from the GpBR as illustrated in Figure 5-7. The Pr-*neo* confers *B. subtilis* neomycin resistance due to full expression of the *neo* gene product. Meanwhile, a CI gene product encoded by the *cI* gene of *E. coli* bacteriophage lambda binds to the Pr promoter sequence and shuts off the promoter activity. The *cI* gene, if present and constitutively expressed in the BGM vector, renders the *B. subtilis* sensitive to neomycin. Absence of the *cI* gene restores the Pr promoter activity and makes the strain resistant to neomycin and vise versa. This small transcriptional gene-circuit worked nicely as a counter selection system known as *cI*-Pr [30, 31] and has proven useful for cloning any DNA lacking an appropriate selection marker in the BGM vector [32–35]. More importantly, the reusable *cI*-Pr system allows repeated integration of DNA segments in the same BGM vector.

5.4.2 Quality Required as Donor DNA

Target DNA for positional cloning has to be as intact as possible [33, 35]. DNA in solution is normally fragmented into small pieces, typically dozens of kilo base pairs on average, caused by physical shearing during isolation step from cells and organelles. The relatively large DNA prepared in agarose gel matrix plug provides a more intact form with minimal breakage [12]. However, DNA inside the gel matrix is

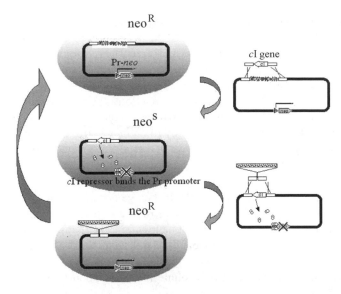

Figure 5-7 Reusable counter selection marker by a *cl*-Pr system. Presence (middle left) or absence (top left) of a repressor gene (*cl*) suppresses or induces antibiotic resistance gene (*neo*) under the Pr promoter. Replacement of the *cl* gene by foreign DNA confers resistance to neomycin (bottom left) and permits another *cl*-Pr selection. Details are described in Section 5.4.1 and [30, 31, 36].

not a good substrate for *B. subtilis* transformation. Therefore, as special protocol and gentle handling are required, we modified some DNA isolation protocols for BGM cloning. For example, a modified genome isolation method improved to yield high-quality DNA applied to the cyanobacterium *Synechocystis* PCC6803, produced DNA fragments on average of a couple of 100 kb as indicated in Figure 5-2. This quality DNA supplied sufficient length for target cloning of 30–70 kb segments [33]. Thus, the target size has to be determined by quality of prepared DNA in BGM cloning.

Ironically in contrast to the above size stipulations, the presence of huge untargeted DNA may result in adverse effects. For example, targeting 50 kb DNA of the 3500 kb *Synechocystis* genome sacrifices efficiency of cloning due to the inhibitory action by the remaining untargeted portion of the cyanobacterial genome present in the DNA preparation. This shortfall is accounted for the limited number of competent complexes formed on the competent cell surface; approximately 50 that take up and guide the DNA inside the cell [19]. Competitive inhibition by irrelevant DNA during *B. subtilis* transformation has been frequently observed, always resulting in marked reduction of the number of correct BGM recombinants [34], albeit is not entirely detrimental to the present BGM protocols.

5.4.3 Repeated Target Positional Cloning

As one can expect, no sooner was the success of target positional cloning, than target cloning of another adjacent segment results in elongation of the target region in the

BGM vector. One should recall that the cI-Pr dependent positional target cloning system is reusable and cloned DNA in the BGM always accompanies two halves of the pBR sequences at both ends. Furthermore, the second target cloning is performed so as to adjoin to the first BGM, as logically extended, and should permit additional positional target cloning by sliding to an adjacent region of the given BGM [36].

The LPS in target positional cloning may function as primer sequences in the PCR amplification method. If one views the two LPS sequences and intervening target DNA, analogous to the head, tail, and body of an inchworm, donor DNA acts as an inchworm walking via integration as shown in Figure 5-8. The inchworm lands first in the tail and head because only head and tail are present between the GpBR. The body part finally lands in the BGM at the completion of the cloning. Addition of a new set of LPS, new "head" aligned with the new "tail" being converted from the "head" in the previous inchworm. This sliding alteration of the two LPS guides the adjacent secondary target DNA so as to be positioned without any gaps. The DNA segment in the BGM vector is then elongated leaving the third inchworm available. Repeated application by renewing "head" and "tail" sets alternately results in progressive elongation of the target DNA until encountering certain constraints inherent in the BGM vector (see Section 5.7.2).

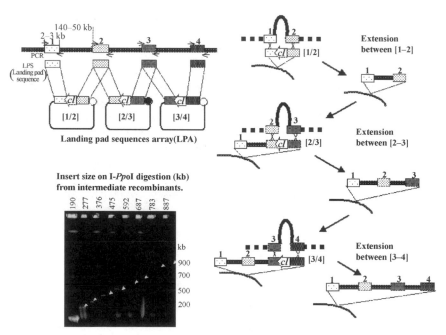

Figure 5-8 Inchworm like elongation in megacloning of the *Synechocystis* genome. Sets of two LPS (LPS array) are prepared separately. Internal sequence between LPS is integrated by the mechanism shown in the right. One cycle of inchworm starts by installation of LPS array and ends up with the incorporation of *Synechocystis* DNA. Details are provided in Sections 5.4.3 and 5.4.4. Discontinuous elongation of the *Synechocystis* genome region 2 (Fig. 5-9 is viewed by I-*Ppo*I fragment size increase. I-*Ppo*I site resides at both the ends of the GpBR in all BGM recombinants (Fig. 5-5).

5.4.4 Megacloning of the *Synechocystis* PCC6803 Genome

The method to elongate a continuous DNA by many inchworms in the BGM vector is termed as megacloning [36], invoking that the cloned DNA segment is in the mega base pair (Mb) size range. The main goal was to use the *B. subtilis* genome as a platform to clone and manipulate DNAs above 500 kb, seemingly the upper clonable limit with *E. coli* plasmids. The whole *Synechocystis* genome cloning stages are summarized in Figure 5-9. The work started in 1997, just 1 year after the whole genome sequence data were published [37]. The project evolved over the past years during which most works have been spent on refining all protocols. First, it should be mentioned as to why *Synechocystis* PCC6803 [37] was chosen as the target for megacloning. There are several reasons noted at that time: The preparation of LPS by PCR-mediated amplification requires sequence information on the whole genome. The high molecular weight genomic DNA prepared as shown in Figure 5-2 [33] was necessary for technical reasons described above. More importantly, the possible expression of cloned genes in *B. subtilis* hazardous for BGM user had to be avoided [38]. Therefore, the sequenced 3573 kb genome of the unicellular photosynthetic bacterium *Synechocystis*, thought to be nonpathogenic, was indeed the only available choice when this work started in 1997. The alleged goal ended by megacloning the whole genome of a life form, the 3.5 Mb of *Synechocystis*. Details are referred to in the recent publication [36], and results are illustrated in Figures 5-9 and 5-10.

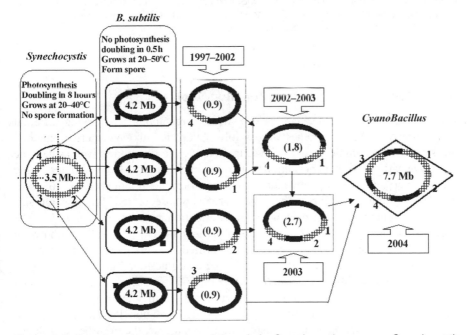

Figure 5-9 Overview of megacloning of the whole *Synechocystis* genome. *Synechocystis* genome was putatively divided in four sectors. Separately megacloned sectors in four BGM vectors were sequentially assembled in the subsequent process. This less straightforward way of the whole cloning was obligatory due to inherent structural constraints of the BGM vector (or the *B. subtilis* genome) as indicated in Section 5.7.2 and Figure 5-23.

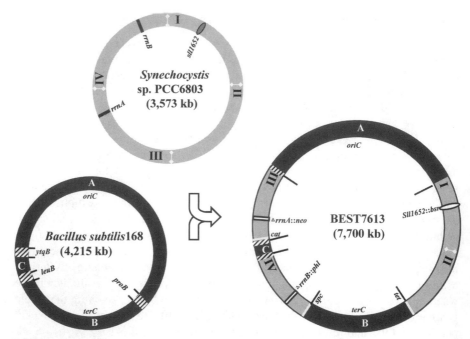

Figure 5-10 Genome anatomy of the *CyanoBacillus*. A chimera (right) of the two genomes from *Synechocystis* (top left) and *B. subtilis* (bottom left) is shown. Certain questions raised by this unprecedented organism are introduced in Section 5.4.6 and Figure 5-12.

5.4.5 Novel Method for Gene Function Analyses

During inchworm walking, various intermediate recombinants carrying different number of *Synechocystis* genes were stably obtained. This characteristic method portrays a novel technique for the investigation of gene function as well as genome function apart from other conventional DNA cloning methods. Two examples noting the discovery of adversely influential genes are worth mentioning here: *sll*1652 and *rrnA* (also equivalent *rrnB*). The gene *sll*1652, whose protein function remains unknown, apparently interfered with the sporulation process of *B. subtilis*. The sporeless phenotype was first found after a particular inchworm walk, as shown in Figure 5-11. The single gene was logically identified by subsequent deletion analysis in the BGM vector. Although the *sll*1652 gene could be found in the conventional *Synechocystis* DNA library made in *B. subtilis*, obvious phenotypic change before and after the presence of a defined region clearly specified the culprit gene. One more functional gene has been similarly suspected in a different region (Itaya M. and Fujita K. unpublished observations). In contrast to the first example, the unsuspected role for *rrnA*/*rrnB* was discovered in reverse manner. Inclusion of *Synechocystis* ribosomal RNA (5S-23S-16S) encoded in the *rrnA* or *rrnB* operon resulted in large deletions from other previously megacloned regions present.

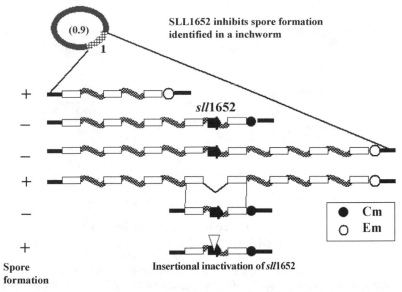

Figure 5-11 Comprehensible discovery of a gene from *Synechocystis*. An apparent sporeless phenotype rooted by a particular inchworm was scrutinized. Genomic manipulation of megacloned recombinants attributed the culprit gene to the *sll*1652 within 40 genes in the inchworm in Section 5.4.5. The sporeless feature was finally confirmed being caused by the gene only.

5.4.6 Questions Raised by Two Genomes in One Cell

Among the many questions arising from the *Bacillus–Synechocystis* chimera, putatively named as *CyanoBacillus*, the potential involvement of ribosomes is currently being intensively investigated. Lack of *Synechocystis* ribosomal RNA is consistent with our recent molecular data that a number of *Synechocystis*-originated transcripts but little translated products in the *CyanoBacillus*. These observations imply that ribosome and associated ribosomal RNA are key switching factors that determine dominant cellular gene networks as illustrated in Figure 5-12. If tremendous amounts of genes, including ribosomal RNA and protein genes are delivered by horizontal gene transfer (HGT), most of them are dormant in one environment but may be retained and utilized as functional and necessary genes when placed under different conditions. The *a priori* consensus for current molecular phylogenic analyses [39] is consistent with the difficulty of finding two natural *rrn* carriers to date [39–42]. The unusual nature of two ribosomes in one cell remains speculative and controversial. Further investigation of *CyanoBacillus* under omics analyses or creation of a second megacloning example should be conducted to access these issues.

One intriguing aspect on chimera structure of *CyanoBacillus* may be related to a proposal that genome fusions and horizontal gene transfer could be deduced from reconstruction of the phylogenetic tree of life. The hypothetical origin of the eukaryotic cell, albeit enigmatic and complex, is that it is the result of a fusion between two diverse prokaryotic genomes [43]. One fusion partner branches from

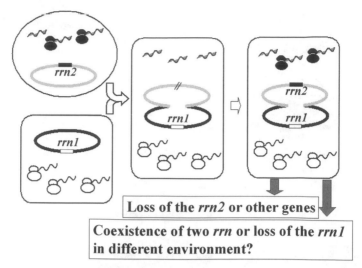

Figure 5-12 Ribosome switches genetic circuit by translational regulation? *Synechocystis* genome possessing *rrn2* (top left) and *B. subtilis* genome *rrn1* (bottom left) do not coexist in *CyanoBacillus* (center) as described in Section 5.4.6 [36]. Many numbers of transcripts from *Synechocystis* genes are observed in *CyanoBacillus* (our unpublished observation). Capturing *rrn2* might be suitable for survival in a different growth environment. The scenario employed has been limited only to complete genome fusion.

deep within an ancient photosynthetic clade, and the other is related to the archaeal prokaryotes. The eubacterial organism is either a proteobacterium, or a member of a large photosynthetic clade that includes the cyanobacteria and the Proteobacteria. This scenario may be investigated by megacloning of two candidate genomes in the BGM vector (see also Section 5.7.3).

5.5 ASSEMBLY OF GENES IN ONE DNA SEGMENT

Location of individual genes in the genome appears not to be determined by a defined set of rules [44–47]. With respect to a set of corelevant genes, typically regarded as operon, the conserved operons in the presently known bacterial genomes are very limited to certain ribosomal proteins and some metabolic pathways. Under the "selfish operon" hypothesis [48], operons are viewed as mobile genetic entities that are constantly disseminated via horizontal gene transfer.

The operon formation rule and the degree of HGT contribution remain controversial [49,50]. Apart from the enigmatic evolutional view for the present-day operons, we can technically modify the present-day gene order by the domino method as briefly mentioned above. Assembly of a number of functionally related genes may be the beginning of a drive to contemplate the ultimate man-made genome with all necessary genes assembled that function as the blueprint for engineered cellular life.

5.5.1   Why Should Genes be Assembled?

Biological processes are series of enzyme-catalyzed biochemical reactions. They include uptake or secretion of materials through the cell surface, production, or degradation of energy-coupled catabolites and metabolites in the cell, construction of cellular membranes/cell walls that are all responsible to sustain life. Biological reactions possess high potential as alternatives to traditional chemical processes for producing valuable molecules. Pioneering attempts have been made to produce materials by introducing a series of relevant genes to carry out their biological process in various hosts. To date, reports are concentrated in the two metabolically and genetically well-understood hosts: *E. coli* and *Saccharomyces cerevisiae* [51–57]. These hosts are eligible for repetitive transformation, circumventing problems associated with increased number of genes required for the biological reaction of interest. Indeed, if all the genes separately prepared from original genomes are included in a single-DNA segment in a row, delivery of such a biological reaction unit to another host becomes more popular and efficient [58]. The DNA manipulation system to construct such recombinant DNA, known as recombinogenic engineering, is limited and significantly dependent on manipulation using *E. coli* [59]. Ordering relevant genes into an all-in-one segment still requires laborious, time-consuming work. This appears to be a bottleneck in outlining a comprehensive blueprint for operon design.

5.5.2   Efficient Assembly of Genes in one DNA Segment using *B. subtilis*

We exploited a method to assemble a number of genes in one DNA segment with very few experimental steps, referred to as an ordered gene assembly in *B. subtilis* (OGAB) [60]. The method, as illustrated in Figure 5-13, stems from the unique *B. subtilis* DNA uptake characteristics. In short, multimeric forms of DNA are favorable substrates to be obtained as plasmid through a unique *B. subtilis* transformation. This consequence is due to the mode of DNA incorporation by competent *B. subtilis*. One may recall that ssDNA is taken up through the transformation apparatus [19] (Fig. 5-3). Only longer than one plasmid unit length results in replication of the full-length complementary strand, and therefore ssDNA of monomer unit length can never be converted to circular dsDNA plasmid. The DNA possessing tandemly repeated unit length is a good substrate for plasmid establishment via *B. subtilis* transformation.

5.5.3   Assembly of Various Numbers of DNA Segments in a Plasmid

How can tandemly repeated unit-length DNA (truDNA) be prepared, and how can multiple DNA fragments be assembled by BGM vector? Our solution was derived by making use of staggered dsDNA ends and modified conditions for T4-DNA ligase. In brief, all the component DNA segments have to possess protruding sequences that specifically connect only once to a singular complimentary end of another segment. As shown in Figure 5-14, for example, the variable three-base sequence within an

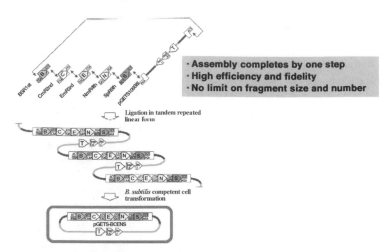

Figure 5-13 The DNA block assembly. Principle of one-step assembly of multimeric DNA fragments in *B. subtilis* plasmid is shown. The six-gene alignment comes from the result indicated in Figure 5-16. Details are stated in Section 5.5.2.

endonuclease *Sfi*I recognition sequence GGCCNNNN/NGGCC, indicated also in Figure 5-14, is suited for this aim. Linear multimer ligation products are formed preferentially in the presence of polyethylene glycol (PEG6000) and high salt concentration (sodium chloride, 150 mM). As shown in Figure 5-15, this condition dramatically suppresses formation of circular products and directs remarkable truDNA

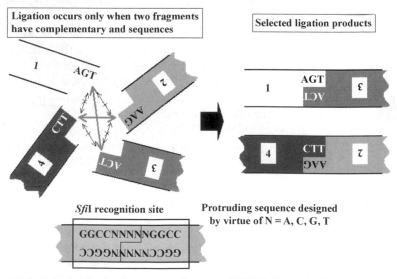

Figure 5-14 Selective ligation determines alignment of DNA blocks. Multiple protruding ends are created by restriction endonuclease such as *Sfi*I including variable nucleotide sequence within its recognition sequence. Flexible design of order and orientation of six gene blocks is possible.

Different ligation products prepared *in vitro*

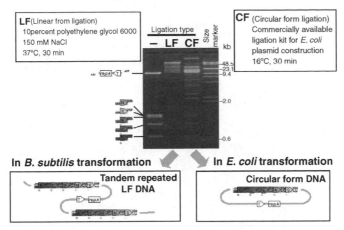

Figure 5-15 Conditions suitable for OGAB substrate on ligation. Multiple DNA blocks in linear form (LF) are preferentially produced in the presence of high molecular weight polymer and at a higher salt concentration [60]. Equivalent mole ratio for all the gene blocks is critical to form LF [60]. circular form (CF) DNA mostly formed in the regular ligation conditions appear as mixtures not suitable even for *E. coli* as indicated in Figure 5-16.

production. It should be emphasized that adjustment to equal moles of all the DNA segments is vital for ligation efficiency of truDNA. Assembly of up to six antibiotic resistance genes, comprising the multiple antibiotic resistance gene cassettes in an 18 kb plasmid, exhibited surprisingly high efficiency and fidelity compared with no equivalent plasmid obtained from *E. coli* transformation as indicated in Figure 5-16.

5.5.4 Application to Assemble Functionally Related Genes

In principle, the OGAB-mediated gene assembly has no limit on the number of included genes and final DNA size. In addition, order and orientation of these DNA fragments can be easily altered by sequence design of the protruding ends. Two recent achievements leading to what we believe is an initial framework for "operon designing" are described here. Two selected metabolic pathways, a pigment biosynthesis and a fungicidal substance production, require five and seven enzyme-coding genes, respectively. The pigment, a carotenoid, and the genes required for its production are basically the same as those used in the domino method (Fig. 5-6b). pACCAR25(ΔcrtX), a natural version, containing six genes under a unique promoter [27], produces the pigment in *E. coli*, albeit the order of genes is not aligned with that of biochemical reactions *in vivo*. Five out of the six genes that catalyze the five biochemical reactions from FPP to give zeaxanthin were prepared as DNA pieces and aligned in parallel with that of the biochemical reactions in a single-DNA segment by OGAB method. Due to the OGAB plasmid that carries dual replication origins for *B. subtilis* and *E. coli*, the recombinant construct when expressed in *E. coli* resulted in

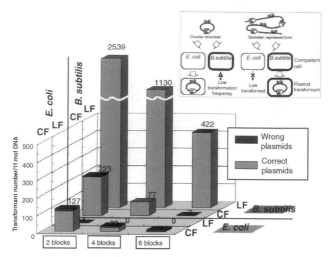

Figure 5-16 Powerful gene assembly by OGAB method. LF preferentially yields correct OGAB plasmid in *B. subtilis*, whereas CF dose not produce the expected ones in *E. coli* as the number of gene blocks increases. Simplified schematic view is shown in the insert.

surprisingly higher zeaxanthin than the original natural construct [61]. This simple and clear result showed that the expression level of a gene-circuit readily varies with the gene order and orientation and raises questions on how to approach the design of such operons, unless the natural operon is available.

We are learning lessons in approaching another biological process where extremely large proteins play roles to synthesize plipastatin, a lipopeptide fungicidal substance whose peptide portion is a nonribosomal peptide (NRP) produced by *B. subtilis* shown in Figure 5-17 [62]. The peptide portion of plipastatin is synthesized by five NRP synthases. These five large enzymes, 289, 290, 287, 407, and 145 kDa, are encoded in 38 kb-long genomic *ppsABCDE* operon in Figure 5-17. Difficulties in manipulating the huge operon and, more importantly, the resultant large transcript are expected to allow for novel insight into gene alignment and associated mRNA design.

For plipastatin production in *B. subtilis,* at least two more genes, *sfp* and *degQ,* are required in addition to the five genes in the *ppsABCDE* operon [62]. The *sfp* (0.9 kb) gene encodes 4′-phosphopantetheinyl transferase that catalyses transfer of 4′-phosphopantetheinyl to apo-peptide synthetases to convert the peptide to the holoenzyme form. The *degQ* (0.6 kb) gene that encodes a polypeptide composed of 42 amino acids shown to be a possible regulator for the *ppsABCDE* operon expression. We used the term "gene block" to define that the DNA fragments can be significantly variable both in number of genes present and in size. It was demonstrated that the OGAB method permits efficient assembly of three gene blocks ranging from 0.6 to 38 kb as predicted [62]. The two examples of zeaxanthin [61] and plipastatin [62] encourage us to enrich our disciplines toward rational operon design that was previously considered quite difficult.

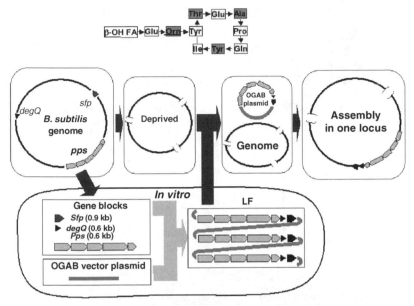

Figure 5-17   A design of operon for plipastatin bioprocess. Molecular structure of plipastatin, an antifungal agent, is shown on top [61]. Combined structure of lipid and peptides containing amino acid with D-configuration and ornithin is synthesized from components by three relevant gene blocks [62]. *B. subtilis* plipastatine gene originally dispersed in three locations are assembled by OGAB and function as stable operon in the genome.

5.5.5   Integration of OGAB Construct in the BGM Vector

The OGAB method yields assembled DNA segment in plasmid form due to its formation mechanism in *B. subtilis* (see the insert in Figure 5-16). The plasmid normally possesses a specific sequence used for initiation of DNA replication (*ori*) independent from the chromosomal counterpart (*oriC*). This physically independent character in replication as well as segregation during cell division ensures properties advantageous for gene cloning; for example, rapid extraction of cloned plasmid. However, in retrospect, the plasmid format sacrifices genetic stability compared with genes integrated in the genome. Thus, function of the newly made operons should be characterized as integrated form in the BGM vector. Attempts to integrate them in the BGM vector are currently pending.

5.6   HOW TO PURIFY RECOMBINANT GENOMES

Much focus has been placed on the DNA-cloning process in the BGM vector. These assembled DNA molecules then have to be delivered to another host system to diversify applications. In particular, biomaterial production, which is regulated by a number of metabolic genes, is advantageous if efficiently expressed in lower costing hosts. Engineering of organelle genomes, mitochondria and chloroplast, is much

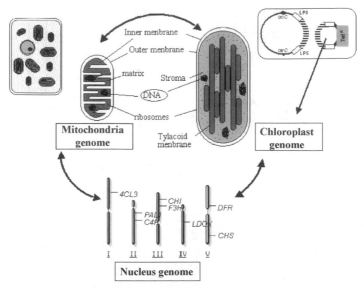

BGM approach toward organella and nucleus genomes

Figure 5-18 Organelle genomes and nucleus genome. Nucleus and two organelle genomes, mitochondria and chloroplast, are schematically viewed. The eight genes relevant for antocyanine bioprocess described in Figure 5-6 are shown. Chloroplast genomes from rice and tobacco are in circular form retrieved out of the BGM vector described in Section 5.6.3.

likely to establish an exciting field because organelle genomes are shown to interfere and exchange genes among genomes *in vivo*, such as transfer with plant nuclear genes [63] shown in Figure 5-18. Three methods for retrieval of the cloned segment from the BGM vector are summarized in Figure 5-19 and stated below.

5.6.1 General Method by Sequence Specific Endonucleases

Digestion by endonuclease of the cloned BGM recombinant and subsequent isolation/ purification of the cloned segment appear to be the most simple and straightforward method. Extreme infrequent recognition sequences, 23-base [ATGACTCTCTTAA/ GGTAGCCAAA] for I-*Ppo*I [64], and 18-base [TAGGGATAA/CAGGGTAAT] for I-*Sce*I [65] have been shown valuable in the BGM vector [16,36]. Owing to I-*Ppo*I site preinstalled at both ends of the GpBR of all the BGM vectors (Fig. 5-5), linearized DNAs produced on I-*Ppo*I digestion are readily isolated from agarose gel resolved by pulsed-field gel electrophoresis and can then be concentrated in liquid form. Because of such simplicity as depicted in Figure 5-19, the method has been of great use in primary analyses as well as pilot preparation of the cloned DNA [23,36].

5.6.2 Dissection of the *B. subtilis* Genome

The second method, genome dissection, largely depends on the *B. subtilis* genetic systems. This method originated from the study on diversity of multiple chromosomes

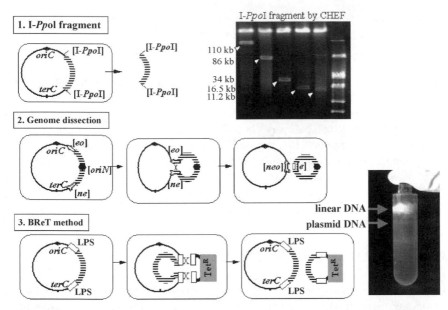

Figure 5-19 Three methods to extract the cloned DNA out of the BGM vector. Three methods are currently BGM vector specific. Examples for DNA resolved by the first method (Section 5.6.1) are shown in top right modified from the data in [90]. The extracted DNA by the second (Section 5.6.2) and third method (Section 5.6.3) is purified on density gradient of cesium chloride formed on ultracentrifugation [21]. Retrieved plasmid DNA is clearly separated from linear DNA mostly comes from sheared genome DNA. The molecular apparatus in the second method is also referred to in Section 5.7.2 and Figure 5-22.

in bacteria illustrated in Figure 5-1. The molecular apparatus briefly described in Figure 5-19 made it possible to physically disconnect long genomic DNA segments. Intrachromosomal homologous recombination between the two DNA repeats separated by 300 kb simply produced the intervening 300 kb as a second chromosome [66,67]. The extreme stability of the reported 300 kb second chromosome, termed the subgenome in the original report [66], exhibited extreme genetic stability via unknown mechanisms. The observation was inconsistent with the somewhat lower stability of the original plasmid, pLS32, whose replicational origin sequence (*oriN*) was required for subgenome replication. Later, presence of one of the essential genes reported by Kobayashi et al. [68] in the subgenome was shown to account for the genetic stability. In spite of the potential ability to make available circular subgenomes larger than 300 kb DNA by the method, broad application remains provisional due to its somewhat complicated procedure (Itaya M. and Fujita K., unpublished observations). Use of the conjugational transfer plasmid may make the delivery process more convenient [69].

5.6.3   Retrieval by Copying the Segment of the *B. subtilis* Genome

Compared with the somewhat elaborate genome dissection method, the third method proceeds through a yet more complicated genetic process referred to as *Bacillus* recombinational transfer (BReT) [70]. Indeed, with the BReT system, copying a DNA segment from the genome and pasting it into the incomplete plasmid, causes the DNA transfer from the genome to plasmid in an apparent reverse direction as that of the domino or inchworm cloning method. The recipient plasmid should possess two landing pad sequences (LPS). If they are the two half pBR sequences, the intervening/cloned DNA segment between the two GpBR halves bridges the gap by landing to the plasmid. As illustrated in simplified manner in Figure 5-19, the complete circular form given only by the BReT pathway should be selected by plasmid-linked markers. Standard extraction protocol for plasmid DNA has resulted in purification of complete recombinant genomes of lambda [70] and organelle genomes from mitochondria and chloroplast [23] as indicated in Figure 5-18. These are the first recombinant genomes shown to be convertible to another host [23]. The BReT system has been widely used to retrieve certain *B. subtilis* genome regions due to technical simplicity [71,72].

5.7   WHAT IS BACTERIAL GENOME?

The ongoing in-depth genome sequencing analyses have unveiled a number of factors in gene composition, location, orientation, accessory sequences, promoters, and so on. Numerous examples deduced from whole genome sequencing results unveiled that HGT plays a significant role in generating subpopulations even to-day [9,29,39,43–47,50]. Is the concept underlying the present bacterial genomes strengthened or amended if more number of whole genome sequence data are added? Evolutional processes responsible for the mitochondrial genome development from the ancestor alpha purple bacteria [1] and chloroplast development from photosynthetic cyanobacteria [2] might represent the largest HGT in history. Gene capturing suggested in some cases [73] has prompted us to mimic the process in using laboratory expertise and aiming at plausible phenotypic conversion upon HGT. A more extensive review on detailed diversity among the sequence-known genomes for subspecies and/or variant genomes will be avoided here, instead focus will be on our experimental approaches, such as genome laundering to create stable mosaic subspecies [74,75] (Figs 5-20 and 5-21) and inversion mutations to expand genome structural variants (Fig. 5-22). It should be mentioned that these two works started before the *B. subtilis* whole genome sequence were determined in 1997 [11]. Although little conclusive gene function based interpretation was made, I believe our primary attempt to convert nascent concepts of plastic genomes to the real examples of gene assembly is demonstrated.

5.7.1   Genome Conversion: Stable Mosaic Genomes by HGT

Obvious traits among closely related strains, where one presents whilst another lacks, may be explained by gain or loss of relevant genes. One of our earlier studies on HGT [74], monitoring terminal phenotypes for *Natto*-production bioprocess, unveiled both predicted and novel constraints underlying the gene content per genome. We used *Natto*, a traditional Japanese food made from boiled soybean. On the soybean surface, a growing strain of *Bacillus natto* produces viscous biofilm-like materials [74].

B. natto whose genome has not been sequenced exhibits high similarity to that of *B. subtilis* with respect to a physical map-based genome comparison [76]. Ability to ferment soybean is only possessed by *B. natto,* and several relevant *Natto* genes have been identified from *B. natto*. Because of the similar context of promoters and identical ribosomal RNA sequences between *B. subtilis* and *B. natto,* these *Natto*-relevant genes must be lost from or deficient in *B. subtilis*. Therefore, consecutive replacement or displacement of *B. subtilis* genome regions by *B. natto* genome DNA incorporated through homologous recombination was performed. The genomic DNA replacement, termed as genome laundering [74], may permit assembly of the lost or deficient *Natto*-genes in the genome of nonproducer *B. subtilis* as illustrated in Figure 5-20. *B. natto* DNA randomly launders the multiple *B. subtilis* genome loci via fragments of approximately 50 kb on average—equivalent to about 50 genes. When the genome laundering process was repeated, the *B. subtilis* genome gradually and discontinuously becomes mosaic as a greater number of genes are introduced. Results summarized in Table 5.1 clearly show that as degree of mosaic increases, phenotypes specific to *B. subtilis* are subdued in parallel with the appearance of *Natto* characteristics. The highest mosaic strain nicknamed as *Natsuko7* exhibiting the most *Natto* traits carries approximately 350 kb of *B. natto*-originated DNA segment, 8 percent of the total 4215 kb *B. subtilis* 168 genome [74]. The estimated degree of DNA heterogeneity was later found not consistent with the expressed proteins profile. As shown in Figure 5-21, the apparent similarity of proteins expressed in *Natusko*7 to those of *B. natto* is biased far greater than the 8 percent estimated from the DNA. *Natusko*7 thus might be classified as *B. natto*. This observation postulates that the global gene networks are

Natto genes of the *B. natto* were assembled in the *B. subtilis* genome with concomitant ability of *natto* production

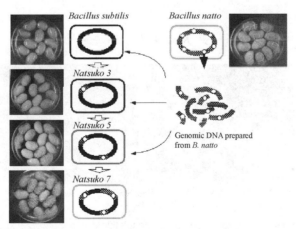

Figure 5-20 Mosaic genomes yielded by HGT. Genome parts from the *B. natto* transferred to and incorporated by replacement in the *B. subtilis* genome. Gradual and stepwise accumulation of *Natto* relevant genes converts non-*Natto* producer to the producer in proportion to the degree of mosaic indicated in Table 5.1. See the details in Section 5.7.1.

Table 5-1 *Natsuko*: Stable intermediate strain retaining both parental traits.
Traits (1,2) or (3–6) are specific to *B. subtilis* or *B. natto*.

Strain	(1)	(2)	(3)	(4)	(5)	(6)
B. subtilis	+++	+++	—	—	—	—
*Natsuko*1	+++	+++	+	—	—	—
*Natsuko*2	+++	+++	++	—	—	—
*Natsuko*3	+++	+++	++	—	—	—
*Natsuko*4	+++	+++	++	+	—	—
*Natsuko*5	+++	++	++	+++	+++	++
*Natsuko*6	++	++	+++	+++	+++	+++
*Natsuko*7	+	+	+++	++++	++++	+++
B. natto	—	—	++++	++++	++++	++++

(1) Ability to develop competent.
(2) Growth on Spizizen plate.
(3) Viscosity of colonies on GSP plate at 42°C for 24 h.
(4) Natto fermentation.
(5) Protease secretion.
(6) Natto fragrance 24 h at 42°C.

physiologically and metabolically dominated by the activity of translation even between closely related species. In this sense, *Natusko*7 may represent the earliest attempt to combine traits form the two separate genomes in one cell. But it does not constitute a recombinant genome because the genome size as well as junction region remain largely indiscriminatory. Rather, the *Natsuko* series strains listed in Table 5.1 invoked an idea for relevant DNA/gene assembly and may be relevant in "the first truly engineered bacterial genome."

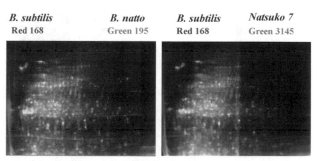

Expressed proteins from *B. subtilis* and *B. natto* are different regardless of sequence similarity

Figure 5-21 Similarity in expressed protein population. Separately labeled proteins prepared from two strains are mixed and run. In proteins resolved by two-dimensional electrophoresis, green spots display major expression in *B. natto* (left) and *Natuko*7 (right). Similar protein profile strongly indicates the highest *Natto*-producer, *Natsuko*7, virtually *B. natto* in spite of the expected degree of DNA converted. See the details in Section 5.7.1.

5.7.2   Rearrangements of Genome Structure by Inversion

Structure of the bacterial genome, a backbone for a set of genes, appears to be stably maintained after a number of replication cycles and cell divisions. The primary structural constraint is clearly exhibited by the diverse modes of replicons as shown in Figure 5-1. A secondary constraint is shown by the symmetry of the genome structure. Two replication arms divided by the *oriC-terC* axis, opposite the location of an initiation and termination locus, are well defined for certain bacterial genomes [44–46].

This structural symmetry together with gene alignment, orientation, and location seems conserved [44].

Bacterial genome plasticity, proposed and argued by pioneers many years ago, does not necessarily dictate apparent phenotypic changes. Thus, in-depth analysis must wait for development of reliable and easily accessible analytical methods such as physical map construction, and comparable sequence determination for closely related genomes. In line with these acknowledgements, the *B. subtilis* genome, an essential component of the BGM vector, is also considered to be plastic. Given analogies to plasmid vectors, genome vectors should surely be tolerant to structural disorder associated with DNA cloning and subsequent maintenance.

Among the naturally occurring events that induce considerable structural disorder listed in Figure 5-22, only inversion does not produce obvious genome size change; accordingly, resulting in little fluctuation of the gene set per genome. Yet, inversion may alter gene order and/or orientations, as well as relative gene locations in the genome. Experimental generation of systematic inversion mutants in the past decades has supported the concept of bacterial genome plasticity [77–79]. Focusing on the *B. subtilis* genome, there is evidence that supports both the stability [15,16,76] and the plasticity of the 4.2 Mb primary sequence [77–79].

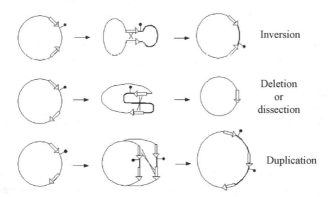

Rearrangements caused by intrachromosomal recombination

Figure 5-22   To induce large disorder of the genome structure. Two identical sequences separately embedded in the different loci of the genome induce inversion, deletion/dissection, or duplication by intrachromosomal homologous recombination. Resulted rearrangement is sequence orientation dependent. The second method to recover the cloned DNA (Fig. 5-19) is aided by the middle protocol here. Inversion mutagenesis is described in Section 5.7.2.

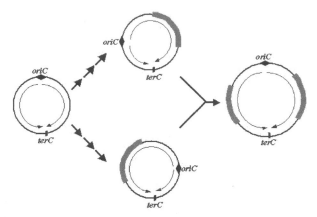

Figure 5-23 Requirement of asymmetry of the bacterial genome for rapid growth is true. Asymmetry around *oriC-terC* axis observed in certain bacterial genome is obstinate to ensure growth. Growth reduction by large DNA insertion is compensated by DNA insertion into opposite half to restore symmetry. This was proven in *Synechocystis* genome megacloning [36] also mentioned in Figure 5-9.

From our attempts to invert 28 various regions of the *B. subtilis* genome, some general conclusions can be drawn. Regions from the smallest 300 kb to the largest 1900 kb, covering any region of the 4215 kb *B. subtilis* 168 genome, have been shown to be invertible without losing mutant viability (Toda T. and Itaya M., unpublished observations). All the inversion mutants induced by the molecular apparatus commonly used to create subgenomes in Figure 5-19 stably grew. The ability to form spores, one of the phenotypes strongly associated with *B. subtilis,* was maintained in all cases. In contrast, competency, another intrinsic feature important to megacloning, shows reduction specific to the inversion of particular regions. Significant growth reduction was also observed in mutants possessing dramatic asymmetry around the *oriC-terC* axis, and therefore restoration of growth rate seems solely dependant on symmetry. This working hypothesis has been clearly evidenced by growth recovery in inversion mutants that restore symmetry (Kuroki A. and Itaya M., unpublished observations) and relocation of the origin of replication so as to make *oric-terC* axis normal (Tomita, S. and Itaya, M., unpublished observation). This was highlighted in a very practical case where elongation of *Synechocystis* DNA above 1000 kb was stalled during megacloning as described in Section 5.4.3 [36]. Additional megacloning in the opposite arm alleviated the asymmetry and allowed for the complete cloning as illustrated in Figure 5-23. The symmetry rule seemed decisively dominant in spite of genome size [36]. On the contrary, the yet undetermined *oriC-terC* loci of *Synechocystis* PCC6803 create a particular concern, if this strain is eligible to harbor the 4.2 Mb *B. subtilis* genome in a similar fashion as shown in Figures 5-9 and 5-10.

5.7.3   The Largest Bacterial Genome

What is the upper size limit of a bacterial genome? The largest evidenced bacterial genomes are from the *Streptomycete* genus with 9.7 Mb [4]. Considering *B. subtilis,* where the cognate 4.2 Mb became a 7.7 Mb hybrid-genome, the following question arises: Can this *CyanoBacillus* accept another 4.8 Mb of DNA to yield a 12.5 Mb hybrid-genome? This size is comparable to the smallest natural eukaryote, *S. cerevisae*, with a 12.5 Mb genome [83]. Why would someone want to do this? Is it just for pure scientific curiosity or some industrial application of these organisms with double or triple the amount of normal DNA? Two points should be addressed here. The long-standing question of the plausible size boundary to discriminate eukaryotes from prokaryotes will be examined experimentally [83]. The large extra DNA serves as a broad palette for various applications, for example, the recurrent use of existing genes through gene duplication or lateral transfer is the most common evolutionary mechanism to generate new protein-coding genes in bacteria [84].

5.7.4 Minimal Set of Genes for Life

The availability of a large number of complete genome sequences raises the question of how many genes are essential for cellular life [9,10,85]. Attempts to reconstruct the core of the protein-coding gene set for hypothetical minimal bacterial cell performed by computational as well as experimental analyses are not detailed here. The main features of such a minimal gene set that must be present in the hypothetical minimal cell would be informational genes (genes involved in transcription, translation, and other related processes) and operational genes (genes involved in cellular metabolic processes such as amino acid biosynthesis, cell envelope and lipid synthesis, and so on).

With regard to terminology "minimal set of genes (MSG)" is not equivalent to minimal genome. This is simply because MSG is a sum of the number of genes elucidated and listed, but a minimal genome requires a real body in which the complete nucleotide sequences must be contained. Apart from the determination that the MSG estimated by *in silico* analyses varies from 206 to 254 [85] gene in number, it is informative to discuss here how to experimentally construct the real minimal genome DNA body. The approach taken may be either top–down or bottom–up. The generation of mutation in all of the nonessential ORFs may be the beginning of the former case, as *B. subltilis* is shown to possess the 270 essential genes required for fixed growth conditions [68,86]. I focus on the latter case only by suggesting how to extract all the essential genes, less than 300 in number, and combine them together in a single-DNA sequence, preferably circular. One can imagine that it is sufficient that given DNA is anchored in a cellular bag, and the bag is exposed in nutritionally sufficient media. But the task is not that simple. Certainly, even limited in DNA structure, millions of questions arise in *de novo* assembly of more than 200 genes or gene blocks relating to the biological meaning when examined *in vivo*. For example, in the case where only 10 genes should be aligned in one DNA segment, the OGAB or possibly the domino method experimentally permits the designed DNA assembly. However, with lacking of information included in Figure 5-6a, no one writes the *de novo* nucleotide sequence.

Intermediate approaches that may be productive and timesaving may start from genomes with a nearly full set of essential genes present but lacking some of those predicted from the *in silico* experimentation. The *Buchnella* species [85], symbionts in certain aphids, may provide several starting points en route to a full set of essential genes. Given the *Buchnella* genome, being several hundred kilo base pairs, it is plausible that by megacloning method or domino method or other combinatorial use, it may serve as the initial body for addition of a certain number of relevant yet lacking genes. By this scenario, modified *Buchnella* genome can be clearly provided in *B. subtilis* cell after genome dissection method or BReT method. The approach sounds at present controversial and remained to be experimentally investigated.

5.8 MISCELLANEOUS USE OF THE BGM VECTOR

5.8.1 Extension of Target DNA to Present DNA Libraries

Our primary assumption of no sequence discrimination during cloning of increased DNA size was strengthened by successful results demonstrated to date. The source of natural DNAs targeted to our BGM vector has been expanding rapidly. Also, the use of currently available cloned DNA resources is progressing. The most likely candidates are DNAs stocked in the prominent *E. coli* cloning vectors, the bacterial artificial chromosome (BAC) [87]. In spite of the nomenclature "chromosome," the structures do not actually resemble true bacterial chromosomes, but are really a large plasmid given the use of the origin of replication from the large stable conjugative plasmid F [87]. Indeed, due to the potential to clone DNA far larger than those considered at the time of introduction in 1991 and its technological simplicity, BACs have greatly attributed to DNA library construction, from mouse to the human genomes, and are widely available. Those DNAs cloned into an *E. coli* BAC vector and directly transferred to the BGM would be another DNA resource or method for DNA library generation. A BGM vector specialized for this aim has been developed [35]. Changing the familiar GpBR sequences to genomic BAC (gBAC) sequences permit direct transfer of a BAC insert ranging up to 200 kb as illustrated in Figure 5-24. It should be emphasized that the DNAs cloned into the BGM vector, in general, automatically serve as a long-term preservation system; owing to the ability for *B. subtilis* to form spores as indicated in Figure 5-24 [35].

5.8.2 Mouse Gneomic DNA in BGM

It is often difficult within BAC libraries to find a single BAC clone that covers an entire genomic gene of interest that possesses a number of introns of various lengths and controlling elements particular to the genes of higher eukaryotes. Connecting together two or more overlapping BAC inserts into a single clone can provide a full-length copy of the gene of interest. Indeed, highly efficient homologous recombination systems have been exploited in *E. coli* that allow for modifications of large BACs without the use of restriction enzymes and ligases, a process called recombineering [88,89]. We have demonstrated similar molecular processing of mouse genomic DNA in the BGM

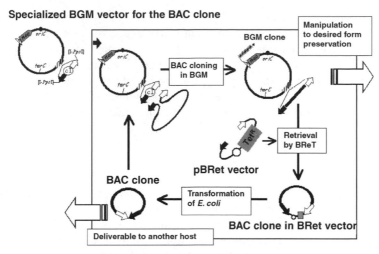

Figure 5-24   Application of the BGM vector to effective transfer from BAC library. A BGM vector providing BAC vector sequence as LPS, open and closed arrow, instead of pBR322 was made (left). Effectiveness of the present BAC library to more versatile use, including long-term preservation as spore form, appears successful [35]. These are mentioned in Section 5.8.1.

vector [34,90]. Joining of the two overlapping BACs to result in a 350 kb continuous mouse DNA segment in the BGM vector (Kaneko S. and Itaya M., unpublished data). Both BAC and BGM methods are similar given that homologous recombination is employed as the basic molecular mechanism. However, I believe our BGM system has super flexible handling to postcloning manipulation.

5.8.3   Sequence Fidelity of Recombinant Genomes

As mentioned above, genomes larger than 100 kb in purified form in solution are not only subjected to physical shearing but also sensitive to contaminating nucleases. The former case may invoke the problem associated with geographical preservation [29], and the latter are technically significant considering laboratory use [21]. From our experience in handling large DNA, most nucleases are removed via careful purification steps, for example, washing two times as a part of the standard protocol wash, wearing gloves, avoiding physical contact with unwrapped instrument, preservation in the presence of nuclease inhibitors such as chelating chemicals EDTA, etc.. Highly concentrated DNA seems resistant to residual nucleases probably due to competitive inhibition.

One fundamental concern for the rGenome is fidelity of the nucleotide sequences of recombinant progenies. Of course deterioration via mutations under given growth conditions must be selected during DNA replication, mutation inevitably accumulates even with the inherent sophisticated repair systems to repair mismatch of bases and removal and replacement of misincorporated nucleotides in the natural genome. Two *E. coli* K-12 derivatives separately cultivated over 40 year revealed nucleotide changes of 10^{-7} per base per year (or 1 in every 10^7 bases) [91]. Hence, we reach a very serious dilemma during construction of the designed rGenomes. Unsuspected mutations may freely accumulate in proportion to both the increase of the DNA size and number of cell divisions during prolonged cultivation. As mentioned in Section 5.4, freshly prepared mtGenome and cpGenome exhibited no obvious nucleotide alteration [23]. One may evade this intrinsic problem by resequencing the DNA whose time and cost turns relevant recently [5].

5.9 SUMMARY

DNA cloning is one of the most basic and inevitable tools to investigate genes and gene functions in biological sciences. Cloning strategy has changed depending on DNA sources, cloning vehicles, and cloning hosts. Innovation of the PCR method allows amplification of unlimited regions of DNA from various species, as well as reducing the required amount and purity of the template DNA. Recently, chemically synthesizing DNA without template DNA above dozens of kilo base pair in size has been shown to be possible and may soon be widely available. The BGM vector is a powerful tool for cloning the present genomes as a whole and for conferring subsequent modifications and reconstruction according to the purpose of the users. General schemes from cloning, pinpoint manipulation, and retrieval protocols are briefly described in Figures 5-2 and 5-19, and examples of progressive elongation of fragments with or without overlapped regions are described in Section 5.3, or in Section 5.4, respectively.

These two examples show reconstruction in the BGM vector of significantly large and continuous target DNA or genomes. Another method to assemble a number of DNA fragments in unit DNA blocks is to form gene clusters, so-called man-made operon, and is described in Section 5.5. Genomes from bacteriophage lambda (48.5 kb), mouse mitochondrial genome (16.3 kb), rice chloroplast genomes (135 kb), mouse genomic genes (up to 350 kb), and the whole genome of a photosynthetic bacterium (3500 kb) were cloned in the BGM vector; this is far above results obtained by conventional gene cloning and technology. The new cloning concept, readily applicable to any sequenced DNA, confers an experimental basis for not only provisional organelle genome engineering but also emerging recombinogenic engineering. Furthermore, the method could be used to assemble sequence-designed DNAs that are made from scratch.

The ultimate goal based on these technologies should be of course to unveil the nature underlying this still somewhat fragile iceberg; as well as to show how to plan and conduct new material production systems beneficial for humans and the earth. Efforts are being undertaken to make all protocols more conventional, hoping that the BGM vector users work just as comfortably as listening to background music.

5.10 FUTURE PERSPECTIVES

Life activities are controlled by the complex and dynamic actions of thousands of genes encoded by their genome. With the sequencing of many genomes, the key problem has shifted from identifying genes to knowing what the genes do; we need a framework for expressing that knowledge. Further comprehensive and systematic investigations regarding other factors listed in Figure 5-6 are required to draw at least some foundational rules on the design of the gene cluster. Gene clusters starting from an even small gene content exhibit various, and far different in some cases, levels of metabolites that may influence terminal phenotypes of the cell where a number of other gene networks may be perturbed. Gene cluster designing and molecular construction for certain basic metabolic pathways such as glycolysis have begun in Keio University (Tsuge et al., unpublished). It should be addressed here that the experimental basis for OGAB and/or domino is offering to provide a substantially large number of isomers with the same gene content. Although it seems a long-way off for creation of genome variations and establishment of subsequent assays for selection, this challenge would be crucial in bottoming up in facilitating the designing of life. The idea of making the appropriate *B. subtilis* genome into a linear chromosome vector similar to the eukaryotic stable vectors, such as YAC, remains to be initiated.

Production of designed genomes and consequently cellular life is one of the most challenging tasks in systems biology and synthetic biology, which still remains a nascent field. As deduced from the impact caused by gene cloning and recombinant genetics in the last century, once technologies show any potential, many start thinking how to use a technology to accomplish their goal. In this context, the present achievement shows that whole bacterial genomes have now become a target for cloning that was previously not an available option. I personally think the present two-genome cell is just a start for teaching many readers that "genome cloning or recombinant genomes" is becoming a reality. I personally believe that modification/manipulation of the recombinant Genomes of *Synechocystis*, would be much faster than genetic conversion of *Synechocystis* itself in traditional manner.

I always recall the aim of recombinant genomes research clearly addressed 35 years ago by late Dr Fujio Egami, primary director of the institute to which the author previously worked for 20 years. He said "The primary aim of life reorganization is to bring about fulfilled and comfortable lives for present and future human. Thus, there is no need to make the same cells as those currently available. As a consequence, what we need is to provide simpler and more beneficial cells for us."

The recombinant genome technology will markedly change the style of research not only limited to biomedical and biomaterial production, but also yielding a significant scientific, economic, and cultural impact. Cloning DNA to some extent mimics HGT. Finally, I would like to say that I make it a rule to keep learning from the greatest tutor, *Nature*.

ACKNOWLEDGEMENTS

I thank Drs Hoshino T., Ikeuchi M., Takeuchi T., Tanaka T., Yanagawa H., and Yoshikawa H. for their useful comments and many discussions. I also thank my colleagues, Drs Kaneko S., Ohashi Y., Oshima H., Tsuge K., and many other colla-borators for their daily works and number of discussions. Many technical staffs and students of graduate as well as undergraduate in my laboratories at Mitsubishi Kagaku Institute of Life Sciences and Keio University are specially acknowledged for their great contributions to improve the BGM vector. I particularly thank Prof. Tomita M., Director, Institute of Advanced Biosciences of Keio, for his continuous encouragement and opportunities for genome designing works. Finally, special thanks are addressed to Ms Fujita. K. whose contribution year to genome designing project is the longest.

REFERENCES

1. Gray MW, Burger G, Lang BF. Mitochondrial evolution. *Science* 1999;283:1476–1481.
2. Bendich A. Circular chloroplast chromosomes: the grand illusion. *Plant Cell* 2004; 161:1661–1666.
3. Endy D. Foundations for engineering biology. *Nature* 2005;438:449–453.
4. http://www.ncbi.nlm.nih.gov
5. Margulies M, Egholm M, Altman WE, Attiya S, Bader JS, LA, Bemben. et al. Genome sequencing in microfabricated high-density picolitre reactors. *Nature* 2005;437: 376–380.
6. Tyson GW, Chapman J, Hugenholtz P, Allen EE, Ram RJ, Richardson PM, Solovyev VV, Rubin EM, Rokhsar DS, Banfield JF. Community structure and metabolism through reconstruction of microbial genomes from the environment. *Nature* 2004;428:37–43.
7. Venter JC. et al. Environmental genome shotgun sequencing of the Sargasso sea. *Science* 2004;304:66–74.
8. Cello J, Paul AV, Wimmer E. Chemical synthesis of poliovirus cDNA: generation of infectious virus in the absence of natural template. *Science* 2002;297:1016–1018.
9. Kooning EV. Minimal set of genes. *Nat Rev Microbiol* 2003;2:127–136.
10. Gil R, Latorre A, Moya A. Bacterial endosymbionts of insects: insights from comparative genomics. *Environ Microbiol* 2004;6:1109–1122.
11. Kunst F. et al. The complete genome sequence of the Gram-positive bacterium *Bacillus subtilis*. *Nature* 1997;390:249–256.
12. Itaya M, Tanaka T. Complete physical map of the *Bacillus subtilis* 168 chromosome constructed by a gene-directed mutagenesis method. *J Mol Biol* 1991;220:631–648.
13. Itaya M. Toward a bacterial genome technology: integration of the *Escherichia coli* prophage lambda genome into the *Bacillus subtilis* 168 chromosome. *Mol Gen Genet* 1995;248:9–16.
14. Itaya M, Tanaka T. Gene-directed mutagenesis on the chromosome of *Bacillus subtilis*. *Mol Gen Genet* 1990;223:268–272.
15. Itaya M. Integration of repeated sequences (pBR322) in the *Bacillus subtilis* 168 chromosome without affecting the genome structure. *Mol Gen Genet* 1993;241: 287–297.
16. Itaya M. Stability and asymmetric replication of the *Bacillus subtilis* 168 chromosome. *J Bacteriol* 1993;175:741–749.

17. Itaya M, Shiroishi T, Nagata T, Fujita K, Tsuge K. Efficient cloning and engineering of giant DNAs in a novel *Bacillus subtilis* genome vector. *J Biochem* 2000;128:869–875.

18. Spizizen J. Transformation of biochemically deficient strains of *Bacillus subtilis* by deoxyribonucleate. *Proc Natl Acad Sci USA* 1958;44:1072–1078.

19. Dubnau DA. DNA uptake in bacteria. *Annu Rev Microbiol* 1999;53:217–244.

20. Chen I. Christie PJ, Dubnau DA. The ins and outs of DNA transfer in bacteria. *Science* 2005;310:1456–1460.

21. Sambrook J, Fritsch EF, Maniatis T. *Molecular Cloning: A Laboratory Manual.* Cold Spring Harbor, NY: Cold Spring Harbor Laboratory Press, 1989.

22. Sanger F, Coulson AR, Hong GF, Hill DF, Peterson GB. Nucleotide sequence of bacteriophage lambda DNA. *J Mol Biol* 1982;162:729–773.

23. Itaya M. Fujita K, Kuroki A, Tsuge K. Bottom-up genome assembly using the *Bacillus subtilis* genome vector. *Nature Methods* 2008;5:41–43.

24. Yoon YG, Koob MD. Efficient cloning and engineering of entire mitochondrial genomes in *Escherichia coli* and transfer into transcriptionally active mitochondria. *Nucleic Acids Res* 2003;31:1407–1415.

25. Winkel-Shirley B. Flavonoid biosynthesis. A colorful model for genetics, biochemistry, cell biology, and biotechnology. *Plant Physiol* 2001;126:485–493.

26. Barkovich R, Liao JC. Metabolic engineering of isoprenoids. *Metab Eng* 2001;3:27–39.

27. Misawa N, Shimada H. Metabolic engineering for the production of carotenoids in non-carotenogenic bacteria and yeast. *J Biotech* 1998;59:169–181.

28. Yoshikawa H, personal communications.

29. Lorenz MG, Wackernagel W. Bacterial gene transfer by natural genetic transformation in the environment. *Microbiol Rev* 1994;58:563–602.

30. Itaya M. Effective cloning of unmarked DNA fragments in the *Bacillus subtilis* 168 genome. *Biosci Biotech Biochem* 1999;63:602–604.

31. Uotsu-Tomita R, Kaneko S, Tsuge K, Itaya M. Insertion of unmarked sequences in multiple loci of the *Bacillus subtilis* 168 genome: an efficient selection method. *Biosci Biotechnol Biochem* 2005;69:1036–1039.

32. Itaya M, Nagata T, Shiroishi T, Fujita K, Tsuge K. Efficient cloning and engineering of giant DNAs in a novel *Bacillus subtilis* genome vector. *J Biochem* 2000;128:869–875.

33. Itaya M, Fujita K, Koizumi M, Ikeuchi M, Tsuge K. Stable positional cloning of long continuous DNA in the *Bacillus subtilis* genome vector. *J Biochem* 2003;134:513–519.

34. Yonemura I, Nakada K, Sato A, Hayashi J-I, Fujita K, Kaneko S, Itaya M. Direct cloning of full-length mouse mitochondrial DNA via *Bacillus subtilis* genome vector. *Gene* 2007;391:171–177.

35. Kaneko S, Akioka M, Tsuge K, Itaya M. DNA shuttling between plasmid vectors and a genome vector: systematic conversion and preservation of DNA libraries using the *Bacillus subtilis* genome (BGM) vector. *J Mol Biol* 2005;349:1036–1044.

36. Itaya M, Tsuge K, Koizumi M, Fujita K. Combining two genomes in one cell: stable cloning of the *Synechocystis* PCC6803 genome in the *Bacillus subtilis* 168 genome. *Proc Natl Acad Sci USA* 2005;102:15971–15976.

37. Kaneko T, Sato S, Kotani H, Tanaka A, Asamizu E, Nakamura Y, Miyajima N, Hirosawa M, Sugiura M, Sasamoto S. et al. Sequence analysis of the genome of the unicellular cyanobacterium *Synechocystis* sp. strain PCC6803. II, Sequence determination of the entire genome and assignment of potential protein-coding regions. *DNA Res* 1996;3:109–136.

38. Nelson KE, Paulsen IT, Heidelberg JF Fraser CM. Status of genome projects for nonpathogenic bacteria and archaea. *Nat Biotechnol* 2000;18:1049–1054.

39. Doolittle WF. Phylogenetic classification and the universal tree. *Science* 1999;284: 2124–2129.

40. Berkum PB, Terefework Z, Paulin L, Suomalainen S, Lindstro K, Eardly BD. Discordant Phylogenies within the *rrn* Loci of Rhizobia. *J Bacteriol* 2003;185:2988–2998.

41. Yap WH, Zhang Z, Wang Y. Distinct types of rRNA operons exist in the genome of the actinomycete *Thermomonospora chromogena* and evidence for horizontal transfer of an entire rRNA operon. *J Bacteriol* 1999;181:5201–5209.

42. Evguenieva-Hackenberg E. Bacterial ribosomal RNA in pieces. *Mol Microbiol* 2005;57: 318–325.

43. Riviera MC, Lake J. The ring of life provides evidence for a genome fusion origin of eukaryote. *Nature* 2004;431:152–155.

44. Rocha EP. Is there a role for replication fork asymmetry in the distribution of genes in bacterial genomes? *Trends Microbiol* 2002;10:393–395.

45. Rocha EP, Danchin A. Essentiality, not expressiveness, drives gene-strand bias in bacteria. *Nat Genet* 2003;34:377–378.

46. Rocha EP, Danchin A, Viari A. Universal replication biases in bacteria. *Mol Microbiol* 1999;32:11–16.

47. Rocha EP, Guerdoux-Jamet P, Moszer I, Viari A, Danchin A. Implication of gene distribution in the bacterial chromosome for the bacterial cell factory. *J Biotechnol* 2000;78:209–219.

48. Lawrence JG. Shared strategies in gene organization among prokaryotes and eukaryotes. *Cell* 2002;110:407–413.

49. Price MN. et al. Operon formation is driven by co-regulation and not by horizontal gene transfer. *Genome Res* 2003;15:809.

50. Omelchenko MV, Makarova KS, Wolf YI, Rogozin IB, Koonin EV. Evolution of mosaic operons by horizontal gene transfer and gene displacement *in situ*. *Genome Biol* 2003;4:R55.

51. Khosla C, Keasling JD. Metabolic engineering for drug discovery and development. *Nat Rev Drug Discov* 2003;2:1019–1025.

52. Wang CW, Oh MK, Liao JC. Direct evolution of metabolically engineered *Escherichia coli* for carotenoid production. *Biotechnol Prog* 2000;16:922–926.

53. Kao CM, Katz L, Khosla C. Engineered biosynthesis of a complete macrolactone in a heterologous host. *Science* 1994;265:509–512.

54. Pfeifer BA, Admiraal S, Gramajo H, Cane DE, Khosla C. Biosynthesis of complex polyketides in a metabolically engineered strain of *E. coli*. *Science* 2001;291:1790–1792.

55. Szczebara FM, Chandelier C, Villeret C, Masurel A, Bourot S, Duport C, Blanchard S, Groisillier A, Testet E, Costaglioli P. et al. Total biosynthesis of hydrocortisone from a simple carbon source in yeast. *Nat Biotechnol* 2003;21:143–149.

56. Alper H, Miyaoku K, Stephanopoulos G. Construction of lycopene-overproducing *E. coli* strains by combining systematic and combinatorial gene knockout targets. *Nat Biotechnol* 2005;23:612–616.

57. Watanabe K, Rude MA, Walsh CT, Khosla C. Engineered biosynthesis of an ansamycin polyketide precursor in *Escherichia coli*. *Proc Natl Acad Sci USA* 2003;100: 9774–9778.

58. Wenzel SC, Gross F, Zhang Y, Fu J, Stewart AF, Müller R. Heterologous expression of a Myxobacterial natural products assembly line in Pseudomonads via Red/ET recombineering. *Chem Biol* 2005;12:349–356.

59. Kumar P, Khosla C, Tang Y. Manipulation and analysis of polyketide synthases. *Methods Enzymol* 2005;388:269–293.

60. Tsuge K, Matsui K, Itaya M. One step assembly of multiple DNA fragments with designed order and orientation in *Bacillus subtilis* plasmid. *Nucleic Acids Res* 2003;31(21): e-133.

61. Nishizaki T, Tsuge K, Itaya M, Doi N, Yanagawa H. Metabolic engineering of carotenoid biosynthesis in *Escherichia coli* by ordered gene assembly in *Bacillus subtilis* (OGAB). *Appl Environ Microbiol* 2007;73:1355–1361.

62. Tsuge K, Matsui K, Itaya M. Production of the non-ribosomal peptide plipastatin in *Bacillus subtilis* regulated by 3 relevant gene blocks assembled in a single movable DNA segment. *J Biotechnol* 2007;129:592–603.

63. Notsu Y, Masood S, Nishikawa T, Kubo N, Akiduki G, Nakazono M, Hirai A, Kadowaki K. The complete sequence of the rice (*Oryza sativa* L,) mitochondrial genome: frequent DNA sequence acquisition and loss during the evolution of flowering plants. *Mol Gen Genomics* 2002;268:434–445.

64. Argast GM, Stephens KM, Emond MJ, Monnat RJ. I-*Ppo*I and I-*Cre*I homing site sequence degeneracy determined by random mutagenesis and sequential *in vitro* enrichment. *J Mol Biol* 1998;280:345–353.

65. Moure CM, Gimble FS, Quiocho FA. The crystal structure of the gene targeting homing endonuclease I-*Sce*I reveals the origins of its target site specificity. *J Mol Biol* 2003;334: 685–695.

66. Itaya M, Tanaka T. Experimental surgery to create subgenomes of *Bacillus subtilis* 168. *Proc Natl Acad Sci USA* 1997;94:5378–5382.

67. Itaya M, Tanaka T. Fate of unstable *Bacillus subtilis* subgenome: reintegration and amplification in the main genome. *FEBS Lett* 1999;448:235–238.

68. Kobayashi K, Ehrich SD, Aibertini A, Amati G, Andersen KK. et al. Essential *Bacillus subtilis* gene. *Proc Natl Acad Sci USA* 2003;100:4678–4683.

69. Itaya M, Sakaya N, Matsunaga S, Fujita K, Kaneko S. Conjugational transfer kinetics for *Bacillus subtilis* in liquid culture. *Biosci Biotechnol Biochem* 2006;70:740–742.

70. Tsuge K, Itaya M. Recombinational transfer of 100-kb genomic DNA to plasmid in *Bacillus subtilis* 168. *J Bacteriol* 2001;183:5453–5456.

71. Tomita S, Tsuge K, Kikuchi Y, Itaya M. Targeted isolation of a designated region of the *Bacillus subtilis* 168 genome by Recombinational transfer. *Appl Environ Microbiol* 2004;70:2508–2513.

72. Tomita S, Tsuge K, Kikuchi Y, Itaya M. Regional dependent efficiency for recombinational transfer of the *Bacillus subtilis* 168 genome. *Biosci Biotech Biochem* 2004;68:1382–1384.

73. Heidelberg JF, Eisen JA, Nelson WC, Clayton RA, Gwinn ML, Dodson RJ, Haft DH, Hickey EK, Peterson JD, Umayam L. et al. DNA sequence of both chromosomes of the cholera pathogen *Vibrio cholerae*. *Nature* 2000;406:477–483.

74. Itaya M, Matsui K. Conversion of *Bacillus subtilis*. 168: natto producing *Bacillus subtilis* with mosaic genomes. *Biosci Biotech Biochem* 1999;63:2034–2037.

75. Itaya M, Tanaka T, Predicted and unsuspected alterations of the genome structure of genetically defined *Bacillus subtilis* 168 strains. *Biosci Biotech Biochem* 1997;61:56–64.

76. Qiu D, Fujita K, Sakuma Y, Tanaka T, Ohashi Y, Ohshima H, Tomita M, Itaya M. Comparative analysis of the structure of complete physical maps of four *Bacillus subtilis* (natto) genome. *Appl Environ Microbiol* 2004;70:6247–6256.

77. Rebollo JE, François V, Louarn JM. Detection and possible role of two large nondivisible zones on the *Escherichia coli* chromosome. *Proc Natl Acad Sci USA* 1988;85:9391–9395.

78. Segall A, Mahan MJ, Roth JR. Rearrangement of the bacterial chromosome: forbidden inversions. *Science* 1988;241:1314–1318.

79. Campo N, Dias MJ, Daveran-Mingot ML, Ritzenthaler P. Le Bourgeois P, chromosomal constraints in Gram-positive bacteria revealed by artificial inversions. *Mol Microbiol* 2004;51:511–522.

80. Itaya M. First evidence for homologous recombination-mediated DNA inversion of the *Bacillus subtilis* 168 chromosome. *Biosci Biotech Biochem* 1994;58:1836–1841.

81. Toda T, Itaya M. I-*Ceu*I recognition sites in the *rrn* operons of the *Bacillus subtilis* 168 chromosome: inherent landmarks for genome analysis. *Microbiology* 1995;141: 1937–1945.

82. Toda T, Tanaka T, Itaya M. A method to invert DNA segments of the *Bacillus subtilis* 168 genome by recombination between two homologous sequences. *Biosci Biotech Biochem* 1996;60:773–778.

83. Bendich A, Drlica K. Prokaryotic and eukaryotic chromosomes: what's the difference? *Bioessays* 2000;22:481–486.

84. Gevers D, Vandepoele K, Simillon C. Van de Peer Y, Gene duplication and biased functional retention of paralogs in bacterial genomes. *Trends Microbiol* 2004;12:148–154.

85. Gil R, Silva FJ, Pereto J, Moya A. Determination of the core of a minimal bacterial gene set. *Microbiol Mol Biol Rev* 2004;68:518–537.

86. Itaya M. An estimation of minimal genome size required for life. *FEBS Lett* 1995;362:257–260.

87. Shizuya H, Birren B, Kim UJ, Mancino V, Slepak T, Tachiiri Y, Simon M. Cloning and stable maintenance of 300-kilobase-pair fragments of human DNA in *Escherichia coli* using an F- factor-based vector. *Proc Natl Acad Sci USA* 1992;89:8794–8797.

88. Copeland NG, Jenkins NA, Court DL. Recombineering: a powerful new tool for mouse functional genomics. *Nat Rev Genet* 2001;2:769–779.

89. Kotzamanis G, Huxley C. Recombining overlapping BACs into a single large BAC. *BMC Biotechnol* 2004;4:I.

90. Kaneko S, Tsuge K, Takeuchi T, Itaya M. Conversion of submegasized DNA to desired structures using a novel *Bacillus subtilis* genome vector. *Nucleic Acids Res* 2003;31(18): e-112.

91. Itoh T, Okayama T, Hashimoto H, Takeda J, Davis RW, Mori H, Gojobori T. A low rate of nucleotide change in *Escherichia coli* K-12 estimated from a comparison of the genome sequences between two different substrains. *FEBS Lett* 1999;450:72–76.

6

IN SILICO GENOME-SCALE METABOLIC MODELS: THE CONSTRAINT-BASED APPROACH AND ITS APPLICATIONS

Andrew R. Joyce[1] and Bernhard Ø. Palsson[2]

[1]*Bioinformatics Program, University of California, San Diego, 9500 Gilman Drive, La Jolla, California 92093*
[2]*Department of Bioengineering, University of California, San Diego, 9500 Gilman Drive, La Jolla, California 92093*

6.1 INTRODUCTION TO MODELING USING THE CONSTRAINT-BASED APPROACH

The development of high-throughput experimental techniques in recent years has led to an explosion of genome-scale data sets for a variety of organisms. Considerable efforts have yielded complete genomic sequences for dozens of organisms [1] from which gene annotation provides a list of individual cellular components. Microarray technology affords researchers the ability to probe gene expression patterns of cells and tissues on a genome scale. Genome-wide location analysis, also known as ChIP-chip [2], provides transcription factor binding site information for the entire cell. Furthermore, advances in the fields of fluxomics [3] and proteomics add to the vast quantity of data currently available to researchers. Integration of these data sets to extract the most relevant information to formulate a comprehensive view of

Systems Biology and Synthetic Biology   Edited by Pengcheng Fu and Sven Panke
Copyright © 2009 John Wiley & Sons, Inc.

biological systems is a major challenge the biological research community [4] currently facing. Achieving this task will require comprehensive models of cellular processes.

A prudent approach to gain biological understanding from these complex data sets involves the development of mathematical modeling, simulation, and analysis techniques [5]. For many years, researchers have developed and analyzed models of biological systems via simulation, but these efforts often have been hampered by lack of complete or reliable data. Some examples of the modeling philosophies and approaches that have been pursued include deterministic kinetic modeling [6,7], stochastic modeling [8,9], and Boolean modeling [10]. Many of these approaches are hindered by requiring knowledge of unknown parameters that are difficult to determine experimentally. Furthermore, the above approaches typically require substantial computational power, thus, limiting the scale of the models that can be developed.

In recent years, however, great strides have been made in developing and using genome-scale metabolic models of a number of organisms using another modeling technique that is not subject to the above limitations. This approach, known as constraint-based modeling [11–15], has been employed to generate genome scale for organisms from all three major branches of the tree of life. While bacterial models dominate this growing collection, a model from archaea has recently appeared, and several eukaryotic models are also available (see Table 6-1 for an overview of existing constraint-based models).

In complimentary efforts, many analytical tools have been developed to use these models in computational investigations of model organisms (reviewed in Ref. [12]). One method in particular, known as flux balance analysis (FBA) [16,17], is a powerful mathematical approach that uses optimization by linear programming (LP) to study the properties of metabolic networks under various conditions. When using FBA, the investigator chooses a property to optimize, such as biomass production in microbial models, and then calculates the optimal flux distribution(s) that lead to this result. Therefore, FBA is useful for computationally assessing the ability of an organism to grow on a particular substrate or in a particular environment and can also be used to assess the effect of metabolic gene deletions under various growth conditions. Given that these types of analyses rely on computer simulation, computational results must be confirmed at the bench through experimental means. However, by first investigating these situations at the computer work station, researchers can be directed to the most interesting and scientifically meaningful experiments to perform, thus limiting the amount of time spent conducting experiments of less scientific value.

In this chapter, we provide an introduction to the principles that underlie constraint-based modeling and FBA of biological systems. We give a brief, but practical example to introduce the method and concepts directly. Furthermore, we discuss both the utility and potential shortcomings of these models by reviewing several published studies that use these models to assess gene essentiality, which is simply defined as the study of organism viability despite harboring single or multiple gene knockouts. Finally, we briefly discuss additional analytical techniques and interesting applications of constraint-based modeling as well as their future implications.

Table 6-1   Currently available constraint-based models

Organism	Total Genes	Model Genes	Model Metabolites	Model Reactions	Reference
Bacteria					
Bacillus subtilis	4225	614	637	754	[114]
E. coli	4405	904	625	931	[61]
		720	438	627	[63]
Geobacter sulfurreducens	3530	588	541	523	[67]
Haemophilus influenzae	1775	296	343	488	[79]
		400	451	461	[82]
Heliobacter pylori	1632	341	485	476	[65]
		291	340	388	[64]
Lactococcus lactis	2310	358	422	621	[115]
Mannheimia succinciproducens	2463	335	352	373	[116]
Staphylococcus aureus	2702	619	571	641	[66]
Streptomyces coelicolor	8042	700	500	700	[68]
Archaea					
Methanosarcina barkeri	5072	692	558	619	[69]
Eukarya					
Mus musculus	28,287	1156	872	1220	[75]
S. cerevisiae	6183	750	646	1149	[71]
		672	636	1038	[72]
		708	584	1175	[70]
Human cardiac mitochondria	615[a]	298	230	189	[56]
Human red blood cell	NA	NA	39	32	[76]

This table summarizes model statistics for the models developed and published to date. *E. coli*, *Escherichia coli*; *S. cerevisiae*, *Saccharomyces cerevisiae*; NA, not applicable.

[a]This number is based on the protein species identified in a proteomics study of the human cardiac mitochondria from which the components of the reconstruction were derived [117].

6.2   BUILDING A CONSTRAINT-BASED MODEL

This section outlines the general procedure (Fig 6-1) followed in constructing a constraint-based model with a slant toward metabolic network. Furthermore, we introduce FBA as an example of a useful analytical method that can be used in conjunction with these models. This model building and analysis approach can be divided into approximately four successive steps:

1. Network reconstruction
2. Stoichiometric (S) matrix compilation
3. Identification and assignment of appropriate constraints to molecular components

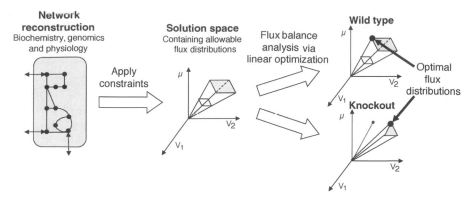

Figure 6-1 Constraint-based modeling. Application of constraints to a reconstructed metabolic network leads to a defined solution space that specifies a cell's allowable metabolic phenotypes. Flux balance analysis uses linear optimization to find solutions in the space that maximize or minimize a given objective. The effects of gene knockouts on the solution space and metabolic capabilities can be assessed by simulating a gene knockout and comparing its ability to grow *in silico* relative to wild type.

    4. Optimal flux distribution determination and assessment of gene essentiality via flux balance analysis

We will consider each of the above components in turn. In addition, a simple example will be provided in Section 6.2.5 to illustrate directly the concepts described herein.

6.2.1   Network Reconstruction

The first step in constraint-based modeling, known as network reconstruction, involves generating a model that describes the system of interest. This process can be decomposed into three parts typically performed simultaneously during model construction. These components, known as data collection, metabolic reaction list generation, and gene–protein reaction relationship (GPR) determination, are detailed in this section.

6.2.1.1   Data Collection   Perhaps the most critical component of the constraint-based modeling approach involves data collection relevant to the system of interest. Not long ago, this was among the most challenging steps as researchers had very limited access to amounts of biochemical data. However, the success of recent genome sequencing and annotation projects, advances in high-throughput technologies, and the extensive development of online database resources have improved matters dramatically.

After identifying the system or organism of interest, relevant data sources must be identified to begin the compiling of appropriate metabolites, biochemical reactions, and associated genes to be included in the model. The three primary types of resources are the biochemical literature, high-throughput data, and integrative database resources.

Biochemical Literature   Direct biochemical information found in the primary literature usually contains the highest quality data for use in reconstructing biochemical networks. Important details, such as precise reaction stoichiometry and reaction reversibility, are often directly available. Given that scrutinizing each study individually is time consuming and tedious task, biochemical textbooks and review articles should be utilized when available, relying on the primary literature used to resolve conflicts. Furthermore, many volumes devoted to individual organisms and organelles, such as *Escherichia coli* [18] and the mitochondria [19], are increasingly available and are typically excellent resources.

High-Throughput Data   Genomic and proteomic data are useful sources of information for identifying relevant metabolic network components. In recent years, the complete genome sequence of hundreds of organisms has been determined and many more sequencing projects are underway [20]. This collection is dominated by microbial and viral sequences, but several highly publicized higher eukaryotic sequences are also available [21–24]. Furthermore, extensive bioinformatics-based annotation efforts continue to make great strides toward automatically identifying all coding regions contained within the sequence [25–27]. To illustrate a common approach to gene functional annotation, consider the case in which a biochemical reaction is known to occur in the organism, but whose corresponding gene(s) are unknown. Sequence alignment tools such as BLAST and FASTA [28] can be utilized to assign putative functions based on similarity to orthologous genes and proteins of known function in other sequenced organisms. However, it should be noted that putative assignments represent functional hypotheses and are subject to revision upon direct biochemical characterization. As one final note on genome annotation, interesting efforts are also underway to automatically reconstruct networks based on annotated sequence information alone [29]. However, these automated approaches are limited in that they can be only as good as the genome annotation from which they are derived. Therefore, considerable quality control efforts should be conducted prior to extensive use of these networks.

The proteome of a biological system defines the full complement, localization, and abundance of proteins. Although these data are generally difficult to obtain, data for some subcellular components and bacteria are available [30,31]. Proteomic data are of particular importance in eukaryotic systems modeling in which care must be taken to assign reactions to their appropriate subcellular compartment or organelle. Similarly, when modeling a system under a single condition, these data are important in identifying active components.

In addition to the primary literature, genomic and proteomic data repositories can be accessed via the Internet as can the additional resources discussed in the next section. Some popular resources are provided in Table 6-2.

Integrative Database Resources   In recent years, significant efforts have been devoted to developing comprehensive databases that integrate many information sources including those data types previously described. Of particular interest are resources that have incorporated these disparate data sources into metabolic

Table 6-2 Online data resources

Data Type	Resource	Description	URL
Genomic	Genomes OnLine Database (GOLD)	Repository of completed and ongoing genome projects	http://www.genomesonline.org
	The Institute for Genomic Research (TIGR)	Curated databases for microbial, plant, and human genome projects	http://www.tigr.org
	National Center for Biotechnology Information (NCBI)	Curated databases of DNA sequences as well as other data	http://www.ncbi.nlm.nih.gov
Transcriptomic	Gene Expression Omnibus (GEO)	Microarray and SAGE-based genome-wide expression profiles	http://www.ncbi.nlm.nih.gov/geo
	Stanford Microarray Database (SMD)	Microarray-based genome-wide expression data	http://genome-www5.stanford.edu/
Proteomic	Expert Protein Analysis System (ExPASy)	Protein sequence, structure, and 2D-PAGE data.	http://au.expasy.org
	BRENDA	Enzyme functional data.	http://www.brenda.uni-koeln.de/
	Open Proteomics Database (OPD)	Mass-spectrometry-based proteomics data	http://bioinformatics.icmb.utexas.edu/OPD
Protein–DNA Interaction	Biomolecular Network Database (BIND)	Published protein–DNA interactions	http://www.bind.ca/Action/
	Encyclopedia of DNA Elements (ENCODE)	Database of functional elements in human DNA	http://genome.ucsc.edu/encode/
Protein–protein interaction	Munich Information Center for Protein Sequences (MIPS)	Links to protein–protein interaction data and resources	http://mips.gsf.de/proj/ppi
	Database of Interacting Proteins (DIP)	Published protein–protein interactions	http://dip.doe-mbi.ucla.edu

Category	Database	Description	URL
Subcellular location	Yeast GFP-fusion Localization Database	Genome-scale protein localization data for yeast	http://yeastgfp.ucsf.edu
Phenotype	A Systematic Annotation Package (ASAP) for community analysis of genomes	Single-gene-deletion phenotype microarray data for *E. coli*	http://www.genome.wisc.edu/tools/asap.htm
	General Repository for Interaction Datasets (GRID)	Synthetic lethal interactions in yeast	http://biodata.mshri.on.ca/grid
Pathway	Kyoto Encyclopedia of Genes and Genomes (KEGG)	Pathway maps for many biological processes	http://www.genome.ad.jp/kegg/
	BioCarta	Interactive graphic models of molecular and cellular pathways	http://www.biocarta.com/genes/index.asp
Organism specific	EcoCyc	Encyclopedia of *E. coli* K12 genes and metabolism	http://www.ecocyc.org
	Saccharomyces Genome Database (SGD)	Scientific database of the molecular biology and genetics of *S. cerevisiae*	http://www.yeastgenome.org
	BioCyc	A collection of 205 pathway/genome databases for individual organisms.	http://www.biocyc.org

This table details some of the databases that store and distribute genome-scale data, gene ontological information, and organism specific data. It should also be noted that this table is by no means comprehensive in its content, but rather provides a reasonably broad sample of the data and resources that are readily accessible to researchers today. 2D-PAGE, two-dimensional polyacrylamide-gel electrophoresis; GFP, green fluorescent protein; SAGE, serial analysis of gene expression.

pathway maps. Among these resource types, Kyoto Encyclopedia of Genes and Genomes (KEGG) [32] is perhaps the most extensive and well known. Pathway maps for numerous metabolic processes are available. KEGG also provides information regarding orthologous genes for a variety of organisms, thus greatly enhancing the power of this resource. Additional organism-specific database resources are also available. EcoCyc [33] incorporates gene and regulatory information as well as enzyme-reaction pathways particular to *E. coli*. The Comprehensive Yeast Genome Database (CYGD) [34] and *Saccharomyces* Genome Database (SGD) [35] are other examples of *Saccharomyces cerevisiae*-specific comprehensive resources. Finally, the BioCyc resource [36,37] contains automated annotation-derived pathway/genome databases for 205 individual organisms.

An additional important wealth of information can be found in resources that provide functional information for individual genes and gene products. These ontology-based tools strive to describe how gene products behave in a cellular context. The most well-known resource is Gene Ontology Consortium (GO) [38,39] that contains information for a variety of organisms. In recent years, organism-specific ontologies, such as GenProtEC [40] for *E. coli*, also have appeared. In sum, these online resources are valuable that they typically integrate information regarding individual genes and proteins as well as information regarding their regulation and participation in enzymatic reactions in a single location.

6.2.1.2 *Metabolic Reaction List Generation*

The next step in defining a constraint-based model requires clearly specifying the reactions to be included based on the metabolite and enzyme information collected in the previous step. A metabolic reaction can be viewed simply as substrate(s) conversion to product(s), often by enzyme-mediated catalysis. In light of this notion, each reaction in a metabolic network must adhere to the fundamental laws of physics and chemistry; therefore, reactions must be balanced in terms of charge and elemental composition. For example, the depiction of the first step of glycolysis in Figure 6-2a is neither elementally nor charge balanced. However, inclusion of hydrogen in Figure 6-2b balances the reaction in both regards.

Biological boundaries also must be considered when defining reaction lists. Metabolic networks are comprised of both intracellular and extracellular reactions. For example, the reactions of glycolysis and the tricarboxylic acid (TCA) cycle take

(a)

$$C_6H_{12}O_6 + ATP^{3-} \xrightarrow{\text{hexokinase}} C_6H_{11}O_6PO_3{}^{2-} + ADP^{2-}$$

(b)

$$C_6H_{12}O_6 + ATP^{3-} \xrightarrow{\text{hexokinase}} C_6H_{11}O_6PO_3{}^{2-} + ADP^{2-} + H^+$$

Figure 6-2 Charge and elementally balanced reactions. (a) This depiction of the hexokinase-mediated conversion of glucose to glucose-6-phosphate is neither elementally, nor charge balanced. (b) Inclusion of hydrogen both elementally and charge balances the reaction.

place intracellularly in the cytosol. However, glucose must be transported into the cell via an extracellular reaction in which a glucose transporter takes up extracellular glucose. An additional boundary consideration must be recognized particularly when modeling eukaryotic cells. Given that certain metabolic reactions take place in the cytosol and others take place in various organelles, reactions must be compartmentalized properly. Data is now being generated in which proteins are tagged, for example, with green fluorescent protein (GFP) or recognized by antibodies and localized to subcellular compartments or organelles [41–43]. Furthermore, computational tools have also been developed to predict subcellular location of proteins in eukaryotes [44].

Finally, reaction reversibility must be defined. Certain metabolic reactions can proceed in both directions. Thermodynamically, this permits reaction fluxes to take on both positive and negative values. The KEGG and BRENDA online resources (Table 6-2) are two useful resources that catalog enzyme reversibility.

6.2.1.3 Determining Gene–Protein Reaction Relationships Upon completing the reaction list, the protein or protein complexes that facilitate each metabolite substrate to product conversion must be determined. Each subunit protein from a complex must be assigned to the same reaction. Additionally, some reactions can be catalyzed by different enzymes. Collectively, each enzyme that fits this criterion is known as an isozyme for a particular reaction. Accordingly, isozymes must all be assigned to the same appropriate reaction. Biochemical textbooks often provide the general name of the enzyme(s) responsible; however, the precise gene and associated gene product specific for the model organism of interest must be identified. The database resources detailed in Section 6.2.1.1 and Table 6-2 assist this process. In particular, KEGG and GO provide considerable enzyme-reaction information for a variety of organisms. Furthermore, protein–protein interaction data sets, for example, those derived from yeast two-hybrid experiments [45], may be useful resource for defining enzymatic complexes in less defined situations. One must take care in using these data because of their high false-positive rate and questionable reproducibility [46,47].

6.2.2 Defining the Stoichiometric Matrix

The compiled reaction list can be represented mathematically in the form of a S matrix. The S matrix is formed from the stoichiometric coefficients of the reactions that participate in the defined reaction network. It has $m \times n$ dimensions, where m is the number of metabolites and n is the number of reactions. Therefore, the S matrix is organized such that every column corresponds to a reaction and every row corresponds to a metabolite. The S matrix describes how many reactions a compound participates in, and thus, how reactions are interconnected. Accordingly, each network that is reconstructed in this way effectively represents a two-dimensional annotation of the genome [11,48].

Figure 6-3 shows how a simple two-reaction system can be represented as an S matrix. In this example, v_1 and v_2 denote reaction fluxes and are associated with

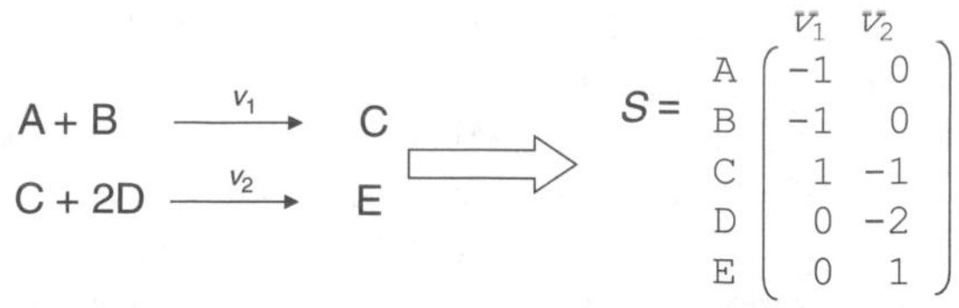

$$
\begin{array}{cc}
A + B & \xrightarrow{\ v_1\ } & C \\
C + 2D & \xrightarrow{\ v_2\ } & E
\end{array}
\qquad\Longrightarrow\qquad
S =
\begin{array}{c}
\\ A \\ B \\ C \\ D \\ E
\end{array}
\begin{pmatrix}
v_1 & v_2 \\
-1 & 0 \\
-1 & 0 \\
1 & -1 \\
0 & -2 \\
0 & 1
\end{pmatrix}
$$

Figure 6-3 Generating the S matrix. The reaction list on the left is mathematically represented by the S matrix on the right. As a convention, each row represents a metabolite and each column represents a reaction in the network. Additionally, input or reactant metabolites have negative stoichiometric coefficients and outputs or products have positive stoichiometric coefficients. Metabolites that do not participate in a given reaction are assigned a zero value.

individual proteins or protein complexes that catalyze the reactions. In the S matrix representation, each row denotes an individual metabolite while each column corresponds to an individual reaction. Element S_{ij} represents the stoichiometric coefficient of the metabolite associated with row i in the reaction associated with column j. Furthermore, notice that substrates are assigned negative coefficients and products are given positive coefficients. Also, for those reactions in which a metabolite does not participate, the corresponding S matrix element is assigned a zero value.

6.2.3 Identifying and Applying Constraints

Having developed a mathematical representation of a metabolic network in the form of the S matrix, the next step requires that any constraints be identified and imposed on the model. Cells are subject to a variety of constraints from environmental, physiochemical, evolutionary, and regulatory sources [12,14]. In and of itself, the S matrix is a constraint in that it defines the mass and charge balance requirements for all possible metabolic reactions available to the cell. These stoichiometric constraints establish a geometric solution space that, in principle, contains all possible metabolic behaviors.

Additional constraints can be identified and imposed on the model, which has the effect of further limiting the metabolic behavior solution space. Maximum enzyme capacity, V_{max}, which can be determined experimentally for some reactions is one example and can be imposed by limiting the flux through any associated reactions to that maximum value. Furthermore, the uptake rates of certain metabolites can be determined experimentally and used to restrict metabolite uptake to the appropriate levels when mathematically analyzing the metabolic model. Additional types of constraints have also been applied including thermodynamic limitations [49], internal metabolic flux determinations [13], and transcriptional regulation [50–53]. This latter topic will receive considerable detailed treatment in Section 6.4.3.

With respect to computationally assessing gene essentiality, a similar strategy to setting the maximum enzyme capacity can be utilized. By simply restricting the flux through reactions associated with the protein of interest to zero, a gene knockout can be simulated. Flux balance analysis then can be used to examine the simulated knockout properties relative to wild type, as outlined in the next section.

6.2.4   Assessing the Model Using Flux Balance Analysis

Flux balance analysis is a powerful computational method that relies on optimization techniques by linear programming [54] to investigate the production capabilities and systemic properties of a metabolic network. By defining an objective, such as biomass production, ATP production, or by-product secretion, linear optimization may be used to find an optimal flux distribution for the network model that maximizes the stated objective. This section briefly introduces some main concepts that underlie FBA, with an emphasis on how FBA can be utilized to assess gene essentiality in a metabolic network.

6.2.4.1 *Linear Optimization*   As stated previously, the solution space defined by constraint-based models can be explored via optimization by linear programming. The LP problem corresponding to the search for the optimal flux distribution determination through a metabolic network can be formulated as follows:

$$\text{Maximize} \quad Z = c^{\mathrm{T}} v$$
$$\text{Subject to} \quad S \cdot v = 0$$
$$\alpha_i \leq v_i \leq \beta_i \quad \text{for all reactions } i$$

In the above representation, Z represents the objective function, and c is a vector of weights on the fluxes v. The weights are used to define the properties of the particular solution that is sought. The latter statements represent the flux constraints for the metabolic network. S is the matrix defined in the previous section and contains the mass and charge balanced representation of the system. Furthermore, each reaction flux v_i in the system is subject to lower and upper bound constraints, represented in α_i and β_i, respectively.

The solution to this problem yields not only a value for Z but also results in an optimal flux distribution (v) that allows the highest flux through the chosen objective function, Z. Furthermore, computational assessment of gene essentiality is performed easily within this framework. By setting the upper and lower flux bound constraints to zero for the reaction(s) corresponding to the gene(s) of interest, a simulated gene deletion strain may be created. Examining the results of simulations run before and after knocking out a gene lead to gene essentiality predictions.

Problems of this type can be formulated and solved readily by commercial software packages, such as Matlab (The MathWorks, Inc., Natick, MA), Mathematica (Wolfram Research, Inc., Champaign, IL), LINDO (LINDO Systems, Inc., Chicago, IL), and tools available through the General Algebraic Modeling System (GAMS Development Corporation, Washington, DC). Section 6.2.5 presents a simple, hypothetical example solved using Matlab. It should also be noted that these types of analyses yield a single answer; however, it is possible that multiple equivalent flux distributions that yield a maximal biomass function value for a given network and simulation conditions. This topic has been explored using mixed integer linear programming (MILP) techniques with genome-scale metabolic models [55,56], but is beyond the scope of this chapter and will not be discussed further.

6.2.4.2 *Constraints*   As previously stated, the S matrix constrains the system by defining the mass and charge balance constraints for all possible metabolic reactions within the system. In mathematical terms, the S matrix is a linear transformation of the reaction flux vector,

$$v = (v_1, v_2, \ldots, v_n)$$

to a vector of time derivatives of metabolic concentrations

$$\mathbf{x} = (x_1, x_2, \ldots, x_n)$$

such that

$$\frac{d\mathbf{x}}{dt} = \mathbf{S} \cdot v$$

Therefore, a particular flux distribution v represents the flux levels through each reaction in the network. Since the time constants that describe metabolic transients are fast (on the order of tens of seconds or less), whereas the time constants for cell growth are comparatively long (on the order of hours to days) the behavior of cellular components can be considered as existing in a quasi steady state. This assumption leads to the reduction of the previous equation to

$$\mathbf{S} \cdot v = 0$$

By focusing only on the steady-state condition, assumptions regarding reaction kinetics are not needed. Furthermore, based on this premise, it is possible to determine all chemically balanced metabolic routes through the metabolic network.

The second constraint set is imposed on the individual reaction flux values. The constraints defined by

$$\alpha_i \le v_i \le \beta_i \quad \text{for all reactions } i$$

specify lower and upper flux bounds for each reaction. If all model reactions are irreversible, α equals to 0. Similarly, if the enzyme capacity, V_{max}, is experimentally defined, setting β to the known experimental value limits the allowable reaction flux through the enzyme. In contrast, a gene knockout is simulated by setting $\beta_i = 0$ for gene i (see Section 6.2.5 and Box 6-1). If no constraints on flux values through reaction v_i can be identified, then α_i and β_i are set to -∞ and +∞, respectively, to allow for all possible flux values. In practice, ∞ is typically represented as an arbitrarily large number that will exceed any feasible internal flux (for an example, see Section 6.2.5).

A brief consideration should also be given to specifying input and output constraints on the system. When analyzing metabolic models in the context of assessing cellular growth capabilities, input constraints effectively define the environmental conditions being considered. For example, organisms have various elemental requirements that must be provided in the environment in order to support growth.

Some organisms that lack certain biosynthetic processes are auxotrophic for certain biomolecules, such as amino acids, and these compounds also must be provided in the environment. From an FBA standpoint, these issues mean that input sources must be specified in the form of input flux constraints specified in v. For example, if one desires to simulate rich medium conditions, flux constraints are specified such that all biomolecules that can be served as inputs to the system, in other words all compounds that are available extracellularly, are left unconstrained and can flow freely into the system. In contrast, when modeling minimal medium conditions (for an example of a large-scale analysis performed of *E. coli* growth simulations on minimal media, see Ref. [57]) only those inputs required for cell growth, or biomass formation in the formalism being considered here, are allowed to flow into the system with all other input fluxes constrained to zero. It should also be noted that certain output flux constraints may need to be set appropriately in order to allow for the simulated secretion of biomolecules that may "accumulate" in the process of forming biomass. A simple example of this is allowing for lactate and acetate secretion when modeling fermentative growth of microbes.

6.2.4.3 *The Objective Function*   Given that multiple possible flux distributions exist for any given network, linear optimization is used to identify a particular solution that maximizes or minimizes a defined objective function. Commonly used objective functions include production of ATP or production of a secreted by-product. When assessing the growth capabilities of a microbe using its associated metabolic model, growth rate, as defined by the weighted consumption of metabolites needed to make biomass, is maximized. The general analysis strategy asks the question "is the metabolic reaction network able to support growth under the specified growth conditions?" Therefore, biomass generation in this modeling framework is represented as a reaction flux that drains intermediate metabolites, such as ATP, NADPH, pyruvate, and amino acids, in appropriate ratios (defined in the vector c of the biomass function Z) to support growth. As a convention, the biomass function is typically written to reflect the needs of the cell in order to make 1 g of cellular dry weight, and has been experimentally determined for *E. coli* [58]. In sum, the choice of biomass as an objective function, cell growth, depicted as a nonzero value for Z, will only occur if all the components in the biomass function can be provided for by the network in the correct relative amounts.

6.2.5 A Simple FBA Example

In order to demonstrate the concepts previously introduced, this section presents a specific example using a simple system. Figure 6-4a shows a hypothetical four metabolite (A,B,C,D), eight reaction ($v_1,v_2,v_3,v_4,v_5,v_6,b_1,b_2$) network. By convention, each internal reaction is associated with a flux v_i whereas reactions that span the system boundary are denoted with flux b_i. Furthermore, external metabolites A and D are denoted with subscript "o" to distinguish them from their corresponding internal metabolite. However, external metabolites need not be explicitly considered in the stoichiometric network representation.

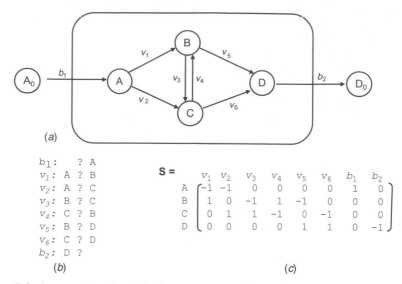

(a)

b_1 :   ? A
v_1 : A ? B
v_2 : A ? C
v_3 : B ? C
v_4 : C ? B
v_5 : B ? D
v_6 : C ? D
b_2 : D ?

(b)

$$S = \begin{array}{c} \\ A \\ B \\ C \\ D \end{array} \begin{array}{cccccccc} v_1 & v_2 & v_3 & v_4 & v_5 & v_6 & b_1 & b_2 \\ \left[\begin{array}{cccccccc} -1 & -1 & 0 & 0 & 0 & 0 & 1 & 0 \\ 1 & 0 & -1 & 1 & -1 & 0 & 0 & 0 \\ 0 & 1 & 1 & -1 & 0 & -1 & 0 & 0 \\ 0 & 0 & 0 & 0 & 1 & 1 & 0 & -1 \end{array} \right] \end{array}$$

(c)

Figure 6-4 An example system. (a) A four-metabolite, eight-reaction system is first decomposed into individual reactions in (b), and then represented mathematically in the S matrix depicted in (c). By convention, internal reactions are denoted by v_i and reactions that span the system boundary are denoted by b_i. External metabolites A_0 and D_0 need not be explicitly represented explicitly within this framework as they are outside the system under consideration.

Figure 6-4b outlines the reaction list associated with the system. Notice that the conversion of metabolite B to C is reversible. Rather than treating this as a single reaction, for simplicity this reaction is decoupled into two separate reactions with individual corresponding fluxes.

The S matrix for this system is detailed in Figure 6-4c. Again, notice how this representation follows directly from the reaction list. Metabolite substrates and products are represented with negative and positive coefficients, respectively. Recall that LP problems take on the following form:

$$\text{Maximize} \quad Z = c^{\mathrm{T}} v$$

$$\text{Subject to} \quad S \cdot v = 0$$

$$\alpha \leq v_i \leq \beta \quad \text{for all reactions } i$$

For example, if the metabolite D output is to be maximized, corresponding to maximizing the flux through b_2 the objective function is defined as follows:

$$Z = \begin{pmatrix} 0 & 0 & 0 & 0 & 0 & 0 & 0 & 1 \end{pmatrix} \cdot \begin{pmatrix} v_1 & v_2 & v_3 & v_4 & v_5 & v_6 & b_1 & b_2 \end{pmatrix}^{\mathrm{T}}$$

Furthermore, in addition to the mass and charge, balance constraints imposed by the S matrix, lower (α), and upper (β) bound vectors must be specified for the reaction vector v. Since all reactions in this network are irreversible, which constrains all fluxes to be positive, the lower bound vector α is set to zero.

$$\alpha = \begin{pmatrix} 0 & 0 & 0 & 0 & 0 & 0 & 0 & 0 \end{pmatrix}^{\mathrm{T}}$$

Upper bound values specified in vector β can be chosen to incorporate experimentally determined maximal enzyme capacities, also known as V_{max} values, or some arbitrarily chosen values to explore network properties. An acceptable example vector is

$$\beta = (2 \quad 10 \quad 4 \quad 6 \quad 10 \quad 8 \quad 100 \quad 100)^{T}$$

The latter two upper bound values for the respective input and output fluxes are set to an arbitrarily large number in this case to reflect an effectively unlimited capacity. Given the constraints on the internal fluxes, however, the actual values of these fluxes in the calculated optimal flux distribution will never approach these values.

Utilizing the information compiled above, the Matlab function **linprog()** can be used to solve for a steady-state flux distribution that maximizes for the output of metabolite D under wild-type conditions, as detailed in Box 6-1. It should be noted that the default Matlab optimization solver is only suitable for problems of this and slightly

---

BOX 6-1 FBA USING MATLAB

Here, we use Matlab to solve an FBA problem for three cases using the system shown in Fig. 6-4. The **linprog()** function accepts six arguments and returns two values in the following form:

$$[v, Z] = \text{linprog}(c, \text{Aeq}, \text{beq}, S, b, \alpha, \beta)$$

This solves the following LP problem:

$$\text{Minimize} \quad Z = c^{T} \times v$$
$$\text{Subject to} \quad \text{Aeq} \times v \leq \text{beq}$$
$$S \cdot v = b$$
$$a \leq v \leq b$$

Since the system does not have inequality constraints other than flux vector bounds, Aeq is set equal to the identity matrix and beq to β, so that

$$\text{Aeq} \cdot v \leq \text{beq}$$

is equivalent to

$$v \leq b$$

The code to solve the wild type problem (Case 6-1) of interest in Matlab's framework follows, using the **linprog()** function and α and β as defined in the text:

```
>> S = [-1 (1 0 0 0 0 1 0; 1 0 -1 1 -1 0 0 0; 0 1 1 -1 0 -1 0 0; 0 0 0 0 1 1 0 -1];
>> b = [0 0 0 0]';
>> alpha = [0 0 0 0 0 0 0 0 ]';
```

```
>> beta = [2 10 4 6 10 8 100 100]';
>> c = [0 0 0 0 0 0 0 1];
>> Aeq = eye (8);
>> [v, z] = linprog (-c, Aeq, beta, S, b, alpha, beta)Optimization
terminated successfully.

v = 2.0000 10.0000 0.1822 3.9137 5.7315 6.2685 12.0000 12.0000
Z = -12.0000
```

Note that since **linprog()** defaults to solving a minimization problem we use the negative of the optimization weight vector c. Use the Matlab Help for more details on **linprog()**.

Case 6-1: Wild Type

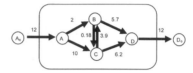

Case 6-2, shown below, solves the same problem but this time after knocking out reaction v_5 by modifying the β vector stored in the beta variable:

```
>> beta = [2 10 4 6 10 0 100 100]';
```

Case 6-2: Growth Impaired v_6 Knockout

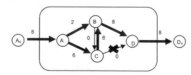

Finally, in Case 6-3 depicted below, by again modifying the beta variable a "lethal" deletion strain can be simulated by knocking out both v_5 and v_6:

```
>> beta = [2 10 4 6 0 0 100 100]';
```

Case 6-3: Lethal v_5 and v_6 Double Knockout

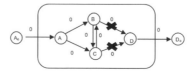

larger magnitude. Typical biological problems that involve many more variables and constraints require more sophisticated optimization software such as the packages available through LINDO Systems, Inc. and GAMS. A thorough discussion of the algorithmic details that underlie solving FBA and other LP problems is beyond the scope of this text. For further details, see Refs [54,59,60] .

Having used the above information to simulate the wild-type case, the upper bound β vector can be modified to simulate a gene deletion. For example, if we want to examine the effects of deleting the enzyme responsible for the conversion of metabolite C to D, flux v_6 is restricted to zero simulation.

$$\beta = (2 \quad 10 \quad 4 \quad 6 \quad 10 \quad 0 \quad 100 \quad 100)^T$$

Similarly, a v_5, v_6 double mutant is simulated using the following vector:

$$\beta = (2 \quad 10 \quad 4 \quad 6 \quad 0 \quad 0 \quad 100 \quad 100)^T$$

6.3 COMPUTATIONAL CHALLENGES

This chapter presents the basic steps required to reconstruct and analyze genome-scale metabolic networks. These model systems quickly grow in size and scale, introducing computational challenges that need to be addressed. As noted previously, with large-scale models it becomes necessary to use a robust computational platform designed specifically for sophisticated optimization problems, such as those developed by LINDO Systems, Inc. and available through GAMS.

Furthermore, data management becomes difficult as models scale up in size. For example, the most current published E. coli model contains 904 genes and 931 unique biochemical reactions [61]. Analyzing a genome-scale model within the framework proposed in Sections 6.2.4 and 6.2.5 is possible, but would be slow, cumbersome, and error prone. In recent years, an integrative data management and analysis software platform called SimPheny™ (Genomatica, San Diego, CA) has been developed specifically to address the data management and computational challenges inherent in building large-scale cellular models. This versatile platform provides network visualization, database, and various analytical tools that greatly facilitate the construction and study of genome-scale cellular models.

Currently, more than a dozen genome-scale metabolic models have been published and are available (Table 6-1) for further research and analysis. Most of these models represent bacteria and range from the important model organism E. coli [61–63] to pathogenic microbes such as H. pylori [64,65], and S. aureus [66]. Furthermore, recently developed models of G. sulfurreducens [67] and S. coelicolor [68] are potentially important for their facilitation of studies that probe these organisms' respective potential bioenergetic and therapeutics-producing properties.

Representative constraint-based models have also appeared from the other two major branches of the tree of life. The recently developed metabolic reconstruction of M. barkeri [69], an interesting methanogen with bioenergetic potential, represents the first constraint-based model of an archaea that has been used to aid in the analysis of experimental data from this relatively obscure group of organisms. Furthermore, several eukaryotic models also have been developed. The metabolic models of the baker's or brewer's yeast *S. cerevisiae* [70-72] are second only to the *E. coli* models in terms of relative maturity and have been used in a variety of studies designed to assess network properties (for recent examples, see Refs [73,74]). Metabolic models of higher order systems are also becoming available such as a model of mouse (*Mus muculus* [75]), human cardiac mitochondria [56], and red blood cell [76].

As more of these genome-scale models are developed, the issue of making their contents available to the broader research community is of primary concern. Given their inherent complexity, there is a need for a standardized format in which their contents can be represented in order to circumvent potential problems associated with the current typical means of distribution of models via nonstandard flat file or spreadsheet format. In an effort to mitigate this deficiency, for example, the Systems Biology Markup Language (SBML) [77], has been developed to provide a uniform framework in which models can be represented, and the recently initiated MIRIAM ("minimum information requested in the annotation of biochemical models") project [78] and affiliated databases have appeared to provide greater transparency as to the contents, and potential deficiencies of models. The adoption of these or similar standards will be important to the advancement of the field and in promoting its general utility in biological research.

6.3.1 Predicting Gene Essentiality

One application of constraint-based modeling in conjunction with FBA that has been particularly successful in computationally assessing metabolic networks is in studies of gene essentiality. Recent studies have used genome-scale constraint-based models to assess gene essentiality for several organisms under various growth conditions. Each study simulated gene deletions by constraining the flux through the associated reaction(s) to zero as described in Section 6.2.5 and Box 6.1. In this section, we will review the results from studies performed using models of *E. coli* [53,63], *H. Influenzae* [79], *H. pylori* [64,65], and *S. cerevisiae* [70-72] as a platform on which we can highlight some of the benefits and limitations of genome-scale metabolic models.

6.3.1.1 Escherichia coli   The bacterium *E. coli* is historically the most studied and perhaps the best characterized model organism to date, and is of important industrial, genetic, and pathologic importance. Thus, *E. coli* is among the most suitable organisms for metabolic reconstruction and constraints based analysis. Accordingly, constraint-based models of *E. coli* have been under development since 1990 (for a historical review of the *E. coli* constraint-based model development [62]). Prior to the complete determination of its genome sequence in 1997 [80], *E. coli* models were limited by data availability and thus included only 300 reactions. The success of genome sequencing efforts, coupled with other high-throughput technological advances led to a dramatic increase in size and scope of available models, yielding the first genome-scale models. The most current model of *E. coli* K-12 MG1655 metabolism includes 904 genes, 931 unique biochemical reactions, and 625 metabolites [61] and will soon crest the 1000 gene, 1000 reaction mark (A. Feist, B. Palsson, personal communication).

Following the completion of the first genome-scale model, gene essentiality was examined by investigating the effects of single-gene deletions on the metabolic capability of *E. coli* [63]. Gene deletions were simulated by restricting flux through the corresponding enzymatic reaction to zero. Each individual gene involved in the central metabolic pathways (glycolysis, pentose phosphate pathway, TCA, and respiration processes) was subjected to deletion under an environment of aerobic growth on minimal glucose medium.

Eleven (*rpiAB, pgk, acnAB, gltA, icdA, tktAB, gapAC*) of the simulated deletion strains failed to grow, and an additional 12 (*atp, fba, pfkAB, tpiA, eno, gpmAB, nuo, ackAB, pta*) deletion strains were impaired in their growth characteristics relative to wild type. When grown on glucose, these genes are involved in the three-carbon stage of glycolysis, three reactions of the TCA cycle, and several points within the pentose phosphate pathway.

This study also simulated gene deletion effects on *E. coli* when grown on other minimal medium formulations. Of 79 cases tested, 68 (86 percent) of the *in silico* predictions matched experimental observations. Most of the mischaracterized growth predictions were failure to predict no growth. A later study showed that the incorporation of transcriptional regulatory information can improve performance of the *E. coli* model [53]. Furthermore, the most recent model of *E. coli* is enhanced by containing elementally and charge balanced reactions, as well as GPR associations [61]. As models are further enhanced and a more detailed knowledge and representation of the biomass formulation is acquired, the predictive performance of these models will continue to improve.

6.3.1.2 Haemophilus influenzae

H. influenzae is a Gram-negative pathogen adapted to living in the upper respiratory mucosa, causing ear infections as well as acute- and chronic-respiratory infections primarily in children. Prior to development of the first vaccine in 1985, *H. influenzae* type b was the leading cause of bacterial meningitis in children less than 5 years of age. Based upon the annotated genome sequence and known biochemical information, the metabolic network of this microbe was reconstructed [79,81]. Four hundred of the approximately 1743 open reading frames (ORFs) predicted to exist in *H. influenzae* were included in the model. By including 49 additional reactions based on general metabolic information on related prokaryotes, the final network consists of 461 reactions acting on 367 internal and 84 external metabolites.

In addition to studying various systems properties of the network, gene essentiality was assessed in *H. influenzae* by examining the effects of simulated gene deletions on growth characteristics. Gene deletions were simulated by constraining the flux through the corresponding enzyme catalyzed reaction to zero. One study examined the single, double, and triple deletion of a set of 36 enzymes involved in central intermediary metabolism [81]. For each simulation, the ability of *H. influenzae* to exhibit *in silico* growth was assessed on a defined media, with fructose and glutamate being the two key substrates. Under these conditions, 12 genes (*eno, fba, fbp, pts, gapA, gpmA, pgi, pgk, ppc, rpiA, tktA, tpiA*) were essential for growth and an additional 10 enzyme deletion strains (*cudABCD, atp, ndh, ackA, pta, gnd, pgl, zwf, talB, rpe*) exhibited impaired growth. This result suggests that *H. influenzae*'s metabolic network is less robust against single central metabolic gene deletions than *E. coli* [63]. Examination of all possible double mutants for evidence of so-called synthetic lethal interactions [82] within this set of enzymes revealed only 7 of 361 lethal gene pairs where each single deletion mutant was viable. Similarly, only 7 of 5270 lethal gene triple knockouts were observed where the double deletion of any two of the gene products does not result in a null phenotype.

A related study also examined simulated gene deletion effects on an expanded 42 enzyme set under two different growth conditions [79]. When simulating growth on a minimal media with fructose as the primary carbon source, the 11 single-gene deletions (*fba, fbp, tpiA, gapA, pgk, pgmA, eno, rpiA, tktA, prsA, ppc*) failed to grow. In order to study these same single-gene mutants under more relevant *in vivo* conditions, simulations were performed in similar fashion, this time using media supplemented with a number of carbon sources likely to be found in the host, mucosal environment. These include fructose, glucose, glycerol, galactose, fucose, ribose, and sialic acid. Under these conditions six mutant strains (*gapA, pgk, pgmA, eno, ppc*) again failed to grow. While these predictions require experimental verification using methods described elsewhere in this volume, these studies show the utility of computational studies in directing the researcher to the most interesting targets.

6.3.1.3 Helicobacter pylori

H. pylori is a human bacterial pathogen that colonizes the gastric mucosa. Infection results in acute inflammation and damage to epithelial cells, ultimately progressing to a number of disease states, including gastritis, peptic ulceration, and gastric cancer. In comparison with the examples of *E. coli, H. influenzae, and S. cerevisiae*, provided elsewhere in this chapter, a little experimental data is available to complement the known genome sequence of *H. pylori*. Recent work shows that even in the absence of extensive experimental data, detailed metabolic models of great utility can be developed, using *H. pylori* as an example [64]. Relying primarily on annotated genome sequence, a model comprised of 388 enzymatic reactions, corresponding to 291 of 1590 known ORFs and 403 metabolites, was developed for *H. pylori*.

Similar to the other modeling efforts described in this chapter, systemic properties of this model were examined, as was gene essentiality by simulating gene deletions. The effect of the loss of enzymatic function corresponding to a gene deletion was assessed under four different simulated growth conditions. The growth conditions include a previously determined minimal medium required to support *in silico* growth, minimal medium supplemented with glucose, other carbon sources, and amino acids.

Each of the 34 reactions in the central intermediary metabolic network was individually eliminated by constraining the flux through the reaction to zero. Of these simulated gene deletions, only four (*aceB, ppa, prsA, and tpi*) failed to exhibit simulated growth under all four conditions. These particular knockouts affect malate synthase activity, the pyrophosphate to inorganic phosphate conversion, synthesis of nucleotides and deoxynucleotides, and impaired glycolytic ability.

The predictive performance of the model was also assessed via comparison with available experimental gene deletion data for *H. pylori* [64]. The model accurately predicted the growth ability for 10–17 gene deletions. Of the seven incorrect predictions, six were predicted to be nonessential when experimental evidence showing their essentiality. This discrepancy could signal deficiencies in the model. For example, given that the model is not complete, all of the relevant information may not be available to accurately predict the given case. In contrast, it could also be because of the experimental conditions not corresponding exactly with simulated conditions. In any case, these discrepancies identify keen areas of interest to probe with further experiments in an effort to ultimately improve and enhance the model.

An updated *H. pylori* was also recently developed and published [65]. This expanded model includes 341 genes and 476 intracellular metabolic reactions. A single-gene deletion analysis was carried out in which the growth capability of all 341 knockouts was assessed by constraining the flux through the associated reaction(s) to zero for FBA simulations of growth on both minimal and rich medium. More than 70 percent of the 72 predictions for which experimental data was available were in congruence. Furthermore, this result represents an improvement over the previous version of the *H. pylori* reconstruction [64]. A simulated double-deletion, or synthetic lethal, screen was also carried out using this network by constraining the flux for reactions associated with all pairwise combinations of genes in the model to zero. Of the more than 22,000 combinations that were tested, 47 pairs involving 64 unique genes were found to be lethal. While no corresponding experimental data exists for validation, this effort is still quite useful in that it can direct researchers to the potentially more interesting portions of the network for experimental investigation in the absence of a labor-intensive high-throughput screen.

6.3.1.4 *Saccharomyces cerevisiae*

In recent years, using the vast quantities of data available for the baker's yeast, *S. cerevisiae*, researchers have developed a genome-scale reconstructed metabolic model using the constraint based approach [70,83]. A total of 708 metabolism related ORFs were accounted for in the reconstructed network, corresponding to 1035 metabolic reactions. An additional 140 reactions were included based on biochemical evidence without direct knowledge of a responsible enzyme, ultimately yielding a reconstructed network containing 1175 metabolic reactions and 584 metabolites. This model has been shown in most cases to predict growth characteristics consistent with observed phenotypic functions [83].

Gene essentiality was assessed by using this large-scale model of *S. cerevisiae* to computationally evaluate the effect of 599 single-gene deletions on viability [84]. In this study, growth of yeast was simulated under aerobic conditions and on complete medium containing glucose, the 20 essential amino acids, and nucleic acids. Ammonia, phosphate, and sulfate were also supplied. Gene deletions were simulated by constraining the flux through the corresponding reactions to zero and optimizing for growth, as in previous studies in *E. coli* [85], and performance was gauged through comparison with experimental data.

The model performed remarkably well, accurately predicting the effect of 90 percent of these mutant alleles. In concurrence with experimental observation [86], a small fraction of these deletion strains exhibit impaired growth and fewer are still lethal. It should be noted, however, that the model had the most difficulty in correctly predicting null mutant phenotypes. This can be attributed to incomplete biochemical information, inadequate biomass equation definition, and gene regulatory effects. Addressing each of these issues in future work will likely improve the model's predictive capability. Future studies might also include the examination of lethal double mutants, also known as synthetic lethality [83,87], as these may provide better insight than single deletion mutants into gene essentiality and network robustness in *S. cerevisiae*.

6.3.2 Model Performance Assessment

Validating model predictions is a critical component in constraint-based model analysis. Growth phenotype data, available for a number of knockout strains and organisms, can be acquired from biochemical literature [88] and online databases, including ASAP [89] for *E. coli*, as well as CYGD and SGD for *S. cerevisiae*. As noted in the previous section, experimental growth phenotype data is available to assess directly the predictive power of the model for three of the four organisms listed previously, and shows that correct predictions were made in approximately 60, 86, and 83 percent of cases for *H. pylori* [64], *E. coli* [53], and *S. cerevisiae* [71], respectively. These comparisons serve two important functions: Validation of the general predictive potential of the model and identification of areas that require refinement. In this sense, constraint-based models are particularly useful in experimental design by directing research to the most or least poorly understood biological components. The next section details how to interpret incorrect model predictions and their likely causes.

6.3.3 Troubleshooting Incorrect Predictions

In the studies discussed in Section 6.3.1, the model predictions when compared to experimental findings failed most often by falsely predicting growth when the gene deletion leads to a lethal phenotype *in vivo*. This trend indicates that the most common cause of false predictions is because of the lack of information included in the network. For example, certain important pathways not related to metabolism in which the deleted gene participates may not be represented. In addition, the objective function may not be defined properly by failing to include the production of a compound required for growth. This case was shown to account for many false predictions when using a yeast metabolic model to account for strain lethality [72] when a few relatively minor changes to the biomass function dramatically improved the model's predictive capability. Alternatively, the gene deletion may lead to the production of a toxic by-product that ultimately kills the cell, a result for which this approach cannot account. Furthermore, certain isozymes are known to be dominant whereas metabolic models typically assign equal ability to each isozyme. The model would predict viable growth for the dominant isozyme deletion whereas *in vivo*, the minor isozyme(s) would not sufficiently rescue the strain from the lethal phenotype perhaps due to lower gene expression or enzymatic activity.

An additional major error source stems from the lack of regulatory information incorporated into the previously described models. Including transcription factor, metabolic gene interactions, using a Boolean logic approach, enhance the accuracy of constraint-based model predictions [53]. Regulatory information is available in the primary literature, in addition to online resources such as EcoCyc and RegulonDB [90]. Furthermore, these interactions can be derived from ChIP-chip analysis of transcription factors and corresponding gene expression microarray data [91]. A more detailed treatment of this latter topic is presented in Section 6.4.3.

Incorrect predictions are less often due to false predictions of lethality. These uncommon cases often suggest the presence of previously unidentified enzyme activities, which if added to the model, would lead to accurate predictions. They may also reflect improper biomass function definition, but in a different sense from the situation described above. For example, rather than failing to include compounds required for growth, it is also possible that certain compounds are included in the biomass function erroneously, and may actually not be essential to support biological growth. In any case, inaccurate [12,92] predictions are most often attributed to a paucity of information available for inclusion in the model and not simply a failure of the technique, thus validating the general strategy of constraint-based modeling.

6.3.4 Additional Analytical Tools

A rapidly growing collection of analytical methods have been developed for use in conjunction with constraint-based models [12], some of which we briefly introduce in this section. Although many of the examples in this chapter focus on the use of constraint-based models to assess gene essentiality, these models can also be used to predict behavior of viable gene deletions. For example, FBA uses LP to identify the optimal metabolic state of the mutant strain. In contrast, Minimization of Metabolic Adjustment (MOMA) uses quadratic programming (QP) to identify optimal solutions that minimize the flux distribution distance between a wild type and simulated gene deletion strain [93,94]. Experimental data seems to confirm the MOMA assumption that knockout strains utilize the metabolic network similar to wild type [93]. It remains to be determined if this is true in all situations or if the network optimizes for growth over time following gene deletion.

A more recent method known as regulatory on/off minimization (ROOM) [95] is another constraint-based analysis technique that uses a mixed integer linear programming strategy to predict the metabolic state of an organism following a gene deletion by minimizing the number of flux changes that occur with respect to wild type. In other words, this algorithm aims to identify flux distributions that are qualitatively the most similar to wild-type in terms of the number and types of reactions that are utilized. While MOMA seems to better predict the initial metabolic adjustment that occurs following the genetic perturbation, ROOM, like FBA, better predicts the later, stabilized growth phenotype.

Constraint-based modeling also has applications in the metabolic engineering field. Identifying optimal metabolic behavior of mutant strains using a bilevel optimization framework has been employed by OptKnock [96]. This metabolic engineering strategy uses genome-scale metabolic models and a dual-level, nested optimization structure to predict which gene deletion(s) will lead to a desired biochemical production while retaining viable growth characteristics. This technique establishes a framework for microbial strain design and improvement and has the potential for significant impact. These and other analytical techniques and applications that rely on constraint-based modeling will be discussed in detail in Chapter 11.

Additional methods have been developed to specifically assess the systemic or topological properties of these networks [12]. Extreme pathway analysis [97] represents one such technique that utilizes convex analysis of the S matrix to define a cone that circumscribes all allowable steady-state solutions within the space defined by the S matrix and its associated constraints (see Fig. 6-1 for a conceptual representation of the this space, also known as the "solution space"). Accordingly, all possible routes through the network can be described by nonnegative combinations of the generated extreme pathways. This technique and analysis of the extreme pathways themselves have been fruitful in a variety of studies (for examples, see Refs [98–100]) and can be readily calculated for reasonably sized networks using available software [79,101].

6.4 FUTURE DIRECTIONS FOR CONSTRAINT-BASED MODELING

Thus far, constraint-based models have had their primary success in assessing the metabolic capabilities of cells, but fail to account for many other important aspects of cellular biology. In the past several years, however, several efforts have been initiated to apply the constraint-based modeling and analysis techniques to other cellular processes. Below we briefly describe relatively recent work that is setting the stage for including RNA and protein synthesis [102] as well as other processes governed by cell signaling [103] and transcriptional regulatory networks (TRNs) into genome-scale, constraint-based models of the cell.

6.4.1 Modeling of RNA and Protein Synthesis

RNA and protein synthesis represent two of the primary energy drains on the cell [58] and are of obvious vital importance in that these processes give rise to many of the active components responsible for cellular activities. Existing constraint-based genome-scale metabolic models do not explicitly account for these processes, rather they are included as abstract, lumped sum quantities of monomeric amino acid, and nucleotide triphosphate demand required to support cellular growth [104]. The specific values for these quantities are determined from measurements of biomass constituents [58] and are independent of the genome sequence. In order to meet this deficiency in the field, a scalable, constraint-based framework was developed to capture the metabolic requirements for gene expression and protein synthesis directly from the genome sequence [102].

The general strategy stems from the observation that RNA and protein synthesis can be broken down into constitutive biochemical reactions that underlie the processing of these polymers. As illustrated in Figure 6-5, the expression of a given gene and the synthesis of the protein that it encodes can be modeled by six essential biochemical reactions. These reactions include transcription initiation, transcription elongation, mRNA degradation, translation initiation, translation elongation, and tRNA charging. Biochemical equations representing each of these processes can be compiled (Fig. 6-5b) and used to formulate an associated S matrix (Fig. 6-5c).

Many of the previously introduced analytical tools can then be used to computationally assess the properties of the S matrix. For example, by choosing protein production as the objective, FBA can be used to determine how much the protein that the RNA and protein synthesis machinery within the cell can produce for a given set of environmental conditions and resources [102]. One can also incorporate promoter strength, transcription elongation, and translational initiation constraints on the system if such information is known or can be approximated. Extreme pathway analysis can also be used to assess the capabilities of these systems and their characteristic states [102]. Thus far, however, this framework and analysis methods have only been applied to small biological systems, namely the malate dehydrogenase (*mdh*) gene and the *lac* operon [102]. Accordingly, the limitations associated with studying large-scale systems in this manner remain to be assessed, although an ongoing study of the *E. coli* RNA and protein synthesis network (I. Thiele and B. Palsson, personal communication) is certain to be illuminating.

6.4.2 Modeling of Cell Signaling Networks

The signal transduction pathways that comprise cell signaling networks are responsible for many critical processes. Signaling events operate both on relatively quick timescales, such as those that cause posttranslational protein changes, and long timescales, such as cell cycle control, cell proliferation and migration, as well as apoptosis. Cell signaling networks are often highly connected and complex involving many molecular players. In an effort to quantitatively characterize their properties, researchers are beginning to reconstruct these networks and apply mathematical methods to analyze them.

One approach to computationally analyzing cell signaling networks relies on many of the same constraint-based modeling principles discussed earlier in this chapter for metabolic networks [103,105]. The key insight is to treat signaling pathways as a series of biochemical transformations starting with an input (the signal) and resulting in an output (posttranslational protein modification, apoptosis, etc.). Accordingly, just as in modeling metabolic networks the first steps of this process focus on network reconstruction. One must first identify the components in the signaling network of interest and the interactions that occur between them. In contrast to modeling of metabolic networks where enzymes and metabolites are the primary players, signaling networks typically include receptors and their corresponding receptor ligands,

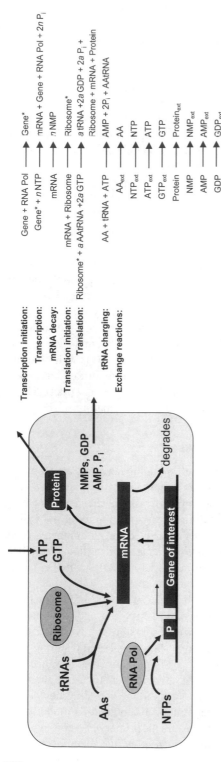

Transcription initiation: Gene + RNA Pol ⟶ Gene*

Transcription: Gene* + n NTP ⟶ mRNA + Gene + RNA Pol + $2n$ P_i

mRNA decay: mRNA ⟶ n NMP

Translation initiation: mRNA + Ribosome ⟶ Ribosome*

Translation:
Ribosome* + a AAtRNA +$2a$ GTP ⟶ a tRNA +$2a$ GDP + $2a$ P_i + Ribosome + mRNA + Protein

tRNA charging: AA + tRNA + ATP ⟶ AMP + $2P_i$ + AAtRNA

Exchange reactions:
AA_{ext} ⟶ AA
NTP_{ext} ⟶ NTP
ATP_{ext} ⟶ ATP
GTP_{ext} ⟶ GTP
Protein ⟶ $Protein_{ext}$
NMP ⟶ NMP_{ext}
AMP ⟶ AMP_{ext}
GDP ⟶ GDP_{ext}
P_i ⟶ $P_{i\,ext}$

(b)

Figure 6-5 Constraint-based modeling of RNA and protein synthesis. (a) A hypothetical system that represents the RNA and protein synthesis network associated with the transcription and translation of a single gene is depicted. The processes of transcription initiation, transcription, mRNA decay, translation initiation, translation, and tRNA charging are depicted. Also shown are some of the exchange fluxes required to balance the system. (b) A biochemical reaction list for the included processes and appropriate exchange reactions can be compiled. Note that the precise stoichiometry can and should be included in each reaction definition. In this system the gene and associated protein length can vary. Accordingly, variables for the number of bases (n) and number of amino acids (a) are included in the reaction stoichiometry. (c) The S matrix can then be formulated based on the reaction list. System components are represented in respective rows and each column denotes individual system reactions. AA, amino acid; AAtRNA, charged tRNA; Gene*, gene undergoing transcription; NMP, nucleotide monophosphate; Pi, inorganic phosphate; Ribosome*, actively translating ribosome; RNA Pol, RNA polymerase.

	Transcription Init.	Transcription	mRNA decay	Translation Init.	Translation	tRNA Charging	AA influx	NTP influx	ATP influx	GTP influx	Protein efflux	NMP efflux	AMP efflux	GDP efflux	P_i efflux
Gene	-1	1	0	0	0	0	0	0	0	0	0	0	0	0	0
Gene*	1	-1	0	0	0	0	0	0	0	0	0	0	0	0	0
RNA Pol	-1	1	0	0	0	0	0	0	0	0	0	0	0	0	0
NTP	0	-n	0	0	0	0	0	-1	0	0	0	0	0	0	0
mRNA	0	1	-1	-1	1	0	0	0	0	0	0	0	0	0	0
P_i	0	2n	0	0	2a	2	0	0	0	0	0	0	0	0	1
NMP	0	0	n	0	0	0	0	0	0	0	0	1	0	0	0
Ribosome	0	0	0	-1	1	0	0	0	0	0	0	0	0	0	0
Ribosome*	0	0	0	1	-1	0	0	0	0	0	0	0	0	0	0
AAtRNA	0	0	0	0	-a	1	0	0	0	0	0	0	0	0	0
Protein	0	0	0	0	1	0	0	0	0	0	1	0	0	0	0
GTP	0	0	0	0	-2a	0	0	0	0	-1	0	0	0	0	0
GDP	0	0	0	0	2a	0	0	0	0	0	0	0	0	1	0
AA	0	0	0	0	0	-1	-1	0	0	0	0	0	0	0	0
tRNA	0	0	0	0	a	-1	0	0	0	0	0	0	0	0	0
ATP	0	0	0	0	0	-1	0	0	-1	0	0	0	0	0	0
AMP	0	0	0	0	0	1	0	0	0	0	0	0	1	0	0

(c)

Figure 6-5 (Continued)

metabolites such as ATP and ADP, as well as intracellular signal transducing proteins. Also, these networks often include transcription factors, transcription factor binding sites, and the resulting target genes.

The data from which components and their interactions are derived have been traditionally difficult to obtain due to the often laborious effort involved in mapping signaling pathways using standard molecular biology techniques. However, recently developed high-throughput, genome-scale techniques are mitigating this issue. For example, whole genome sequencing and annotation identifies the possible network components, ChIP-chip assays identify protein-DNA interactions, and yeast two-hybrid assays identify protein–protein interactions. As previously noted, Table 6-2 summarizes many useful online resources that contain publicly accessible data. Several strategies for mapping signaling pathways and networks have been developed in recent years by integrating these and other high-throughput data [4]. These methods have been employed to map DNA damage response as well as developmental pathways [4] among others.

Having identified the components and interactions that occur between them, a list of biochemical reactions that describes the cell signaling network can be listed. A stoichiometric matrix is then derived from this list (Fig. 6-6) in very much the same

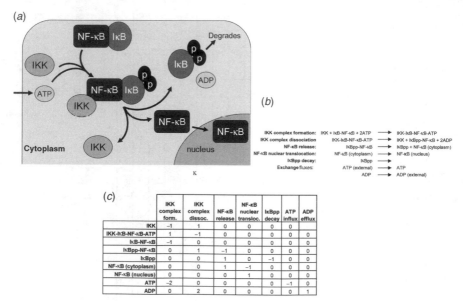

Figure 6-6 Constraint-based modeling of cell signaling networks. (a) A schematic that includes a portion of the nuclear factor (NF)-κB signaling-related network is depicted. (b) A reaction list that corresponds to the schematic in (a) is detailed. Reactions are included for the interaction of IκB kinase (IKK) with the inhibitor of NF-κB (IκB)-NF-κB complex. The subsequent phosphorylation of IκB and release of NF-κB are also shown in addition to the degradation of phosphorylated IκB (IκBpp) and NF-κB translocation to the nucleus, and exchange fluxes required for the system. (c) The associated *S* matrix is compiled based on the reaction list. System components are depicted in each respective row, and reactions are represented in each column.

manner as previously described for metabolic as well as RNA and protein synthesis networks. It is important to note that each state of a component must be explicitly accounted for in the network. For example, a protein must be differentially represented in separate phosphorylated and unphosphorylated forms [105,106].

This stoichiometric framework explicitly defines the underlying network reactions in a chemically consistent form. Accordingly, network properties can be readily and quantitatively assessed using previously introduced analytical tools. Extreme pathway analysis, in particular, is an immensely useful tool for characterizing cell signaling networks. Using existing software [101,107], one can enumerate the extreme pathways using the stoichiometric matrix from the reconstructed cell signaling network.

All routes through the cell signaling network can be described by nonnegative linear combinations of the extreme pathways. Accordingly, network cross talk, signaling redundancy, correlated reaction sets, as well as reaction participation and likely relative importance are all properties that can be derived from this analysis. Network cross talk refers to an analysis of how disjoint, overlapping, or identical inputs can lead to disjoint, overlapping, or identical outputs within a signaling network and are derived from pairwise comparisons of individual extreme pathways. Strictly speaking, signaling redundancy is the multiplicity of routes through a network by which identical inputs lead to identical outputs, but it can be further delineated into considerations of input and redundancy alone. Correlated reaction sets are a collection of reactions that is always either present or absent in all of the extreme pathways. In other words, these sets of reactions represent functional modules that act together in a given network, although the reactions themselves may not necessarily be adjacent on the reaction map. Finally, reaction participation is the percentage of pathways in which a given signaling reaction is used. This relatively simple calculation can indicate important biological insights. For example, reactions with high participation values are likely to be critical for network functionality while low participation values indicate more specialized portions of the network.

Thus far, the stoichiometric approach to modeling signaling networks has only been applied to a prototypic network [105] and the human B-cell JAK (Janus activated kinase)–STAT (signal transducer and activator of transcription) signaling network [106]. While the prototypic network study served simply as proof of concept, the work on the JAK–STAT network showed that the constraint-based approach can be used to analyze real biological systems and yield quantitative insights into its properties. Accordingly, as more signaling networks are delineated and reconstructed, this approach will likely be of great utility.

6.4.3 Modeling of Transcriptional Regulatory Networks

With the huge success of whole genome sequencing efforts and the appearance of hundreds of genome sequences, there is an increased interest in understanding how the genes within a given genome are regulated through complex TRNs). Consequently, efforts are underway to define and catalog the set of regulatory rules for model organisms. Due to the large number of regulated genes and associated regulatory proteins as well as their extensive interconnectivity, there is a significant need for a structured framework to integrate regulatory rules and interrogate TRN functions in a systematic fashion.

Previous work has integrated models of regulatory and metabolic networks to analyze and predict the effect of transcriptional regulation on cellular metabolism at the genome scale [50,52,53,108]. These studies developed and utilized a framework in which regulatory rules are represented as Boolean logic rules that control the expression of enzyme-encoding genes that ultimately facilitate metabolic reactions within a constraint-based metabolic model of the type described previously within this chapter. The regulatory rules are defined such that metabolic enzyme genes are determined to be present or absent based on the presence or absence of extracellular and intracellular metabolites. If an enzyme-encoding gene is determined to be absent then the flux through that enzyme is set to zero in the metabolic model, which in effect adds a temporary constraint on the system. In effect, this is equivalent to carrying out FBA on the network following a gene deletion.

Using an iterative computational scheme in which time, t, is divided into small steps (usually on the order of minutes), a dynamic profile of growth can be simulated. At $t = 0$ the metabolic model is used to predict the optimal flux distribution for the network using FBA, as described in Section 6.2.4. The resulting flux distribution is used as initial conditions from which the Boolean transcriptional regulatory rules are evaluated. The rule evaluations specify the transcriptional status of enzymes for the next time step. As noted above, if the transcriptional state of an enzyme-encoding gene results in the absence of the corresponding enzyme the reaction flux(es) mediated by it are set to zero for the FBA carried out on the system for the next time step. This process of iterative Boolean rule evaluation and FBA calculation continues for the user-defined time span [50,52].

This type of integrated analysis of metabolic and regulatory networks has been performed for both small prototypic systems [52] as well as for a genome-scale model of *E. coli* [50] and more recently in yeast [108]. In the study of *E. coli*, this analysis was performed in conjunction with dual perturbation growth experiments coupled with genome-wide expression analysis. This systematic approach to reconstructing and interrogating the integrated network of *E. coli* led to the identification of many novel regulatory rules, and an expanded characterization of the genome-scale TRN, based on a model-driven analysis of multiple high-throughput data sets. Furthermore, a recent study has also used this model in a large-scale simulation project to study all potential network states and found them to be organized primarily based upon terminal electron acceptor availability [57]. However, one shortcoming of this framework is that it does not facilitate a detailed analysis of transcriptional regulatory network properties.

In an effort to address this limitation, a structured and self-contained representation of TRNs that can be quantitatively interrogated has been developed relying on the principles of the constraint-based approach [109]. This strategy, which effectively connects environmental cues to transcriptional responses, is conceptually similar to the previously described constraint-based approach to modeling cell signaling networks. The first step in the process involves defining the components of the system and interactions between them based on legacy data from traditional molecular biology studies or from recently generated high-throughput data. In particular, ChIP-chip data provides direct information regarding transcription factor-target gene relationships and genome-wide expression profiling data can yield insight as to the type of regulation. For example, when examined using conditions in which a given transcription factor is known to be active, the upregulation of a gene identified to be a target of the given transcription factor indicates that the transcription factor serves as an activator, whereas downregulation of the target gene suggests that the transcription factor acts as a repressor for that particular target gene. The environmental cues or stimuli to which the transcription factors respond as well as any required cofactors such as cyclic AMP (cAMP) must also be identified and included in this representation.

Having gathered this type of information that describes the regulatory system of interest, the next step is to write quasi stoichiometric, biochemical equations that describe the regulatory logic for each interaction in the network (Fig. 6-7b). The quasi stoichiometric nature of these equations is not required of course, but rather is used due to the general lack of specific chemical detail for most regulatory interactions. As the specific stoichiometry of regulatory interactions becomes available [110], however, higher levels of detail can be readily incorporated into this framework. As is the case in reconstructing cell signaling networks, it is important to reiterate that each state of a component must be explicitly accounted for in the network. For example, for regulatory networks, this case is encountered when transcription factors interact with cofactors to form activating or inhibitory complexes.

One peculiarity of this methodology is that it requires the inclusion of the converse of regulatory rules in addition to the regulatory rules themselves. The *converse* of the regulatory rules—the regulatory reactions that lead to the inhibition of gene transcription in our sample system—is necessary to reflect the lack of protein production for a given set of environmental cues. Many regulatory rules are inhibitory, such that the expression of a protein depends on the absence of a given metabolite or protein product. Additional reactions that include the converse of the regulatory rules and the absence of metabolites and protein products where appropriate must be included in the system. Also, note that regulatory rules of the Boolean type "OR" require two separate reactions to indicate that there are two independent ways in which the target gene can be transcribed.

A matrix can then be compiled from this list of biochemical reactions (Fig. 6-7c) in much the same way as was done for the other network types described previously in this chapter. Each row of the matrix describes a component of the system and each column represents regulatory events, or reactions. As a reminder, notice that each metabolite is represented in both present and absent forms, as is each transcription factor. Furthermore, the quasi stoichiometric formalism needs to be supplemented by *exchange* reactions that balance the entry of external cues or stimuli into the system as well as the production of proteins and their exit from the system. These exchange reactions describe the role of external cues and stimuli as inputs to the regulatory system and the role of the proteins as outputs of the transcriptional regulatory system. Therefore, columns representing the exchange of external stimuli as well as protein products are incorporated.

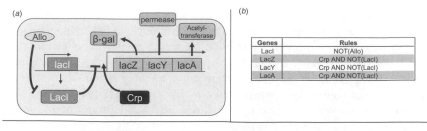

(b)

Genes	Rules
Lacl	NOT(Allo)
LacZ	Crp AND NOT(Lacl)
LacY	Crp AND NOT(Lacl)
LacA	Crp AND NOT(Lacl)

(c)

	vlacl	vLacl	vNOTLacl	vlacOperon	vLacOperon	Allo	Crp	Lacl	LacA	LacZ	LacY
lacl	−1	1	0	0	0	0	0	0	0	0	0
lacl*	1	−1	0	0	0	0	0	0	0	0	0
Lacl	0	1	−1	0	0	0	0	−1	0	0	0
Allo	0	0	−1	0	0	−1	0	0	0	0	0
NOT_Lacl	0	0	1	−1	0	0	0	0	0	0	0
Crp	0	0	0	−1	0	0	−1	0	0	0	0
lacZYA	0	0	0	−1	1	0	0	0	0	0	0
lacZYA*	0	0	0	1	−1	0	0	0	0	0	0
LacZ	0	0	0	0	1	0	0	0	0	−1	0
LacY	0	0	0	0	1	0	0	0	0	0	−1
LacA	0	0	0	0	1	0	0	0	−1	0	0

(d)

$$r^T = $$

vlacl	vLacl	vNOTLacl	vlacOperon	vLacOperon	Allo	Crp	Lacl	LacZ	LacY	LacA
1	1	0	0	0	0	0	1	0	0	0
1	1	1	1	1	−1	−1	0	1	1	1

(e)

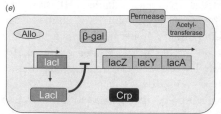

Extreme pathway 1:
lac operon not expressed

(f)

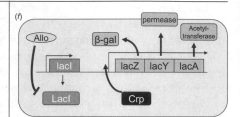

Extreme pathway 2:
lac operon expressed

Figure 6-7 Constraint-based modeling and analysis of transcriptional regulatory networks. (a) The *lac* operon regulatory system is depicted and defined to include the *lac* operon genes (*lacZ, lacY, lacA*), the inhibitor gene *lacl*, the activator *Crp*, and the inducer allolactose (*Allo*). (b) A reaction list that summarizes the Boolean rules that capture the regulatory logic of the system is shown. (c) The *R* matrix that corresponds to the regulatory rule list from (b) is depicted with each row corresponding to system components and each column specifying regulatory reactions in a quasi stoichiometric formalism. Accordingly, a "−1" represents a "consumed" component, whereas a "+1" represents a "produced" component. (d) The two extreme pathways for this system are listed in **r** with the corresponding reaction labels listed as well for reference. A nonzero value indicates that the corresponding reaction is active. The negative coefficients in the second extreme pathway reflect that *Allo* and *Crp* can be thought of as conceptually flowing into the system. (e) Pathway 1 is graphically illustrated and reflects the conditions for the Lacl-mediated inhibition of the *lac* operon. (f) The graphical depiction of Pathway 2 shows the activation of the *lac* operon (i.e., inhibition of Lacl by allolactose, thus allowing for derepression and Crp-activated expression of *lacZYA*). *r*[T], the transpose of the extreme pathway vectors reported in *r* (depicted in this way simply out of space considerations).

With a regulatory matrix in hand, many of the analytical tools previously discussed can be applied to assess its properties. For example, extreme pathway analysis generates a set of vectors that encompasses all possible expression states of the network. Recall that all possible regulatory pathways, and thus expression states can be described as a nonnegative linear combination of these extreme pathways. Consequently, extreme pathway analysis represents an *in silico* technique for evaluating global characteristics of gene expression. Furthermore, the pervasiveness of signal inputs, percentage of environments in which a given gene is expressed, numbers of genes coordinately expressed, and correlated gene sets represent the type of data that can be readily generated based on extreme pathway analysis for a transcriptional regulatory network.

To illustrate some of these regulatory matrix ideas, we briefly consider the *lac* operon in *E. coli*. For the purpose of this investigation, the system is defined to include the *lac* operon (*lacZYA*) and the proteins each gene encodes, the inhibitor of the operon (*lacI*), an activator of the operon (Crp), and the intracellular inducer molecule allolactose, which inhibits the LacI inhibitor thus activating *lacZYA* transcription (Fig. 6-7a) by way of derepression.

Having defined the system (Fig. 6-7a) and Boolean rules that specify the regulatory logic of this small transcriptional regulatory network (Fig. 6-7b), the system can be formulated and the associated R matrix constructed (Fig. 6-7c). For the purposes of this analysis, each gene/operon is depicted within the matrix twice: *lacI* and *lacI**, as well as *lacZYA* and *lacZYA**. The former entity represents the open form, whereas the latter, asterisk-marked entity represents the actively transcribed form of the gene. This level of detail is not required in formulating R as the actively transcribed form of the gene is only a transient entity between transcription and translation. Rather, this is meant to show concretely that such mechanistic detail about ORFs and other network relationships can be readily incorporated into the current formalism as the data becomes available.

Extreme pathway analysis on this system yields two vectors, denoted by r (Fig. 6-7d). Each entry in the vectors represents the activity of a reaction in the expression state, or pathway. For reaction names prefaced with a "v," a 1 indicates that the reaction is active, and a 0 indicates that it is inactive. In the remaining reactions that specify flow across the system boundary, a 1 indicates flow out of the system (for example, a protein is produced), a -1 indicates flow into the system, and a 0 indicates that the associated component is neither produced nor consumed. Note that the entries are not quantitative but denote an active connection, and further, that a series of connections leads to a "causal path." The first vector represents the LacI-mediated inhibition of the *lac* operon. The second vector defines the inhibition of LacI by allolactose, thus resulting in derepression and Crp-activated expression of *lacZYA*. These two vectors thus represent the two expression states of the *lac* operon system, as depicted graphically in Figure 6-7e and f.

Thus far, this approach has only been applied to the small *lac* operon system described above and a larger 25 gene prototypic network [111]. While this proof of concept study validates the utility of this approach for small systems, potential complications associated with scaling this approach up to genome-scale systems remain to be determined. Nonetheless, transcriptional regulatory network matrix reconstructions for model organisms will likely be important not only in studies of regulatory network properties but also in guiding experimental programs based upon results from these analyses.

6.4.4 The Next Big Challenge

The constraint-based approach has proven immensely successful for modeling metabolic models and, as described in this section, is showing promise for RNA and protein synthesis, cell signaling, and transcriptional regulatory networks. However, as the field currently stands, each respective framework produces models that exist as independent entities. Arguably, the ultimate goal of systems biology is to integrate data from disparate sources and generate comprehensive models that reflect biological reality for entire cells. Therefore, these modeling strategies present an opportunity to take a significant step forward in realizing this aim through integrative modeling efforts.

To elaborate, the interconnectivity between these distinct networks is clear. For example, a simplistic, but illustrative conceptual picture (Fig. 6-8) can be envisioned in

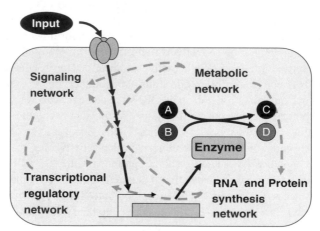

Figure 6-8 The next big challenge: model integration. This chapter has illustrated the utility of constraint-based modeling and analysis in computationally representing many cellular processes. To date, however, these models have been developed and analyzed in isolation despite the fact that these systems are all interrelated, as shown in this conceptual figure. For example, cellular signals or inputs are recognized by the cell signaling network, which in turn stimulate regulatory processes. These regulatory processes mediate RNA and protein synthesis ultimately leading to the production of enzymes that perform metabolic processes that result in cell growth or maintenance. The dashed arrows highlight the interconnectivity of these networks in the form of shared molecular components or feedback mechanisms. In principle, the constraint-based formalism can be used as a platform to capture these systems into a single picture. Accordingly, one of the next major challenges facing the field is to integrate these models of disparate cellular processes, thus pushing toward one of the field of systems biology's foundational goals: To computationally represent and analyze models of entire cells and biological systems.

which system inputs are recognized by cell signaling networks that in turn stimulate regulatory processes. These regulatory processes mediate RNA and protein synthesis ultimately leading to the production of enzymes that perform metabolic processes and lead to cell growth or maintenance. Additional connectivity between the systems also exists in the form of feedback processes and shared currency metabolites such as ATP and GTP, for example. Thus, in principle, the stoichiometric and pseudo-stoichiometric representations of the networks described in this chapter could be integrated into a unified model of the cell. While there are certainly computational challenges that will need to be overcome in order to facilitate the development and analysis of such a model, this notion seems feasible and is sure to be tackled in the near future. Representing additional cellular processes, such as differentiation, and accounting for multicellularity await novel research efforts and represent open problems to be addressed in the more distant future.

6.5 CONCLUSIONS

Despite the challenges outlined in the previous section associated with pushing the field forward, constraint-based modeling and its associated analyses are (and will remain as) powerful tools that facilitate system-level modeling [11,53,103] and analysis of biological networks [57,99,111,112]. Furthermore, these model-based studies can be used to help researchers prioritize experimental projects and save considerable time at the bench. Beyond its utility as a tool for basic biological research and in metabolic engineering applications [97,113], this computational approach also has potential medical relevance. For example, in pathogenic microbial models, each gene that is predicted to be essential by constraint-based modeling and analysis represents a potential drug target that could be used to develop effective therapeutics in the future. As more genome-scale models are developed and existing models enhanced, additional applications—a broad range of fields—will likely become apparent. Consequently, the flexibility of constraint-based models will continue to be exploited to drive the exploration of countless exciting biological questions in the future.

REFERENCES

1. Wheeler DL, et al. Database resources of the National Center for Biotechnology Information: update. *Nucleic Acids Res* 2004;32(Database issue): D35–D40.
2. Wyrick JJ, Young RA. Deciphering gene expression regulatory networks. *Curr Opin Genet Dev* 2002;12(2): 130–136.
3. Sanford K, et al. Genomics to fluxomics and physiomics—pathway engineering. *Curr Opin Microbiol* 2002;5(3): 318–322.
4. Joyce AR, Palsson BO. The model organism as a system: integrating omics' data sets. *Nat Rev Mol Cell Biol* 2006;7(3):198–210.
5. Arkin AP. Synthetic cell biology. *Curr Opin Biotechnol* 2001;12(6): 638–644.
6. Tomita M,et al. E-CELL: software environment for whole-cell simulation. *Bioinformatics* 1999;15(1): 72–84.

7. Hoffmann A, et al. The IkappaB-NF-kappaB signaling module: temporal control and selective gene activation. *Science* 2002;298(5596): 1241–1245.

8. Elowitz MB, et al. Stochastic gene expression in a single cell. *Science* 2002;297(5584): 1183–1186.

9. Arkin A, Ross J, McAdams HH. Stochastic kinetic analysis of developmental pathway bifurcation in phage lambda-infected *Escherichia coli* cells. *Genetics* 1998;149(4): 1633–1648.

10. Sarkar A, Franza BR. A logical analysis of the process of T cell activation: different consequences depending on the state of CD28 engagement. *J Theor Biol* 2004;226(4): 455–466.

11. Reed JL, et al. Towards multidimensional genome annotation. *Nat Rev Genet* 2006;7(2): 130–141.

12. Price ND, Reed JL, Palsson BO. Genome-scale models of microbial cells: evaluating the consequences of constraints. *Nat Rev Microbiol* 2004;2(11): 886–897.

13. Edwards JS, Covert M, Palsson B. Metabolic modelling of microbes: the flux-balance approach. *Environ Microbiol* 2002;4(3):133–140.

14. Covert MW, Famili I, Palsson BO. Identifying constraints that govern cell behavior: a key to converting conceptual to computational models in biology? *Biotechnol Bioeng* 2003;84 (7): 763–772.

15. Price ND, et al. Genome-scale microbial *in silico* models: the constraints-based approach. *Trends Biotechnol* 2003;21(4): 162–169.

16. Varma A, Palsson BO. Stoichiometric flux balance models quantitatively predict growth and metabolic by-product secretion in wild-type *Escherichia coli* W3110. *Appl Environ Microbiol* 1994;60(10): 3724–3731.

17. Kauffman KJ, Prakash P, Edwards JS. Advances in flux balance analysis. *Curr Opin Biotechnol* 2003;14(5): 491–496.

18. Neidhardt FC, Curtiss R. *Escherichia coli* and *Salmonella*: cellular and molecular biology, 2nd ed. Washington, DC: ASM Press, 2 v. xx, 1996, p. 2822.

19. Scheffler IE. *Mitochondria*. New York: Wiley-Liss, 1999,xiv p. 367.

20. Liolios K, et al. The Genomes On Line Database (GOLD) v.2: a monitor of genome projects worldwide. *Nucleic Acids Res* 2006;34(Database issue): D332–D334.

21. Consortium CSAA. Initial sequence of the chimpanzee genome and comparison with the human genome. *Nature* 2005;437(7055): 69–87.

22. Gibbs RA,et al. Genome sequence of the Brown Norway rat yields insights into mammalian evolution. *Nature* 2004;428(6982): 493–521.

23. Istrail S, et al. Whole-genome shotgun assembly and comparison of human genome assemblies. *Proc Natl Acad Sci USA* 2004;101(7): 1916–1921.

24. Kirkness EF, et al. The dog genome: survey sequencing and comparative analysis. *Science* 2003;301(5641): 1898–1903.

25. Stein L. Genome annotation: from sequence to biology. *Nat Rev Genet* 2001;2(7): 493–503.

26. Brent MR. Genome annotation past, present, and future: how to define an ORF at each locus. *Genome Res* 2005;15(12): 1777–1786.

27. Mao X, et al. Automated genome annotation and pathway identification using the KEGG Orthology (KO) as a controlled vocabulary. *Bioinformatics* 2005;21(19): 3787–3793.

28. Chen Z. Assessing sequence comparison methods with the average precision criterion. *Bioinformatics* 2003;19(18): 2456–2460.

29. Karp PD, Paley S, Romero P. The Pathway Tools software. *Bioinformatics* 2002;18(Suppl 1): S225–S232.

30. Cash P. Proteomics of bacterial pathogens. *Adv Biochem Eng Biotechnol* 2003;8393–115.

31. Taylor SW, Fahy E, Ghosh SS. Global organellar proteomics. *Trends Biotechnol* 2003;21 (2): 82–88.

32. Kanehisa M, et al. The KEGG resource for deciphering the genome. *Nucleic Acids Res* 2004;32(Database issue): D277–D280.

33. Karp PD, et al. The EcoCyc Database. *Nucleic Acids Res* 2002;30(1): 56–58.

34. Mewes HW, et al. MIPS: analysis and annotation of proteins from whole genomes. *Nucleic Acids Res* 2004;32(Database issue): D41–D44.

35. Christie KR, et al. Saccharomyces Genome Database (SGD) provides tools to identify and analyze sequences from *Saccharomyces cerevisiae* and related sequences from other organisms. *Nucleic Acids Res* 2004;32(Database issue): D311–D314.

36. Caspi R, et al. MetaCyc: a multiorganism database of metabolic pathways and enzymes. *Nucleic Acids Res* 2006;34(Database issue): D511–D516.

37. Karp PD, et al. Expansion of the BioCyc collection of pathway/genome databases to 160 genomes. *Nucleic Acids Res* 2005;33(19): 6083–6089.

38. Harris MA, et al. The Gene Ontology (GO) database and informatics resource. *Nucleic Acids Res* 2004;32(Database issue): D258–D261.

39. The Gene Ontology (GO) project in 2006 Nucleic Acids Res 2006;34(Database issue): D322–D326.

40. Serres MH, Goswami S, Riley M. GenProtEC: an updated and improved analysis of functions of *Escherichia coli* K-12 proteins. *Nucleic Acids Res* 2004;32(Database issue): D300–D302.

41. Coulton G. Are histochemistry and cytochemistry 'Omics'? *J Mol Histol* 2004;35(6): 603–613.

42. Arita M, Robert M, Tomita M. All systems go: launching cell simulation fueled by integrated experimental biology data. *Curr Opin Biotechnol* 2005,16(3): 344–349.

43. Huh WK, et al. Global analysis of protein localization in budding yeast. *Nature* 2003;425 (6959): 686–691.

44. Guda C Subramaniam S TARGET: a new method for predicting protein subcellular localization in eukaryotes. *Bioinformatics* 2005.

45. Fields S. High-throughput two-hybrid analysis. The promise and the peril. *Febs J* 2005;272(21): 5391–5399.

46. Deeds EJ, Ashenberg O, Shakhnovich EI. A simple physical model for scaling in protein–protein interaction networks. *Proc Natl Acad Sci USA* 2006;103 (2): 311–316.

47. Sprinzak E, Sattath S, Margalit H. How reliable are experimental protein–protein interaction data? *J Mol Biol* 2003;327(5): 919–923.

48. Palsson B. Two-dimensional annotation of genomes. *Nat Biotechnol* 2004;22(10): 1218–1219.

49. Beard DA, Liang SD, Qian H. Energy balance for analysis of complex metabolic networks. *Biophys J* 2002;83(1): 79–86.

50. Covert MW, ct al. Integrating high-throughput and computational data elucidates bacterial networks. *Nature* 2004;429(6987): 92–96.

51. Covert MW, Palsson BO. Constraints-based models: regulation of gene expression reduces the steady-state solution space. *J Theor Biol* 2003;221(3): 309–325.

52. Covert MW, Schilling CH, Palsson B. Regulation of gene expression in flux balance models of metabolism. *J Theor Biol* 2001;213(1): 73–88.

53. Covert MW, Palsson BO. Transcriptional regulation in constraints-based metabolic models of *Escherichia coli*. *J Biol Chem* 2002;277(31): 28058–28064.

54. Chvatal V. *Linear Programming*. New York: W.H. Freeman and Company, 1983.

55. Reed JL, Palsson BO. Genome-scale *in silico* models of *E. coli* have multiple equivalent phenotypic states: assessment of correlated reaction subsets that comprise network states. *Genome Res* 2004;14(9): 1797–1805.

56. Vo TD, Greenberg HJ, Palsson BO. Reconstruction and functional characterization of the human mitochondrial metabolic network based on proteomic and biochemical data. *J Biol Chem* 2004;279(38): 39532–39540.

57. Barrett CL, et al. The global transcriptional regulatory network for metabolism in *Escherichia coli* exhibits few dominant functional states. *Proc Natl Acad Sci USA* 2005;102(52): 19103–19108.

58. Neidhardt FC, Ingraham JL, Schaechter M. *Physiology of the Bacterial Cell*. Sunderland, MA: Sinauer Associates, Inc., 1990.

59. Hillier FS, Lieberman GJ. *Introduction to Mathematical Programming*. New York: McGraw-Hill, xv, 1990,p. 649.

60. Williams HP, NetLibrary Inc. *Model Building in Mathematical Programming*, 4th ed. New York: Wiley, 1999,p. 368.

61. Reed JL, et al. An expanded genome-scale model of *Escherichia coli* K-12 (iJR904 GSM/GPR). *Genome Biol* 2003;4(9): R54.

62. Reed JL, Palsson BO. Thirteen years of building constraint-based *in silico* models of *Escherichia coli*. *J Bacteriol* 2003;185(9): 2692–2699.

63. Edwards JS, Palsson BO. The *Escherichia coli* MG1655 *in silico* metabolic genotype: its definition, characteristics, and capabilities. *Proc Natl Acad Sci USA* 2000;97(10): 5528–5533.

64. Schilling CH, et al. Genome-scale metabolic model of *Helicobacter pylori* 26695. *J Bacteriol* 2002;184(16): 4582–4593.

65. Thiele I, et al. An expanded metabolic reconstruction of *Helicobacter pylori* (iIT341 GSM/GPR): an *in silico* genome-scale characterization of single and double deletion mutants. *J Bacteriol* 2005;187(16): 5818–5830.

66. Becker SA, Palsson BO. Genome-scale reconstruction of the metabolic network in *Staphylococcus aureus* N315: an initial draft to the two-dimensional annotation. *BMC Microbiol* 2005;5(1): 8.

67. Mahadevan R, et al. Characterization of metabolism in the Fe(III)-reducing organism *Geobacter sulfurreducens* by constraint-based modeling. *Appl Environ Microbiol* 2006;72(2): 1558–1568.

68. Borodina I, Krabben P, Nielsen J. Genome-scale analysis of *Streptomyces coelicolor* A3 (2) metabolism. *Genome Res* 2005;15(6): 820–829.

69. Feist AM, et al. Modeling methanogenesis with a genome-scale metabolic reconstruction of *Methanosarcina barkeri*. 2006;2(1): msb4100046-E1–msb4100046-E14.

70. Forster J, et al. Genome-scale reconstruction of the *Saccharomyces cerevisiae* metabolic network. *Genome Res* 2003;13(2): 244–253.

71. Duarte ND, Herrgard MJ, Palsson BO. Reconstruction and Validation of *Saccharomyces cerevisiae* iND750, a Fully Compartmentalized Genome-scale Metabolic Model. *Genome Res* 2004;14(7): 1298–1309.

72. Kuepfer L, Sauer U, Blank LM. Metabolic functions of duplicate genes in *Saccharomyces cerevisiae*. *Genome Res* 2005;15(10): 1421–1430.

73. Almaas E, Oltvai ZN, Barabasi AL. The activity reaction core and plasticity of metabolic networks. *PLoS Comput Biol* 2005;1(7): e68

74. Segre D, et al. Modular epistasis in yeast metabolism. *Nat Genet* 2005;37(1): 77–83.

75. Sheikh K, Forster J, Nielsen LK. Modeling hybridoma cell metabolism using a generic genome-scale metabolic model of *Mus musculus*. *Biotechnol Prog* 2005;21(1): 112–121.

76. Wiback SJ, Palsson BO. Extreme pathway analysis of human red blood cell metabolism. *Biophys J* 2002;83(2): 808–818.

77. Hucka M, et al. The systems biology markup language (SBML): a medium for representation and exchange of biochemical network models. *Bioinformatics* 2003;19 (4): 524–531.

78. Novere NL, et al. Minimum information requested in the annotation of biochemical models (MIRIAM). *Nat Biotechnol* 2005;23(12): 1509–1515.

79. Schilling CH, Palsson BO. Assessment of the metabolic capabilities of *Haemophilus influenzae* Rd through a genome-scale pathway analysis. *J Theor Biol* 2000;203(3): 249–283.

80. Blattner FR, et al. The complete genome sequence of *Escherichia coli* K-12. *Science* 1997;277(5331): 1453–1474.

81. Edwards JS, Palsson BO. Systems properties of the *Haemophilus influenzae* Rd metabolic genotype. *J Biol Chem* 1999;274(25): 17410–17416.

82. Tong AH,et al. Global mapping of the yeast genetic interaction network. *Science* 2004;303(5659): 808–813.

83. Famili I, et al. *Saccharomyces cerevisiae* phenotypes can be predicted by using constraint-based analysis of a genome-scale reconstructed metabolic network. *Proc Natl Acad Sci USA* 2003;100(23): 13134–13139.

84. Forster J, et al. Large-scale evaluation of *in silico* gene deletions in *Saccharomyces cerevisiae*. *Omics* 2003;7(2): 193–202.

85. Edwards JS, Palsson BO. Metabolic flux balance analysis and the *in silico* analysis of *Escherichia coli* K-12 gene deletions. *BMC Bioinformatics* 2000;1(1): 1.

86. Giaever G, et al. Functional profiling of the *Saccharomyces cerevisiae* genome. *Nature* 2002;418(6896): 387–391.

87. Hartwell L, Genetics. Robust interactions. *Science* 2004;303(5659): 774–775.

88. Baba T, et al. Construction of *Escherichia coli* K-12 in-frame, single-gene knockout mutants: the Keio collection. 2006;2 (1): pmsb4100050-E1–msb4100050-E11.

89. Glasner JD, et al. ASAP, a systematic annotation package for community analysis of genomes. *Nucleic Acids Res* 2003;31(1): 147–151.

90. Salgado H, et al. RegulonDB (version 4.0): transcriptional regulation, operon organization and growth conditions in *Escherichia coli* K-12. *Nucleic Acids Res* 2004;32(Database issue): D303–D306.

91. Covert MW, et al. Integrating high-throughput data and computational models leads to *E. coli* network elucidation. *Nature* 2004;429(6987): 92–96.

92. Palsson BO. *Systems Biology: Properties of Reconstructed Networks.* Cambridge University Press, 2006.

93. Segre D, Vitkup D, Church GM. Analysis of optimality in natural and perturbed metabolic networks. *Proc Natl Acad Sci USA* 2002;99(23): 15112–15117.

94. Segrc D, et al. From annotated genomes to metabolic flux models and kinetic parameter fitting. *Omics* 2003;7(3): 301–316.

95. Shlomi T, Berkman O, Ruppin E. Regulatory on/off minimization of metabolic flux changes after genetic perturbations. *Proc Natl Acad Sci USA* 2005;102(21): 7695–7700.

96. Burgard AP, Pharkya P, Maranas CD. Optknock: a bilevel programming framework for identifying gene knockout strategies for microbial strain optimization. *Biotechnol Bioeng* 2003;84(6): 647–657.

97. Schilling CH, Letscher D, Palsson BO. Theory for the systemic definition of metabolic pathways and their use in interpreting metabolic function from a pathway-oriented perspective. *J Theor Biol* 2000;203(3): 229–248.

98. Price ND, et al. Analysis of metabolic capabilities using singular value decomposition of extreme pathway matrices. *Biophys J* 2003;84(2): 794–804.

99. Papin JA, Price ND, Palsson BO. Extreme pathway lengths and reaction participation in genome-scale metabolic networks. *Genome Res* 2002;12(12): 1889–1900.

100. Papin JA, et al. The genome-scale metabolic extreme pathway structure in *Haemophilus influenzae* shows significant network redundancy. *J Theor Biol* 2002;215(1): 67–82.

101. Bell SL, Palsson BO. Expa: a program for calculating extreme pathways in biochemical reaction networks. *Bioinformatics* 21(8): 2005;1739–1740.

102. Allen TE, Palsson BO. Sequence-based analysis of metabolic demands for protein synthesis in prokaryotes. *J Theor Biol* 2003;220(1): 1–18.

103. Papin JA, et al. Reconstruction of cellular signalling networks and analysis of their properties. *Nat Rev Mol Cell Biol* 2005;6(2): 99–111.

104. Varma A, Palsson BO. Metabolic capabilities of *Escherichia coli*: II. Optimal growth patterns. *Journal of Theoretical Biology* 1993;165(4): 503–522.

105. Papin JA, Palsson BO. Topological analysis of mass-balanced signaling networks: a framework to obtain network properties including crosstalk. *J Theor Biol* 2004;227(2): 283–297.

106. Papin JA, Palsson BO. The JAK–STAT signaling network in the human B-cell: an extreme signaling pathway analysis. *Biophys J* 2004;87(1): 37–46.

107. Schilling CH, Palsson BO. Assessment of the Metabolic Capabilities of *Haemophilus influenzae* Rd through a Genome-scale Pathway Analysis. *J Theor Biol* 2000;203(3): 249–283.

108. Herrgard MJ, et al. Integrated Analysis of Regulatory and Metabolic Networks Reveals Novel Regulatory Mechanisms in *Saccharomyces cerevisiae*. *Genome Res* 2006;16(5): 627–635.

109. Gianchandani EP, et al. Matrix formalism to describe functional states of transcriptional regulatory systems. *PLoS Comput Biol* 2006;2(8): e101.

110. von Hippel PH. Biochemistry. Completing the view of transcriptional regulation. *Science* 2004;305(5682): 350–352.

111. Price ND, et al. Analysis of metabolic capabilities using singular value decomposition of extreme pathway matrices. *Biophys J* 2003;84(2 Pt 1): 794–804.

112. Price ND, Papin JA, Palsson BO. Determination of redundancy and systems properties of the metabolic network of *Helicobacter pylori* using genome-scale extreme pathway analysis. *Genome Res* 2002;12(5): 760–769.

113. Fong SS, et al. *In silico* design and adaptive evolution of *Escherichia coli* for production of lactic acid. *Biotechnol Bioeng* 2005;91(5): 643–648.

114. Oh YK, et al. Genome-scale reconstruction of metabolic network in *Bacillus subtilis* based on high-throughput phenotyping and gene essentiality data. *J Biol Chem* 2007;282 (39): 28791–28799.

115. Oliveira AP, Nielsen J, Forster J. Modeling *Lactococcus lactis* using a genome-scale flux model. *BMC Microbiol* 2005; 539.

116. Hong SH et al. The genome sequence of the capnophilic rumen bacterium *Mannheimia succiniciproducens*. *Nat Biotechnol*, 2004;22(10): 1275–1281.

117. Taylor SW, et al. Characterization of the human heart mitochondrial proteome. *Nat Biotechnol* 2003;21(3): 281–286.

7

MATHEMATICAL MODELING OF GENETIC REGULATORY NETWORKS: STRESS RESPONSES IN *Escherichia coli*

Delphine Ropers[1], Hidde de Jong[1], and Johannes Geiselmann[2]

[1]*INRIA Grenoble-Rhône-Alpes, 655 avenue de l'Europe, Montbonnot, 38334 Saint Ismier Cedex, France*
[2]*Laboratoire Adaptation et Pathogénie des Microorganismes, Université Joseph Fourier, UMR CNRS 5163, Bâtiment Jean Roget, Faculté de Médecine-Pharmacie, Domaine de la Mercie, La Tronche 38700, France*

7.1 INTRODUCTION

Living organisms have to adapt and respond to an ever-changing environment. The genes of the organism are the basis of both immediate responses to these changes and long-term evolutionary adaptation. In fact, the functional capabilities of an organism are the result of complex interactions between the gene products encoded by its genome, and cellular functions are therefore tightly linked to the regulation of gene expression.

We call genetic regulatory network the set of genes of an organism and the molecular components controlling gene expression. This control is generally exerted by proteins

Systems Biology and Synthetic Biology   Edited by Pengcheng Fu and Sven Panke
Copyright © 2009 John Wiley & Sons, Inc.

regulating the different stages of gene expression, that is, transcription, mRNA stability, translation, and protein degradation. The regulators not only include proteins but also signaling molecules allosterically modulating the activity of proteins and small RNAs regulating gene expression [1]. The ultimate effect of a regulator of gene expression is to modulate the concentration or activity of a gene product. The characteristic timescale of interactions within the genetic regulatory network is therefore generally determined by the speed of transcription and translation, ranging from several minutes for prokaryotes to several hours for higher eukaryotes.

The genetic regulatory network is connected to other cellular networks, such as signal-transduction cascades and metabolic pathways. The interactions within these networks are typically much faster than gene regulation: The average metabolic enzyme carries out a reaction cycle within milliseconds to microseconds, and most signal-transduction events also involve rapid covalent modifications, such as phosphorylation. Even though a complete description of the functioning of an organism will have to include all these networks and their interactions, the genetic regulatory network occupies a central position. Modifications of gene expression are at the very basis of developmental decisions and the response to a particular environment in the short term (adaptation) and long term (evolution). Moreover, due to differences in characteristic timescales, metabolic and signal-transduction pathways can often be seen as mediating indirect interactions on the genetic level. For instance, if a metabolite produced by a particular enzyme affects the activity of a transcriptional regulator, we can hide the molecular details of the (fast) enzymatic reactions and simply say that the gene coding for the enzyme indirectly regulates the activity of the regulator [2].

Genetic regulatory networks are the product of evolutionary processes that are better described as tinkering than engineering, in the words of François Jacob [3]. In fact, evolution does not work according to a preconceived plan, but achieves efficient performance by exploiting contingent events. It does not build an organism from scratch for a well-defined purpose, but modifies and reorganizes what is already available in order to meet arising environmental challenges. Notwithstanding these differences, many aspects of the structure and dynamics of biological systems can be compared with the principles governing man-made, engineered systems [4]. For instance, biological systems can be seen as being put together from reusable parts, assembled into modules, in much the same way as are man-made systems. Moreover, the question which aspects of the structure of a system allow it to reliably function over a range of environmental conditions, in the presence of noise, can be asked of biological and man-made systems alike. Not surprisingly therefore, living organisms have been fruitfully compared to airplanes [5] and genetic regulatory networks to electronic circuits [6]. In addition, the analogies between a signal-transduction pathway and a transistor radio have inspired some insightful comments on current biological research [7].

From the observation that the functioning of biological systems might be understood in much the same way as man-made systems, it is only a small step to applying traditional engineering methods to the study of cellular networks. This is one of the main inspirations of the emerging field of systems biology [8], and it also underlies our contribution. More specifically, the aim of this chapter is to review different approaches toward the mathematical modeling of genetic regulatory networks, from both a theoretical and a practical point of view. Much emphasis will be put on a point that is familiar to engineers, but is often forgotten when it comes to biological modeling: A model is not a faithful copy of reality, but a simplified representation adapted to a particular type of biological questions. Instead of a single modeling approach, we therefore need a multiplicity of approaches, each capturing a different aspect of the biological system under study.

In addition to the observation that adopting an engineering approach might lead to fresh insights into the functioning of biological processes, the parallels between biological and man-made systems can be pushed further by applying engineering methods to the design of genetic regulatory networks and their actual implementation in living cells. Such networks could be useful in a variety of ways as a test bed for the study of naturally occurring networks or as a vector for biotechnological and medical applications. The rise of synthetic biology [9,10] is the second major theme of this book and is addressed in a number of other contributions, such as the chapter by Fussenegger. In this chapter, we will describe modeling approaches that can be used not only for the analysis of genetic regulatory networks but also for their design. Although, in our examples, we focus on applications in the field of systems biology, many of the theoretical and practical considerations carry over to network design in synthetic biology.

In order to illustrate the different kinds of modeling approaches, as well as the kinds of questions that can be addressed by each of these, we focus in this chapter on one particular model system: *Escherichia coli*. Although its ecological niche is the human colon, this enterobacterium has turned out to be an excellent model system for biological research as it is capable of persisting in diverse environments, easy to manipulate in the laboratory, and evolutionarily close to many pathogenic bacteria. In this chapter, we will show what the different modeling approaches have taught us about the stress responses of *E. coli*, that is, the adaptation of the bacterium to a variety of stresses, such as a lack of essential nutrients, overcrowding, and temperature shocks [11,12]. The capability to respond to challenges arising from its environment is essential to the survival of the bacterium in the short and long term.

The stress responses of *E. coli* are controlled at the molecular level by a genetic regulatory network integrating various environmental signals. The network involves the interplay of numerous signal-transduction cascades, metabolic pathways, and gene expression interactions, which together control the reorganization of the bacterial physiology and metabolism in response to a given stress [11,12]. Although many of the molecular components of the networks have been identified, currently not much is known about how the interactions between these components give rise to the cellular response to external stresses. It is clear that, when dealing with networks of this size and complexity, intuitive reasoning about the dynamical behavior of the system quickly becomes infeasible or fraught with error. This motivates the use of modeling and simulation approaches to better understand the survival strategies of the bacterium.

More generally, due to the enormous amount of information that has been accumulated about cellular interaction networks [13], *E. coli* has become a system of choice for modeling and simulation studies. The first models on the molecular level of its response to particular nutrient shifts have appeared already in the early seventies (e.g., see Ref. [14]), while pioneering attempts to develop whole-cell models of *E. coli* adaptation appeared more than 20 years ago [15]. Recently, an International *E. coli* Alliance has been founded aiming at the coordination of modeling efforts so as to create an *in silico* cell corresponding to the bacterium [16,17].

In the remainder of this chapter, three different kinds of modeling formalisms are discussed: graphs, ordinary differential equations, and stochastic master equations. We summarize the mathematical basis of the formalisms as well as their application to the analysis of various *E. coli* stress–response networks. In particular, we investigate how these formalisms have helped address questions on (i) the structural decomposition of the stress–response network into modules and motifs, (ii) the existence of steady states and the dynamic response of the stress–response network to external perturbations, and (iii) the emergence of robust network behavior in the presence of intracellular and extracellular noise. In the concluding discussion we consider which questions are suitably addressed by each of the modeling formalisms and emphasize the point of model pluralism. For further information, the reader may wish to consult other reviews on the modeling of genetic regulatory networks [18–22].

7.2 GRAPH MODELS

7.2.1 Model Formalism and Analysis Techniques

Probably the most straightforward way to model a genetic regulatory network is to view it as a graph. Formally, a *graph* is defined as a tuple (V, E), with V indicating a set of *vertices*, and $E \subseteq V \times V$ indicating a set of *edges* [23] as follows:

$$G = (V, E) \tag{7-1}$$

The edges represent the relation between vertices and may be directed or undirected. A *directed* edge is a pair $(i, j) \in E$ of vertices, where i denotes the head, and j denotes the tail of the edge. (i, j) is an *undirected* edge if the order of the vertices is of no importance. The vertices of a graph correspond to genes or other elements of interest in the cell, while the edges denote interactions among the genes. In the case of directed graphs, edges point from regulating to regulated genes, for example, from genes encoding transcription factors to the targets of the transcription factors. The graph representation of a genetic regulatory network can be generalized in several ways. For instance, the vertices and edges could be labeled, by adding information about genes and their interactions. Defining a directed edge as (i, j, s), with s equal to $+$ or $-$, allows one to indicate whether i is activated or inhibited by j, respectively.

An example of a simple directed graph model is shown in Figure 7-1. It consists of three genes, connected by labeled interactions that indicate whether a gene positively or negatively regulates the expression of its target. Many of the pictures of biological networks found in the literature can be mapped to some sort of graph representation. Two particularly impressive examples are the mammalian cell-cycle control

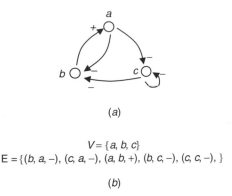

(a)

$$V = \{a, b, c\}$$
$$E = \{(b, a, -), (c, a, -), (a, b, +), (b, c, -), (c, c, -), \}$$

(b)

Figure 7-1 (a) Directed, labeled graph representing a genetic regulatory network composed of three genes (a, b, and c) and their mutual regulatory interactions. The symbols + and − indicate whether the regulator gene activates or inhibits its target. (b) Formal definition of the graph in (a).

network [24] and the network regulating endomesoderm specification in the sea urchin [25].

The representation of a genetic regulatory network as a graph allows the analysis of its *topological* properties by means of graph–theoretical techniques [26,27]. The global connectivity properties of the network can, for instance, be described by the average degree and the degree distribution of the vertices. The *degree k* of a vertex indicates the number of edges to which it is connected (if necessary, incoming and outgoing edges can be distinguished). $\langle k \rangle$ denotes the *average degree* and $P(k)$ denotes the *degree distribution* of the graph. These properties give an indication of the complexity of the graph and allow different types of graphs, and therefore networks, to be distinguished (Fig. 7-2). In classical *random graphs* (Fig. 7-2a), also called Erdős-Rényi graphs, the probability that a given vertex has k edges follows a Poisson distribution $P(k)$. That is, the vertices typically have $\langle k \rangle$ edges, and the vertices having significantly more or less edges than $\langle k \rangle$ are extremely rare, as shown in part (c) of the figure. By contrast, in *scale-free graphs* (Fig. 7-2b), the vertex degrees obey a power-law distribution $P(k) \sim k^{-\gamma}$, shown in part (d) of the figure. Scale-free graphs are inhomogeneous, in the sense that most of the vertices have few edges, whereas some vertices, called *hubs*, have many edges and hold the graph together.

For values of the degree exponent γ between 2 and 3, scale-free graphs have a number of surprising properties. First, the average length of the path between two vertices of the graph is proportional to log log |V|, where |V| denotes the number of vertices of the graph [26,27]. This is even shorter than the average path length in random graphs, which scales as log |V| and confers them the *small-world property* [28]. The small-world property implies that local perturbations can quickly spread out through the entire network. Second, the presence of hubs makes scale-free graphs robust against accidental failures [29–31]. Whereas randomly removing a certain number of vertices disintegrates a random graph, in a scale-free graph this mainly affects the numerous low-degree vertices, the absence of which does not decompose the graph. Finally, a scale-free topology may also confer robustness to the dynamical

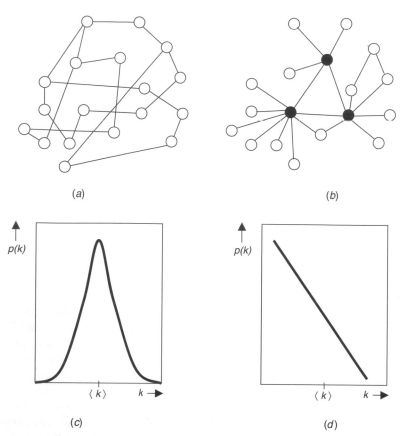

Figure 7-2 Schematic illustration of the architecture of (a) random and (b) scale-free undirected graphs [170]. The degree distribution follows (c) a Poisson distribution in random graphs and (d) a power-law distribution in scale-free graphs. k denotes the degree of a vertex and $P(k)$ denotes the degree distribution. The filled vertices in (b) are hubs.

properties of the network [32], suggesting that the latter are relatively insensitive to the precise values of the parameters (Section 3.1).

The relation between the scale-freeness of a graph and such fundamental properties of living systems as reactivity and robustness makes this type of graph interesting as a model of genetic regulatory networks. In recent years quite some evidence has accumulated, showing that genetic regulatory networks, and many other biological and nonbiological networks, are indeed scale-free [30,31,33–39]. The results should be interpreted with some care though. Because current data on regulatory interactions are incomplete, only subnetworks of the actual networks can be analyzed, which may have a different degree distribution [40,41]. Moreover, the particular graph representation chosen to model the network may bias the results, as shown by Arita for the *E. coli* metabolic network [42]. Further, in the case of genetic regulatory networks, graph models are usually restricted to direct transcriptional regulatory interactions, thus ignoring indirect interactions that are mediated by metabolites binding to transcription regulators [43].

Another example of the use of graph analysis is to study the structural decomposition of a graph into subgraphs. Here, we focus on two kinds of subgraphs: modules and motifs. Informally speaking, a *module* is a (possibly hierarchically structured) cluster of vertices, such that the vertices within a module are strongly connected, while the connections between modules are much looser. The formalization of this intuition can be achieved by means of different graph–theoretical concepts, for instance clustering coefficients [44], shortest path distances [45], and edge betweenness, denoting the number of shortest paths between pairs of vertices that run through an edge [46]. On a different level of granularity, *motifs* are small subgraphs, consisting of a few vertices only, which frequently recur in the graph [47–49]. More precisely, motifs are defined with respect to a statistical background consisting of a randomized version of the graph: a small subgraph is called a motif if it occurs significantly more often in the original graph than in the randomized graphs.

The interest of the structural decomposition of a graph into modules and motifs is that the latter may correspond to a particular function of the genetic regulatory networks. Some results validating this intuition will be presented below, in the case of the *E. coli* transcriptional regulatory network [50,51]. One should be careful in interpreting the results of such graph analyses though. As mentioned above, currently available data are incomplete and specific modeling choices may introduce a bias. Moreover, this point needs to be emphasized, the relation between topological concepts like modules and motifs on the one hand, and the functioning of biological systems on the other, is far from straightforward. Consider the example of a module.

Even if some of the genes in the module are known to play a role in a particular biological function, this may not be sufficient for concluding that the module is responsible for the function. For instance, some of the interactions between the genes in the network may not be operative at all under the physiological conditions where the function is called upon. One could counter this objection to some extent by integrating other kinds of data in the process of module identification, such as transcriptome data, phylogenetic profiles, and biological sequences (e.g., see Refs. [52–59]). This certainly allows for more refined answers to the question *which* genes and interactions are relevant for a particular biological function. However, by itself it does not explain *how* the function emerges from the genes and interactions in the module. In order to deal with the latter questions, we need dynamical models of the kind discussed in later sections of this chapter.

In summary, graph models of genetic regulatory networks allow one to address questions concerning the network topology, giving insights into global structural properties like the edge distribution. In addition, they enable the identification of local substructures like modules and motifs that may be related to the functioning of the biological system. Graph models are applicable to genome-scale models, and the computer tools to support the analysis exist, such as mfinder [60] and TopNet [41]. However, in order to clarify the relation between the topological properties of graph models and the functioning of genetic regulatory networks, more powerful dynamical models are necessary. The capabilities and limitations of graph models will be illustrated in the next section.

7.2.2 Modules and Motifs in the Transcriptional Regulatory Network of *E. coli*

Transcription factors are key components in the control of the *E. coli* stress responses, in that they link the sensing of environmental changes to the reorganization of the pattern of gene expression, and thus to the control of metabolic pathways. Depending on the environmental conditions, different sets of transcription factors are used by the bacterium. We would therefore like to ask such questions as: Can we relate the stress adaptation capabilities of *E. coli* to the topological organization of its transcriptional regulatory network? More precisely, are the different sets of genes organized in modules, for example, can we define a carbon utilization module and a nitrogen assimilation module? And more generally, how can we define such modules and detect them from a graph model of the regulatory dependencies of the genes of the bacterium?

The topological analysis of the *E. coli* network has been much facilitated by the rich store of information about the components of the network and their interactions, which are published in the literature or stored in databases, like RegulonDB [61] or EcoCyc [62]. Hence, several studies aiming at the analysis of structural properties of genetic regulatory networks by means of the approaches mentioned in the previous section have exploited the information on the *E. coli* network stored in RegulonDB [33,49,63–65].

An example of the search for modules is the study by Resendis-Antonio and collaborators [65]. Considering only genes for which experimental evidence on their involvement in regulatory interactions is available, the authors analyzed a network composed of 55 transcription factors controlling the expression of 747 genes. The relations between the genes in the network of transcription factors were determined by computing the shortest path distance for every pair of genes. Based on this information, eight topological modules were identified using a clustering approach. Further analysis revealed that the modules are composed of functionally related genes, for example, involved in (i) respiration, (ii) stress response, and (iii) chemotaxis, motility, and biofilm formation. The largest module (iv) gathers genes involved in the assimilation of the various carbon sources. The remaining modules are composed of genes involved in various cellular responses, like (v) sulphur assimilation, (vi) nitrogen metabolism, (vii) fermentative conditions, and (viii) chromosome replication.

The topological analysis of Resendis-Antonio *et al.* suggests that the *E. coli* network possesses a modular structure, where each module consists of genes that perform one or more tasks in response to particular environmental conditions. For instance, the *nac* and *asnC* genes, coding for transcription factors known to be involved in the control of nitrogen assimilation, are found inside the same module [65]. Further, the carbon assimilation module includes transcription factor genes like *crp*, *araC*, *malF*, *fruR*, which regulate the utilization of carbon sources. Interestingly, the latter module can be further decomposed into submodules, each submodule being specialized in the use of a different carbon source. One can guess that this supplementary internal organization makes *E. coli* cells able to easily grow on various carbon sources and switch from the use of one carbon source to another. Other analyses have found a similar modular structure of the *E. coli* transcriptional regulatory network, though using different approaches and arriving at a different number of modules [59,63].

Several questions regarding the *E. coli* network structure remain open. For instance, how is the global coordination of cellular responses to be explained? Most often, a stress response does not involve a single module but rather a combination of modules. For instance, *E. coli* uses its motility, controlled by module (iii), to seek optimal oxygen concentrations, required for the respiration task performed by module (i). It seems obvious that accomplishing this function requires a connection between the two modules, which agrees with the fact that, generally, the modules are not clearly separated from the rest of the network but tend to overlap [65]. Can these interconnections be characterized by means of certain topological properties?

To address this question, we need to take into account the local topology of networks, defined in terms of motifs. Using information from RegulonDB and the literature, Shen-Orr et al. [49] have analyzed the transcriptional regulatory network of *E. coli*. They found that in this network, consisting of 855 genes and 1330 regulatory interactions, three different motifs occur more frequently than expected: the *feedforward loop*, in which a transcription factor regulates a second transcription factor and both regulate together a target gene; the *single-input* motif, in which a group of genes is controlled by a single transcription factor; and the *dense overlapping regulons*, in which genes and the transcription factors controlling their expression form a highly overlapping structure. The feedforward loop is the motif occurring most frequently (40 times) in the *E. coli* network. This has been subsequently confirmed by means of an extended version of the same network, in which an even higher number of feedforward loop motifs were found [63]. The different motifs are not equally distributed in the network of *E. coli*. In the above-mentioned study, Resendis-Antonio and collaborators found that the feedforward loop motifs are mainly located inside modules (71 percent of the cases), whereas the *bifan* motif (which forms the basic building block of the above-mentioned dense overlapping regulons) is the main motif connecting modules (65 percent of the cases).

What is the advantage for the cell of conserving certain network motifs? Do they have a functional role, in addition to their structural role? The group of Alon demonstrated both theoretically and experimentally the information-processing task carried out by the coherent feedforward loop. Using a differential equation model of the feedforward loop motif, they showed that its role might be to filter out fluctuations in input stimuli and allow a rapid response when the stimuli disappear [66,67]. Consider the coherent feedforward loop motif in Figure 7-3, where the transcription factors X and Y together activate the gene *z*. When X is active and above a threshold concentration, the input signal activating X is transmitted to the output Z through a direct path from X and an indirect path from X through Y. Hence, a transient signal is not transmitted to Z, since it does not allow the concentration of Y to reach a threshold level high enough to stimulate the expression of gene *z* (Fig. 7-3). On the contrary, a persistent input signal enables the concentration of Y to rise and eventually

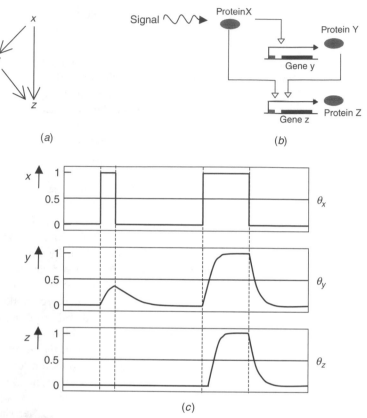

Figure 7-3 (a) Coherent feedforward loop motif in a graph representation. (b) Feedforward loop in a genetic regulatory network, where it is assumed that both X and Y are necessary for expression of z. (c) Dynamic properties of the feedforward loop [49]; x, y, and z denote the concentrations of X, Y, and Z, respectively, while θ_x, θ_y, and θ_z denote their threshold levels. The input signal activates X.

allows Z to pass its threshold level. The functioning of the feedforward loop motif is asymmetric, since the inactivation of X leads to the rapid downregulation of z. The above predictions have been experimentally verified for the L-arabinose utilization system in *E. coli* using reporter genes [67]. In this feedforward loop motif, CRP corresponds to the general transcription factor X and AraC to the specific transcription factor Y, while z is the operon *araBAD*.

The study by the group of Alon assigns a function to a pattern of interactions, the coherent feedforward loop, which is overrepresented in the transcriptional regulatory network of *E. coli*. However, the relation between structure and function is not straightforward, given that motifs do not usually occur in isolation, but rather overlap to generate *motif clusters* [33]. Does the function of a motif change when it is embedded within a network and interacting with many other components? The group of Alon partially answered this question in a subsequent study on the *incoherent feedforward loop*, that is, a feedforward loop in which the transcription factor X activates genes y and z, while Y represses z [68]. The motif was experimentally shown to perform response acceleration, as predicted by a differential equation model [66], despite the fact that it participates in additional interactions that were not included in the model. However, it is not sure that this will turn out to be true in general.

7.3 ORDINARY DIFFERENTIAL EQUATION MODELS

7.3.1 Model Formalism and Analysis Techniques

As concluded in the previous section, a better comprehension of the relation between the structure and functioning of a regulatory system requires the use of dynamical models. *Ordinary differential equations* [69] are probably the most-widespread formalism for modeling the dynamical behavior of cellular interaction networks. In this formalism, the concentrations of gene products (mRNAs or proteins) are represented by continuous, time-dependent variables, $x(t)$, $t \in T$ and T being a closed time interval ($T \subseteq \mathbf{R}_{\geq 0}$). The variables take their values from the set of nonnegative real numbers ($x: T \rightarrow \mathbf{R}_{\geq 0}$), reflecting the constraint that a concentration cannot be negative.

The regulatory interactions between genes are modeled by a system of ordinary differential equations having the following general form:

$$\mathrm{d}x_i/\mathrm{d}t = f_i(x), \quad i \in \{1, \dots, n\}, \tag{7-2}$$

where $x = (x_1, \dots, x_n)'$ is the vector of concentration variables of the system, and the usually highly nonlinear function $f_i: \mathbf{R}_{\geq 0}{}^n \rightarrow \mathbf{R}_{\geq 0}$ represents the regulatory interactions. The above system of equations describes how the temporal derivative of the concentration variables depends on the values of the concentration variables themselves. Several variants of Equation 7-2 can be imagined [22]. For instance, by taking into account input variables u, it becomes possible to express the dependence of the temporal derivative on external factors, such as the presence of nutriments. In order to account for the delays resulting from the time it takes to complete transcription, translation, and the other stages of the synthesis and the transport of proteins, Equation 7-2 could be changed into a system of delay differential equations.

In Figure 7-4, an example of a simple genetic regulatory network and its associated differential equation model is shown, based on early work by Goodwin [70,71]. The end product of a metabolic pathway coinhibits the expression of a gene coding for an enzyme that catalyzes a reaction step in the pathway. This gives rise to a negative feedback loop involving the mRNA concentration x_1, the enzyme concentration x_2, and the metabolite concentration x_3. The equations each express a balance between the increase and decrease of the molecular concentration per unit time due to the occurrence of the various reactions. More precisely, the equations describe the rate of synthesis of the enzyme ($k_2 x_1$) and the metabolite ($k_3 x_2$), as well as the rate of synthesis of mRNA ($k_1 r(x_3)$). The nonlinear, sigmoidal Hill function r expresses that the rate of synthesis of mRNA depends in a cooperative way on the concentration of the metabolite, which binds and thereby activates a repressor of the gene (Fig. 7-4b). The terms $-g_1 x_1$, $-g_2 x_2$, and $-g_3 x_3$ indicate that the concentrations x_1, x_2, and x_3 decrease through degradation and growth dilution, at a rate proportional to the concentrations themselves.

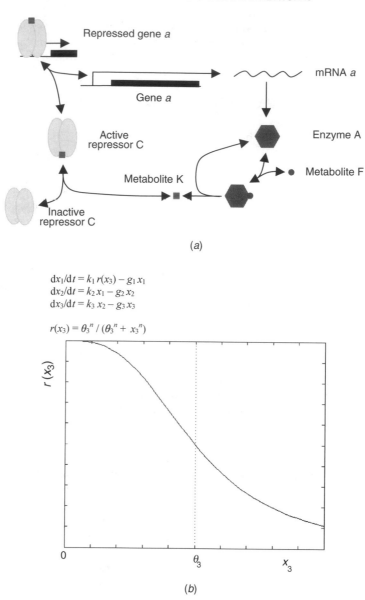

$$dx_1/dt = k_1\, r(x_3) - g_1\, x_1$$
$$dx_2/dt = k_2\, x_1 - g_2\, x_2$$
$$dx_3/dt = k_3\, x_2 - g_3\, x_3$$

$$r(x_3) = \theta_3^{\,n} / (\theta_3^{\,n} + x_3^{\,n})$$

(b)

Figure 7-4 (a) Simple example of gene regulation involving end-product inhibition and (b) the corresponding differential equation model. A is an enzyme and C a repressor protein, while K and F are metabolites. x_1, x_2, and x_3 represent the concentrations of mRNA a, protein A, and metabolite K, respectively, k_1, k_2, k_3 are production constants, g_1, g_2, g_3 degradation constants, and $r: R_{\geq 0} \rightarrow R_{\geq 0}$ is a decreasing Hill function ranging from 0 to 1, with threshold parameter θ and exponent n. All parameter values are positive and $n > 1$, in order to obtain the sigmoidal shape characteristic of cooperative interactions.

A first dynamical property that can be studied by means of ordinary differential equation models is the *asymptotic behavior* of the system, notably the occurrence of equilibrium points and limit cycles, as well as their stability and basin of attraction. The equilibrium points and limit cycles may correspond to functional modes of the systems, for instance a particular growth stage or a particular response of the cell to an external stress. The equilibrium points are simply determined by setting every dx_i/dt given in Equation 7-2 to 0 and solving for x_i. In the example of the end-product inhibition network, we thus obtain a single equilibrium point [72]. This follows from Equation 7-2 by noting that at equilibrium $x_1 = (k_1/g_1)\, r(k_2 k_3 x_1/g_2 g_3)$, and bearing in mind that r is a monotonically decreasing function (Fig. 7-5a).

The stability of the equilibrium point x^* can be determined by linearizing the system of differential equations given in Equation 7-2 around x^*, computing the characteristic equation, and solving for the eigenvalues. The sign of the (real part of the) eigenvalues then determines the stability of the system [69,73]. The characteristic equation for the end-product inhibition network is given by $(\lambda + g_1)\,(\lambda + g_2)\,(\lambda + g_3) - k_1 k_2 k_3\, \partial r(x_3^*)/\partial x_3 = 0$, where $x^* = (x_1^*, x_2^*, x_3^*)'$. The equation can be rewritten as a third-order polynomial whose roots λ are the eigenvalues. Depending on the exact numerical values of the parameters, different configurations of eigenvalues are found, notably (i) three negative real eigenvalues, (ii) a negative real eigenvalue and two conjugate complex eigenvalues with negative real part, or (iii) a negative real eigenvalue and two conjugate complex eigenvalues with positive real part. In the former two cases, the equilibrium point is asymptotically stable, meaning that after a (small and temporary) perturbation the system will eventually return to the equilibrium point. In contrast, in the third case the equilibrium point is unstable: a perturbation will cause the system to diverge from the equilibrium point and approach a stable limit cycle, corresponding to sustained oscillations in the protein concentrations. Figure 7-5b illustrates case (ii) for arbitrary but not unrealistic parameter values.

A second dynamical property of interest is the *transient behavior* of the system. The transient behavior provides information on the manner in which the genetic regulatory network controls the response of the system to an external perturbation, for example, by switching from one functional mode to another. In order to predict the transient behavior, we need to compute the solutions of the system of ordinary differential equations 7-2. Since the models of most genetic regulatory networks of practical interest are nonlinear, it is usually not possible to find an analytical solution. This means that in all but the simplest cases we have to resort to *numerical simulations* [74], which yield approximations of the exact solutions. The solutions obtained by simulation can be visualized by plotting their trajectories in the phase space, for two or three-dimensional systems, or by simply plotting the solutions as a function of time. This is illustrated in Figure 7-5b and c for the model of the end-product inhibition network. The plots show how the system adapts to a perturbation from its steady state, by returning to this state through damped oscillations.

The analysis of the feedback inhibition network shows that it is a homeostatic system, with a tendency to maintain a stable steady state or stable oscillations. The negative feedback loop, arising from the (indirect) inhibition of the expression of the gene by its own product, tends to compensate for a transient perturbation. Examples of

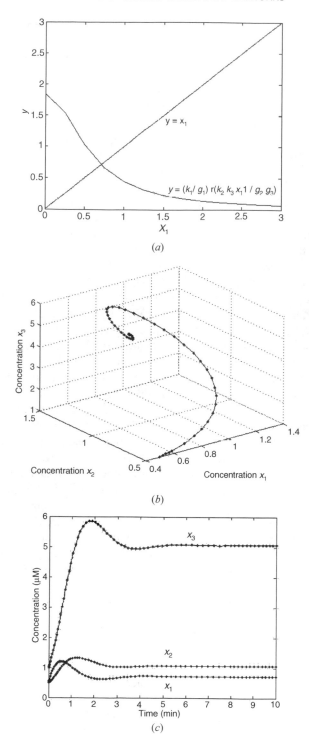

(a)

(b)

(c)

negative feedback loops abound in biological systems and play an important role in gene expression, metabolism, and signal transduction (e.g., see Refs. [75–77]). More generally, Thomas conjectured that negative feedback loops are a prerequisite for homeostasis [78,79]. In a similar vein, he proposed that positive feedback loops are a necessary condition for the occurrence of multiple steady states, corresponding to different functional modes of the system. Several proofs of the latter conjecture have been given under increasingly general conditions [80–84]. These results illustrate the potential of mathematical models to highlight fundamental relations between the topology and the dynamics of regulatory networks.

The qualitative dynamics of the end-product inhibition network, the stability of the equilibrium point and the occurrence of a stable limit cycle, are determined by the values for parameters in the differential equation model in Figure 7-4. For large ranges of parameter values, the qualitative dynamics of the system remains invariant, that is, the qualitative dynamics is *robust* to fluctuations in the parameter values. This robustness is an essential property of living systems, which have to cope with continuous fluctuations in physiological and environmental conditions as well as with genetic variability. Following pioneering work by Savageau (1971), the study of the robustness of dynamical properties of regulatory networks to changes in the parameter values has been an active research area, demonstrating robust behavior of the chemotaxis system of *E. coli* [85,86], the development of the *Drosophila* embryo [87,88], the *Xenopus* cell cycle [89], and the circadian clock of *Drosophila* [90,91]. In the case of synthetic networks, the ability of the system to reliably function in the presence of noise is an important design objective. Control theory provides a range of methods that could be used to assess the robustness of naturally occurring networks and improve the robustness of synthetic networks (e.g., see Refs. [91–94]).

Although differential equation models allow making precise, quantitative predictions on the dynamics of large and complex genetic regulatory networks, they may be difficult to apply in practice. Most regulatory networks of interest are large and complex, possibly involving hundreds of genes, proteins, and other molecules. If these networks were to be modeled in the same way as the simple autoinhibition network in Figure 7-4, we would obtain huge models that cannot be analyzed other than by massive numerical simulations. Apart from the fact that such simulations may be difficult to carry out, given that numerical values for the parameters are often not available (see below), it is not sure that the generation of time-course predictions of hundreds of molecular components will be of much help in gaining a better understanding of the functioning of the system. This has stimulated an interest in strategies for model simplification, often based on indications that the networks have a modular structure (Section 7.2). In order to study large and complex networks, it may be more

◄─────────────────────────────────

Figure 7-5 (a) The differential equation model of the end-product inhibition network in Fig. 7-4 has a single equilibrium point. The stability of this equilibrium point varies with the parameter values. (b) Asymptotically stable equilibrium point with a trajectory spiraling toward this point. (c) Time-series representation of the solution. The parameter values are as follows: $k_1 = 4.6\,\mu$M/min, $k_2 = 1.8$/min, $k_3 = 10$/min, $g_1 = 2.5$/min, $g_2 = 1.2$/min, $g_3 = 2.1$/min, $\theta = 4\,\mu$M, and $n = 2$. The simulations have been carried out with Matlab.

judicious to first analyze the modules individually, and only afterward the question of how they are woven together, preferably using simpler and more abstract models for this second step. The definition of network modules may be based on topological criteria, not unlike those used for graph models but usually more directly relevant to the dynamics of the system, such as the feedback structure of the network [95–97]. Another way to define modules is based on the distinction between rapid and slow processes in the system, for example, allowing the separation of metabolism and gene expression in separate modules [98,99].

Even after model simplification, for most networks we will be left with large and complex models. Their analysis requires quantitative information on the values of kinetic constants and molecular concentrations, but unfortunately this information is only rarely available, especially when modeling systems on the forefront of experimental research. Several ways to deal with this problem have been proposed in the literature. First of all, pushing the robustness argument further, one could argue that important dynamical properties of actual regulatory networks do not so much depend on particular molecular mechanisms or precise values for the parameters, but rather on the network topology. A second strategy is to try to estimate the parameter values from experimental data [100]. The use of these techniques has been shown to work well on small to medium-sized systems, in cases where the interactions are well described by linear or quasilinear functions (e.g., see Refs. [101–104]). A third way out would be turn to simplified models, having a particular mathematical form that simplifies their analysis [105,106]. Examples of such models are the piecewise-linear differential equation models proposed by Glass and Kaufmann [107] or the logical models proposed by Kauffman [108] and Thomas [109,110].

In conclusion, differential equation models allow questions related to the transient or asymptotic dynamics of genetic regulatory networks to be answered. Many examples of their application exist, some of which will be discussed later in the context of the *E. coli* stress response. Techniques for the mathematical analysis and numerical solution of differential equation models are standard engineering tools, and a large variety of computer programs are available, ranging from general-purpose mathematical problem solvers like Matlab to tools specifically adapted to the analysis of cellular interaction networks such as Copasi [111], ProMoT/DIVA [112], Virtual Cell [113], and XPPAUT [114].[1] Due to the size and complexity of networks of practical interest as well as the lack of precise, quantitative information on the molecular mechanisms and kinetic constants, standard techniques for numerical analysis may be difficult to apply in practice. Several strategies to cope with this problem have been proposed, some of which will be illustrated in the next section.

7.3.2 Response of *E. coli* to Carbon-Source Availability

As seen in Section 2.2, the transcriptional regulatory network of *E. coli* contains a carbon assimilation module, allowing the bacterium to use a large range of carbon

---

[1] The SBML format [115] allows models to be exchanged between different computer tools. A list of computer tools compatible with the SBML format is available at http://www.sbml.org.

sources under a variety of conditions. For instance, when several carbon sources are available, the bacteria choose the "best" nutrient, meaning the nutrient sustaining fastest growth. Hence, if *E. coli* is presented with two carbon sources, for example glucose and lactose, it starts using glucose until this preferred nutrient is depleted from the medium. Growth then temporarily arrests while the bacterium modifies its pattern of gene expression so as to produce the enzymes necessary for the uptake and metabolism of lactose. This physiological response is referred to as *diauxic growth*. When all carbon sources in the growth medium have become depleted, *E. coli* bacteria are no longer able to sustain fast growth rates and enter into a *stationary phase* of growth, characterized by no net change of the size of the bacterial population. In response to carbon starvation, the bacteria completely modify their physiology to cope with the absence of nutrients. This implies that they conserve energy by shutting off most biosynthetic functions and protect their DNA from potential damage, while at the same time maintaining a minimal metabolism in order to explore potential alternative nutrient sources and "be ready" as soon as nutrients become available again.

Given the numerous tasks ensured by the carbon assimilation module, several questions arise regarding its functioning: How does the module coordinate the different responses of *E. coli* cells to carbon-source availability? How does the reorganization of gene expression and metabolism emerge from the interactions between the many components making up the regulatory network of *E. coli*? Can we develop comprehensive dynamic models of these interactions that account for the bacterial responses to carbon-source availability?

In the remainder of this section, we give two examples of differential equation models, describing, respectively, the diauxic growth of *E. coli* and its response to carbon starvation. Even though *E. coli* is a well-studied system, the development of the models has been limited by the lack of quantitative information on most of the molecular concentrations and the kinetic parameters characterizing the interactions inside the carbon assimilation module. To overcome these constraints, different strategies were chosen, one based on the estimation of parameter values from experimental data [116] and the other on a more abstract description of the network [117].

The group of Gilles has developed a dynamical model describing the successive assimilation of different carbon sources in *E. coli* (glucose, lactose, etc.), leading to diauxic growth [116]. The central part of the regulatory network controlling this process is a large, membrane-bound enzyme complex called *Phospho-Transferase System (PTS)*. The PTS transfers a phosphate onto the carbohydrates (e.g., glucose), which makes the transport irreversible and prepares the carbohydrate for metabolic breakdown and conversion into cellular energy. In the absence of glucose, the same complex activates another membrane-bound enzyme, adenylate cyclase (Cya). Cya produces a signaling molecule, cyclic adenosine mono-phosphate (cAMP), which in turn binds a transcription factor, CRP (cAMP receptor protein) and enables the latter to activate or inhibit transcription. The promoter of the lactose operon is one of the targets activated by cAMP–CRP. The same promoter is also under the negative control of the *lac* repressor. This transcription factor is inactivated by a metabolite, allolactose, which is produced in the presence of lactose. This allows derepression of the transcription of the operon and the subsequent use of lactose as a carbon source.

The model by Bettenbrock and colleagues is the last in a series of detailed models of the carbon assimilation module developed by the group of Gilles [118–121]. While other models of the same system are available in the literature (e.g., see Refs. 122–124), the Bettenbrock model provides the most comprehensive picture to date. It describes the PTS and its interactions with several uptake systems and metabolic pathways, accounting for the growth of *E. coli* on different carbohydrates. The network is composed of different types of interactions, involving genetic regulatory interactions, metabolic reactions, and reactions involved in the signal-transduction pathway. These were modeled by ordinary differential equations of the form described in Section 3.1, using kinetic rate laws appropriate for each type of interaction, and algebraic equations expressing conservation relations among the different molecular components of the system. In total, the model is composed of 50 differential equations and 14 algebraic equations.

Even though the network controlling diauxic growth is a well-characterized system, it was not always possible to include parameter values reported in the literature in the model, as they are often obtained under different experimental conditions and with different strains. To circumvent this problem, Bettenbrock and colleagues have carried out their own experiments, measuring the concentration of the various metabolites over time and have used the resulting data to estimate the value of the model parameters by means of the ProMoT/Diva environment [112]. In this way, some fifty uncertain or unknown parameter values could be obtained.

By means of the resulting numerical model, *E. coli* growth on various carbohydrates was simulated (see Fig. 7-6a, for example). The confrontation of these predictions with time-series measurements performed under the experimental conditions corresponding to the simulations revealed a number of contradictions that required model revision. For instance, the model could not account for the behavior of the system during disturbed batch experiments, consisting of the exponential growth of the cell on a carbon source (glycerol or lactose), followed by the application of a pulse of glucose. Although the simulations showed glucose uptake, as observed experimentally, the process was predicted to proceed too fast (Fig. 7-6b). The inclusion into the model of the regulation of the *pts* operon allowed a much better fit of the experimental data with the model predictions. In this instance, the model not only confirmed what is currently known about the accumulation of carbon sources by *E. coli* but also provided novel explanations of the role of certain network components in the process. Hence, the cAMP metabolite appears to play a key role in the short-term adaptation to a new carbon source during diauxic growth, whereas the complex cAMP–CRP seems to be more important for long-term adaptation.

The model of Bettenbrock and colleagues provides a detailed and rigorous description of the molecular events underlying diauxic growth. However, the model does not address the functioning of the carbon assimilation module in the broader context of the genetic regulatory network of *E. coli*. For instance, it is known that the PTS is closely connected to some major transcription regulators of the bacterium called *global regulators*. These transcription factors control the expression of large sets of genes in response to environmental stimuli [125,126]. More precisely, the PTS transfers information on the lack of carbon source to the global regulators, which reorganize gene expression and allow the bacteria to stop exponential growth and enter

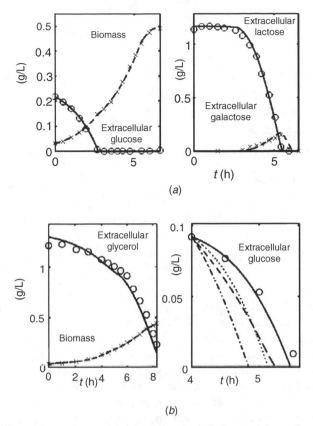

Figure 7-6 Differential equation model of the carbon assimilation module: confrontation of model predictions and experimental measurements [116]. The circles denote measurements, and the lines denote simulation results. The biomass and the extracellular concentrations of carbohydrates are mentioned on the curves. (a) Diauxic growth on glucose and lactose. Galactose is a product of lactose metabolism. (b) Disturbed batch experiment with application of a pulse of glucose on bacteria growing on glycerol. Dashed lines denote simulation results from different versions of the model that do not take into account regulation of the *pts* operon expression. Simulation of the model including this additional regulatory interaction results in the continuous line.

stationary phase upon carbon starvation. How does the growth adaptation of *E. coli* emerge from the interactions between the global regulators in response to a carbon starvation signal transmitted by the PTS?

In order to address these questions, we have developed an initial, simple model of the network of global regulators, including six genes believed to play a key role in the carbon starvation response (Fig. 7-7) [117]. The network includes genes that are targets of the PTS (the global regulator *crp* and the adenylate cyclase *cya*), genes involved in the control of metabolism (the global regulator *fis*), cellular growth (the *rrn* genes coding for stable RNAs), and DNA supercoiling, an important modulator of gene expression (the topoisomerase *topA* and the gyrase *gyrAB*). Although these genes have been the focus of intensive study over the last few decades, the development of a model

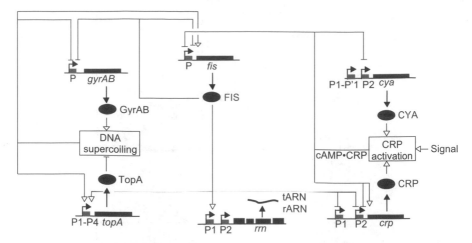

Figure 7-7 Key genes, proteins, and regulatory interactions making up the network involved in the response of *E. coli* bacteria to carbon-source availability [117].

of the network is limited by the lack of quantitative information about the concentrations of the network components and the parameters characterizing their interactions.

To overcome the lack of quantitative information, we have used a qualitative modeling and simulation method to analyze the network [127,128]. This method is based on piecewise-linear differential equations of the regulatory interactions and employs inequality constraints on the parameters to make predictions of the qualitative dynamics of the system. The piecewise-linear models of genetic regulatory networks are based on the use of step-function approximations of the sigmoidal functions describing the regulatory interactions (Fig. 7-4b). This approximation simplifies the analysis of the dynamics in that it allows the phase space to be subdivided into hyperrectangular regions where the system behaves in a qualitatively homogeneous way. The continuous dynamics of the system in the phase space can be discretized into a state transition graph, that is, a graph composed of states corresponding to the phase-space regions and transitions between these states. The state transition graph describes the possible qualitative behaviors of the system and allows the attractors of the system and their reachability to be determined.

Based on the qualitative simulations, two regulatory feedback loops were hypothesized to play a key role in the response of *E. coli* cells to carbon starvation. A positive feedback loop, involving *fis* and *crp*, seems to function as a switch controlling the transition of *E. coli* cells between the exponential and the stationary growth phase in response to a carbon starvation signal transmitted by the PTS. The other loop is a negative feedback loop, a homeostatic mechanism involving *fis* and DNA supercoiling, which regulates the resumption of cellular growth when a carbon source is available again, causing damped oscillations in certain protein concentrations. The qualitative simulations provide a description of the ordering of qualitative events (such as the upregulation and downregulation of key genes), which can be tested by monitoring gene expression over time, for instance through the use of gene reporter systems.

The assimilation of carbon sources by *E. coli*, in particular lactose, has been the subject of a large number of modeling studies (e.g., see Refs. 14,129–133). However, differential equation models have also been used to model the response of *E. coli* to other stresses, such as the response to a heat shock [96], bacteriophage infection [6,134,135], phosphate starvation [136], or the SOS response [103,136].

A common assumption underlying these models is that individual bacteria, under identical conditions, respond to a stress in exactly the same way. However, it is known that, whereas most bacteria enter a nongrowth state in response to carbon starvation, some of them continue to grow and divide. These different behaviors of individual cells arise from the stochasticity of the underlying processes that is not accounted for by the differential equation models. The next section will elaborate this point and introduce a modeling approach capable of dealing with the stochastic aspects of gene expression.

7.4 STOCHASTIC MASTER EQUATION MODELS

7.4.1 Model Formalism and Analysis Techniques

Ordinary differential equations provide a deterministic view on genetic regulatory networks, in the sense that, for given parameter values and initial conditions, Equation 7-2 has a unique solution and consequently predicts a single behavior of the system. Real biological systems, however, are not deterministic since noise arises inside and outside the system, due to fluctuations in the synthesis and degradation of proteins—strengthened by the low number of molecules of each species—and fluctuations in the environmental conditions [138–140]. As a consequence, genetically identical cells evolving under the same conditions may display different phenotypic characteristics [141,142]. In order to capture the stochastic aspects of cellular processes on the molecular level, different types of models can be used [140,143,144]. Here we focus on *stochastic master equations*, which give a detailed description of the biochemical reactions occurring in a cell.

Instead of continuous concentrations x_i, the variables in a stochastic master equation denote discrete numbers of molecules $X_i \in N$. For each different species in the system—proteins, RNA, DNA, or metabolites—a separate variable X_i is introduced. The continuous rates of change $f_i(x)$ in ordinary differential equations are replaced by discrete reaction events occurring with a certain probability per time interval. We can write the following equation for the time evolution of the system:

$$p[X(t+\Delta t) = V, t+\Delta t] = p[X(t) = V, t](1 - \sum_{j=1,\ldots,m} \alpha_j \Delta t)$$
$$+ \sum_{j=1,\ldots,m} p[X(t) = V - \nu_j, t]\beta_j \Delta t, \qquad (7\text{-}3)$$

where $X = (X_1,\ldots,X_n)'$, m is the number of reactions that can occur in the system, $\alpha_j \Delta t$ is the probability that reaction j will occur in the time interval $[t, t + \Delta t]$ given that $X(t) = V$, and $\beta_j \Delta t$ is the probability that reaction j will bring the system from a state $X(t) = V - \nu_j$ to a state $X(t + \Delta t) = V$ in $[t, t + \Delta t]$, where ν_j represents the stoichiometry of the reaction. In other words, Equation 7-3 expresses that the probability of having

V molecules at time $t + \Delta t$ equals the sum of the probability of having already V molecules at t with no reaction occurring on $[t, t + \Delta t]$, and the probability of having $V-\nu_j$ molecules at t and reaction j occurring on $[t, t + \Delta t]$. Rearranging Equation 7-3 and taking the limit $\Delta t \rightarrow 0$ yield the stochastic master equation (see [145] and [146] for details):

$$\partial p[X(t) = V, t]/\partial t = \sum_{j=1,\ldots,m} p[X(t) = V - \nu_j, t]\beta_j - p[X(t) = V, t]\alpha_j. \quad (7\text{-}4)$$

Compare this equation with the ordinary differential equation given in Equation 7-2. Whereas the latter specifies how the state of the system evolves over time, Equation 7-4 describes how the probability that the system is in a certain state evolves over time. Notice that the variables in Equation 7-4 can be reformulated as concentrations by dividing the number of molecules X by the cell volume.

Figure 7-8 gives an example of a negative feedback loop that is even simpler than the one shown in Figure 7-4. It consists of a single gene a coding for a protein A that forms a dimer capable of binding to the promoter region of a, thus inhibiting the expression of the gene. The reactions involving the different molecular species of the system are shown in the figure. For instance, the dimerization of the repressor is represented by the reaction $A + A \rightarrow A_2$. Even for this simple system, the stochastic master Equation 7-3 cannot be solved analytically. Under certain conditions, however, it can be approximated by stochastic differential equations, so-called Langevin equations, which consist of a differential equation Equation 7-2 extended with a noise term [140,146,147]. The conditions under which the approximation is valid may not always be possible to satisfy in the case of genetic regulatory networks.

An alternative way to proceed would be to disregard the stochastic master equation altogether and directly simulate the time evolution of the regulatory system. This idea underlies the stochastic simulation approach developed by Gillespie [145]. Basically, the stochastic simulation algorithm (i) determines *when* the next reaction occurs and of *which* type it will be, given that the system is in a state $X(t) = V$ at t, (ii) revises the state of the system in accordance with this reaction, and (iii) continues at the resulting next state. The stochastic variables τ and ρ are introduced, which represent the time that has passed until the next reaction occurs and the type of reaction, respectively. At each state a value for τ and ρ is randomly chosen from a set of values whose joint probability density function $p[\tau, \rho]$ has been derived from the same principles as those underlying the master equation 7-4. This guarantees that when a large number of stochastic simulations are carried out, the computed distribution for X at t will approach the distribution implied by the master equation.

It is obvious that stochastic simulation is a computationally intensive process, especially when dealing with species involving a large number of molecules and/or with reactions occurring at high frequency. Examples are metabolic reactions, which may occur millions of times on the timescale of one generation of a bacterial cell [148]. Another reason is that a large number of different molecular species may need to be taken into account, for instance when a protein has a large number of phosphorylation or methylation states, each of which participates in different reactions and therefore needs to be treated as a separate species [149,150]. Various improvements of the

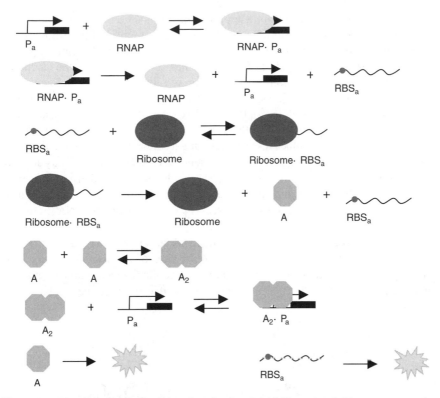

Figure 7-8 List of biochemical reactions in a simple autoinhibition network. The gene *a* encodes a repressor protein A that forms a dimer A_2. Gene expression consists of a transcription step (involving an RNA polymerase that binds to the promoter P_a on the DNA) and a translation step (involving a ribosome that binds to the ribosome binding site RBS_a on the mRNA). The promoter region P_a contains a binding site for A_2, which allows the protein to inhibit gene expression. Both the protein A and mRNA *a* are degraded.

original Gillespie algorithm have been proposed, directed at reducing the computational complexity of the procedure. For instance, Gibson and Bruck [151] have proposed a modification that reduces the number of random numbers to be generated. Whereas this improved algorithm remains exact, in the sense that it yields results consistent with the stochastic master equation, other algorithms address the performance problems by exploiting approximations that lower the accuracy but improve the computational complexity. A popular approximation is the *τ-leap method*, which chooses the time *τ* between two states such that the algorithm "leaps over" a large number of frequently occurring reactions [152,153]. This speeds up the simulation in that only a single random number needs to be generated for the latter reactions. Another approximation is to explicitly distinguish fast and slow reactions and to use composite reaction mechanisms based on quasi-steady-state approximations for the fast reactions [154] or simulate the latter by means of ordinary or stochastic differential equations [155].

In Figure 7-9 we show the results of applying the original Gillespie algorithm to the autoregulatory feedback network of Figure 7-8, for a maximum of 8000 reaction steps. In comparison with the simulations of the deterministic ordinary differential equation models in Figure 7-5, the number of A_2 molecules fluctuates due to the stochastic nature of the underlying reaction events. The figure illustrates that expression of the gene occurs in bursts [138,142], associated with the binding of RNA polymerase to the promoter, which initiates the transcription of mRNA molecules, in turn translated into proteins. It can be seen in the figure that, as the number of A_2 molecules increases, transcription initiation becomes less frequent due to the occupation of the promoter region by the repressor protein. As a consequence, A_2 reaches a stationary level of about 65 molecules. Not surprisingly, a much higher level is reached in a variant of the above model in which autoregulation has been disabled, for instance due to a mutation in the promoter region that prevents the repressor from binding to the DNA (figure not shown).

A network with the same autoregulatory feedback structure as in Figure 7-5, as well as its mutant variant, has been designed and constructed on a plasmid by Becskei and Serrano [156]. Measurements of the repressor protein concentration in the two networks, by means of a fluorescent reporter, show that the negative feedback loop

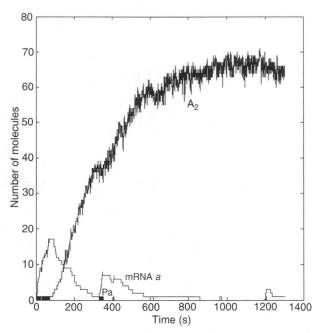

Figure 7-9 Example of a stochastic simulation of the autoinhibition network shown in Fig. 7-8, using a Matlab implementation of the Gillespie algorithm. The figure shows the temporal evolution of the number of molecules of three molecular species (A_2, mRNA a, and promoter P_a occupied by RNA polymerase) over 8000 steps. The values of the kinetic constants used in the simulation have been adapted from [141] and [148].

has the effect of decreasing fluctuations in the concentration. This illustrates how topological properties of the regulatory network may reduce the effect of noise arising from the stochasticity of the reaction events [138,140]. Besides mechanisms to *reduce* the effect of noise, the network may also include mechanisms to *amplify* fluctuations so as to increase differences between individual cells in a population. For instance, Isaacs et al. [157] have constructed an autoregulatory feedback network with an activator rather than a repressor protein. The positive feedback loop leads to bistability with states of high and low expression of the gene. Due to stochasticity in gene expression, cells may switch from a high to low expression state, giving rise to a bimodal distribution of protein concentrations in the cell population (see also Refs. [158,159]). The resulting population heterogeneity may have important pheno-typic consequences, as discussed in the next section for *E. coli* cells. Pedraza and van Oudenaarden [160] have shown that noise attenuation and amplification can also arise from other mechanisms, for instance gene cascades propagating noise through the network.

In summary, stochastic models of genetic regulatory networks focus on aspects of gene expression that are not taken into account by the deterministic models discussed in Section 7.3. In particular, stochastic models allow the effects of noise on the dynamical behavior of the cell to be studied, by analyzing the way in which fluctuations are filtered out or exploited by means of different molecular mechanisms. A number of computer tools for the stochastic simulation of molecular reaction systems are available, for instance Copasi [111], STOCKS [148], and StochSim [161]. Although stochastic simulation results in closer approximations of the molecular reality than can possibly be obtained by means of the other model types reviewed in this chapter, it is also more difficult to put in practice. Apart from the fact that it requires detailed knowledge of the reactions occurring in the system, notably the value of the kinetic parameters that specify the probability density function $p[\tau, \rho]$ [145], stochas-tic simulation is a computationally intensive process. In many cases, conventional deterministic models may provide an adequate description of the dynamics of genetic regulatory networks [90].

7.4.2   Effects of Noise in the Carbon Assimilation of *E. coli*

We know experimentally that, even when genetically completely identical, not all individuals of a bacterial population behave in the same way. For example, a long time ago already, it has been observed that in a population of *E. coli* cells the activity of the *lac* operon is not homogeneous [162,163]. That is, under conditions favoring the use of lactose, the *lac* operon is expressed in most but not all cells. An intuitive explanation of this phenomenon relies on the observation that certain kinds of molecules are present at very low numbers in the cell. For example, only about ten copies of the *lac* repressor protein are present in an *E. coli* cell. If these proteins are distributed randomly during cell division, about one cell in a thousand will not contain any *lac* repressor just after cell division. This would lead to derepression of the *lac* operon even in the absence of lactose in the growth medium. What are the consequences of such stochastic phenomena on the behavior of cells and their progeny? Has the cell developed compensatory mechanisms to cope with the fluctuations, or are they propagated throughout the entire network?

To address these questions, several studies have analyzed the role of stochasticity in the control of carbon assimilation in E. coli [148,164,165]. We will focus here on the stochastic model of the growth of bacteria on various carbon sources (glucose, lactose, and glycerol) developed by Puchałka and Kierzek [165]. Using the approach presented in Section 4.1, they have described the PTS and the metabolic pathways involved in the assimilation of these three carbohydrates. In particular, a list of more than 80 molecular species and 120 reactions has been compiled, as well as kinetic parameter values characterizing these reactions.

Stochastic simulation of such a large biochemical reaction system is extremely computationally intensive, given that the interactions in the carbon assimilation module take place on quite different timescales. For instance, the breakdown of carbohydrates and signal transduction by the PTS are fast reactions (less than 1 sec) involving large numbers of molecules, whereas the regulation of gene expression is a slow process (several minutes) involving a very small number of molecules. As discussed in Section 4.1, this prevents the simulation of individual reaction events by means of the basic Gillespie algorithm. Puchałka and Kierzek have therefore used a variant, the *maximal time-step method*, which dynamically partitions the reactions into fast and slow reactions. Whereas the slow reactions are simulated using the Gillespie algorithm, the fast reactions are treated by the τ-leap method (Section 4.1; [165]).

Stochastic simulation of the assimilation of various carbon sources by E. coli reproduced expected and well-known phenomena, like the use of glucose as a preferred nutrient. In addition, the simulations showed that stochastic fluctuations in reactions involving a small number of molecules may propagate through the network and influence the time course of other processes in the system, even metabolic pathways processing large numbers of molecules. For instance, during the transition from glucose to a mixture of lactose and glycerol, random delays in the expression of the *lac* operon may favor the use of glycerol (even though lactose is the preferred nutrient). This results in an almost complete shutdown of the glycolytic pathway, fuelled by glucose and lactose but not by glycerol. A striking effect of these time delays is the heterogeneity in the induction of the *lac* operon within the cell population switching from glucose to a mixture of lactose and glycerol (Fig. 7-10). Moreover, this heterogeneity in the use of carbon sources is conserved throughout consecutive cell divisions.

The model of Puchałka and Kierzek is probably the most extensive stochastic model to date, describing an integrated network of gene expression regulation, signal transduction, and metabolism. There exist a few other examples of stochastic models in E. coli, notably the model of the lysis-lysogeney decision following λ phage infection [141] and the model of the regulation of the *pap* operon in a pathogenic strain of E. coli [166]. However, as noted in Section 4.1, the development of such models requires precise knowledge about the molecular mechanisms underlying the biological processes, and their analysis involves high computational costs. As a consequence, the development of stochastic models has been limited to rather small, well-characterized systems.

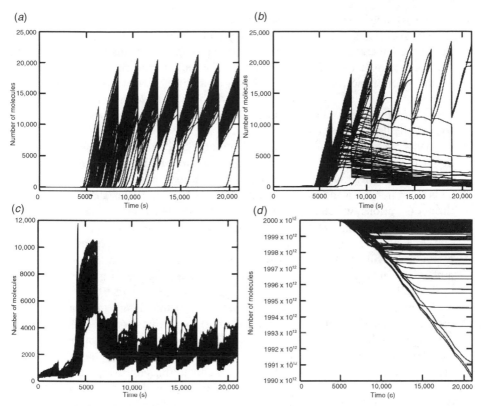

Figure 7-10   Stochastic simulation of *E. coli* growth on a mixture of glucose, lactose, and glycerol, using a model of the carbon assimilation module [164]. Glucose is depleted during the first 5000 s of the simulation. Plots A, B, C, and D show the time-course predictions for LacZ (expressed from the lactose operon), GlpF (expressed from the glycerol operon), cAMP, and external glycerol, respectively. The sawtooth patterns arise from the occurrence of a cell division every 2100 s, during which the molecules present in the cell are distributed over two daughter cells. Plots B and D reveal that there exists a small subpopulation of cells expressing proteins allowing the consumption of glycerol, instead of lactose, after the depletion of glucose.

7.5   DISCUSSION

In this chapter, we have reviewed three different approaches toward the modeling of genetic regulatory networks, based on graphs, ordinary differential equations, and stochastic master equations, respectively. The approaches make different modeling assumptions. Whereas graph models provide a static description of the network, formalizing the structure of interactions between genes, proteins, and other network components, ordinary differential equation and stochastic master equation models describe the dynamic behavior of the system. However, they do so in quite different ways. Differential equations are deterministic models, whereas master equations take into account the stochastic nature of the underlying biochemical reaction processes.

When going from graphs to stochastic master equations, the models take into account increasingly more aspects of physical reality. The counterpart is that this makes them increasingly more difficult to treat in practice. In many cases, even for a well-studied model system like *E. coli*, the information required for building a model on the level of individual reactions is not available. Moreover, the computational burden of simulating such detailed systems is high and does not currently allow its application to large reaction systems, although technical improvements are expected to push the limit further up [167].

In this context, it is crucial to stress that more detailed models are not necessarily *better* models. A model is by its very nature a simplified representation, based on assumptions that ignore certain aspects of reality so as to better bring out others. This is even true for the most detailed models discussed in this chapter, stochastic master equations, which implicitly assume that the reaction volumes are spatially homogeneous, an assumption that is not generally true [168]. In the end, what counts is whether a certain type of model, and thus certain types of simplifying assumptions, are adequate for answering the biological questions at hand. As some authors put it, the art of modeling consists in choosing the "right model for the job" [169].

The discussion of *E. coli* stress response models in this chapter has confirmed that different kinds of models are appropriate for different kinds of biological questions. The graph models are able to answer questions about the structure of the transcriptional regulatory network of the bacterium, such as the manner in which the network is composed of building blocks like modules and motifs. However, in order to study the dynamics of the building blocks, for instance the carbon assimilation module, one has to resort to ordinary differential equations and stochastic master equations. The former are well adapted for studying the steady states of a regulatory module and the way in which the system may evolve from one steady state to another in response to a perturbation. The latter are especially appropriate for questions about the way the network deals with noise arising from intracellular and extracellular processes, which in the case of noise amplification may give rise to heterogeneous phenotypes in a genetically identical population. This was illustrated by the differential induction of the *lac* operon in *E. coli* cells when glucose in the medium is depleted.[2]

The most effective strategy for studying a complex biological system therefore relies on *model plurality,* using different kinds of models that look at the system from different angles. Instead of building one large supermodel, describing the entire system on the most detailed level possible, it is more fruitful to build a hierarchy of models, accounting for different aspects of the system on different levels of abstraction.

---

[2] Of course, there is no one-to-one correspondence between biological questions and model formalisms. For instance, the robustness of a dynamic property of the system to fluctuations in the environment can be studied by means of a stochastic model of the biochemical reaction system, but also by varying parameter values in a differential equation model.

ACKNOWLEDGMENTS

The authors acknowledge financial support from the ACI IMPBio initiative of the French Ministry of Research (project BacAttract) and the NEST Adventure and Pathfinder programs of the European Commission (projects Hygeia and Cobios).

REFERENCES

1. Majdalani N, Vanderpool CK, Gottesman S. Bacterial small RNA regulators. *Crit Rev Biochem Mol Biol* 2005;40(2):93–113.

2. Brazhnik P, de la Fuente A, Mendes P. Gene networks: how to put the function in genomics. *Trends Biotechnol* 2002;20(11):467–472.

3. Jacob F. Evolution and thinkering. *Science* 1977;196(4295):1161–1166.

4. Alon U. Biological networks: the tinkerer as engineer. *Science* 2003;301(5641):1866.

5. Csete ME, Doyle JC. Reverse engineering of biological complexity. *Science* 2002;295 (5560):1664–1669.

6. McAdams HH, Shapiro L. Circuit simulation of genetic networks. *Science* 1995;269 (5224):650–656.

7. Lazebnik Y. Can a biologist fix a radio? Or, what I learned while studying apoptosis. *Biochemistry (Mosc)* 2004;69(12):1403–1406.

8. Kitano H. Systems biology: a brief overview. *Science* 2002;295(5560):1662–1664.

9. Endy D. Foundations for engineering biology. *Nature* 2005;438(7067):449–453.

10. Sprinzak D, Elowitz MB. Reconstruction of genetic circuits. *Nature* 2005;438(7067): 443–448.

11. Hengge-Aronis R. Signal transduction and regulatory mechanisms involved in control of the σ^S (RpoS) subunit of RNA polymerase. *Microbiol Mol Biol Rev* 2002;66(3): 373–395.

12. Wick LM, Egli T. Molecular components of physiological stress responses in *Escherichia coli*. *Adv Biochem Eng/Biotechnol* 2004;89:1–45.

13. Neidhardt FC, et al., editors. *Escherichia coli and Salmonella: Cellular and Molecular Biology*, 2nd ed. Washington, DC: ASM Press, 1996.

14. Babloyantz A, Sanglier M. Chemical instabilities of 'all-or-none' type in β-galactosidase induction and active transport. *FEBS Lett* 1972;23(3):364–366.

15. Domach MM, et al. Computer model for glucose-limited growth of a single cell of *Escherichia coli* B/r-A. *Biotechnol Bioeng* 1984;26(3):203–216.

16. Holden C. Alliance launched to model *E. coli*. *Science* 2002;297(5586):1459–1460.

17. Wanner BL, Finney A, Hucks M. Modeling the *E. coli* cell: the need for computing, cooperation, and consortia. *Top Curr Genet* 2005;13:163–189.

18. Bolouri H, Davidson EH. Modeling transcriptional regulatory networks. *BioEssays* 2002;24(12):1118–1129.

19. de Jong H. Modeling and simulation of genetic regulatory systems: a literature review. *J Comput Biol* 2002;9(1):67–103.

20. Hasty J, et al. Computational studies of gene regulatory networks: *in numero* molecular biology. *Nat Rev Genet* 2001;2(4):268–279.

21. McAdams HH, Arkin A. Simulation of prokaryotic genetic circuits. *Annu Rev Biophys Biomol Struct* 1998;27:199–224.

22. Smolen P, Baxter DA, Byrne JH. Modeling transcriptional control in gene networks: methods, recent results, and future directions. *Bull Math Biol* 2000;62(2):247–292.

23. Berge C. *The Theory of Graphs*. Mineola, NY: Dover Publications, 2001.

24. Kohn KW. Molecular interaction map of the mammalian cell cycle control and DNA repair systems. *Mol Biol Cell* 1999;10:2703–2734.

25. Davidson EH, et al. A genomic regulatory network for development. *Science* 2002;295 (5560):1669–1678.

26. Barabási AL, Oltvai ZN. Network biology: understanding the cell's functional organization. *Nat Rev Genet* 2004;5(2):101–113.

27. Newman MEJ. The structure and function of complex networks. *SIAM Rev* 2003;45(2): 167–256.

28. Watts DJ, Strogatz SH. Collective dynamics of 'small-world' networks. *Nature* 1998;393 (6684):440–442.

29. Albert R, Jeong H, Barabási AL. Error and attack tolerance of complex networks. *Nature* 2000;406(6794):378–382.

30. Jeong H, et al. Lethality and centrality in protein networks. *Nature* 2001;411(6833): 41–42.

31. Jeong H, et al. The large-scale organization of metabolic networks. *Nature* 2000;407 (6804):651–654.

32. Aldana M, Cluzel P. A natural class of robust networks. *Proc Natl Acad Sci USA* 2003;100 (15):8710–8714.

33. Dobrin R, et al. Aggregation of topological motifs in the *Escherichia coli* transcriptional regulatory network. *BMC Bioinform* 2004;5(1):10.

34. Featherstone DE, Broadie K. Wrestling with pleiotropy: genomic and topological analysis of the yeast gene expression network. *BioEssays* 2002;24(3):267–274.

35. Guelzim N, et al. Topological and causal structure of the yeast transcriptional regulatory network. *Nat Genet* 2002;31(1):60–63.

36. Lee TI, et al. Transcriptional regulatory networks in *Saccharomyces cerevisiae*. *Science* 2002;298(5594):799–804.

37. Maslov S, Sneppen K. Specificity and stability in topology of protein networks. *Science* 2002;296(5569):910–913.

38. Tong AH, et al. Global mapping of the yeast genetic interaction network. *Science* 2004;303(5659):808–813.

39. Wagner A, Fell DA. The small world inside large metabolic networks. *Proc R Soc Lond, B*, 2001;268(1478):1803–1810.

40. Stumpf MPH, Wiuf C, May RM. Subnets of scale-free networks are not scale-free: sampling properties of networks. *Proc Natl Acad Sci USA* 2005;102(12):4221–4224.

41. Yu H, et al. TopNet: a tool for comparing biological sub-networks, correlating protein properties with topological statistics. *Nucleic Acids Res* 2004;32(1):328–337.

42. Arita M. The metabolic world of *Escherichia coli* is not small. *Proc Natl Acad Sci USA* 2004;101(6):1543–1547.

43. Alm E, Arkin AP. Biological networks. *Curr Opin Struct Biol* 2003;13(2):193–202.

44. Ravasz E, et al. Hierarchical organization of modularity in metabolic networks. *Science* 2002;297(5586):1551–1555.

45. Rives AW, Galitski T. Modular organization of cellular networks. *Proc Natl Acad Sci USA* 2003;100(3):1128–1133.

46. Girvan M, Newman M. Community structure in social and biological networks. *Proc Natl Acad Sci USA* 2002;99(12):7821–7826.

47. Milo R, et al. Superfamilies of evolved and designed networks. *Science* 2004;303(5663): 1538–1542.

48. Milo R, et al. Network motifs: simple building blocks of complex networks. *Science* 2002;298(5594):824–827.

49. Shen-Orr SS, et al. Network motifs in the transcriptional regulation network of *Escherichia coli*. *Nat Genet* 2002;31(1):64–68.

50. Wuchty S, Oltvai ZN, Barabási A-L. Evolutionary conservation of motif constituents in the yeast protein interaction network. *Nat Genet* 2003;35(2):118–119.

51. Conant GC, Wagner A. Convergent evolution of gene circuits. *Nat Genet* 2003;34(3): 264–266.

52. Bar-Joseph Z, et al. Computational discovery of gene modules and regulatory networks. *Nat Biotechnol* 2003;21:1337–1342.

53. Ihmels J, et al. Revealing modular organization in the yeast transcriptional network. *Nat Genet* 2002;31(4):370–377.

54. Luscombe NM, et al. Genomic analysis of regulatory network dynamics reveals large topological changes. *Nature* 2004;431(7006):308–311.

55. Segal E, et al. Module networks: identifying regulatory modules and their condition-specific regulators from gene expression data. *Nat Genet* 2003;34(2):166–176.

56. Snel B, Bork P, Huynen MA. The identification of functional modules from the genomic association of genes. *Proc Natl Acad Sci USA* 2002;99(9):5890–5895.

57. Tanay A, et al. Revealing modularity and organization in the yeast molecular network by integrated analysis of highly heterogeneous genomewide data. *Proc Natl Acad Sci USA* 2004;101(9):2981–2986.

58. Mering CV, et al. Genome evolution reveals biochemical networks and functional modules. *Proc Natl Acad Sci USA* 2003;100:15428–15433.

59. Wu H, et al. Prediction of functional modules based on comparative genome analysis and gene ontology application. *Nucleic Acids Res* 2005;33(9):2822–2837.

60. Kashtan N, et al. Efficient sampling algorithm for estimating subgraph concentrations and detecting network motifs. *Bioinformatics* 2004;20(11):1746–1758.

61. Salgado H, et al. RegulonDB (version 5.0): *Escherichia coli* K-12 transcriptional regulatory network, operon organization, and growth conditions. *Nucleic Acids Res* 2006;34:D394

62. Keseler IM, et al. EcoCyc: a comprehensive database resource for *Escherichia coli*. *Nucleic Acids Res* 2005;33:D334–D337

63. Ma H-W, et al. An extended transcriptional regulatory network of *Escherichia coli* and analysis of its hierarchical structure and network motifs. *Nucleic Acids Res* 2004;32(22): 6643–6649.

64. Madan Babu M, Teichmann SA. Evolution of transcription factors and the gene regulatory network in *Escherichia coli*. *Nucleic Acids Res* 2003;31(4):1234–1244.

65. Resendis-Antonio O, et al. Modular analysis of the transcriptional regulatory network of *E. coli*. *Trends Genet* 2004;21(1):16–20.

66. Mangan S, Alon U. Structure and function of the feed-forward loop network motif. *Proc Natl Acad Sci USA* 2003;100(21):11980–11985.

67. Mangan S, Zaslaver A, Alon U. The coherent feedforward loop serves as a sign-sensitive delay element in transcription networks. *J Mol Biol* 2003;334(2):197–204.

68. Mangan S, et al. The incoherent feed-forward loop accelerates the response time of the *gal* system of *Escherichia coli*. *J Mol Biol* 2006;356(5):1073–1081.

69. Boyce WE, DiPrima RC. *Elementary Differential Equations and Boundary Value Problems*, 5th ed. John Wiley & Sons, 1992.

70. Goodwin BC. Oscillatory behavior in enzymatic control processes. *Adv Enzyme Regul* 1965;425–438.

71. Goodwin BC. *Temporal Organization in Cells*. New York: Academic Press, 1963.

72. Tyson JJ, Othmer HG. The dynamics of feedback control circuits in biochemical pathways. *Prog Theor Biol* 1978;5:1–62.

73. Hirsch MW, Smale S, editors. *Differential Equations, Dynamical Systems, and Linear Algebra*. Academic Press, 1974.

74. Lambert JD. *Numerical Methods for Ordinary Differential Equations*. Chichester, UK: Wiley, 1991.

75. Freeman M. Feedback control of intercellular signalling in development. *Nature* 2000;408(6818):313–319.

76. Perrimon N, McMahon AP. Negative feedback mechanisms and their roles during pattern formation. *Cell* 1999;97(1):13–16.

77. Thieffry D, et al. From specific gene regulation to genomic networks: a global analysis of transcriptional regulation in *Escherichia coli*. *BioEssays* 1998;20(5):433–440.

78. Thomas R. On the relation between the logical structure of systems and their ability to generate multiple steady states or sustained oscillations. In: Lacolle B et al., editors. *Numerical Methods in the Study of Critical Phenomena*. Berlin: Springer-Verlag, 1981, pp. 180–193.

79. Thomas R, d'Ari R. *Biological Feedback*. Boca Raton, FL: CRC Press, 1990.

80. Cinquin O, Demongeot J. Positive and negative feedback: striking a balance between necessary antagonists. *J Theor Biol* 2002;216(2):229–241.

81. Gouzé J-L. Positive and negative circuits in dynamical systems. *J Biol Syst* 1998;6(1): 11–15.

82. Plahte E, Mestl T, Omholt SW. Feedback loops, stability and multistationarity in dynamical systems. *J Biol Syst* 1995;3(2):409–413.

83. Snoussi EH. Necessary conditions for multistationarity and stable periodicity. *J Biol Syst* 1998;6(1):3–9.

84. Soulé C. Graphic requirements for multistationarity. *ComPlexUs* 2003;1(3):123–133.

85. Alon U, et al. Robustness in bacterial chemotaxis. *Nature* 1999;397(6715):168–171.

86. Barkai N, Leibler S. Robustness in simple biochemical networks. *Nature* 1997;387 (6636):913–917.

87. Barkai N, Shilo B-Z. Modeling pattern formation: counting to two in the Drosophila Egg. *Curr Biol* 2002;12(14):R493

88. von Dassow G, et al. The segment polarity network is a robust developmental module. *Nature* 2000;406(6792):188–192.

89. Morohashi M, et al. Robustness as a measure of plausibility in models of biochemical networks. *J Theor Biol* 2002;216(1):19–30.

90. Gonze D, Halloy J, Goldbeter A. Robustness of circadian rhythms with respect to molecular noise. *Proc Natl Acad Sci USA* 2002;99(2):673–678.

91. Stelling J, et al. Robustness of cellular functions. *Cell* 2004;118(6):675–686.

92. Ma L, Iglesias PA. Quantifying robustness of biochemical network models. *BMC Bioinform* 2002;3:38.

93. Alves R, Savageau MA. Comparing systemic properties of ensembles of biological networks by graphical and statistical methods. *Bioinformatics* 2000;16(6):527–533.

94. Yi T-M, et al. Robust perfect adaptation in bacterial chemotaxis through integral feedback control. *Proc Natl Acad Sci USA* 2000;97(9):4649–4653.

95. Bhalla US, Iyengar R. Emergent properties of networks of biological signaling pathways. *Science* 1999;283(5400):381–387.

96. El-Samad H, et al. Surviving heat shock: control strategies for robustness and performance. *Proc Natl Acad Sci USA* 2005;102(8):2736–2741.

97. Saez-Rodriguez J, et al. Modular analysis of signal transduction networks. *IEEE Control Systems Magazine* 2004;24(4):35–52.

98. Heinrich R, Schuster S. *The Regulation of Cellular Systems*. New York: Chapman & Hall, 1996.

99. Pécou E. Splitting the dynamics of large biochemical interaction networks. *J Theor Biol* 2005;232(3):375–384.

100. Ljung L. *System Identification: Theory for the User*. Upper Saddle River, NJ: Prentice Hall, 1999.

101. Gardner TS, et al. Inferring genetic networks and identifying compound mode of action via expression profiling. *Science* 2003;301(5629):102–105.

102. Lemeille S, Latifi A, Geiselmann J. Inferring the connectivity of a regulatory network from mRNA quantification in *Synechocystis* PCC6803. *Nucleic Acids Res* 2005;33(10): 3381–3389.

103. Ronen M et al. Assigning numbers to the arrows: parameterizing a gene regulation network by using accurate expression kinetics. *Proc Natl Acad Sci USA* 2002;99(16): 10555–10560.

104. van Someren EP, Wessels LFA, Reinders MJT. Linear modeling of genetic networks from experimental data. *Proceedings of the Eight International Conference on Intelligent Systems for Molecular Biology, ISMB 2000*, Menlo Park, CA: AAAI Press, 2000.

105. de Jong H, Ropers D. Qualitative approaches towards the analysis of genetic regulatory networks. In: Stelling J, editor. *System Modeling in Cellular Biology: From Concepts to Nuts and Bolts*. Cambridge, MA: MIT Press, 2006, pp. 125–148.

106. Gagneur J, Casari G. From molecular networks to qualitative cell behavior. *FEBS Lett* 2005;579(8):1867–1871.

107. Glass L, Kauffman SA. The logical analysis of continuous non-linear biochemical control networks. *J Theor Biol* 1973;39(1):103–129.

108. Kauffman SA. Homeostasis and differentiation in random genetic control networks. *Nature* 1969;224:177–178.

109. Thomas R. Regulatory networks seen as asynchronous automata: a logical description. *J Theor Biol* 1991;153:1–23.

110. Thomas R. Boolean formalization of genetic control circuits. *J Theor Biol* 1973;42:563–585.

111. Hoops S, et al. COPASI—a COmplex PAthway SImulator. *Bioinformatics* 2006;22(24): 3067–3077.

112. Ginkel M, et al. Modular modelling of cellular systems with ProMoT/Diva. *Bioinformatics* 2003;19:1169–1176.

113. Slepchenko BM, et al. Quantitative cell biology with the virtual cell. *Trends Cell Biol* 2003;13(11):570–576.

114. Ermentrout B. *Simulating, Analyzing, and Animating Dynamical Systems: A Guide to XPPAUT for Researchers and Students*. Philadelphia: SIAM, 2002.

115. Hucka M, et al. The systems biology markup language (SBML): a medium for representation and exchange of biochemical network models. *Bioinformatics* 2003;19 (4):524–531.

116. Bettenbrock K, et al. A quantitative approach to catabolite repression in *Escherichia coli*. *J Biol Chem* 2006;281(5):2578–2584.

117. Ropers D, et al. Qualitative simulation of the carbon starvation response in *Escherichia coli*. *BioSystems* 2006;84(2):124–152.

118. Kremling A, et al. The organization of metabolic reaction networks: III. Application for diauxic growth on glucose and lactose. *Metab Eng* 2001;3(4):362–379.

119. Kremling A, et al. Time hierarchies in the *Escherichia coli* carbohydrate uptake and metabolism. *BioSystems* 2003;73(1):57–72.

120. Sauter T, Gilles ED. Modeling and experimental validation of the signal transduction via the *Escherichia coli* sucrose phosphotransferase system. *J Biotechnol* 2004;110(2): 181–199.

121. Wang J, et al. Modeling of inducer exclusion and catabolite repression based on a PTS-dependent sucrose and non-PTS-dependent glycerol transport systems in *Escherichia coli* K-12 and its experimental verification. *J Biotechnol* 2001;92(2):133–158.

122. Chassagnole C, et al. Dynamic modeling of the central carbon metabolism of *Escherichia coli*. *Biotechnol Prog* 2002;79:53–73.

123. Covert MW, Palsson BO. Constraints-based models: regulation of gene expression reduces the steady-state solution space. *J Theor Biol* 2003;221(3):309–326.

124. Wong P, Gladney S, Keasling JD. Mathematical model of the lac operon: inducer exclusion, catabolite repression, and diauxic growth on glucose and lactose. *Biotechnol Prog* 1997;13:132–143.

125. Hengge-Aronis R. Regulation of gene expression during entry into stationary phase. In: Umbarger HE, editor. *Escherichia coli* and Salmonella: Cellular and Molecular Biology. Washington, DC: ASM Press, 1996, pp. 1497–1512.

126. Martinez-Antonio A, Collado-Vides J. Identifying global regulators in transcriptional regulatory networks in bacteria. *Curr Opin Microbiol* 2003;6(5):482–489.

127. de Jong H, et al. Qualitative simulation of genetic regulatory networks using piecewise-linear models. *Bull Math Biol* 2004;66(2):301–340.

128. Batt G, et al. Validation of qualitative models of genetic regulatory networks by model checking: analysis of the nutritional stress response in *Escherichia coli. Bioinformatics* 2005;21(Suppl. 1):i19–i28

129. Laffend L, Schuler ML. Structured model of genetic control via the lac promoter in *E. coli. Biotechnol Bioeng* 1994;43:399–410.

130. Ozbudak EM, et al. Multistability in the lactose utilization network of *Escherichia coli. Nature* 2004;427:737–740.

131. Santillán M, Mackey MC. Influence of catabolite repression and inducer exclusion on the bistable behavior of the lac operon. *Biophys J* 2004;86(3):1282–1292.

132. Yildirim N, Mackey MC. Feedback regulation in the lactose operon: a mathematical modeling study and comparison with experimental data. *Biophys J* 2003;84(5): 2841–2851.

133. Sanglier M, Nicolis G. Sustained oscillations and threshold phemomena in an operon control circuit. *Biophys Chem* 1976;4:113–121.

134. Endy D, et al. Computation, prediction, and experimental tests of fitness for bacteriophage T7 mutants with permuted genomes. *Proc Natl Aca Sci USA* 2000;97(10):5375–5380.

135. Ranquet C, et al. Control of bacteriophage Mu lysogenic repression. *J Mol Biol* 2005;353 (1):186–195.

136. Van Dien SJ, Keasling JD. A dynamic model of the *Escherichia coli* phosphate-starvation response. *J Theor Biol* 1998;190(1):37–49.

137. Gardner TS, et al. Inferring genetic networks and identifying compound mode of action via expression profiling. *Science* 2003; 301(5629):102–105.

138. Kaern M, et al. Stochasticity in gene expression: from theories to phenotypes. *Nat Rev Genet* 2005;6(6):451–464.

139. McAdams HH, Arkin A. It's a noisy business! Genetic regulation at the nanomolar scale. *Trends Genet* 1999;15(2):65–69.

140. Rao CV, Wolf DM, Arkin AP. Control, exploitation and tolerance of intracellular noise. *Nature* 2002;420(6912):231–237.

141. Arkin A, Ross J, McAdams HA. Stochastic kinetic analysis of developmental pathway bifurcation in phage lambda-infected *Escherichia coli* cells. *Genetics* 1998;149(4): 1633–1648.

142. McAdams HH, Arkin A. Stochastic mechanisms in gene expression. *Proc Natl Acad Sci USA* 1997;94(3):814–819.

143. Gillespie DT, Petzold LR. Numerical simulation for biochemical kinetics. In: Szallasi VPZ, Stelling J, editors. *System Modeling in Cellular Biology: From Concepts to Nuts and Bolts*, Cambridge, MA: MIT Press, 2006.

144. Paulsson J, Elf J. Stochastic modelling of intracellular kinetics. In: Szallasi VPZ, Stelling J, editors. *System Modeling in Cellular Biology: From Concepts to Nuts and Bolts*. Cambridge, MA: MIT Press, 2006.

145. Gillespie DT. Exact stochastic simulation of coupled chemical reactions. *J Phys Chem* 1977;81(25):2340–2361.

146. van Kampen NG. *Stochastic Processes in Physics and Chemistry.* Amsterdam: Elsevier, 1997.

147. Gillespie DT. The chemical Langevin equation. *J Chem Phys* 2000;113(1):297–306.

148. Kierzek AM. STOCKS: STOChastic Kinetic Simulations of biochemical systems with Gillespie algorithm. *Bioinformatics* 2002;18(3):470–481.

149. Morton-Firth CJ, Bray D. Predicting temporal fluctuations in an intracellular signalling pathway. *J Theor Biol* 1998;192(1):117–128.

150. Morton-Firth CJ, Shimizu TS, Bray D. A free-energy-based stochastic simulation of the tar receptor complex. *J Mol Biol* 1999;286:1059–1074.

151. Gibson MA, Bruck J. Efficient exact stochastic simulation of chemical systems with many species and many channels. *J Phys Chem A* 2000;104(9):1876–1889.

152. Gillespie DT. Approximate accelerated stochastic simulation of chemically reacting systems. *J Chem Phys* 2001;115(4):1716–1733.

153. Gillespie DT, Petzold LR. Improved leap-size selection for accelerated stochastic simulation. *J Chem Phys* 2003;119(16):8229–8234.

154. Rao CV, Arkin AP. Stochastic chemical kinetics and the quasi steady-state assumption: application to the Gillespie algorithm. *J Chem Phys* 2003;118(11):4999–5010.

155. Haseltine EL, Rawlings JB. Approximate simulation of coupled fast and slow reactions for stochastic chemical kinetics. *J Chem Phys* 2002;117(15):6959–6969.

156. Becskei A, Serrano L. Engineering stability in gene networks by autoregulation. *Nature* 2000;405(6786):590–591.

157. Isaacs FJ, et al. Prediction and measurement of an autoregulatory genetic module. *Proc Natl Acad Sci USA* 2003;100(13):7714–7719.

158. Becskei A, Séraphin B, Serrano L. Positive feedback in eukaryotic gene networks: cell differentiation by graded to binary response conversion. *EMBO J* 2001;20(10):2528–2535.

159. Hasty J, et al. Noise-based switches and amplifiers for gene expression. *Proc Natl Acad Sci USA* 2000;97(5):2075–2080.

160. Pedraza JM, van Oudenaarden A. Noise propagation in gene networks. *Science* 2005;307 (5717):1965–1969.

161. Le Novère N, Shimizu TS. STOCHSIM: modelling of stochastic biomolecular processes. *Bioinformatics* 2001;17(6):575–576.

162. Benzer S. Induced synthesis of enzymes in bacteria analyzed at the cellular level. *Acta Biochim Biophys* 1953;11:383–395.

163. Novick A, Weiner M. Enzyme induction as an all-or-none phenomenon. *Proc Natl Acad Sci USA* 1957;43(7):553–566.

164. Carrier TA, Keasling JD. Investigating autocatalytic gene expression systems through mechanistic modeling. *J Theor Biol* 1999;201:25–36.

165. Puchałka J, Kierzek AM. Bridging the gap between stiochastic and deterministic regimes in the kinetic simulations of the biochemical reaction networks. *Biophys J* 2004;86(3): 1357–1372.

166. Jarboe LR, Beckwith D, Liao JC. Stochastic modeling of the phase-variable *pap* operon regulation in uropathogenic *Escherichia coli*. *Biotechnol Bioeng* 2004;88(2):189–203.

167. Endy D, Brent R. Modelling cellular behaviour. *Nature* 2000;409(6818):391–396.

168. Kruse K, Elf J. Kinetics in spatially-extended systems. In: Szallasi VPZ, Stelling J, editors. *System modeling in cellular biology: From concepts to nuts and Bolts*, Cambridge, MA: MIT Press, 2006.

169. Falkenhainer B, Forbus KD. Compositional modeling: finding the right model for the job. *Artif Intell* 1991;51(1–3):95–143.

170. Bray D. Molecular networks: the top–down view. *Science* 2003;301(5641):1864–1865.

8

SYNTHETIC LIFE: ETHOBRICKS FOR A NEW BIOLOGY

Jordi Vallverdú[1] and Claes Gustafsson[2]

[1]*Philosophy Department, Universitat Autònoma de Barcelona,*
E-08193 Bellaterra (BCN), Catalonia
[2]*DNA2.0, Inc., 1430 O'Brien Drive, Menlo Park, California 94025*

8.1 INTRODUCTION

The purpose of this study is to bring the current state of gene synthesis and synthetic biology into the context of normative ethics, the branch of philosophy that classifies actions as good or bad. What are the demarcation lines between existing and new normative ethics as the technology evolves? What tools are needed by philosophers to capture and merge or resolve conflicting synthetic biology norms in a multicultural society, and how can we extend the conversation between philosophy and technology on these issues?

We here have no intent to review all the exciting new applications of synthetic biology. This information is already available in several outstanding reviews [1,2]. We further do not attempt to cover the significant regulatory and intellectual property issues brought to light by the rapidly escalating technology. Such discussions can be found elsewhere [3]. Our sole aim with this publication is to invite an open discussion on what, if any, consequences the rapidly increasing ability to build new genetic information has on our current normative ethics.

Systems Biology and Synthetic Biology   Edited by Pengcheng Fu and Sven Panke
Copyright © 2009 John Wiley & Sons, Inc.

8.2 HISTORICAL AND CONCEPTUAL BACKGROUND

In early nineteenth century, scientists of the era believed that compounds from living organisms could not be synthesized and that they possessed a nonphysical inner energy that could self-propagate (vitalism). Compounds derived from living organisms were labeled "organic" and most analytical efforts were instead focused on the inorganic types of compounds: metals, salts, and other nonbio materials. This vitalism myth was forever shattered in 1828 when Friedrich Wöhler synthesized urea, an organic molecule. This revelation changed much of chemistry from a discovery-based science to an engineering field devoted to the construction of novel organic molecules. Today there are almost no limits to the kind of organic molecules a good synthetic chemist can synthesize. The acceptance that organic materials can be made and modified to fit the need of mankind in conjunction with our significant understanding of organic chemistry has had far-reaching consequences in today's society, culture, and economy.

With the current advent of molecular biology, genomics, and most recently synthetic biology, we are again breaking through an imaginary barrier as now we have the ability to modify, edit, and create new biological entities by directly altering the biological source code—DNA. We are no longer limited to creating chimeras of naturally existing information, as is the case with "classic" genetic engineering. Instead, as the formal rules and grammar of biological information are gradually deconvoluted and gene synthesis technology improves, we now are able to create designed genetic templates for nonexisting proteins, replicative units, metabolic pathways, and, entire organisms (Table 8-1).

Synthetic biology is now emerging at the interface between chemistry, molecular biology, engineering, and computer science. The discipline is often suggested to be the "other half" of systems biology (Fig. 8-1) [4]. While systems biology is focused on cataloging all parts of biology, synthetic biology aims instead at building novel genetic circuitry and processes from scratch based on new or existing biological parts [5].

Our increasing ability to efficiently create any genetic information imaginable will transform life sciences into an engineering discipline just as what happened to organic chemistry more than a century ago.

Table 8-1 Historical milestones in creating increasingly larger and more complex synthetic DNA

First synthetic gene	1970	Yeast tRNA Ala (207 bp)	[8]
First synthetic peptide coding gene	1977	Human growth hormone (56 bp)	[54]
First synthetic protein coding gene	1981	Alpha interferon (514 bp)	[55]
First synthetic bacterial replicating unit	1995	Plasmid (2.7 kb)	[56]
Identification of minimal genome	1999	265–350 protein-coding genes of *Mycoplasma genitalium* are essential	[57]
First synthetic enzymatic pathway	1999	>50 erythromycin analogues	[58]
First synthetic genome	2002	poliovirus (7.4 kb)	[59]
First synthetic metabolic operon	2004	PKS gene cluster (32 kb)	[60]
First synthetic prokaryotic chromosome	2008	*Mycoplasma genitalium* genome	[61]

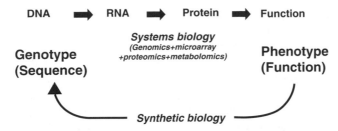

Figure 8-1 Systems biology, synthetic biology, and information flow. Information from genomics, proteomics, and metabolomics can be analyzed to create models describing biological activities. Systems biology develops methods to predict behavior of networks of these elements. Synthetic biology closes the loop by reversing the direction of information flow, creating novel genes, proteins, and organisms based on data from naturally occurring systems. This allows testing of predictive algorithms and the creation of new and useful biology. Figure previously published [62] and reprinted with permission from Elsevier.

8.3 DNA AS INFORMATION CARRIER

DNAs are discrete entities of coded information just as letters in the alphabet, musical annotation, and bits of computer code are all coded information. However, unique to genetic information and computational code is that the medium is also the message. The information carrier itself (be it ACGT or 100110) has the ability to perform a defined function (metabolize glucose or query a database) without any manual intervention. Similar to computational source code, the genetic code of synthetic biology captures an algorithm for a function and converts it to a step-by-step protocol.

DNA, as any type of coded information, can be both written and read. Reading is done by DNA sequencing and writing by gene synthesis. Most of the molecular biology over the last decades has focused on reading and analyzing naturally existing DNA sequences, as evident in the massive DNA sequencing effort of the human genome [6,7]. In contrast, writing new genetic information in the form of synthetic biology has only recently become commonplace. Although the first synthetic gene was made as early as 1970 [8], gene synthesis as a standard process to create completely synthetic genetic information only appeared over the last few years.

Today we are not only able to decipher the coding sequence of ourselves and of all other living organisms, but we are also in a position where we have the technology and knowledge to write synthetic genetic code that can operate new biological entities.

8.4 CREATING NEW BIOLOGICAL ENTITIES

A synthetic gene can be made to be an identical copy of a naturally existing gene sequence or it can be made to be a gene that has never existed before and not even be remotely similar to anything previously seen. Or it can be anything in between.

Currently, the majority of ongoing synthetic biology efforts are directed toward making minor deviations from existing genetic information. Typical synthetic biology applications today include efficiently making genetic constructs using "classic" genetic engineering that would be very labor intensive [9–12]. Even though the genes are completely synthetic, the coded information itself is identical or very similar to existing natural genetic information. This type of "synthetic–natural" genes is the application that today drives much of the technical development of gene synthesis.

Due to the degeneracy of the genetic code, a nucleotide sequence can be designed to be only ~60 percent identical to an existing gene, while the encoded protein is still an exact copy of a natural protein. This degenerate property of genetic information can be utilized to create synthetic genes for applications such as making genes for expression in foreign hosts [13–16] or for making RNAi-resistant mRNA transcripts [17]. Here the synthetic genes encode the natural protein information, but the corresponding DNA sequence has been recoded and can be drastically different on the DNA level to encode additional properties that are not found in the natural DNA information.

The availability and acceptance of "synthetic–natural" and recoded synthetic genes are important stepping stones on the path toward completely synthetic biological systems. The commercial demand for "synthetic–natural" and recoded synthetic genes today drives the technology to make all sorts of disparate synthetic genes faster at lower cost and without mutations. This type of synthetic genes are also critical in that they teach us the ground rules for what changes can and cannot be made in the genetic information.

Synthetic genes have also been used to create biological entities that have no precedence in the existing biological environment. These new entities can be completely new DNA or protein structures [18,19], genetic networks [20] with fascinating applications that include molecular computers [21], programmed pattern formation [22], an unbeatable Tic-Tac-Toe player [23], and even a bacteria that take pictures [24].

As our understanding of the rules and grammar of biological information increases, we expect synthetic biology constructs to deviate more and more from naturally existing sequences. But what does this new information imply? If a nanosized octahedron made from synthetic DNA encodes a replicon and can multiply in a surrogate host just as a virus, does it mean it is alive? Is the octahedron a complex organic molecule or is it an organism? And if a molecular DNA computer is built into a self-contained synthetic eukaryotic cell, is that a living calculator? The answers to these questions belong not only to science but also to ethics.

8.5 INTRODUCTION TO NORMATIVE ETHICS IN A GLOBAL COMMUNITY

From the very beginning of human societies, it was necessary to have a set of rules that could guide and regulate human behavior and actions. Such set of rules would define good as well as bad human actions in the context of society. A subset of these rules over time formed the basis for the judicial system. Most, if not all, of these early rules had a supranatural foundation, that is, the rules were perceived to have been defined by a God or set of Gods. Later, in the Ancient Greece (fourth century BC), philosophers like Plato and Aristotle designed rational normative ethics, a tradition followed later by the likes of Spinoza (and his geometrical style ethics), Kant (with his categorical imperative), or Moore (with his critics to the naturalistic fallacy and his stunning and absurd claim about "the direct object of Ethics is knowledge and not practice," *Principia Ethica*, 1903, Chapter 1, Section 14). Both positions, supranatural and rationalist, can be labeled as *foundationalist* approaches, because they find a clear foundation for their beliefs or ideas. For them, there is one and only one truth. Consequently, their moral codes are based on that absolute truth.

At the end of nineteenth and beginning of the twentieth centuries, several philosophers made new approaches to the ethical analysis: Nietzsche killed God and started an *Übermensch*'s ethics based on new myths, whereas Wittgenstein delimitated the possible rational, linguistic, and, therefore, thinkable spaces, excluding ethics from rational debate; to quote, "Ethics, if it is anything, is supernatural and our words will only express facts; as a teacup will only hold a teacup full of water and if I were to pour out a gallon over it" [25]. Wittgenstein continues to pragmatically point out the obvious in an ethical crossroad, "the absolutely right road would be the road which everybody on seeing it would, with logical necessity, have to go, or be ashamed for not going."

This new way of thinking led to the development of an antifoundamentalist ethics by philosophers like Eduardo Rabossi or Richard Rorty. Rabossi, for example, thought that the human rights phenomenon rendered human rights foundationalism outmoded and irrelevant [26]. From this perspective can be understood the claims of underdeveloped countries arguing about the imposition of Christian– Occidental values as if they were universal truths [27–29].

As monolithic religious/rationalist-based ethics is today losing its monopoly in contemporary developed societies, an intense blend of cultures and opinions is increasingly making its presence heard. This change is reflected in the increasingly global perspective of a multicultural society.

An additional ethical concept that has emerged over the last few years is the concept of risk society [30]. Risk society is often described as a systematic way of dealing with hazards and insecurities induced and introduced by modernization itself. Risk society specifically attempts to address how all human beings are connected by ecological and industrial risks, including everything from global warming to electromagnetic radiation from cell phones.

Although the globalized world now requires common global solutions for everything from economic markets to law enforcement, the moral pluralism is instead expanding and common ethics frameworks are diminishing on the contemporary ethical arena. This contradiction has been described as the "collapse of consensus" [31] and is making the efforts to find common solutions and compromises increasingly difficult to achieve.

The facts and promises of biological engineering and synthetic biology create conceptual problems about future decisions because they involve completely new and previously unexpected ways of changing reality. For that reason, opinions like those of Cho et al. [32], "Without prior discussion of ethical issues, the general public cannot develop a framework or common language to discuss acceptable issues of a new biomedical technology, or even whether it should be used at all," are premature as the general public does not share common normative ethics. Few words have in fact so disparate and ethically pregnant meaning as "life." The meaning of the word varies widely in the global transnational community based on respective individual beliefs and historical and casuistic background.

Can there be a common ethical space or must we resign to a complete ethical anarchy? This is not a problem specific for the synthetics biology community but for any ethical problem affecting a global society. If it is true that bioethics is a specialized part of common ethics, with its own topics of interest, we must also consider that it belongs to the debate about sense and meaning among general ethics.

It could be argued that the general public should not decide on ethical aspects of synthetic biology, since several studies about risk perception and scientific literacy show that most citizens of our societies have a distorted, at best, or false ideas about science and the concept of risk [33–35]. The counter argument however is simple: A democratic society must rely on democratic principles as the foundation for ethical framework of any normative behavior.

8.6 EMOTIONS AS THE BASE FOR SYNTHETIC BIOETHICS?

The interest in the ethical aspects of synthetic biology is not only due to the moral implications of this kind of research but also due to the cognitive implications of ethical values for scientific practices. As previously discussed [36], theoreticians of science studies consider nonepistemic values (specifically ethical and moral values) as alien to the scientist's process of making rational decisions [37]. However, neuroscience and the emerging field of neuroethics propose that epistemic values are inherent to natural science [38–40]. At the same time, analysis of the emotional aspects of human reasoning suggests that most of the moral actions imply emotional attitudes and responses [41–43]. From an anthropocentric perspective, information is not just information, a state, but an active quantity of data meaningful for action. We are not perfect rational robots, nor strange Mr. Spock without emotions [44]. Minds have not been evolutionarily designed to just capture the neutral realities but to interpret them in a framework based on previous experiences and in the context of other related pieces of information. Only by categorizing and sorting new information into an existing framework can we start to interact with the captured information through behavior and actions. Emotions shape thoughts and how to relate to and acknowledge new information. And as we now are able to create new biological data, we need to find meanings for those pieces of information to build a framework for our understanding of living entities and biological systems.

Results from neuroethics are suggesting that nonepistemic values, formerly considered alien to the scientific praxis, are instead anchored in the scientist's neuronal processes and are determining their actions. Several investigations in neuroimaging have shown the central role of emotions in the formation of rational judgments [45–47] and in how moral dilemmas initiate cerebral activity in the areas associated with emotion and moral cognition. These emotions are also socially distributed among human communities [48,49].

It can thus be concluded that rational decisions are cognitively processed through emotional elements that can be explained from insights derived from new advances in neuroethics. Therefore, any kind of bioethics that can be thought must be developed from the sentimental frame. The relationships between ethics and sentimentality were initially developed by Rorty [50], at the same time when contemporary discoveries about the limits of rationality (beyond the formal incongruence of the logical classic research, as was demonstrated by Gödel and Russell) and the basic role of emotions in human thinking were made, led contemporary efforts on ethics toward an ethics based on emotions with strong limitations of fundamentalist positions.

8.7 ETHOBRICKS

From the perspective of the collapse of the consensus in contemporary ethics and after the historical failures to achieve a universal ethical code, we here propose a simple project: Create an ongoing ethical frame that can offer answers to the synthetic biology community, a kind of ethical Nash equilibrium based on simple and shared ethobricks. This would be an open and collaborative project (like an universal *wiki*), in which different social agents (scientists, artists, civil society organizations, and so on) define basic ethical pieces for configuring the action's puzzle [5].

The current ethical frame for synthetic biology was defined through the social contemporary circumstances at the time of inception. We can see the early pioneers in this effort when reading the personal writings of the Dolly sheep's creators [52] or we can see it in the Critical Art Ensemble's conflictive artistic works (http://www.critical-art.net). Specially interesting is the use of synthetic biology and genetic engineering techniques by Eduardo Kac, an artist, on his *Move 36*, an open reflection on the limits between artificial and human intelligence through the visual results of the incorporation of a synthetic gene on a new plant (http://www.ekac.org/move36.html).

Currently we have not defined meanings for those genetic information realities, and we are afraid of what to do with it. This is a good starting point: We feel uncomfortable with our actual ideas and the language with which we have developed them, and we are looking for a new way to understand (and modify or create) that biological reality. Before genetic engineering and synthetic biology, humans created names for existing semantic genetic meanings. Chimeras such as mermaids and centaurs were part of the imaginary realm and not a real world. Ethics is an integral part of scientific decision making (e.g., stem cells). There were clear roads in the scientific framework. But our capacity to create new biological meanings require that we create new names and new ethical frameworks to shape the future of living systems, including ourselves, humans. For this we have no references, because those realities were not previously thinkable by whatever philosophical or religious ideas we could consider. As Benner and Sismour state, "A synthetic goal forces scientists to cross uncharted ground to encounter and solve problems that are not easily encountered through analysis. This drives the emergence of new paradigms in ways that analysis cannot easily do" [53]. This is the situation for synthetic biology and, as consequence, for synthetic bioethics.

The origin and justification of the proposed value do not matter. Dividing the ethical concepts into minimal building ethobricks alleviates the need to discern between the truthfulness of different religious, traditional, and cultural beliefs. Ethobricks can be used to define the consensus among beliefs and how to apply ethics to the scientific question asked. Ethobricks would be, then, small ethical blocs with which we could regulate our present and future relationship with synthetic biology research.

With that approach, we can create a common ethical space while avoiding discussions of the foundational basis of ethics. Instead of an ethics of confrontation between deep truths, ethobricks means a "common sense" ethics based on basic emotions, flexible and adaptable to continuous changes. Once synthetic biology community reaches a stable set of ethobricks, it will be transformed into an ethical core with an external belt of concepts under day-to-day supervision. All, core and external ethical belt, are provisional but accepted ways to regulate action.

Like biobricks, basic biological synthetic pieces with which to create new life forms or processes, ethobricks are basic ethical points of departure for a common ethical background. Very important for our approach is the consideration of synthetic biology problems as radically new questions about life for which we have no answers (otherwise, it would not be a problem!). And a crucial problem: We know that people who have values based on divine beliefs or supposed universal principles cannot convince all others the truth of their beliefs (based on divine truths as well as on "rational" ones). Absolute ethics is only possible from absolute beliefs. This is an exclusionary project that separates between those who have the (ethical) truth and those who have not.

Multicultural and democratic societies must develop ethical agreement tools to be able to have coordinated responses to the contemporary ethical dilemmas such as synthetic biology.

8.8 APPLYING ETHOBRICKS

Our project on ethobricks is not just as an engineering code of ethics for practitioners. Bioethics is part of ethics, and we must always remember that ethics is a practice and not an intellectual or mere regulatory process. It is a way of life if considered personally, but a merged moral state if considered socially.

The question is how to find common ways of life and practice? From our perspective, it would be illusory to pretend to find a unique and absolute moral code for all humans. Instead, we must negotiate provisional but reasonably stable codes for deciding our actions. There are many application fields of ethics into synthetic biology, such as bioterrorism, biosafety, patents, life definition, right to manipulation, and control over research (http://openwetware.org/wiki/Synthetic_Society#Synthetic_Society. 2FUnderstanding.2C_Perception_.26_Ethics). These concepts will change in different societies due to the transformations of their sciences, industry and technology.

Therefore, we should try to find provisional ways to develop our activities as scientists, as citizens, as artists, or whatever role we have in our societies.

Ethobricks cannot be regulated by a metaorganization, because it would imply that there is an upper lever from which certain experts know the truth about the discussed facts. Ethobricks is instead an open and continuous project that is gradually implemented on legal traditions. However, there is not only one channel of debate, because it would exclude most of the interested participants from the debate. The equilibrium of an independent ethobrick is achieved if, for continuous space of time, there is not a deep debate about its ethical values. Accordingly, our approach does not imply a different way to define ethics on synthetic biology issues, but it requires a commitment to avoid absolute values. One may think that this leads to a weak ethical frame, but it is the only possible path forward inside true democratic societies.

Can we define the first ethobrick? Certainly yes, *trust*. We can and should trust the fact that all implied participants in this ethical debate seek the best for them and their societies (the basic *pleasure of happiness*). The point is, then, to harmonize for *common* ethical spaces in this increasingly globalized world.

BOX 8-1 MAKING SYNTHETIC LIFE

Current list of examples where genomes have been synthesized *de novo*. In all cases but the T7 phage below, the synthetic genomes are very similar to the natural counterpart. For the T7 phage genome, only a quarter of the genome was synthesized and subsequently combined with the remaining natural three quarters. The synthetic quarter of the T7 genome was redesigned and is significantly different from the natural T7 genome counterpart.

- In August 2002, Dr Wimmer (SUNY) announced that his research team had assembled an infectious poliovirus (7.4 kb) *de novo* using DNA sequence of the viral genome available from GenBank [59].
- In 2003, Dr Smith and colleagues at the Venter Institute developed a two-week selection-based method for the *de novo* synthesis of a phage genome, the 5.4 kb bacteriophage φX174 [63].
- In 2005, Dr Tumpey and colleagues at the U.S. Centers for Disease Control in Atlanta synthesized the 1918 pandemic flu virus genome (13.5 kb) and showed they were infectious in mice [64].
- In 2005, Drew Endy and coworkers at MIT redesigned and synthesized 12 kb of the 40 kb T7 phage genome to make the virus simpler to model and more amenable to manipulation [65].
- In 2008, the J. Craig Venter Institute published the first synthetic prokaryotic chromosome (~600 kb). The DNA was synthesized by three commercial gene synthesis companies (DNA2.0, Blue Heron Bio, and GeneArt) and stitched together by scientists at the J. Craig Venter Institute [61].

ACKNOWLEDGMENTS

The research of Dr Vallverdú has been developed under the main activities of the TECNOCOG research group (UAB, HUM2005-01552), and two special grants funded by Fundació Víctor Grífols i Lucas.

Dr Gustafsson is financially supported by DNA2.0, a synthetic biology company.

We are also grateful to the Synthetic Biology 3.0 (Zurich, Switzerland) for its trust in our research, which we presented during the conference, and to the Royal Society for collecting our preliminary results at http://royalsociety.org/downloaddoc.asp?id=4773, inside the cell, for views about synthetic biology (Nov 22, 2007, http://royalsociety.org/document.asp?tip=0&id=7290).

REFERENCES

1. Forster AC, Church GM. Synthetic biology projects *in vitro*. *Genome Res* 2007;17:1–6.
2. Pleiss J. The promise of synthetic biology. *Appl Microbiol Biotechnol* 2006;73:735–739.
3. Serrano L. *Synthetic Biology: Applying Engineering to Biology*. Brussels, Belgium: European Commission Directorate-General for Research, 2005.
4. Breithaupt H. The engineer's approach to biology. *EMBO J* 2006;7:21–23.
5. Endy D. Foundations for engineering biology. *Nature* 2005;438:449–453.
6. Consortium IHGS. Initial sequencing and analysis of the human genome. *Nature* 2001;409:860–921.
7. Venter JC, Adams MD, Myers EW, Li PW, Mural RJ, Sutton GG, Smith HO. The sequence of the human genome. *Science* 2001;291:1304–1351.
8. Agarwal KL, Büchi H, Caruthers MH, Gupta N, Khorana HG, Kleppe K, Kumar A, Ohtsuka E, Rajbhandary UL, Van De Sande JH, Sgaramella V, Weber H, Yamada T. Total synthesis of the gene for an alanine transfer ribonucleic acid from yeast. *Nature* 1970;227:27–34.
9. Falkowska E, Durso RJ, Gardner JP, Cormier EG, Arrigale RA, Ogawa RN, Donovan GP, Maddon PJ, Olson WC, Dragic T. L-SIGN (CD209L) isoforms differently mediate trans-infection of hepatoma cells by hepatitis C virus pseudoparticles. *J Gen Virol* 2006;87:2571–2576.
10. Meyn MA, Wilson MB, Abdi FA, Fahey N, Schiavone AP, Wu J, Hochrein JM, Engen JR, Smithgall TE. Src family kinases phosphorylate the Bcr-Abl SH3–SH2 region and modulate Bcr-Abl transforming activity. *J Biol Chem* 2006;281: 30907–30916.
11. Smith WW, Pei Z, Jiang H, Moore DJ, Liang Y, West AB, Dawson VL, Dawson TM, Ross CA. Leucine-rich repeat kinase 2 (LRRK2) interacts with parkin, and mutant LRRK2 induces neuronal degeneration. *Proc Natl Acad Sci USA* 2005;102:18676–18681.
12. Harvey SL, Charlet A, Haas W, Gygi SP, Kellogg DR. Cdk1-dependent regulation of the mitotic inhibitor Wee1. *Cell* 2005;122:407–420.
13. Galonic DP, Vaillancourt FH, Walsh CT. Halogenation of unactivated carbon centers in natural product biosynthesis: trichlorination of leucine during barbamide biosynthesis. *J Am Chem Soc* 2006;128:3900–3901.

14. Maroney AC, Marugan JJ, Mezzasalma TM, Barnakov AN, Garrabrant TA, Weaner LE, Jones WJ, Barnakova LA, Koblish HK, Todd MJ, Masucci JA, Deckman IC, Galemmo RAJ, Johnson DL. Dihydroquinone ansamycins: toward resolving the conflict between low *in vitro* affinity and high cellular potency of geldanamycin derivatives. *Biochemistry* 2006;45:5678–5685.

15. Keravala A, Groth AC, Jarrahian S, Thyagarajan B, Hoyt JJ, Kirby PJ, Calos MP. A diversity of serine phage integrases mediate site-specific recombination in mammalian cells. *Mol Genet Genomics* 2006;276:135–146.

16. Gustafsson C, Govindarajan S, Minshull J. Codon bias and heterologous protein expression. *Trends Biotechnol* 2004;22:346–353.

17. Kumar D, Gustafsson C, Klessig DF. Validation of RNAi silencing specificity using synthetic genes: salicylic acid-binding protein 2 is required for plant innate immunity. *Plant J* 2006;45:863–868.

18. Shih WM, Quispe JD, Joyce GF. A 1.7-kilobase single-stranded DNA that folds into a nanoscale octahedron. *Nature* 2004;427:618–621.

19. Kuhlman B, Dantas G, Ireton GC, Varani G, Stoddard BL, Baker D. Design of a novel globular protein fold with atomic-level accuracy. *Science* 2003;302:1364–1368.

20. Guido NJ, Wang X, Adalsteinsson D, McMillen D, Hasty J, Cantor CR, Elston TC, Collins JJ. A bottom-up approach to gene regulation. *Nature* 2006;439:856–860.

21. Benenson Y, Paz-Elizur T, Adar R, Keinan E, Livneh Z, Shapiro E. Programmable and autonomous computing machine made of biomolecules. *Nature* 2001;414:430–434.

22. Basu S, Gerchman Y, Collins CH, Arnold FH, Weiss R. A synthetic multicellular system for programmed pattern formation. *Nature* 2005;434:1130–1135.

23. Stojanovic MN, Stefanovic D. A deoxyribozyme-based molecular automaton. *Nat Biotechnol* 2003;21:1069–1074.

24. Levskaya A, Chevalier AA, Tabor JJ, Simpson ZB, Lavery LA, Levy M, Davidson EA, Scouras A, Ellington AD, Marcotte EM, Voigt CA. Engineering *E. coli* to see light. *Nature* 2005;438:441–442.

25. Wittgenstein L, Lecture on Ethics, 1929.

26. Levisohn JA. On Richard Rorty's ethical anti-foundationalism. *Harvard Rev Phil* 1993; Spring:48–58.

27. Information Office of the State Council of the People's Republic of China. Human Rights in China. Beijing, 1991[http://www.china.org.cn/e-white/7/7-L.htm].

28. Davis M, editor. *Human Rights and Chinese Values*. Oxford: OUP, 1995.

29. Li X. "Asian Values" and the Universality of Human Rights. *Report from the Institute for Philosophy and Public Policy* **16** (2), Spring, 1996.

30. Beck U. From industrial society to the risk society: questions of survival. *Theor, Cult, Soc* 1992;9:97–123.

31. Engelhardt TH, editor. *Global Bioethics: The Collapse of Consensus*. Salem, MA: M6M Scrivener Press, 2006.

32. Cho MK, Magnus D, Caplan AL, McGee D, Ethics of Genomics Group. Ethical considerations in synthesizing a minimal genome. *Science* 1990;286:2087–2090.

33. European Commission, Special Eurobarometer 238/Wave 64.1–TNS Opinion & Social "Risk Issues", http://www.efsa.europa.eu/EFSA/General/comm_report_eurobarometer_en2,3.pdf, 2006.

34. Slovic P. Perception of risk. *Science* 1987;236:280–285.

35. Durant JR, Evans GA, Thomas GP. The public understanding of science. *Nature* 1989;340:11–14.

36. Delgado M, Vallverdú J. Valores en controversias: la investigación con células madre. *Revista Iberoamericana de Ciencia, Tecnología y Sociedad* 2007;3(9):9–31.

37. Thagard P. The passionate scientist: emotion in scientific cognition. In: Carruthers P, Stich S, Siegal M, editors. *The Cognitive Basis of Science.* Cambridge: Cambridge University Press, 2002.

38. Casebeer W. Moral cognition and its neural constituents. *Nat Neurosci* 2003;4:841–846.

39. Farah M. Neuroethics: the practical and the philosophical. *Trends Cogn Sci* 2005;9(1): 34–40.

40. Illes J. editor. *Neuroethics: Defining the Issues in Theory, Practice, and Policy.* New York: Oxford University Press, 2006.

41. Ramachandran VS. *A Brief Tour of Human Consciousness.* New York: Pi Press, Pearson Education, 2004.

42. Denton D. *Emotions: The Dawning of Consciousness.* Oxford: Oxford University Press, 2006.

43. Damasio AR. *Descartes' Error: Emotion, Reason, and the Human Brain.* New York: Putnam, 1994.

44. Pinker S. *How the Mind Works.* New York: W.W. Norton & Company, 1997.

45. Phan L, Wager T, Taylor SF, Liberzon I. Functional neuroanatomy of emotion: a meta-analysis of emotion activation studies in PET and fMRI. *NeuroImage* 2002; 16:331–348.

46. Canli IT, Amin Z. Neuroimaging of emotion and personality: scientific evidence and ethical considerations. *Brain Cogn* 2002;50:414–431.

47. Ochsner K, Bunge SA, Gross JJ, Gabrielo JDE. Rethinking feelings: an fMRI study of the cognitive regulation of emotion. *Science* 2002;293:2108.

48. Rimé B, Mesquita B, Boca S, Philippot P. Beyond the emotional event: six studies on the social sharing of emotions. *Cognition Emotion* 1991;5:435–465.

49. Rimé B, Finkenauer C, Luminet O, Zech E, Philippot P. Social sharing of emotion: new evidence and new questions. *European Rev Soc Psychol* 1998;9:145–189.

50. Rorty R. Human rights, rationality, and sentimentality. In: *On Human Rights: The Oxford Amnesty Lectures.* London: Basic Books, 1993.

51. Vallverdú J. Bioethical art. Genome sense construction through artistic interactions. *Aesthethika: Int J Culture, Subjectivity, Aesthetics* 2006;2(2):7–16.

52. Wilmut I, Campbell K, Tudge C. *The Second Creation: Dolly and the Age of Biological Control.* Cambridge, MA: Harvard University Press, 2001.

53. Benner SA, Sismour AM. Synthetic biology. *Nat Rev Genet* 2005;6:533–543.

54. Itakura K, Hirose T, Crea R, Riggs AD, Heyneker HL, Bolivar F, Boyer HW. Expression in *Escherichia coli* of a chemically synthesized gene for the hormone somatostatin. *Science* 1977;198:1056–1063.

55. Edge MD, Green AR, Heathcliffe GR, Meacock PA, Schuch W, Scanlon DB, Atkinson TC, Newton CR, Markham AF. Total synthesis of a human leukocyte interferon gene. *Nature* 1981;292:756–762.

56. Stemmer WP, Crameri A, Ha KD, Brennan TM, Heyneker HL. Single-step assembly of a gene and entire plasmid from large numbers of oligodeoxyribonucleotides. *Gene* 1995;164:49–53.

57. Hutchison CA, Peterson SN, Gill SR, R.T. C, White O, Fraser CM, Smith HO, Venter JC. Global transposon mutagenesis and a minimal *Mycoplasma* genome. *Science* 1999;286:2165–2169.

58. McDaniel R, Thamchaipenet A, Gustafsson C, Fu H, Betlach M, Betlach M, Ashley G. Multiple genetic modifications of the erythromycin polyketide synthase to produce a library of novel "unnatural" natural products. *Proc Natl Acad Sci USA* 1999;96:1846–1851.

59. Cello J, Paul AV, Wimmer E. Chemical synthesis of poliovirus cDNA: generation of infectious virus in the absence of natural template. *Science* 2002;297:1016–1018.

60. Kodumal SJ, Patel KG, Reid R, Menzella HG, Welch M, Santi DV. Total synthesis of long DNA sequences: synthesis of a contiguous 32-kb polyketide synthase gene cluster. *Proc Natl Acad Sci USA* 2004;101:15573–15578.

61. Gibson DG, Benders GA, Andrews-Pfannkoch C, Denisova EA, Baden-Tillson H, Zaveri J, Stockwell TB, Brownley A, Thomas DW, Algire MA, Merryman C, Young L, Noskov VN, Glass JI, Venter JC, Hutchison CA 3rd, Smith HO. Complete chemical synthesis, assembly, and cloning of a *Mycoplasma genitalium* genome. *Science* 2008;319(5867):1215–1220.

62. Minshull J, Ness JE, Gustafsson C, Govindarajan S. Predicting enzyme function from protein sequence. *Curr Opin Chem Biol* 2005;9:202–209.

63. Smith HO, Hutchison CA III, Pfannkoch C, Venter JC. Generating a synthetic genome by whole genome assembly: phiX174 bacteriophage from synthetic oligonucleotides. *Proc Natl Acad Sci USA* 2003;100:15440–15445.

64. Tumpey TM, Basler CF, Aguilar PV, Zeng H, Solórzano A, Swayne DE, Cox NJ, Katz JM, Taubenberger JK, Palese P, García-Sastre A. Characterization of the reconstructed 1918 Spanish influenza pandemic virus. *Science* 2005;310:77–80.

65. Chan LY, Kosuri S, Endy D. Refactoring bacteriophage T7. *Mol Sys Biol* 2005;1(1):1–10

9

YEAST AS A PROTOTYPE FOR SYSTEMS BIOLOGY

Goutham Vemuri[1] and Jens Nielsen[2]

[1]*Center for Microbial Biotechnology, Denmark Technical University, Biocentrum-DTU, Lyngby DK2800, Denmark*
[2]*Department of Chemical and Biological Engineering, Chalmers University of Technology, Kemivägen 10, SE-412 96, Gothenberg, Sweden*

9.1 INTRODUCTION

The bakers yeast, *Saccharomyces cerevisiae*, is arguably one of the earliest microorganisms to be domesticated for early biotechnological applications such as brewing and baking. Gradually, this yeast has established itself as the primary model eukaryote in the field of genetics and molecular biology. Currently, *S. cerevisiae* is also the most commonly used eukaryote for bioprocess applications, owing to its flexibility in aerobic and anaerobic modes of metabolism and its amenability to genetic manipulations. With the advent of the high-throughput omics technology, it is not surprising that *S. cerevisiae* served as the platform for deciphering the molecular details of chromosomal activity such as DNA replication and transcription as well as physiological activity such as translation and metabolism at a global level. The wealth of information on the cellular components and their structural components has greatly facilitated the progress of yeast biotechnology and metabolic engineering. Several aspects of *S. cerevisiae* fundamental metabolism have been modified and improved to meet the end bioprocess objectives, but more complex aspects such as expanding the range of consumable substrates and the mechanism of glucose repression still remain obscure. The primary reason impeding progress is the lack of knowledge on how the different

Systems Biology and Synthetic Biology   Edited by Pengcheng Fu and Sven Panke
Copyright © 2009 John Wiley & Sons, Inc.

cellular components, such as genes, proteins, and metabolites interact with each other to impart the phenotype. The lack of detailed knowledge of regulation of cellular processes is impeding progress in metabolic engineering, despite rapid progress of technology in genomics, transcriptomics, proteomics, and metabolomics areas.

The cellular processes comprise of the transfer of mass in the metabolic pathways and the transfer of information in the regulatory and signal transduction pathways. The regulation of these processes as a result of environmental or genetic changes is the key step in imparting the phenotype. The information flow commences at the genome level with the DNA. For a given microbial strain, the sequence of the genome remains unvarying. The variation that begins with the primary step in the flow of information is transcription, the process of making mRNA. From this step forward, the abundance of the cellular components highly depends on the environment. For example, in the presence of high glucose concentrations, those genes whose products are required for rapid glucose consumption are transcribed to a greater extent. The next step in the information pipeline is the translation of mRNA into proteins. Proteins are the structural as well as functional entities, carrying out all the cellular functions such as adaptation, regulation, and even catalysis. The transfer of mass from the substrate to the product occurs depending on the protein availability and, therefore, proteins are the link between the information transfer and mass transfer. A simplified schematic of these two cellular pipelines is depicted in Figure 9-1.

Until the late twentieth century, the focus of most traditional enquiries was limited to in-depth analysis of only a small number of cellular components (usually genes, proteins, or signaling pathways) in relative isolation from the remaining system. Although this reductionist approach has been extremely useful in providing detailed description of the individual cellular components, it is to be noted that these components do not function in isolation in the system. Therefore, their biological role has to be elucidated in the context of the remaining components in the system. This line of thinking is the inspiration for modern systems biology, and will be the main focus of this chapter. This chapter begins with a historical perspective of the research that led to the current notion of systems biology and the experimental and computational tools available. The chapter focuses on the development and applications of systems biology in the context of *S. cerevisiae*.

9.2 INTEGRATIVE PHYSIOLOGY AS THE BIRTH OF SYSTEMS BIOLOGY

Although systems biology has entered the popular lexicon only after the millennium, the idea is not new. The natural confluence of systems science with biology and the representation of biological entities as systems were described as early as 1929 [18] in Walter Cannon's homeostasis theory that described the human body as dynamic control system. In 1963, Jacob and Monod followed this line by implementing the concept of control theory to the operation and regulation of the *lac* operon [109]. Subsequently, the concept of a holistic "systems approach," was developed, and it was in 1968 that the term "systems biology," was first used by Mesarovic [107] to indicate

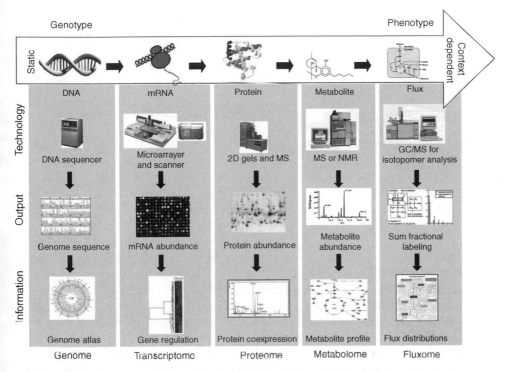

Figure 9-1 The transfer of information from DNA that defined the genotype to metabolic fluxes, which quantify the phenotype. The genotype of a strain is a static entity, but the expression of subsequent components is context dependent. The technology used to quantify the various components in the information pipeline is depicted schematically along with the typical output. It requires careful analysis to extract useful biological information out of the data. Integration of the high-throughput data from these different stages of hierarchy in a context-dependent manner will reveal the interactions between the various components, leading to a holistic understanding of how the phenotype is linked to the genotype.

the application of the techniques of systems scientists (who were conventional control engineers, physicists, and mathematicians) to experimental biology. This was one of the early invitations for biologists to study vital biological phenomenon from a systems perspective. Development continued into the 1970s when researchers developed biochemical systems theory and metabolic control theory to create simplified mathematical models of biological systems, enticing nonbiologists to work with biological systems. This concept was immediately picked up by several researchers who reported many exemplary applications of the control systems theory to life sciences in the following decade. For example, in 1969, Yates pointed out the similarities in the conventional mechanical and electrical control systems and adrenal glucocorticoid control system in humans [178]. Goldbeter and Segel developed the kinetic theory of enzyme action in microorganisms in 1977 [52]. This concept was further developed by Iberall in 1977 using the laws of irreversible thermodynamics to describe the hierarchy in physiological systems. The results from this paper were subsequently used to describe several other regulatory phenomena in living systems [71]. Gradually, these concepts were developed to describe modeling of the structure, control, and optimality of metabolic networks [61].

Along with systems theory, cybernetics played a key role in drawing parallels between the information transfer in electronic systems and biological systems. Moreover, with the development of a formal framework for studying the design of biological networks in terms of error correction, feedback and feedforward control loops, and other circuit concepts, the confluence of the two fields became even more obvious. Together with systems theory and cybernetics, another field that contributed to early systems biology was the field of reaction engineering. The focus of reaction engineering is on the properties of complex reaction networks while monitoring the individual reactants. Although in classical reaction engineering it is possible to calculate important thermodynamic and kinetic parameters, the description of biological systems is far from such quantification. As the application of the concepts of control systems and reaction engineering in biology gained popularity, there was also a simultaneous progress in the development of experimental techniques and high-throughput methods, particularly the ability to sequence complete genomes. The availability of complete genome sequences provides abundance of information, but without any rules pertaining to how the cell processes the genomic information. In addition, the RNA microarray-based expression technology expedited the progress in understanding the transfer of information from the genome to proteins. The paradigm shift in the approach to study biological systems from a reductionist approach to an integrative approach provides a new meaning to the integrative approaches developed thus far and has given birth to the modern meaning of systems biology. Systems biology, in a very general way, can be defined as the integration of genomic, proteomic, transcriptomic, and metabolomic data using computational methods for a holistic understanding of systemic functions. In the context of metabolic engineering, this definition can be interpreted as the unification of information from the flow of mass and energy (in metabolic pathways) and the flow of information from the DNA (transcriptional regulatory pathways and signal transduction pathways). Understanding how the flow of information and the flow of mass occur in tandem is the fundamental tenet of systems biology.

9.3 SYSTEMS BIOLOGY AS AN ENGINEERING DISCIPLINE

We are currently at a crossroads in proceeding with the study of biology. The paradigm shift in the study of biology from a descriptive science to a well-defined quantitative discipline reflects the need to incorporate the theories and principles developed in other disciplines, particularly engineering sciences. The Human Genome Project validated the discovery-driven approach to systems biology for augmentation of the previous purely hypothesis-driven paradigm. With the completion of the human genome and the genomes of various other species, we are now introduced to a number of genes we have never even known existed before. At the same time we are also troubled with the disturbingly finite size of this gene list, and we quickly learned that the diversity of the genes could not approximate the diversity of functions within an organism. The key to this discrepancy is in the combinatorial use of the gene products to impart the diversity. This section will bring out the engineering concepts that are highly applicable in the progress of systems biology.

Systems are central to engineering. The traditional concepts of analysis, synthesis, and design that form the core of the engineering discipline are unified in the systems approach, as shown in Figure 9-2. The system is first decomposed into well-defined subsystems and each subsystem is analyzed for its components and functionality. This defines the analysis component, represented by the left arm in the figure. The knowledge gleaned from the components of the subsystems is assembled into larger and larger subsystems, until the complete system is synthesized. The methodology described here is also known as the bottom-up approach. As applied to biological systems, all the information of individual genes, proteins, metabolites, and so on is gathered, followed by assembling these components in the context of the observed phenotype. Therefore, each level of information processing shown in Figure 9-1 serves as one subsystem. Such a model designed by the bottom-up approach should be capable of describing exactly how the cell functions in response to a certain genetic or environmental alteration. Other less commonly used approach in systems analysis are

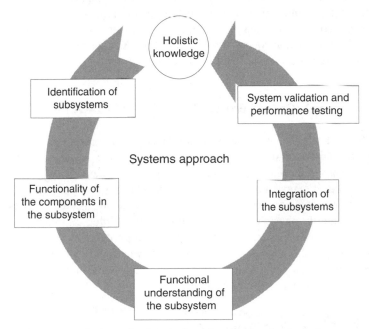

Figure 9-2 One iteration in the cycle of analysis and synthesis using the bottom-up approach to systems biology. As in other disciplines, the system is first defined, followed by the identification of subsystems it comprises of. The components that make up the subsystem are studied in detail for their functionality to understand their role in the context of the whole system. The knowledge from these components is assembled to synthesize bigger subsystems until the complete system is put together and functionally evaluated. This approach reflects the implementation of the classic engineering principles of analysis and synthesis in the context of biological systems.

the top–down approach where the rules are defined for all the individual components identified by the analysis, allowing them to freely interact with each other.

There are three fundamental concepts that an engineer uses to understand a system: emergence, robustness, and modularity. An inherent property of complex systems is that they are larger than the sum of their individual parts, a property known as "emergent property.," The properties of a cell cannot be deduced based on the properties of DNA, RNA, or proteins. It takes a holistic understanding using systems level analyses for a comprehensive understanding of these emergent properties. The robustness of a mechanical or an electronic system is judged on its ability to maintain its functionality despite perturbations. Similarly, a biological system maintains its phenotypic robustness in the event of environmental and genetic perturbations and is a strong determinant in evolution. The feedback and feedforward control loops that comprise a biological or nonbiological system impart robustness to these systems. The third concept central to systems is their modularity. An engineer would define modularity as a subsystem, as shown in Figure 9-2. It is a collection of components that perform a distinct function through interactions and has clear inputs, control processes, and outputs. In biology, modularity refers to a set of components that have close interactions and share a common function. An example of a module of a subsystem in a biological system is the respiratory chain, which is composed of several genes, proteins, cofactors, and regulators that work together in the transport of electrons to oxygen with concomitant energy generation. From an evolutionary perspective, modularity contributed to robustness by restricting the change (malfunction) to the subsystem, thereby decreasing the severity of system failure.

9.4 HIGH-THROUGHPUT EXPERIMENTAL TECHNIQUES

Although the development of the systemic concepts and applying them to biology appealed to a large community of researchers, the lack of experimental techniques to verify the results of the analogy limited the progress of systems biology in 1980s and 1990s. The recent exponential increase in the availability of biological information in the form of genome sequences, RNA, and protein abundance and metabolic flux analysis to quantify physiology transformed systems biology into one of the most exciting scientific developments. Having provided an overview of the concept of systems biology and its evolution to the present-day notion, we devote this section to the experimental techniques that contributed to the advancement of the field from integrative physiology to systems biology.

9.4.1 Yeast Genome Sequencing

The main catalyst behind the rapid progress is the ability to sequence complete genomes. Since the completion of sequencing of the genome of the first independently living organism, *Haemophilus influenzae*, genome sequencing became a routine procedure with the genomes of several microorganisms, including *S. cerevisiae* becoming available. The landmark invention that triggered this explosion was the invention of nucleotide sequencing method by Frederick Sanger [140] and subsequent automation of the process. Sequencing of the yeast genome was the offshoot of a broad international consortium, acting upon a consensus reached in 1988 [1], that was committed to working on a 15-year massive effort to sequence the human genome, supported by a $3 billion funding. The recommendation of this consensus was that the genome sequences of some other eukaryotes should be determined alongside the human genome. The "model," eukaryotic genomes specifically chosen were those of yeast (*S. cerevisiae*), a nematode worm (*Caenorhabditis elegans*), and a fruitfly (*Drosophila melanogaster*). New and faster DNA sequencers were developed, following this initiative and sequencing individual genes became a routine process in yeast. Another landmark result in eukaryotic sequencing was the determination of the complete sequence of a whole chromosome (chromosome II in *S. cerevisiae*) in 1992 [117]. Subsequently, there were several reports of sequencing large fragments of the individual chromosomes in *S. cerevisiae*, which provided the foundation for completion of the chromosome sequencing. Figure 9-3 shows the time line when

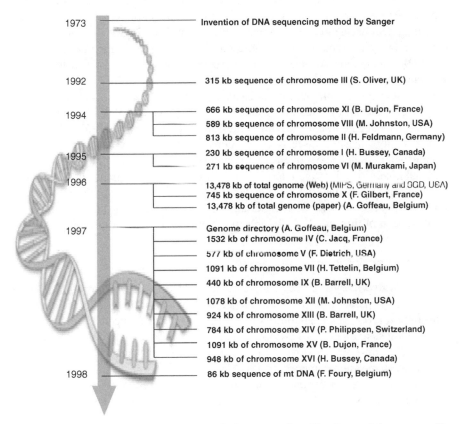

Figure 9-3 The time line of significant events in the sequencing of the *S. cerevisiae* genome. The sequencing of the individual chromosomes paved the path for determining the complete genome sequence. The final draft of the genome sequence was completed in 1996. This is a collaborative effort that required several laboratories across the globe.

the sequences of the other chromosomes became available. The availability of the individual chromosome sequences finally led to determining the nucleotide sequence of the first eukaryotic genome, *S. cerevisiae*, in 1996 [50]. Thus, the yeast genome became the pioneer eukaryotic genome and yeast research community was the primary beneficiary of this knowledge of the complete sequence.

A dramatic transformation of yeast research ensued that presaged similar transformations in the approach to research on other model organisms, as their sequences, became available. This transformation began with technical improvements that accelerated research involving DNA cloning and recombinant techniques. The same consortium that played a key role in yeast genome sequencing undertook another major project of producing deletion mutants of every yeast open reading frame (ORF) [48,173], which led to the development of a whole class of genome-scale genetic methods. The estimated number of ORFs in *S. cerevisiae* is 6034, spread over the 16 chromosomes. A comparative analysis of the complete genome of yeast with the genomes of other model organisms and humans validated the conservation of sequence and function in evolution, despite the difference in the size and number of genes. This observation of "grand unification," has particularly useful ramifications in functional genomics. It became clear that a similarity in sequence is an important factor in assuming functional similarity. Therefore, comparative genomics permits the elucidation of a gene or protein in one organism to be applied to the same in another "lesser," known organism. Since yeast is still the most tractable eukaryotic system, much of the annotation of basic cellular functions in other eukaryotes, including humans, have functional identity in yeast. The availability of the entire genome sequence permits the asking of new kinds of research questions that can be answered only when one has truly comprehensive information about an organism. As previously mentioned, the sequence is the static entity and stable part in the organism under all conditions. It is the interplay between flow of information from the DNA sequence and flow of material in the metabolic network that imparts the variation. The subsequent sections will describe the high-throughput methods that have made the quantification of this variation possible and contributed to systems biology.

9.4.2 Transcriptomics

9.4.2.1 Microarrays The genome basically defines the phenotypic space an organism can operate within and all phenotypic changes ultimately originate at the transcription level. The availability of genome sequences induced the development of technologies to quantify transcriptional activity on a genome scale and to identify the nature of information flow at the transcription level. Once the entire genome sequence became known, it became possible, for the first time, to study expression of all the genes at once; earlier one could study genes only a few at a time. The very idea of what constitutes "specificity," has been changed by the ability to study expression of all the genes without exception. It is now a routine procedure to simultaneously measure the abundance of mRNA species of every ORF in the genome using two-dye spotted arrays [141] or GeneChips [102] in an organism that respond to a specific stimulus or stress. One of the earliest genome-wide transcription characterizations

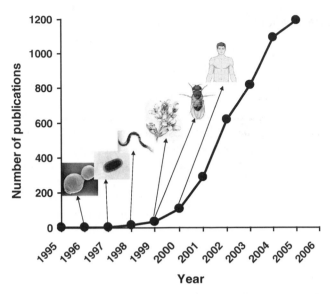

Figure 9-4 The number of publications in peer-reviewed journals that contained the words/ phrases "microarray," "global gene expression," or "oligonucleotide array" in the title or in the abstract. The exponential increase in the interest in quantifying global gene expression stems from the availability of the genomes of the various model organisms: *S. cerevisiae* (1996), *E. coli* (1997), *C. elegans* (1998), *Arabidopsis thaliana* and *Drosophila melanogaster* (1999), and, finally, human (2000). This technology has arguably revolutionized the concept of systems biology.

was to study the expression changes of all the genes in *S. cerevisiae* in response to metabolic shift from growth on glucose to diauxic shift to growth on ethanol [28]. Identifying similarities in the transcriptional profile, the role of many previously uncharacterized genes was predicted, based on the assumption that coexpressed genes are coregulated. Since then, there has been tremendous interest in quantifying gene expression in response to various conditions and, consequently, the number of publications using gene expression microarrays has exponentially increased over the past 10 years (Fig. 9-4).

Prior to the availability of complete sequences, cDNA clones from cDNA banks were PCR amplified and robotically printed onto glass slides, which were used to study gene expression [103,141]. On the other hand, in the photolithography technique, which is popularized by Affymetrix, synthetic linkers are adhered to a glass surface using photosensitive groups, and a light mask is used to direct light to specific areas on the glass to remove the exposed groups. A new mask is used to direct coupling at other sites, and the process is repeated until the desired sequence and length of the oligonucleotide is synthesized. This method, which is very similar to the production of computer chips, is very efficient in high-throughput generation of identical arrays. However, the method is quite expensive in the design phase and the DNA array generated using this method is not flexible when the new genes need to be added. A completely new array needs to be redesigned if new features are to be added. A slightly modified version of this method that resolves this issue is the ink-jet printing of

60mer oligonucleotides [10,68]. This method can generate new arrays or modify the gene content by reprogramming the synthesis of the new set of oligonucleotide sequences. The availability of the genome sequences for several model organisms has facilitated several researches to PCR-amplified genes (either a selected few or the entire list of open reading frames) from chromosomal DNA or design oligonucleotides to develop arrays that are very specifically suited for their purpose. Transcriptional profiling by global gene expression technology is a paradigm of the convergence of several technologies, such as DNA sequencing and amplification, synthesis of oligonucleotides, and fluorescence biochemistry. Transcriptional profiling is based on the fundamental base pairing ability of the nucleotides. The conventional terminology is to refer to robotically printed sets of PCR products or conventionally synthesized oligonucleotides on glass slides as microarrays [30,141], whereas high-density arrays of oligonucleotides that are synthesized *in situ* using photolithography are referred to as GeneChips [101,102], although here we refer to both as microarrays. Numerous reviews have been published that describe the methodologies and analytics behind these methods.

For *S. cerevisiae*, extensive applications of microarrays have been reported, and there are many examples of analysis of genome-wide responses to several environmental and genetic perturbations. These initial transcription applications relied on existing knowledge to confirm some of the results as a means of validating new discoveries. For example, the application of microarrays to the classical study of aging and cell cycle identified several previously known genes in addition to discovering several new ones. Although the cell division cycle in yeast is known to regulate the expression of several histone genes [63], the transcriptional changes in the genome were followed in synchronized yeast cells during various stages of the cell cycle [21,147]. About 7 percent of the genome oscillated with the cell cycle, and every chromosome contained at least one cell cycle-dependent gene. By correlating the expression of the oscillating gene with the stage of the cell cycle, hundreds of transcripts were discovered that exhibited rhythmic expression trends exhibiting close periodicity to the cell cycle. Based on the cell cycle stage, these genes were grouped into different clusters, and analyzing the upstream sequences of genes from the same cluster revealed binding sites for several known as well as unknown transcription factors, indicating the involvement of additional transcription factors in regulating gene expression during the cell cycle. Considering that a large number of human proteins have high homology to yeast proteins, this research could have important applications in understanding human aging.

Microarrays have been extremely useful in understanding regulation and metabolism in yeast. For example, studying the transcriptional responses of *S. cerevisiae* to growth limitation by carbon, nitrogen, phosphorus, sulfur, or oxygen enabled the identification of gene clusters that are involved in sensing nutrient limitation and trigger alternate pathways to minimize stress [11,150]. The stress response induced by nutrient limitation during steady-state growth is apparently different from that observed in normal stationary phase cultures, and that some aspect of starvation, possibly a component of stress response, may therefore be required for triggering metabolic reprogramming associated with diauxic shift as demonstrated by transcription profiling [14,138]. Another important feature of yeast metabolism that was characterized by microarrays is that of glucose repression. Yeast preferentially metabolizes glucose while repressing the genes required for the uptake of several other carbon sources. It also induces the expression of genes required for glucose metabolism such as the glucose transporters and those in glycolysis. Although some of the key players in this complex signal transduction cascade were known for a long time, the true complexity involved in this process is just beginning to be gauged. The central components in the glucose repression pathway are *MIG1*, a DNA-binding transcriptional repressor, and its homologue, *MIG2*, a protein kinase, *SNF1* and its associated regulators such as *SNF4*, and a protein phosphatase, *GLC7* and its regulatory subunit *REG1*. The binding of *MIG1* upstream of many of the genes seems to be the most prevalent mechanism by which the repression pathway acts. In the presence of glucose, Mig1 is localized in the nucleus where it is dephosphorylated and represses gene expression. Upon removing glucose, Snf1 phosphorylates Mig1 and transports it out of the nucleus, resulting in derepression of the genes sensitive to glucose. It is believed that AMP (or even more likely the AMP:ATP or ADP:ATP ratio) signals the phosphorylation/dephosphorylation of Mig1. Several excellent reviews have been written detailing the state-of-the-art knowledge about this mechanism [87,134,155]. The glucose induction pathway is triggered by a completely different mechanism, which induces the *HXT* and *HXK* genes for glucose uptake and phosphorylation, respectively. In this pathway, the key components are a transcriptional repressor, *RGT1*, a protein complex, SCF (SCF complexes are named for their constituent proteins: Skp1, Cdc53 and Cdc34, and an F-box-containing protein), and membrane-bound glucose sensors, *SNF3* and *RGT2*. Upon sensing glucose, Rgt1 binds to the glucose sensors and generates a signal that causes the SCF complex to inactivate the Rgt1 repressor, thereby enabling glucose uptake and metabolism. In the absence of glucose, Rgt1 binds to the HXT promoters and represses their transcription. In this process, Grr1 (glucose repression resistant) plays a key role through ubiquitinating the proteins involved in the signal transduction pathway. The expression of *GRR1* is independent of the carbon source and both the mRNA and protein are constitutively expressed in *S. cerevisiae* in low amounts, thus supporting the role of Grr1p being a regulatory protein [121,122].

Microarrays played a key role in elucidating the regulatory role of *GRR1* in glucose induction. Upon comparing the transcription profiles in Δ*grr1* with its isogenic control strain, we observed large transcriptional changes spread out over different parts of the metabolism [172]. Several genes of the TCA cycle, respiration, and oxidative phosphorylation were induced while many transporters and amino acid biosynthetic genes were repressed in Δ*grr1*. Since Grr1 has also been implicated to play a key role in regulating glucose transport, profiling the transcriptional response of the hexose transporters in *S. cerevisiae* indicated strong repression of the low-affinity transporters (*HXT1*, *HXT3*) and one high-affinity transporter (*HXT4*) while inducing another high-affinity transporter (*HXT8*) and *HXT16*, a hexose permease. These results indicate differential regulation of even the different high-affinity transporters. Analysis of the sequence upstream of these genes revealed the binding sites for the transcription regulators, suggesting a key role for Rgt1 in the repression mechanism. Similarly, upon the identification of a second homologue for *MIG1*, YER028, microarray experiments revealed its glucose-dependent transcription repressing nature [105]. Subsequent DNA binding assays revealed that the binding affinities of *MIG1*, *MIG2*, and YER028 are different, although they recognize the same binding sequence. Transcription profiling also revealed that about 50 percent of the genes that responded to *MIG1* or *MIG2* were of unknown function. High-throughput experiments such as these could help identify genes that could serve as indicators to sense nutrient limitation and so on for inverse metabolic engineering applications.

One such example of transcriptome-guided inverse metabolic engineering was that of designing a strain with enhanced galactose uptake capability [119]. Overcoming glucose control over galactose metabolism has industrial interest in prompt utilization of galactose that is present in lignocelluloses and beet molasses along with glucose. The *GAL* system that contains genes responsible for the uptake of galactose is subjected to dual regulation of glucose repression and galactose induction. Galactose induces the *GAL* system by an ATP-dependent mechanism where the transducer protein Gal3 interacts with Gal80 [149,177]. In forming a complex with Gal80, Gal4 binds to the activator sequences in the *GAL* system, expressing the structural genes, *GAL2*, *GAL1*, *GAL7*, and *GAL10* [174]. These genes code for galactose permease, galactokinase, galactose-1-phosphate uridylyltransferase, and UDP-glucose 4-epimerase, respectively (Fig. 9-5) and are responsible for galactose uptake and its

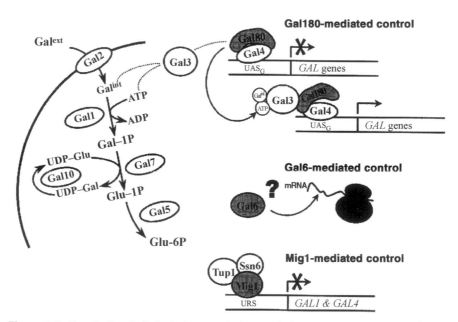

Figure 9-5 Despite the similarity between galactose and glucose, yeast consumes galactose almost three times slower than it can consume glucose. The uptake and metabolism of galactose in the pathway is shown up to its entry in the glycolysis, where its subsequent metabolism is identical to that of glucose. Using inverse metabolic engineering strategies and global transcription analysis, we increased the rate of galactose consumption.

subsequent conversion to glucose-1-phosphate in the Leloir pathway. Therefore, deleting Gal6, Gal80, and Mig1 increased galactose uptake rate by about 40 percent and upon comparing the transcription profile in this strain and in a Gal4 overexpressed strain with a reference strain, there was no clear reason for the enhanced galactose uptake rates. Besides *GAL4*, *GAL6*, and *GAL80*, only the *PGM2* transcript, encoding the major isoform of phosphoglucomutase, exhibited a statistically significant change (of about 1.5 fold) and overexpressing *PGM2* resulted in a 70 percent increase in galactose uptake rate [15]. This case study presents a microarray-guided approach for inverse metabolic engineering, where the targets for metabolic engineering are identified by screening various strains.

Engineering the redox metabolism is an attractive target for metabolic engineering applications since it plays a crucial role in determining growth efficiency and product formation. Despite the importance of redox in metabolism, little knowledge exists about the transcriptional changes that emulate following a redox perturbation. As a fundamental study to identify redox-sensitive genes, a *S. cerevisiae* strain with co-factor modifications in the glutamate generation pathway was compared with the reference strain [16]. Gdh1 (encoding glutamate dehydrogenase) is one of the principal NADPH-consuming pathways in biomass synthesis, consuming more than half of NADPH generated. The sustenance of the Δ*gdh1* strain is ensured by substituting this pathway with the glutamate synthase reaction (GS-GOGAT), encoded by *GDH2*, which uses NADH as the cofactor (Fig. 9-6). Therefore, the switching of Gdh1 with Gdh2 perturbs the redox balance without disturbing biomass generation. Not surprisingly, comparing the transcript levels between the mutated strain with those in the

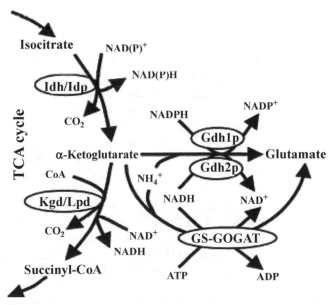

Figure 9-6 Engineering cofactor utilization by replacing the *GDH1* gene with *GDH2* in *S. cerevisiae* [16].

reference revealed that several genes responsible for the regeneration of NADPH have altered expression. This study indicates a possible redox-dependent regulation among these genes, as revealed by the gene expression analysis.

The current use and application of microarray technology is tremendously valuable to systems biology. After more than 10 years of its conception, the rate-limiting step in microarray technology is not in the technical aspects, but rather in the data handling. Currently, only a small fraction of the data generated in a microarray experiment is being used to make inferences for functional testing. This significantly undermines the potential of microarrays in the ability to investigate the genomic response when only a handful of genes undergo further study. The next technological leap for microarray technology lies not in the study of model organisms, but in the interrogation and analyses of uncharacterized or even mixed cell samples whose complete genome may not be available.

9.4.2.2 *Serial Analysis of Gene Expression*

The genome sequences of all eukaryotes are large and contain enormous number of genes, the functions of most of which are yet to be elucidated. In the transcription stage of information transfer, each ORF synthesizes widely varying number of mRNA species. Techniques based on subtractive hybridization and differential display have been useful in identifying the differences among transcripts. However, these methods provide only a partial picture and may miss transcripts expressed at low levels. Oligonucleotide arrays were useful in comparing the expression of thousands of genes in a variety of tissues, including small cell populations, but they are limited to analyzing only previously identified transcripts. In contrast serial analysis of gene expression (SAGE) allows quantification and simultaneous analysis of a large number of transcripts without the prior knowledge of the genes [160]. The SAGE technique can be used in a variety of applications, including analysis of the effect of drugs on tissues, identification of disease-related genes, and elucidation of disease pathways. This method produces 9–10 base sequences or "tags," that uniquely identify one mRNA species. These unique tags are concatenated serially into long DNA sequences for high-throughput sequencing. The frequency of each tag in the sequence quantifies the abundance of the corresponding transcripts. The resulting sequence data are analyzed to identify each gene expressed in the cell and the levels at which each gene is expressed. This information forms a library that can be used to analyze the differences in gene expression between cells. The frequency of each SAGE tag in the cloned multimers directly reflects the transcript abundance. Therefore, SAGE results in an accurate picture of gene expression at both the qualitative and the quantitative levels.

The transcriptome, as defined above, was described for the first time in *S. cerevisiae* [161]. To maximize the representation of the genes involved in normal growth and cell cycle, SAGE libraries were generated from three stages of cell cycle: exponential phase, S-phase arrested, and G2/M phase arrested. The number of SAGE tags required to define the yeast transcriptome depends on the desired confidence to detect low-abundance mRNA species. Employing this method to determine the complete set of yeast genes expressed under a given set of conditions (the transcriptome), 4665 genes (approximately 75 percent of the predicted protein-coding ORFs) were detected, with most genes being expressed at low level. Among the highly expressed genes were those corresponding to well-defined metabolic functions and energy generation under the three stages [161]. Using the SAGE method, metabolic modules that are subject to suppressed translation under normal conditions (translation on demand) were identified in *S. cerevisiae* [8]. This study, which investigated the relation of transcription, translation, and protein turnover on a genome scale, demonstrated significant posttranscriptional control of protein levels for a number of different compartments and functional modules in eukaryotes using *S. cerevisiae* as a model organism, a concept that is missed when exclusively focusing on transcript levels.

Like other genome-wide analyses, SAGE analysis of the yeast transcriptome also has several limitations. For example, a small number of transcripts that lack the appropriate site for tagging could not be detected by this method. Second, there is a basal level of frequency only above which the transcripts could be detected. Despite these limitations, the SAGE method has established itself as a more accurate method to quantify global and local snapshots of gene expression.

9.4.2.3 *Chromatin Immunoprecipitation*

Genome sequencing and microarrays have provided the ability to simultaneously quantify the expression of the entire genome to study transcription. The transcription of genes highly depends on the environment, and the level of expression of a particular gene is controlled by transcription factors (TFs), which bind to the specific DNA sequences upstream of the gene either inducing it or repressing it. Almost every gene in eukaryotic cell is regulated by several positive and negative TFs that recognize the specific binding sites. TF–DNA interaction in a living cell is a complex process with most TFs interacting with other sequence-specific binding proteins and general transcription machinery. These protein–protein interactions (PPI) may affect DNA binding characteristics of the TF of interest. A comprehensive understanding of where enzymes and their regulatory proteins interact with the genome *in vivo* would greatly increase our understanding of the mechanism and logic of critical cellular events. Detailed *in vitro* studies of DNA–protein interactions have provided, and will continue to provide, useful information; it is clear that studies of TF–DNA interactions are critical to understanding the cause and effect relationship between transcription and environment. However, these traditional methods of investigation have failed to create high-resolution, genome-wide maps of the interaction between a DNA binding protein and DNA. For example, the DNA binding properties of a protein determined by *in vitro* oligo selection or gel–shift assays are often poor predictors of a factor's actual binding targets *in vivo*. The study of TF–DNA interactions has undergone a major revolution by overcoming this limitation, owing to the development of combining chromatin immunoprecipitation (ChIP) with DNA microarray analysis (ChIP–chip analysis).

The first ChIP-to-chip experiments were reported at more or less the same time by the Young and Brown groups [77,131]. Both studied TF–chromatin binding in *S. cerevisiae*. Yeast is a good model system for TF–DNA interaction studies for many reasons, one of which is that its genome is much smaller than that of mammals, allowing genome-wide microarrays. DNA fragments from cells grown under controlled experimental conditions that are bound to the transcriptional regulators are recovered by a ChIP assay using an antibody specific to the protein of interest and are hybridized to DNA microarrays that contain the complete set of intergeneic regions. The strength of hybridization intensity signal of a particular gene reflects binding of the transcriptional regulator to the promoter site of that gene. Ranging from yeast to cultured mammalian cells, there is surprisingly little variation in published ChIP–chip protocols. This second generation application of microarrays reveals the network of genes that are bound by one or more transcriptional regulators and presents a very powerful experimental methodology into revealing the first step in transcriptional regulation by identifying gene sets that are bound by the same transcription regulators.

The ChIP–chip technique was first applied successfully to identify binding sites for individual transcription factors in *S. cerevisiae* [77,96,131]. Later, also in yeast, a c-Myc epitope protein tagging system was used to map the genome-wide positions of 106 transcription factors [93]. Other applications including the study of DNA replication, recombination, and chromatin structure have also been reported in *S. cerevisiae* providing a wealth of information on the transcriptional regulation governing these mechanisms. In these experiments, microarrays containing ~1 kb PCR products representing ORFs, intergeneic regions, or both were used in conjunction with a two-color experimental scheme. The PCR products in these arrays were "tiled," across the genome, meaning the PCR products were directly adjacent to one another along the genome, with little or no DNA sequence between arrayed elements. The relatively compact and nonrepetitive nature of the simple genome harbored by yeast made such an approach feasible.

Based on known regulatory information gleaned from biochemistry, gene expression, and ChIP results, it was demonstrated that the strength of interactions between transcription factors and genes is context dependent in *S. cerevisiae* [104]. Studying the changes in gene expression patterns in response to changes in cell cycle, sporulation, diauxic shift, DNA damage, and stress, it was concluded that a few transcription factors are always involved in regulation whereas others depend on the stimulus, thus constantly reprogramming the regulatory network. Only a few target genes are expressed under a specific condition. One of the ramifications of this conclusion, based on over 7000 interactions between genes and transcription factors in *S. cerevisiae*, is that one must use caution when extrapolating the interactions and regulatory mechanisms identified under condition to another.

Recently, an *in vitro* DNA microarray technology for genome-scale characterization of the sequence specificities of DNA–protein interactions was reported based on the ChIP–chip protocol [110]. This technology, known as the protein binding microarray (PBM), allows rapid determination of *in vitro* binding specificities of individual transcription factors by assaying the sequence-specific binding of those individual transcription factors directly to double-stranded DNA microarrays spotted with a large number of potential DNA binding sites. A DNA binding protein of interest is expressed with an epitope tag, purified, and then bound directly to a double-stranded DNA microarray. The PBM is then washed to remove any nonspecifically bound protein and labeled with a fluorophore-conjugated antibody specific for the epitope tag. The PBM technology was used to compare the binding site specificities of the three yeast TFs, Abf1, Rap1, and Mig1 *in vitro* and *in vivo*. The PBM-derived binding site sequences are reportedly more accurate in identifying *in vivo* binding sites. In addition to previously identified targets, Abf1, Rap1, and Mig1 have been reported to bind to several new target intergeneic regions, many of which were upstream of previously uncharacterized open reading frames. Comparative sequence analysis indicated that many of these newly identified sites are highly conserved across five sequenced *sensu stricto* yeast species and, therefore, are probably functional *in vivo* binding sites that may be used in a condition-specific manner [110].

Although the ChIP–chip method can only map the probable protein–DNA interaction loci within 1–2 kb resolution, it also fails to distinguish between positive and negative regulation. The development of the ChIP–chip assay has provided an extraordinarily powerful tool for the analysis of DNA–protein interactions in living cells or tissues on a global scale. In the near future, further advances in microarray construction and the increased availability of useful antibodies will increase the utility of this approach even more. Genomic profiling of transcription factor binding sites, histone modifications, and so on will almost certainly emerge as a central tool in understanding the systems biology of gene regulation in eukaryotic cells. In addition, studies of the genomic distribution of nuclear proteins that are not sequence-specific DNA binders, such as general transcription machinery, the proteasome and its component pieces, DNA replication and repair complexes, and so on will shed new light on fundamental aspects of basic genome function and maintenance. Already, the realization that the majority of transcription factors examined to date are localized outside the promoter sequences has contributed significantly to our growing realization of the importance of abundant noncoding small RNAs in the cell.

9.4.3 Proteomics

Even though global changes in gene expression provide deep insights into understanding transcriptional control, proteins have to be recruited to perform the process since they are the actual functional units. Therefore, knowledge of protein abundance reveals the extent to which regulatory proteins and transcription binding factors participate in the resulting change in gene expression profile. Since gene function is heavily associated with proteins, analysis of proteins will divulge more information on protein function and the pathways they act on. Moreover, although proteins are the end products of gene transcription, there is no one-to-one correspondence between the number of proteins and the number of genes. Therefore, mere transcriptome analysis does not reflect the functional profile at the protein level. This section will outline the emerging quantitative proteomic techniques that are often first developed and tested in *S. cerevisiae*. The focus will primarily be on the two major proteomic technologies that are commonly in use, 2D gel electrophoresis and liquid chromatography coupled to mass spectrometry (LC–MS). The applications of these technologies to investigate protein expression levels of yeast grown under different growth conditions and its implications on systems biology are also discussed.

9.4.3.1 2D Gel Electrophoresis and LC–MS   The most common trend in analyzing proteomes employs two-dimensional gel electrophoresis to separate proteins, followed by mass spectrometry to identify proteins. On a 2D gel, proteins are separated using isoelectric focusing (separation based on isoelectric point of proteins) in the first dimension and sodium dodecyl sulfate polyacrylamide gel electrophoresis (separation based on molecular mass of proteins) in the second dimension. The separated proteins can be visualized using a variety of staining methods such as Coomassie blue dye, silver staining, or fluorescent dyes. Generally, in the first dimension, the proteins are brought on a strip that contains an immobilized pH gradient. By applying an electric field over this strip, the proteins will migrate over the strip until they reach the pH area on the strip where they will be neutral. Each protein therefore will be separated and focused on the strip at the position of its isoelectric point. In the second dimension, proteins are separated on their size/mass. On the resulting two-dimensional gel each protein is present at a position that reveals its approximate pI and mass. Although the concept of 2D gel electrophoresis was introduced more than 30 years ago, its application to proteomics has really taken off since the development of MS-based techniques that enabled high-throughput protein identification.

As with other high-throughput technologies where *S. cerevisiae* was one of the first organisms in which these methods were tested and were subsequently used to conduct genome-level interrogations, some of the early proteomic studies in this context were performed in *S. cerevisiae*. These early large-scale separation and visualization of protein resulted in yeast reference maps, which can be used to locate and identify proteins. The digitalized image maps from these experiments were established by which annotated proteins can be localized and identified directly from the image. For example, the SWISS-2D PAGE yeast database at http://www.expasy.org/ch2d/2-index.html presents the 2D protein pattern of yeast in the pH range 4–9 with 101 spots identified and localized in this area so far. The yeast protein map at http://www.ibgc.u-bordeaux2.fr/YPM/ contains a protein pattern of pH 4–7 with 410 proteins identified. Depending on the protein staining method, approximately 1000 proteins can be visualized on such gels. Also, subproteome reference maps of, for example, yeast mitochondria, have been generated [116,146]. Similar 2D reference maps have been constructed for important industrial yeast strains, such as an ale-fermenting strain, a wine strain, and a lager-brewing strain. These annotated reference maps are useful tools for yeast researchers because they can be used for 2D gel comparisons; however, because of poor gel-to-gel reproducibility and strain variation, protein spot identities should always be confirmed using MS.

Although the conventional trend in analyzing proteomes using two-dimensional gel electrophoresis has had a good turnover of information, the greatest drawback in this method is that it is heavily biased toward proteins expressed at high concentrations [55]. It is also extremely labor intensive and is often hampered by poor gel-to-gel reproducibility. Different staining methods have been developed to improve the accuracy and the sensitivity of protein detection and quantification [125,157], yet proteins expressed at low concentrations may not be detected accurately. Therefore, mass spectrometers are used to detect and identify proteins on a 2D gel. Nowadays, an ordinary mass spectrometer can precisely determine the masses even of large proteins (approximately 1 Da precision at 50 kDa). Since determining only the mass of a protein does not give a direct clue about its identity, two MS techniques are commonly used for protein identification. In the first method, a peptide fingerprint of a protein is recorded, usually by matrix-assisted laser desorption/ionization time-of-flight (MALDI-TOF) mass spectrometry. In the second, slightly more complicated method, short amino acid sequences, the so-called sequence tags, are determined by tandem mass spectrometry.

In the first of the two approaches, the protein spot to be identified is cut out of the gel and digested (in-gel) with a protease, most often trypsin. The resulting peptide mixture is eluted from the gel and analyzed by MALDI–TOF mass spectrometry or alternatively by using electrospray ionization mass spectrometry. Collectively, these peptide masses form a fingerprint, which is indicative for the protein concerned. This fingerprint is then compared to theoretically expected tryptic peptide masses for each protein entry in the database. Generally, peptide finger printing is still the most rapid and efficient method for protein identification. As identification occurs via consultation of protein and genome databases, it may be apparent that their increasing comprehensiveness greatly aids in protein identification. In the second method, the peptide of interest is fragmented in the MS and mass analysis of the resulting fragments allows determining the amino acid sequence from the peptide. Although these fragmentation patterns maybe quite complicated, they generally allow the determination of partial sequences. With this partial sequence, possibly in combination with the peptide fingerprint already obtained, the chance of a unique hit in the database is considerably enhanced. With one or two of these short sequence tags (often no more than five amino acids), it is often possible to unambiguously identify a protein. These strategies, as with other technologies, come with inherent drawbacks. For example, only those proteins that can be visualized on a 2D gel can be analyzed. The 2D gels are incapable of handling large proteins and in general have bad reproducibility and are extremely cumbersome.

In recent years, proteomics methods that employ LC–MS have proven to provide strong alternatives. LC-based technologies have several advantages compared with 2D gel-based techniques. LC–MS, which can be automated, combines high-speed, high-resolution, and high-sensitivity separation of extremely complex peptide mixtures. Several 2D gel independent LC–MS–MS approaches have been introduced to overcome some of the inherent disadvantages of 2D gels. In one approach the proteins in the total proteome are only separated and resolved by molecular mass using 1D gels. Subsequently, this 1D gel is cut into pieces, all proteins in such a band are digested, and the mixture of peptides is analyzed by LC–MS and/or LC–MS–MS [129]. This approach provides an intermediate form between analyzing the very complex large peptide mixture obtained when digesting all proteins of a lysate and the single protein digested when using a 2D gel. As advantages over the 2D gels, a 1D gel-based approach is less elaborate. In addition, very large and basic proteins are easier to handle using just one-dimensional gels. In a third approach, the whole cell lysate is digested chemically or by a protease. This generates a very complex set of peptides, beyond the separation capacity of 1D separation techniques. For the analysis of such complex mixtures, several multidimensional separation techniques have been introduced. An example of an innovative online 2D chromatographic approach is the MudPIT, the multidimensional protein identification technology [169]. In this approach, the complete cellular protein mixture is protealyzed and the resulting peptide mixture is then separated and analyzed using online 2D chromatography directly coupled to tandem MS, which enables the identification of proteins by peptide sequencing. In one of the first applications, MudPIT was used to analyze the yeast proteome and a total of 1484 could be identified [169]. The resulting data set identified proteins from all subcellular compartments, with wide-ranging isoelectric points and molecular weights. Moreover, low abundance proteins, such as transcription factors and protein kinases, as well as hydrophobic membrane proteins, were detected. More recently, the MudPIT method was improved by adding an additional reversed phase column to the biphasic column, resulting in an online 3D LC method [171]. Using this method, it was possible to identify 3109 yeast proteins, which is the most comprehensive proteome coverage reported to date.

Alongside the continuing efforts to develop reliable methods to quantify the proteome, an important advancement in our understanding of the function is the global identification of protein localization in the cell [47,69]. Information about the localization of a protein reveals its function, activation state, and its potential interactions with other proteins particularly in eukaryotic cell, which is compartmentalized. For example, in *S. cerevisiae*, 82 new proteins were discovered in the nucleolus and were predicted to be involved in ribosomal function and, in general, the localization results had 80 percent agreement with the data in the Saccharomyces Genome Database [69]. This study confirmed previously known protein–protein interactions in addition to identifying new ones such as those between cell structure and morphology. Localization of proteins depends on the cell signaling events and their state of activation, which depends on the environmental conditions. Such intercompartmental translocation of proteins triggers new signals. Among the various methods used to study protein localization, variants of GFP are commonly used to tag the protein for visualization using a light microscope [23,69]. The dynamic nature of protein synthesis and consequent modifications, identification, and quantification of proteins alone may not be sufficient. It is also necessary to identify complex formation *in vivo* to obtain a systems view of cellular functioning.

9.4.3.2 Two Hybrid System

An important goal of systems biology is the identification of functional interactions between different cellular components. Since microarrays and 2D gels cannot contribute to the knowledge of protein interaction, protein–protein interactions play a crucial role in elucidating the nature of these mechanisms. Recently, innovative methods for a comprehensive analysis of protein interaction events and signaling pathways have been implemented to provide additional information such as the high-throughput yeast two-hybrid (Y2H) system. The yeast system provided the perfect platform for this assay since it had the advantages of speed, sensitivity, and simplicity in addressing an important biological question when the identification of an interacting protein following its purification was difficult. The Y2H system detects interaction of two proteins by their ability to reconstitute the activity of a split transcription factor, thus allowing the use of a simple growth selection in yeast to identify new interactions. Although the test case for this assay was only a single example of yeast proteins previously known to interact, the results led to the suggestion that the approach might be applicable to the identification of new interactions via a search of a library of activation domain-tagged proteins. Subsequently, the Y2H also proved to be very applicable to study protein interactions in any organisms, although certain types of proteins such as membrane-bound or extracellular proteins were less amenable to this method. The Y2H assay was subsequently adapted to detect protein–DNA, protein–RNA, or protein–small molecule interactions as well as protein–protein interactions that depend on posttranslational modifications, that occur in compartments of the cell other than the nucleus, or that yield signals other than transcription of a reporter gene.

Since its introduction about 15 years ago [35], the assay largely has been applied to single proteins, successfully uncovering thousands of novel protein partners. In the last few years, however, two-hybrid experiments have been scaled up to the proteome scale to identify the complement of all the proteins found in an organism. In the first array-based Y2H of the whole proteome, 192 "bait," proteins were used to survey interactions with 6000 yeast "prey," proteins, resulting in 281 distinct protein pairs [156]. Using a similar strategy with more "bait," proteins to search the yeast genome for protein interactions, 4549 interactions were deduced, out of which a subset of 841 protein pairs were classified as "core," interactions, that is, highly reliable [75,76].

Despite its routine use, the classical Y2H suffers from the appearance of a large number of false positives, even though arrays and other confirmation experiments help to identify them. Two hybrid systems in other organisms such as bacteria or mouse have not been used for large-scale screens, making it difficult to identify if the reproducibility issue is specific to Y2H or if it is a general trait in all such assays.

9.4.3.3 *Protein Arrays*

Considering the pivotal functional role proteins play in defining the phenotype, it is important to quantify protein abundance as well as activity. In the lines of DNA microarrays, protein arrays are rapidly becoming powerful high-throughput tools to identify proteins, monitor their expression, and elucidate their function and interactions within them and, more importantly, the posttranslational changes that they undergo. Several properties of proteins make building protein microarrays more challenging than building their DNA counterparts. First, unlike the simple hybridization chemistry of nucleic acids, proteins demonstrate a staggering variety of chemistries, affinities, and specificities. Moreover, proteins may require multimerization, partnership with other proteins, or posttranslational modification to demonstrate activity or binding. Second, there is no equivalent amplification process like PCR that can generate large quantities of protein. Third, expression and purification of proteins is a tedious task and does not guarantee the functional integrity of the protein. Finally, many proteins are notoriously unstable, which raises concerns about microarray shelf life. Despite these challenges, the development of protein microarrays has begun to achieve some recent success. Currently, protein arrays come in two main formats. The first, abundance-based microarrays, seeks to measure the abundance of specific biomolecules using analyte-specific reagents such as antibodies. The second, function-based microarrays, examines protein function in high-throughput by printing a collection of target proteins on the array surface and assessing their interactions and biochemical activities. Although the applications of these arrays widely differ, they all function on the underlying principle of detecting interaction partners. Abundance-based microarrays include antibody microarrays and reverse protein microarrays. In antibody microarray, antibodies are immobilized and purified proteins and complex mixtures are screened for antibody characterization as well as to quantify protein abundance. Fractionated proteins or protein mixtures are immobilized in reverse protein microarrays and single antibodies are the target screen partners. Function-based microarrays include the standard protein microarrays, where the immobilized component is the protein itself and proteins, antibodies, DNA, or other chemicals are used as the screening partners in functional characterization of the immobilized proteins and to identify their interaction partners. By far the greatest obstacle in developing function-based protein microarrays is the construction of a comprehensive expression clone library from which a large number of distinct protein samples can be produced. In building a clone library, it is desirable to construct recombinant genes where fusion proteins can be produced for the purpose of affinity purification and/or slide surface attachment. Cloning the genes of interest with an inducible promoter allows individual proteins to be expressed in high abundance. High-throughput purification can be accomplished with the addition of C- or N-terminal tags, such as glutathione-s-transferase or the IgG binding domain of protein A. The incorporation of fusion tags also facilitates the verification of clone inserts by sequencing across the vector–insert junction. It is highly desirable to transform the expression vector into a homologous or related cell type, ensuring the proper delivery of the protein product to the secretory pathway and hence correct folding and posttranslational modification of each recombinant protein.

Using these protein microarrays for the first time, the binding activities of three known pairs of interacting proteins was investigated in *S. cerevisiae* [106]. One protein of each pair was printed in quadruplicate onto aldehyde slides, and the arrays were probed with the labeled partners. The most important outcome of this research was that the researchers were able to quantify the concentrations of the bound and solution phase proteins necessary to carry out the experiments. Thus, these experiments demonstrated the feasibility of arraying proteins in a standard microarray format and at feature densities comparable with those of DNA arrays. In a subsequent study, a yeast high-density (13,000 samples per array) proteome microarray was developed that contained full-length, purified expression products of over 93 percent of the organism's complement of 6280 protein coding genes [183]. A total of 5800 ORFs were cloned as glutathione-s-transferase::His6 fusions, and expressed in their native cells under a Gal-inducible promoter. This work represented the first systematic cloning and purification of an entire eukaryotic proteome as well as the first large-scale functional protein array comprising discrete functional proteins. Several different experiments were performed with the arrays, including a calmodulin binding survey to assess protein–protein interactions and a large-scale screen for phospholipid binding specificity [182]. More recently, these proteome chips were used to study global protein phosphorylation in yeast [128], and this study identified over 4000 phosphorylation events involving 1325 proteins from a wide range of biochemical functions and cellular roles. It was also found that these interactions even occur across different compartments, and have helped construct the first draft of a phosphorylation map for *S. cerevisiae*. These results are expected to provide valuable insights into the mechanisms and role of protein phosphorylation in many eukaryotes since several of these proteins are highly conserved.

In spite of these advances, the fundamental aspect that currently limits the advancement of proteomics (in contrast to genomics) is the lack of protein amplification mechanisms analogous to PCR. Therefore, only those proteins that are produced naturally in large quantities or by recombinant techniques can be analyzed. Nevertheless, protein microarrays have shown considerable promise in determining protein–protein, protein–lipid, protein–ligand, and enzyme–substrate interactions. Protein microarrays also have great potential in drug development and clinical diagnostics. We can expect protein microarrays for other organisms as well as for membrane proteins in the near future. Although there is no established proteomics technology to detect all the desired aspects of proteins, aggressive research in the area of proteomics reflects the pivotal role that proteins play in executing metabolic control. It is expected that proteomics will continue to be in the forefront of systems biology research.

9.4.4 Metabolomics

The cells control the concentrations of their intracellular metabolites very rigidly. There is normally a very low tolerance on the allowable variation in the metabolite concentrations for a given physiological state. Conversely stated, a change in the concentration of a metabolite beyond the tolerance level induces a change in the cell physiology. Since they are the intermediates of biochemical reactions, metabolites play a pivotal role in maintaining the connectivity in the metabolic network. Certain metabolites such as ATP or NADH, which are involved in a large number of reactions in the metabolic network, are capable of bringing about significant changes in large parts of the metabolism [115]. The level of the metabolites is a complex function of enzymatic properties and regulatory processes at different levels of information hierarchy. Therefore, similar to the transcriptome and proteome, the metabolome (global set of all the intracellular metabolites) also presents a snapshot of the physiological state of the cell and measuring the changes in the concentrations of intracellular metabolites would reveal an aspect of regulation (such as allosteric inhibition/activation, metabolite–DNA binding, and so on), which cannot be studied by any other omic approaches described. Indeed, metabolome profile presents a closer snapshot of metabolism than the transcriptome or the proteome, because the information flow at this level is the closest to the phenotype (Fig. 9-1). Metabolome profiling also presents a more complete representation of metabolism by defining the thermodynamic equilibrium of a reaction. Therefore, metabolite profiling is now considered an important part of systems biology, playing a complementary role to genomics and proteomics [153,154,170]. However, this field is still in its infancy, mostly due to the lack of analytical techniques. In comparison to more than 6000 protein-coding genes in *S. cerevisiae* [50], there are only about 600 metabolites in *S. cerevisiae* [118]. Thus, even though the goal of any metabolome experiment is to quantify the level of all intracellular metabolites in a cell, tissue, or an organism, there is no single analytical method that can measure all metabolites.

Although the technology to quantify and study the genome (consisting of 4 nucleotides as building blocks) and the proteome (consisting of 22 amino acids as building blocks) is developed based on the similarity in their structure, the metabolomics technology is vastly more complex owing to the highly diverse building blocks, ranging from carbohydrates and organic acids to volatile alcohols and ketones. Consequently, it is virtually impossible to simultaneously determine the complete metabolome with current technologies. Nevertheless, the phrase "metabolome analysis," is used to describe the experimental approaches employed to quantify or detect metabolites. Currently, it is possible to quantify about 50 metabolites (Fig. 9-7). Although metabolite profiling has long been applied for medical and diagnostic purposes as well as for phenotypic characterization, particularly in plants, it is only recently that efforts toward the development of high-throughput analyses are being undertaken [33,34,153]. Mass spectrometry and nuclear magnetic resonance (NMR) are the most frequently used methods of detection in the analysis of the metabolome. The NMR is very useful in determining the structure of unknown compounds, but comes with the drawback of expensive instrumentation. In addition, NMR has the advantages that it is nondestructive to samples and provides rich information on the structures of molecules in complex mixtures. On the other hand, MS is considerably more sensitive and comes with the identification of unknown and unexpected compounds. The combination of separating the metabolites using a gas chromatogram or liquid chromatogram coupled with the MS is transpiring to be the most promising technique for metabolite profiling, thus far. The reader is directed to a very comprehensive review for detailed description and analysis on the different analytical methods employed to identify and quantify metabolites [164]. The issues related with different sampling methods and subsequent processing of the samples, particularly from yeast cultures, are described in another paper [162].

We reported a novel derivatization method for metabolome analysis of yeast that enabled us to measure several metabolites in the central carbon metabolites as well as in the amino acid biosynthesis pathways. Using this methodology, we compared responses of the metabolite profile in a Δ*gdh1* (NADPH-dependent glutamate dehydrogenase) and *GDH2* (NADH-dependent glutamate dehydrogenase) overexpressed mutant and its isogenic reference yeast strains under aerobic and anaerobic conditions [165]. During aerobic growth, the level of all the TCA cycle intermediates increased in the mutant compared with the wild type, indicating a higher TCA cycle flux in this mutant (Fig. 9-8). An increased level of 2-oxoglutarate reflects an alteration in ammonium metabolism due to the thermodynamically less favorable glutamate synthesis using NADH as the cofactor. Moreover, an elevated level of all amino acids was observed, indicating a wide change in amino acid metabolism. More recently, we reported the identification of a pathway for glycine catabolism and glyoxylate biosynthesis in *S. cerevisiae* using metabolite profiling and combining it with pathway

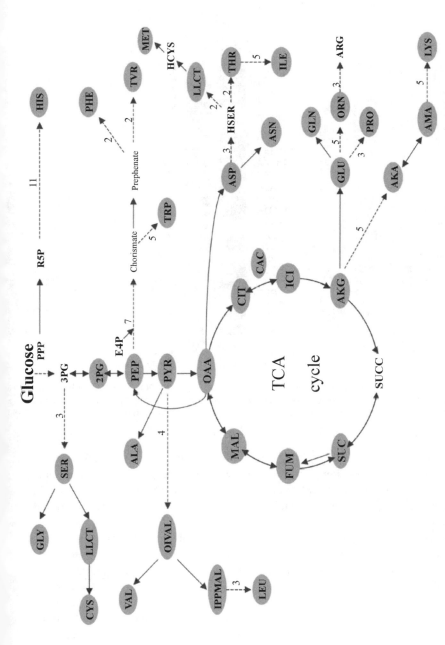

Figure 9-7 Metabolites that can be quantified by the current state-of-the-art methods in metabolomics [162]. In addition to most of the amino acids, several sugar phosphates could also be quantified. However, there are several other metabolites that could be identified (but not quantified) on the chromatograms.

311

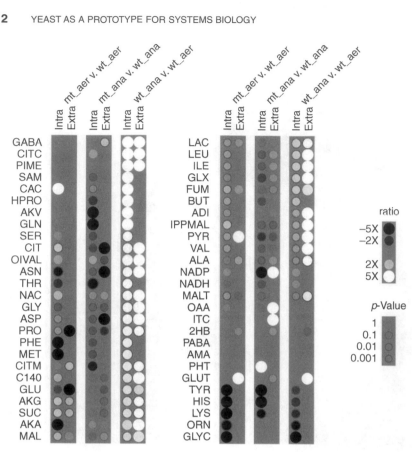

Figure 9-8   Using the current metabolite profiling technology, we determined the differential levels of various central carbon metabolites [166] under various conditions, validating the method as well as laying the groundwork for an integrated transcription–metabolome studies in *S. cerevisiae*.

analysis [163]. Metabolic footprinting ability opens a new avenue in yeast systems biology research by providing results that neither gene expression nor proteome analysis could. Second, we demonstrated that the levels of specific metabolites could be quantified using this method, enabling the targeted and quantitative microbial metabolome analysis.

These examples demonstrate the immense potential of metabolite profiling in providing supplemental information to transcriptome and proteome analysis. However, there are a number of challenges for this nascent field. The fundamental problem arises due to the rapid turnover time of metabolite (in the order of 2–3 s), which makes it extremely difficult to capture a reliable snapshot of a metabolite profile. Second, the analytical methods for identifying and quantifying these metabolites are still in its infancy. Third, there is no robust data analysis methodology to integrate metabolite profile in the context of genome and proteome and interpret the physiological significance of an observed change in the metabolite level. Finally, the lack of standards in this field results in poor reproducibility. Recently, metabolomics ontology and experimental reporting standards have been proposed by the Metabolomics Society [100] to facilitate the establishment of credibility to the large amount of data that is being generated. Despite these challenges, there is growing belief in the scientific community that metabolomics holds the promise to expedite the progress of systems biology.

9.5 COMPUTATIONAL APPROACHES IN SYSTEMS BIOLOGY

The rapid advancement in the experimental approaches in measuring genome, transcriptome, proteome, and metabolome, as described in previous sections, gives rise to enormous data. It has now become clear that further discovery and progress in biological research will be limited not by the availability of data but by the lack of the right tools to analyze and interpret these data. Systems biology calls for the development of mathematical principles to integrate these high-throughput data. From its humble origins in control engineering and general systems theory, major challenges in the dynamic pathway modeling have been addressed and goals realized by (1) characterizing model structures that could realize the given stimulus–response relationship, (2) determining values for model parameters from experimental data and simulations, and (3) predicting the consequences of perturbations by introduction/removal of feedback or feedforward control loops. The new fields of genomics, proteomics, transcriptomics, and metabolomics are extremely essential not only to divulge information in the different levels of material processing in the cell but also to serve as precursors for realizing the larger objective of phenotypic characterization. Computational biology now plays a predominant role in the discovery process through automated genome reconstruction, flux balance analysis and metabolic networks, protein structure determination, and elucidation of regulatory networks. The synthesis arm of the systems biology cycle as depicted in Figure 9-1 heavily relies on the robust, reliable computational biology aspects. Since the ultimate cellular phenotype is the result of coordinated activity of multiple gene products and environmental factors, understanding the connectivity and interaction among these elements is pivotal.

9.5.1 Constraint-Based Genome-Scale Models

The key role of computational approaches in systems biology has been acknowledged and accepted. The development of mathematical models that can simulate the cellular phenotype by integrating high-throughput data forms the foundation of systems biology approaches. We view systems biology as an iterative process where mathematical models are built and developed based on the experimental data available. The goal of these models depends on the questions one is trying to address. For example, to describe and understand the metabolic dynamics, one uses a detailed kinetic model, or to study the mechanism of signaling cascades in a regulatory network, one formulates differential equations to describe the inputs to each module in the cascade and its response. The predictions and simulations from these models are then validated by experimental methods to complete one iteration in the process of understanding cellular properties. Any discrepancy between the model predictions and experimental observations will be addressed by incorporating the new experimental results in the model to start a new iteration. Since it is virtually impossible to determine the kinetics of all the steps in the system, these models have an obvious limitation. Among the several classes of mathematical models available to analyze cellular behavior, the constraint-based linear models are the only kind that can incorporate extensive biochemical data, genomic sequence data, and information from metabolic pathway databases into the context of simulating and predicting cellular metabolism and phenotype.

The development of linear metabolic models begins at the identification and functional annotation of the ORFs in the genome sequence. The Saccharomyces Genome Database (http://www.yeastgenome.org), Munich Information for Protein Sequences (http://mips.gsf.de/), and Kyoto Encyclopedia of Genes and Genomes (http://www.genome.jp/kegg) are the most commonly used databases to search for genes, their products, and their functional role in metabolism for *S. cerevisiae*. A comprehensive list of all the metabolic genes constitutes the genotype and an *in silico* representation of this subset of genes is the basis for creating of *in silico* strains. The gene products derived from the genes in the metabolic genotype carry out all the enzymatic reactions and transport processes that occur within the cell (Fig. 9-9). For example, the *in silico* representation of *S. cerevisiae* includes the genes involved in central metabolism, amino acid metabolism, nucleotide metabolism, fatty acid and lipid metabolism, carbohydrate assimilation, vitamin and cofactor biosynthesis, energy and redox generation, and macromolecule production (i.e., peptidoglycan, glycogen, and nucleotides). The reactions that are mediated by each of the gene(s) are represented as a linear equation, and all the stoichiometric coefficients from all the reactions are collected in the stoichiometric matrix, S, and the velocity (flux) of each reaction is collected in the velocity matrix v, which has the same number of rows as the number of reactions. Any exchange fluxes that are involved in material transfer across the systemic boundary are represented by the matrix, b. Under any given condition, the *in silico* metabolic network can then be represented as $S \cdot v = b$ [158]. The construction of the metabolic network is covered in greater detail in other chapters of this book. The fundamental drawback of these models is that they operate strictly on the stoichiometry and do not consider thermodynamic constraints and kinetics and, therefore, cannot resolve the directionality of the reaction. This inherent drawback is partly addressed by imposing constraints on the fluxes and defining their directionality and degree of utilization as $\alpha_i \leq v_i \leq \beta_i$, where α_i and β_i are the lower and upper bounds, respectively, of the *i*th reaction. The values of the fluxes are estimated by imposing an objective function, often maximizing for biomass production rate, which is also expressed as an equation that is conceptualized by including the individual biomass precursors that contribute to the synthesis of biomass. Since typically the S matrix is underdetermined (the number of unknown fluxes is far greater than the number of measured parameters), linear programming is the most commonly used procedure to estimate the unknown fluxes. Due to the existence of an infinite number of solutions in the feasible space and even the presence of several solutions of the flux vector that fulfils the objective

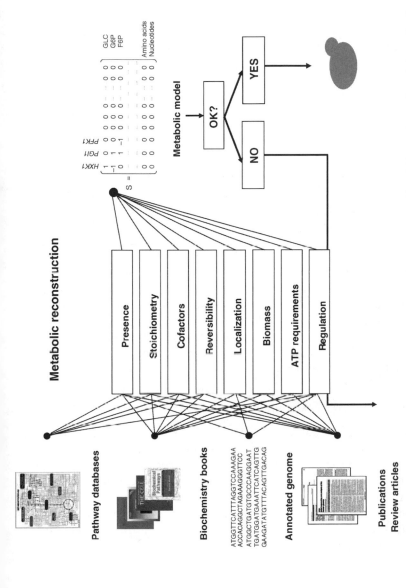

Figure 9-9 A schematic methodology involved in constructing the genome-scale stoichiometric matrices. Information from various databases, knowledge from fundamental biochemistry, functional annotation of the genome of the organism, and information from various review and research articles are gleaned to assess the presence of a certain pathway and its stoichiometry. Additional information such as localization and reversibility are also incorporated in the matrix. Simulating the metabolism by flux balance analysis using this matrix and validating with experimental data will decide further fine-tuning of the model or its implementation to determine metabolic capabilities.

function, it is very likely that the estimated fluxes may not accurately represent biological reality [29]. Nevertheless, these models are extremely useful in characterizing the metabolic capabilities of the cell. The first genome-scale metabolic model of a eukaryote ever constructed was that of *S. cerevisiae* [39], which includes over 800 reactions and 500 metabolites. The details of this model can be viewed at http://www.cpb.dtu.dk/models/yeastmodel.html. Using this model, several aspects of *S. cerevisiae* metabolism such as biomass yields under various carbon sources, gene lethality and synthetic lethality, pathway utilization, and general network properties such as connectivity were successfully studied [39].

It is commonly observed that the performance of the *in silico* cell, such as the rates and yields of biomass and product formation, is far below the predicted theoretical maxima, particularly for strains that have undergone the first iteration of metabolic engineering. This phenomenon has detrimental impact on the utility of *S. cerevisiae* as a cell factory for commercial applications. This is due to the extensive adaptive mechanism of the cell to counteract any mutation. The stoichiometric models cannot predict this sluggish performance of the cell but rather provide the maximum cellular capabilities under the conditions of mutation. The discrepancy arises due to the assumption that maximizing biomass formation drives flux distribution. Even in response to a mutation, this approach assumes that the metabolic network could readjust to maintain optimal flux toward biomass. Recent evidence suggests that this could be achieved by selecting for fast growth [70]. However, in most industrial strain improvement scenarios, cells are subject to natural selection and a modification of the flux balance models using constraint-based linear programming approach is recently described to predict the sluggish metabolic phenotype [144]. This approach, known as minimization of metabolic adjustment (MOMA), assumes minimal response of the metabolic network to gene perturbations and suggests that the metabolic network has an inherent inertia to change and prefers to remain as close as possible to the original steady state (of the wild-type genotype).

9.5.2 Metabolic Pathway Analysis

The cell has multiple pathways at its disposal to attain its natural objective of survival. In the process of engineering these metabolic pathways, we attempt to manipulate these pathways to eliminate the ineffective ones or enhance the performance of the rate-limiting ones. Whether the goal is to delete pathways or overexpress them, it is necessary to develop an understanding of how the cell meets its metabolic objectives. This is the goal of metabolic pathway analysis. It is an integral analytical part in the discovery of meaningful routes in the metabolic networks, constructed as described in the previous section. By virtue of the complexity in the wide array of feasible metabolic pathways, it is not always intuitive which set of pathways are employed in reaching the cellular objective. The most commonly used mathematical tool that is used to analyze the set of all feasible pathways for robustness and efficiency is by elementary flux modes (EFMs) [142]. A flux mode is a steady-state flux distribution in which the proportions of the fluxes are fixed. If this steady-state solution is non-decomposable, then it is classified as elementary. In other words, an elementary flux mode is the minimal set of enzymes that could operate at steady state with all the reversible reactions assumed to proceed in the appropriate direction. Therefore, this concept assumes three conditions in determining an EFM: a pseudo-steady-state condition, a nondecomposability condition, and a feasibility condition.

Elementary flux modes are idealized representations of metabolism and it is very likely that any one EFM cannot represent biological reality. Instead, the real flux distribution is a linear combination of several EFMs, each of which has a fraction of contribution to the final flux. When *S. cerevisiae* grows on glucose, all the pathways that use other substrates are downregulated, even in the presence of a mixture of substrates. This is clearly demonstrated using DNA microarrays during the diauxic shift where 183 genes are induced and 203 genes are repressed at least fourfold [28]. This reflects a marked shift in the utilization of different metabolic modes, which are the likely superpositions of other EFMs. In fact, a very strong correlation was observed between the EFMs, as determined for yeast grown on different carbon sources, and the transcript measurements from microarray experiments [17]. Above all, the method of determining the control-effective fluxes to calculate the theoretical transcript values and correlating them with the experimentally derived transcript ratios demonstrates the importance of flexibility in metabolic networks. In this regard, the EFMs have a greater applicability over flux balance analysis. Moreover, since there is no objective function in this kind of analysis, unlike the flux balance analysis where the objective function is usually maximization of biomass formation, the system is not forced to behave in a particular manner. Since the metabolic reaction system is allowed full flexibility, it is free to choose all the possible routes toward product formation. Metabolic pathways analysis has also been used to assign function to orphan genes in *S. cerevisiae* based on convex analysis of its simplified metabolic network by combining metabolome analysis with metabolic pathway analysis [40]. Based on this analysis, a change in the pathway structure of deletion mutants could be combined with the different metabolite profile for that mutant to disclose the functionality of an orphan gene.

In many situations, the biosynthesis of a product is feasible by multiple routes and it is interesting to identify the pathways that give maximal yield. The optimal flux distributions, as predicted by the flux balance analysis, may not always be obtainable, thereby making it necessary to determine the suboptimal solutions using EFMs. The concept of EFM can also be used to predict the effects of an insertion or a deletion of a pathway, resulting in a pathway with new functional capabilities. This method allows a comparison of sets of admissible routes for product formation in wild-type cell and its engineered mutant. By comparing the elementary modes in the complete system with those in a deficient system, it can be shown whether or not an essential biological substance can still potentially be synthesized, via a bypass in the network system. Elementary flux modes essentially capture all the possible flux distributions (optimal as well as suboptimal) in the metabolic network as defined by the stoichiometric matrix, unlike flux balance analysis, which returns only one optimal solution. An important aspect of EFM that cannot be determined using the flux balance analysis is that of futile cycles. Futile cycles play an important role in regulation in eukaryotes and it is extremely important to identify them to avoid wasteful expenditure of cellular energy. Although such cycles are difficult to identify in large networks, they can be detected by calculating elementary modes, which include both cyclic and noncyclic metabolic pathways. This method is valuable for comprehending the complex architecture of cell physiology and together with other theoretical tools such as metabolic control theory, it can help to engineer living cells in a directed and rational way.

9.5.3   Gene and Regulatory Networks

The information pipeline in cells is extremely efficient and can robustly respond to multiple environmental and genetic signals. The mechanisms by which cells are able to achieve this are still not clear due to complex regulatory circuitry in the cell. To uncover the mechanisms that dictate the information processing, a modular approach is the most common approach [60,92]. The high degree of complexity involved in cellular response can be simplified by considering the large-scale genetic networks as composed of modules of simpler components that are interconnected through input and output signals, analogous to electrical circuits [130]. The analogy between genetic circuits and electrical circuits extends beyond just the superficial level. Just as electrical engineers construct circuits, genetic network engineers make use of the biological equivalents of inverters and transistors to manipulate living organisms by connecting these modules into gene regulatory networks that can control cellular function. Two landmark studies published in 2000 [31,43] clearly illustrate this concept, in which one describes a genetic circuit engineered into *Escherichia coli* cells that oscillates asynchronously with regard to the cell division cycle [31] and the other describes a toggle-switch circuit that can be switched between two stable states by transient external signals [43]. In both studies, the circuits' qualitative performance is consistent with the predictions of relatively simple differential equation models that characterize the dynamics of production, degradation, and genetic regulation.

The interactions between the functional modules in the gene regulatory networks involve proteins, DNA, RNA, and small molecules. For example, a simple module consists of a promoter, the genes expressed from that promoter, and the regulatory proteins that affect the expression of the promoter. The idea behind formulating gene networks and subnetworks is essentially to identify those genes that are commonly bound by the same transcription factor. Since the output of a microarray experiment is the end result of the interplay between transcription factors and genes, this aspect has been the focus of recent data analysis methods. Since one gene is under the control of multiple transcription factors, the amount of control from each transcription factor is not easy to quantify. Associating transcription with binding information for 106 transcription factors, Bar-Joseph et al. clustered coexpressed genes to reconstruct regulatory networks in *S. cerevisiae* [6]. They identified established interactions as well as discovered new interactions that they used to construct regulatory models. Liao et al. developed a similar approach called network component analysis to quantify the strength of interactions between genes and transcription factors [95]. The interactions were modeled as a two-layered network with transcription factors consisting of the first layer and the genes in the next layer and the interactions between the two layers as edges. Implementing this technique for glucose to acetate diauxic shift in *E. coli*, 16 transcription factors were found to be significantly involved in the transition. The biggest advantage of this method is that it does not assume independence or orthogonality of genes, unlike independent component analysis or the principal component analysis, respectively. Although these reports demonstrated the use of gene expression microarrays to study the regulation of specific pathways at the transcriptional level, they still do not account for regulatory effects brought about by proteins and metabolites interacting with DNA, and therefore such an approach would not be feasible in higher organisms with a greater level of complexity. As pointed out by Nielsen, the percentage of genes that are encoded for nonmetabolic functions (particularly for regulatory functions) increases with increasing cellular complexity [115]. To reveal regulatory phenomena based only on the changes in gene expression, detailed information about interactions between genes and their transcription factor proteins must be elucidated. Recently, it was demonstrated that a stochastic simulation algorithm can be efficiently implemented by using field programmable gate array devices to build a microelectronic circuit that simulates the kinetics of biochemical networks [139]. Such devices, built as an array of simple configurable logic blocks embedded in a programmable interconnection matrix, are ideally suited to implement highly parallel architectures comparable in complexity to biochemical networks. The parallel architecture of this logic-based programming can simulate the basic reaction steps in biological networks and since they can be scaled up efficiently, simulations of realistic biological systems should be possible.

We are still far from completely understanding the wiring of the regulatory circuit in a system, and the challenge lies in designing selection schemes that can be used to drive cells containing artificially engineered gene circuits for a robust, reliable, and noise-resistant behavior. The current paradigm for engineering regulatory circuits is to use computational methods to incorporate the desired changes in the cell. The engineered cells usually exhibit weak compliance with the desired objective and by using a directed evolution selection screen, more compliant mutants could be produced. The engineering of regulatory networks has immense applications in the production of industrial or medically important chemicals such as proteins and antibiotics and in the design of cells to perform complex multistage tasks such as conversions in bioremediations or cell-specific activity for gene therapy. A variety of relatively simple but useful types of biological circuits similar to switches, transducers, signal processors, sensors, and actuators are already being developed from the existing knowledge of the cellular components.

9.5.4 Protein–Protein Interactions

Proteins are the functional units in the cell and carry out most of the information processing such as intracellular communication, signal transduction, and even gene regulation via interaction with other proteins. Identification of protein–protein interactions on a proteome-wide scale is currently one of the main challenges of systems biology. Although the genome sequencing projects have identified the comprehensive set of genes and proteomic studies identified the protein abundance in several species, there are no thorough methods to identify interactions between proteins comprehensively. The current methods to identify PPIs include genetic and biochemical screens, which identify few interactions at a time but when applied in combination produce highly reliable results. Using a combination of expression profile reliability index to estimate biologically relevant fraction of protein interactions and paralogous verification method to score the interactions, over 8000 pairwise PPIs were detected in *S. cerevisiae* [25]. These interactions identified by such small-scale screens represent only a small fraction of the biologically significant interactions in yeast [53]. To identify the PPIs on a proteome scale, several high-throughput experimental methods have been developed such as the yeast two-hybrid assay described earlier [156], tandem affinity precipitation [44], and high-throughput mass spectrometry protein complex identification [64]. Although all these methods have the capability to detect thousands of interactions, their reliability is limited due to the high occurrence of false positives and false negatives. Besides these approaches, there also exist other approaches to infer PPIs based on indirect evidence, such as synthetic lethality [152], correlated expression of gene pairs [27], or identifying structural domains and subcellular localization [114]. Despite the relatively lower confidence of the interactions predicted by these methods, they are still very popular in elucidating PPIs. Recent reports indicated that integrating data from different levels of information hierarchy with these high-throughput methods significantly improve the reliability of the inferred interactions [51,80,180]. The computational aspect of predicting the interactions (e.g., between protein A and protein B) is usually based on the general criteria such as (1) they should have appropriate domains to facilitate interactions, (2) the expression levels of genes A and B should correlate, and (3) proteins A and B should be localized in the same compartment.

Using Y2H assays, proteins were assigned to functional classes on the basis of their network of physical interactions as determined by minimizing the number of protein interactions among different functional categories [159]. Such a functional assignment is proteome wide and is determined by the global connectivity pattern of the protein network. Using this approach, multiple functional assignments could be possible for a given protein, depending on its interaction with other proteins. This analysis is based on the concept that interacting proteins may belong to at least one common functional class, and thus knowledge of the functional classification of a subset of the proteins involved in the network may lead to an accurate prediction of the functional classification of the remaining subset of uncharacterized proteins. The idea behind this approach is to assign function to unclassified proteins based on its position in the interaction network, also known as the "majority rule," assignment [143]. The majority rule derives from the empirical observation that 70–80 percent of interacting protein pairs share at least one function. In most cases, only a few unclassified proteins interact with more than one protein of known function and often the interacting proteins with known functions do not generally share functionalities. In this respect, the majority rule assignment is inconclusive because the analysis does not include the links among proteins of unknown function. Therefore, much of the information contained in a reconstructed protein–protein interaction network is not used. A major concern in implementing network-based predictive methods is the topological accuracy of the protein interaction network. It is known that protein–protein interaction data obtained from two-hybrid experiments contain a certain number of false positive and negative results, as discussed earlier. These errors could compromise on the quality of the predictions by incorporating spurious connections into the network (false or missing edges).

9.6 INTEGRATING THE HIGH-THROUGHPUT DATA

The primary step in understanding any biological entity from a systems perspective is to identify its structural organization, such as the gene interactions and biochemical networks, followed by the dynamic interactions between them. Characterization of biological networks requires detailed maps elucidating proteins, RNAs, promoters, and other macromolecules. Toward this broad goal, metabolic networks [81], regulatory networks [93], and protein interaction networks [156] have already begun to be established. These maps are commonly represented as a static set of nodes to represent the components (RNA, proteins, macromolecules, transcription factors, etc.) of the network and edges to represent the interactions (activation/inhibition or induction/repression, etc.) between them. Human minds are incapable of inferring the emergent properties of a system from thousands of data points, but we have evolved to intelligently interpret an enormous amount of visual information. The data are therefore transferred to visualization programs. This is the initiation point for the formulation of detailed graphical or mathematical models, which are then refined by hypothesis-driven, iterative systems perturbations and data integration (Fig. 9-10). For example, using a bipartite graphical visualization, Patil and Nielsen showed similarities in metabolic network patterns and transcriptional responses that led to the identification of "reporter metabolites," in *S. cerevisiae*, which represent the hub of regulatory action [124]. Similarly, topological analysis of metabolism in 43 organisms revealed hierarchical modularity in the network organization [130]. Using the path of shortest length in graph-theory approach, Said et al. identified that the toxicity-modulating proteins in *S. cerevisiae* have more interactions with other proteins, leading to a greater degree of metabolic adaptation upon modulating the functioning of these proteins [137]. This result has direct implications on many human degenerative disorders such as cancer and even aging. The authors demonstrate that the protein interaction network is much more complex than the metabolic network, consistent with the knowledge that signaling pathways and regulatory networks have more complex organizational structure than the metabolic network. Although only protein interactions were studied, deeper regulatory aspects could have been revealed by also including protein interactions with DNA, particularly since the study focused on the recovery of *S. cerevisiae* from DNA-damaging agents. As opposed to the representation of biological networks as graphs that reflect only the static properties of system, de Lichtenberg et al. have recently reported the dynamics of protein interactions during the yeast cell cycle [24]. They used previously published gene expression data from different stages of the cell cycle [21,147] and integrating it with a network of

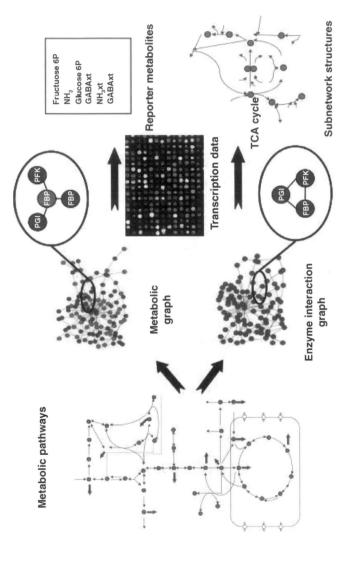

Figure 9-10 The data integration methodology we have developed identifies reporter metabolites, which indicate the hubs of transcriptional regulation and the subnetwork structures [124]. Gene expression data from a particular experiment then is used to identify highly regulated metabolites (reporter metabolites) and significantly correlated subnetworks in the enzyme interaction graph.

physically interacting proteins from public databases such as MIPS discovered that most of the protein complexes are comprised of both constitutively and just-in-time expressed proteins. Currently, the mathematical models that represent cellular components and their interactions compromise either on the specificity or lack the sensitivity. This is due to several reasons, such as a limitation in biological information available and lack of mathematical rules to integrate the available information. Learning how the structure changes in response to various conditions and, more importantly, what makes the system respond in this fashion will enable identifying precise targets for metabolic engineering [86]. Established protocols are not immediately available to guide the merger of global information from various omes indicated in Figure 9-1. Ideker et al. [72] compared the global changes in the expression of mRNA and proteins in *S. cerevisiae* in response to a series of perturbations in the GAL regulatory system. They used the yeast galactose metabolic model as a prototype and studied the global responses to genetic and environmental perturbations. The key feature of this study that is missing from the previous comparisons was that the authors also considered protein interactions with other proteins and with DNA in their model. Not surprisingly, the expression of those genes that are linked by physical interactions exhibited a higher degree of correlation with corresponding protein levels. Information about protein–protein interactions in *S. cerevisiae* [143,156] facilitates the integration of the resulting mRNA and protein responses with known physical interactions to discover and/or refine gene functions. Since it is the proteins that actually execute the genetic program, mapping global interactions between proteins or "interactome," in single-celled [156] and multicellular [94] organisms is particularly valuable in revealing the signal transduction pathways, which play an integral part in overall regulation. These reports on transcriptome–proteome–interactome analysis communicate a unified theme, suggesting strong posttranscriptional as well as posttranslational control of metabolism.

Ihmels et al. [73] developed an integrated analysis methodology, called signature algorithm for *S. cerevisiae*, which analyzes patterns in gene expression changes over a large number of data sets with varying conditions to establish proximity between genes in terms of their expression under various conditions. Although this work did not incorporate changes in the metabolic profile as that of Ideker et al. [72] did, physiological changes were used to provide functionalities to genes, based on similarity profiles. The premise of organizing genes into transcription modules is that genes that are expressed similarly under a large variety of conditions are more likely to be coregulated than those clustered based on fewer conditions. This method was then used to study various cellular functions as well as the global transcription program. For example, applying this method to a *S. cerevisiae* data set, genes with previously unknown (or speculated) function such as YGR067C, YGL186C, and YJL1200C were identified with the regulation of the glyoxylate shunt, purine transport, and lysine biosynthesis, respectively [74]. An interesting discovery made by Ihmels et al. [74] was that only 63 percent of the isozyme pairs were not coregulated. An experimental validation of one such prediction of isozymes not being coregulated was that of the two glutamate dehydrogenases, encoded by GDH1 and GDH3. In a completely independent work, these isozymes were demonstrated to be nonredundant and their expression is carbon dependent [26]. This result agrees very nicely with the work of Kafri et al. [83] on identifying the nature of backup functions that genes perform. They argue that genes that are similarly expressed do not back up each other in the event of a mutation but rather through a transcriptional reprogramming mechanism that *S. cerevisiae* has evolved; paralogues for the mutated genes are activated only when the gene in question is inactivated. Although the authors did not discuss this aspect, this result might provide some clues to the nature of silent mutations. Hundreds of components in the cell are organized into modules and dynamically interact with one another. The consequent phenotype is a reflection of these dynamic interactions. Although there is no clear boundary between these modules, the probability of interaction of a component with k other components, $p(k)$, has been shown to decrease according to the power law $k^{-2.2}$ [81]. However, few widely connected components such as ATP connect a large portion of the metabolism and result in an integrated module-free metabolic network. This dilemma has been resolved by demonstrating that metabolic networks are organized in highly connected modules that operate in conjunction with each other in a hierarchical manner [130]. Elucidating the principles that govern the nature and function of these individual modules may be possible with help from engineering, life sciences, and computer applications.

One of the several examples of such an integrative approach is that of identifying overlooked genes in *S. cerevisiae* [91]. Although the sequence information is extremely valuable, its ultimate utility lies in its accuracy and the completeness with which it is annotated. The yeast genome was sequenced and published to have 6274 genes, based on eukaryotic gene finding algorithms [108]. In the integrated approach, Kumar et al. [91] identified candidate genes by large-scale insertional mutagenesis using a modified transposon as a simple gene trap. The expression of each candidate gene is independently verified by microarray analysis. Only those gene sequences detected by both gene trapping and microarray analysis are classified as potential candidates. In this manner, they identified 137 previously overlooked genes in yeast, a majority of which are either short or overlap a previously annotated gene on the opposite strand. In yet another example of high-throughput data integration, the gene expression profiling and protein–protein interaction maps were integrated to compare the interactions between proteins encoded by genes that belong to common expression-profiling clusters with those between proteins encoded by genes that belong to different clusters [45]. The clusters derived from transcription profiling experiments were organized in a matrix, with each element of the matrix representing all pairwise combinations of genes either in a single cluster (diagonal or intracluster squares) or between two different clusters (nondiagonal or intercluster squares). This kind of a correlation approach suggested that the interactome data could help identify expression clusters with greater biological relevance. This study provides evidence that genes with similar expression profiles are more likely to encode interacting proteins and establishes a platform to integrate other functional genomic and proteomic data, both in yeast as well as in higher organisms.

The fundamental tenet of systems biology is capturing and integrating global data sets from biological systems from as many hierarchical levels as necessary. These include the static DNA sequences, context-dependent mass flow measurements in the form of RNA and protein quantifications, regulatory measurements such as protein–protein or protein–DNA interactions, and information flow measurements such as signaling pathways. The data collected from these measurements are transferred to a database where it is warehoused and analyzed for emergent properties systemic properties. The visualization methods described earlier permit a means to integrate the phenotypic features of the system directly to protein and gene regulatory networks. Cycles of iteration will result in a more accurate model to explain the subsystem or even the complete system (Fig. 9-2). Once the model has achieved sufficient level of accuracy and detail, it will allow biologists to accomplish tasks that remained elusive until now: predict the systemic response to a perturbation and redesign the regulatory networks to create new emergent systems. The second aspect of the systems biology will be addressed in greater detail in the next section. Therefore, fundamentally, systems biology is a hypothesis-driven, global, iterative, integrative, and dynamic branch in biological engineering.

9.7 SYNTHETIC BIOLOGY: STATE OF THE ART

Synthetic biology is a new and emerging direction that engineering of biological systems has taken. It is the synthesis of complex, biologically inspired systems that exhibit novel functionality, which do not exist naturally. This engineering perspective may be applied at all levels of hierarchy of biological structures. Therefore, synthetic biology is the design of biological systems in a rational and systematic way. The realization that the way to understand the cellular complexity requires a lot more than just compiling a "parts list,," as provided by the genome sequencing, for example, has precipitated into the origins of synthetic biology. Elucidating the interaction between the parts is central to systems biology and is providing the necessary conceptual tools needed for synthetic biology. This nascent offspring of systems biology will share a symbiotic relationship with the fundamental sciences to expand on the biological control mechanisms using engineering approaches. These approaches include, but are not restricted to, the design and synthesis of novel genes and proteins, modifying the genetic code, altering regulatory mechanisms and signal sensing and enzymatic reactions, constructing multicomponent modules that impart complex phenotype, and even generating engineered cells.

The field of synthetic biology involves taking existing biological pieces, transforming them into micromachines, and creating artificial systems that mimic the properties of living systems. By creating systems that mimic what nature has created, scientists can discover the basic principles that rule living systems, manipulate these systems, and eventually find treatments for many diseases plaguing humanity. Today's synthetic biologists are looking to channel genetic engineering from a hit-or-miss field of discovery to the type of discipline used by engineers to build bridges, computers, and buildings. This approach can translate into more specific anticancer therapies and antiviral drugs, as well as more efficient drug delivery systems that will have a significant impact on the health care industry.

9.7.1   Systems Biology and Synthetic Biology

Systems biology merely provides the analytical framework within which synthetic biology develops. The fundamental difference between systems biology and synthetic biology is that quite unlike systems biology, synthetic biology is not a discovery science (Fig. 9-11). It is a new way of constructing biology by adapting natural biological mechanisms to the requirements of an engineering approach. Similar to the mundane origins of systems biology described in Section 9.2, the first contribution of synthetic biology as defined above was made in 1964, when the first functional synthetic gene was made by a research team led by Khorana [84] as part of their work on elucidation of the genetic code. This gene, encoding tyrosine transfer RNA, was built from basic chemicals and was successfully tested in bacteria. Subsequently, this technology was automated and was used in making primers for polymerase chain reactions [111] and sequencing [140]. Since the simulation tools and models that are developed in systems biology could be used in synthetic biology, it is considered the design counterpart of systems biology. The design process demands sophisticated technology to target large number of components in addition to the high-throughput approaches. Therefore, synthetic biology will take some time before it matures to the status that systems biology is currently enjoying.

9.7.2   Synthetic Biological Circuits and Cascades

The discovery of signaling pathways controlling fundamental physiology [79] led to the application of nonlinear dynamics to understand gene regulation analogous to electric circuits and the development of the concept of a regulatory network. However, in the pregenomic era, the lack of sufficient experimental techniques precluded further expansion in this field. However, the recent explosion in the development of quantitative experimental methodology sparked interest in the elucidation of biological circuits and, more recently, introduction of synthetic circuits in biological systems. The simplest circuit is a transcriptional cascade, where genes are arranged in series and each gene product regulates the expression of one or more targets downstream in the series (Fig. 9-12). Although this concept has been optimized to perfection in natural systems using over evolution, in synthetic biology the networks are assembled from components that may not be related to each other. Therefore, the main obstacle in engineering synthetic circuits is to match the impedance of the individual elements such that they are kinetically functional in the context of the desired objective. There are two methods to optimize a synthetic circuit. The first one employs sensitivity analysis where randomly chosen kinetic rates are assigned to the functionality of the components and the contribution of each element's kinetics to the overall system behavior can be determined from analyzing the data from a large number of runs [32]. These data can be subsequently used to manipulate or fine-tune the system to achieve the end goal. The second approach is by directed evolution, which does not require detailed knowledge of the component kinetics. Directed evolution is most commonly achieved by subjecting a given component (usually a gene) in the circuit to a random mutation followed by a screening process to select the mutants that meet the

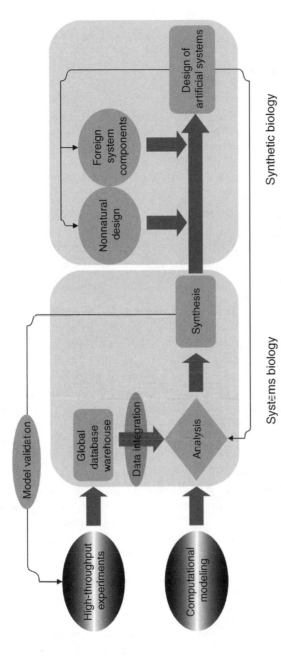

Figure 9-11 We view synthetic biology as a specific case of the synthetic arm of systems biology. Information from the global X-omes is integrated with the aid of computational modeling to understand the system performance in systems biology. This will lead to new understanding of the system functioning, which can be modified by rewiring the system in a nonnatural way or replacing some of the systemic components with foreign components to impart new features that lead to novel functions. This approach has tremendous applications in industrial as well as medical biotechnology.

327

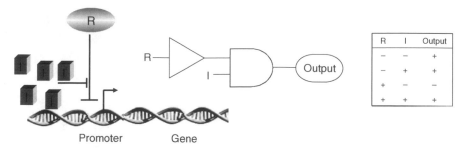

Figure 9-12 A simple depiction of how a gene regulatory network can be represented as an analogous electric circuit. The table on the right shows the conditions when the circuit will have an output. The circuit indicates how the output can be generated even in the presence of a repressor protein, making the gene circuit function like a switch.

desired criteria. This technique has been demonstrated in the optimization of a transcriptional cascade in *E. coli* [179] and cell-to-cell communication elements in *Vibrio fishcheri* [22]. The former strategy of rational design of synthetic circuits by computational approaches works well when properties dictating the component activity are well established, as is often the case with ribosome binding sites and operators. Directed evolution is more useful when the mechanism of the elements is not well known.

Signaling cascades are useful in elucidating the fundamental mechanisms of information flow in regulatory networks, and are usually characterized by having a steady-state output that is a monotonic function of the input. The steady-state behavior of most signaling cascades is similar to the digital logic with an ultrasensitive step-like dosage–response function, as illustrated in the case of mitogen-activated protein kinase in *S. cerevisiae* [67]. The response properties of transcription cascades also possess similar response characteristics. The dynamic and steady-state analysis of synthetic transcriptional cascades comprising one, two, and three repression stages has shown an ultrasensitive response to stimulus, and the sensitivity of the cascade increases as more elements are added to the cascade [66]. Synthetic transcriptional cascades have also been useful in studying noise propagation and in quantifying the contribution of intrinsic and extrinsic factors to phenotypic variations [126,135]. Most of the synthetic signaling cascades have been studied in prokaryotes, but long cascades are more common in eukaryotes, and therefore are more complicated. In contrary to the prokaryotic transcriptional cascades, eukaryotes exhibit a nonmonotonous response to stimulus, particularly in the presence of feedforward loops in the cascade.

9.7.3 Challenges for Synthetic Design

The construction of a functional synthetic network required assembling diverse genetic elements and getting them to work together. This process involves combining disparate components and tuning of biological parameters such as kinetic constants. Moreover, characterization of the circuit may not be valid under all conditions. To overcome some of these problems, several strategies have been suggested. First, the use of tunable elements such as transcription factors and promoters allow external control over some of these parameters. Second, the host cell in which the synthetic circuit has been integrated could be subjected to directed evolution in the laboratory and selected for optimized parameters. Another strategy is to implement a robust circuit design that is inherently insensitive to any kind of stimulus. These strategies have their basis in natural selection and are extremely useful in incorporating synthetic circuits in biological systems.

Another aspect of designing synthetic circuits is that of computational modeling. Simulation of synthetic models is essential for both the analysis of natural systems and also for engineering synthetic ones. Some of the problems that complicate the straightforward application of mathematical modeling to synthetic circuits include parameter sensitivity, lack of mathematical principles to model the complex biological circuits, and the inherent difficulty in distinguishing signal from noise in the circuits. On the positive side, synthetic circuits are simpler and, therefore, are better characterized than their natural counterparts; they will serve as ideal test systems to study their natural counterparts.

Currently, synthetic biology offers the ability to study cellular regulation and behavior using *de novo* networks. However, in the future, synthetic biology is expected to greatly contribute to the progress of medicine, biotechnology, and other areas of biology. The true potential of synthetic biology will be realized when the synthetic regulatory cascades mentioned above are interfaced with sensory inputs and biological response outputs. The inputs permit noninvasive monitoring of external environmental conditions and internal cell state and the outputs enable the engineered circuitry to control metabolism, cell cycle, and so on. Although the ability to program cell behaviors is still in its infancy, it is clear that the power to freely manipulate the set of instructions governing the behavior of organisms will have a tremendous impact on our quality of life and our ability to interact with and control the physical world surrounding us. One important difference between established quantitative engineering disciplines and synthetic biology is that state-of-the-art biological modeling tools still do not offer the same level of precision and predictive power. The future of synthetic biology looks very promising, with two goals clearly becoming obvious: understanding natural circuits by mimicking the natural systems and discovering what alternate nonnatural circuit designs are possible given the biological components. These hold the promise for immense potential in industrial as well as medical biotechnology.

9.8 COORDINATED RESEARCH IN YEAST SYSTEMS BIOLOGY

We are currently witnessing a transition in the approach to yeast physiology from traditional macroscopic procedures to a molecular approach and from a reductionist approach to an integrated approach (Fig. 9-2). Research in the field of systems biology and engineering is primarily driven by its end use and the quest for fundamental understanding. Truly comprehensive approaches to systems biology lie at the confluence of pure basic research and use-inspired basic research. Since such comprehensive approaches seem to be the future trend in studying physiology, it is necessary to establish a common platform to enable effective information exchange between different research groups. The generation of high-throughput global data that will be used in the integrated methodologies will prove to be an expensive venture and will undeniably require extensive knowledge about computer modeling, physiology, and metabolism, as well as excellent technical skills in measuring gene and protein expression and metabolic flux analysis. Although the current trend of generating high-throughput data is increasingly popular, we believe that there is extremely useful information that could still be extracted from the data that are already generated. Such a multidisciplinary approach paves the way to establishing strong symbiotic research collaborations. In this vein, there is also an increased government funding for systems biology. Notably, the U.S. National Institute of Health's roadmap for medical research provides $2.1 billion in funding over 5 years with heavy emphasis on systems biology, computational biology, and interdisciplinary programs. On a smaller scale, the U.S. National Science Foundation has launched a funding initiative entitled, "Quantitative Systems Biology FY 2004.," Also in the United Kingdom, BBSRC and EPSRC have launched a focused research program on systems biology resulting in the establishment of six national research centers tackling different aspects of systems biology. These and similar initiatives worldwide are catalyzing a renaissance in systems biology with special emphasis on producing a new generation of researchers trained in their core discipline and in complementary fields as well. We will focus on the European efforts in performing coordinated research in yeast systems biology.

9.8.1 European Functional Analysis

The European Functional Analysis (EUROFAN) network precipitated from the Yeast Genome Sequencing Network, which played a key role in yeast genome sequencing efforts. The goal of EUROFAN, which was established 2 years after the yeast genome sequence was published, is to provide a central repository of yeast mutants and characterize their transcriptome and proteome profiles to elucidate the biological function of novel genes revealed by the yeast genome sequence. The systemic functional analysis of the yeast genome is not intended to replace the regular biological enquiries that are conducted to answer specific questions. There are established approaches that permit the study of biological significance of every gene with increasing specificity. The approach implemented by EUROFAN is very efficient since it is not necessary to perform the global analyses on all the single gene disruption mutants. The ultimate idea of this project was to distribute the novel genes among the various laboratories in Europe, where additional information about its physiologic significance is evaluated. EUROFAN has now a well-curated database of gene function for most of the novel genes, an effort still in progress, and also serves as a genetic archive and stock center comprising yeast strains containing specific deletion mutants containing the individual genes as well as disruption cassettes allowing their manipulation in any laboratory or industrial yeast. This resource partly runs under the patronage of the Yeast Industrial Platform (YIP), which ensures rapid and efficient technology transfer to maintain European leadership in industrial yeast research.

The EUROFAN Project B0 has characterized more than 700 novel genes with respect to growth and morphology of deletion strains at several conditions of media and temperature. Project B1 has carried out quantitative phenotypic analysis of 564 deletion mutants with respect to 31 inhibitory chemicals and temperature shift [9]. Other EUROFAN projects have focused on other postgenomic technologies or have characterized the deletion mutants for phenotypes according to the specialties of the participating laboratories. The EUROFAN projects represent a major source of phenotypic data for the novel nonessential genes targeted by the European consortium. Details regarding the EUROFAN reports can be searched from the MIPS site (http://mips.gsf.de/proj/eurofan) and the deletion mutants, plasmids containing individual genes, and disruption cassettes are available at EUROSCARF (http://www.uni-frankfurt.de/FB/mikro/euroscarf/index-htlm) or Research Genetics (http://www.resgen.com). The MIPS primary gene query page has a link to the gene-specific EUROFAN data but the results of the EUROFAN functional analyses have not yet been linked to any yeast genome database and consequently, there is no single downloadable compendium of the EUROFAN data. However, the results from the B0 project have been curated into YPD. A resource such as EUROFAN laid the foundation for thorough high-throughput research using yeast to serve as a model as well as a tool.

9.8.2 Yeast Systems Biology Network

Systems biology is, by definition, multidisciplinary. It requires close collaboration of various laboratories specializing in experimental as well as theoretical disciplines to exploit the variety of methods to describe complex interactions in the yeast system. Thus far, it has not been possible for any single lab to possess the economic capability that is required to perform a variety of high-throughput experiments at the genome, transcriptome, proteome, metabolome, and the fluxome levels as well as the computational capability required to integrate data from these experiments to qualify for a true systems approach. Therefore, it is only through the coordination of activities in different labs such a systems approach can be realized. Despite the extensive government and private funding that the yeast systems biologists are enjoying worldwide, there is no concerted multilaboratory effort to coordinate and pool the individual competences toward studying yeast. The recognition for a unified effort for a symbiotic collaboration in yeast systems biology precipitated in launching the Yeast Systems Biology Network (YSBN) at the XXI International Conference on Yeast Genetics and Molecular Biology in Göthenburg, Sweden [65]. This alliance is expected to provide a platform for fostering collaboration between experimental yeast biologists and theoretical modelers in the "systems community.," The integrating platform for the alliance will be an internet-based functionality that generates a global virtual research community. The wider vision of the YSBN as part of both the yeast research and the emerging systems biology community is to work toward a comprehensive understanding of the function of the yeast cell, which will continue to serve as a paradigm for all eukaryotic cells. The other objectives of the YSBN are developing experimental and computational methods to iteratively improve the integration of experimental data in computational models and using the experimental approaches, in turn, to validate the predictions. This will facilitate easy data sharing for establishing standards for generating and documenting high-throughput data. Another important goal is to establish competence centers at regional as well as global levels for training of students and researchers. It is also necessary to spread the awareness of yeast systems biology among the general public and at the level of school education. These activities, in turn, are expected to increase the visibility of the YSBN to attract funding and financial support for yeast systems biology. An update on the progress of this international collaboration and its future activities and conferences are recently published [112].

9.9   SOCIETAL IMPACT OF YEAST SYSTEMS BIOLOGY

Systems biology, with its interdisciplinary approach to devising computational models of complex biological systems, may very well hold the key to unlocking the true value of the genome. There are vast commercial opportunities available for the pharmaceutical, human health, biotechnology, diagnostics, and agribusiness industries within systems biology. Market projections made by Research and Markets (http://www.researchandmarkets.com) for systems biology products and services are expected to grow at an annual compound rate of 66 percent to $785 million by 2008. *S. cerevisiae* has long served as a model eukaryote by virtue of the plethora of tools with which it can be manipulated genetically. In this section, we will illustrate some cases where similarities between yeast genes and human genes have been exploited to understand the mechanism of disease to improve human health and drug discovery in the pharmaceutical industry. We will also provide some case studies where novel metabolic engineering strategies in yeast have aided the bioprocess industry.

9.9.1   Human Health

The identification of several of the orthologues of human disease genes in this yeast has made it indispensable tool as a prototype system for medical research. Importantly, genetic dissections of yeast physiology serendipitously led to significant advances in our understanding of several human diseases, most notably cancer, through the exemplary studies on the regulation of the cell cycle performed by Hartwell et al. [59]. More recently, however, the genetic and biochemical tools available in yeast have been recruited for the purpose of directly examining the molecular basis and to aid in the treatment of several human diseases. High-throughput screening methods using the technology described earlier have been used to identify novel pharmacological targets produced in yeast or, through the two-hybrid screen, to obtain protein partners of medically relevant gene products. Moreover, the heterologous expression of proteins in yeast that lead to human disease has been used to uncover physiological responses to these proteins; yeast also encode homologues of several disease-causing proteins. In particular, the expression of specific proteins in yeast that fail to adopt their proper conformations or whose conformation lead to a pathological state in humans has helped us to understand how "conformational diseases," arise and how eukaryotic cells respond to malconformed polypeptides [20,90].

9.9.1.1 *Mitochondrial Disorders*   Despite the extensive research conducted on the structure of mammalian respiratory complexes, our knowledge of mitochondrial biogenesis in humans relies on yeast genetics and biochemistry. Human cDNAs have been isolated based on their homology with newly discovered yeast genes and have been used to rescue yeast mutants deficient in the corresponding genes. This approach has led to isolation of human genes involved in mitochondrial protein import, expression, biogenesis, and assembly of the respiratory complexes. Although the complete sequence of the human 16 kb mitochondrial DNA circle was published in 1981 [3], the mitochondrial gene sequence in *S. cerevisiae* was achieved only in 1998 as a complement to the nuclear genome. In humans as well as in yeast, only a few polypeptides of the respiratory complexes and ATP synthases are mitochondrially encoded with the vast majority of the mitochondrial proteins encoded in the nucleus and imported into the mitochondria by sophisticated machinery. Among the diseases of mitochondrial origin, cystic fibrosis is the most common lethal, inherited disease in North America and Europe, the common problems being breathing disorders, pancreatic dysfunctioning, and male infertility. Although over 900 mutations have been identified in the gene encoding, the cystic fibrosis transmembrane conductance regulator (CFTR), a phenylalanine in the 508 position of the protein, accounts for more than 70 percent of all the disease-causing mutations as a result of poor folding (http://www.genet.sickkids.on.ca/cftr/). The mutant form of CFTR localizes in the endoplasmic reticulum instead of the plasma membrane. When expressed in yeast, CFTR expressed in the endoplasmic reticulum was degraded; however, the degradation was attenuated when expressed in yeast containing a rapid-acting thermosensitive allele of a cytosolic Hsp70 chaperone [181]. These results indicate that Hsp70 facilitates CFTR degradation. Moreover, based on the genome sequence, a new essential mitochondrial metabolic pathway was discovered in yeast that appears as a promising model to study human iron–sulfur clusters [85], since this pathway is conserved in the human mitochondria as well [97].

9.9.1.2 *Nutrient Sensing and Metabolic Response*   All organisms appear to have the nutrient sensing mechanism that can rapidly detect changes in the concentration of available nutrients, adjust flux through metabolic pathways, and networks accordingly. In single-celled organisms, certain nutrients can regulate their own uptake, synthesis, and utilization. By contrast, higher eukaryotes sense nutrient availability primarily through endocrine and neuronal signals (e.g., insulin, glucagon, epinephrine, and so on). However, research performed in the last decade has shown that many types of mammalian cells can directly sense changes in the levels of a variety of nutrients and transduce this sensory information into changes in flux through metabolic pathways. These signal transduction pathways appear to operate both independently from and coordinately with the hormonal pathways. Since several of these pathways are conserved from the unicellular yeasts to mammals, they must have originally evolved independent of hormonal control. This conservation has proven extremely useful in delineating these pathways. In this section, we will discuss the sensing and response to macronutrients, particularly glucose, with particular focus on the modulation of cellular energy and aging.

Yeast has also been the prototype in evaluating the onset and modulation of cellular energy metabolism with respect to glucose homeostasis and, therefore, plays an important role in elucidating the mechanisms of metabolic syndrome. The metabolic syndrome is characterized by insulin resistance, hyperinsulinemia, dyslipidaemia, and a predisposition to type-2 diabetes, hypertension, premature atherosclerosis, and other diseases such as nonalcoholic fatty liver. Patients with this syndrome are usually overtly obese or have more subtle manifestations of increased adiposity, such as an increase in visceral fat. This syndrome has reached an epidemic level in our modern society due to a number of environmental factors, in particular overnutrition and inactivity. A major collaborative effort between basic researchers, clinicians, dieticians, health care authorities, and the pharmaceutical industry is required to halt progression of this devastating clustering of diseases. The AMP-activated protein kinase (AMP kinase) plays a key role in the modulation of cellular energy metabolism by phosphorylating key metabolic enzymes in response to increased AMP levels (Fig. 9-13). AMP levels rise during states of low energy charge (i.e., reduced ATP/AMP ratios) that occur in a variety of normal processes such as exercise and possibly also in some pathological states such as diabetes. Activated AMP kinase phosphorylates key enzymes in both biosynthetic and oxidative pathways and differentially modulates their activities to promote a reestablishment of normal ATP/AMP ratios. Besides maintaining the energy balance within the cells, AMP kinase also plays a key role in sensing intracellular ATP levels. The discovery of naturally occurring mutations in AMP kinase that cause cardiac hypertrophy provides direct evidence that AMP kinase has a fundamental role in maintaining normal human

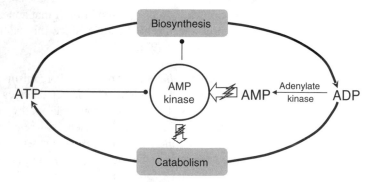

Figure 9-13   The AMP-activated protein kinase in yeast serves as a sensor of cellular energetic state. If the rate of ATP consumption exceeds its production (rate of biosynthesis exceeds catabolism), the concentration of ADP will increase, stimulating adenylate kinase to convert ADP to AMP. The rise in the level of AMP along with the reduction in ATP levels activates AMP kinase, which then switches off ATP-consuming processes and stimulates catabolism. The exact mechanisms involved in the activation of AMP kinase and its subsequent action are not yet known.

physiology. Moreover, the recent discovery of an upstream kinase in the AMP kinase cascade could implicate the role of AMP kinase in cancer development [57].

AMP kinase is a heterotrimeric complex with a catalytic α-subunit and two regulatory β- and γ-subunits, and homologues of all these three subunits have been identified in all eukaryotes [58]. The identification of these subunits in yeast, catalytic α-subunit (*SNF1* in yeast), the regulatory γ-subunit (*SNF4*), and the scaffolding β-subunit (three partly redundant proteins in yeast: *GAL83, SIP1*, and *SIP2*), provided *S. cerevisiae* as an ideal platform to elucidate the regulation and control of AMP kinase in humans. This conservation suggests an essential role of this complex in the functioning of the kinase [58]. Detailed studies on *S. cerevisiae SNF1* complex revealed an intimate role of this complex in transcriptional activation of many genes that are sensitive to glucose repression [19]. Growth on sucrose requires the expression of invertase, whereas growth on nonfermentable carbon sources requires expression of mitochondrial genes needed for oxidative metabolism. The expression of all of these genes is repressed by glucose, and the *SNF1* and *SNF4* genes are required for their derepression. One mechanism by which this is mediated is the phosphorylation of the repressor protein Mig1 by the *SNF1* complex (Fig. 9-14). Phosphorylation causes Mig1 to bind to a nuclear export protein that promotes its removal from the nucleus [82]. Therefore, the primary role of AMP kinase in yeast appears to be in the regulation of energy metabolism by repressing ATP-consuming processes and stimulating ATP generation via control of glucose uptake and its catabolism. Upon activation, AMP kinase controls many metabolic processes, ranging from stimulating fatty acid oxidation and glucose uptake to inhibiting protein, fatty acid, glycogen, and cholesterol synthesis. Its central role as a metabolic glucose sensor is illustrated by

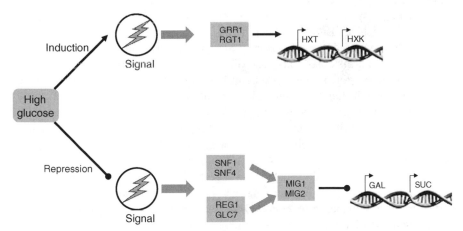

Figure 9-14 An extremely simplified schematic depicted the induction and repression mechanisms that are triggered by the presence of high glucose concentrations. The genes responsible for glucose metabolism (e.g., hexose transporters and kinases) are induced, whereas those responsible for the metabolism of other sugars are repressed (e.g., galactose and sucrose). Currently, there are several unknown steps involved in both the pathways and we believe that at least some of these uncertainties could be solved by employing systems biology techniques.

recent studies showing that mice lacking one of the AMP kinase isoforms have abnormal glucose tolerance and are insulin resistant [167]. Upon the discovery that AMP kinase is the major target of the antidiabetes drugs metformin and rosiglitazone, there has been tremendous interest in understanding the kinetics and action of this enzyme [136].

The fact that several of the nutrient-sensing pathways are conserved between humans and yeasts has proven extremely useful in studying the control and utilization of these pathways. Another aspect of medical research where *S. cerevisiae* has been used as the model system is in the elucidation of the mechanism of aging. Traditionally, rodents have been used to study these phenomena, analogous to the human processes. Over the past 75 years, many studies have shown that caloric restriction extends life span in a wide variety of species, from invertebrates to rodents to mammals. So far, no long-term studies have been completed in primates or conducted in humans because of the sheer length of any proposed study (perhaps a century or more for human studies!). With the recent explosion in yeast biology, coupled with the identification of the cell cycle regulators that share high homology with the human genes, yeasts are taking over as the ideal system to study aging. Moreover, the short life span of yeast makes them the convenient and preferred hosts over rodents.

Aging in budding yeast is measured by the number of mother cell divisions before senescence. Genetic studies have linked aging in *S. cerevisiae* to the *Sir* (silent information regulator) genes, which mediate genomic silencing at telomeres, mating-type loci, and the repeated ribosomal DNA (rDNA) [54]. Sir2 determines life span in a dose-dependent manner by creating silenced rDNA chromatin, thereby repressing recombination and the generation of toxic rDNA circles. This protein also functions in a meiotic checkpoint that monitors the fidelity of chromosome segregation [98]. Glucose enters yeast cells via highly regulated glucose-sensing transporters (HXT) and is then phosphorylated by hexokinases (Hxk1, Hxk2, and Glk1) to generate glucose-6-phosphate. Limiting the glucose availability by mutating *HXK2* also significantly extended the life span [98]. This is brought about by the yeast NAD^+-dependent histone deacetylase Sir2 and it is shown to be required for life span extension by glucose restriction and low-intensity stress [2,99]. The function of Sir2 enzymes in longevity and cell survival appears to be conserved in higher organisms as well. Currently, it is not clear how calorie restriction stimulates Sir2 activity, whether by feedback regulation of nicotinamide, an inhibitory product of Sir2 itself, or by increasing either NAD^+ or the NAD^+:NADH ratio. Although it is possible to affect Sir2 activity by genetically manipulating NAD^+ metabolic pathways, it is not known whether NAD^+ is a bona fide regulator of Sir2 in normal cells. Sir2 represses transcription by removing acetyl groups from lysines of histone tails and certain transcription factors (e.g., FOXO and p53) [62]. These findings have led to the intriguing possibility that Sir2 acts as a metabolic sensor, via its NAD^+ dependence, that links caloric intake to a transcriptional program that modulates life span. The fact that Sir2 requires the central metabolic cofactor NAD^+ to catalyze protein deacetylation is surprising, since from the chemical perspective, deacetylation does not require the destruction of a high-energy cofactor. There is also no indication that the breakdown of NAD^+ during the deacetylation reaction is coupled to any form of protein

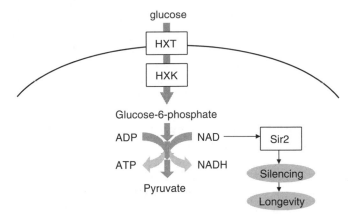

Figure 9-15 In another example of nutrient sensing and metabolic response in yeast, the Sir2 protein plays a key role in the aging and longevity via calorie restriction. This protein requires NAD for its activity is believed to serve as a cellular redox sensor. When the NAD levels are high (low sugar uptake or calorie restricted conditions), Sir2 is activated that extends life span. This protein is also believed to have partial control on glucose uptake, depending on the intracellular NAD levels.

conformational change or other work. Instead, the NAD^+ requirement may serve to link the activity of Sir2 to the metabolic status of the cell. Mutation of the NAD^+-salvage pathway in yeast lowers the NAD^+ concentration and prevents the life span extension conferred by caloric restriction [99]. This is similar to what was seen for *Sir2* mutants and led to the suggestion that Sir2 activity might depend on the intracellular concentration of some component of the NAD^+ pathway (Fig. 9-15). Support for the idea that Sir2 acts as a sensor of the NAD^+/NADH ratio (or the concentration of some other component that would be influenced by this ratio) comes from a study on mammalian skeletal muscle cell differentiation [42]. These studies provide strong evidence that Sir2 might be functioning as a metabolic or a redox sensor. However, the difficulty in measuring the *in vivo* NAD^+/NADH ratio and the threshold value it triggers in the activation of Sir2 is inhibiting further insight into its sensory and regulatory role.

9.9.1.3 *Mechanism of Cancer*   Most human cancers are the consequence of some form of genome instability, and therefore maintaining the stability of the genome is critical to cell survival and normal cell growth. In general, these aberrations occur either due to increased rate of chromosome instability or due to increased rates of point mutations and frameshift mutations [89]. Mismatch repair is the process by which incorrectly paired nucleotides in DNA are recognized and repaired. Our understanding of mismatch repair in eukaryotes relevant to cancer research mostly comes from studies completed in *S. cerevisiae* and, to a lesser extent, in higher eukaryotes. This section will deal with some of the recent insights into these issues that have emerged from recent genetic studies in *S. cerevisiae*.

S. cerevisiae contains at least two genes (*MSH2* and *MSH1*), which function in mismatch repair in the nucleus and mitochondria, respectively. It was identified that mutations in *MSH2* caused high spontaneous mutation rate, a defect in the repair of base pair mismatches with 1–4 nucleotide insertion/deletions, along with a modulation of genetic recombination. This is consistent with the view that *MSH2* functions in the major mismatch repair pathway in *S. cerevisiae* [36]. The human MSH2 protein (hMSH2) was identified as a minor component of a protein fraction that was purified by virtue of its mismatch binding activity, providing evidence that hMSH2 protein also recognizes mispaired bases [123]. A considerable amount of evidence has accumulated indicating that mutations in this gene are the primary cause of hereditary nonpolyposis colon cancer (commonly known as colorectal cancer) in humans [37]. Colorectal cancer is the disorder where rapid cell proliferation occurs in the lining of the large intestine, and these aberrant cells invade other tissues. This disorder most often begins as a benign polyp, which subsequently develops into malignant cancer. Cancer in the colon is the second largest cause of cancer-related deaths in the United States, and if discovered in the early states, it is treatable. Even when the abnormal cell proliferation spreads into nearby lymph nodes, surgical treatment followed by chemotherapy has been demonstrated to be highly successful (information from Colorectal Cancer Alliance Website, http://www.ccalliance.org/). Mapping studies have shown that *hMSH2* mapped to the chromosome 2 colon cancer locus, and analysis of chromosome 2-linked colon cancer families revealed germline *msh2* mutations that cosegregate with colon cancer in these families [88]. By combining the approaches used to define the yeast and human *MSH* genes with methods for identifying the yeast *MLH1* gene in a database of cDNA sequences, the human *MLH1* (*hMLH1*) gene has isolated and demonstrated to map to the chromosome 3p colon cancer locus, and mutational analysis indicate cosegregation of *hMLH1* mutants with the colon cancer locus, providing evidence that inheriting *hMLH1* mutations also causes colon cancer [145].

In the case of cancers caused by mutations in mismatch repair genes, genome instability arises due to elevated mutation rate, although the cause behind this is not clearly understood. However, very little is known about the molecular mechanisms underlying the genome rearrangements, their suppression mechanisms, and the possible defects in the suppression mechanisms that could potentially lead to many cancers. The utility of *S. cerevisiae* to study genome rearrangements began more than 20 years ago, when an extra copy of a DNA sequence was inserted at a site on an unrelated chromosome, followed by selection for recombination [148]. This resulted in chromosomal translocations due to mitotic recombination, similar to those seen in leukemia. The checkpoints shown in the S-phase of the cell cycle were originally identified to promote cell cycle delay or arrest in response to DNA damage, providing the cell an opportunity to repair the damage (Fig. 9-16) [38]. The sensitivity of the checkpoint-defective mutants to killing by DNA-damaging agents suggested that these checkpoints might function in suppressing genome instability. A survey of the *S. cerevisiae* genome for these checkpoints revealed that mutations that disrupt the replication checkpoint (*RFC5-1*, *DPB11-1*, *MEC1*, *DDC2*, and *DUN1*) significantly increase the rate of genome rearrangements [113]. In contrast, mutations in the genes required for the classical G1 and G2 DNA damage checkpoints and the mitotic spindle checkpoints had little effect, suggesting that the DNA replication checkpoint in the S-phase plays a critical role in suppression of spontaneous genome instability.

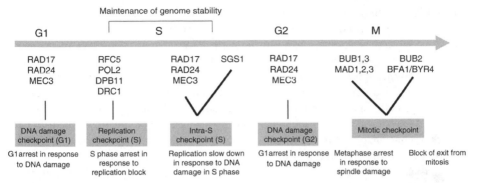

Figure 9-16   Different stages of the cell cycle and the checkpoints for DNA damage, replication, and mitosis. The proteins that are believed to detect the faults at each checkpoint are indicated below the cell cycle stage. The effect of activating the checkpoint is shown below the proteins in a box. This figure is redrawn from Ref. [89].

Therefore, replication errors appear to be the cause for genome rearrangements. The function of the replication checkpoint in suppressing genome instability likely includes regulating cell cycle progression in response to replication errors, modulating DNA repair functions, ensuring the establishment of sister chromatid cohesion, and maintaining stalled replication forks in a state that allows them to restart DNA synthesis. All of the genome rearrangements seen when this checkpoint was inactivated involved deletion of a chromosome end coupled with *de novo* addition of a new telomere [89]. Although data-driven systems biology in its purest form has generated progress mainly in the area of basic research, the more general concept of combining global data of multiple types is already making significant contributions, especially in the areas of drug discovery and development.

9.9.2   Drug Discovery

The extraordinary advances in biological research over the last decade have failed to translate into successful applications in drug discovery. Indeed, a recent analysis reported a decline in the productivity of pharmaceutical R&D, despite a 13 percent annual growth in investment in biomedical research from industry and government [12]. Moreover, the pharmaceutical industry will lose nearly $80 billion in revenue by 2008 due to patent expiration, and the current drug pipeline will replace only a small fraction of this value [4]. The bottleneck in the drug development technology lies in our inability to visualize the complexity of biological systems. Three major issues are associated with identifying effective new drugs: first, discovery of a relevant drug target; second, identification of a drug that will appropriately perturb the target; and third, assessment of the possible side effects and pharmaceutical properties of the drug before its deployment in clinical trials. Systems biology offers powerful new approaches for dealing with these problems.

In the long run, systems biology approach to drug discovery holds the promise to have a profound impact on medical practice, allowing a detailed evaluation of underlying predisposition to disease, diagnosis of disease, and the progression of disease. However in the near future, as a consequence of vigorous biomedical research, systems biology will provide powerful means for validating new drug targets, improving the success with which pharmaceuticals are identified. Farther into the future, the same approaches will drive the development of early diagnostics, enabling disease stratification, individualized therapy, and ultimately preventive drugs, based on both genetic and environmental considerations. Although systems biology as currently envisioned does not have a direct impact on the chemistry of identifying drugs or pharmacological challenges of drug metabolism, it may provide rapid and useful assays for these in the future.

Yeast can contribute to the drug discovery pipeline at an early state in identifying potential drug targets and evaluating the physiological outcome of modulating the activities of these targets. Although there are obvious limitations to using a micro-organism to identify potential human drug targets, several yeast proteins share a significant part of their primary amino acid sequence with at least one known or predicted human protein (around 2700 at BLAST with e-value less than 10^{-10} and around 1100 at BLAST e-value $<10^{-50}$). Among these are several hundred with sequence similarity to proteins implicated in human disease [7,13]. A large number of familiar drugs used against human targets specifically inhibit the orthologous proteins in yeast, providing a strong case for the use of yeast physiology to identify and study potential human drug targets. Among the conserved proteins that are uncharacterized, functional studies in yeast will shed light upon possible utility of the human counterparts as drug targets. Most of the proteins conserved between yeast and humans are involved in basic cellular processes such as small-molecule metabolism, protein synthesis, cell division, DNA synthesis and repair, secretion, and so on. Hence, target identification in yeast has proven especially relevant for cancer, which at the simplest level is a disorder of proliferation control caused by accumulated mutations. Many of the common mutations in human cancers including genetic and physical interactions between the mutated genes/proteins can be modeled in yeast, greatly simplifying and accelerating directed study. The concept of "synthetic lethality," a phenomenon where a combination of two innocuous genetic mutations renders the cell inviable, has shown great promise in identifying targets for anticancer therapy. Screening for mutation pairs that display synthetic lethality could lead to identifying drug targets that could selectively inhibit pro-liferation only in cells carrying a cancer-causing mutation. Such "gene-therapy," applications are presumably less detrimental than chemical or radiation therapy. Due to the obvious combinatorial problem associated with the experimental analysis of all the ordered pairwise mutations (even with 6000 genes), an automated system for creating and analyzing all pairwise combinations between a single mutant and all of the around 5000 viable single-gene deletion mutants has recently been described [151,152].

The increasing cases of fungal infections, particularly among immunity-compromised persons (those with AIDS and transplant patients), the need for safer and more effective antifungals is widely recognized. Although *Candida albicans* and *Aspergilli* have been used for the development of antifungals, *S. cerevisiae* presents a ready-made model system, particularly for azole-based antifungals. The discovery of pathogenic strains of *S. cerevisiae* that display invasive filamentous growth [49] or biofilm formation [132] provides excellent opportunity to examine the association between gene function and hyphal growth and infective capacity and biofilm formation, potentially leading to the identification of new antifungals. Screening for antifungals begins with a specific target with a known mechanism of action, since they could be used as templates for combinatorial modifications. An ideal antifungal should be required for the growth of yeast and should have minimal or no activity in humans (and therefore, not be conserved in humans). Among the 1100 essential genes in yeast, 350 do not have orthologues in humans and a subset of these genes would make an ideal target to screen for antifungals. However, some of the most successful antifungal compounds have properties far from the ideal criteria. For example, morpholines inhibit the Erg2 protein in the ergosterol pathway. Deletion of the *ERG2* gene is not lethal, and it shares sequence similarity with human sigma receptor protein [5]. The standard method used to screen for antifungals in particular and drug targets in general is the Y2H, which has been described earlier.

9.9.3 Food and Chemical Technology

White biotechnology (or industrial biotechnology) is an emerging field that specifically caters to the needs of the chemical and environmental industry [41]. It relies largely on using living cells like yeast as cell factories for sustainable production of biochemicals, biomaterials, and biofuels from renewable resources. A recent study conducted by McKinsey and Co predicts immense growth potential for white biotechnology in the future (http://www.mckinsey.com/clientservice/chemicals/pdf/BioVision_Booklet_final.pdf), with some of the large chemical companies such as BASF and DSM already replacing their chemical processes with cleaner, more efficient bioprocesses. An important component in developing yeast as a cell factory for an economically viable, efficient bioprocess is to optimize its metabolic network and systems biology has propelled the field of white biotechnology to new heights. In addition to the traditional use of yeast for baking purposes and ethanol production, it is also the system of choice for producing a variety of recombinant proteins such as insulin and various vaccines. There are a number of advantages of using yeast as a cell factory such as

- the availability of complete genome sequence
- its generally regarded as safe (GRAS) status
- well-defined cellular architecture
- established genetic manipulation techniques
- ease of scale-up of yeast bioprocesses
- availability of metabolic models

Besides its conventional applications in the brewing industry and as bakers yeast, *S. cerevisiae* is now used for a number of other industrial applications (Table 9.1).

Table 9-1   Industrial applications of bakers yeast

	Nonproprietary Name	Trade Name	Company	Reference
Pharmaceuticals	Hepatitis surface antigen	Ambirix	GlaxoSmithKline	[46,168]
		Comvax	Merck	[46,168]
		HBVAXPRO	Aventis Pharma	[46,168]
		Infanrix-Penta	GlaxoSmithKline	[46,168]
		Pediarix	GlaxoSmithKline	[46,168]
		Procomvax	Aventis-Pasteur	[46,168]
		Twinrix	GlaxoSmithKline	[46,168]
	Insulin	Actrapid	NovoNordisk	[46,168]
		Novolog	NovoNordisk	[46,168]
		Levemir	NovoNordisk	[46,168]
	Hirudin/Desirudin	Refludan	Aventis	[46,168]
	Urate oxidase	Elitex	Sanofi-Synthelabo	[46,168]
Fine chemicals	Epicedrol	—	—	[78]
	Lycopene	—	—	[175]
	β-carotene	—	BASF, Roche	[175]
	Artemesinin	—	Amyris Biotechnologies	[133]
	Flavanones	—	—	[176]
	Ascorbic acid	—	—	[56]
Bulk chemicals	Glycerol	—	—	[120]
	Lactic acid	—	—	[127]

9.10 PERSPECTIVE

Systems biology offers an opportunity to study how the phenotype is generated from the genotype and with it a glimpse of how evolution has crafted the phenotype. One aspect of systems biology is the development of techniques to examine broadly the level of protein, RNA, and DNA on a gene-by-gene basis and even the posttranslational modification and localization of proteins. In a very short time we have witnessed the development of high-throughput biology, forcing us to consider cellular processes *in vivo*. Even though much of the data is noisy and today partially inconsistent and incomplete, this has been a radical shift in the way we address problems one interaction at a time. When coupled with gene deletions by RNAi and classical methods and with the use of chemical tools tailored to proteins and protein domains, these high-throughput techniques become still more powerful. It is evident that a wide range of experimental approaches are being developed for use in *S. cerevisiae* that will allow functional genomics to build up an integrative view of the workings of a simple eukaryotic cell. This should enable a deeper understanding of more complex eukaryotes, both by the identification of orthologous genes in the different species and also by the expression of foreign coding sequences in yeast for complementation or two-hybrid analyses. However, many of these techniques are sufficiently general that once they have been tried and tested in the experimentally tractable yeast system, they should be directly applicable to the study of the functional genomics of higher organisms.

ACKNOWLEDGMENT

The authors wish to thank Michael L. Nielsen for his comments and suggestions on the contents of this chapter.

REFERENCES

1. Albert B, Botstein D, Brenner S, Cantor CR, Doolittle RF, Hood L, McKusick VA, Nathans D, Olson MV, Orkin S. Mapping and Sequencing the Human Genome, 1988.

2. Anderson RM, Latorre-Esteves M, Neves AR, Lavu S, Medvedik O, Taylor C, Howitz KT, Santos H, Sinclair DA. Yeast life-span extension by calorie restriction is independent of NAD fluctuation. *Science* 2003;302:2124–2126.

3. Anderson S, Bankier AT, Barrell BG, de Bruijn MH, Coulson AR, Drouin J, Eperon IC, Nierlich DP, Roe BA, Sanger F, Schreier PH, Smith AJ, Staden R, Young IG. Sequence and organization of the human mitochondrial genome. *Nature* 1981;290: 457–465.

4. Anonymous Pharmaceutical Development, http://www.newsrx.com 2004.

5. Bammert GF, Fostel JM. Genome-wide expression patterns in *Saccharomyces cerevisiae*: comparison of drug treatments and genetic alterations affecting biosynthesis of ergosterol. *Antimicrob Agents Chemother* 2000;44:1255–1265.

6. Bar-Joseph Z, Gerber GK, Lee TI, Rinaldi NJ, Yoo JY, Robert F, Gordon DB, Fraenkel E, Jaakkola TS, Young RA, Gifford DK. Computational discovery of gene modules and regulatory networks. *Nat Biotechnol* 2003;21:1337–1342.

7. Bassett DE Jr, Boguski MS, Spencer F, Reeves R, Kim S, Weaver T, Hieter P. Genome cross-referencing and XREFdb: implications for the identification and analysis of genes mutated in human disease. *Nat Genet* 1997;15:339–344.

8. Beyer A, Hollunder J, Nasheuer HP, Wilhelm T. Post-transcriptional expression regulation in the yeast *Saccharomyces cerevisiae* on a genomic scale. *Mol Cell Proteomics* 2004;3:1083–1092.

9. Bianchi MM, Ngo S, Vandenbol M, Sartori G, Morlupi A, Ricci C, Stefani S, Morlino GB, Hilger F, Carignani G, Slonimski PP, Frontali L. Large-scale phenotypic analysis reveals identical contributions to cell functions of known and unknown yeast genes. *Yeast* 2001;18:1397–1412.

10. Blanchard AP, Hood L. Sequence to array: probing the genome's secrets. *Nat Biotechnol* 1996;14:1649.

11. Boer VM, de Winde JH, Pronk JT, Piper MD. The genome-wide transcriptional responses of *Saccharomyces cerevisiae* grown on glucose in aerobic chemostat cultures limited for carbon, nitrogen, phosphorus, or sulfur. *J Biol Chem* 2003;278:3265–3274.

12. Booth B, Zemmel R. Prospects for productivity. *Nat Rev Drug Discov* 2004;3:451–456.

13. Botstein D, Chervitz SA, Cherry JM. Yeast as a model organism. *Science* 1997;277: 1259–1260.

14. Brauer MJ, Saldanha AJ, Dolinski K, Botstein D. Homeostatic adjustment and metabolic remodeling in glucose-limited yeast cultures. *Mol Biol Cell* 2005;16:2503–2517.

15. Bro C, Knudsen S, Regenberg B, Olsson L, Nielsen J. Improvement of galactose uptake in *Saccharomyces cerevisiae* through overexpression of phosphoglucomutase: example of transcript analysis as a tool in inverse metabolic engineering. *Appl Environ Microbiol* 2005;71:6465–6472.

16. Bro C, Regenberg B, Nielsen J. Genome-wide transcriptional response of a *Saccharomyces cerevisiae* strain with an altered redox metabolism. *Biotechnol Bioeng* 2004;85: 269–276.

17. Cakir T, Kirdar B, Ulgen KO. Metabolic pathway analysis of yeast strengthens the bridge between transcriptomics and metabolic networks. *Biotechnol Bioeng* 2004;86:251–260.

18. Cannon WB. *Bodily Changes to Pain, Hunger, Fear and Rage*, 2nd ed. New York and London: D. Appleton and Co, 1929.

19. Carlson M. Glucose repression in yeast. *Curr Opin Microbiol* 1999;2:202–207.

20. Carrell RW, Lomas DA. Conformational disease. *Lancet* 1997;350:134–138.

21. Cho RJ, Campbell MJ, Winzeler EA, Steinmetz L, Conway A, Wodicka L, Wolfsberg TG, Gabrielian AE, Landsman D, Lockhart DJ, Davis RW. A genome-wide transcriptional analysis of the mitotic cell cycle. *Mol Cell* 1998;2:65–73.

22. Collins CH, Arnold FH, Leadbetter JR. Directed evolution of Vibrio fischeri LuxR for increased sensitivity to a broad spectrum of acyl-homoserine lactones. *Mol Microbiol* 2005;55:712–723.

23. Davis TN. Protein localization in proteomics. *Curr Opin Chem Biol* 2004;8:49–53.

24. de Lichtenberg U, Jensen LJ, Brunak S, Bork P. Dynamic complex formation during the yeast cell cycle. *Science* 2005;307:724–727.

25. Deane CM, Salwinski L, Xenarios I, Eisenberg D. Protein interactions: two methods for assessment of the reliability of high-throughput observations. *Mol Cell Proteomics* 2002;1:349–356.

26. DeLuna A, Avendano A, Riego L, Gonzalez A. NADP-glutamate dehydrogenase isoenzymes of *Saccharomyces cerevisiae*. Purification, kinetic properties, and physiological roles. *J Biol Chem* 2001;276:43775–43783.

27. Deng M, Sun F, Chen T. Assessment of the reliability of protein–protein interactions and protein function prediction. *Pac Symp Biocomput* 2003;140–151.

28. DeRisi JL, Iyer VR, Brown PO. Exploring the metabolic and genetic control of gene expression on a genomic scale. *Science* 1997;278:680–686.

29. Edwards JS, Ramakrishna R, Palsson BO. Characterizing the metabolic phenotype: a phenotype phase plane analysis. *Biotechnol Bioeng* 2002;77:27–36.

30. Eisen MB, Brown PO. DNA arrays for analysis of gene expression. *Methods Enzymol* 1999;303:179–205.

31. Elowitz MB, Leibler S. A synthetic oscillatory network of transcriptional regulators. *Nature* 2000;403:335–338.

32. Feng XJ, Hooshangi S, Chen D, Li G, Weiss R, Rabitz H. Optimizing genetic circuits by global sensitivity analysis. *Biophys J* 2004;87:2195–2202.

33. Fiehn O. Metabolomics: the link between genotypes and phenotypes. *Plant Mol Biol* 2002;48:155–171.

34. Fiehn O, Kopka J, Dormann P, Altmann T, Trethewey RN, Willmitzer L. Metabolite profiling for plant functional genomics. *Nat Biotechnol* 2000;18:1157–1161.

35. Fields S, Song O. A novel genetic system to detect protein–protein interactions. *Nature* 1989;340:245–246.

36. Fishel R, Kolodner RD. Identification of mismatch repair genes and their role in the development of cancer. *Curr Opin Genet Dev* 1995;5:382–395.

37. Fishel R, Lescoe MK, Rao MR, Copeland NG, Jenkins NA, Garber J, Kane M, Kolodner R. The human mutator gene homolog MSH2 and its association with hereditary non-polyposis colon cancer. *Cell* 1993;75:1027–1038.

38. Foiani M, Pellicioli A, Lopes M, Lucca C, Ferrari M, Liberi G, Muzi Falconi M, Plevani P. DNA damage checkpoints and DNA replication controls in *Saccharomyces cerevisiae*. *Mutat Res* 2000;451:187–196.

39. Forster J, Famili I, Fu P, Palsson BO, Nielsen J. Genome-scale reconstruction of the *Saccharomyces cerevisiae* metabolic network. *Genome Res* 2003;13:244–253.

40. Forster J, Gombert AK, Nielsen J. A functional genomics approach using metabolomics and *in silico* pathway analysis. *Biotechnol Bioeng* 2002;79:703–712.

41. Frazzetto G. White biotechnology. *EMBO Rep* 2003;4:835–837.

42. Fulco M, Schiltz RL, Iezzi S, King MT, Zhao P, Kashiwaya Y, Hoffman E, Veech RL, Sartorelli V. Sir2 regulates skeletal muscle differentiation as a potential sensor of the redox state. *Mol Cell* 2003;12:51–62.

43. Gardner TS, Cantor CR, Collins JJ. Construction of a genetic toggle switch in *Escherichia coli. Nature* 2000;403:339–342.

44. Gavin AC, Bosche M, Krause R, Grandi P, Marzioch M, Bauer A, Schultz J, Rick JM, Michon AM, Cruciat CM, Remor M, Hofert C, Schelder M, Brajenovic M, Ruffner H, Merino A, Klein K, Hudak M, Dickson D, Rudi T, Gnau V, Bauch A, Bastuck S, Huhse B, Leutwein C, Heurtier MA, Copley RR, Edelmann A, Querfurth E, Rybin V, Drewes G, Raida M, Bouwmeester T, Bork P, Seraphin B, Kuster B, Neubauer G, Superti-Furga G. Functional organization of the yeast proteome by systematic analysis of protein complexes. *Nature* 2002;415:141–147.

45. Ge H, Liu Z, Church GM, Vidal M. Correlation between transcriptome and interactome mapping data from *Saccharomyces cerevisiae. Nat Genet* 2001;29:482–486.

46. Gerngross TU. Advances in the production of human therapeutic proteins in yeasts and filamentous fungi. *Nat Biotechnol* 2004;22:1409–1414.

47. Ghaemmaghami S, Huh WK, Bower K, Howson RW, Belle A, Dephoure N, O'shea EK, Weissman JS. Global analysis of protein expression in yeast. *Nature* 2003;425:737–741.

48. Giaever G, Chu AM, Ni L, Connelly C, Riles L, Veronneau S, Dow S, Lucau-Danila A, Anderson K, Andre B, Arkin AP, Astromoff A, El-Bakkoury M, Bangham R, Benito R, Brachat S, Campanaro S, Curtiss M, Davis K, Deutschbauer A, Entian KD, Flaherty P, Foury F, Garfinkel DJ, Gerstein M, Gotte D, Guldener U, Hegemann JH, Hempel S, Herman Z, Jaramillo DF, Kelly DE, Kelly SL, Kotter P, LaBonte D, Lamb DC, Lan N, Liang H, Liao H, Liu L, Luo C, Lussier M, Mao R, Menard P, Ooi SL, Revuelta JL, Roberts CJ, Rose M, Ross-Macdonald P, Scherens B, Schimmack G, Shafer B, Shoemaker DD, Sookhai-Mahadeo S, Storms RK, Strathern JN, Valle G, Voet M, Volckaert G, Wang CY, Ward TR, Wilhelmy J, Winzeler EA, Yang Y, Yen G, Youngman E, Yu K, Bussey H, Boeke

JD, Snyder M, Philippsen P, Davis RW, Johnston M. Functional profiling of the *Saccharomyces cerevisiae* genome. *Nature* 2002;418:387–391.

49. Gimeno CJ, Ljungdahl PO, Styles CA, Fink GR. Unipolar cell divisions in the yeast *S. cerevisiae* lead to filamentous growth: regulation by starvation and RAS. *Cell* 1992;68:1077–1090.

50. Goffeau A, Barrell BG, Bussey H, Davis RW, Dujon B, Feldmann H, Galibert F, Hoheisel JD, Jacq C, Johnston M, Louis EJ, Mewes HW, Murakami Y, Philippsen P, Tettelin H, Oliver SG. Life with 6000 genes. *Science* 1996;274:546, 563–546, 567.

51. Goldberg DS, Roth FP. Assessing experimentally derived interactions in a small world. *Proc Natl Acad Sci USA* 2003;100:4372–4376.

52. Goldbeter A, Segel LA. Unified mechanism for relay and oscillation of cyclic AMP in *Dictyostelium discoideum. Proc Natl Acad Sci USA* 1977;74:1543–1547.

53. Grigoriev A. On the number of protein–protein interactions in the yeast proteome. *Nucleic Acids Res* 2003;31:4157–4161.

54. Guarente L. Diverse and dynamic functions of the Sir silencing complex. *Nat Genet* 1999;23:281–285.

55. Gygi SP, Corthals GL, Zhang Y, Rochon Y, Aebersold R. Evaluation of two-dimensional gel electrophoresis-based proteome analysis technology. *Proc Natl Acad Sci USA* 2000;97:9390–9395.

56. Hancock RD, Galpin JR, Viola R. Biosynthesis of L-ascorbic acid (vitamin C) by *Saccharomyces cerevisiae. FEMS Microbiol Lett* 2000;186:245–250.

57. Hardie DG. The AMP-activated protein kinase cascade: the key sensor of cellular energy status. *Endocrinology* 2003;144:5179–5183.

58. Hardie DG, Carling D, Carlson M. The AMP-activated/SNF1 protein kinase subfamily: metabolic sensors of the eukaryotic cell? *Annu Rev Biochem* 1998;67:821–855.

59. Hartwell LH, Culotti J, Pringle JR, Reid BJ. Genetic control of the cell division cycle in yeast. *Science* 1974;183:46–51.

60. Hartwell LH, Hopfield JJ, Leibler S, Murray AW. From molecular to modular cell biology. *Nature* 1999;402:C47–C52.

61. Heinrich R, Schuster S. The modelling of metabolic systems. Structure, control and optimality. *Biosystems* 1998;47:61–77.

62. Hekimi S, Guarente L. Genetics and the specificity of the aging process. *Science* 2003;299:1351–1354.

63. Hereford LM, Osley MA, Ludwig TR, McLaughlin CS. Cell-cycle regulation of yeast histone mRNA. *Cell* 1981;24:367–375.

64. Ho Y, Gruhler A, Heilbut A, Bader GD, Moore L, Adams SL, Millar A, Taylor P, Bennett K, Boutilier K, Yang L, Wolting C, Donaldson I, Schandorff S, Shewnarane J, Vo M, Taggart J, Goudreault M, Muskat B, Alfarano C, Dewar D, Lin Z, Michalickova K, Willems AR, Sassi H, Nielsen PA, Rasmussen KJ, Andersen JR, Johansen LE, Hansen LH, Jespersen H, Podtelejnikov A, Nielsen E, Crawford J, Poulsen V, Sorensen BD, Matthiesen J, Hendrickson RC, Gleeson F, Pawson T, Moran MF, Durocher D, Mann M, Hogue CW, Figeys D, Tyers M. Systematic identification of protein complexes in *Saccharomyces cerevisiae* by mass spectrometry. *Nature* 2002;415:180–183.

65. Hohmann S. The Yeast Systems Biology Network: mating communities. *Curr Opin Biotechnol* 2005;16:356–360.

66. Hooshangi S, Thiberge S, Weiss R. Ultrasensitivity and noise propagation in a synthetic transcriptional cascade. *Proc Natl Acad Sci USA* 2005;102:3581–3586.

67. Huang CY, Ferrell JE Jr. Ultrasensitivity in the mitogen-activated protein kinase cascade. *Proc Natl Acad Sci USA* 1996;93:10078–10083.

68. Hughes TR, Mao M, Jones AR, Burchard J, Marton MJ, Shannon KW, Lefkowitz SM, Ziman M, Schelter JM, Meyer MR, Kobayashi S, Davis C, Dai H, He YD, Stephaniants SB, Cavet G, Walker WL, West A, Coffey E, Shoemaker DD, Stoughton R, Blanchard AP, Friend SH, Linsley PS. Expression profiling using microarrays fabricated by an ink-jet oligonucleotide synthesizer. *Nat Biotechnol* 2001;19:342–347.

69. Huh WK, Falvo JV, Gerke LC, Carroll AS, Howson RW, Weissman JS, O'shea EK. Global analysis of protein localization in budding yeast. *Nature* 2003;425:686–691.

70. Ibarra RU, Edwards JS, Palsson BO. *Escherichia coli* K-12 undergoes adaptive evolution to achieve *in silico* predicted optimal growth. *Nature* 2002;420:186–189.

71. Iberall AS. A field and circuit thermodynamics for integrative physiology. I. Introduction to the general notions. *Am J Physiol* 1977;233:R171–R180.

72. Ideker T, Thorsson V, Ranish JA, Christmas R, Buhler J, Eng JK, Bumgarner R, Goodlett DR, Aebersold R, Hood L. Integrated genomic and proteomic analyses of a systematically perturbed metabolic network. *Science* 2001;292:929–934.

73. Ihmels J, Friedlander G, Bergmann S, Sarig O, Ziv Y, Barkai N. Revealing modular organization in the yeast transcriptional network. *Nat Genet* 2002;31:370–377.

74. Ihmels J, Levy R, Barkai N. Principles of transcriptional control in the metabolic network of *Saccharomyces cerevisiae*. *Nat Biotechnol* 2004;22:86–92.

75. Ito T, Chiba T, Yoshida M. Exploring the protein interactome using comprehensive two-hybrid projects. *Trends Biotechnol* 2001;19:S23–S27.

76. Ito T, Ota K, Kubota H, Yamaguchi Y, Chiba T, Sakuraba K, Yoshida M. Roles for the two-hybrid system in exploration of the yeast protein interactome. *Mol Cell Proteomics* 2002;1:561–566.

77. Iyer VR, Horak CE, Scafe CS, Botstein D, Snyder M, Brown PO. Genomic binding sites of the yeast cell-cycle transcription factors SBF and MBF. *Nature* 2001;409:533–538.

78. Jackson BE, Hart-Wells EA, Matsuda SP. Metabolic engineering to produce sesquiterpenes in yeast. *Org Lett* 2003;5:1629–1632.

79. Jacob F, Monod J. Genetic regulatory mechanisms in the synthesis of proteins. *J Mol Biol* 1961;3:318–356.

80. Jansen R, Yu H, Greenbaum D, Kluger Y, Krogan NJ, Chung S, Emili A, Snyder M, Greenblatt JF, Gerstein M. A Bayesian networks approach for predicting protein–protein interactions from genomic data. *Science* 2003;302:449–453.

81. Jeong H, Tombor B, Albert R, Oltvai ZN, Barabasi AL. The large-scale organization of metabolic networks. *Nature* 2000;407:651–654.

82. Johnston M. Feasting, fasting and fermenting. Glucose sensing in yeast and other cells. *Trends Genet* 1999;15:29–33.

83. Kafri R, Bar-Even A, Pilpel Y. Transcription control reprogramming in genetic backup circuits. *Nat Genet* 2005;37:295–299.

84. Khorana HG. Polynucleotide synthesis and the genetic code. *Fed Proc* 1965;24:1473–1487.

85. Kispal G, Csere P, Prohl C, Lill R. The mitochondrial proteins Atm1p and Nfs1p are essential for biogenesis of cytosolic Fe/S proteins. *EMBO J* 1999;18:3981–3989.

86. Kitano H. Computational systems biology. *Nature* 2002;420:206–210.

87. Klein CJ, Olsson L, Nielsen J. Glucose control in *Saccharomyces cerevisiae*: the role of Mig1 in metabolic functions. *Microbiology* 1998;144(Part 1):13–24.

88. Kolodner RD, Hall NR, Lipford J, Kane MF, Rao MR, Morrison P, Wirth L, Finan PJ, Burn J, Chapman P. Human mismatch repair genes and their association with hereditary non-polyposis colon cancer. *Cold Spring Harb Symp Quant Biol* 1994;59:331–338.

89. Kolodner RD, Putnam CD, Myung K. Maintenance of genome stability in *Saccharomyces cerevisiae*. *Science* 2002;297:552–557.

90. Kopito RR, Ron D. Conformational disease. *Nat Cell Biol* 2000;2:E207–E209.

91. Kumar A, Harrison PM, Cheung KH, Lan N, Echols N, Bertone P, Miller P, Gerstein MB, Snyder M. An integrated approach for finding overlooked genes in yeast. *Nat Biotechnol* 2002;20:58–63.

92. Lauffenburger DA. Cell signaling pathways as control modules: complexity for simplicity? *Proc Natl Acad Sci USA* 2000;97:5031–5033.

93. Lee TI, Rinaldi NJ, Robert F, Odom DT, Bar-Joseph Z, Gerber GK, Hannett NM, Harbison CT, Thompson CM, Simon I, Zeitlinger J, Jennings EG, Murray HL, Gordon DB, Ren B, Wyrick JJ, Tagne JB, Volkert TL, Fraenkel E, Gifford DK, Young RA. Transcriptional regulatory networks in *Saccharomyces cerevisiae*. *Science* 2002;298:799–804.

94. Li S, Armstrong CM, Bertin N, Ge H, Milstein S, Boxem M, Vidalain PO, Han JD, Chesneau A, Hao T, Goldberg DS, Li N, Martinez M, Rual JF, Lamesch P, Xu L, Tewari M, Wong SL, Zhang LV, Berriz GF, Jacotot L, Vaglio P, Reboul J, Hirozane-Kishikawa T, Li Q, Gabel HW, Elewa A, Baumgartner B, Rose DJ, Yu H, Bosak S, Sequerra R, Fraser A, Mango SE, Saxton WM, Strome S, Van Den Heuvel S, Piano F, Vandenhaute J, Sardet C, Gerstein M, Doucette-Stamm L, Gunsalus KC, Harper JW, Cusick ME, Roth FP, Hill DE, Vidal M. A map of the interactome network of the metazoan *C. elegans*. *Science* 2004;303:540–543.

95. Liao JC, Boscolo R, Yang YL, Tran LM, Sabatti C, Roychowdhury VP. Network component analysis: reconstruction of regulatory signals in biological systems. *Proc Natl Acad Sci USA* 2003;100:15522–15527.

96. Lieb JD, Liu X, Botstein D, Brown PO. Promoter-specific binding of Rap1 revealed by genome-wide maps of protein–DNA association. *Nat Genet* 2001;28:327–334.

97. Lill R, Muhlenhoff U. Iron–sulfur–protein biogenesis in eukaryotes. *Trends Biochem Sci* 2005;30:133–141.

98. Lin SJ, Defossez PA, Guarente L. Requirement of NAD and SIR2 for life-span extension by calorie restriction in *Saccharomyces cerevisiae*. *Science* 2000;289:2126–2128.

99. Lin SJ, Ford E, Haigis M, Liszt G, Guarente L. Calorie restriction extends yeast life span by lowering the level of NADH. *Genes Dev* 2004;18:12–16.

100. Lindon JC, Nicholson JK, Holmes E, Keun HC, Craig A, Pearce JT, Bruce SJ, Hardy N, Sansone SA, Antti H, Jonsson P, Daykin C, Navarange M, Beger RD, Verheij ER, Amberg A, Baunsgaard D, Cantor GH, Lehman-McKeeman L, Earll M, Wold S, Johansson E, Haselden JN, Kramer K, Thomas C, Lindberg J, Schuppe-Koistinen I, Wilson ID, Reily MD, Robertson DG, Senn H, Krotzky A, Kochhar S, Powell J, van der Ouderaa F, Plumb R, Schaefer H, Spraul M. Summary recommendations for standardization and reporting of metabolic analyses. *Nat Biotechnol* 2005;23:833–838.

101. Lipshutz RJ, Morris D, Chee M, Hubbell E, Kozal MJ, Shah N, Shen N, Yang R, Fodor SP. Using oligonucleotide probe arrays to access genetic diversity. *Biotechniques* 1995; 19:442–447.

102. Lockhart DJ, Dong H, Byrne MC, Follettie MT, Gallo MV, Chee MS, Mittmann M, Wang C, Kobayashi M, Horton H, Brown EL. Expression monitoring by hybridization to high-density oligonucleotide arrays. *Nat Biotechnol* 1996;14:1675–1680.

103. Lou XJ, Schena M, Horrigan FT, Lawn RM, Davis RW. Expression monitoring using cDNA microarrays. A general protocol. *Methods Mol Biol* 2001;175:323–340.

104. Luscombe NM, Babu MM, Yu H, Snyder M, Teichmann SA, Gerstein M. Genomic analysis of regulatory network dynamics reveals large topological changes. *Nature* 2004;431:308–312.

105. Lutfiyya LL, Iyer VR, DeRisi J, DeVit MJ, Brown PO, Johnston M. Characterization of three related glucose repressors and genes they regulate in *Saccharomyces cerevisiae*. *Genetics* 1998;150:1377–1391.

106. MacBeath G, Schreiber SL. Printing proteins as microarrays for high-throughput function determination. *Science* 2000;289:1760–1763.

107. Mesarovic MD, Systems theory and biology: view of a theoretician. *Systems Theory and Biology*. New York: Springer, 1968, pp. 59–87.

108. Mewes HW, Albermann K, Bahr M, Frishman D, Gleissner A, Hani J, Heumann K, Kleine K, Maierl A, Oliver SG, Pfeiffer F, Zollner A. Overview of the yeast genome. *Nature* 1997;387:7–65.

109. Monod J, Changeux JP, Jacob F. Allosteric proteins and cellular control systems. *J Mol Biol* 1963;6:306–329.

110. Mukherjee S, Berger MF, Jona G, Wang XS, Muzzey D, Snyder M, Young RA, Bulyk ML. Rapid analysis of the DNA-binding specificities of transcription factors with DNA microarrays. *Nat Genet* 2004;36:1331–1339.

111. Mullis K, Faloona F, Scharf S, Saiki R, Horn G, Erlich H. Specific enzymatic amplification of DNA *in vitro*: the polymerase chain reaction. *Cold Spring Harb Symp Quant Biol* 1986; 51(Part 1):263–273.

112. Mustacchi R, Hohmann S, Nielsen J. Yeast systems biology to unravel the network of life. *Yeast* 2006;23:227–238.

113. Myung K, Datta A, Kolodner RD. Suppression of spontaneous chromosomal re-arrangements by S phase checkpoint functions in *Saccharomyces cerevisiae*. *Cell* 2001;104:397–408.

114. Ng SK, Zhang Z, Tan SH, Lin K. InterDom: a database of putative interacting protein domains for validating predicted protein interactions and complexes. *Nucleic Acids Res* 2003;31:251–254.

115. Nielsen J. It is all about metabolic fluxes. *J Bacteriol* 2003;185:7031–7035.

116. Ohlmeier S, Kastaniotis AJ, Hiltunen JK, Bergmann U. The yeast mitochondrial proteome, a study of fermentative and respiratory growth. *J Biol Chem* 2004;279:3956–3979.

117. Oliver SG, van der Aart QJ, Agostoni-Carbone ML, Aigle M, Alberghina L, Alexandraki D, Antoine G, Anwar R, Ballesta JP, Benit P. The complete DNA sequence of yeast chromosome III. *Nature* 1992;357:38–46.

118. Oliver SG, Winson MK, Kell DB, Baganz F. Systematic functional analysis of the yeast genome. *Trends Biotechnol* 1998;16:373–378.

119. Ostergaard S, Olsson L, Johnston M, Nielsen J. Increasing galactose consumption by *Saccharomyces cerevisiae* through metabolic engineering of the GAL gene regulatory network. *Nat Biotechnol* 2000;18:1283–1286.

120. Overkamp KM, Bakker BM, Kotter P, Luttik MA, van Dijken JP, Pronk JT. Metabolic engineering of glycerol production in *Saccharomyces cerevisiae*. *Appl Environ Microbiol* 2002;68:2814–2821.

121. Ozcan S, Johnston M. Three different regulatory mechanisms enable yeast hexose transporter (HXT) genes to be induced by different levels of glucose. *Mol Cell Biol* 1995;15:1564–1572.

122. Ozcan S, Johnston M. Function and regulation of yeast hexose transporters. *Microbiol Mol Biol Rev* 1999;63:554–569.

123. Palombo F, Hughes M, Jiricny J, Truong O, Hsuan J. Mismatch repair and cancer. *Nature* 1994;367:417.

124. Patil KR, Nielsen J. Uncovering transcriptional regulation of metabolism by using metabolic network topology. *Proc Natl Acad Sci USA* 2005;102(8): 2685–2689.

125. Patton WF. Detection technologies in proteome analysis. *J Chromatogr B Analyt Technol Biomed Life Sci* 2002;771:3–31.

126. Pedraza JM, van Oudenaarden A. Noise propagation in gene networks. *Science* 2005;307:1965–1969.

127. Porro D, Brambilla L, Ranzi BM, Martegani E, Alberghina L. Development of metabolically engineered *Saccharomyces cerevisiae* cells for the production of lactic acid. *Biotechnol Prog* 1995;11:294–298.

128. Ptacek J, Devgan G, Michaud G, Zhu H, Zhu X, Fasolo J, Guo H, Jona G, Breitkreutz A, Sopko R, McCartney RR, Schmidt MC, Rachidi N, Lee SJ, Mah AS, Meng L, Stark MJ, Stern DF, De Virgilio C, Tyers M, Andrews B, Gerstein M, Schweitzer B, Predki PF, Snyder M. Global analysis of protein phosphorylation in yeast. *Nature* 2005;438:679–684.

129. Rappsilber J, Siniossoglou S, Hurt EC, Mann M. A generic strategy to analyze the spatial organization of multi-protein complexes by cross-linking and mass spectrometry. *Anal Chem* 2000;72:267–275.

130. Ravasz E, Somera AL, Mongru DA, Oltvai ZN, Barabasi AL. Hierarchical organization of modularity in metabolic networks. *Science* 2002;297:1551–1555.

131. Ren B, Robert F, Wyrick JJ, Aparicio O, Jennings EG, Simon I, Zeitlinger J, Schreiber J, Hannett N, Kanin E, Volkert TL, Wilson CJ, Bell SP, Young RA. Genome-wide location and function of DNA binding proteins. *Science* 2000;290:2306–2309.

132. Reynolds TB, Fink GR. Bakers' yeast, a model for fungal biofilm formation. *Science* 2001;291:878–881.

133. Ro DK, Paradise EM, Ouellet M, Fisher KJ, Newman KL, Ndungu JM, Ho KA, Eachus RA, Ham TS, Kirby J, Chang MC, Withers ST, Shiba Y, Sarpong R, Keasling JD. Production of the antimalarial drug precursor artemisinic acid in engineered yeast. *Nature* 2006;440:940–943.

134. Ronne H. Glucose repression in fungi. *Trends Genet* 1995;11:12–17.

135. Rosenfeld N, Young JW, Alon U, Swain PS, Elowitz MB. Gene regulation at the single-cell level. *Science* 2005;307:1962–1965.

136. Rutter GA, Da, Silva X, Leclerc I. Roles of 5'-AMP-activated protein kinase (AMPK) in mammalian glucose homoeostasis. *Biochem J* 2003;375:1–16.

137. Said MR, Begley TJ, Oppenheim AV, Lauffenburger DA, Samson LD. Global network analysis of phenotypic effects: protein networks and toxicity modulation in *Saccharomyces cerevisiae*. *Proc Natl Acad Sci USA* 2004;101:18006–18011.

138. Saldanha AJ, Brauer MJ, Botstein D. Nutritional homeostasis in batch and steady-state culture of yeast. *Mol Biol Cell* 2004;15:4089–4104.

139. Salwinski L, Eisenberg D. *In silico* simulation of biological network dynamics. *Nat Biotechnol* 2004;22:1017–1019.

140. Sanger F, Donelson JE, Coulson AR, Kossel H, Fischer D. Use of DNA polymerase I primed by a synthetic oligonucleotide to determine a nucleotide sequence in phage fl DNA. *Proc Natl Acad Sci USA* 1973;70:1209–1213.

141. Schena M, Shalon D, Davis RW, Brown PO. Quantitative monitoring of gene expression patterns with a complementary DNA microarray. *Science* 1995;270:467–470.

142. Schuster S, Dandekar T, Fell DA. Detection of elementary flux modes in biochemical networks: a promising tool for pathway analysis and metabolic engineering. *Trends Biotechnol* 1999;17:53–60.

143. Schwikowski B, Uetz P, Fields S. A network of protein–protein interactions in yeast. *Nat Biotechnol* 2000;18:1257–1261.

144. Segre D, Vitkup D, Church GM. Analysis of optimality in natural and perturbed metabolic networks. *Proc Natl Acad Sci USA* 2002;99:15112–15117.

145. Shimodaira H, Filosi N, Shibata H, Suzuki T, Radice P, Kanamaru R, Friend SH, Kolodner RD, Ishioka C. Functional analysis of human MLH1 mutations in *Saccharomyces cerevisiae*. *Nat Genet* 1998;19:384–389.

146. Sickmann A, Reinders J, Wagner Y, Joppich C, Zahedi R, Meyer HE, Schonfisch B, Perschil I, Chacinska A, Guiard B, Rehling P, Pfanner N, Meisinger C. The proteome of *Saccharomyces cerevisiae* mitochondria. *Proc Natl Acad Sci USA* 2003;100:13207–13212.

147. Spellman PT, Sherlock G, Zhang MQ, Iyer VR, Anders K, Eisen MB, Brown PO, Botstein D, Futcher B. Comprehensive identification of cell cycle-regulated genes of the yeast *Saccharomyces cerevisiae* by microarray hybridization. *Mol Biol Cell* 1998;9: 3273–3297.

148. Sugawara N, Szostak JW. Construction of specific chromosomal rearrangements in yeast. *Methods Enzymol* 1983;101:269–278.

149. Suzuki-Fujimoto T, Fukuma M, Yano KI, Sakurai H, Vonika A, Johnston SA, Fukasawa T. Analysis of the galactose signal transduction pathway in *Saccharomyces cerevisiae*: interaction between Gal3p and Gal80p. *Mol Cell Biol* 1996;16:2504–2508.

150. Tai SL, Boer VM, ran-Lapujade P, Walsh MC, de Winde JH, Daran JM, Pronk JT. Two-dimensional transcriptome analysis in chemostat cultures: combinatorial effects of oxygen availability and macronutrient limitation in *Saccharomyces cerevisiae*. *J Biol Chem* 2004;280:437–447.

151. Tong AH, Boone C. Synthetic genetic array analysis in *Saccharomyces cerevisiae*. *Methods Mol Biol* 2006;313:171–192.

152. Tong AH, Evangelista M, Parsons AB, Xu H, Bader GD, Page N, Robinson M, Raghibizadeh S, Hogue CW, Bussey H, Andrews B, Tyers M, Boone C. Systematic genetic analysis with ordered arrays of yeast deletion mutants. *Science* 2001;294: 2364–2368.

153. Trethewey RN. Gene discovery via metabolic profiling. *Curr Opin Biotechnol* 2001; 12:135–138.

154. Trethewey RN, Krotzky AJ, Willmitzer L. Metabolic profiling: a Rosetta Stone for genomics? *Curr Opin Plant Biol* 1999;2:83–85.

155. Trumbly RJ. Glucose repression in the yeast *Saccharomyces cerevisiae*. *Mol Microbiol* 1992;6:15–21.

156. Uetz P, Giot L, Cagney G, Mansfield TA, Judson RS, Knight JR, Lockshon D, Narayan V, Srinivasan M, Pochart P, Qureshi-Emili A, Li Y, Godwin B, Conover D, Kalbfleisch T, Vijayadamodar G, Yang M, Johnston M, Fields S, Rothberg JM. A comprehensive analysis of protein–protein interactions in *Saccharomyces cerevisiae*. *Nature* 2000;403:623–627.

157. Unlu M. Difference gel electrophoresis. *Biochem Soc Trans* 1999;27:547–549.

158. Varma A, Palsson BO. Stoichiometric flux balance models quantitatively predict growth and metabolic by-product secretion in wild-type *Escherichia coli* W3110. *Appl Environ Microbiol* 1994;60:3724–3731.

159. Vazquez A, Flammini A, Maritan A, Vespignani A. Global protein function prediction from protein–protein interaction networks. *Nat Biotechnol* 2003;21:697–700.

160. Velculescu VE, Zhang L, Vogelstein B, Kinzler KW. Serial analysis of gene expression. *Science* 1995;270:484–487.

161. Velculescu VE, Zhang L, Zhou W, Vogelstein J, Basrai MA, Bassett DE Jr, Hieter P, Vogelstein B, Kinzler KW. Characterization of the yeast transcriptome. *Cell* 1997;88:243–251.

162. Villas-Boas SG, Hojer-Pedersen J, Akesson M, Smedsgaard J, Nielsen J. Global metabolite analysis of yeast: evaluation of sample preparation methods. *Yeast* 2005;22: 1155–1169.

163. Villas-Boas SG, Kesson M, Nielsen J. Biosynthesis of glyoxylate from glycine in *Saccharomyces cerevisiae*. *FEMS Yeast Res* 2005;5:703–709.

164. Villas-Boas SG, Mas S, Akesson M, Smedsgaard J, Nielsen J. Mass spectrometry in metabolome analysis. *Mass Spectrom Rev* 2004;24(5): 613–646.

165. Villas-Boas SG, Moxley JF, Akesson M, Stephanopoulos G, Nielsen J. High-throughput metabolic state analysis: the missing link in integrated functional genomics of yeasts. *Biochem J* 2005;388(Part 2):669–677.

166. Villas-Boas SG, Moxley JF, Akesson M, Stephanopoulos G, Nielsen J. High-throughput metabolic state analysis: the missing link in integrated functional genomics of yeasts. *Biochem J* 2005;388:669–677.

167. Viollet B, Andreelli F, Jorgensen SB, Perrin C, Flamez D, Mu J, Wojtaszewski JF, Schuit FC, Birnbaum M, Richter E, Burcelin R, Vaulont S. Physiological role of AMP-activated protein kinase (AMPK): insights from knockout mouse models. *Biochem Soc Trans* 2003;31:216–219.

168. Walsh G. Biopharmaceuticals: recent approvals and likely directions. *Trends Biotechnol* 2005;23:553–558.

169. Washburn MP, Wolters D, Yates JR III. Large-scale analysis of the yeast proteome by multidimensional protein identification technology. *Nat Biotechnol* 2001;19: 242–247.

170. Weckwerth W. Metabolomics in systems biology. *Annu Rev Plant Biol* 2003;54: 669–689.

171. Wei J, Sun J, Yu W, Jones A, Oeller P, Keller M, Woodnutt G, Short JM. Global proteome discovery using an online three-dimensional LC–MS/MS. *J Proteome Res* 2005;4:801–808.

172. Westergaard SL, Bro C, Olsson L, Nielsen J. Elucidation of the role of Grr1p in glucose sensing by *Saccharomyces cerevisiae* through genome-wide transcription analysis. *FEMS Yeast Res* 2004;5:193–204.

173. Winzeler EA, Shoemaker DD, Astromoff A, Liang H, Anderson K, Andre B, Bangham R, Benito R, Boeke JD, Bussey H, Chu AM, Connelly C, Davis K, Dietrich F, Dow SW, El-Bakkoury M, Foury F, Friend SH, Gentalen E, Giaever G, Hegemann JH, Jones T, Laub M, Liao H, Liebundguth N, Lockhart DJ, Lucau-Danila A, Lussier M, M'Rabet N, Menard P, Mittmann M, Pai C, Rebischung C, Revuelta JL, Riles L, Roberts CJ, Ross-Macdonald P, Scherens B, Snyder M, Sookhai-Mahadeo S, Storms RK, Veronneau S, Voet M, Volckaert G, Ward TR, Wysocki R, Yen GS, Yu K, Zimmermann K, Philippsen P, Johnston M, Davis RW. Functional characterization of the *S. cerevisiae* genome by gene deletion and parallel analysis. *Science* 1999;285:901–906.

174. Wu Y, Reece RJ, Ptashne M. Quantitation of putative activator–target affinities predicts transcriptional activating potentials. *EMBO J* 1996;15:3951–3963.

175. Yamano S, Ishii T, Nakagawa M, Ikenaga H, Misawa N. Metabolic engineering for production of beta-carotene and lycopene in *Saccharomyces cerevisiae*. *Biosci Biotechnol Biochem* 1994;58:1112–1114.

176. Yan Y, Kohli A, Koffas MA. Biosynthesis of natural flavanones in *Saccharomyces cerevisiae*. *Appl Environ Microbiol* 2005;71:5610–5613.

177. Yano K, Fukasawa T. Galactose-dependent reversible interaction of Gal3p with Gal80p in the induction pathway of Gal4p-activated genes of *Saccharomyces cerevisiae*. *Proc Natl Acad Sci USA* 1997;94:1721–1726.

178. Yates FE, Brennan RD, Urquhart J. Application of control systems theory to physiology. Adrenal glucocorticoid control system. *Fed Proc* 1969;28:71–83.

179. Yokobayashi Y, Weiss R, Arnold FH. Directed evolution of a genetic circuit. *Proc Natl Acad Sci USA* 2002;99:16587–16591.

180. Zhang LV, Wong SL, King OD, Roth FP. Predicting co-complexed protein pairs using genomic and proteomic data integration. *BMC Bioinformatics* 2004;5:38.

181. Zhang Y, Nijbroek G, Sullivan ML, McCracken AA, Watkins SC, Michaelis S, Brodsky JL. Hsp70 molecular chaperone facilitates endoplasmic reticulum-associated protein degradation of cystic fibrosis transmembrane conductance regulator in yeast. *Mol Biol Cell* 2001;12:1303–1314.

182. Zhu H, Bilgin M, Bangham R, Hall D, Casamayor A, Bertone P, Lan N, Jansen R, Bidlingmaier S, Houfek T, Mitchell T, Miller P, Dean RA, Gerstein M, Snyder M. Global analysis of protein activities using proteome chips. *Science* 2001;293:2101–2105.

183. Zhu H, Klemic JF, Chang S, Bertone P, Casamayor A, Klemic KG, Smith D, Gerstein M, Reed MA, Snyder M. Analysis of yeast protein kinases using protein chips. *Nat Genet* 2000;26:283–289.

10

CONSTRUCTION AND APPLICATIONS OF GENOME-SCALE *IN SILICO* METABOLIC MODELS FOR STRAIN IMPROVEMENT

Sang Yup Lee[1,2], Jin Sik Kim[1], Hongseok Yun[1], Tae Yong Kim[1], Seung Bum Sohn[1], and Hyun Uk Kim[1]

[1]*Department of Chemical and Biomolecular Engineering (BK21 Program), Center for Systems and Synthetic Biotechnology, Institute for the BioCentury, KAIST, Daejeon 305-701, Korea*
[2]*Department of Bio and Brain Engineering, BioProcess Engineering Research Center and Bioinformatics Research Center, KAIST, Daejeon 305-701, Korea*

10.1 INTRODUCTION TO SYSTEMS BIOTECHNOLOGY

Since the first genome sequence of a microorganism was finished in 1995, a number of projects for sequencing microbial genomes have been completed [1]. Currently, the complete sequences of more than 300 genomes are available in various databases [2]. The processes of sequencing and annotating microbial genomes have now become more routine, which resulted in the continued introduction of complete genome sequences of new microorganisms to the life science and biotechnology community. In addition, breakthroughs in studying biological systems at transcriptomic, proteomic, and other omic levels have enabled the researchers to generate and analyze high-throughput data

Systems Biology and Synthetic Biology   Edited by Pengcheng Fu and Sven Panke
Copyright © 2009 John Wiley & Sons, Inc.

for the better characterization of the organisms of interest [3]. Furthermore, computational (*in silico*) tools for modeling and simulation of biological systems on large or genome scale have been developed and used for deciphering the characteristics of metabolic, regulatory, and signaling networks [4]. With such advances in experimental and computational techniques, microorganisms can be systematically engineered to be suitable for various industrial applications that fall into a new paradigm of research called "systems biotechnology" [3].

Systems biotechnology aims at improving the biotechnological processes by systems-level optimization of cellular metabolism, regulations and signaling circuits, and mid- to down-stream processes altogether [3]. Understanding basic genotype–phenotype relationship in an organism is important, but it is not sufficient to understand and control the entire behavior of the organism. For this reason, high-throughput technologies have been indispensable tools as they allow the expression of genes to be monitored on global scale at transcriptional and translational level. One of the high-throughput techniques that has helped make this progress is transcriptomics, which allows the analysis of mRNA expression levels of the entire genes using DNA microarray. Proteomics allows analysis of the protein contents in an organism or a given sample. Metabolomics and fluxomics, which quantitatively profile the metabolites and fluxes, respectively, in the cell, also occupy an important portion of the omics research to carry out systems biotechnology research. By combining all the information generated from these omics disciplines, it will be possible to model an organism at the systems level (although not complete yet) and perform a systematic analysis of large-scale data using bioinformatics for a better understanding of how that system works and how it can be best adjusted for our applications [5,6].

Analysis of the *in silico* metabolic network can be used as a powerful approach for the identification of drug targets and targets for the improvement of microbial performance suitable for industrial applications such as production of useful materials [7–9]. *In silico* model is a mathematical representation of the biological system in interest and allows researchers to perform experiments on a computer to predict physiological behaviors much faster and economically than the actual experiments. Recently, various approaches for the construction of reliable metabolic network model have been suggested [1,10].

In this chapter, we describe the recent developments and trends in systems biotechnology research based on the *in silico* genome-scale metabolic models. Various strategies are described for the reconstruction of genome-scale metabolic network. Thereafter, we will review their applications with specific examples from the metabolic engineering perspectives. Readers are recommended to read Chapter 7 in parallel, which presents the state-of-the-art review on building the constraints-based metabolic models and their use in flux balance analysis (FBA).

10.2   DATABASES AND TOOLS FOR THE RECONSTRUCTION OF METABOLIC NETWORKS

From the last decade, unprecedentedly large amounts of information have been accumulated from experiments in genomics and other omics research projects.

As a result, many different databases and related applications have been developed for researchers to use to extract suitable information for the analysis of pathways and the reconstruction of genome-scale metabolic networks. The databases and applications commonly used for the systems biotechnology research are listed in Table 10-1. These databases are mainly used for the retrieval and analysis of sequences, protein analysis, functional annotation of genes and sequences, metabolic pathways, and other information needed for the reconstruction of metabolic networks. Here we shall focus on the effective construction and analysis of the *in silico* genome-scale metabolic networks using the information present in the databases.

Databases such as DDBJ, EMBL, and NCBI Entrez contain information regarding the DNA, RNA, and protein sequences and other related information [11–13]. Along with these databases, the controlled vocabularies, such as Gene Ontology (GO), are used for standardizing the results of genome annotations. Other databases contain information on metabolic networks such as reactions and network maps, tools for comparative analysis, and various information on enzymes, metabolites, and other biomolecules. For example, the automatic annotation tools import raw genome sequences and find proper open reading frames (ORFs) and gene candidates by applying gene finding algorithms. The databases for protein profiles and motifs are very helpful in enhancing the quality of genome annotation and in predicting the detailed functions of proteins by taking advantage of the conserved domains found in the proteins [14–16].

Reconstruction of metabolic pathways is mostly based on the information from metabolic databases [17–21]. Most of these databases provide graphical references or metabolic maps for users to find the metabolic information such as gene names, enzyme commission (EC) numbers, and reactions that are highly interlinked within the frame of metabolic network. KEGG is one of the most widely used metabolic resources and provides various data on the genomes, pathways, compounds, and controlled vocabularies. The pathway maps supplied by KEGG can be used as a backbone for the reconstruction of the networks. The BioSilico database integrates components of heterogeneous metabolic databases such as LIGAND, ENZYME, and BioCyc for easy querying and comparison of metabolic information present in multiple databases [18].

10.3 *IN SILICO* MODELING AND SIMULATION OF GENOME-SCALE METABOLIC NETWORK

The first step for the reconstruction of genome-scale metabolic model is the analysis of genome information in the databases. The availability of the annotation results from the completely sequenced genomes for many organisms makes it possible to reconstruct the *in silico* models on a genome scale. Thus, the automatic annotation process, which uses the reference databases and relevant information to identify potential ORFs, is the first step for the reconstruction of *in silico* metabolic model. However, as many shortcomings become obvious in the reconstruction process [22], the automatic annotation process appears to be insufficient and various complementary processes

Table 10-1 Databases and tools useful for the reconstruction of genome-scale metabolic network

Database	Availability	Brief Description
Resources of sequences and genomic information		
DDBJ [11]	http://www.ddbj.nig.ac.jp/	DNA Database of Japan
EMBL [12]	http://www.ebi.ac.uk/embl/	Europe's primary nucleotide sequence resource
Entrez [13]	http://www.ncbi.nlm.nih.gov/sites/gquery	The integrated, text-based search and retrieval system used at NCBI
COG [30]	http://www.ncbi.nlm.nih.gov/COG/	Clusters of Orthologous Groups
Controlled vocabularies and ontology		
GO [86]	http://www.geneontology.org/	A controlled vocabulary to describe gene and gene product attributes in any organism
KO [20]	http://www.genome.jp/kegg/ko.html	KEGG Orthology
Protein sequences, motifs, and profiles		
InterPro [14]	http://www.ebi.ac.uk/interpro/	A database of protein families, domains, and functional sites
PROSITE [16]	http://www.expasy.org/prosite/	A database of protein families and domains
Metabolic databases and tools		
BioCyc [17]	http://biocyc.org/	A collection of pathway/genome databases
BioSilico [18]	http://biosilico.kaist.ac.kr/	Integrated metabolic databases
BRENDA [19]	http://www.brenda-enjymes.info/	The comprehensive enzyme information system
KEGG [20]	http://www.genome.ad.jp/kegg/	Kyoto Encyclopedia of Genes and Genomes
Pathway tools [25]	http://bioinformatics.ai.sri.com/ptools/	A software system for pathway analysis of genomes and for creating Pathway/Genome Databases (PGDBs)
PATIKA [21]	http://www.patika.org/	Pathway Analysis Tools for Integration and Knowledge Acquisition
Gene annotation and comparative genomics tools		
Glimmer [88]	http://www.cbcb.umd.edu/software/glimmer/	A system for finding genes in microbial DNA, especially the genomes of bacteria and archaea.

Table 10-1   (*Continued*)

Database	Availability	Brief Description
ERGO [90]	http://ergo.integratedgenomics. com/	Accommodation of data integration, providing the tools to support comparative analysis of genomes
STRING [35]	http://string.embl.de/	Search tool for the retrieval of interacting genes/proteins

are required for the validation of the constructed metabolic models. Recently developed genome-scale *in silico* metabolic models are listed in Table 10-2.

10.3.1   Reconstruction Using the Known Pathways and Enzymes

The common method for the reconstruction of genome-scale metabolic network has been the utilization of information obtained from the previously constructed biochemical pathways, related sequences, and proteins [1,10,23]. Such information is mostly derived from the sequence-based search. The major advantage of metabolic reconstruction using the sequence-based comparison is that proper function of the genes can be quickly assigned. However, the presence of multiple relationships between genes and metabolic reactions can cause an erroneous assignment of genes on the metabolic map [9]. For example, imprecise annotations may occur for the homologues within the metabolic network, which hampers the accurate assignment of specific metabolic functions to the ORFs. Therefore, advanced curating methods have been introduced to eliminate the limitation of sequence-based annotation method [10].

Currently, a number of databases and tools have been developed for systems biotechnology research. Among them, several resources have been developed to represent the biochemical reactions and pathways on a two-dimensional space. Representative resources are KEGG [20] and BioCyc [17,24]. These are the most easily accessible and widely used databases on genes, enzymes, metabolites, and biochemical reactions. In addition to these tools, numerous databases and tools for the analysis of metabolic pathways have been released (Table 10-1). The utilization of these resources helps to gather the information on biochemical reactions and their location on the metabolic map. For example, the PathoLogic software, part of Pathway Tools that also contains MetaCyc database, automatically reconstructs the metabolic pathways of any organism only if the annotation file is available as an input [25]. The core algorithm in this process is that the software matches the enzyme in the annotation file (input file) to the ones defined in the MetaCyc database by EC number or enzyme name. Then, the software graphically displays the metabolic pathways and the associated components including reactions, enzymes, substrates, and products. The initial version of the automatically reconstructed metabolic network can be used as a basic framework and can be upgraded by manual curation.

Table 10-2 Recently developed genome-scale *in silico* models

Organism	Year	Genome Size (kbp)	Metabolites (ea)	Reactions (ea)	Referred Strains or Species for Biomass Composition
Corynebacterium glutamicum [91]	2009	3,309	411	446	*C. glutamicum* ATCC 13032, *C. glutamicum* ATCC17965, *C. glutamicum* CGL2005, *C. glutamicum* CGL2022, *E. coli*
Pseudomonas putida iJP815 [92]	2008	6,182	886	877	*E. coli*
P. putida iJN746 [93]	2008	6,182	710	950	*E. coli, P. putida*
Clostridium acetobutylicum [94]	2008	4,132	422	552	*S. aureus* 292, *S. aureus* 49/1974, *S. aureus* 6571, *S. aureus* 8325, *S. aureus* ATCC 12600, *S. aureus* DSM 20233, *S. aureus* Duncan, *S. aureus* H1AA, *Staphylococcus aureus* NCTC, *S. aureus* U-71, *S. aureus* oxford 209p *E. coli*
C. acetobutylicum [95]	2008	4,132	479	502	*C. acetobutylicum* ATCC 824, *B. subtilis*
Acinetobacter baylyi [96]	2008	3,583	701	875	*A. calcoaceticus, Micrococcus cerificans* HO1-N, *A. sp.* MJT/F5/199A
S. cerevisiae iIN800 [44]	2008	12,069	1013	1446	*S. cerevisiae*
P. aeruginosa iMO1056 [97]	2008	6,264	760	883	*P. aeruginosa* PAO1, *E. coli*
Rhizobium etli iOR363 [98]	2008	6,159	371	387	No biomass equation
Neisseria meningitidis [99]	2007	2,272	471	496	*N. meningitidis* HB-1
E. coli iAF1260 [65]	2007	4,639	1039	2077	*E. coli*
Bacillus subtilis iYO844 [100]	2007	4,214	988	1020	*B. subtilis* RB50:pRF69
Mycobacterium tuberculosis iJN661 [101]	2007	4,412	828	939	*M. tuberculosis* H37Rv, *M. bovis*
M tuberculosis [102]	2007	4,412	739	849	*M. tuberculosis* H37Rv, *M. bovis* BCG
Geobacter sulfurreducens [103]	2006	3,814	541	523	*G. sulfurreducens* ATCC 51573, *E. coli*
M. succinciproducens MBEL55E [104]	2005	2,314	519	686	*M. succinciproducens, E. coli*

Model [ref]	Year				Organisms
L. plantarum WCFS1 [38]	2005	3,308	658	762	*L. plantarum* WCFS1
S. aureus N315 [42]	2005	2,813	712	774	*S. aureus* 292, *S. aureus* 49/1974, *S. aureus* 6571, *S. aureus* 8325, *S. aureus* ATCC 12600, *S. aureus* DSM 20233, *S. aureus* Duncan, *S. aureus* H1AA, *Staphylococcus aureus* NCTC, *S. aureus* U-71, *S. aureus* oxford 209p *E. coli*
					B. subtilis
S. aureus N315 [45]	2005	2,813	571	641	*L. lactis* NCDO 2118, *L. lactis* subsp cremoris, *L. lactis* subsp. cremoris NCDO763. *E. coli*
Lactococcus lactis [46]	2005	2,365	422	621	
Helicobacter pylori [75]	2005	1,668	485	476	*H. pylori* NCTC 11638, *E. coli*
S. coelicolor A3(2) [22]	2005	8,667	500	971	*S. antibioticus* RIA-594, *S. clavuligerus, S. antibioticus, S. chrysomallus, S. roseoflavus, S. roseoflavus* var. *Roseofungini, E. Coli, S. typhimurium*
M. succinciproducens MBEL55E [59]	2004	2,314	352	373	*M. succinciproducens, E. Coli*
S. cerevisiae iND750 [105]	2004	12,069	646	1,149	*S. cerevisiae*
S. cerevisiae iFF708 [77]	2003	12,069	584	842	*S. cerevisiae*
E. coli K-12 iJR904 GSM/GPR [106]	2003	4,639	625	931	*E. coli*
H. pylori [107]	2002	1,668	340	388	*H. pylori* NCTC 11638, *E. coli*
E. coli K-12 iJE660a GSM [43]	2000	4,639	438	627	*E. coli*

10.3.2   Reconstruction Using Controlled Vocabulary

The interactions among the molecules in the metabolic and regulatory networks are known to be highly complex and incompletely understood [21]. To understand this, abstractions on different levels are used to analyze the cellular processes more effectively and to deal with the complex network structure more easily. The abstractions can be utilized to construct and analyze the graphical representation of metabolic pathways [26].

Ontologies for the standardization of the vocabularies were used for the automatic annotation analyses in many projects. The sequence similarity can be directly related to the potential protein functions by utilizing the ontology [27]. However, the limitation of gene ontology is that it cannot be directly connected to cellular metabolism. This is compensated by the application of metabolism-based orthology concept such as KEGG orthology to the annotation process [28]. When the proper KEGG orthology term can be assigned to a gene, the associated metabolic pathways can be found by tracing back the hierarchical structure of KEGG orthology [28].

10.3.3   Completion of Reconstruction Using Phylogenetic Profiles and Contexts

As the amount of sequence data increases explosively, the noise in the data also increases; accumulation of incomplete and/or wrong sequences causes obvious problems during bioinformatic analyses [29]. Annotation and analysis based only on these resources can generate wrong results and result in incorrect interpretations. This limitation can be overcome by employing controlled vocabulary and large-scale phylogenetic trees. Bacteria share many functional components with a high degree of conservation in the components. As mentioned in Section 10.3.1, Clusters of Orthologous Groups (COGs) use the grouping of previously annotated genes based on the sequence homology [30]. There are many ways to construct phylogenetic trees [29,31]. Different from the sequence-based analysis, the genome-scale phylogenetic profiles use various components of the genome such as the metabolic profiles and the distribution of gene contents [32]. Especially, the highly conserved components such as transporters and proteins involved in signaling and carbon source utilization can be used to find the proper orthologous genes [33,34]. The molecular interactions and network can be identified by using databases for protein–protein interaction and metabolic context such as STRING [35]. Similar to STRING, the SEED genome annotation system is based on the fundamental principle that the value of genome analysis increases with the number of genomes available as a context for comparative analysis [36].

Various bioinformatic methods, such as genome context analysis that includes chromosomal gene clustering, protein fusions, occurrence profiles, and shared regulatory sites, can be employed to obtain further information [37]. For example, a draft *in silico* metabolic model of *Lactobacillus plantarum* showed that succinyl-CoA is involved in a reaction related to methionine biosynthesis. However, after the phylogenetic studies and pathway analysis of the *L. plantarum* metabolic network, it was concluded that succinyl-CoA is not produced due to the operation of a branched tricarboxylic acid (TCA) cycle and that the actual substrate is most likely acetyl-CoA [38].

In addition to these methods, several integrated programs have been developed. The recent version of PathoLogic provides the function called "Pathway Hole Filler," which employs genome context analysis to fill in missing genes using the candidate sequences from the database, and subsequently a Bayes classifier to evaluate the probability of how likely the candidate has the desired function for the missing genes in the newly reconstructed metabolic network [39].

10.3.4 Completion of Reconstruction Using the Information from Various Sources

When the metabolic network reconstruction is complete, it should be able to describe and predict various phenotypic characteristics of the organism reasonably well under different genotypic and environmental conditions. However, some metabolic data are missing, inconsistent and insufficient to fully represent the physiology of a particular organism. In particular, reaction reversibility, substrate specificity, isoenzyme functions, cofactor specificity, and absence of certain pathways can make reconstruction process difficult.

Updated and new knowledge on the metabolic pathways and their components can be obtained by a thorough examination of literature. For example, the initial reconstruction of metabolic model of *Streptomyces coelicolor* A3(2) suggested that valine dehydrogenase (E.C. 1.4.1.8) is an NADP-dependent enzyme. However, after thorough examination of literature, it was found to use NAD as the preferred cofactor [22,40]. In the case of *Staphylococcus aureus* N315, literature indicates that acetate can be transported by acetate permease [41]. This transport reaction was then added to the reconstruction model to allow proper representation of observed physiological behavior *in vivo* [42].

When all the possible inconsistencies are considered, the reconstructed model should be validated and tested to see whether mathematical methods, such as convex analysis and linear programming, can effectively represent the physiology of the organism under the various genetic and environmental conditions. If the results reasonably represent what are observed in actual experiments, the reconstruction of genome-scale metabolic model is said to be done. However, it should be emphasized that metabolic reconstruction is not truly complete but has to be upgraded continuously as new information and knowledge on metabolic pathways and their participating components are discovered.

10.3.5 Simulation of Genome-Scale *In Silico* Metabolic Network

Once the genome-scale metabolic network is constructed from the genomic and other related information, computer-based experiments such as quantitative flux analysis, network topology analysis, and simulation can be performed to characterize the metabolic network under various conditions. There are two main strategies of quantitative *in silico* simulation of metabolic systems: static analysis and dynamic analysis.

Metabolic flux analysis (MFA), which utilizes stoichiometric matrices, has been employed for the large-scale analyses of metabolism (see Box 10-1). MFA calculates the intracellular flux distribution with an assumption of steady-state condition and does

BOX 10-1 VARIOUS MODELING APPROACHES

Metabolic Flux Analysis

MFA is a mathematical analysis of metabolic pathways in which metabolic fluxes are calculated by constructing a stoichiometric model of the biochemical reactions along with mass balances on intracellular metabolites [85]. Given a metabolic system, the mass conservation around metabolites can be expressed as

$$\frac{dc}{dt} = S \cdot v - b$$

where c is the concentration vector of metabolites, S is the $m \times n$ stoichiometric matrix in which m is the number of metabolites and n is the number of reactions, and v is the n-dimensional vector of intracellular fluxes. b is the concentration vector of metabolites that are diluted owing to biomass growth. Assuming the pseudo-steady or stationary state based on rapid turnover of most metabolites and dilution effects that are relatively small compared with the fluxes, we can simplify the kinetic model into a static representation. Unlike the dynamic approach, static model only considers the network's connectivity and capacity as time-invariant properties of the metabolic system.

$$S \cdot v = 0$$

The metabolic network can be classified as determined, overdetermined, and underdetermined systems if the degrees of freedom are zero, negative, and positive, respectively.

Different approaches are undertaken depending on the degree of freedom of the system. In general, two general methodologies of MFA have been practiced most widely: isotopomer balance analysis and flux balance analysis. The notable difference between these two methods is that the former is usually employed for overdetermined system whereas the latter is applicable to underdetermined system. For isotopomer balance analysis, ^{13}C carbon labeling measurements produce the flux data that can help solve the overdetermined system; it has been shown that the combination of information gathered from such isotopomer measurements using NMR and GC/MS and metabolite balancing enabled refined analysis of the metabolic fluxes. However, it should be mentioned that isotopomer analysis has so far been used for the analysis of small-scale metabolic networks because of the complicated mathematical formulation and limited availability of parameters. FBA allows determination of intracellular fluxes even for a large underdetermined system through linear optimization. Even though the accuracy of FBA can be thought as not as good as that achievable with isotopomer analysis, it generally gives satisfactory flux distribution under various genotypic and environmental conditions. Many successful examples are available in the literature, which report the use of FBA in various applications (see the text and Chapter 7).

Minimization of Metabolic Adjustment

MOMA [69] is based on the same constraints as the FBA. However, quadratic programming (QP) is used instead of linear programming to formalize the MOMA. The goal is to minimize the Euclidian distance from a wild-type flux distribution as follows:

$$\text{Minimize } (v-w)^T(v-w)$$

$$\text{Subject to } S \cdot v = 0, v_{min} \leq v \leq v_{max}$$

$$v_j = 0, \quad j \in R$$

where w is the wild-type flux distribution and R is a set of reactions related to the deleted genes.

Regulatory On/Off Minimization

ROOM [70] is based on the same constraints as FBA. The goal is to minimize the number of significant flux changes. A range $[w^l, w^u]$ around the vector w is defined for nonsignificant flux change. The mixed integer linear programming (MILP) can be formulated as

$$\text{Minimize } \sum_{i=1}^{m} y_i$$

$$\text{Subject to } S \cdot v = 0$$

$$v - y(v_{max} - w^u) \leq w^u$$

$$v - y(v_{min} - w^l) \geq w^l$$

$$v_j = 0, \quad j \in R, \quad y_i \in \{0, 1\}$$

$$w^u = w + \delta|w| + \varepsilon, w^l = w - \delta|w| - \varepsilon$$

where, for each flux i, $1 \leq i \leq m$, $y_i = 1$ for a significant flux change in v_i, and $y_i = 0$ otherwise.

Optknock

The bilevel optimization framework, OptKnock, was introduced to propose reactions to be eliminated from the *E. coli* network for maximizing the production of simple compounds such as succinate, lactate, and 1,3-propane-diol [71]. This is accomplished by calculating solutions that simultaneously optimize two objective functions, biomass formation and secretion of a target biochemical. This bilevel optimization algorithm is based on the fact that the overproduction of target biochemical can be achieved by altering the structure of the metabolic network through gene deletion such that the stoichiometry of the perturbed network forces production of the target metabolite while normal biomass precursors are generated.

$$\text{Maximize } v_{\text{biochemical}}(\text{over } y_j)$$

$$\text{Maximize } v_{\text{biomass}}(\text{over } v_j)$$

$$\text{subject to } \sum_{j=1}^{M} S_{ij}v_j = 0, \quad \forall i \in N$$

$$v_{\text{pts}} + v_{\text{glk}} = v_{\text{glucose uptake}}$$

$$v_{\text{ATP}} \geq v_{\text{ATP maintenance}} \qquad \qquad \theta$$

$$v_{\text{biomass}} \geq v_{\text{biomass}}^{\text{target}}$$

$$v_j^{\text{min}}y_j \leq v_j \leq v_j^{\text{max}}y_j, \quad \forall j \in M$$

$$\sum_{j=M}(1-y_j) \leq K$$

$$y_j \in \{0,1\}, \quad \forall j \in M$$

where S_{ij} is the coefficient of metabolite i in reaction j, biomass formation is quantified as an aggregate reaction flux, v_{biomass}, draining biomass components in their appropriate biological ratios, and $v_{\text{ATP maintenance}}$ is the non-growth-associated minimum ATP requirement. The uptake rate of glucose $v_{\text{glucose\_uptake}}$ is fixed and encompasses both the phosphotransferase system, v_{pts}, and glucokinase reaction, v_{glk}. K is the number of allowable reactions to be eliminated. Binary variable, y_j, is one if a particular reaction is active, and zero otherwise. An active reaction has an upper bound, v_j^{max}, and a lower bound, v_j^{min}, obtained by maximizing and minimizing each flux subject to the constraints.

not require rate equations and kinetic parameters. The result is a flux map showing the distribution of anabolic and catabolic fluxes within the metabolic network. Among the various applications of MFA, two general ones are as follows. The first application field is to characterize the cell's physiology under genetic and environmental perturbations. MFA has been used to characterize the effects of acute metabolic perturbation, especially, gene deletion in the organism. It was also performed under the combination of rich and minimal media and aerobic and anaerobic conditions to predict which reactions are essential for the growth of the organism under these conditions [4,42–46]. The second application field is to improve the production of various products including commodity chemicals by overexpression/deletion of key metabolic pathways that already exist in the host organism or by introducing new routes of metabolism. Of course, MFA is used to identify the candidate target genes to be manipulated.

The MFA solution provides a snapshot of a certain pathway in a defined state, but is insufficient to predict the dynamic behavior of metabolism. Recently, this approach was extended to allow the prediction of dynamic behavior. Dynamic simulation of genome-scale network model can be performed using the differential equations

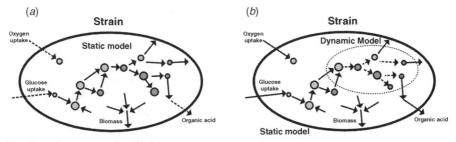

Figure 10-1   Concepts of dynamic flux balance analysis (a) and hybrid dynamic/static simulation (b). Fluxes represented by dashed arrows are given by kinetic equations.

representing the dynamic mass balances incorporating the reaction rate equations [47]. One of the major difficulties in the dynamic simulation of metabolic network is the lack of accurate kinetic equations and parameters for the reactions in the metabolic network. The parameters also tend to vary as the environmental conditions change. Therefore, the dynamic simulation of large-scale network requires many assumptions and is generally restricted to the small-scale network model. However, much effort has been devoted to solve this problem; a static simulation method was combined with a dynamic method, which is called dynamic flux balance analysis (DFBA) [48]. This method was developed to incorporate extracellular metabolite dynamics and substrate uptake kinetics within the flux balance analysis for extracellular glucose, acetate, and liquid- and gas-phase oxygen. Simple Michaelis–Menten kinetics and mass transfer kinetics are used to model the glucose uptake rate, oxygen uptake rate, and the acetate secretion rate (Fig. 10-1). When applied to the analysis of diauxic growth of *Escherichia coli* on glucose [48], the results from DFBA were qualitatively similar to the experimental observations. Yugi et al. [47] improved the DFBA method by introducing the dynamic methods (kinetics) to the rate-limiting steps of the metabolic reactions and the static methods (FBA) to the remaining reactions. In this method, the reactions expressed in the form of the static model require no prior information about kinetic equations and parameters or about the initial concentrations of metabolites (Fig. 10-1). This method was successfully used for the simulation of the erythrocyte model [47].

The hybrid method reduced the cost for the development of large-scale *in silico* models as well as the number of experiments for the identification of kinetic properties for dynamic simulation.

There are several software programs available for performing analyses and simulations of genome-scale *in silico* metabolic network. MetaFluxNet is a software package for the modeling and simulation of metabolic reaction networks focusing on MFA [49,50] (Table 10-3). It also provides the management of metabolic information and supports the systems biology markup language (SBML) and the metabolic flux analysis markup language (MFAML) [50] for the exchange of metabolic models. Simpheny (Genomatica, San Diego, CA) is a commercial software program for the construction and simulation of *in silico* genome-scale metabolic models [51]. Simpheny allows construction of *in silico* cells from their molecular components and simulation of the complete biochemical reaction network of a cell. Simpheny can be used for the prediction of various phenotypic characteristics based on FBA. General

Table 10-3 Useful softwares for the analyses of genome-scale
in silico **metabolic network**

Application	Web Site Address	Reference	Note (Usability)
MetaFluxNet	http://mbel.kaist.ac.kr/lab/mfn/	[49,50]	MetaFluxNet is a powerful software package for the *in silico* modeling and simulation of metabolic network using metabolic flux analysis. It supports for the generation and management of metabolic model using MFAML
GEPASI	http://www.gepasi.org/	[52]	GEPASI is a dynamic modeling software to construct and optimize network models with kinetic parameters
COPASI	http://www.copasi.org	[53]	COPASI provides tools for metabolic model generation, time course simulation, and metabolic control analysis. COPASI improved the GEPASI
BioSPICE	http://biospice.sourceforge.net/	[54,55]	BioSPICE provides an integration framework/workbench to integrate various tools according to their purpose

Pathway Simulator (GEPASI) has been widely used for the dynamic simulation and metabolic control analysis (MCA) of the metabolic network [52]. GEPASI contains several predefined kinetic models for the easy construction of dynamic simulation model. The Complex Pathway Simulator (COPASI) is an application for the simulation and analysis of biological networks [53]. It was developed based on the dynamic simulation tool, GEPASI. COPASI provides various tools including model generator, stochastic simulation tool, metabolic control analysis, and elementary mode analysis. It also supports the SBML format for the effective description of parameters of the kinetic equations. BioSPICE is an integrated system of the systems biology workbench (SBW) that allows the sharing of computational codes in various tools. It can be used to develop metabolic and genetic models using the common software framework [54,55].

10.4 ITERATIVE *IN SILICO* MODEL DEVELOPMENT

Since the *in silico* metabolic network cannot truly represent the real cell, it needs to be improved by iterative process. This process involves creating the metabolic network model, obtaining experimental data, comparing the predicted outcomes with experimental data, and resolving inconsistencies in the results to update the model. Hypotheses based on the results of *in silico* analysis can be tested by experiments, from which the model can be updated and improved based on the experimental results.

Both biochemical and genetic engineering experiments as well as computational tests are parts of the iterative process. As additional data for an organism become available, such as gene expression data and metabolic profiles, new biological information will be discovered to further refine and improve the *in silico* model.

In silico microbial models have been found to correctly predict experimentally observed behaviors of microbes *in vitro* 70–80 percent of the time [43,56,57]. Despite the relatively good agreement between the model predictions and actual experimental results, it is the 20 percent "failure" rate that is of most interest to us. These *in silico* "failures" point to areas of the model in which current knowledge on the organism is lacking (such as unknown pathways in the reconstruction, unaccounted-for regulatory interactions, etc.). These gaps in information must be filled in through new biological discovery. It is through the iterative process of model construction, testing, validation, and revision that new information on the organism can be discovered for filling in those gaps that will refine and improve the *in silico* model of the organism. By this iterative process, the most comprehensive and predictive *in silico* model of the organism can be built.

The current strategy for this process is to involve both experiments and the mathematical modeling/simulation in a feedback and iterative fashion (Fig. 10-2). The feedback approach is based on the prediction of genetic and metabolic modifications that can be compared with the experimental results, leading to a more rational strategy for the reconstruction of *in silico* model. Palsson et al. [58] showed that FBA could be used to predict what the eventual effects of genetic modifications would be on the global host cell physiology. The ability of a constraints-based model of *E. coli* describing genetic modifications was examined by subjecting them to adaptive evolution under different growth conditions.

Lee et al. [59] used this iterative approach by integrating genome and fluxome information in the characterization of a relatively less studied bacterium *Mannheimia succiniciproducens*. The genome was used to construct the genome-scale *in silico* metabolic map of *M. succiniciproducens*, and flux analysis was used to calculate the succinic acid yields and flux distributions under various conditions. It was found from the genome-scale flux analysis that carboxylation of phosphoenolpyruvate to oxaloacetate by PEP carboxykinase is the most important anaplerotic pathway leading to the efficient production of succinic acid by the reductive tricarboxylic acid cycle and menaquinone system [59]. In this iterative process, the proteome reference map of *M. succiniciproducens* was established by 2-DE coupled with mass spectrometry [60], and the results obtained were used to fine-tune the *in silico* metabolic network. The *in silico* metabolic network thus improved can be used to design new experiments for flux profiling and consequently for characterizing the metabolic characteristics under various environmental conditions.

10.5 METABOLIC ENGINEERING BASED ON THE *IN SILICO* MODEL FOR THE ENHANCED PRODUCTION OF VARIOUS BIOPRODUCTS

After the valid model is constructed, many *in silico* experiments can be carried out to quantify flux distributions under the numerous conditions of interest. These *in silico*

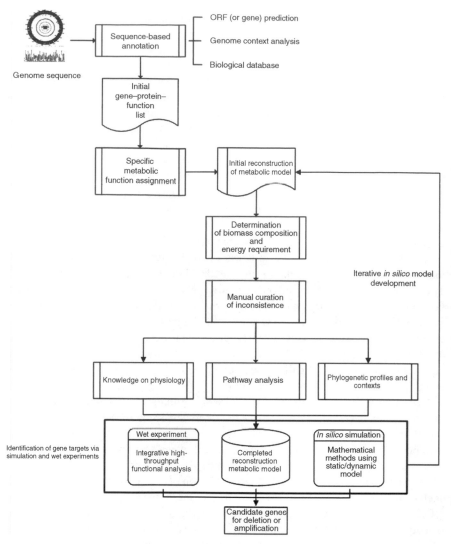

Figure 10-2 Flow chart for identifying gene targets by combining computational modeling/ simulation and high-throughput experimental analyses. The outcomes of these analyses evolve during the iterations to allow identification of new gene targets.

experiments make it possible to decipher the metabolic and physiological changes of the cells under various genetic and/or environmental conditions, and consequently establish a more rational metabolic engineering strategy to achieve desired goals. Furthermore, plausible targets for genetic modifications can be identified to improve the strain's performance through the comparative study of responses observed under various genetic and environmental perturbations (Fig. 10-3).

Currently, there are a large number of microorganisms that are used industrially for the production of bioproducts. Although these microorganisms do produce the desired

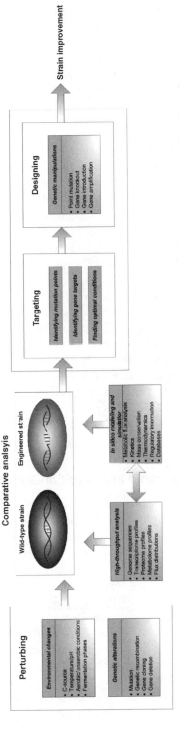

Figure 10-3 Procedure for the improvement of strains. Environmental and genetic changes significantly affect the cellular physiology. Wild-type and engineered strains can be comparatively analyzed by combining high-throughput technologies and *in silico* modeling and simulation. This comparative analysis would lead to identification of new gene targets to be introduced, gene knockout or amplification targets, and optimal production conditions. This newly generated knowledge can then be used to design experiments for genetic manipulations to bring about the desired phenotypes.

bioproducts, they do not naturally produce them to the concentrations and productivity high enough for commercialization. Additionally, the biological networks of microorganisms are robust enough to resist many changes introduced to them. Therefore, many combinatorial experiments including genetic manipulations (gene amplification and knockout), regulatory modification, and cultivation experiments need to be carried out to understand the metabolic characteristics and improve the phenotype to a desired level good enough for industrial applications. Here, *in silico* metabolic modeling and simulation can be used to overcome the impossibility of carrying out these many combinatorial experiments.

In silico organisms have been constructed to generate more knowledge about the cell and tackle the aforementioned problems. *E. coli*, the most well known and widely used bacterium, has been used for the production of a wide variety of bioproducts ranging from primary and secondary metabolites to biopolymers [61–64]. The *in silico* *E. coli* metabolic network has been expanded to contain up to 2077 reactions with 1039 metabolites [65]. However, baker's yeast, *Saccharomyces cerevisiae*, has been a model organism for understanding cellular physiology and compartmentalized intracellular biochemical behavior of a eukaryotic cell. Genome-based yeast model has the biochemical network of 1446 biochemical reactions and 1013 metabolites covering cytosolic and mitochondrial and transport reactions [44].

Obviously, gene manipulation is a very essential tool for strain improvement for the production of industrially valuable bioproducts. However, it is not possible to try every possible combination of gene targets as it is very time consuming and laborious. This is where FBA comes into play. FBA has most widely been exploited to quantitatively analyze the metabolic system thanks to its capability to predict the phenotypic behavior under various genetic and/or environmental conditions, and its applicability to genome-scale metabolic models [66,67]. Herein, strategies for the identification of gene knockout and addition targets as well as the combinatorial deletion, amplification, and regulation are described.

10.5.1 Identifying Gene Knockout and Addition Targets

Identifying the target genes for metabolic engineering to enhance the production of certain products is not always easy because of the large number of genes to be considered in the organism. Also, there is no guarantee that the identified single or even multiple target genes will enhance the production of the desired product due to the robustness of the biological network against changes to be made. At the initial stage, the potential target genes can be found through comparative analysis. The main obstacle to obtaining a rational solution to the problem of introducing genetic modifications is the lack of a reliable, global, metabolic model that captures stoichiometric, kinetic, and regulatory effects of the modifications on metabolite interconversions and metabolic flux distributions through the cellular reaction network. As a result, strain improvement has conventionally been achieved by random approaches whereby the target genes to be knocked out or amplified were intuitively selected rather than systematically. Consequently, the unexpected outcomes were often obtained. However, the genome-scale *in silico* metabolic model has changed a paradigm by enabling systemic approaches for strain improvement. Such genome-scale model has been simulated by means of linear optimization with a particular objective function such as maximization of cellular growth rate or production rate of certain metabolite of industrial value. Although the optimal value obtained by linear programming does not exactly describe the actual state of the cellular physiology, this methodology is still worthy to consider as it provides an overall picture of the cell metabolism, particularly carbon and energy distribution.

Raman et al. [68] employed FBA to search drug targets from the mycolic acid pathway of *Mycobacterium tuberculosis*, an important human pathogen. Mycolic acid constitutes the protective layer of this pathogen, and the inhibition of its biosynthesis has been the drug target due to its essentiality in cell growth, survival, and pathogenicity. Based on this biochemical background, a comprehensive model of mycolic acid biosynthetic system was built, and FBA was performed to identify essential genes by systematically knocking out the genes. Those genes that, when knocked out, resulted in a zero value for the objective function, the maximization of mycolic acid production in this case, were considered as drug targets as the pathogens cannot survive without mycolic acids. Candidate drug targets were further screened by homology search of these genes against the human genome to ensure that the host system does not possess the similar genes, which may be unexpectedly targeted by the drug, leading to adverse effects. This study is a nice example of how *in silico* analyses can be applied to the drug development process.

Since FBA does not account for the physiological changes caused by genotypic mutation, the simulation results may deviate from the experimental data. This has led to the development of a new algorithm called minimization of metabolic adjustment (MOMA) (see Box 10-1). This method attempts to determine more realistic flux distributions in knockout mutants by minimizing the changes in the flux distribution of the mutant with respect to the wild type instead of maximizing the biomass formation in the mutant [69]. This method takes into account that the mutant strain is not optimized for the production of metabolites because it has not had a chance to fine-tune its new metabolic network through evolution. This framework can be used to identify target genes to be knocked out to present a phenotype that is closest to the wild type (Fig. 10-4). It was found that this suboptimal profile actually lies between the wild-type and the mutant optimals. In one study using MOMA, *in silico* single- and multiple-gene knockout experiments were performed to systematically identify the gene targets and, ultimately, increase the lycopene yield [61]. This strategy can be used to guide the choice of gene knockout targets. This method yielded a triple knockout mutant that produced less than 40 percent more lycopene compared with an engineered overproducing *E. coli* strain. This study demonstrates the value of system optimization using MOMA for the strain improvement.

Another method that is similar to MOMA is the regulatory on/off minimization (ROOM) method (see Box 10-1). MOMA is based on the minimization of the changes in the metabolic fluxes in the mutant strain from the wild type. There may be one or two fluxes that require huge changes to compensate for the effects the mutation puts on the system. ROOM, however, minimizes all the fluxes with respect to the wild type regardless of any other factors. This method is based on the assumption that the system chooses to minimize its adaptation cost through regulation of the fluxes to maintain the

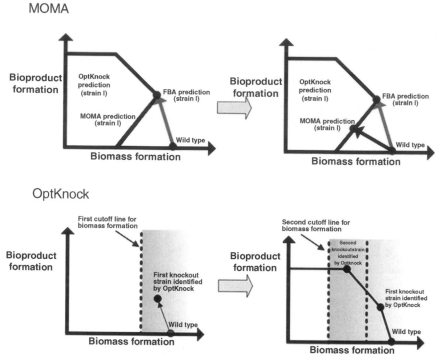

Figure 10-4 Graphical representation of the principles of MOMA and OptKnock. MOMA utilizes quadratic programming to find a metabolic state, in which artificially generated mutants try to minimize the redistribution of intracellular fluxes compared to the optimal flux distribution of wild type. Consequently, MOMA identifies a suboptimal metabolic state of the mutant that lies somewhere between the optimal state of the mutant and wild type in the altered solution space. This approach is based on the assumption that artificially generated mutant cannot immediately redistribute its fluxes toward the optimal growth rate since it has not undergone evolutionary pressure for an enough period of time as wild type had. OptKnock is a framework that suggests gene(s) to be knocked out for the enhanced production of bioproducts by considering both cell growth rate and objective metabolite production rate. This approach only considers the optimal production rate of the strain whose biomass formation rate is greater than the predetermined cutoff line. As a result, it will suggest a mutant genotype that allows faster growth only when it simultaneously produces a metabolite at faster rate.

wild-type stoichiometric and thermodynamic constraints. While both methods do not maximize biomass for the mutant strain, ROOM, by constraining the fluxes to "run in parallel" to the wild type, implicitly gives results under the maximum growth rates. ROOM was found to give similar or better prediction compared with FBA or MOMA in knockout experiments eight out of nine times [70].

The OptKnock method is another approach of identifying the knockout targets. This method identifies genes to be knocked out for bioproduct overproduction while considering the cell's needs as well (Fig. 10-4) (see Box 10-1). This approach was applied to lactate production in *E. coli* under anaerobic conditions where lactate production was maximized as an objective function in addition to the biomass objective function. This resulted in a coupling of lactate production with the formation of biomass [71]. The OptKnock method was also employed by the same research group to optimize the production of amino acids. Additional constraints, such as ammonia and oxygen transport by the cell, were introduced to eliminate alternative solutions. The OptKnock method is especially well suited for the study of amino acid production system because the metabolic reactions for amino acid production are highly regulated. Although OptKnock does not consider regulatory networks, it is satisfactorily acceptable because it considers the global effects of any changes made. It is because of this global consideration on the cell that less intuitive strategies need to be formulated by using this method [72]. In the production of amino acids, OptKnock suggested a number of knockout strategies that could enhance the production of various amino acids. The results of the study showed an increase in the amino acid production compared to the current strains used in industry. For example, an alanine yield of 91.5 percent from glucose could be achieved, which is much higher than that (45–55 percent) currently achieved in industry [73].

Lee et al. [74] compared the metabolism of *M. succiniciproducens*, a succinic acid overproducer, with that of *E. coli* to engineer an *E. coli* strain to overproduce succinic acid (Fig. 10-5). Several candidate genes for deletion were identified in *E. coli*. From

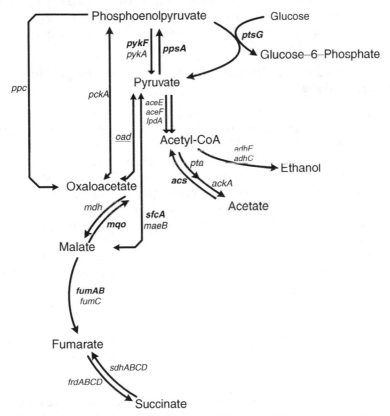

Figure 10-5 Comparison of the central metabolic pathways related to succinic acid formation in *M. succiniciproducens* and *E. coli*. Underlined genes represent those present only in *M. succiniciproducens* while those in boldface represent genes only present in *E. coli*.

this comparative genome analysis and using flux analysis to reverse engineer the metabolic network, succinic acid overproduction by *E. coli* could be achieved. During this process, our understanding on the general fermentative metabolic pathways in *E. coli* could be broadened. The fluxes to pyruvate and other acids were found to be the knockout targets for redirecting metabolic pathways toward enhanced succinic acid production. This example shows the effectiveness of combining comparative genomics, metabolic flux prediction, gene knockout, and fermentation toward strain development.

In contrast to gene knockout, FBA has also been used to identify genes to be amplified for enhanced metabolite production. FBA on poly(3-hydroxybutyrate) (PHB) producing *E. coli* predicted that the Entner–Doudoroff (ED) pathway, which was known to be inactive under normal culture conditions, was active during the production of PHB from glucose [64] (Fig. 10-6). This prediction was validated by actual experiments with a mutant *E. coli* strain defective in the activity of 2-keto-3-deoxy-6-phosphogluconate aldolase (Eda), a key enzyme in the ED pathway. Low PHB accumulation in the *eda* mutant strain compared to its parent strain could be restored when the *eda* gene was overexpressed in the *eda* mutant *E. coli* strain [64]. Also, the overexpression of the target genes (*fba* and *tpiA*) identified by FBA allowed enhanced production of PHB [75]. Therefore, MFA allows not only the knockout targets but also amplification targets to achieve enhanced metabolite production.

10.5.2 Combining the Deletion, Amplification, and Regulation of the Target Genes

The *in silico* genome-scale metabolic model can be used to further enhance the production of useful materials by combining the strategies of gene deletion, amplification, and regulation. Bro et al. [76] employed the genome-scale *in silico* model for the metabolic engineering of *S. cerevisiae* to improve the ethanol production. To increase the ethanol yield and reduce the yield of glycerol, an unnecessary by-product, a number of strategies were simulated using the previously reconstructed genome-scale model of *S. cerevisiae* [77]. Before they actually perform the simulations with the model, a few modifications were made to the model including incorporation of the necessary reactions. For example, those reactions catalyzed by xylose reductase and xylitol dehydrogenase for xylose metabolism were added to reflect actual experimental conditions as microorganisms were cultivated on the mixture of glucose and xylose. They then performed a gene insertion analysis by adding reactions one at a time from a pool of 3800 biochemical reactions that are derived from the LIGAND database [23]. The results of simulation by linear programming were scored based on the improvement of growth and ethanol yield and decreased glycerol yield. Consequently, the best-scored strategy, which predicted to improve the ethanol yield by 10 percent, but completely block the glycerol formation, was chosen for the actual experiment. According to the suggested strategy, they constructed a *S. cerevisiae* mutant, in which NADP-dependent glyceraldehydes-3-phosphate dehydrogenase (GAPN) was overexpressed, and achieved a 40 percent reduced glycerol yield with 3 percent increase in ethanol yield without affecting the specific growth rate. In a later study, the increased ethanol yield was also achieved with a GAPN expressing strain containing

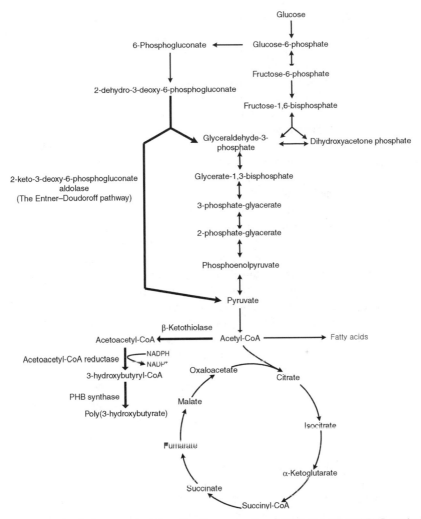

Figure 10-6 Metabolic network of *E. coli* for the production of PHB. The ED and PHB producing pathways are indicated with thick arrows. The ED pathway had been known to be inactive under the normal growth of *E. coli* using glucose as a carbon source. However, the simulation results of the *E. coli* model by FBA showed that the ED pathway is active. Consequently, overexpression of the corresponding enzyme, 2-keto-3-deoxy-6-phosphogluconate aldolase, in *E. coli* led to the improved production yield of PHB, and thus validated the simulation results of FBA.

xylose reductase and xylitol dehydrogenase, cultured on the mixture of glucose and xylose. This study is another good example of genome-scale *in silico* model for the hypothesis-driven metabolic engineering capable of predicting various strategies with acceptable accuracies. Moreover, it provides deeper insight into the metabolic characteristics because it shows how cofactors are linked with one another in different parts of the metabolic network. All benefits would lead to the more efficient way of a desired strain development.

Random mutagenesis, such as transposon mutagenesis and overexpression libraries, takes opposite approach to the systematical *in silico* analysis by randomly mutating the host organism and, thereby, producing a wide range of mutants. Since a large number of mutants must be screened for their improved phenotypes, smart screening system is essential. This approach is particularly beneficial and complementary to the *in silico* analysis because it can create mutants that cannot be predicted with the current techniques [78]. When coupled with global *in silico* metabolic analysis, this method becomes powerful for identifying targets for strain improvement [61]. Furthermore, it allows the dissection of critical subnetworks within the cell and a deeper understanding of that network, such as regulatory networks. By investigating how product formation correlates with these regulatory networks, putative molecular interactions may be inferred and examined in subsequent perturbations.

The metabolism of a living organism is controlled by not only mass balances but also various regulatory mechanisms such as transcriptional, translational, and allosteric regulations. By the incorporation of regulatory mechanisms, conditional activation and inactivation of metabolic networks can be mimicked, and optimal metabolic distributions can be obtained for different environmental conditions. So far, this has been achieved by incorporating transcriptional regulation into an *E. coli* MFA model based on Boolean logic [79–81]. With the inclusion of the transcriptional regulatory mechanisms, the accuracy of the MFA results increased to match experimental data better. Although this Boolean rule has been successfully combined with stoichiometric analysis, there is an inherent limitation in this method as the gene expression is somewhat stochastic and is not distinctive on-and-off type phenomenon in the real biological system [82,83]. In this context, probabilistic graphical models have been employed to model such regulatory networks, but its integration with a genome-scale metabolic model remains to be an open problem [83,84].

10.6 CONCLUSIONS AND FUTURE PROSPECTS

In this chapter, we have described the processes for the reconstruction of genome-scale *in silico* metabolic network using the genomic information and the applications of these models. These genome-scale metabolic networks are being applied to various fields. When combined with metabolic engineering, the genome-scale network can be utilized as a fundamental platform to identify key steps of bioproduct production under different conditions.

However, the construction of the metabolic network is by no means complete. Limitations on the network from the insufficient knowledge on the genetic characteristics of the genome create missing information such as gaps in the network. To complement the incompleteness of the model, experimental data should be sufficiently supported. In the post genomic era, the high-throughput omics technologies including transcriptomics, proteomics, fluxomics using ^{13}C-labeling flux analysis, and metabolomics can be efficiently used to validate the genome-scale model at least in a qualitative manner. For instance, simulation results (fluxes of biochemical reactions) of a genome-scale model can be compared with the transcriptome data to confirm whether the corresponding genes are expressed in the transcriptome profile. Likewise, proteome and metabolome profiles can be compared with the simulation results and used to generate further constraints. At present, these constraints are rather on-and-off type (e.g., the flux is set to be zero if there is no transcription of the gene encoding the enzyme carrying out that reaction). It is expected that an efficient algorithm will be developed for integrating the changing levels of various omics data in a quantitative manner during MFA. Such upgrade of modeling and simulation based on the integration of omics data will reveal the metabolic and regulatory characteristics more realistically and help designing strategies for the future experiments aiming at strain improvement. Eventually, all these efforts will lead to the development of virtual cell factory that can be used to tailor-design strains that are capable of producing various useful materials for human life.

ACKNOWLEDGMENTS

Our work described in this chapter was supported by a grant from the Korean Ministry of Education, Science and Technology (Korean Systems Biology Research Grant, M10309020000-03B5002-00000) and by the Brain Korea 21 Project. Furthermore supports through the LG Chem Chair Professorship, IBM SUR program, and Microsoft are greatly appreciated.

REFERENCES

1. Bansal AK. Bioinformatics in microbial biotechnology—a mini review. *Microb Cell Fact* 2005;4:19.
2. Liolios K, Tavernarakis N, Hugenholtz P, Kyrpides NC. The Genomes OnLine Database (GOLD) v.2: a monitor of genome projects worldwide. *Nucleic Acids Res* 2006;34:D332–D334.
3. Lee SY, Lee DY, Kim TY. Systems biotechnology for strain improvement. *Trends Biotechnol* 2005;23;349–358.
4. Borodina I, Nielsen J. From genomes to *in silico* cells via metabolic networks. *Curr Opin Biotechnol* 2005;16:350–355.
5. Ishii N, Robert M, Nakayama Y, Kanai A, Tomita M. Toward large-scale modeling of the microbial cell for computer simulation. *J Biotechnol* 2004;113:281–294.
6. Hocquette JF. Where are we in genomics? *J Physiol Pharmacol* 2005;56(Suppl 3):37–70.
7. Koffas M, Stephanopoulos G. Strain improvement by metabolic engineering: lysine production as a case study for systems biology. *Curr Opin Biotechnol* 2005;16:361–366.
8. Rajasethupathy P, Vayttaden SJ, Bhalla US. Systems modeling: a pathway to drug discovery. *Curr Opin Chem Biol* 2005;9:400–406.
9. Smid EJ, Molenaar D, Hugenholtz J, de Vos WM, Teusink B. Functional ingredient production: application of global metabolic models. *Curr Opin Biotechnol* 2005;16:190–197.
10. Francke C, Siezen RJ, Teusink B. Reconstructing the metabolic network of a bacterium from its genome. *Trends Microbiol* 2005;13:550–558.

11. Tateno Y, Saitou N, Okubo K, Sugawara H, Gojobori T. DDBJ in collaboration with mass-sequencing teams on annotation. *Nucleic Acids Res* 2005;33:D25–D28.

12. Kanz C, Aldebert P, Althorpe N, Baker W, Baldwin A, Bates K, Browne P, van den Broek A, Castro M, Cochrane G, Duggan K, Eberhardt R, Faruque N, Gamble J, Diez FG, Harte N, Kulikova T, Lin Q, Lombard V, Lopez R, Mancuso R, McHale M, Nardone F, Silventoinen V, Sobhany S, Stoehr P, Tuli MA, Tzouvara K, Vaughan R, Wu D, Zhu W, Apweiler R. The EMBL Nucleotide Sequence Database. *Nucleic Acids Res* 2005;33:D29–D33.

13. Maglott D, Ostell J, Pruitt KD, Tatusova T. Entrez Gene: gene-centered information at NCBI. *Nucleic Acids Res* 2005;33:D54–D58.

14. Mulder NJ, Apweiler R, Attwood TK, Bairoch A, Bateman A, Binns D, Bradley P, Bork P, Bucher P, Cerutti L, Copley R, Courcelle E, Das U, Durbin R, Fleischmann W, Gough J, Haft D, Harte N, Hulo N, Kahn D, Kanapin A, Krestyaninova M, Lonsdale D, Lopez R, Letunic I, Madera M, Maslen J, McDowall J, Mitchell A, Nikolskaya AN, Orchard S, Pagni M, Ponting CP, Quevillon E, Selengut J, Sigrist CJ, Silventoinen V, Studholme DJ, Vaughan R, Wu CH. InterPro, progress and status in 2005. *Nucleic Acids Res* 2005;33: D201–D205.

15. Bateman A, Coin L, Durbin R, Finn RD, Hollich V, Griffiths-Jones S, Khanna A, Marshall M, Moxon S, Sonnhammer EL, Studholme DJ, Yeats C, Eddy SR. The Pfam protein families database. *Nucleic Acids Res* 2004;32:D138–D141.

16. Hulo N, Sigrist CJ, Le Saux V, Langendijk-Genevaux PS, Bordoli L, Gattiker A, De Castro E, Bucher P, Bairoch A. Recent improvements to the PROSITE database. *Nucleic Acids Res* 2004;32:D134–D137.

17. Krummenacker M, Paley S, Mueller L, Yan T, Karp PD. Querying and computing with BioCyc databases. *Bioinformatics* 2005;21:3454–3455.

18. Hou BK, Kim JS, Jun JH, Lee DY, Kim YW, Chae S, Roh M, In YH, Lee SY. BioSilico: an integrated metabolic database system. *Bioinformatics* 2004;20:3270–3272.

19. Schomburg I, Chang A, Ebeling C, Gremse M, Heldt C, Huhn G, Schomburg D. BRENDA, the enzyme database: updates and major new developments. *Nucleic Acids Res* 2004;32: D431–D433.

20. Kanehisa M, Goto S, Kawashima S, Okuno Y, Hattori M. The KEGG resource for deciphering the genome. *Nucleic Acids Res* 2004;32:D277–D280.

21. Demir E, Babur O, Dogrusoz U, Gursoy A, Nisanci G, Cetin-Atalay R, Ozturk M. PATIKA: an integrated visual environment for collaborative construction and analysis of cellular pathways. *Bioinformatics* 2002;18:996–1003.

22. Borodina I, Krabben P, Nielsen J. Genome-scale analysis of *Streptomyces coelicolor* A3(2) metabolism. *Genome Res* 2005;15:820–829.

23. Goto S, Okuno Y, Hattori M, Nishioka T, Kanehisa M. LIGAND: database of chemical compounds and reactions in biological pathways. *Nucleic Acids Res* 2002;30:402–404.

24. Karp PD, Ouzounis CA, Moore-Kochlacs C, Goldovsky L, Kaipa P, Ahren D, Tsoka S, Darzentas N, Kunin V, Lopez-Bigas N. Expansion of the BioCyc collection of pathway/ genome databases to 160 genomes. *Nucleic Acids Res* 2005;33:6083–6089.

25. Karp PD, Paley S, Romero P. The Pathway Tools software. *Bioinformatics* 2002;18(Suppl 1): S225–S232.

26. Demir E, Babur O, Dogrusoz U, Gursoy A, Ayaz A, Gulesir G, Nisanci G, Cetin-Atalay R. An ontology for collaborative construction and analysis of cellular pathways. *Bioinformatics* 2004;20:349–356.

27. Zehetner G. OntoBlast function: from sequence similarities directly to potential functional annotations by ontology terms. *Nucleic Acids Res* 2003;31:3799–3803.

28. Mao X, Cai T, Olyarchuk JG, Wei L. Automated genome annotation and pathway identification using the KEGG Orthology (KO) as a controlled vocabulary. *Bioinformatics* 2005;21:3787–3793.

29. Snel B, Huynen MA, Dutilh BE. Genome trees and the nature of genome evolution. *Annu Rev Microbiol* 2005;59:191–209.

30. Tatusov RL, Natale DA, Garkavtsev IV, Tatusova TA, Shankavaram UT, Rao BS, Kiryutin B, Galperin MY, Fedorova ND, Koonin EV. The COG database: new developments in phylogenetic classification of proteins from complete genomes. *Nucleic Acids Res* 2001;29:22–28.

31. Snel B, Bork P, Huynen MA. Genome phylogeny based on gene content. *Nat Genet* 1999;21:108–110.

32. Hong SH, Kim TY, Lee SY. Phylogenetic analysis based on genome-scale metabolic pathway reaction content. *Appl Microbiol Biotechnol* 2004;65:203–210.

33. Eisen JA, Wu M. Phylogenetic analysis and gene functional predictions: phylogenomics in action. *Theor Popul Biol* 2002;61:481–487.

34. Holder M, Lewis PO. Phylogeny estimation: traditional and Bayesian approaches. *Nat Rev Genet* 2003;4:275–284.

35. von Mering C, Jensen LJ, Snel B, Hooper SD, Krupp M, Foglierini M, Jouffre N, Huynen MA, Bork P. STRING: known and predicted protein-protein associations, integrated and transferred across organisms. *Nucleic Acids Res* 2005;33:D433–D437.

36. Overbeck R, Begley T, Butler RM, Choudhuri JV, Chuang HY, Cohoon M, de Crecy-Lagard V, Diaz N, Disz T, Edwards R, Fonstein M, Frank ED, Gerdes S, Glass EM, Goesmann A, Hanson A, Iwata-Reuyl D, Jensen R, Jamshidi N, Krause L, Kubal M, Larsen N, Linke B, McHardy AC, Meyer F, Neuweger H, Olsen G, Olson R, Osterman A, Portnoy V, Pusch GD, Rodionov DA, Ruckert C, Steiner J, Stevens R, Thiele I, Vassieva O, Ye Y, Zagnitko O, Vonstein V. The subsystems approach to genome annotation and its use in the project to annotate 1000 genomes. *Nucleic Acids Res* 2005;33:5691–5702.

37. Osterman A, Overbeek R. Missing genes in metabolic pathways: a comparative genomics approach. *Curr Opin Chem Biol* 2003;7:238–251.

38. Teusink B, van Enckevort FH, Francke C, Wiersma A, Wegkamp A, Smid EJ, Siezen RJ. *In silico* reconstruction of the metabolic pathways of *Lactobacillus plantarum*: comparing predictions of nutrient requirements with those from growth experiments. *Appl Environ Microbiol* 2005;71:7253–7262.

39. Green ML, Karp PD. A Bayesian method for identifying missing enzymes in predicted metabolic pathway databases. *BMC Bioinformatics* 2004;5:76.

40. Navarrete RM, Vara JA, Hutchinson CR. Purification of an inducible L-valine dehydrogenase of *Streptomyces coelicolor* A3(2). *J Gen Microbiol* 1990;136:273–281.

41. Somerville GA, Said-Salim B, Wickman JM, Raffel SJ, Kreiswirth BN, Musser JM. Correlation of acetate catabolism and growth yield in *Staphylococcus aureus*: implications for host-pathogen interactions. *Infect Immun* 2003;71:4724–4732.

42. Heinemann M, Kummel A, Ruinatscha R, Panke S. *In silico* genome-scale reconstruction and validation of the *Staphylococcus aureus* metabolic network. *Biotechnol Bioeng* 2005;92:850–864.

43. Edwards JS, Palsson BO. The *Escherichia coli* MG1655 *in silico* metabolic genotype: its definition, characteristics, and capabilities. *Proc Natl Acad Sci USA* 2000;97:5528–5533.

44. Nookaew I, Jewett MC, Meechai A, Thammarongtham C, Laoteng K, Cheevadhanark S, Nielsen J, Bhumiratana S. The genome-scale metabolic model ilN800 of *Saccharomyces cerevisiae* and its validation: a scaffold to query lipid metabolism. *BMC syst Biol* 2008;2:71.

45. Becker SA, Palsson BO. Genome-scale reconstruction of the metabolic network in *Staphylococcus aureus* N315: an initial draft to the two-dimensional annotation. *BMC Microbiol* 2005;5:8.

46. Oliveira AP, Nielsen J, Forster J. Modeling *Lactococcus lactis* using a genome-scale flux model. *BMC Microbiol* 2005;5:39.

47. Yugi K, Nakayama Y, Kinoshita A, Tomita M. Hybrid dynamic/static method for large-scale simulation of metabolism. *Theor Biol Med Model* 2005;2:42.

48. Mahadevan R, Edwards JS, Doyle FJ, 3rd, Dynamic flux balance analysis of diauxic growth in *Escherichia coli*. *Biophys J* 2002;83:1331–1340.

49. Lee DY, Yun H, Park S, Lee SY. MetaFluxNet: the management of metabolic reaction information and quantitative metabolic flux analysis. *Bioinformatics* 2003;19:2144–2146.

50. Lee SY, Woo HM, Lee D-Y, Choi HS, Kim TY, Yun H. Systems-level analysis of genome-scale microbial metabolisms under the integrated software environment. *Biotechnol Bioproc Eng* 2005;10:425–431.

51. Mahadevan R, Burgard AP, Famili I, Dien SV, Schilling CH. Applications of metabolic modeling to drive bioprocess development for the production of value-added chemicals. *Biotechnol Bioproc Eng* 2005;10:408–417.

52. Mendes P. GEPASI: a software package for modelling the dynamics, steady states and control of biochemical and other systems. *Comput Appl Biosci* 1993;9:563–571.

53. Vallabhajosyula RR, Chickarmane V, Sauro HM. Conservation analysis of large biochemical networks. *Bioinformatics* 2006;22:346–353.

54. Sauro HM, Hucka M, Finney A, Wellock C, Bolouri H, Doyle J, Kitano H. Next generation simulation tools: the Systems Biology Workbench and BioSPICE integration. *Omics* 2003;7:355–372.

55. Garvey TD, Lincoln P, Pedersen CJ, Martin D, Johnson M. BioSPICE: access to the most current computational tools for biologists. *Omics* 2003;7:411–420.

56. Famili I, Forster J, Nielsen J, Palsson BO. *Saccharomyces cerevisiae* phenotypes can be predicted by using constraint-based analysis of a genome-scale reconstructed metabolic network. *Proc Natl Acad Sci USA* 2003;100:13134–13139.

57. Thiele I, Vo TD, Price ND, Palsson BO. Expanded metabolic reconstruction of *Helicobacter pylori* (iIT341 GSM/GPR): an *in silico* genome-scale characterization of single- and double-deletion mutants. *J Bacteriol* 2005;187:5818–5830.

58. Palsson BO, Ibarra RU, Edwards JS. *Escherichia coli* K-12 undergoes adaptive evolution to achieve *in silico* predicted optimal growth. *Nature* 2002;420:186–189.

59. Lee SY, Hong SH, Kim JS, In YH, Choi SS, Rih JK, Kim CH, Jeong H, Hur CG, Kim JJ. The genome sequence of the capnophilic rumen bacterium *Mannheimia succiniciproducens*. *Nat Biotechnol* 2004;22:1275–1281.

60. Lee JW, Lee SY, Song H, Yoo JS. The proteome of *Mannheimia succiniciproducens*, a capnophilic rumen bacterium. *Proteomics* 2006;6:3550–3566.

61. Alper H, Miyaoku K, Stephanopoulos G. Construction of lycopene-overproducing *E. coli* strains by combining systematic and combinatorial gene knockout targets. *Nat Biotechnol* 2005;23:612–616.

62. Lee SY, Hong SH, Moon SY. *In silico* metabolic pathway analysis and design: succinic acid production by metabolically engineered *Escherichia coli* as an example. *Genome Inform* 2002;13:214–223.

63. Hong SH, Lee SY. Metabolic flux distribution in metabolically engineered *Escherichia coli* strain producing succinic acid. *J Microbiol Biotechnol* 2000;10:496–501.

64. Hong SH, Park SJ, Moon SY, Park JP, Lee SY. *In silico* prediction and validation of the importance of the Entner–Doudoroff pathway in poly(3-hydroxybutyrate) production by metabolically engineered *Escherichia coli*. *Biotechnol Bioeng* 2003;83:854–863.

65. Feist Am, Henry CS, Reed JL, Krummenacker M, Joyce AR, Karp PD, Broadbelt LJ, Hatzimanikatis V, Paisson BØ. A genome-scale metabolic reconstruction for *Escherichia coli* K-12 MG1655 that accounts for 1260 ORFs and thermodynamic information. *Mol Syst Biol* 2007;3:121.

66. Patil KR, Akesson M, Nielsen J. Use of genome-scale microbial models for metabolic engineering. *Curr Opin Biotechnol* 2004;15:64–69.

67. Varma A, Palsson BO. Stoichiometric flux balance models quantitatively predict growth and metabolic by-product secretion in wild-type *Escherichia coli* W3110. *Appl Environ Microbiol* 1994;60:3724–3731.

68. Raman K, Rajagopalan P, Chandra N. Flux balance analysis of mycolic acid pathway: targets for anti-tubercular drugs. *PLoS Comput Biol* 2005;1:e46.

69. Segre D, Vitkup D, Church GM. Analysis of optimality in natural and perturbed metabolic networks. *Proc Natl Acad Sci USA* 2002;99:15112–15117.

70. Shlomi T, Berkman O, Ruppin E. Regulatory on/off minimization of metabolic flux changes after genetic perturbations. *Proc Natl Acad Sci USA* 2005;102:7695–7700.

71. Fong SS, Burgard AP, Herring CD, Knight EM, Blattner FR, Maranas CD, Palsson BO. *In silico* design and adaptive evolution of *Escherichia coli* for production of lactic acid. *Biotechnol Bioeng* 2005;91:643–648.

72. Pharkya P, Burgard AP, Maranas CD. Exploring the overproduction of amino acids using the bilevel optimization framework OptKnock. *Biotechnol Bioeng* 2003;84:887–899.

73. Ikeda M. Amino acid production processes. *Adv Biochem Eng Biotechnol* 2003;79:1–35.

74. Lee SJ, Lee DY, Kim TY, Kim BH, Lee J, Lee SY. Metabolic engineering of *Escherichia coli* for enhanced production of succinic acid, based on genome comparison and *in silico* gene knockout simulation. *Appl Environ Microbiol* 2005;71:7880–7887.

75. Park SJ. Metabolic engineering for the production of poly(3-hydroxyalkanoate) in recombinant *Escherichia coli*, Dissertation, Korea Advanced Institute of Science and Technology, Daejeon, Korea, 2003.

76. Bro C, Regenberg B, Forster J, Nielsen J. *In silico* aided metabolic engineering of *Saccharomyces cerevisiae* for improved bioethanol production. *Metab Eng* 2006;8:102–111.

77. Forster J, Famili I, Fu P, Palsson BO, Nielsen J. Genome-scale reconstruction of the *Saccharomyces cerevisiae* metabolic network. *Genome Res* 2003;13:244–253.

78. Ohnishi J, Mitsuhashi S, Hayashi M, Ando S, Yokoi H, Ochiai K, Ikeda M. A novel methodology employing Corynebacterium glutamicum genome information to generate a new L-lysine-producing mutant. *Appl Microbiol Biotechnol* 2002;58:217–223.

79. Covert MW, Palsson BO. Transcriptional regulation in constraints-based metabolic models of *Escherichia coli*. *J Biol Chem* 2002;277:28058–28064.

80. Covert MW, Knight EM, Reed JL, Herrgard MJ, Palsson BO. Integrating high-throughput and computational data elucidates bacterial networks. *Nature* 2004;429:92–96.

81. Herrgard MJ, Lee BS, Portnoy V, Palsson BO. Integrated analysis of regulatory and metabolic networks reveals novel regulatory mechanisms in *Saccharomyces cerevisiae*. *Genome Res* 2006;16:627–635.

82. Kaern M, Elston TC, Blake WJ, Collins JJ. Stochasticity in gene expression: from theories to phenotypes. *Nat Rev Genet* 2005;6:451–464.

83. Sun N, Zhao H. Genomic approaches in dissecting complex biological pathways. *Pharmacogenomics* 2004;5:163–179.

84. Friedman N. Inferring cellular networks using probabilistic graphical models. *Science* 2004;303:799–805.

85. Lee SY, Papoutsakis ET. *Metabolic Engineering*. New York: Marcel Dekker, 1999.

86. Harris MA, Clark J, Ireland A, Lomax J, Ashburner M, Foulger R, Eilbeck K, Lewis S, Marshall B, Mungall C, Richter J, Rubin GM, Blake JA, Bult C, Dolan M, Drabkin H, Eppig JT, Hill DP, Ni L, Ringwald M, Balakrishnan R, Cherry JM, Christie KR, Costanzo MC, Dwight SS, Engel S, Fisk DG, Hirschman JE, Hong EL, Nash RS, Sethuraman A, Theesfeld CL, Botstein D, Dolinski K, Feierbach B, Berardini T, Mundodi S, Rhee SY, Apweiler R, Barrell D, Camon E, Dimmer E, Lee V, Chisholm R, Gaudet P, Kibbe W, Kishore R, Schwarz EM, Sternberg P, Gwinn M, Hannick L, Wortman J, Berriman M, Wood V, de la Cruz N, Tonellato P, Jaiswal P, Seigfried T, White R. The Gene Ontology (GO) database and informatics resource. *Nucleic Acids Res* 2004;32:D258–D261.

87. Hoersch S, Leroy C, Brown NP, Andrade MA, Sander C. The GeneQuiz Web server: protein functional analysis through the Web. *Trends Biochem Sci* 2000;25:33–35.

88. Delcher AL, Harmon D, Kasif S, White O, Salzberg SL. Improved microbial gene identification with GLIMMER. *Nucleic Acids Res* 1999;27:4636–4641.

89. Riley ML, Schmidt T, Wagner C, Mewes HW, Frishman D. *The PEDANT genome database in 2005. Nucleic Acids Res* 2005;33:D308–D310.

90. Overbeek R, Larsen N, Walunas T, D'Souza M, Pusch G, Selkov E, Jr., Liolios K, Joukov V, Kaznadzey D, Anderson I, Bhattacharyya A, Burd H, Gardner W, Hanke P, Kapatral V, Mikhailova N, Vasieva O, Osterman A, Vonstein V, Fonstein M, Ivanova N, Kyrpides N. The ERGO genome analysis and discovery system. *Nucleic Acids Res* 2003;31:164–171.

91. Kjeldsen KR, Nielsen J. *In silico* genome-scale reconstruction and validation of the *Corynebacterium glutamicum* metabolic network. *Biotechnol Bioeng* 2009;102:583–597.

92. Puchałka J, Oberhardt MA, Godinho M, Bielecka A, Regenhardt D, Timmis KN, Papin JA, Martins dos Santos VA. Genome-scale reconstruction and analysis of the *Pseudomonas putida* KT2440 metabolic network facilitates applications in biotechnology. *PLoS Comput Biol* 2008;4:e1000210.

93. Nogales J, Palsson BØ, Thiele I. A genome-scale metabolic reconstruction of *Pseudomonas putida* KT2440: iJN746 as a cell factory. *BMC Syst Biol* 2008;2:79.

94. Senger RS, Papoutsakis ET. Genome-scale model for *Clostridium acetobutylicum*: Part I. Metabolic network resolution and analysis. *Biotechnol Bioeng* 2008;101:1036–1052.

95. Lee J, Yun H, Feist AM, Palsson BØ, Lee SY. Genome-scale reconstruction and *in silico* analysis of the *Clostridium acetobutylicum* ATCC 824 metabolic network. *Appl Microbiol Biotechnol* 2008;80:849–862.

96. Durot M, Le Fèvre F, de Berardinis V, Kreimeyer A, Vallenet D, Combe C, Smidtas S, Salanoubat M, Weissenbach J, Schachter V. Iterative reconstruction of a global metabolic model of *Acinetobacter baylyi* ADP1 using high-throughput growth phenotype and gene essentiality data. *BMC Syst Biol* 2008;72:85.

97. Oberhardt MA, Puchałka J, Fryer KE, Martins dos Santos VA, Papin JA. Genome-scale metabolic network analysis of the opportunistic pathogen *Pseudomonas aeruginosa* PAO1. *J Bacteriol* 2008;190:2790–2803.

98. Resendis-Antonio O, Reed JL, Encarnación S, Collado-Vides J, Palsson BØ. Metabolic reconstruction and modeling of nitrogen fixation in *Rhizobium etli*. *PLoS Comput Biol* 2007;3:1887–1895.

99. Baart GJ, Zomer B, de Haan A, van der Pol LA, Beuvery EC, Tramper J, Martens DE. Modeling *Neisseria meningitidis* metabolism: from genome to metabolic fluxes. *Genome Biol* 2007;8:R136.

100. Oh YK, Palsson BO, Park SM, Schilling CH, Mahadevan R. Genome-scale reconstruction of metabolic network in *Bacillus subtilis* based on high-throughput phenotyping and gene essentiality data. *J Biol Chem* 2007;282:28791–28799.

101. Jamshidi N, Palsson BØ. Investigating the metabolic capabilities of *Mycobacterium tuberculosis* H37Rv using the *in silico* strain iNJ661 and proposing alternative drug targets. *BMC Syst Biol* 2007;1:26.

102. Beste DJ, Hooper T, Stewart G, Bonde B, Avignone-Rossa C, Bushell ME, Wheeler P, Klamt S, Kierzek AM, McFadden J. GSMN-TB: a web-based genome-scale network model of *Mycobacterium tuberculosis* metabolism. *Genome Biol* 2007;8:R89.

103. Mahadevan R, Bond DR, Butler JE, Esteve-Nuñez A, Coppi MV, Palsson BO, Schilling CH, Lovley DR. Characterization of metabolism in the Fe(III)-reducing organism *Geobacter sulfurreducens* by constraint-based modeling. *Appl Environ Microbiol* 2006;72:1558–1568.

104. Kim TY, Kim HU, Park JM, Song HH, Kim JS, Lee SY. Genome-scale analysis of *Mannheimia succiniciproducens* metabolism. *Biotechnol Bioeng* 2007;97:657–671.

105. Duarte NC, Herrgard MJ, Palsson BO. Reconstruction and validation of *Saccharomyces cerevisiae* iND750, a fully compartmentalized genome-scale metabolic model. *Genome Res* 2004;14:1298–1309.

106. Reed JL, Vo TD, Schilling CH, Palsson BO. An expanded genome-scale model of *Escherichia coli* K-12 (iJR904 GSM/GPR). *Genome Biol* 2003;4:R54.

107. Schilling CH, Covert MW, Famili I, Church GM, Edwards JS, Palsson BO. Genome-scale metabolic model of *Helicobacter pylori* 26695. *J Bacteriol* 2002;184:4582–4593.

11

SYNTHETIC BIOLOGY: PUTTING ENGINEERING INTO BIOENGINEERING

Matthias Heinemann[1] and Sven Panke[2]

[1]ETH Zurich, Institute of Molecular Systems Biology, Wolfgang-Pauli-Str. 16, 8093 Zurich, Switzerland
[2]ETH Zurich, Institute of Process Engineering, Universitätsstr. 6, 8092 Zurich, Switzerland

11.1 HISTORY AND PERSPECTIVES OF SYNTHETIC BIOLOGY

The field of synthetic biology has recently received tremendous attention. Nevertheless, to most researchers it remains somewhat elusive what synthetic biology really is. Is it a new discipline? Or is it just a new phrase for old stuff? Is it similar to the contemporary field of systems biology as the phonetic similarity might suggest?

Briefly, no single mature concept of synthetic biology exists yet, which makes a short historic view on the early occurrences of the term and the different proposed conceptual backgrounds for synthetic biology a potentially good point to start. As we will see, there are a number of different strands of origin for synthetic biology. In a further step, we illustrate some perspectives of and requirements for synthetic biology.

11.1.1 History

To the best of our knowledge, the first user of the term "synthetic biology" was Stéphane Leduc (1853–1939) at the Medical School in Nates, France, who had an

Systems Biology and Synthetic Biology Edited by Pengcheng Fu and Sven Panke
Copyright © 2009 John Wiley & Sons, Inc.

interest in defining life and to create lifelike forms from chemicals. In his book "La biologie synthétique" published in 1912 Leduc covered a multitude of experiments with inanimate substances that seem to mimic various animate structures—crystal growth, mineral formations, electrolytic and colloidal solutions that react and develop similarly as cellular structures, tissues, and nuclei. The ultimate aim was to present the reader with new ideas about the nature and definition of life, the physicochemical basis for biological activity, evolution, and morphogenesis. Leduc thought that the appearance of forms resembling plants produced by osmotic effects in concentrated colloidal mixtures of inorganic salts had something to tell us about the emergence of life. Although he did not claim that these forms were actually living, even during his lifetime Leduc became completely marginalized and the passion for this topic died out in the early 1930s with the rise of cell physiology, biochemistry, and genetics.

It took then more than 60 years until the term "synthetic biology" was used for the second time. In 1978, the Nobel Prize in Physiology and Medicine was awarded to Werner Arber, Daniel Nathans, and Hamilton O. Smith for their discovery of restriction enzymes and their application to molecular genetics. In an editorial comment of the journal *Gene*, Waclaw Szybalski and Ann Skalka wrote: "The work on restriction nucleases not only permits us easily to construct recombinant DNA molecules and to analyze individual genes but also has led us into the new era of *synthetic biology* where not only existing genes are described and analyzed but also new gene arrangements can be constructed and evaluated" [1].

Maybe prompted by this comment, "synthetic biology" headed a *Nature* review on a book that discussed recombinant DNA technology in 1979 [2] and a review article published by Barbara Hobom in 1980 in *Medizinische Klinik* that covered the corresponding new possibilities [3]. The subsequent time of public debate on possible accompanying biohazards led to an article on "social responsibility in an age of synthetic biology," published in the journal *Environment* [4]. Finally, in an article published in 1986 again in a German journal (*Verhandlungen der Deutschen Gesellschaft für Innere Medizin*), Gerd Hobom reviewed the recent advances in gene technology and stated that biology had left the status of a purely descriptive scientific discipline and was now heading toward a synthetic discipline—synthetic biology. He compared the new possibilities of gene technology, that is, the possibility to recombine genes from different organisms with the development of organic chemistry, where 150 years before there had been a transition from mere description and analysis of naturally occurring chemical compounds to the directed synthesis of novel chemicals. Correspondingly, he stated that the new technologies could also be viewed as tools to create simple biological systems for further analysis [5].

While the term "synthetic biology" had been primarily used to address the new capability of recombining existing genes so far, the synthesis of new genes came into focus in 1988 at a conference organized by Steven Benner in Interlaken, Switzerland. Benner, a chemist at the University of Florida, titled this conference "Redesigning the molecules of life" after the originally intended title "Redesigning life" was considered too provocative in the light of the ongoing recombinant DNA debates [6,7]. Benner's goals were, and still are, to generate molecules by chemical synthesis that reproduce the complex behavior of living systems, including self-reproduction and Darwinian-like evolution, thereby contributing to our understanding of the chemistry behind life. At the time of this conference, although the term synthetic biology was not explicitly used for the ongoing endeavors, the notion of synthetic biology in the sense of designing artificial DNA molecules was around. It took another 22 years until this notion of chemically designing molecules for manipulating living systems was labeled synthetic biology: At the annual meeting of the American Chemical Society (ACS) in San Francisco, Benner's colleague Eric T. Kool, professor of chemistry at Stanford University, described his work of designing nonnatural, synthetic molecules that nevertheless function in biological systems as synthetic biology [8].

Besides these chemical research-driven activities, another strand of synthetic biology was initiated around the year 2000, when several groups mainly from the biophysics community published on designing and engineering genetic circuits [9,10]. The driving force of these activities was the idea that new insights into the functioning of circuits could be obtained by their *de novo* reconstruction. Taking this a step further led to an engineering perspective of synthetic biology, aiming at the rational construction of biological parts, devices, or systems that have new and not necessarily natural functionality and can be employed for useful purposes.

These issues featured very prominently in the "The First International Meeting on Synthetic Biology," which took place in June 2004 at the Massachusetts Institute of Technology in Cambridge, USA. We consider this meeting as the inaugural event of the discipline. In addition to the work on designing genetic circuits, research from various other areas such as protein engineering, metabolic engineering, and biological chemistry was presented.

Since then, the term synthetic biology has reaped tremendous popularity, which is reflected by the significant boost in the number of mentioning of the term "synthetic biology" in scientific publications over the recent years (Fig. 11-1). As synthetic biology has gained momentum, various research communities have embraced the term, and most likely, many other disciplines will follow suit.

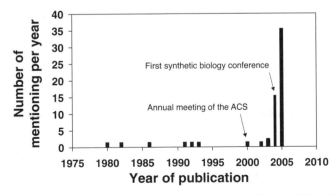

Figure 11-1 Number of mentionings of the term "synthetic biology"" in scientific publications over the recent years.

11.1.2 Perspectives of Synthetic Biology

As seen before, the term synthetic biology was used in different research communities rather independently. Today, basically, as a result of the outlined historic development, one can distinguish between two different perspectives of synthetic biology, the science and the engineering perspective. We will first sketch both the perspectives and then concentrate on the engineering perspective and illustrate this in detail (cf. Table 11-1 and [11]).

The scientific perspective of synthetic biology is mainly discovery and understanding driven. Biologists are interested in learning more about how natural living systems work by rebuilding biological systems and functions (i.e., real physical instances) from scratch according to the current understanding and to test these rebuilt systems or functions, very much in the spirit of "reverse engineering" or "reverse systems biology" (cf. [12]). Chemists involved in synthetic biology try to synthesize new, nonnatural "biochemicals," such as alternate self-replicating macromolecules, to ultimately study the origin of life. Thus, the chemistry-oriented branch of synthetic biology represents a specific field of chemical research striving to analyze and understand our living world, which is an extension of the concept of "biomimetic chemistry".

Synthetic biology can also be viewed from an engineering perspective. Biological systems or their parts are used in processing chemicals, energy, information, and materials. Unfortunately, the engineers' efforts in this area (e.g., in the areas of metabolic or protein engineering) are only decorated with a few success stories, reflecting today's limited ability to engineer biology in a directed and successful manner. In the engineering perspective, synthetic biology aims at overcoming the existing fundamental inabilities by developing foundational technologies to ultimately enable a systematic forward engineering of biology for improved and novel applications. In this perspective,

Table 11-1 Different perspectives of synthetic biology

Synthetic Biology View From the Different Sides ...	Biology	Chemistry	Engineering
Respective goals	Rebuilding represents a vehicle to test our understanding of complex systems.	Creating new biochemicals to study the origin of life.	Designing new biological systems in a forward engineering manner for useful purposes.
Synthetic biology seen as	A research tool	A specific research area	A discipline
Also known as	Reverse engineering	Organic chemistry, biological chemistry	Biological engineering
In the tradition of	Biology	Biomimetric chemistry	Biochemical engineering, metabolic engineering, protein engineering

"synthetic biology" would be synonymous to "biological engineering" and describe another field of engineering next to, for example, mechanical or electrical engineering, with which it would share a common set of methodologies.

Despite these fundamental differences, a common denominator exists in the described areas of synthetic biology: All branches are similar in so far as they deal with the designing and building of biological components, functions, and systems. In each branch, however, the final purpose for doing so is different.

11.1.3 Synthetic Biology from the Engineering Perspective

Biology, as a scientific discipline, has traditionally focused on studying single events or mechanisms in a more or less isolated manner, but in great detail (Fig. 11-2). Examples are the detailed investigation on the mechanism of a specific enzyme reaction or the in-depth analysis of a single gene's function.

This is now complemented by the new field of systems biology, which targets at a system-level understanding of whole biological systems [13]. Armed with detailed mechanistic knowledge on a multitude of isolated phenomena, this new approach aims at a holistic understanding of biological systems with all the interactions between different cellular processes. It is powered by the recognition that biology cannot be understood by looking at its parts alone but requires an understanding of its systemic characteristics and also by the advent of powerful measurement techniques ("omics techniques") that enable this type of research.

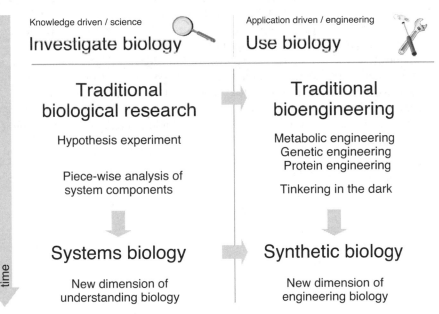

Knowledge driven / science
Investigate biology

Application driven / engineering
Use biology

Traditional biological research

Hypothesis experiment

Piece-wise analysis of system components

Traditional bioengineering

Metabolic engineering
Genetic engineering
Protein engineering

Tinkering in the dark

Systems biology

New dimension of understanding biology

Synthetic biology

New dimension of engineering biology

time

Figure 11-2 An overview of certain aspects of the scientific and applied side of biology at different times.

In all the previous periods of biological research from ancient times to the era of recombinant DNA technology, the knowledge acquired was exploited by engineers for practical applications, from dairy products and beer to metabolic and protein engineering. However, the modifications of the biological material (i.e., strains, enzymes, etc.) to achieve improved properties involved (and still involves) a great degree of uncertainty. Indeed, the desired output of a manipulation is rarely obtained in a straightforward manner, but requires a prolonged trial-and-error period ("tinkering in the dark"). Here, it is important to note that this is in stark contrast to the work in other (nonbiology-related) engineering disciplines, such as mechanical or civil engineering.

Now, that systems biology promises a new quality of understanding and, at least, an intellectual framework to understand biology from first principles, just as good as we understand mechanics or thermodynamics, we can start to think of designing biological systems, and ultimately we will want to do it in a way as we design other functional objects such as cars or bridges. In other words, at least in specific areas biology has matured enough to start thinking of designing biological parts in a forward engineering manner. Such forward engineering design of biological functions or systems we would call synthetic biology. In summary, one could argue that the scientific discipline of systems biology paves the way for the engineering discipline of synthetic biology aiming at the design of new and improved biological functions.

The following thought experiment might be helpful to grasp the difference between systems and synthetic biology:

Assume that a car was something derived from nature that had been optimized by evolution—like a biological cell. Furthermore, assume that our knowledge about this biological car would be very limited. The systems biologist would start investigating the car. He would discover that there is an engine and a gearing system, and that the engine is linked to the gearing system, which causes the wheels to turn, and eventually he would understand how this biological system, the car, works.

In turn, the synthetic biologist would use the knowledge acquired about the gearing system, engine, and so on and would dismantle these parts, would try to optimize (redesign), for example, the engine, to standardize the parts so that they can be used for other cars, but also for other systems, work on the corresponding interfaces, and finally reassemble the parts of the car in a new manner to build something new, for example, a moon rocket.

11.2   WHAT IS REAL ENGINEERING?

In the last section, we have used the term engineering several times and have also mentioned that "engineering" (as in metabolic engineering) is not necessarily equal to "engineering" (as in mechanical engineering). Looking at a classical engineering project, we will try to derive the characteristic features of a "true" engineering work.

11.2.1   An Engineering Example

Imagine the manufacturing of a new car. First, properly skilled mechanical engineers are needed, who were trained to know that an engine, a gearing system, wheels and so on are required, and how these parts are interconnected. With this knowledge, the engineers make use of computer software (e.g., for computer-aided design), do calculations, and finally come up with a design of a new car. According to their plans, parts (e.g., headlights) are then manufactured. Previously introduced standards (e.g., ISO standards) that are respected during the design process ensure that the different parts will later fit together, even though they might have been produced by different companies. Once manufactured, the produced parts will in most cases first be stored in warehouses until they are used for assembly. Note that design engineers not only develop the plans for the fabrication of the single parts but also elaborate ways for assembling these parts (e.g., in which order) so that finally the designed car becomes a reality as a result of a structured design process.

11.2.2 Key Features of Engineering Endeavors

From this short illustration of a typical engineering project, we can derive several characteristic features of true engineering endeavors whose relevance to biological engineering is worth exploring: (1) forward engineering design on the basis of know-how, (2) abstraction, (3) standardization of components and conditions, and (4) decoupling of system design from system fabrication. Some of the ideas presented in the following were taken from a recent review by Drew Endy [11].

11.2.2.1 Forward Engineering Design on the Basis of Know-How   In nonbiology-related areas, engineers can usually draw on a sound knowledge base. Phenomena relevant for design projects in chemical, mechanical, electrical, or civil engineering are in most cases understood from first principles or at least up to a level that makes forward engineering design possible. The sound mastering of thermodynamics and reaction kinetics (chemical engineering), mechanics (mechanical engineering), physics (electrical engineering), or statics (civil engineering) can serve as an example. In each of these areas, the existing in-depth understanding permits computer-based design of new systems by going through iterations between computer models and simulations (but in most cases not including experimentation). By this procedure, extensive testing of new design variants can be performed *in silico*, which in most cases is more time and cost efficient and also much safer than an actual realization and real-life testing. In other words, sound knowledge acts as a basis for real engineering and enables forward engineering design with a predictable outcome.

11.2.2.2 Component and Device Abstraction   Engineering endeavors are typically characterized by hierarchies of abstractions. As already indicated in the illustration of the car fabrication, the different parts of the car are set up at different hierarchical levels. The car (overarching top level) contains one specific part—the engine, which at a lower level consists of a number of cylinders, which again can be decomposed into several other parts, such as seal rings, and so on. Generally, there are parts that cannot be decomposed into smaller parts (such as screws); there are parts that consist of several other parts (such as headlights) and finally the whole system (car) that has been built of various parts.

This hierarchical structure has several practical advantages: First, the introduction of system boundaries basically hides information and is thus a way to manage complexity. People, who assemble headlights into the chassis of a car, do not need to know the exact bodywork of the rest of the car. In other words, abstraction is useful as it allows individuals to work independently at each level of the hierarchy. Furthermore, abstraction provides an organization or a methodology of how to combine the parts and consequently supports the engineering of systems with many integrated components.

11.2.2.3 *Standardization of Components and Conditions*
To efficiently make use of an introduced abstraction hierarchy that principally allows for plug-and-play of the different components, the connection and interfaces between the different parts need to be defined, that is, standardization is required. In classical engineering disciplines, standardization is provided by institutions such as the International Organization for Standardization, a federation of national standards bodies, providing standards for almost every sector of business, industry, and technology. Defined standards for components and conditions ensure that connections or interfaces between components fit, even when they are fabricated by different companies. Beyond, as the metrics of tools (e.g., screw drivers) are also subject to standardization, this guarantees that the tools match the according part (e.g., screw).

11.2.2.4 *Decoupling of Design from Fabrication*
Another typical feature of true engineering is the decoupling of the design process from the actual fabrication of novel devices or systems. There are people designing new devices or systems (i.e., developing the plans of how to fabricate them) and other people (construction people, craftsmen) actually realize these plans (i.e., actually building or fabricating the devices or systems according to the specified design). This separation between design and fabrication is realized since both tasks (design and construction) require a distinct set of skills and expertise, which is typically not provided by the same individuals in a mature engineering field.

Nevertheless, this decoupling of design and fabrication requires the design engineers to have a sound knowledge about how things are actually produced and how parts are assembled together. In other words: The design for an object is useless if no way exists to fabricate it. Or, the design for parts of a car is useless if no concept is provided of how these parts can be assembled together (e.g., in which order—compare the planning of an assembly line). Of course, this includes the respecting the importance of standards ranging from a common language needed between the two interacting sets of people (the designers and the craftsmen) to the fact that it is necessary that the craftsmen's tools fit the designed parts.

11.3 VISION FOR SYNTHETIC BIOLOGY

11.3.1 A Little Bit of Science Fiction

To imagine what synthetic biology could become in the future, we just have to replace the car with a biological cell and have to employ the outlined features of classical

Towards a cell-level biofactory

1. Design catalyst strain
2. Construct catalyst strain (synthesis of DNA segments and assembly)
3. Produce catalyst strain (cultivation on rich medium, and then transition to a catalytic machinery)
4. Produce chemical compound with catalyst strain (from inexpensive starting material)

Figure 11-3 Steps toward a cell-based biofactory.

engineering disciplines. Some of the ideas presented in the following were taken from a recent article published in *The Scientist* [14]. Imagine, for example, a cell-level biofactory, easy to produce by cultivation, that replaces a 50-step chemical synthesis route, for example, to a complex oligosaccharide drug molecule (such as the antithrombotic pentasaccharide Arixtra [15]).

The steps from the design to the application of the catalyst are outlined in Figure 11-3. Ultimately, the design of this cellular catalyst would be straightforward, solely computer-based, and would draw on readily available parts that simply would have to be combined in a plug-and-play manner. Then, after the design and the genetic construction of the strain, comprising synthesis of the required genetic segments and assembly in a bacterial strain, the cell-based biofactory would be amplified by cultivation and finally would be used for production of the complex oligosaccharide starting from the inexpensive substrate.

At a first glance, one could gain the impression that this concept is very much in line with the classical approach of metabolic engineering. However, we will see that much more is required than the overexpression or knockout of a few genes. For example, the *de novo* design and construction of new biofunctional systems will involve building of novel proteins, genetic circuits, and metabolic networks.

Most likely, we would start our endeavors of designing this novel catalyst with an organism with a reduced, possibly redesigned genome (serving as a sort of chassis on which we can expand in a rational fashion), to which we would add additional functionality in the form of nucleotide sequences including the required regulatory, gene coding and other functional regions. The organism with the minimized set of protein-coding genes would most probably have only a rudimentary set of metabolic capabilities (to eliminate interference with the inserted pathways), would have lost all elements that contribute to genome dynamics (such as transposons and insertion elements), and would, in general, be reduced to the specific functions that are required for the well-characterized behavior under predefined manufacturing conditions.

To this organism with a minimized genome, we would then add the set of *de novo* synthesized genes that provide the capability to synthesize the desired oligosaccharide starting from a cheap substrate such as glucose. The amount of proteins could be carefully controlled by adjusting the corresponding elements on the DNA, such as promoter and ribosome binding strength. The genes of the pathway might have been assembled from templates from different species and then adapted for the expression in the chosen host or they might be the result of a rational protein engineering effort that has conveyed the desired functionality to a specific protein. As energy and reducing power must be provided for this synthesis, preferably in a carefully stoichiometrically balanced fashion to prevent the production of side reactions, we additionally would have to include reactions that fulfill these tasks. For this, we would ultimately employ carefully characterized and readily available DNA modules that have been used for these tasks frequently before.

To prevent unnecessary metabolic burdens in early process phases, the conversion of the host cell to the actual catalyst would be inducible and comprehensive—for example, to such an extent, that growth and production could be completely uncoupled but cellular functions could be rescued for maintenance on the pathway. One fundamental prerequisite here would be that we are able to indeed decouple specific cellular functions from the remaining cellular activities. One way to achieve this might be the deliberate introduction of mutually independent functionalities (orthogonality), such as ribosomes that interact with novel ribosome binding sites (see also below) or enzymes that depend on novel coenzymes.

But these mutually independent functionalities would hopefully also extend to the dynamic properties of the designed pathways. Natural enzymes are frequently adapted to the needs of the cell to maintain homeostasis and operate with metabolite concentrations within the μM–mM range. Consequently, for an improved version of the catalyst that produces high product titers, we need to identify the relevant allosteric inhibitions (some of which might be still be unknown since they only become apparent at concentrations higher than the typical intracellular ones) and remove these inhibitions by redesigning the respective enzymes.

The novel properties of this cellular production machine will hopefully ultimately be designed and optimized at the computer. These designed components would then be converted into the respective amino acid sequences and finally translated into a nucleotide sequence, from which the desired functionality can be expressed. Finally, this designed DNA sequence will then be chemically synthesized, assembled with other parts and introduced in the chassis—the organism with the minimized genome.

It is important to note that such cell-based biofactory for fine chemical manufacturing is just an example of many conceivable synthetic biology projects and applications. From the degree of mastership that is required to execute such a project, it becomes clear that the implications are much more far-reaching and can be extended to any other area for which biotechnology has been or will be considered. To be clear, today, every ongoing synthetic biology project only scratches at a project like the one illustrated above.

11.3.2 Potential Fields of Application

The illustration of the cell-based biofactory represents an example where a synthetically devised organism could execute various functions that allow producing a chemical compound, a drug, or even maybe energy in the form of hydrogen from agricultural waste. The recent and envisioned breakthroughs in biology and technology, however, do not only present unprecedented opportunities that could restructure and revolutionize the manufacture of chemicals, pharmaceuticals, and energy, but also may offer unique ways to enable carbon sequestration and environmental remediation. In addition, several medical applications can be envisioned as well as projects

Table 11-2   An overview of potential areas of application for synthetic biology together with illustrating examples

Area	Examples
Production of pharmaceuticals, chemicals, energy	Develop a bacterial or fungal cell that can be programmed to produce complex hydrocarbon precursors (e.g., oils, plastics), to produce hydrogen or ethanol, to convert waste into energy, to convert sunlight into hydrogen
Chemical/biological threat detection and decontamination	Develop a bacterial cell that can be programmed to fix any desirable amount of atmospheric CO_2 Develop a bacterial cell that swims to the threat and decontaminates it
Medical applications	Develop bacteria that can parasitize cancer cells Develop circuits that guards against cancer, which if activated self-deconstructs the cell
Analytics and diagnostics, sensors and actuators	Develop bacteria, fungi, or plants that can be programmed to monitor environmental state, but that never survive mutation, whose DNA are not subject to horizontal gene transfer (both coming and going) Develop proteins that can sense any kind of harmful chemical compound (e.g., TNT)

stemming from the field of processing of information. Here, Table 11-2 provides an overview of potential areas of application, together with some illustrating examples.

The examples foreseen for areas not related to chemical synthesis underline that indeed the term "metabolic engineering" is too narrow to describe the new discipline of synthetic biology, as this term is always used in the context of chemical production by means of manipulated biological cells. In contrast, it is envisioned that synthetic biology will employ organisms and biological systems more broadly to solve real-world problems, and thus it has an enormous potential for human health, renewable energy, and the environment.

In fact, a few companies have already been founded with the goal of harvesting some of the early benefits in the area of synthetic biology. Possibly, the most prominent companies are the two U.S. companies Synthetic Genomics and Codon Devices, both founded in 2005. Synthetic Genomics wants to develop and commercialize genome reconstruction and synthesis technologies and particularly engages in the area of ethanol and hydrogen production. Codon Devices aim to develop a technology platform that is expected to accurately synthesize kilobase-to-megabase-long DNA sequences. The company's early commercial focus is on providing engineered devices for molecular biology research and biotherapeutics.

11.4 REQUIREMENTS FOR SYNTHETIC BIOLOGY

After portraying our vision of synthetic biology and after suggesting some ideas of what the discipline might be able to deliver in the future, we now examine the actual requirements to make this vision come true.

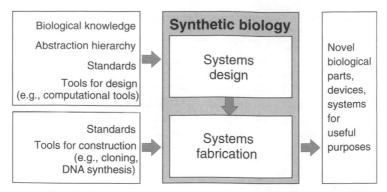

Figure 11-4 Synthetic biology encompasses systems design and fabrication. Each part has its specific prerequisites and inputs. Ultimately, synthetic biology delivers novel biological entities with improved functionality.

Synthetic biology encompasses the redesign of existing, natural biological systems for useful purposes and in the long run the design and construction of new biological parts, devices, and systems. Generally, future synthetic biology endeavors can be subdivided into two distinct divisions: systems design and systems fabrication (cf. Fig. 11-4).

11.4.1 Design

The design division of synthetic biology deals with the forward engineering (re)design of biological parts, devices, or systems. In the following, we will discuss the general requirements for the design as well as the current limitations and problems in the respective areas: (1) knowledge, (2) computational design and (3) standardization, modularity and orthogonality.

11.4.1.1 Knowledge   As outlined in one of the previous sections, every established engineering classical discipline (such as mechanical or civil engineering) can draw on a sound body of knowledge, ideally ranging down to the first principles. In biology, we have not yet reached such level of in-depth understanding and consequently, true biological engineering has not been possible until now. The recent advances in the postgenomic research, however, provide hope that sooner or later we will be able to draw on a body of knowledge that allows for such a directed engineering of biology. Here, especially the concerted efforts of systems biology provide novel degree of comprehension, so that systems biology could be considered as a driving force for synthetic biology.

11.4.1.2 Computational Design   As a further requisite for synthetic biology, computational tools are necessary that enable the computer-based (re)design of biological parts, devices, or systems and form the synthetic biologist's computer-aided design (CAD) software package equivalent, in analogy to the design tools available in the areas of mechanical or civil engineering. Such a design tool would need to integrate the detailed available knowledge into a user-friendly program and thus would bring this knowledge out of the realm of research into the realm of engineering. Also in the field of synthetic biology, the designs tools would be based on mathematical models that realistically reproduce biological behavior. Using such software, the design engineer would try to improve the behavior of a system *in silico* by adjusting design parameters—a task that can also be fulfilled by automatic optimization procedures targeting a selected objective function. Simulation capabilities implemented in the design tool would finally allow computational testing of design variants. Still, to obtain such design tools much research work is necessary as first rough mathematical models (e.g., describing gene transcription and translation or kinetics of metabolic or signaling pathways) are only now becoming available [16]. Moreover, to be useful for a forward engineering design, the employed mathematical models need to have predictive power. Beyond, in cases where only small numbers of molecules are involved (such as in gene transcription and translation, where transcription factors and mRNA molecules only occur in very low discrete numbers), the models need to be able to even reproduce the inherent stochasticity of such processes. This is imperative, as it was shown that stochasticity in combination with certain system architectures can— on a stochastic basis for decision making—result in different system states [17]. Thus, a robust design of new devices and systems must be able to exclude such eventualities.

Another important area for further computational design efforts is the field of protein design: Nowadays, *de novo* protein structure prediction from a linear amino acid sequence can only be achieved for small protein domains [18] and quantitative prediction of enzymatic activity and selectivity from 3D protein structures, in general, is not yet feasible—although significant progress has been made in this direction. For example, by selecting an enhanced "dead-end elimination" algorithm, an efficient computational procedure could be established to first convert *E. coli*'s periplasmic ribose binding protein into a set of proteins with completely novel substrate binding specificities [19] and later into a protein with triose phosphate isomerase activity [20]. The novel enzyme had a catalytic constant in the order of $0.1 \, s^{-1}$, which is quite a remarkable achievement for a computational design. Nevertheless, there remains a long way to go until true forward engineering of proteins "at discretion" becomes possible.

Our still limited abilities in protein design only highlight another important limitation in our current synthetic biology designs: the lack of detailed knowledge of many important systems parameters, which synthetic biology shares with systems biology. When trying to implement specific behaviors, it is usually possible to identify parts that qualitatively have the required behavior (positive or negative regulation, composite promoters, etc.). However, to organize these parts into a system with the desired behavior, we also need the right dynamics—in other words, we need specific DNA–protein or protein–protein binding constants, Hill coefficients, protein degradation rates, and so on. However, the quantitative characterization of many systems is by far not comprehensive enough, and even if the corresponding parameters were measured, they might not fit the system we intend to desire. Here is another big field for protein design in synthetic biology: the rational modification of specific binding properties.

Alternatively, one has to resort to evolutionary methods to obtain the required modifications. This was shown by rescuing a nonfunctional inverter gene network in which LacI was supposed to repress the synthesis of CI repressor, but did so only partially, so that CI was constantly synthesized [21]. Possible solutions here were decreasing the leakiness of *cI* repression or reduction of the repressive effect of CI itself. Screening a library of mutated DNA fragments covering the RBS of the *cI* gene and the gene itself delivered clones with a variety of mutations that, in general, seemed to weaken the repression effect of CI, such as reducing the dimerization of CI molecules or reducing DNA–protein interactions. In other words, directed evolution was an excellent tool to adapt the system parameters such that the inverter characteristic could be rescued.

11.4.1.3 *Standardization, Modularity, and Orthogonality* Another requisite for the design in a true engineering sense is the availability of standards. Parts, devices, or systems stored in a database (or warehouse) need to have standardized interfaces so that a design engineer can make use of these as modules for his design in a plug-and-play manner. A first synthetic biology warehouse, the MIT Registry of Standard Biological Parts, has been established at the Massachusetts Institute of Technology, Cambridge, United States of America (http://parts.mit.edu/). It uses the standardized vector format of "idempotent vectors" that lends itself easily to assemble and allows interoperability of assembled sequences (https://dspace.mit.edu/bitstream/1721.1/21168/1/biobricks.pdf). Alternatively, the NOMAD technology has been suggested [22]. Here, vectors are designed in such a way that a DNA insertion into an assembly vector recreates exactly the same restriction site architecture of the assembly vector alone by exploiting restriction sites with compatible but noncleavable ends or by exploiting type II restriction enzymes. Both techniques allow multiple rounds of insertion on either side of an insert. However, despite its success, initiatives such as the MIT Registry have to be backed up in the future by more sophisticated design tools and a large capacity to validate and document such standard parts.

Next, it is not clear whether we already measure the most useful quantitites for engineering and, if we can agree on the set of quantitites, how to measure them so that measurements can be reproduced and contributed by multiple labs. Even such an apparently simple concept like promoter strength is poorly defined. Frequently, it is reported in terms of protein activities. This, however, is an aggregate quantity that integrates the number of messages produced per time from a promoter, the (again aggregated) efficiency of initiation of translation, the efficiency of the translation itself including codon usage effects, the amount of protein that has correctly folded into a functional form and the current steady-state of protein production and degradation. Consequently, the corresponding experiments across labs are usually difficult to compare. Alternatively, the promoter strength could be very narrowly defined as "PoPS" (polymerases per second) that would quantitatively describe the number of RNA polymerase molecules that pass a specific point on the DNA per time (http://parts.mit.edu/r/parts/htdocs/AbstractionHierarchy/index.cgi). Such a quantity would lend itself easily to comparison of many promoters, as it is much more narrowly defined. However, it is currently unclear how to measure such a parameter directly.

This idea of standardization of parts is inevitably linked to the concept of modularity and functional self-containment of parts. After all, it will be almost impossible and highly undesirable to recreate the interdependency that is characteristic of today's living systems. Rather, from an engineering perspective, it is much more desirable to draw on a limited number of well-characterized and optimized parts and devices that do not interact with each other besides the interactions that have been introduced on purpose. This provokes, of course, the question whether such modularity—composability "at discretion"—is possible in biology.

Current evidence is that it is, at least for many instances. Even though the current failure rate when assembling modular parts on DNA level—promoters, ribosome binding sites, coding sequences, terminators, and so on—for example, from the MIT registry is still significant, this reflects more of a deficiency in implementing all available knowledge into the design process than of a fundamental problem. The composability of genetic elements is, after all, the underlying dogma of recombinant DNA-technology.

Recent work on RNA molecules has extended the concept of modularity to gene regulation. For example, it is possible to design modular RNA aptamer domains for small molecules that can be coupled to antisense effector domains to regulate translation in response to the presence or absence of small molecule effectors [23]. Aptamer domains responsive to different small molecules have been used successfully with the same antisense domain and vice versa, demonstrating the modularity of the concept. Taking to the extreme, this means that there is a new series of modular tools available that can interfere in a programmable and responsive manner with gene expression on RNA level of many different genes.

However, the modularity concept also appears to work on protein level: In particular, the domain architecture of many regulatory proteins plays here very much in favor of such approaches. One specific example is the design of polydactyl zinc finger DNA binding proteins [24]. Combinations of zinc finger domains provide the sequence specificity of the DNA binding domain (DBD) by recognizing essentially a subset of three or four nucleotides per zinc finger domain. Such proteins display modularity in two ways: typically, the DBD is functionally independent from the effector domain and zinc finger domains are functionally relatively independent of each other—so that by selecting a set of specific zinc fingers *in silico*, one can specify arbitrary sequence specificity into a novel DBD, which can then be coupled to a novel effector domain. This concept, though not yet truly universal, has already delivered some spectacular successes in designing DNA-binding domains that recognized up to 18 nucleotides [25–28].

Similar functional reprogramming could also be achieved on the level of protein–protein interactions with signaling proteins. For example, the domains of the eukaryotic neuronal N-WASP protein, involved in actin polymerization, are very amenable to recombination, including with domains from other proteins. These recombinations lead, for example, to proteins that execute novel logical behavior [29].

Alternatively, nature provides examples where signaling proteins have left the task of providing specificity within signaling pathways to scaffold proteins. These recruit a kinase and the kinase's substrate and assemble them in close proximity for phosphorylation [30]. Reorganizing such scaffolds along common kinases allows recombining signaling pathways, so that osmotolerance could be converted to a function of mating pathway induction.

A final point that will be essential for successful design is its orthogonality (mutual independence). When implemented, the design has to be executed by the cell in the intended fashion, which implies that the interactions of the design with the cell occur only at the anticipated points. Given the high degree of interdependency of cellular functions and one common reaction space for all the different interactions (the cytoplasm), this is an absolutely nontrivial task. For example, it will be vital to eliminate cross talk of gene regulatory elements such as regulatory proteins, particularly when considering the design of large artificial networks. A promising approach is presented here by the concept of engineered riboregulators [31], where *cis*-acting repressing parts of the mRNA and trans-acting activating RNA molecules combine to regulate gene expression in a fashion that could be widely extended to many genes. Importantly, when four sets of such crRNA–taRNA combinations were tested for cross talk, none could be detected, advocating well for the exploitation of this technology in large artificial networks.

An alternative approach would be to reserve subsets of specific functions in the cell only for the execution of a design. For example, rather than feeding mRNAs of genes that are part of the design into the common cellular ribosome pool, proper engineering of the mRNA–ribosome pair can reserve a subset of ribosomes specifically for the translation of one mRNA [32] and thus isolate the translation of the target mRNA from the rest.

Taking this concept a step further, it should be possible to introduce not only new specificity with existing molecular species, but also to introduce new molecular species. Again, this might allow whole new sets of unique interactions that exist quasi "in parallel" to traditional cellular functions. The topic of alternative chemical structures with self-replicative properties is explored further in Chapter 13 of this book by Holliger and Loakes, so we are not treating it here.

11.4.2 Fabrication

The fabrication division of synthetic biology is responsible for the actual realization of the design engineers' plans, so that finally a new biological part, device, or system turns into reality. Once the design engineer has delineated a novel functional module and has converted (or, in other words, coded) this design into a sequence of nucleotide bases, it is necessary to physically produce this strand of DNA, possibly to assemble it with other already existing oligonucleotide segments and finally to introduce it in an organism (ideally with a minimal genome), which will then express the implemented functions.

So far, our ability to extensively modify chromosomal DNA was restricted by the possibilities of the traditional (and laborious) molecular biology techniques, such as traditional cutting and pasting of DNA, site directed mutagenesis, PCR and error-prone versions of it, and so on. Actually, we rather modified or combined existing natural DNA sequences than constructed DNA from scratch. However, this repertoire of competencies appears much too limited and time-consuming for the envisioned novel functions.

11.4.2.1 DNA Synthesis Recently, however, tremendous improvements in the speed, accuracy, and price of *de novo* chemical synthesis of DNA have been made, so that these limitations will be eliminated soon (see also Chapter 12). Significant efforts are currently being made in the development of DNA synthesis technology, so that there is reasonable hope that existing technical challenges will be rapidly overcome, and consequently the designers can outsource the DNA preparation task of the fabrication division and concentrate on the actual design task. Bulk DNA synthesis capacity appears to have doubled approximately every 18 months for the last ten years; the commercial price of synthesis of long fragments of DNA (>500 bp) has decreased by a factor of $\sim$2 over the past years [33]. Right now, we can witness the change from classical DNA synthesis technology to novel forms, such as reactors based on microfluidic concepts and photochemical methods. At present, DNA *de novo* synthesis is performed by assembling overlapping short (25–70 bp long), chemically synthesized oligonucleotides into longer DNA fragments in a PCR-based assembly process [34] and has already led to the complete reconstruction of some smaller phage genomes such as the polio virus [35,36]. These efforts are typically accompanied by enzymatic efforts to reduce the error rate [36,37]. But many of, at least, the chemical steps involved can now be reproduced in miniaturized forms in microfluidics chips, where exploiting the small scale should lead to not only a reduction in the materials costs, but also in the opportunity to provide optimized reaction conditions and thus reduced error frequencies [38]. Taking the concept even further, oligonucleotide synthesis can also be miniaturized on photoprogrammable chips. Coupled with error detection by hybridization, exceptionally low error rates in the order of one mistake in every 1400 bp are possible [39].

11.4.2.2 Chassis and Cloning of Giant DNA Finally, once we have synthesized novel strands of DNA, we need to integrate them into an organism. This splits into two aspects—the organism and the actual introduction.

We have already discussed the desire for reduced genomes in model organisms. Taking this to the extreme, growth in the presence of a rich but synthetic and defined medium requires as few as 206 genes, basically comprising the DNA replication, transcriptional, and translational machinery, rudimentary DNA repair functions, protein processing and degradation, cell division, and rudimentary metabolic and energy functions [40]. Toward this theoretical goal, one can either substantially reduce the relatively large genomes of established model systems and exploit the abundance of molecular biology tools for these model organisms, or reduce very small genomes of other organisms in exchange of the requirement to develop novel molecular biology tools.

With respect to the latter, nonpathogenic *Mesoplasma florum* with very attractive cultivation properties and a genome size of 793 kb is currently being established as a chassis. Its genomic sequence has become recently available and molecular biology methods have been developed (http://www.broad.mit.edu/annotation/genome/

mesoplasma_florum.2/Info.html). A similar approach is followed with *Mycoplasma genitalium*, which was already carefully investigated by transposon mutagenesis for nonessential genes [41].

Regarding the former, a prominent example is *Escherichia coli* whose genome has been reduced in various projects by 6 percent [42], 8 percent [43], or 15 percent [44], respectively, without any noticeable effect on the investigated physiological properties and by 30 percent resulting in defects in cell replication [45]. *Bacillus subtilis'* genome has been reduced by 8 percent, again with only minor effects on physiology [46], confirming the observation that under controlled laboratory conditions a substantial part of a bacterium's genome is indeed dispensable.

The complementation of such minimized genomes will inevitably involve the handling of giant DNA. This requires novel methods from storing via faithfully amplifying to insertion in stable fashion into an organism. First steps in this direction have been made recently with the megacloning technique that allowed insertion of a 3.5 Mb *Synecocystis* genome (a photosynthetic bacterium) into the 4.2-megabase genome of *Bacillus subtilis* [47].

11.5 DESIGN AND APPLICATION

There are two areas in which the ideas of synthetic biology, in our view, have already been implemented to a substantial extent—the design of artificial genetic networks and the design of novel production pathways for chemicals. The first topic is intensively covered in Chapter 15 of this book by Greber and Fussenegger, so we will concentrate on the latter subject.

In the production of novel pathways, the benefits of *de novo* DNA design are particularly apparent. Here, suitable designs allow a significant acceleration of the construction process, for example, when, codon usage is from the very beginning optimized for each novel gene and novel DNA elements are suitably structured, for example, by flanking restriction sites, so that the adaptation of the DNA element to novel insights is very simple. This has played a major role in expressing polyketide synthases in *E. coli*, but also in recombining their domains them in such a way that novel polyketides could be produced [48,49].

Another project that catches very much the spirit of synthetic biology is the attempt to construct from scratch a cheap terpenoid production pathway in *E. coli* leading to artemisinic acid, a precursor to the antimalaria drug artemisinin. This goal requires essentially the building of an entire new pathway in a suitable production organism, which in this case is *E. coli* or *Saccharomyces cerevisiae*. A pathway from acetyl-CoA to amorphadiene was created in *E. coli* by splicing the genes of the mevalonate pathway of *S. cerevisiae* into artificial operons and *de novo* synthesizing the amorphadiene synthase gene from the plant *Artemisa annua* [50]. The remaining step from amorphadiene to artemisinic acid required an *A. annua* cytochrome P450 monooxygenase that catalyzes the remaining three oxidation steps and could so far only be functionally expressed in *S. cerevisiae*, so the pathway from the *S. cerevisae* metabolite farnesylpyrophosphate to amorpadiene and then to artemisinic acid was reconstituted in an *S. cervisiae* mutant engineered for farnesylpyrophosphate overproduction [51].

Although the design of novel biological systems is only beginning, all ingredients of the engineering approach are visible: the role of *de novo* DNA synthesis/fabrication, the design of well-behaved parts on DNA and protein level, the organization of parts into the next functional level of devices and the corresponding abstractions, and the attempt to introduce standardization, even though for the time being only on parts level. With the design of ever more complex systems, the need to emphasize these elements will undoubtedly increase.

11.6 SAFETY AND SECURITY ASPECTS

The reports on the resynthesis of the genomes of the polio virus [35] and the 1918 influenza virus [52] graphically illustrate that large scale *de novo* DNA synthesis might also be used for activities that raise concerns about the safety of synthetic biology, which also have been picked up by the community of synthetic biology researchers (http://openwetware.org/wiki/Synthetic_Biology:SB2.0/Biosecurity_resolutions). If biology is becoming indeed engineerable and we acquire indeed the capabilities to manufacture even more complex systems according to our specifications, then the question that arises is how we are going to manage the safety and security aspects of this potential technological revolution successfully. The answer can be given on two levels—organizational and technical.

From a European perspective—which is the perspective of the two authors of this chapter—the organizational issues appear to be well taken care of. At the moment, synthetic biology does not contain fundamentally new scientific issues that require a reevaluation of the current safety or security standards. Rather, it tries to exploit selected existing concepts to accelerate the progress in the application of biological sciences. Furthermore, even though the agenda of synthetic biology is ambitious and promising, it is for the time being exactly that—an agenda, not a reality. So even if the goal might be the design of complex systems, our current capabilities are much more modest. In addition, the applications that one might have in mind for synthetic biology—for example, a more efficient production of chemicals—will typically lead to strains that are much less fit to survive in natural environments than their nonengineered counterparts. With these arguments in mind, our view is that there are the rules that apply to genetically modified organisms and just as well apply to the field of synthetic biology. Such experimental work is typically regulated in considerable detail and ethics commissions to evaluate researches that might touch upon the questions of fundamental ethical importance are in place. Where synthetic biology has links to technology that is already working reliably—for example, large-scale *de novo* DNA synthesis, see the cases above—regulations (such as analysis of ordered large DNA sequences) are in place that should prevent the abuse of these technologies in those areas where these regulations can be effectively enforced. Beyond that, it is important to note that even the capacity to produce a viral genome within tolerated error margins does not produce viruses that could be applied. For the time being, such efforts would face the same problems that all contemporary biological weapons face (storage, distribution, application) and which makes them difficult to manage.

The improvement of safety or security on technical grounds—the question whether synthetic biology could provide techniques that render the field inherently safer—is so far less obvious. Although some suggestions have been made as to how "synthetic biology-engineered systems" could be prevented from interacting with the natural environment—for example, relying on unnatural amino acids that are not available in the environment, introducing unnatural codon amino acid assignments and the corresponding codon sequence in genes, so that important genes could only be translated in correspondingly synthesized organisms—these strategies have to be tested first to confirm that the introduction of such alternate "codes of life" into the environment does not have unexpected consequences. But this example should suffice to illustrate that if there are concerns that synthetic biology might pose a novel safety or security risk, it is also likely that the accelerated development capabilities that would go along with the successful progress of the field would deliver clues on how to address such issues.

In summary, if synthetic biology indeed turns out to be the revolutionary approach that we envision it to be, there might be safety and security questions that we will need to address. For the time being and the foreseeable future, the potential dangers appear to be well within the grasp of existing safety regulations.

11.7 CONCLUSION

Engineering requires a sound knowledge base and exploits a number of distinguishing features such as forward engineering design, abstraction and standardization of components and conditions, and the decoupling of system design from system fabrication. While systems biology hopefully will allow the consolidation of the knowledge base to a sufficient extent, the implementation of these engineering-specific methodological elements into the application of biological systems is in our view the most powerful aspect of the new discipline of synthetic biology. Adoption of these elements would lead to a much accelerated design process that, at some point in the not-so-distant future, will generate biological designs with a very high chance of success and predictability. Crucial elements in the implication of these elements are on the design side suitable computer-based design tools, the successful establishment of standards, the success of the concepts of modularity of parts and orthogonality. On the fabrication side, further progress in the accuracy and the efficiency of large-scale *de novo* DNA synthesis and assembly and the providing of suitably engineered chassis will be the key.

11.8 TEACHING MATERIAL LINKS

- http://www.syntheticbiology.org—synthetic biology community homepage
- http://en.wikipedia.org/wiki/Synthetic_biology—Wikipedia, the free encyclopedia, about synthetic biology

- http://parts.mit.edu/—Registry of Standard Biological parts
- http://www.igem.org/—iGEM (International Genetically Engineered Machine competition) is an international arena where student teams compete to design and assemble engineered machines using advanced genetic components and technologies.

REFERENCES

1. Szybalski W, Skalka A. Nobel prizes and restriction enzymes. *Gene* 1978;4:181–182.
2. Roblin R. Synthetic biology. *Nature* 1979;282:171–172.
3. Hobom B. Genchirurgie: an der schwelle zur synthetischen biologie. *Medizinische Klinik* 1980;75:834–841.
4. Krimsky S. Social responsibility in an age of synthetic biology. *Environment* 1982; 24: 2–11.
5. Hobom G. Genetik: von der analyse des vererbungsvorganes zur synthetischen biologie. *Verhandlungen der Deutschen Gesellschaft für Innere Medizin* 1986;92:100–105.
6. Presnell SR, Benner SA. The design of synthetic genes. *Nucleic Acids Res* 1988; 16:1693–1702.
7. Benner SA. *Redesigning the Molecules of Life*. Heidelberg: Springer, 1988.
8. Rawls R. Synthetic biology makes its debut. *Chem Eng News* 2000;49–53.
9. Elowitz MB, Leibler S. A synthetic oscillatory network of transcriptional regulators. *Nature* 2000;403:335–338.
10. Gardner TS, Cantor CR, Collins JJ. Construction of a genetic toggle switch in *Escherichia coli. Nature* 2000;403:339–342.
11. Endy D. Foundations for engineering biology. *Nature* 2005;438:449–453.
12. Pawson T, Linding R. Synthetic modular systems: reverse engineering of signal transduction. *FEBS Lett* 2005;579:1808–1814.
13. Kitano H. Systems biology: a brief overview. *Science* 2002;295:1662–1664.
14. Silver P, Way J. Cells by design. *Scientist* 2004;18:30–31.
15. Petitou M, Duchaussoy P, Driguez PA, Jaurand G, Herault JP, Lormeau JC, van Boeckel CAA, Herbert JM. First synthetic carbohydrates with the full anticoagulant properties of heparin. *Angew Chem Int Edit* 1998;37:3009–3014.
16. Endy D, Brent R. Modelling cellular behaviour. *Nature* 2001;409:391–395.
17. Kærn M, Elston TC, Blake WJ, Collins JJ. Stochasticity in gene expression: from theories to phenotypes. *Nat Rev Genet* 2005;6:451–464.
18. Bradley P, Misura KMS, Baker D. Toward high-resolution *de novo* structure prediction for small proteins. *Science* 2005;309:1868–1871.
19. Looger LL, Dwyer MA, Smith JJ, Hellinga HW. Computational design of receptor and sensor proteins with novel functions. *Nature* 2003;423.
20. Dwyer MA, Looger LL, Hellinga HW. Computational design of a biologically active enzyme. *Science* 2004;304:1967–1971.
21. Yokobayashi Y, Weiss R, Arnold FH. Directed evolution of a genetic circuit. *Proc Natl Acad Sci USA* 2002;99:16587–16591.

22. Rebatchouk D, Daraselia N, Narita JO. NOMAD: a versatile strategy for *in vitro* DNA manipulation applied to promoter analysis and vector design. *Proc Natl Acad Sci USA* 1996;93:10891–10896.

23. Bayer TS, Smolke CD. Programmable ligand-controlled riboregulators of eukaryotic gene expression. *Nat Biotechnol* 2005;23:337–343.

24. Blancafort P, Segal DJ, Barbas CF, III. Designing transcription factor architectures for drug discovery. *Mol Pharmacol* 2004;66:1361–1371.

25. Kim CA, Berg JM. A 2.2 angstrom resolution crystal structure of a designed zinc finger protein bound to DNA. *Nat Struct Biol* 1996;3:940–945.

26. Sera T, Uranga C. Rational design of artificial zinc-finger proteins using a nondegenerate recognition code table. *Biochemistry* 2002;41:7074–7081.

27. Segal D, Dreier B, Beerli R, Barbas Cr. Toward controlling gene expression at will: selection and design of zinc-finger domains recognizing each of the 5′-GNN-3′ DNA target sequences. *Proc Natl Acad Sci USA* 1999;96:2758–2763.

28. Dreier B, Beerli RR, Segal DJ, Flippin JD, Barbas Cr. Development of zinc finger domains for recognition of the 5′-ANN-3′ family of DNA sequences and their use in the construction of artificial transcription factors. *J Biol Chem* 2001;276:29466–29478.

29. Dueber JE, Yeh BJ, Chak K, Lim WA. Reprogramming control of an allosteric signaling switch through modular recombination. *Science* 2003;301:1904–1908.

30. Park S-Y, Zarrinpar A, Lim WA. Rewiring MAP kinase pathways using alternative scaffold assembly mechanisms. *Science* 2003;299:1061–1064

31. Isaacs FJ, Dwyer DJ, Ding C, Pervouchine DD, Cantor CR, Collins JJ. Engineered riboregulators enable post-transcriptional control of gene expression. *Nat Biotechnol* 2004;22:841–847.

32. Rackham O, Chin JW. A network of orthogonal ribosome-mRNA pairs. *Nat Chem Biol* 2005;1:159–166.

33. Carlson R. The pace and proliferation of biological technologies. *Biosecurity and Bioterrorism* 2003;1:203–214.

34. Stemmer WP, Crameri A, Ha KD, Brennan TM, Heyneker HL. Single-step assembly of a gene and entire plasmid from large numbers of oligodeoxyribonucleotides. *Gene* 1995; 164:49–53.

35. Cello J, Paul AV, Wimmer E. Chemical synthesis of poliovirus cDNA: Generation of infectious virus in the absence of natural template. *Science* 2002;297:1016–1018.

36. Smith HO, Hutchison CA III, Pfannkoch C, Venter JC. Generating a synthetic genome by whole genome assembly: phiX174 bacteriophage from synthetic oligonucleotides. *Proc Natl Acad Sci USA* 2003;100:15440–15445.

37. Young L, Dong QH. Two-step total gene synthesis method. *Nucleic Acids Res* 2004;32: e59.

38. Zhou X, Cai S, Hong A, You Q, Yu P, Sheng N, Srivannavit O, Muranjan S, Rouillard JM, Xia Y, et al. Microfluidic PicoArray synthesis of oligodeoxynucleotides and simultaneous assembling of multiple DNA sequences. *Nucleic Acids Res* 2004;32:5409–5417.

39. Tian J, Gong H, Sheng N, Zhou X, Gulari E, Gao X, Church G. Accurate multiplex gene synthesis from programmable DNA microchips. *Nature* 2004;432:1050–1054.

40. Gil R, Silva FJ, Pereto J, Moya A. Determination of the core of a minimal bacterial gene set. *MMBR* 2004;68:518–537.

41. Glass JI, Assad-Garcia N, Alperovich N, Yooseph S, Lewis MR, Maruf M, Hutchison CA III, Smith HO, Venter JC. Essential genes of a minimal bacterium. *Proc Natl Acad Sci USA* 2006;103:425–430.

42. Yu BJ, Sung BH, Koob MD, Lee CH, Lee JH, Lee WS, Kim MS, Kim SC. Minimization of the *Escherichia coli* genome using a Tn5-targeted Cre/loxP excision system. *Nat Biotechnol* 2002;20:1018–1023.

43. Kolisnychenko V, Plunkett G, Herring CD, Feher T, Posfai J, Blattner FR, Posfai G. Engineering a reduced *Escherichia coli* genome. *Genome Res* 2002;12:640–647.

44. Posfai G, Plunkett G III, Feher T, Frisch D, Keil GM, Umenhoffer K, Kolisnychenko V, Stahl B, Sharma SS, de Arruda M, et al. Emergent properties of reduced-genome *Escherichia coli*. *Science* 2006;312:1044–1046.

45. Hashimoto M, Ichimura T, Mizoguchi H, Tanaka K, Fujimitsu K, Keyamura K, Ote T, Yamakawa T, Yamazaki Y, Mori H, et al. Cell size and nucleoid organization of engineered *Escherichia coli* cells with a reduced genome. *Mol Microbiol* 2005;55:137–149.

46. Westers H, Dorenbos R, van Dijl JM, Kabel J, Flanagan T, Devine KM, Jude F, Seror SJ, Beekman AC, Darmon E, et al. Genome engineering reveals large dispensable regions in *Bacillus subtilis*. *Mol Biol Evol* 2003;20:2076–2090.

47. Itaya M, Tsuge K, Koizumi M, Fujita K. Combining two genomes in one cell: Stable cloning of the *Synechocystis* PCC6803 genome in the *Bacillus subtilis* 168 genome. *Proc Natl Acad Sci USA* 2005;102:15971–15976.

48. Kodumal SJ, Patel KG, Reid R, Menzella HG, Welch M, Santi DV. Total synthesis of long DNA sequences: synthesis of a contiguous 32-kb polyketide synthase gene cluster. *Proc Natl Acad Sci USA* 2004;101:15573–15578.

49. Menzella HG, Reid R, Carney JR, Chandran SS, Reisinger SJ, Patel KG, Hopwood DA, Santi DV. Combinatorial polyketide biosynthesis by *de novo* design and rearrangement of modular polyketide synthase genes. *Nat Biotechnol* 2005;23:1171–1176.

50. Martin VJJ, Pitera DJ, Withers ST, Newman JD, Keasling JD. Engineering a mevalonate pathway in *Escherichia coli* for production of terpenoids. *Nat Biotechnol* 2003, 21:796–802.

51. Ro DK, Paradise EM, Ouellet M, Fisher KJ, Newman KL, Ndungu JM, Ho KA, Eachus RA, Ham TS, Kirby J, et al. Production of the antimalarial drug precursor artemisinic acid in engineered yeast. *Nature* 2006;440:940–943.

52. Tumpey TM, Basler CF, Aguilar PV, Zeng H, Solórzano A, Swayne DE, Cox NJ, Katz JM, Taubenberger JK, Palese P, et al. Characterization of the reconstructed 1918 Spanish influenza pandemic virus. *Science* 2005;310:77–80.

12

RATIONALES OF GENE DESIGN AND *DE NOVO* GENE CONSTRUCTION

Marcus Graf[1], Thomas Schoedl[1], and Ralf Wagner[1,2]

[1]GENEART AG, Josef-Engert-Str. 11, Regensburg 93053, Germany
[2]University of Regensburg, Molecular Microbiology and Gene Therapy,
Franz-Josef-Strauss Allee 11, Regensburg 93053, Germany

12.1 INTRODUCTION INTO THE FIELD

As per definition, synthetic biology combines science and engineering in order to build novel biological functions and systems. Genetic engineering paved the way for the development of this new field of research; for instance, in 1978, when the Nobel prize in physiology or medicine was awarded to Werner Arber, Daniel Nathans, and Hamilton O. Smith for the discovery of restriction enzymes and their application to molecular genetics, one could already read in an editorial comment in *Gene* that "...The work on restriction nucleases not only permits us easily to construct recombinant DNA molecules and to analyze individual genes but also has led us into the new era of synthetic biology where not only existing genes are described and analyzed but also new gene arrangements can be constructed and evaluated" [1]. Another cornerstone in genetic engineering and synthetic biology was developed when Genentech scientists and their academic partners in 1977 generated the first example of recombinant expression of a human protein

Systems Biology and Synthetic Biology   Edited by Pengcheng Fu and Sven Panke

(somatostatin) in *Escherichia coli*. However, not many scientists are today aware of the fact that the group of Boyer, Itakura, and colleagues were not only the first to invent heterologous recombinant expression but they also did so without using a natural gene. At that time, 9 years before the polymerase chain reaction (PCR) was introduced into genetic engineering, it was easier rationally to design the 14 amino acid long somatostatin gene and synthesize it with methods of organic chemistry than to clone it using the natural template. Since then, only very limited sequence information on the *E. coli* genome was available, codons preferentially used by the MS2 phage were assumed to be beneficial for recombinant gene expression in *E. coli*. However, most of the presently used genetically engineered biotherapeutics are based on natural genes cloned by reverse transcribing message RNAs (mRNAs) into complementary DNAs (cDNAs) and subsequent cloning using restriction enzymes and other DNA modifying enzymes. From 1986 on, in particular PCR-based cloning methods started soon dominating the field either by direct amplification of genomic information or using cDNA libraries as templates for amplification and subsequent cloning and recombinant expression.

As more genetic information became available, the clearer it got that the coding region of a gene comprises more information than just the primary amino acid sequence. The genetic code provides various codon options for each of the 20 amino acids that contribute to the primary sequence of a protein except for methionine and tryptophane, which are encoded by only one codon each. However, the codon options are used in an unequal frequency in different species showing a clear tendency for certain codons, which lead to the "genome hypothesis" postulated in 1980 by Grantham et al. [2]. They analyzed 90 genes of 7 different species and found a nonrandom codon choice pattern, which seemed to be specific for the analyzed species and therefore established the first codon usage table named "codon catalog" [2]. Just 1 year later with more genes analyzed the same group correlated a certain, species-specific subset of codons with mRNA expressivity, that is, the amount of protein made by a particular messenger transcript [3]. Also Ikemura, who synthesized the first human gene to be expressed in *E. coli* found a strong positive correlation between the transfer RNA (tRNA) abundance and the choice of codons in all *E. coli* genes encoding proteins that had been sequenced completely at that time [4]. The positive correlation of codon usage and the amount of available tRNAs in a cell was also confirmed in other unicellular (*Saccharomyces cerevisiae, Salmonella typhimurium*) and multicellular (human, rat, plant, chicken, fish) organisms [5] indicating that species-specific codon bias is a general phenomenon and can be even used to perform phylogenetic analyses (see Fig. 12-1).

After it was generally accepted that species-specific codon bias exists and influences expression rates, it seemed clear that recombinant gene expression can be improved by adapting at least the codon choice of the gene of interest to the preferred codons of the target expression host. In addition to codon choice, various intragenic *cis*-acting elements heavily impact expression yields in heterologous and even autologous cell factories. However, these elements like splice sites, TATA-boxes, or ribosomal entry sites are highly species-specific and have to be taken into account when rationally designing genes.

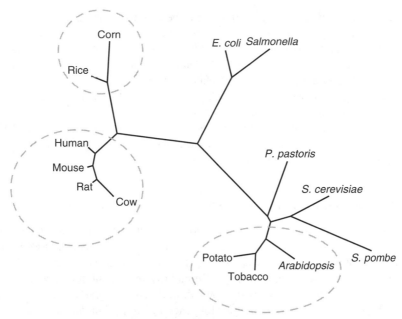

Figure 12-1 Phylogenetic tree based on codon usage. For each species pair the "codon usage distance" was calculated as follows: First, the difference of the frequency of each codon for each amino acid was calculated and then the total sum of frequency differences was computed. The resulting difference matrix was normalized with the highest difference to be set as 1. Phylogenetic branching of the data was performed using the Fitch–Margoliash least squares algorithm by the program "Fitch" (Felsenstein, J. 1989. PHYLIP—Phylogeny Inference Package (Version 3.2). Cladistics 5: 164–166). Codon usage data was obtained from the Kazusa Codon Usage Database (http://www.kazusa.or.jp/codon/). Kindly provided by M. Liss, GENEART (AG).

12.2 RATIONALS OF GENE DESIGN

12.2.1 Codon Usage Adaptation Strategies

The easiest and simplest approach to design an expression host-specific DNA sequence is to reversely translate ("backtranslate") the amino acid sequence into a DNA sequence using the most frequently used codon. A comparably simple approach is to mimic the codon usage distribution that can be derived from the host's codon usage table. As most codon usage tables are generated on the genome-wide codon usage distribution, this approach levels out the natural differentiation between highly and low-expressed genes in terms of their codon usage. Using codon usage tables based on highly expressed genes bares several other unpredictable uncertainties: How many genes are sufficient for the compilation of the codon usage table? Is the different promoter strength of the highly expressed genes taken into account? Especially for multicellular organisms tissue and compartment specificity is often not addressed by those codon usage tables. The following chapters deal with the different codon adaptation and gene optimization strategies that should be applied for different expression host systems. In order to be able to compare the overall codon quality of a given gene several measurements and indices were developed. The most accepted index is the codon adaptation index (CAI) introduced by Sharp and Li in 1987 [6]. The CAI is a measurement for the relative adaptiveness of the codon usage of a gene toward the codon usage of highly expressed genes. The relative adaptiveness of each codon is the ratio of the usage of each codon to that of the most abundant codon for the same amino acid. The CAI index is defined as the geometric mean of these relative adaptiveness values. Nonsynonymous codons and termination codons (dependent on genetic code) are excluded. CAI values are always between 0 (where no optimal codons are used) and 1 (where only optimal codons are used).

12.2.2   Designing Genes for Prokaryotic Expression *In Vivo*

The adverse effect of rare codons in recombinant expression in *E. coli* is known for many years. It was shown for instance that the production of β-galactosidase decreased when rare AGG codons were inserted near the translational initiation site [7]. Rare codons for arginine seem to be the most difficult to express, but other codons such as proline, glycine, or leucine are also problematic. Already very early studies revealed a strong bias in synonymous codon usage for genes encoding abundant proteins such as ribosomal proteins and elongation factors, RNA polymerase subunits, and glycolytic enzymes [3,8,9]. A strong correlation was found between copy numbers of protein and the frequency of codons whose cognate tRNAs were most abundant. Ikemura first discovered that this correlation was strongest in the most highly expressed genes which almost exclusively use optimal, that is, most frequently used codons [4,5,10] and he suggested that this bias in codon usage might both regulate gene expression and should act as an optimal strategy for recombinant gene expression. Although most frequently used codons correspond with the most abundant tRNAs, one has to keep in mind that the total tRNA composition of *E. coli* increases by 50 percent as growth rate increases to maximum [11]. However, the relative but not the absolute tRNA levels stay the same, thus the most frequently used codons still provide the largest tRNA pools feeding the translational machinery.

Therefore, it is now a widely accepted strategy for recombinant gene expression in *E. coli* to change the codons of the gene of interest or alternatively to coexpress certain tRNA-encoding genes in order to supplement tRNA pools with low abundance in *E. coli*. There are three *E. coli* strains commercially available (Novagen, Stratagene) coexpressing arginine encoding tRNA-genes specific for the nonfrequently used codons AGG, AGA, AUA, CUA, CCC, and GGA codons. Although coexpression of selected tRNAs can overcome certain expression problems due to the presence of extreme rare arginine, proline, or glycin codons, best and consistent results will be achieved only by adapting consequently the entire gene to most frequently used *E. coli* codons [12]. For instance, the sequence of the mature human IL-18 gene shows 37 rare *E. coli* codons (fraction <0.1) among 157 amino acids with an overall CAI of 0.58. Expression of nonoptimized IL-18 versus codon-optimized IL-18 was recently analyzed in depth. Although supplementation with rare tRNA genes did increase expression, Li et al. [13] found that a codon-optimized gene with a CAI of 0.84 was

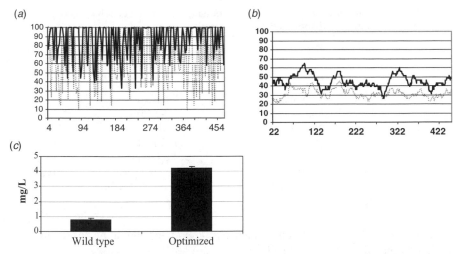

Figure 12-2 Codon, GC content, and expression analysis of human IL-18 expression in *E. coli*. (a) The plot shows the quality of the used codon at the indicated codon position. The quality value of the most frequently used codon was set to 100, the remaining synonymous codons were scaled accordingly (relative adaptiveness, see also Ref. [6]). Dotted line: wild-type gene; bold black line: optimized gene. (b) The plot shows the GC content in a 40 bp window centered at the indicated nucleotide position. Thin line: wild-type gene; bold black line: optimized gene. (c) Expression of wild-type and codon optimized human IL-18 in *E. coli* using BL-21 (DE3) strain 5 hours after induction. Modified from Ref. [6].

more beneficial and increased expression yields by a factor of five (see Fig. 12-2). Although the optimized gene shows elevated GC levels compared to the wild-type gene, which was increased from 35 percent up to 45 percent in average, the authors neither discuss nor find any correlation between elevated expression levels and GC content that could be thus solely coincidental. Interestingly, the authors also showed that the biochemical properties and protein activity of the proteins produced using wild-type gene expression and optimized gene expression were exactly the same.

In addition to codon choice, there also seems to be an influence of mRNA secondary structure on translation efficiency. Nomura et al., for instance, showed that the presence of an 8 bp stem-loop structure preceding the Shine Dalgarno ribosomal entry site may hamper expression and that destabilization of such an element could increase expression dramatically [14]. Moreover, the presence of intragenic *E. coli* ribosomal entry sites, as found in many mammalian genes, may lead to truncated products during heterologous expression and should therefore be avoided throughout the gene [15]. Hatfield et al. found by statistical analysis that certain codon pairs are seemingly overrepresented within the *E. coli* genome and reported a negative effect of such codon pairs on translational efficiency because of a translational pausing effect [16]. However, the negative activity of codon pairs could not be reproduced by another group utilizing the T7 promoter for transcription control [17].

Finally, termination efficiencies vary significantly depending on both the stop codon used and the nucleotide immediately following the stop codon. Efficiencies are ranging from 80 (TAAT) to 7 percent (TGAC) indicating that TAAT is the most

Table 12-1 Selected publications where synthetic genes were successfully used to express proteins in *E. coli* (modified from Ref. [69])

Gene Origin	Protein Name	Improvement	Reference
H. sapiens	IL2	16-fold	[19]
H. sapiens	TnT	10–40-fold	[20]
C. tetani	Fragment C	Fourfold	[21]
S. oleracea	Plastocyanin	1.2-fold	[22]
H. sapiens	Neurofibromin	Threefold	[23]
H. sapiens	M2-2	140-fold	[24]
H. sapiens	IL-6	Threefold	[25]
Pyrococcus abyssi	Phosphopantetheine adenylyltransferase	Below Detection to 15–20 mg/L	[26]

efficient translational termination sequence in *E. coli* and should be chosen to avoid undesired read through products [18].

12.2.3 Designing Genes for Prokaryotic Expression *In Vitro*

Despite coexpression of rare tRNA *in vitro* a more and more popular strategy to avoid expression problems *in vivo* is to choose *in vitro* transcription/translation systems. *In vitro* expression allows upscaling for production, which is, however, still comparably costly and more important, highly parallel screening of different protein variants for functional analysis.

To test whether gene optimization can increase protein yields of *in vitro* expression systems, differently optimized HIV-1 p24-encoding genes and wild-type p24 were translated *in vitro* under the transcriptional control of T7 in a Roche RTS 100 system using *E. coli* lysats (Table 12-1). Expression of different p24-encoding genes correlated nicely with the CAI of the respective genes. Lowest expression yields (770–831 µg/mL) were obtained using the wild-type p24 gene showing a $CAI_{E.\ coli}$ of 0.54. The mammalian-optimized gene with a $CAI_{E.\ coli}$ of 0,68 ($CAI_{H.\ sapiens} = 0.98$) showed a slight increase by 35–50 percent (1061–1170 µg/mL), but only the *E. coli*-optimized gene with a $CAI_{E.\ coli}$ of 0.98 raised protein yields by 3–3.5-fold (2565–2648 µg/mL) (Fig. 12-3). Gene optimization with a particular emphasis on codon choice improvement that results in high $CAI_{E.\ coli}$ values seems therefore beneficial to increase the yields when using *in vitro* expression systems.

Today, several companies offer cell-free expression systems that include chaperons and other proteins positively influencing correct folding of the translated protein. Alternatively, *in vitro* expression can be also performed in cells of higher eukaryotes such as rabbit reticulocytes or wheat germ lysats (Roche, Novartis, etc.).

12.2.4 Designing Genes for Yeast Expression

A very early study revealed in 1982 that an extreme codon bias is seen for the highly expressed *S. cerevisiae* genes alcohol dehydrogenase isozyme I (ADH-I) and glyceraldehyde-3-phosphate dehydrogenase. A proportion of more than 96 percent of

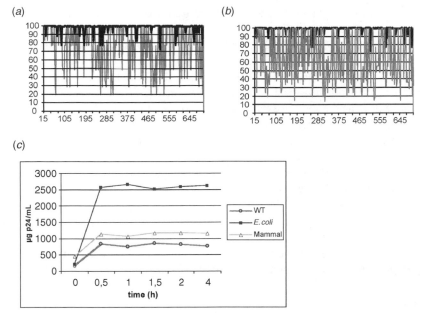

Figure 12-3 *In vitro* expression of differently optimized and wild-type HIV-1 p24 in *E. coli* lysats: (a, b) The plot shows the quality of the used codon at the indicated codon position. The quality value of the most frequently used codon was set to 100, the remaining synonymous codons were scaled accordingly (relative adaptiveness, see also Ref. [6]) gray line: wild-type gene or mammalian optimized in (a) and (b), respectively; bold black line: *E. coli*-optimized gene. (c) Expression of wild-type (WT), mammalian optimized (mammal), and *E. coli* optimized (*E. coli*) HIV-1 p24 in *E. coli* lysats under the transcriptional control of T7 using the Roche RTS system. Cell lysats (50 μL) were harvested at the indicated time points and p24 expression measured by commercial ELISA.

the 1004 amino acid residues are encoded by only 25 of the 61 possible coding triplets [27]. Only a few years later, Sharp et al. were able to show in a compiled analysis on 110 *S. cerevisiae* genes that there are two groups of genes. One group matched to highly expressed genes with a strong codon bias, which was speculated to match abundant tRNAs [28], whereas the other group corresponded to nonhighly expressed genes. More than 10 years later, Percudani et al. [29] were analyzing the now fully sequenced *S. cerevisiae* genome and were able to identify all tRNA-encoding genes. The respective 274 tRNA genes were assigned to 42 classes of distinct codon specificity. The gene copy number for individual tRNA species, which ranges from 1 to 16, correlated well with most frequently used *S. cerevisiae* codons. Moreover, they were able to show that tRNA gene copy numbers as well as codon frequencies nicely correlate with tRNA abundancy of the respective codon, thereby, confirming the previous assumptions of Sharp et al.

Other yeasts like *Pichia pastoris* or *Schizosaccharomyces pombe* also shows a highly biased, and species-specific codon choice (see Fig. 12-1) indicating that codon choice adaptation should also be beneficial for recombinant expression in these different yeasts. However, codon optimization procedures dedicated to obtain very high CAI values (> 0.95) will usually result in a significant decrease in the overall GC content due to the relative low-GC bias of *P. pastoris* preferred codons. It was shown for instance that expression of *P. pastoris* codon-optimized human glucocerebroside fragments can be increased up to 10-fold above wild-type levels, hence almost the same levels (7.5-fold) can be achieved with a gene fragment in which rare codons were not removed, but the overall GC content was increased [30]. Consequently, it was shown by others that it is indeed beneficial not only to adapt codon choice but also to increase the GC content of the encoding gene for optimal expression in *P. pastoris* [31–33]. Therefore, an optimal balance has to be found between optimal codon choice and CAI values on the one hand and overall GC content on the other hand. To test whether this might be also true for *S. cerevisiae*, we tested expression of human MIP-1α-encoding genes with an optimal codon choice (CAI = 1.0), but low overall GC content (36 percent) and a gene with elevated GC content (47 percent), where we had to introduce several nonfrequently GC biased codons resulting in slightly lower CAI value (0.83). Protein yields were compared with gene expression resulting from nonaltered wild-type gene showing a GC content of 58 percent and a CAI of 0.56. As depicted in Figure 12-4 there seems to be no correlation between expression yields and elevated GC content in *S. cerevisiae* since highest expression was achieved with the fully codon-optimized gene showing the

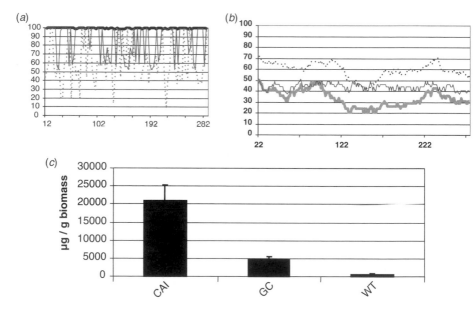

Figure 12-4 Expression of differently optimized and wild-type human MIP-1α in *S. cerevisiae*. (a) The plot shows the quality of the used codon at the indicated codon position. The quality value of the most frequently used codon was set to 100, the remaining synonymous codons were scaled accordingly (relative adaptiveness, see also Ref. [6]). Dotted line: wild-type gene; bold black line: codon-optimized gene; gray line: GC content-optimized gene. (b) The plot shows the GC content in a 40 bp window centered at the indicated nucleotide position. Dotted line: wild-type gene; bold black line: codon-optimized gene; gray line: GC content-optimized gene. (c) Transient expression of wild type (WT), GC content-optimized (GC), and codon-optimized (CAI) human MIP-1α in *S. cerevisiae* from lysed cell pellets using commercial ELISA formats.

lowest overall GC content. Expression yields exceeded GC-rich gene expression by a factor of 4.5 and by a factor of more than 35 compared to wild-type gene expression. We therefore conclude that for best expression results in *S. cerevisiae* codon choice seems to be a more dominant factor than GC content.

However, mRNA nucleotide composition itself has implications on the relative efficiency of protein expression through effects on secondary structures and stability irrespective of codon choice [34,35]. For instance, it seems advantageous to eliminate or avoid putative polyadenylation sites located in AT-rich DNA [36,37] in order to avoid premature polyadenylation.

12.2.5 Designing Genes for Plant Expression

The most prominent example and probably also the first example to express a rationally designed gene in higher plants is the insecticidal cry gene of *Bacillus thuringiensis*. Transgenic plants expressing active protein using wild-type cry genes failed to protect from insects due to poor expression. The use of different promoters, fusion proteins and leader sequences could not significantly increase protein expression either and failed to protect transgenic plants from insects [38,39]. Therefore, it was speculated that codon choice and the presence of sequence motifs not common in plants might hamper transgenic expression rates (see also Fig. 12-5). Table 12-2 indicates the differences in human, *B. thuringiensis*, rice (*Oryza sativa*), and *Arabidopsis thaliana* codon preference. Not only prokaryotic and mammalian codon choices differ greatly, but also the ones of monocots (rice) and dicots (*A. thaliana*), indicating that a general codon usage adaptation to plant genes will not reflect their phylogenetic diversity and consequently will not improve individual gene expression in the respective plant host.

Due to the preference for A or T in the third nucleotide position of preferred dicot codons (see Table 12-2), a simple reverse translation of the amino acid sequence into the respective encoding DNA sequence using the condon most frequently used by dicots will lead to a nonfavorable extremely AT-rich gene. Therefore, gene design for expression enhancement in dicots has to be carefully balanced between improving codon usage on the one hand and increasing GC content on the other hand.

Consequently, Perlak et al. tested differently modified cry-encoding genes for expression in tobacco and tomato plants. Among the plants transformed with the partially modified cry gene they identified a 10-fold higher and among plants transformed with the fully modified gene they identified 100-fold higher level of insect-control protein compared with plants expressing the wild-type gene ([40], see Table 12-3). Besides codon choice, they increased GC content, removed potential polyadenylation signals, and removed putative RNA instability elements, which have been shown previously to destabilize mRNA in other systems [41].

The very same gene design strategies were applied by others [42] to increase cryIC expression in dicots with similar positive results. Another important factor to keep in mind when designing genes for expression in plants is removal or avoidance of putative active splice motifs within the coding region. Rouwendal et al. for instance identified an 84 bp long cryptic intron within the coding region of the wild-type green fluorescent protein (GFP) encoding gene of the jellyfish *Aequorea victoria* after

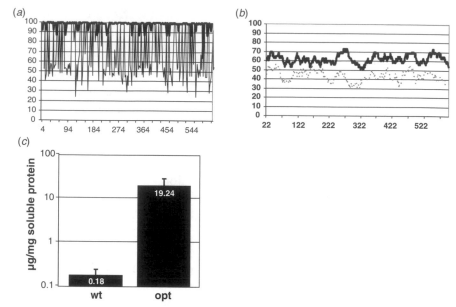

Figure 12-5 Codon, GC content, and expression analysis of (1.3–1.4)-b-glucanase tested for expression in barley. (a) The plot shows the quality of the used codon at the indicated codon position. The quality value of the most frequently used codon was set to 100, the remaining synonymous codons were scaled accordingly (relative adaptiveness, see also Ref. [6]). Dotted line: wild-type gene; bold black line: optimized gene. (b) The plot shows the GC content in a 40 bp window centered at the indicated nucleotide position. Thin line: wild-type gene; bold black line: optimized gene. (c) Expression of recombinant heat-stable (1.3–1.4)-b-glucanase under the control of the D-hordein gene (Hor3) promoter in T1 grains. For the codon-optimized gene, two individual grains were analyzed, and values are given. Modified from Refs [44,45].

transgenic expression in tobacco. Therefore, Rouwendal et al. adapted the codon usage of GFP for expression in tobacco. After codon adaptation where 42 percent of all codons were altered, the overall GC content was improved from 32 to 47 percent and the intragenic cryptic splice sites were altered to be nonfunctional. As expected, several transgenic tobacco lines containing the wild-type GFP gene contained a smaller nonfunctional protein cross-reacting with GFP antiserum, whereas only one protein of the predicted size was found in transgenics expressing the optimized GFP gene. Thus, the authors concluded that the smaller protein was probably encoded by a truncated GFP mRNA created by splicing of an 84 bp cryptic intron present only in the natural GFP-encoding gene [43].

In contrast to dicots, codon adaptation to monocots elevates the GC content automatically due to the preference for G or C in the third nucleotide position of frequently used monocot codons. Gene optimization can therefore be focused on removal of rare codons and other negatively *cis*-acting motifs as discussed above. Jensen et al., for instance, successfully expressed a synthetically engineered (1.3–1.4)-b-glucanase in barley (*Hordeum vulgare*) and were able to elevate expression levels to 107-fold compared to the wild-type gene ([44], see Fig. 12-5).

Table 12-2 Comparison of codon choice of selected amino acids for mammals, prokaryotes, and plants

Amino Acid	Codon	*B. thuringiensis*	*H. sapiens*	*O. sativa*	*A. thaliana*
	GCA	**100**	58	55	61
Ala	GCC	24	**100**	**100**	36
	GCG	41	28	88	32
	GCT	80	65	61	**100**
	AGA	**100**	**100**	61	**100**
	AGG	24	95	91	57
Arg	CGA	39	52	43	34
	CGC	17	90	**100**	20
	CGG	10	**100**	87	26
	CGT	56	38	48	49
	CTA	29	18	32	42
	CTC	8	50	**100**	65
Leu	CTG	10	**100**	82	42
	CTT	38	33	57	**100**
	TTA	**100**	18	25	54
	TTG	23	33	57	85
% GC	Total	37%	53%	56%	45%
% GC	Third	25%	59%	62%	42%

The most frequently used codon of each amino acid was set to 100 and the remaining scaled accordingly [6]. Black bold: most frequently used codon; gray bold: rare codon. GC total, overall GC content within all coding regions; GC third, GC content of the third nucleotide position of all codons.

Table 12-3 Optimization strategy for increased cry expression in tobacco and tomato

	Wild Type	Partially Modified	Fully Modified
Adapted codons	—	10%	60%
GC content	37%	41%	49%
Polyadenylation sites	18	7	1
RNA instability element	13	7	0

Modified from Ref. [40].

Taken together, there is great potential in increasing recombinant expression in transgenic plants using rationally designed genes. Codon choice, RNA instability motifs, and GC content seem to be important factors that have to be taken into account.

12.2.6 Codon Adaptation for Mammalian Expression

Synonymous codon choice also affects gene expression in mammals. In particular when nonmammalian genes are to be expressed in mammalian cells, the substitution of nonfrequently used codons with more common synonyms can significantly

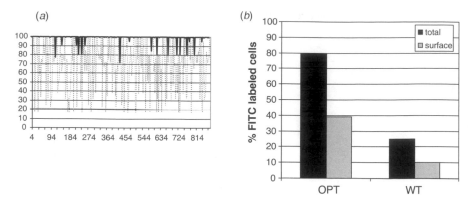

Figure 12-6 Surface and total cellular expression of HHV7 U51 in transiently transfected 293A cells was analyzed by flow cytometry. (a) The plot shows the quality of the used codon at the indicated codon position. The quality value of the most frequently used codon was set to 100, the remaining synonymous codons were scaled accordingly (relative adaptiveness, see also Ref. [6]). Dotted line: wild-type gene; bold black line: codon-optimized gene; gray line: GC-content-optimized gene. (b) Flow cytometry expression analysis of wild type and codon-optimized (OPT) HHV7 U51 in transiently transfected 293A using 2 μg plasmid DNA. Modified from Ref. [50].

increase expression [46–50]. A very prominent example is the widely used jellyfish *Aequorea victoria* green fluorescent protein. An analysis of the GFP encoding sequence showed that the codon usage frequencies of this jellyfish gene are quite different from those prevalent in the human genome [47]. Consequently, after having removed rare codons the synthetic humanized GFP allowed 5–10-fold higher expression rates compared to the wild-type cDNA in transfected mammalian cells. In another example, inhibition of expression of viral genes in mammalian cells could only be overcome by modification of the codon composition or by provision of excess tRNA [49]. A dramatic increase of 10–100-fold in gene expression was achieved when human herpesvirus (HHV) type 6- and HHV type 7-encoding genes were optimized for mammalian expression by adapting codon choice and elevating the overall GC content (see also Fig. 12-6).

More recently, Plotkin et al. [51] discovered systematic differences in synonymous codon usage between genes expressed in different human tissues. They were able to demonstrate that liver-specific genes differ in their codon choice from brain-specific genes, uterus differ from testis genes, etc. Since differences in relative tRNA abundancies in tissues of the same organisms were not reported so far, the authors suggested that codon mediated translational control might be the reason for the observed tissue-specific codon choice. Therefore, even codon choice optimization of mammalian genes (in particular nonhousekeeping genes) for autologous expression in mammalian cells can have a significant impact on the expression rates. For instance, we were able to show that expression of the human granulocyte macrophage colony stimulating factor (GM-CSF) could be increased by a factor of 2.1 by gene optimization and simultaneous removal of rare human codons and negative *cis*-acting RNA instability elements (see Fig. 12-7). In each codon position within the optimized

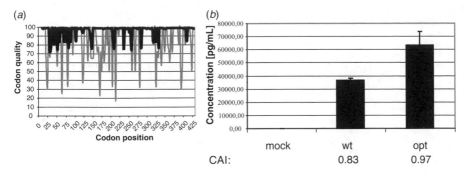

Figure 12-7 (a) The plot shows the quality of the used codon at the indicated codon position. The quality value of the most frequently used codon was set to 100, the remaining synonymous codons were scaled accordingly (relative adaptiveness, see also Ref. [6]). Gray line: wild-type gene; bold black line: optimized gene. (b) Transient transfection of human H1299 cells using wild-type (wt) and optimized (opt) human GM-CSF-encoding genes. Cytokine concentration in the supernatant was measured by commercial ELISA. At least five independent transfection experiments were performed and cytokine concentration measured. CAI, codon adaptation index.

coding region of GM-CSF where not the most frequent human codon was used, certain RNA instability elements have been avoided.

12.2.7 RNA Optimization for Mammalian Expression

Human codon usage bias seems not only to be correlating with available tRNAs but also synonymous codons are placed along coding regions in a way to minimize the number of TA and CG dinucleotides in mammalian genomes [52]. The rarity of CG dinucleotides in mammalian genes is usually ascribed to the tendency of CG to mutate to TG [53], whereas the rarity of TA in coding regions is considered adaptive because UA dinucleotides are preferably cleaved by endonucleases. Moreover, coding regions of mammalian housekeeping genes seem to have an increased GC content compared to low-expressing genes [54]. A correlation of mRNA expression levels with the third nucleotide position GC in codons of mice and rat genes was found by Konu et al., suggesting that higher GC levels may provide the rodent genes with a selective advantage for translational efficiency [55]. Consequently, we were able to show that mRNA stability of AT-rich HIV genes can be increased by orders of magnitude by elevating the GC content of the encoding mRNA from 44.1 to 62.7 percent in average thereby also reducing UA numbers from 96 to 11 (see Fig. 12-8). The increased amounts of available message lead to 100-fold higher expression of the HIV Pr55gag polyprotein compared to the nonoptimized cDNA-based gene expression [56]. This dramatic increase in expression cannot solely be explained by solely improved translational rates, rather then by the increased amounts of available optimized mRNA for cytoplasmatic expression. The optimized and wild-type RNA differ in their biochemical and physiological properties to such a great extent that they even use different nucleocytoplasmatic export pathways. Whereas expression of the wild-type HIV-1 gag genes was blocked using leptomycin B, an antibiotic known to directly interfere with exportin1-mediated nuclear export of mRNAs, the expression of the optimized RNA was not blocked [56].

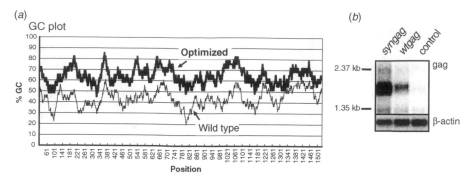

Figure 12-8 (a) The plot shows the GC content in a 40 bp window centered at the indicated nucleotide position. Thin line: wild-type gene; bold black line: optimized gene. (b) Transient transfection of human H1299 cells using wild-type (wtgag) and optimized (syngag) HIV-1 Pr55-encoding genes. Cells were lysed and total RNA was isolated and subjected to Northern Blot analysis. Pr55gag encoding transcripts were detected by a radiolabeled probe and standardized by the amount of detected β-actin RNA (lower panel). The Northern Blot analysis was repeated several times yielding comparable results. Modified from Refs [56,57].

Genes of other non-HIV related pathogens such as malaria or human papilloma virus, or certain proto-oncogenes show similar sequence properties indicating a general mode of action of these *cis*-acting RNA elements and the positive effects of RNA optimization.

When designing genes for optimal expression in mammalian cells, one therefore has to take several parameters into account. Codon choice is one parameter to influence the translation efficiency but, more importantly, the available mRNA levels necessary for translation have a much larger impact on expression yields. RNA optimization, like modification of GC content, avoidance or preferable introduction of UA/CG dinucleotides, removal of *cis*-acting RNA elements negatively influencing expression is the key for stable and high level expression in mammalian cells. Examples for these *cis*-acting RNA elements are adenine-rich elements (ARE), which are found within many cytokines, nuclear transcription factors or proto-oncogenes or within their flanking untranslated regions and are thought to be the most common determinant of RNA instability in mammalian cells [58,59]. Other, probably closely related, *cis*-acting elements are known to destabilize or retain mRNAs and should be avoided [60–63]. Despite these elements and ARE-directed mRNA degradation it is known that isolated cryptic splice sites may retain mRNA within the splicing machinery or lead to truncated transcripts [64].

12.2.8 Other Important Designing Rules for Mammalian Expression

In addition to optimizing the coding region for improved expression yields in mammalian cells, there are other sequence elements known to be beneficial for efficient and stable expression. In mammalian mRNAs, initiation sites usually align to all or part of the sequence GCCRCCAUGG referred to as the Kozak sequence [65]. A strong contribution of G directly downstream of the starting ATG was confirmed in a study that directly monitored the initiation step [66], thus negating the concern that this conservation might simply reflect a preference for certain amino acids in the second position of eukaryotic proteins. Not only the efficiency of the translational initiation is context dependent but also the efficiency of translational termination. In order to avoid undesired read through by-products in mammalian expression, two stop codons are usually recommended to ensure efficient termination. Finally, natural mammalian transcripts always comprise and intron/exon structure and undergo complex splicing and editing before leaving the nucleus for cytoplasmatic translation. Therefore, expression of intronless cDNAs or artificial genes can be improved by adding an efficiently spliced intron 5′ of the coding region [67]. Finally, there is a recent report that by removal/avoidance of certain codon pairs gene expression can be significantly increased [68].

Taken together, gene design for mammalian expression is a sophisticated task since many different parameters have to be optimized in parallel for optimal results. Table 12-3 shows selected examples from literature where gene optimizations led to increased expression results compared to wild-type expression.

12.2.9 Methods of Analyzing Genes

Most standard software offers the possibility to analyze genes, for example, concerning their codon usage or GC content. Tables 12-4 and 12-5 list easy to handle Internet-based software for the analysis of genes with respect to codon usage, prediction of repetitive elements, and splice sites.

Further link compilations can be found on http://bioweb2.pasteur.fr/ and http://www.expasy.org/tools/. Nevertheless all those software tools just concentrate on the evaluation of one single parameter. As shown above, especially when it comes to gene optimization, several parameters have to be considered important. This complicates gene optimization as it becomes a multiparameter and multitask problem: Multiparameter as several constraints like codon choice or GC content have to be accounted for, multitask as several jobs like local and global sequence alignments have to be performed. This can only be achieved by a multiparameter process that simultaneously takes into account all constraints. State-of-the-art multiparameter optimization software (GeneOptimizer™, Geneart AG, Regensburg, Germany) allows for different weighting of the constraints and evaluates the quality of codon combinations concurrently. Approaches that are based on a successive modulation of parameters most likely spoil findings of previous optimization runs in subsequent runs (e. g., first optimization focusing on codon choice, second on GC content).

12.3 *DE NOVO* GENE CONSTRUCTION

12.3.1 Oligonucleotide Synthesis—Creating Your Template

Since rationally designed DNA comprising genes, operons, or even genomes only exist *in silico*, nature cannot provide a natural template for PCR-based amplification. Therefore, the only possible way to get access to such a rationally designed DNA is by *de novo* gene synthesis. Virtually, all gene synthesis methods are based on oligonucleotide synthesis causing many technical as well as cost implications.

Table 12-4 Selected publications where synthetic genes were used to express proteins in mammalian cells (modified from Ref. [69])

Gene Origin	Protein Name	Expression Host	Improvement	Reference
HIV	gp120	*H. sapiens*	Expression: >40-fold	[70]
Aequorea victoria	GFP	*H. sapiens*	Expression: 22-fold	[47]
Listeria monocytogenes	LLO	*M. musculus*	Expression: 100-fold	[71]
BPV1	L1, L2	Mammalian	Expression: 1.000-fold	[49]
HIV	gag	*H. sapiens*	Expression: >322-fold	[72]
HIV	gag	*H. sapiens*	Expression: >10–100-fold	[56]
Plasmodium spp.	EBA-175 region II and MSP-1	*M. musculus*	Expression: 4-fold	[73]
Tn10/HSV	rtTA	*M. musculus*	Expression: >20-fold	[74]
HPV	L1	*H. sapiens*	Expression: 1.000–10.000-fold	[75]
P1 phage	Cre	Mammalian	Expression: 1.6-fold	[76]
Schistosoma mansoni	SmGPCR	*H. sapiens*	Expression: barely detectable versus strong signal	[77]
HV-fold	U51	Mammalian	Expression: 10–100-fold	[50]
HIV	gag, pol, env, nef	*H. sapiens*	Expression: >250×, >250×, >45×, >20×, respectively	[78]
HPV	E5	Mammalian	Expression: 6–9-fold	[79]
HPV	E7	Mammalian	Expression: 20–100-fold	[80]
HIV, SIV	GagPol	*H. sapiens*	Safety in gene therapy	[80]
HIV	Gag	*H. sapiens*	Efficacy, vaccine development	[81]
HIV	GagPolNef	*H. sapiens*	Efficacy, safety for vaccine development	[82]
H. sapiens	Dopamine receptor D2	*H. sapiens*	RNA stability	[52]
Choristoneura fumiferana	GE, GEvy, VR	*H. sapiens*	Increase in reporter induction: 2.7- and 1.7-fold	[83]
Coronavirus	Spike and nucleocapsid proteins	*H. sapiens*	Safe and efficient expression for studying antibody binding	[84]

Table 12-5   Selected web-based software tools for gene analysis

Codon usage distribution and codon usage table compilation

   http://www.kazusa.or.jp/codon/countcodon.html
   http://www.bioinformatics.org/sms2/codon_usage.html
   http://gcua.schoedl.de/
   http://www.bioinformatics.org/sms2/codon_plot.html

Repetitive or secondary structure elements

   http://www.genebee.msu.su/services/rna2_reduced.html
   http://bioweb.pasteur.fr/seqanal/interfaces/dottup.html

Splice site prediction

   http://www.cbs.dtu.dk/biolinks/pserve2.php
   http://genes.mit.edu/GENSCAN.html
   http://www.fruitfly.org/seq_tools/splice.html

Reverse translation

   http://www.bioinformatics.org/sms2/rev_trans.html

The chemical synthesis of short single strand DNA dates back to 1955 and provided the directed chemical synthesis of a dithymidinyl nucleotide [85]. Over the decades oligonucleotide chemistry was improved, but it took until the beginning of the 1980s when the phosphoramidite approach and solid phase synthesis allowed the development of routine oligonucleotide synthesis, as we know it today. The oligonucleotide synthesis starts on a solid phase (controlled pore glass CPG) in 3′ to 5′ direction. The first 3′ nucleotide is already immobilized on the CPG solid phase and a synthesis cycle comprising four steps builds up the desired oligonucleotide. Briefly, the first step is deblocking which removes a protective group of the 5′-end thereby exposing a reactive hydroxyl group at which the next nucleotide (phosphoramidite) is added (coupling step) after activation. An overview of the cycle is depicted in Fig. 12-9.

The capping step prevents all oligonucleotides that were not elongated in the previous coupling step from subsequent coupling steps. Finally, oxidation step stabilizes the phosphate bond of the newly added amidite and the cycle may commence from the beginning until the full-length oligonucleotide is synthesized. During synthesis, there are several steps of this cycle that can lead either to deletions or nucleotide substitutions within the desired oligonucleotide for instance due to inefficient capping, deblocking, or removal of purines in an acidic solution. No matter what gene synthesis method is used to build up long DNA fragments, the sequence identity and the respective error frequency of the used oligonucleotides has a major impact on costs, efforts, and duration of subsequent *de novo* gene production.

12.3.2   *De Novo* Gene Synthesis Methods

The field of gene synthesis was pioneered by Khorana et al. with the groundbreaking work to synthesize a full-length tRNA encoding sequence, which took them several years [86]. Koester et al. synthesized the first protein-encoding gene (angiotensin II)

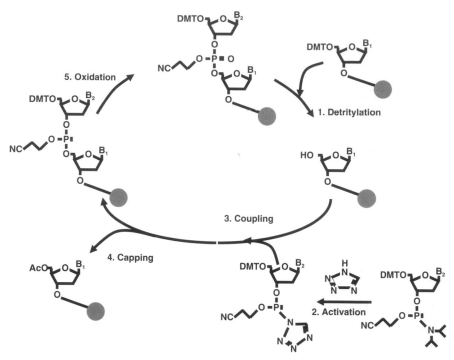

Figure 12-9   Oligonucleotide synthesis cycle description (see text). CPG, controlled pore glass, solid phase. B_1, base 1; B_2, base 2; DMTO, dimethoxytrityl; Ac, acid.

just 5 years later and Itakura et al. revolutionized the field in 1977 with the synthesis of a human gene in just 6 months [87,88]. Although Itakura et al.'s work gave birth to the field of genetic engineering and recombinant DNA technology, gene synthesis was restricted these days by the limited availability of synthetic oligonucleotides. DNA synthesis methods and molecular biology methods needed to coevolve to allow high-throughput gene synthesis, as we know it today from specialized gene synthesis laboratories, some of which are capable to produce several hundreds of genes per month.

In principle, all gene synthesis methods used so far rely on the elongation of hybridized oligonucleotides with long overlaps or the ligation of the phosphorylated oligonucleotides. The latter technique was used already by Koester et al. who phosphorylated oligonucleotides by T4 polynucleotide kinase and joined them using T4 ligase giving rise to a double-stranded DNA consisting of 33 bp. A similar approach was used by Edge et al. to synthesize long DNA fragments, which were assembled and ligated to a 514 bp long human leukocyte interferon encoding synthetic gene in 1981 [89]. After Francis Barany introduced the ligase chain reaction (LCR) using a heat stable ligase [90], it was possible to anneal oligonucleotides for *de novo* gene synthesis at high temperatures making ligation of phosphorylated oligonucleotides a very robust, but yet labor-intensive, time consuming, and expensive gene synthesis strategy [91].

In 1982, 3 years before the PCR was invented by Kary B. Mullis, the gene synthesis pioneer Itakura introduced a primer extension method enabling him to build up short genetic sequences *de novo*. A subsequently filed patent application was granted many years later in 1997 and is thus still active within the United States for many years to come. After the introduction of PCR into genetic engineering in 1985, several PCR-based oligonucleotide assembly methods emerged but all of these are based on one or more primer extension steps with subsequent amplification. This has to be kept in mind when using PCR-based assembly methods and resulting genes for commercial purposes or for clinical testing. By applying PCR-/primer-extension-based methods soon the 1000 bp size barrier was broken in 1990 by the synthesis of a 2.1 kb long fully synthetic plasmid [92]. Stemmer et al. used 132 oligonucleotides in a single primer extension reaction of overlapping complementary oligonucleotides with subsequent PCR amplification to assemble a 2.7 kb sized plasmid [93]. Similar approaches were used by Withers-Martinez et al. in 1999, who assembled an apparently difficult to construct AT-rich malaria gene [94]. In Figure 12-10 a schematic drawing shows the principles of ligase-based and PCR-/primer-extension-based gene synthesis methods.

Surely, one major cost factor in gene synthesis is still the bulk of oligonucleotides needed for *de novo* gene construction. To reduce oligonucleotide costs,

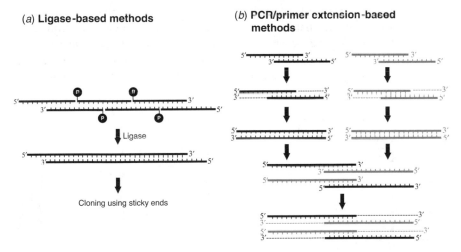

(a) **Ligase-based methods**

(b) **PCR/primer extension-based methods**

Ligase

Cloning using sticky ends

Figure 12-10 Assembly of complementary overlapping oligonucleotides. (a) The inner oligonucleotides need to be phosphorylated (P) and no gaps should separate the oligonucleotides. To allow subsequent cloning, 5' overhangs should be added to the flanking oligonucleotides thereby creating sticky ends which are compatible with the cohesive ends created after restriction digestion of a suitable cloning vector. After annealing the oligonucleotides can be ligated, the ligation product can be purified and cloned in a suitable cloning vector. (b) PCR cycling allows the annealing of overlapping oligonucleotides and their elongation. The assembled sequence may be subsequently amplified using flanking PCR primer oligonucleotides and cloned into a suitable cloning vector.

George Church et al. recently combined photo-programmable microfluids chips with classic PCR-based gene synthesis techniques. Although use of chemicals can be dramatically reduced, the oligonucleotides are present after release into solution only in femtomolar concentrations per sequences that are insufficient to allow bimolecular priming necessary for *de novo* gene construction. Therefore, the oligonucleotides released from the chip need to be reamplified by PCR using universal primers, endonuclease treatment to remove the universal priming region, and final purification. These extensive post synthesis treatments in addition to the setup costs of the oligonucleotide chip add up to the overall costs of this new and visionary gene synthesis method [95].

All gene synthesis methods so far have in common that they totally rely on quality and sequence identity of the oligonucleotides used in the assembly process. Due to the nature of the chemical synthesis of oligonucleotides there will be a certain fraction of oligonucleotides present showing deviations (like deletions, duplications, substitutions) from the desired end product. To test the quality and sequence correctness of oligonucleotides, we analyzed different batches and differently long oligonucleotides from a commercial source for gene synthesis. Oligonucleotides of 24mer, 44mer, and 64mer were used to assemble a 513 bp long fully artificial gene encoding all epitopes of the HIV-1 tat gene. A ligase-based gene assembly technique was used to avoid introduction of PCR-based mutations. After gene assembly, cloning and transformation 161 different clones were established, sequenced, and analyzed (see Fig. 12-11a). Apparently, there are huge deviations in oligonucleotide quality that differ from batch to batch even when the same oligonucleotide sequence is ordered. There seems to be a clear tendency that the longer the oligonucleotides are the higher the likelihood of mutations will be within the final clones (see Fig. 12-11b). This is most probably due to the fact that an increasing number of chemical reactions

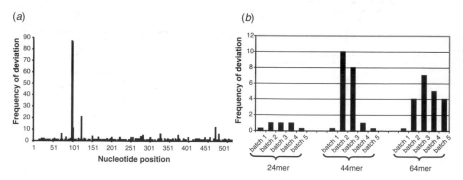

Figure 12-11 (a) Sequence analysis of a 513 bp long synthetic gene. The gene was assembled using 24 different overlapping, phosphorylated oligonucleotides using a ligase-based approach. The synthetic genes were cloned and 161 colonies sequenced. Frequency of deviations was plotted along the sequence. (b) Within the 513 long synthetic gene distinct oligonucleotides were replaced with a 24mer, 44mer and 64mer to test the influence of oligonucleotide length error frequency found in clones. Five different batches of each oligonucleotide sequence were tested and cloned. Twenty clones of each batch were analyzed by sequencing and frequency of deviations analyzed.

take place during synthesis leading to an increase of undesired side products. As deduced from this example it is most likely to get a statistical clustering of mutations within a synthetic gene. In our case, almost 90 percent of all clones showed mutations in one specific oligonucleotide, resulting in a deletion frequency of approximately 55 percent at position 97 and 98, respectively. Consequently, only 2 of 161 sequenced clones showed no deviation from the desired sequence design. Obviously, it would have been cheaper and easier to replace the low-quality oligonucleotide batch with new one resulting in a much higher likelihood to identify a 100 percent correct clone. In addition, gene length has clearly also influence on the chance to identify 100 percent correct clones by sequencing screening. When 50mer oligonucleotides with a given deviation frequency of approximately 10 percent would be used to build up a 500 bp long synthetic gene the probability of identifying a correct clone will be 12 percent (0.90^{20}, 20 oligonucleotides used in assembly). One would therefore need to screen at least 12–24 colonies to identify a 100 percent correct clone. However, to assemble a 1000 bp gene one would need at least 40 oligonucleotides to build up a synthetic gene. Subsequently, only 1.5 percent (0.90^{40}) of all screened clones would be correct. By doubling the gene length the screening effort increased by a factor of 10!

Consequently, one should therefore use only oligonucleotides with highest possible sequence identy, which are unfortunately not easy to access from commercial oligonucleotide suppliers. Alternative strategies were published by several groups trying to eliminate the fraction of mutated genes from the initial oligonucleotide assembly product in order to reduce the sequencing screening workload in gene synthesis. By denaturation and annealing of the crude oligo assembly product, heteroduplexes are most probably formed by DNA products showing mutations and homoduplexes are formed with the highest probability only from such DNA sequences showing no mutations. In a subsequent step, the heteroduplex fraction is removed by either enzymatic treatment, for example, T4 endonuclease VII cleavage [91,96] or by addition of proteins binding specifically to heteroduplex DNA and subsequent homoduplex enrichment [97].

More recently, Cello et al. reported the synthesis of a 7.5 kb cDNA poliovirus by a combination of PCR-based oligonucleotide assembly/PCA (polymerase cycling assembly) and ligation methods [98]. Being one of the first genomes to be fully created synthetically, the generation of an infectious and pathogenic virus based entirely on a *de novo* constructed genome will surely be considered as a hallmark in synthetic biology and even raised further ethical questions regarding biosafety of such synthetic genomes. Using similar approaches, only 2 years later, Kodumal et al. of Kosan Biosciences reported the synthesis of a contiguous 32 kb polyketide synthase gene cluster being the longest synthetic DNA assembled from synthetic oligonucleotides so far [99]. The functionality of the gene cluster was demonstrated by successfully expressing the polyketide synthase and producing its polyketide product in *E. coli*, being the first report of a functionally operon synthesized and assembled *de novo*. Just 2 years later, the same laboratory reported the synthesis of a redesigned polyketide synthase gene cluster expressing significantly more protein that the wild-type cluster did [100]. The last remarkable cornerstone was set here by Craig Venter who managed to synthesise and artificially construct the first bacterial genome of more than 500.000 bp in size [101].

Today, it is therefore not unrealistic to predict that in the years to come more and more synthetic operons, artificial chromosomes, and synthetic genomes will be synthesized being the major enabling technology for the new field of synthetic biology.

REFERENCES

1. Szybalski W, Skalka A. Nobel prizes and restriction enzymes. *Gene* 1978;4:181–182.

2. Grantham R, Gautier C, Gouy M, Mercier R, Pave A. Codon catalog usage and the genome hypothesis. *Nucleic Acids Res* 1980;8:r49–r62.

3. Grantham R, Gautier C, Gouy M, Jacobzone M, Mercier R. Codon catalog usage is a genome strategy modulated for gene expressivity. *Nucleic Acids Res* 1981;9:r43–r74.

4. Ikemura T. Correlation between the abundance of *Escherichia coli* transfer RNAs and the occurrence of the respective codons in its protein genes: a proposal for a synonymous codon choice that is optimal for the *E. coli* translational system. *J Mol Biol* 1981;151: 389–409.

5. Ikemura T. Codon usage and tRNA content in unicellular and multicellular organisms. *Mol Biol Evol* 1985;2:13–34.

6. Sharp PM, Li WH. The codon adaptation index—a measure of directional synonymous codon usage bias, and its potential applications. *Nucleic Acids Res* 1987;15: 1281–1295.

7. Chen GF, Inouye M. Suppression of the negative effect of minor arginine codons on gene expression; preferential usage of minor codons within the first 25 codons of the *Escherichia coli* genes. *Nucleic Acids Res* 1990;18:1465–1473.

8. Grosjean H, Fiers W. Preferential codon usage in prokaryotic genes: the optimal codon–anticodon interaction energy and the selective codon usage in efficiently expressed genes. *Gene* 1982;18:199–209.

9. Gouy M, Gautier C. Codon usage in bacteria: correlation with gene expressivity. *Nucleic Acids Res* 1982;10:7055–7074.

10. Ikemura T. Correlation between the abundance of yeast transfer RNAs and the occurrence of the respective codons in protein genes. Differences in synonymous codon choice patterns of yeast and *Escherichia coli* with reference to the abundance of isoaccepting transfer RNAs. *J Mol Biol* 1982;158:573–597.

11. Emilsson V, Naslund AK, Kurland CG. Growth-rate-dependent accumulation of twelve tRNA species in *Escherichia coli*. *J Mol Biol* 1993;230:483–491.

12. Zhou Z, Schnake P, Xiao L, Lal AA. Enhanced expression of a recombinant malaria candidate vaccine in *Escherichia coli* by codon optimization. *Protein Expr Purif* 2004;34:87–94.

13. Li A, Kato Z, Ohnishi H, Hashimoto K, Matsukuma E, Omoya K, Yamamoto Y, Kondo N. Optimized gene synthesis and high expression of human interleukin-18. *Protein Expr Purif* 2003;32:110–118.

14. Nomura M, Ohsuye K, Mizuno A, Sakuragawa Y, Tanaka S. Influence of messenger RNA secondary structure on translation efficiency. *Nucleic Acids Symp Ser* 1984;(15): 173–176.

15. Ivanov IG, Alexandrova R, Dragulev B, Leclerc D, Saraffova A, Maximova V, Abouhaidar MG. Efficiency of the 5'-terminal sequence (omega) of tobacco mosaic virus RNA for the initiation of eukaryotic gene translation in *Escherichia coli*. *Eur J Biochem* 1992;209:151–156.

16. Irwin B, Heck JD, Hatfield GW. Codon pair utilization biases influence translational elongation step times. *J Biol Chem* 1995;270:22801–22806.

17. Cheng L, Goldman E. Absence of effect of varying Thr-Leu codon pairs on protein synthesis in a T7 system. *Biochemistry* 2001;40:6102–6106.

18. Poole ES, Brown CM, Tate WP. The identity of the base following the stop codon determines the efficiency of *in vivo* translational termination in *Escherichia coli*. *EMBO J* 1995;14:151–158.

19. Williams DP, Regier D, Akiyoshi D, Genbauffe F, Murphy JR. Design, synthesis and expression of a human interleukin-2 gene incorporating the codon usage bias found in highly expressed *Escherichia coli* genes. *Nucleic Acids Res* 1988;16:10453–10467.

20. Hu X, Shi Q, Yang T, Jackowski G. Specific replacement of consecutive AGG codons results in high-level expression of human cardiac troponin T in *Escherichia coli*. *Protein Expr Purif* 1996;7:289–293.

21. Makoff AJ, Oxer MD, Romanos MA, Fairweather NF, Ballantine S. Expression of tetanus toxin fragment C in *E. coli*: high level expression by removing rare codons. *Nucleic Acids Res* 1989;17:10191–10202.

22. Ejdeback M, Young S, Samuelsson A, Karlsson BG. Effects of codon usage and vector host combinations on the expression of spinach plastocyanin in *Escherichia coli*. *Protein Expr Purif* 1997;11:17–25.

23. Hale RS, Thompson G. Codon optimization of the gene encoding a domain from human type 1 neurofibromin protein results in a threefold improvement in expression level in *Escherichia coli*. *Protein Expr Purif* 1998;12:185–188.

24. Johansson AS, Bolton-Grob R, Mannervik B. Use of silent mutations in cDNA encoding human glutathione transferase M2-2 for optimized expression in *Escherichia coli*. *Protein Expr Purif* 1999;17:105–112.

25. Li Y, Chen CX, von Specht BU, Hahn HP. Cloning and hemolysin-mediated secretory expression of a codon-optimized synthetic human interleukin-6 gene in *Escherichia coli*. *Protein Expr Purif* 2002;25:437–447.

26. Nalezkova M, de Groot A, Graf M, Gans P, Blanchard L. Overexpression and purification of *Pyrococcus abyssi* phosphopantetheine adenylyltransferase from an optimized synthetic gene for NMR studies. *Protein Expr Purif* 2005;39:296–306.

27. Bennetzen JL, Hall BD. Codon selection in yeast. *J Biol Chem* 1982;257:3026–3031.

28. Sharp PM, Tuohy TM, Mosurski KR. Codon usage in yeast: cluster analysis clearly differentiates highly and lowly expressed genes. *Nucleic Acids Res* 1986;14:5125–5143.

29. Percudani R, Pavesi A, Ottonello S. Transfer RNA gene redundancy and translational selection in *Saccharomyces cerevisiae*. *J Mol Biol* 1997;268:322–330.

30. Sinclair G, Choy FY. Synonymous codon usage bias and the expression of human glucocerebrosidase in the methylotrophic yeast, *Pichia pastoris*. *Protein Expr Purif* 2002;26:96–105.

31. Woo JH, Liu YY, Mathias A, Stavrou S, Wang Z, Thompson J, Neville DM Jr. Gene optimization is necessary to express a bivalent anti-human anti-T cell immunotoxin in *Pichia pastoris*. *Protein Expr Purif* 2002;25:270–282.

32. Outchkourov NS, Stiekema WJ, Jongsma MA. Optimization of the expression of equistatin in *Pichia pastoris. Protein Expr Purif* 2002;24:18–24.

33. Gurkan C, Ellar DJ. Expression in *Pichia pastoris* and purification of a membrane-acting immunotoxin based on a synthetic gene coding for the *Bacillus thuringiensis* Cyt2Aa1 toxin. *Protein Expr Purif* 2003;29:103–116.

34. Oliveira CC, van den Heuvel JJ, McCarthy JE. Inhibition of translational initiation in *Saccharomyces cerevisiae* by secondary structure: the roles of the stability and position of stem-loops in the mRNA leader. *Mol Microbiol* 1993;9:521–532.

35. Seffens W, Digby D. mRNAs have greater negative folding free energies than shuffled or codon choice randomized sequences. *Nucleic Acids Res* 1999;27:1578–1584.

36. Romanos MA, Makoff AJ, Fairweather NF, Beesley KM, Slater DE, Rayment FB, Payne MM, Clare JJ. Expression of tetanus toxin fragment C in yeast: gene synthesis is required to eliminate fortuitous polyadenylation sites in AT-rich DNA. *Mol Biol* 1991;19:1461–1467.

37. Scorer CA, Buckholz RG, Clare JJ, Romanos MA. The intracellular production and secretion of HIV-1 envelope protein in the methylotrophic yeast *Pichia pastoris. Gene* 1993;136:111–119.

38. Hofte H, Seurinck J, Van Houtven A, Vaeck M. Nucleotide sequence of a gene encoding an insecticidal protein of *Bacillus thuringiensis* var. *tenebrionis* toxic against Coleoptera. *Nucleic Acids Res* 1987;15:7183.

39. Muller BC, Raphael AL, Barton JK. Evidence for altered DNA conformations in the simian virus 40 genome: site-specific DNA cleavage by the chiral complex lambda-tris(4,7-diphenyl-1,10-phenanthroline)cobalt(III). *Proc Natl Acad Sci USA* 1987;84: 1764–1768.

40. Perlak FJ, Fuchs RL, Dean DA, McPherson SL, Fischhoff DA. Modification of the coding sequence enhances plant expression of insect control protein genes. *Proc Natl Acad Sci USA* 1991;88:3324–3328.

41. Shaw G, Kamen R. A conserved AU sequence from the 3′ untranslated region of GM-CSF mRNA mediates selective mRNA degradation. *Cell* 1986;46:659–667.

42. Strizhov N, Keller M, Mathur J, Koncz-Kalman Z, Bosch D, Prudovsky E, Schell J, Sneh B, Koncz C, Zilberstein A. A synthetic cryIC gene, encoding a *Bacillus thuringiensis* delta-endotoxin, confers *Spodoptera* resistance in alfalfa and tobacco. *Proc Natl Acad Sci USA* 1996;93:15012–15017.

43. Rouwendal GJ, Mendes O, Wolbert EJ, Douwe de Boer A. Enhanced expression in tobacco of the gene encoding green fluorescent protein by modification of its codon usage. *Plant Mol Biol* 1997;33:989–999.

44. Jensen LG, Olsen O, Kops O, Wolf N, Thomsen KK, von Wettstein D. Transgenic barley expressing a protein-engineered, thermostable (1,3-1,4)-beta-glucanase during germination. *Proc Natl Acad Sci USA* 1996;93:3487–3491.

45. Horvath H, Huang J, Wong O, Kohl E, Okita T, Kannangara CG, von Wettstein D. The production of recombinant proteins in transgenic barley grains. *Proc Natl Acad Sci USA* 2000;97:1914–1919.

46. Levy JP, Muldoon RR, Zolotukhin S, Link CJ Jr. Retroviral transfer and expression of a humanized, red-shifted green fluorescent protein gene into human tumor cells. *Nat Biotechnol* 1996;14:610–614.

47. Zolotukhin S, Potter M, Hauswirth WW, Guy J, Muzyczka N. A "humanized" green fluorescent protein cDNA adapted for high-level expression in mammalian cells. *J Virol* 1996;70:4646–4654.

48. Wells KD, Foster JA, Moore K, Pursel VG, Wall RJ. Codon optimization, genetic insulation, and an rtTA reporter improve performance of the tetracycline switch. *Transgenic Res* 1999;8:371–381.

49. Zhou J, Liu WJ, Peng SW, Sun XY, Frazer I. Papillomavirus capsid protein expression level depends on the match between codon usage and tRNA availability. *J Virol* 1999; 73:4972–4982.

50. Bradel-Tretheway BG, Zhen Z, Dewhurst S. Effects of codon-optimization on protein expression by the human herpesvirus 6 and 7 U51 open reading frame. *J Virol Methods* 2003;111:145–156.

51. Plotkin JB, Robins H, Levine AJ. Tissue-specific codon usage and the expression of human genes. *Proc Natl Acad Sci USA* 2004;101:12588–12591.

52. Duan J, Antezana MA. Mammalian mutation pressure, synonymous codon choice, and mRNA degradation. *J Mol Evol* 2003;57:694–701.

53. Salser W. Globin mRNA sequences: analysis of base pairing and evolutionary implications. *Cold Spring Harb Symp Quant Biol* 1978;42(Part 2): 985–1002.

54. Lercher MJ, Hurst LD. Can mutation or fixation biases explain the allele frequency distribution of human single nucleotide polymorphisms (SNPs)? *Gene* 2002; 300:53–58.

55. Konu O, Li MD. Correlations between mRNA expression levels and GC contents of coding and untranslated regions of genes in rodents. *J Mol Evol* 2002;54:35–41.

56. Graf M, Bojak A, Deml L, Bieler K, Wolf H, Wagner R. Concerted action of multiple *cis*-acting sequences is required for Rev dependence of late human immunodeficiency virus type 1 gene expression. *J Virol* 2000;74:10822–10826.

57. Graf M, Deml L, Wagner R. Codon-optimized genes that enable increased heterologous expression in mammalian cells and elicit efficient immune responses in mice after vaccination of naked DNA. *Methods Mol Med* 2004;94:197–210.

58. Chen CY, Xu N, Shyu AB. mRNA decay mediated by two distinct AU-rich elements from c-fos and granulocyte-macrophage colony-stimulating factor transcripts: different deadenylation kinetics and uncoupling from translation. *Mol Cell Biol* 1995,15: 5777–5788.

59. Chen CY, Shyu AB. AU-rich elements: characterization and importance in mRNA degradation. *Trends Biochem Sci* 1995;20:465–470.

60. Maldarelli F, Martin MA, Strebel K. Identification of posttranscriptionally active inhibitory sequences in human immunodeficiency virus type 1 RNA: novel level of gene regulation. *J Virol* 1991;65:5732–5743.

61. Nasioulas G, Zolotukhin AS, Tabernero C, Solomin L, Cunningham CP, Pavlakis GN, Felber BK. Elements distinct from human immunodeficiency virus type 1 splice sites are responsible for the Rev dependence of env mRNA. *J Virol* 1994;68:2986–2993.

62. Olsen HS, Cochrane AW, Rosen C. Interaction of cellular factors with intragenic *cis*-acting repressive sequences within the HIV genome. *Virology* 1992;191:709–715.

63. Schwartz S, Felber BK, Pavlakis GN. Distinct RNA sequences in the gag region of human immunodeficiency virus type 1 decrease RNA stability and inhibit expression in the absence of Rev protein. *J Virol* 1992;66:150–159.

64. Chang DD, Sharp PA. Regulation by HIV Rev depends upon recognition of splice sites. *Cell* 1989;59:789–795.

65. Kozak M. At least six nucleotides preceding the AUG initiator codon enhance translation in mammalian cells. *J Mol Biol* 1987;196:947–950.

66. Kozak M. An analysis of 5'-noncoding sequences from 699 vertebrate messenger RNAs. *Nucleic Acids Res* 1987;15:8125–8148.

67. Chapman BS, Thayer RM, Vincent KA, Haigwood NL. Effect of intron A from human cytomegalovirus (Towne) immediate-early gene on heterologous expression in mammalian cells. *Nucleic Acids Res* 1991;19:3979–3986.

68. Trinh R, Gurbaxani B, Morrison SL, Seyfzadeh M. Optimization of codon pair use within the (GGGGS)$_3$ linker sequence results in enhanced protein expression. *Mol Immunol* 2004;40:717–722.

69. Gustafsson C, Govindarajan S, Minshull J. Codon bias and heterologous protein expression. *Trends Biotechnol* 2004;22:346–353.

70. Haas J, Park EC, Seed B. Codon usage limitation in the expression of HIV-1 envelope glycoprotein. *Curr Biol* 1996;6:315–324.

71. Uchijima M, Yoshida A, Nagata T, Koide Y. Optimization of codon usage of plasmid DNA vaccine is required for the effective MHC class I-restricted T cell responses against an intracellular bacterium. *J Immunol* 1998;161:5594–5599.

72. zur Megede J, Chen MC, Doe B, Schaefer M, Greer CE, Selby M, Otten GR, Barnett SW. Increased expression and immunogenicity of sequence-modified human immunodeficiency virus type 1 *gag* gene. *J Virol* 2000;74:2628–2635.

73. Narum DL, Kumar S, Rogers WO, Fuhrmann SR, Liang H, Oakley M, Taye A, Sim BK, Hoffman SL. Codon optimization of gene fragments encoding *Plasmodium falciparum* merozoite proteins enhances DNA vaccine protein expression and immunogenicity in mice. *Infect Immun* 2001;69:7250–7253.

74. Valencik ML, McDonald JA. Codon optimization markedly improves doxycycline regulated gene expression in the mouse heart. *Transgenic Res* 2001;10:269–275.

75. Leder C, Kleinschmidt JA, Wiethe C, Muller M. Enhancement of capsid gene expression: preparing the human papillomavirus type 16 major structural gene L1 for DNA vaccination purposes. *J Virol* 2001;75:9201–9209.

76. Shimshek DR, Kim J, Hubner MR, Spergel DJ, Buchholz F, Casanova E, Stewart AF, Seeburg PH, Sprengel R. Codon-improved Cre recombinase (iCre) expression in the mouse. *Genesis* 2002;32:19–26.

77. Hamdan FF, Mousa A, Ribeiro P. Codon optimization improves heterologous expression of a *Schistosoma mansoni* cDNA in HEK293 cells. *Parasitol Res* 2002;88: 583–586.

78. Gao X, Yo P, Keith A, Ragan TJ, Harris TK. Thermodynamically balanced inside-out (TBIO) PCR-based gene synthesis: a novel method of primer design for high-fidelity assembly of longer gene sequences. *Nucleic Acids Res* 2003;31:e143.

79. Disbrow GL, Sunitha I, Baker CC, Hanover J, Schlegel R. Codon optimization of the HPV-16 E5 gene enhances protein expression. *Virology* 2003;311:105–114.

80. Cid-Arregui A, Juarez V, zur Hausen H. A synthetic E7 gene of human papillomavirus type 16 that yields enhanced expression of the protein in mammalian cells and is useful for DNA immunization studies. *J Virol* 2003;77:4928–4937.

81. Barouch DH, Pau MG, Custers JH, Koudstaal W, Kostense S, Havenga MJ, Truitt DM, Sumida SM, Kishko MG, Arthur JC, Korioth-Schmitz B, Newberg MH, Gorgone DA, Lifton MA, Panicali DL, Nabel GJ, Letvin NL, Goudsmit J. Immunogenicity of recombinant adenovirus serotype 35 vaccine in the presence of pre-existing anti-Ad5 immunity. *J Immunol* 2004;172:6290–6297.

82. Didierlaurent A, Ramirez JC, Gherardi M, Zimmerli SC, Graf M, Orbea HA, Pantaleo G, Wagner R, Esteban M, Kraehenbuhl JP, Sirard JC. Attenuated poxviruses expressing a synthetic HIV protein stimulate HLA-A2-restricted cytotoxic T-cell responses. *Vaccine* 2004;22:3395–3403.

83. Karzenowski D, Potter DW, Padidam M. Inducible control of transgene expression with ecdysone receptor: gene switches with high sensitivity, robust expression, and reduced size. *Biotechniques* 2005;39:191–192, 194, 196.

84. van den Brink EN, Ter Meulen J, Cox F, Jongeneelen MA, Thijsse A, Throsby M, Marissen WE, Rood PM, Bakker AB, Gelderblom HR, Martina BE, Osterhaus AD, Preiser W, Doerr HW, de Kruif J, Goudsmit J. Molecular and biological characterization of human monoclonal antibodies binding to the spike and nucleocapsid proteins of severe acute respiratory syndrome coronavirus. *J Virol* 2005;79:1635–1644.

85. Michelson AM, Todd AR. Synthesis of a dithymidine dinucleotide containing a 3′:5′-internucleotidic linkage. *J Chem Soc* 1955;2632.

86. Agarwal KL, Buchi H, Caruthers MH, Gupta N, Khorana HG, Kleppe K, Kumar A, Ohtsuka E, Rajbhandary UL, Van de Sande JH, Sgaramella V, Weber H, Yamada T. Total synthesis of the gene for an alanine transfer ribonucleic acid from yeast. *Nature* 1970;227:27–34.

87. Koster H, Blocker H, Frank R, Geussenhainer S, Kaiser W. Total synthesis of a structural gene for the human peptide hormone angiotensin II. *Hoppe Seylers Z Physiol Chem* 1975;356:1585–1593.

88. Itakura K, Hirose T, Crea R, Riggs AD, Heyneker HL, Bolivar F, Boyer HW. Expression in *Escherichia coli* of a chemically synthesized gene for the hormone somatostatin. *Science* 1977;198:1056–1063.

89. Edge MD, Green AR, Heathcliffe GR, Meacock PA, Schuch W, Scanlon DB, Atkinson TC, Newton CR, Markham AF. Total synthesis of a human leukocyte interferon gene. *Nature* 1981;292:756–762.

90. Barany F, Gelfand DH. Cloning, overexpression and nucleotide sequence of a thermostable DNA ligase-encoding gene. *Gene* 1991;109:1–11.

91. Young L, Dong Q. Two-step total gene synthesis method. *Nucleic Acids Res* 2004;32:e59.

92. Mandecki W, Hayden MA, Shallcross MA, Stotland E. A totally synthetic plasmid for general cloning, gene expression and mutagenesis in *Escherichia coli*. *Gene* 1990;94:103–107.

93. Stemmer WP, Crameri A, Ha KD, Brennan TM, Heyneker HL. Single-step assembly of a gene and entire plasmid from large numbers of oligodeoxyribonucleotides. *Gene* 1995;164:49–53.

94. Withers-Martinez C, Carpenter EP, Hackett F, Ely B, Sajid M, Grainger M, Blackman MJ. PCR-based gene synthesis as an efficient approach for expression of the A +T-rich malaria genome. *Protein Eng* 1999;12:1113–1120.

95. Tian J, Gong H, Sheng N, Zhou X, Gulari E, Gao X, Church G. Accurate multiplex gene synthesis from programmable DNA microchips. *Nature* 2004;432:1050–1054.

96. Greger B, Kemper B. An apyrimidinic site kinks DNA and triggers incision by endonuclease VII of phage T4. *Nucleic Acids Res* 1998;26:4432–4438.

97. Smith J, Modrich P. Removal of polymerase-produced mutant sequences from PCR products. *Proc Natl Acad Sci USA* 1997;94:6847–6850.

98. Cello J, Paul AV, Wimmer E. Chemical synthesis of poliovirus cDNA: generation of infectious virus in the absence of natural template. *Science* 2002;297:1016–1018.

99. Kodumal SJ, Patel KG, Reid R, Menzella HG, Welch M, Santi DV. Total synthesis of long DNA sequences: synthesis of a contiguous 32-kb polyketide synthase gene cluster. *Proc Natl Acad Sci USA* 2004;101:15573–15578.

100. Menzella HG, Reisinger SJ, Welch M, Kealey JT, Kennedy J, Reid R, Tran CQ, Santi DV. Redesign, synthesis and functional expression of the 6-deoxyerythronolide B polyketide synthase gene cluster. *J Ind Microbiol Biotechnol* 2006;33:22–28.

101. Gibson DG, Benders GA, Andrews-Pfannkoch C, Denisova EA, Baden-Tillson H, Zaveri J, Stockwell TB, Brownley A, Thomas DW, Algire MA, Merryman C, Young L, Noskov VN, Glass JI, Venter JC, Hutchison CA 3rd, Smith HO. Complete chemical synthesis, assembly, and cloning of a Mycoplasma genilalium genome. *Science* 2008; 29:319 (5867): 1215–1220. Epub 2008 Jan.

13

SELF-REPLICATION IN CHEMISTRY AND BIOLOGY

Philipp Holliger and David Loakes

MRC Laboratory of Molecular Biology, Cambridge CB2 2QH, UK

13.1 INTRODUCTION: SELF-REPLICATION, FIDELITY, AND HEREDITY

Self-replication involves the product-directed assembly of components to form a new product; in its simplest form, it involves the joining of just two components. The product acts as a template both to correctly position the two components and to allow for efficient joining of them. The newly formed product can then dissociate to provide a new template for further replication [1,2] (Fig. 13-1).

Important concepts for consideration are fidelity and heredity. Self-replication may be perfect, in which case all products (all "offspring") are identical to the template (the parents), or imperfect, in which case, offspring may differ from their parents. The degree of perfection (or imperfection) of self-replication is called the fidelity, which varies greatly among self-replicating systems. High-fidelity replication denotes a system where only few alterations are introduced into the offspring molecules while low-fidelity replicators will produce a great deal of variation in their offspring. Depending on the system these variations can again be transmitted through the next self-replication cycle. In such a case, the self-replication system will display heredity. Self-replication with heredity is a fundamental property of life and a prerequisite for evolution.

The most widely studied self-replicating systems involve nucleic acids and these are most relevant to extant or plausible primordial biological systems, as nucleic acids are uniquely suited for self-replication, heredity, and evolution. We also briefly discuss chemical replicators based on autocatalytic networks or template-driven replicators

Systems Biology and Synthetic Biology   Edited by Pengcheng Fu and Sven Panke
Copyright © 2009 John Wiley & Sons, Inc.

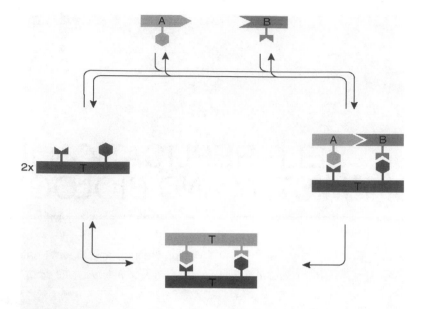

Figure 13-1 Scheme of a simple self-replication system (after [1,2]). A and B, building blocks; T, template.

with limited heredity (e.g., peptides, prions). However, replicators based on digital (e.g., computer viruses) or cultural heredity (e.g., memes), or self-replicating macroscopic machines are beyond the scope of this review, despite striking progress in the latter field [3]. The review will consider primarily recent literature referring to older literature only when necessary, and is not meant to be exhaustive.

13.2 CHEMICAL SYSTEMS CAPABLE OF SELF-REPLICATION

For a number of years chemists have explored a host of molecular systems with autocatalytic and dynamic combinatorial properties. Some of these are capable of templating and catalyzing their own synthesis, that is, catalyze the product-directed synthesis of more products from its constituent parts. Due to the limited complexity of such systems, either self-replication is perfect or alterations (side reactions) are nonhereditary as they would interfere with the self-replication ability. One of the earliest reported is a nucleoside-based system, described by Rebek and colleagues, and involves hydrogen-bonding and stacking interactions in organic media [4–8] (Fig. 13-2).

There have been a number of reports of autocatalytic Diels–Alder reactions. The bicyclic transition state that is formed offers a basis for efficient self-replication, that is, for the transfer of chemical information coded in terms of both regio- and stereoselectivity. In this example, the diene is also chiral, thus allowing for the transfer of diastereomeric information. Following this first published report of

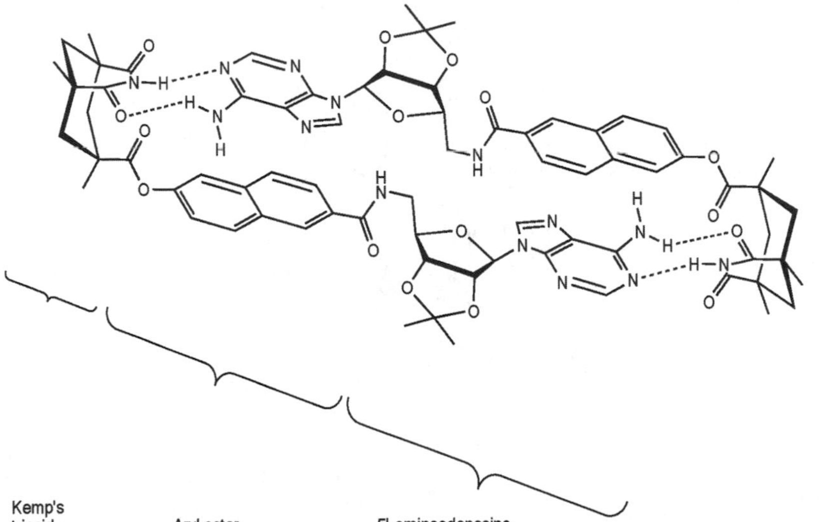

Figure 13-2 One of the earliest self-replicating nucleoside-based systems involves hydrogen-bonding and aryl-stacking interactions in organic media. 5′-Aminoadenosine reacts with the aryl-pentafluorophenyl ester derivative of Kemp's triacid, which then acts as catalyst for further coupling reactions. The reaction involves hydrogen bonding of adenosine to the imine, as blocking of the imine NH group leads to a 10-fold drop in catalytic rate.

the autocatalytic Diels–Alder reaction, it has been demonstrated that using a chiral-starting material only one of the four possible diastereoisomers were formed [9]. Other groups have examined self-replicating Diels–Alder reactions [10,11] including von Kiedrowski [9]. In addition to the advantages of being capable of self-replicating by homochiral autocatalysis and heterochiral cross-catalysis, they are much more efficient replicators than nucleic acids or peptides, giving rise to almost exponential replication.

13.3 PEPTIDE SELF-REPLICATION

Peptides of a certain length fold spontaneously into three-dimensional structures defined by their sequence. These in turn may specifically associate with other peptides in defined oligomeric complexes. Peptide self-replication is generally based on a peptide A acting as a template and promoting the template-directed ligation of two smaller isomorphic peptides. In its simplest form, the two smaller peptides are fragments of A (A′, A″) and ligation thus produces further copies of the parent peptide in a homodimeric complex (Fig. 13-3).

The first self-replicating peptide described was a 32-residue α-helical coiled-coil peptide based on the leucine-zipper domain of the yeast transcription factor GCN4 [12]. It has been shown to promote thioester-mediated amide bond formation

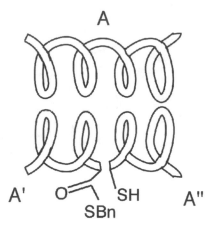

Figure 13-3 Coiled-coil peptide self-replicators are able to promote template-directed ligation of two smaller peptides (A', A'') producing a further copy of the parent peptide (A). Typical coupling reactions occur with carbodiimide or thioester (shown) chemistry.

between a 15- and 17mer fragments in neutral aqueous conditions. It was subsequently shown that this system was also able to distinguish between native and mutant precursor peptides. When mutant peptides are used there is a catalytic step in which mutant progeny are produced, but due to changes in the hydrophobic core of the precursor peptides the correct peptide may be preferentially produced [13,14]. This peptide replicator is also capable of chiral selectivity by efficiently amplifying homochiral products from a racemic mixture of peptide fragments [15] and is capable of discriminating between structures containing only a single chiral mutation. The system demonstrates a dynamic stereochemical editing function whereby heterochiral sequences promote the production of homochiral products. Thus, the peptide replicator system demonstrates the emergence of fidelity of replication.

The self-replicating peptide described by Ghadiri has been computationally analyzed where it was found that the dynamics are governed principally by two reversible hydrophobic interactions between the template and a peptide fragment and between two template molecules [16]. The association of two template molecules was found to be most favorable leading to a build up of the inactive template dimer in the autocatalytic step, thus limiting the self-replication. Analysis of the heterochiral system described by Ghadiri [15] indicated that cross-catalytic processes involving D- and L-species play a significant role. Chiral amplification is mainly due to the formation of *meso*-like species, leading to an enantiomeric excess in the final product [17,18].

Chmielewski has developed a self-replicating peptide that is pH-dependent. The sequence contains glutamic acid side chains such that at physiological pH the peptide is a random coil. However, under acidic conditions the peptide adopts a coiled-coil structure, similar to that developed by Ghadiri, which is then able to promote self-replication [19]. As noted above, the self-replication of peptides is limited as the most stable species is the template dimer. Chmielewski has therefore designed a self-replicating peptide containing a proline residue in place of one of the glutamic acid residues in the pH-dependent replicator. The presence of the proline causes a kink in the coiled-coil structure, which allows for more efficient separation of the template dimer species, thus improving the efficiency of self-replication [20].

Finally, there are two examples of cross-replication between peptides and nucleic acids. In the first example, L-α-amino-γ-nucleobase-butyric acids (NBAs) were substituted into peptides adopting coiled-coil structures to enhance peptide recognition. Templates and fragments were then synthesized containing complementary adenine–thymine or guanine–cytosine sequences at various positions within the peptide. While it was found that the effect of NBAs in the peptide was sequence dependent, it was shown that the increased recognition architecture could be used to design more efficient self-replicating peptides [21]. In the second example, Ellington has examined the ligation of short oligonucleotides by a peptide [22]. Using a 17mer arginine-rich motif (ARM), a 35mer anti-REV RNA aptamer was developed for ligation studies. Aptamer half-molecules bearing a 5′-iodine and a 3′-phosphorothioate could be chemically ligated by cyanogen bromide in the presence of the ARM peptide.

The systems described in this section may also have relevance for the prebiotic synthesis of peptides. It has been demonstrated that amino acids can adsorb to mineral surfaces where they undergo chemical ligation to form random polypeptide species. Together with self-replication, this may provide a process for the selective enrichment of a defined set of peptide sequences. All of the self-replicating peptides described so far adopt α-helical coiled-coil structures, but Ghadiri has speculated that self-replication through β-sheet motifs are also likely [23,24].

Interesting examples of peptide self-replication are provided by prions. These are a number of metastable proteins, which can be converted to a misfolded insoluble form. The insoluble form is capable of catalyzing the conversion of soluble prion protein into the insoluble form. In some cases the insoluble form is infectious and can be transmitted within and across species giving rise to so-called transmissible spongiform encephalopathies (TSEs), of which "mad cow disease" (BSE) is the best known. Intriguingly, prions display heredity in the form of strain and species specific characteristics, which appear to be encoded in the conformation of the prion protein [25]. In yeast, these can provide diverse, heritable phenotypes that are beneficial under certain circumstances. Indeed it has been proposed that prions may act as an epigenetic switch in yeast and fungi or even as a form of molecular long-term memory in the nervous system of *A. californica* [26].

13.4 NUCLEIC ACIDS

In 1953, Watson and Crick published their seminal article on the structure of DNA, which ends with the now famous understatement "It has not escaped our notice that the specific pairing we have postulated immediately suggests a copying mechanism for the genetic material" [27]. Indeed, another 50 years of research into the structure and function of DNA (and RNA) have not only confirmed the proposed semiconservative mode of replication, whereby one strand of DNA acts as a template for synthesis of the opposing strand (and vice versa), but also has revealed that DNA and RNA are singularly suited as molecules for information storage and transmission, for replication and heredity [28]. For one, in nucleic acids, the polyanionic phosphate backbone dominates the physicochemical properties of the molecule (e.g., solubility) to such an extent that changes to neither base composition nor sequence have much effect. In other words, in sharp contrast to, for example, proteins, nucleic acids display similar properties (e.g., solubility) regardless of the sequence, that is, the information encoded within. Furthermore, charge repulsion along the polyphosphate backbone favors an extended conformation of nucleic acid polymers facilitating their templating function in replication and read-out of the hydrogen-bonding pattern at the Watson–Crick face of the bases. Finally, nucleic acid polymorphism is constrained to essentially just two apomorphic classes, A and B (there is also a left-handed helix system, Z, which occurs only under certain conditions and is restricted to alternating purine–pyrimidine (GC) sequences).

13.4.1 Altered Backbones

A- and B-form nucleic acids arise as a result of the restricted spectrum of furanose (ribose or deoxyribose) sugar conformations. This relative inflexibility provides a stable scaffold for the nucleobases and is essential for duplex stability. It is therefore not surprising that many modifications to backbone chemistry have led to nucleic acids that are no longer capable of forming stable duplex structures with either DNA or RNA or themselves. A notable exception is peptide nucleic acids (PNAs), in which the ribofuranose-phosphate backbone of DNA/RNA is replaced by N-(2-aminoethyl)-glycine (Fig. 13-4). PNAs can hybridize specifically and extraordinarily strongly to DNA and RNA making them of significant use in both antisense and antigene strategies [29]. However, longer PNAs can be poorly soluble. Nevertheless, PNA can be used in information transfer to DNA and RNA and it has been proposed that PNA may have been involved in prebiotic evolution [34–36] (see later).

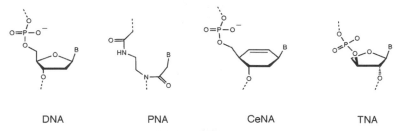

DNA   PNA   CeNA   TNA

Figure 13-4 Various nucleic acid backbone modifications have been examined as alternative genetic systems. These include peptide nucleic acid, cyclohexene nucleic acid (CeNA) and α-L-threose nucleic acid (TNA).

Another example is morpholino nucleosides that are neutral analogues of DNA in which the sugar is substituted by a morpholine ring [30]. They have been shown to bind well with RNA, and have therefore been a subject for investigation in the field of antisense therapy [31]. However, they are poor substrates for enzymes, including RNase-H, and act by a steric-blocking mechanism.

13.4.2 Altered Sugars

Nucleic acid chemists have also synthesized a variety of modifications to the ribofuranose sugars in DNA and RNA in an attempt to modify hybridization properties and study the determinants of Watson–Crick-directed duplex formation. For example, Orgel, Herdewijn, and colleagues have investigated the properties of hexose sugars and have demonstrated nonenzymatic information transfer from nucleic acids derived from 1,5-anhydrohexitol nucleosides (HNA) [32–34] and altritol nucleosides (ANA) [35]. A number of other hexopyranosyl- and pentopyranosyl-nucleoside systems have been studied by Eschenmoser [36]. These systems show a remarkable spectrum of hybridization properties in not only self-pairing systems but also cross-pairing with DNA and RNA.

Cyclohexane- and cyclohexene-nucleic acid systems are conformationally flexible nucleic acid mimics that can hybridize with themselves, DNA and RNA (Fig. 13-4) [37–39]. The self-pairing system is more stable than that with DNA, but is most stable with RNA. While they are yet to be shown to be of use in information transfer they are recognized by some enzymes as they can RNase-H activity when hybridized with RNA, and are therefore of use as antisense agents [40].

Another sugar modification that displays interesting properties is the tetrofuranose α-L-threose nucleic acids (TNA) (Fig. 13-4) [41], which forms specific base pairs with itself, DNA, and RNA. Furthermore, TNA has been shown to be functional in replacing RNA as part of an RNA cleaving ribozyme, albeit with somewhat reduced activity [42]. Finally, TNA templates and TNA triphosphates have been shown to be reasonable substrates for various DNA and RNA polymerises [43–46] and like PNA has been proposed to be involved in prebiotic evolution (see Section 13.6).

13.4.3 Altered Bases

For the reasons discussed above, the introduction of alternative base pairing systems into nucleic acids is much less problematic than alterations to the sugar–phosphate backbone structure. Such systems would not give rise to alternative nucleic acid structures, but may be used to introduce alternative or additional information content into nucleic acids without altering their overall structure. As a result, nucleic acids comprising modified bases are often better substrates for enzymatic replication. The challenge here lies in devising alternative systems, which are both orthogonal to the canonical bases as well as specific in recognition.

There are limited ways in which such novel base-pairing schemes can be devised. Nucleobases can be designed that will display altered recognition based on alternative hydrogen-bonding patterns [47,48], hydrophobic interactions [49], or chelation of a metal ion [50]. Alternatively, specificity and orthogonality can be achieved using size and/or steric effects [51,52].

One attractive strategy for forming an alternative base pair is to use two nucleosides that can form a specific base pair without pairing with any of the natural nucleobases. One of the first such base pair to be described was that between isocytidine (iC) and isoguanine (iG), in which the hydrogen-bonding groups of cytosine and guanosine are inverted. This allows for the formation of a specific base pair with a different donor and acceptor pattern from that of the natural base pairs (Fig. 13-5). The iC–iG pair has been shown to be replicated by both DNA and RNA polymerases [53], including in PCR [54]. One problem associated with this new base pair is that the iG exists in two different tautomeric forms, the minor of which specifically pairs with thymidine, leading to a loss of fidelity in replication reactions. Different strategies have been developed to avoid this. For example, Benner has used 2-thiothymidine instead of thymidine to prevent mispairing with the minor iG tautomer, as the 2-thio-group does not hydrogen bond effectively [55], while Seela has shown that the 7-deaza analogue of iG does form tautomers to a much reduced extent ($>10^3$-fold less) [56].

Benner and coworkers have devised a complete set of alternative hydrogen-bonded base pairs, each of which are held together by three hydrogen bonds and retain the size and geometry of the canonical base pairs [57]. One of the more advanced pairs is

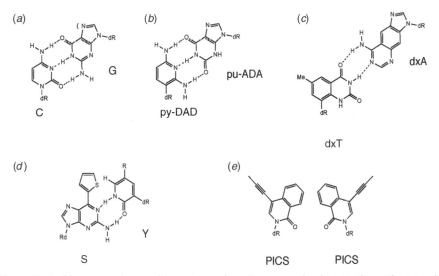

Figure 13-5 Many novel base pairing systems have been examined as an alternative genetic coding system. Specific alternative hydrogen-bonding systems can be used in conjunction with the native base pairs (a) such as py-DAD/pu-ADA (b) and an expanded version of native base pairs (c). Other systems use hydrogen-bonding and steric effects (d) or non-hydrogen-bonding self-pairs (e).

that between 2,4-diaminopyrimidine (py-DAD (py: pyrimidine; D: H-bond donor (e.g., NH_2 group); A: H-bond acceptor (e.g., C=O group)) and xanthine (pu-ADA) (Fig. 13-5) [58,59]. The py-DAD/pu-ADA base pair has been shown to be a substrate for a mutant HIV-1 reverse transcriptase, which replicates the new base pair with good fidelity in the presence of native DNA nucleotides [60].

Kool has examined the effect of size as an alternative genetic base pairing system. In the most striking example, the natural nucleosides have been redesigned with an expanded size by incorporation of a phenyl group between the sugar and the hydrogen-bonding ring [51,61–63]. This size expanded system, termed xDNA (x for expanded) (Fig. 13-5) or yDNA (y for wide), retain the features of regular DNA, such as Watson–Crick base pairing and right-handed helicity, but possess an expanded diameter when in a double helix [64]. When xDNA nucleosides are incorporated into regular duplex DNA there is distortion of the backbone due to the increased size of the base pair (2.4 Å) [65], but a duplex comprised solely of xDNA shows enhanced stability compared to DNA due to enhanced stacking interactions [66,67]. xDNA and yDNA represent two novel genetic systems, possessing many of the features found in regular DNA, but their expanded sizes should make them distinct from DNA.

A further method for developing a new base pair is an analogue that preferentially forms a self-pair. This class of analogue tends to be planar, aromatic, and non-hydrogen bonding, yet they can still be recognized by cellular enzymes, such as polymerases. There are a number of such analogues reported. Romesberg, Schultz, and coworkers have synthesized a number of analogues such as 7-azaindole (7-AI), propynylisocarbostyrile (PICS) (Fig. 13-5) as well as some fluoroaromatic analogues [68] and evaluated them as potential self-pairing nucleosides [69,70]. Various of the analogues prepared are also recognized with reasonable selectivity by DNA polymerases [49,68,71].

One of the most highly developed systems of novel, specific base pairing has been designed by Hirao and Yokoyama, who used steric effects to design various novel base pairs, two of which were found to be compatible with various cellular events. The systems they devised replaced the pyrimidine base with a pyridone and the purine with a C6-modified diaminopurine (Fig. 13-5), and retained hydrogen-bonding capability [72,73]. The pyridone will not form stable base pairs with the natural purines while if the purine base pairs with thymine it will be destabilized by a steric clash between the pyrimidine O4 and the purine C6 modification (Fig. 13-5). These analogues have been shown to form specific base pairs and to be recognized by DNA polymerises [74,75], RNA polymerises [52,76–79] and in translation, allowing the site-specific introduction of an unnatural amino acid *in vitro* [80].

13.5 NUCLEIC ACID SELF-REPLICATION

13.5.1 De Novo Synthesis of Nucleic Acid Polymers

In 1954, physicist George Gamow founded the RNA-tie club with a group of 20 scientists (one for each of the naturally occurring amino acids) who were interested in the function of RNA. One of the original members of this group, Leslie Orgel, has carried out significant studies in the field of self-replicating systems, and aspects of his work are discussed in this chapter. Over the following years the relationship between nucleic acids and proteins became better understood, but scientists such as Orgel started to ask questions about the origins of life and in particular about the prebiotic synthesis of nucleic acids.

While self-replication requires a template molecule to start from, these had to be first generated *de novo* from precursor molecules. Ferris *et al.* [81–84] have investigated the *de novo* synthesis of RNA polynucleotides on common clay minerals such as montmorillonite as a model for prebiotic synthesis. It was shown that mononucleotides activated as phosphoro-imidazolides would react with other nucleotide polyphosphates, for example, triphosphates, to form predominantly 3′, 5′-linked oligonucleotides in the presence of montmorillonite clay with a rate enhancement of 1000-fold compared to the absence of montmorillonite. It has also been shown that oligonucleotide 5′-polyphosphates (including triphosphates) can be formed from polynucleotide monophosphates and sodium trimetaphosphate [85]. Thus, a feasible mechanism for the synthesis of the original RNA polynucleotides has been described. Much of the further work carried out to investigate template-directed self-replication nevertheless makes use of 5′-imidazole-activated nucleotides as they are more reactive derivatives for the synthesis of oligo- and polynucleotides.

13.5.2 Template-Directed Synthesis of Nucleic Acids

Orgel et al. [86] have been involved in a majority of the work in the field of nonenzymatic template-directed synthesis of oligonucleotides. Early work from this group demonstrated that random copolymer RNA templates could be used to replicate RNA in solution without the need for an enzyme or catalyst over several days and at high Mg^{2+} concentrations (Fig. 13-6). These reactions are template-dependent, and under the reaction conditions AT base pairs are formed much less efficiently than GC pairs. Under these conditions, oligonucleotides in the range of 20–30 nucleotides can be produced over a period of 1 week. Analysis of the products demonstrated that there is a mixture of 2′,5′- and 3′,5′-linkages, with the 2′,5′-linkages predominating. This is probably due to the fact that the 2′-hydroxyl group is six to nine times more reactive than the 3′-hydroxyl group [87,88]. Synthesis of DNA using a DNA template and activated deoxynucleotides is much less efficient, and it has been reported that some sequences cannot be copied [89].

Szostak has studied the nonenzymatic template-directed ligation of oligoribonucleotides and shown that there is a dependence for binding to metal ions before ligation can occur [90]. A series of metal ions were assayed and Mn^{2+} and Mg^{2+} ions are most efficient for catalysis while Pb^{2+} and Zn^{2+} ions do not. They also demonstrated that the nonenzymatic ligation proceeds with a preference for 3′–5′ phosphodiester linkages in preference to 2′–5′, though it is dependent on the ligation chemistry (imidazolide or triphosphate) [91]. The preference for 3′–5′ linkages is in contrast to that reported by Orgel, who reported a preference for 2′–5′ linkages for template-directed replication, suggesting that the type of linkage obtained may be sensitive to reaction conditions.

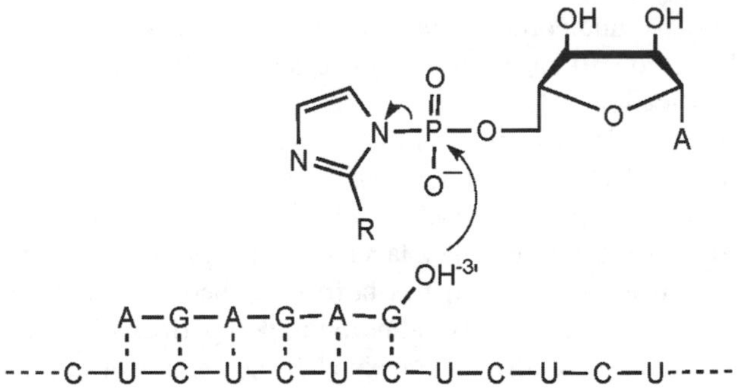

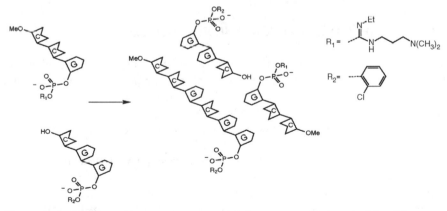

Figure 13-6 Nonenzymatic template-directed synthesis of RNA involving activated nucleoside monophosphates.

The above examples describe nonenzymatic synthesis of the complementary $(-)$ strand of oligonucleotides templated by the $(+)$ strand. For autocatalysis to occur, it is required that the two strands $((+)$ and $(-))$ separate and the $(-)$ strand templates the re-synthesis of the $(+)$ strand. The first example of such a truly self-replicating system was described by von Kiedrowski [92]. In this work, two trideoxynucleotides leading to a hexameric palindromic template were used, each trideoxynucleotide was 3′-protected to prevent elongation beyond a hexamer sequence. Initial coupling was carried out using a water-soluble carbodiimide (EDC) under conditions that led to the hexamer template rather than pyrophosphate dimer. Once formed, the product serves as template for further self-replication, and being palindromic, both $(-)$ and $(+)$ strands are formed in the same reaction (Fig. 13-7). As noted above, a possible product from the EDC-mediated coupling reaction is an oligodeoxynucleotide with an internal pyrophosphate linkage. Such

Figure 13-7 The first example of an autocatalytic system was described by Günter von Kiedrowski involved carbodiimide coupling of two trideoxynucleotides leading to a hexameric palindromic template. Once formed, the product serves as template for further self-replication, and being palindromic both $(-)$ and $(+)$ strands are formed in the same reaction.

modified oligodeoxynucleotides have been examined as substrates for self-replication and shown to still carry out sequence-dependent autocatalysis despite the phosphate modification [93].

Similar autocatalysis was observed for the synthesis of palindromic oligodeoxynucleotides using EDC-mediated formation of a $3'-5'$-phosphoramidate linkage between the two trimer building blocks [94,95]. The kinetics of self-replication has also been studied using fluorescently labeled tetramers by measurement of FRET [96]. Another autocatalytic system has been described by Nicolaou [97] for the synthesis of longer (24 mer) duplex palindromic polypurine/polypyrimidine DNA.

The early work by von Kiedrowski involved the replication of self-complementary sequences while natural replication involves the replication of complementary sequences. Using the previous system of chemical ligation of trimers, a minimal system for the synthesis of complementary replication has been described based on cross-catalytic template-directed synthesis using phosphoramidate linkages [98,99]. Two self-complementary and two complementary templates compete for four common trimeric precursors, and evidence was obtained to show that cross-catalytic self-replication of complementary sequences occurs with an equal efficiency to autocatalysis of the self-complementary sequence.

A common problem with replication by these systems is product inhibition, whereby the product dimer does not efficiently dissociate. As a result of this, there is parabolic rather than exponential amplification, and exponential amplification is a dynamic prerequisite for Darwinian selection. Using a system-denoted SPREAD (surface-promoted replication and exponential amplification of DNA analogues) exponential amplification was achieved by using a step to liberate the daughter strands from the template and cycling the amplification process [100]. More recent work by von Kiedrowski describes the self-assembly of three-dimensional DNA nanoscaffolds as a step toward artificially self-replicating systems on a nanometer scale [101,102] (see Section 13.10).

13.6 RNA SELF-REPLICATION: THE RNA WORLD

The emergence of a polymer (such as RNA) capable of self-replication, mutation, and hence evolution toward more efficient self-replication, represents an attractive and plausible concept for the origin of life. Several strands of evidence support the concept of such an "RNA" world, whereby RNA would serve as both genetic material as well as catalyst, preceding modern biology. These include aspects of modern metabolism (such as nucleotide cofactors, genetic control (self-splicing introns [103], riboswitches [104]), and most strikingly protein synthesis [105–107] that involve RNA and may thus represent relics from the "RNA world." The versatility of RNA to serve as both a receptor and catalyst has been further underlined by the wide range of activities documented in naturally occurring RNA receptors and ribozymes as well as in the ready evolution of novel activities using *in vitro* evolution methods like SELEX [108].

Despite its catalytic and conformational versatility RNA seems a somewhat perverse choice as the primordial genetic material, because it appears to be both difficult to synthesize and extremely unstable under presumed prebiotic conditions. This has led some to propose a "pre-RNA world," which utilized other polymers such as PNA (in which the ribofuranose-phosphate backbone is replaced by an achiral peptide backbone) or TNA (in which the ribose is replaced by a tetrofuranose), which were superseded by RNA at a later stage. Both PNA and TNA can form stable helices with RNA (and DNA) and interpolymer genetic information transfer should thus be possible. Indeed, it has been shown that information can be transferred non-enzymatically between PNA and DNA [109], DNA and PNA [110] and PNA to RNA [111]. Using "Therminator" polymerase, it has been shown that TNA strands up to 80-nucleotides long can be synthesized from a DNA template with good fidelity [46,112]. Orgel and coworkers have also examined other nucleic acid systems and found that nucleosides containing 1,5-anhydrohexital (HNA) can be used in place of ribose to carry out templated nonenzymatic replication [32,33]. The information transfer of HNA to RNA requires the formation of an A-form product and therefore information transfer to DNA is inefficient [34]. The templating of information with hexose sugars is even more efficient when the 1,5-anhydrohexital sugar is replaced by altritol (ANA, HNA that has an additional hydroxyl group) [35]. TNA appears the most attractive pre-RNA polymer as long PNA strands suffer from solubility problems due to the uncharged nature of the polypeptide backbone. Nevertheless, it remains to be seen if TNA displays similar versatility as a receptor and catalyst as RNA.

The case for RNA has recently been further strengthened by the discovery of long RNA polymers in eutectic ice phases [113,114], the stabilization of ribose by borate evaporates [115], the selective uptake of ribose (compared to other aldopentoses) by phospholipid and fatty acid vesicles [116] and the sequestration of enatiomerically pure D-ribose from a prebiotic mixture [117]. The latter is especially significant as the presence of small amounts of L-enantiomers of nucleosides effectively poison chain elongation in templated nonenzymatic RNA synthesis using the natural D-enantiomer [118].

13.6.1 The Search for an RNA Replicase

A cornerstone of the "RNA world" hypothesis is that there exists somewhere in sequence space a ribozyme replicase capable of self-replication. Indeed a number of naturally occurring as well as selected ribozymes display some ability for self-replication, most notably through assembly and enzymatic ligation of oligonucleotides [119]. Indeed, recently a self-replicating ligase ribozyme was described that directed its own assembly from constituent parts, and in an initial phase displayed true exponential growth [120]. This report demonstrates the potential of the approach toward a self-replicating system. However, because of the need to provide presynthesized oligonucleotide substrates and the need to retain substantial base-pairing with the ligase, the ability of such system to evolve is restricted. More complex, multicomponent self-ligation networks [95,121] may allow the inclusion of sufficient molecular diversity for some evolution to proceed.

A more general self-replication capability may be achieved by the use of shorter oligonucleotide substrates, ideally activated nucleotide precursors such as the nucleotide triphosphates (NTPs) utilized by modern polymerases. Intriguingly, both natural as well as evolved ribozymes have been shown to display weak primer extension ability using NTPs as substrates [122]. In ground-breaking work, Bartel and colleagues have evolved the primer extension capability of one such ribozyme, the R18 replicase, to the point where template-directed replication of up 14 nucleotides is possible [123]. As the R18 ribozyme is about 180 nucleotides long, an increase of processivity of a little more than one order of magnitude, should bring true self-replication within reach.

However, self-replication must proceed with a degree of fidelity, as defined by the "error threshold," above which genetic information encoded in the replicase would be irretrievably corrupted. An extensive theoretical framework on error threshold has been developed but it is unclear to what extent these can be applied to the practical case of an RNA replicase ribozyme. For example, the R18 replicase does appear to display fairly substantial template-dependent differences in processivity and fidelity [124,125], making it difficult to assign a meaningful overall mutation rate. While a recent study indicates that ribozymes in general may have an "relaxed error threshold" and thus be able to tolerate higher mutation rates than previously assumed [126], a number of *in vitro* evolution studies suggest that the class I ligase core (on which the R18 replicase is based) is rather resistant to mutation [108,127]. This may indicate that it represents a structure close to an evolutionary optimum, suggesting that at least half of the R18 replicase might be rather sensitive to poor fidelity in self-replication.

13.7 COMPARTMENTALIZATION: TOWARD THE DESIGN OF A SIMPLE CELL

For Darwinian evolution to proceed a putative replicase needs a form of "genetic packaging" such as confinement inside a compartment or at the very least spatial colocalization, for example, on the surface of mineral grains. Without such diffusion-limitation a replicase would fruitlessly replicate unrelated (and most likely inactive) sequences and eventually disappear from the sequence pool. Theoretical studies have also shown that limited diffusion aids replicase evolution by limiting the spread of replication parasites [128]. Physical proximity of a replicase to its "offspring" thus ensures both the growth and spread of the self-replicating entity as well as preventing takeover by fast-replicating "parasites."

13.7.1 Vesicles

Compartmentalization can potentially occur in many forms. An attractive format is vesicles comprising a bilayer of amphiphilic lipids. Such vesicles form spontaneously upon mixing of the constituent lipids with an aqueous solution. Some clay minerals, which promote the synthesis of polynucleotides from activated precursors, have been found to also catalyze the formation of vesicles. Szostak, Luisi, and colleagues [129,130] in particular have shown that vesicles comprising fatty acids as their main

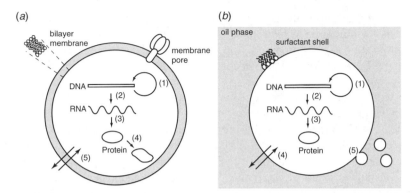

Figure 13-8 Compartmentalization in (a) vesicles and (b) water-in-oil emulsions. Both systems support DNA replication (1), transcription (2) and translation (3), which in the case of vesicles can be made to last for days using a membrane pore (hemolysin) to feed the reaction from surrounding medium. Bilayer membranes as well as W/O are partially permeable (5) to small molecules (e.g., H^+, Mg^{2+}, NTP). Specific molecules can be targeted to emulsion compartments using nanoemulsions (6).

constituents can display both autocatalytic growth as well as multiple cycles of growth and division. The fluid bilayer membrane also allows ready exchange of small molecules across membranes and this can drive competition for limited resources, as vesicles containing a larger amount of an osmotically-active compound (e.g., RNA) grow in size at the expense of others. Permeability is related *inter alia* to the length of the aliphatic chain and is thus in principle, controllable, with longer aliphatic chains leading to progressively less fluid, less permeable membranes [131]. Ribozyme activity [132], DNA as well as RNA replication [131,133], long-lasting transcription and translation [134] and even a two-stage genetic cascade have been demonstrated in vesicles (Fig. 13-8). Vesicles are therefore potentially attractive formats for a synthetic protocell.

From a synthetic biology perspective of engineering a suitable protocell, one of the problems that remain to be solved is that of cell reproduction. While vesicles made from hydrolyzable surfactants can be made to reproduce (whereby hydrolysis generates building blocks to form new vesicles) and vesicle fission and budding can be induced by application of physical and chemical forces [130], such replication is largely independent of vesicle content.

13.7.2 Emulsions

Although unlikely to have been relevant in prebiotic evolution, from a synthetic biology perspective an alternative format for a protocell may be based on emulsions. Emulsions are heterogeneous and, in general, metastable mixtures of two immiscible liquid phases with one of the phases dispersed in the other as droplets of microscopic size. Emulsions may be produced from any suitable combination of immiscible liquids by stirring, homogenization, or through microfluidic methods [135]. For the construction of a protocell so-called "water-in-oil" (W/O) emulsions are preferable, in which the disperse, internal phase forms a suspension of cell-like, aqueous "droplets" within an inert hydrophobic liquid matrix. Nevertheless, a O/W design for a protocell has been proposed [136], in which the genetic material (made from the neutral DNA analogue, PNA) is contained within lipid droplets suspended in an aqueous phase.

As with vesicles, cell-like aqueous compartments formed in W/O emulsions support various enzymatic reactions including coupled *in vitro* transcription and translation [137,138], as well as DNA replication and PCR [139] (Fig. 13-8). The size of aqueous compartments can readily be controlled by varying emulsion composition and mechanical energy input (between 70 nM [140] and 150 µM [141]). Just like vesicles, emulsions are also remarkably permeable to small molecules such as solvated ions and (at high temperatures) even nucleoside triphosphates [139]. Reagents can also be delivered to emulsion compartments in a controlled way using nanoemulsions [142]. However, even after prolonged exposure to high temperatures there appears to be little, if any, exchange of polypeptides or nucleic acids (>30 bp) between compartments [139].

13.7.3 Compartmentalized Evolution

In vitro compartmentalization (IVC) in emulsions allows a stable linkage of genotype and phenotype [137] and this has been exploited for *in vitro* evolution. IVC has allowed the evolution of DNA methylases with altered substrate specificity [143], a super-fast phosphotriesterase [144] as well as novel ribozymes [127,140]. Emulsions can also be used to segregate self-replication reactions. Compartmentalized self-replication (CSR) exploits this for the directed evolution of polymerases [139]. In CSR, polymerases catalyze the replication of their own encoding gene. As a result, adaptive gains by the polymerases translate directly into more genetic "offspring" (i.e., more efficient self-replication). Due to this positive feedback loop, the genes encoding polymerases that are well adapted to the selection conditions (and therefore capable of efficient self-replication) will increase in copy number while genes encoding poorly adapted polymerases will disappear from the gene pool.

CSR has allowed the directed evolution of polymerases with increased thermostability, inhibitor tolerance or a generically expanded substrate spectrum [139,145]. CSR may also be regarded as a simple test bed for self-replication. For example, a classic outcome of *in vitro* replication experiments is an adaptation of the template sequence toward more rapid replication [146]. This typically takes the form of truncation as well as mutation and (in solution) invariably gives rise to (often heavily truncated) "replication parasites," which have lost much of the genetic information encoding the original phenotype but are optimized for replication speed. Such parasites arise frequently, for example in PCR amplifications (primer dimers) or *in vitro* evolution experiments [147]. While template evolution appears to occur in CSR through silent mutations reducing GC content facilitating strand separation and destabilizing secondary structures [139]. However, template truncation was not observed (despite the considerable size of the *Taq* gene (2.5 kb)). Presumably, this is due to both the strong phenotypic selection in CSR as well as the effect of compartmentalization, which limits the spread of parasites to the compartment, where they occur (e.g., see Ref. 128). For a self-replicating RNA replicase, template evolution (e.g., through mutations that destabilize secondary structures) may be a mixed blessing and requires a trade-off between structural stability of the replicase structure itself and its replicability.

A wide range of other forms of compartmentalization (or diffusion limitation) are imaginable that may have played a role in prebiotic replication. These include the surface of fine particulate matter or in porous minerals (e.g., clays such as montmorillonite; see Section 13.5.2), eutectic ice phases [113], or aerosol droplets in the atmosphere [148].

13.8 *IN VIVO* REPLICATION

All of the examples discussed so far involve *ex vivo* designs of self-replicating entities. However, for various applications it may be desirable to consider invading present-day biological systems with self-replicating species. These already exist of course in biology in the form of plasmids, viruses, and so on; however, these interact extensively with the host organism. One concern when building synthetic devices is both their potentially toxic effect on host biology, as well as the possible interference of host cellular functions with the operation of the device.

In order to escape such interference one may ask, if it would be possible to build synthetic self-replicating circuits that are capable of operating independently from the rest of the cell. Such orthogonal episomes would carry their own polymerases for specific replication and transcription "on board" (in analogy to many viruses) but would still be subject to recombination, mutation, and degradation by the host genetic machinery. It might therefore be advantageous to consider synthetic episomes that are orthogonal in composition as well as replication. Such episomes would comprise unnatural nucleic acids (XNA) and thus would be isolated from the host genetic machinery by chemical, steric, or semantic differences.

The requirements for such a system parallel in many ways those for the polymers of a "pre-RNA" world, requiring the ability of cross-talk and mutual interconversion (transliteration) between "XNA" and DNA/RNA (Fig. 13-9). A design for an artificial genetic system might therefore be preferably based on an alternative backbone structure. This has the potential benefits of providing orthogonality, that is, synthetic and functional isolation within the cell (as altered backbone chemistry precludes utilization by the cellular genetic machinery) without altering the coding potential of the nucleic acids. In other words, a genetic entity constructed this way may be built from precursors that are sufficiently different from the natural nucleosides that they cannot be utilized by the pre-existing genetic machinery of the cell (replication/transcription/translation) and therefore do not give rise to toxicity while at the very same time are able to communicate with it. Interestingly, just such a scenario has recently been put forward for the origin of DNA. It proposes that such an orthogonal nucleic acid with a modified backbone (DNA) was "invented" by viruses infecting riboorganisms of the RNA world in order to avoid cellular defenses (e.g., RNAses) [149].

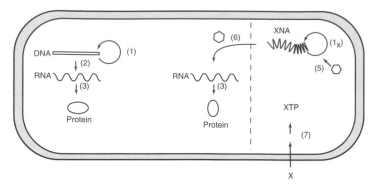

Figure 13-9 Invasion of the natural genetic system by an orthogonal episome made from unnatural nucleic acids (XNA). Because of its orthogonal design it is impervious to cellular degradation, recombination, or mutation mechanisms. A fully functional orthogonal episome must be capable of replication (1) through a dedicated polymerase (5) and transcription into RNA (or DNA) through a second dedicated polymerase (6). Thus, a specific set of proteins (or RNAs) could be encoded on the orthogonal episome to alter the cellular phenotype. Maintenance of the orthogonal episome will also require the engineering of efficient uptake (and possibly phosphorylation) pathways (7) for XNA precursors from the medium.

An alternative strategy, for which there are no known precedents in nature, would be steric orthogonality, which may be based around the expanded DNA (xDNA, yDNA) described by the Kool laboratory at Stanford (see Section 13.4.3). Because the base-pairing is not altered, x, y DNA are still capable of base-pairing with DNA and RNA but forming double helices with an expanded diameter and higher stability due to the increased stacking of the expanded bases [62,66]. Finally, orthogonality might simply be semantic in that genetic information encoded in such a way that it is "meaningless" to the cellular host, unless specific transliterases are provided.

A potentially important advantage of such systems could be safety. As these transgenes would be based around nucleic acid chemistry not present in nature its function and transmission will be entirely dependent and controlled by the supply of orthogonal precursors. Among other things, this will provide a novel and complete control of genetic safety issues as the propagation and inheritance of "foreign" genetic material in a transgenic organism can be simply turned off and the transgene excised by removing the supply of precursors.

13.9 MOLECULAR DEVICES AND AUTOMATA

Unlike any other molecule DNA affords ready control over intermolecular associations. DNA molecules associate according to well-understood rules of complementarity providing a diverse and programmable system, with known structures and a high degree of control over molecular interactions. DNA is thus increasingly recognized as a material of choice for self-assembling "bottom-up" nanostructures and nanodevices [150].

For example, using stable branched structure in conjunction with "sticky end" cohesion has allowed the generation of, for example, DNA cubes and octahedrons as well as two-dimensional DNA arrays some of which can be used for computation [150]. However, the conformational flexibility of DNA also lends itself to the construction of multistate devices with a flexible response to the input conditions. For example, Seeman and colleagues have constructed a nanomechanical device, which exploits the structural transition between the canonical right-handed B-DNA and left-handed Z-DNA in $[CG]_n$-rich sequences in response to high salt concentrations [151]. Thus, the device translates on input signal (ionic strength) into an observable fluorescence signal through fluorescence resonance energy transfer (FRET) differences of the two states. Other devices include DNA tweezers [152], rotary motors [153], walkers [154], and even a translation machine [155].

A different class of DNA devices has been built around cycles of ligation and cleavage with a type II restriction enzyme. Shapiro, Benenson and colleagues showed that this allowed the construction of DNA-based finite-state automata capable of autonomous computation at the molecular level [156]. One such automaton was shown to be able to analyze *in vitro* the levels of several RNAs involved in prostate cancer and compute an appropriate response, that is, release of an antisense molecule [157] offering the prospect of programmable, logical control of biological processes at the molecular scale.

These devices are assembled and sometimes powered by oligonucleotide fragments and thus their topologies are not amenable to replication. However, different topological designs are possible. In a striking example, Joyce and colleagues recently described a single-stranded 1.7-kb DNA sequence that folds into a octahedron in the presence of short DNA oligonucleotides [158]. In conjunction with engineered polymerases [159] this offers the future prospect of replicable nanostructures endowed with expanded chemical capabilities and amenable to iterative cycles of replication, mutation and selection, bringing directed evolution to nanotechnology and material science.

13.10 CONCLUSION

For synthetic biology, self-replication should be a long-term goal for the engineering of material devices. While the construction and implementation of circuits and devices *in vivo* (i.e., in extant biological systems) provides for self-replication as part of the reproduction of the organism, self-replication *ex vivo* or as part of a whole synthetic quasibiotic entity (e.g., a synthetic cell) may have a number of long-term advantages. For one, it may remove some of the unpredictability and instability that can be a consequence of integrating new functionalities into the cellular network [160]. Furthermore, the design and fabrication of synthetic conduits for self-replication promises significant insights and advances in understanding of the transition from prebiotic to biotic matter and the early evolution of life. Finally, self-replication as applied to the emergent technologies such as DNA nanotechnology and molecular automata [161] promises decisive reductions in manufacturing costs as well as bringing the potential of Darwinian evolution to nanosensors and molecular devices.

ACKNOWLEDGMENTS

We would like to thank our colleagues Dan Brown and Jason Chin for comments on the manuscript.

REFERENCES

1. Paul N, Joyce GF. Minimal self-replicating systems. *Curr Opin Chem Biol* 2004;8: 634–639.

2. Szathmary E. The evolution of replicators. *Philos Trans R Soc Lond B Biol Sci* 2000;355: 1669–1676.

3. Zykov V, Mytilinaios E, Adams B, Lipson H. Robotics: self-reproducing machines. *Nature* 2005;435:163–164.

4. Tjivikua T, Ballester P, Rebek J. A self-replicating system. *J Am Chem Soc* 1990;112: 1249–1250.

5. Nowick JS, Feng Q, Tjivikua T, Ballester P, Rebek J. Kinetic studies and modelling of a self-replicating system. *J Am Chem Soc* 1991;113:8831–8839.

6. Wintner EA, Tsao B, Rebek J. Evidence against an alternative mechanism for a self-replicating system. *J Org Chem* 1995;60:7997–8001.

7. Conn MM, Wintner EA, Rebek J. Template effects in new self-replicating molecules. *Angew Chem Int Ed* 1994;33:1577–1579.

8. Pieters RJ, Huc I, Rebek J. Reciprocal template effects in a replication cycle. *Angew Chem Int Ed* 1994;33:1579–1581.

9. Kindermann M, Stahl I, Reimold M, Pankau WM, von Kiedrowski G. Systems chemistry: kinetic and computational analysis of a nearly exponential organic replicator. *Angew Chem Int Ed* 2005;55:6750–6755.

10. Pearson RJ, Kassianidis E, Slawin AMZ, Philp D. Self-replication vs. reactive binary complexes—manipulating recognition-mediated cycloadditions by simple structural modifications. *Org Biomol Chem* 2004;2:3434–3441.

11. Kassianidis E, Pearson RJ, Philp D. Specific autocatalysis in diastereoisomeric replicators. *Org Lett* 2005;7:3833–3836.

12. Lee DH, Granja JR, Martinez JA, Severin K, Ghadiri MR. A self-replicating peptide. *Nature* 1996;382:525–528.

13. Lee DH, Severin K, Yokobayashi Y, Ghadiri MR. Emergence of symbiosis in peptide self-replication through a hypercyclic network. *Nature* 1997;390:591–594.

14. Severin K, Lee DH, Martinez JA, Vieth M, Ghadiri MR. Dynamic error correction in autocatalytic peptide networks. *Angew Chem Int Ed* 1998;37:126–128.

15. Saghatelian A, Yokobayashi Y, Soltani K, Ghadiri MR. A chiroselective peptide replicator. *Nature* 2001;409:797–801.

16. Islas JR, Pimienta V, Micheau J-C, Buhse T. Kinetic analysis of artificial peptide self-replication. Part I: The homochiral case. *Biophys Chem* 2003;103:191–200.

17. Islas JR, Pimienta V, Micheau J-C, Buhse T. Kinetic analysis of artificial peptide self-replication. Part II: The heterochiral case. *Biophys Chem* 2003;103:201–211.

18. Islas JR, Micheau J-C, Buhse T. Kinetic analysis of self-replicating peptides: possibility of chiral amplification in open systems. *Orig Life Evol Biosph* 2004;34:497–512.

19. Yao S, Ghosh I, Zutshi R, Chmielewski J. A pH modulated, self-replicating peptide. *J Am Chem Soc* 1997;119:10559–10560.

20. Li X, Chmielewski J. Peptide self-replication enhanced by a proline kink. *J Am Chem Soc* 2003;125:11820–11821.

21. Matsumura S, Takahashi T, Ueno A, Mihara H. Complementary nucleobase interaction enhances peptide-peptide and self-replicating catalysis. *Chem Eur J* 2003;9: 4829–4837.

22. Levy M, Ellington AD. Peptide-templated nucleic acid ligation. *J Mol Evol* 2003;56: 607–615.

23. Ghadiri MR, Granja JR, Milligan RA, McRee DE, Khazanovich N. Self-assembling organic nanotubes based on a cyclic peptide architecture. *Nature* 1993;366:324–327.

24. Kobayashi K, Granja JR, Ghadiri MR. Beta-sheet peptide architecture: measuring the relative stability of parallel vs antiparallel beta-sheets. *Angew Chem Int Ed* 1995;34:95–98.

25. Peretz D, Williamson RA, Legname G, Matsunaga Y, Vergara J, Burton DR, DeArmond SJ, Prusiner SB, Scott MR. A change in the conformation of prions accompanies the emergence of a new prion strain. *Neuron* 2002;34:921–932.

26. Shorter J, Lindquist S. Prions as adaptive conduits of memory and inheritance. *Nat Rev Genet* 2005;6:435–450.

27. Watson JD, Crick FHC. Molecular structure of nucleic acids: a structure for deoxyribose nucleic acid. *Nature* 1953;171:737–738.

28. Benner SA. Chemistry. Redesigning genetics. *Science* 2004;306:625–626.

29. Nielsen PE. The many faces of PNA. *Lett Peptide Sci* 2003;10:135–147.

30. Stirchak EP, Summerton JE, Weller DD. Uncharged stereoregular nucleic acid analogs: 2. Morpholino nucleoside oligomers with carbamate internucleoside linkages. *Nucleic Acids Res* 1989;17:6129–6141.

31. Summerton J. Morpholino antisense oligomers: the case for an RNase H-independent structural type. *Biochem Biophys Acta* 1999;1489:141–158.

32. Kozlov IA, Politis PK, Pitsch S, Herdewijn P, Orgel LE. A highly enantio-selective hexitol nucleic acid template for nonenzymatic oligoguanylate synthesis. *J Am Chem Soc* 1999;121:1108–1109.

33. Kozlov IA, De Bouvere B, Van Aerschot A, Herdewijn P, Orgel LE. Efficient transfer of information from hexitol nucleic acids to RNA during nonenzymatic oligomerization. *J Am Chem Soc* 1999;121:5856–5859.

34. Kozlov IA, Politis PK, Van Aerschot A, Busson R, Herdewijn P, Orgel LE. Nonenzymatic synthesis of RNA and DNA oligomers on hexitol nucleic acid templates: the importance of the A structure. *J Am Chem Soc* 1999;121:2653–2656.

35. Kozlov IA, Zielinski M, Allart B, Kerremans L, Van Aerschot A, Busson R, Herdewijn P, Orgel LE. Nonenzymatic template-directed reactions on altritol oligomers, preorganised analogues of oligonucleotides. *Chem Eur J* 2000;6:151–155.

36. Eschenmoser A. Chemical etiology of nucleic acid structure. *Science* 1999;284:2118–2124.

37. Maurinsh Y, Rosemeyer H, Esnouf R, Medvedovici A, Wang J, Ceulemans G, Lescrinier E, Hendrix C, Busson R, Sandra P, Seela F, Van Aerschot A, Herdewijn P. Synthesis and pairing properties of oligonucleotides containing 3-hydroxy-4-hydroxymethyl-1-cyclo-hexanyl nucleosides. *Chem Eur J* 1999;5:2139–2150.

38. Lescrinier E, Froeyen M, Herdewijn P. Difference in conformational diversity between nucleic acids with a six-membered 'sugar' unit and natural 'furnaose' nucleic acids. *Nucleic Acids Res* 2003;31:2975–2989.

39. Nauwelaerts K, Lescrinier E, Scelp G, Herdewijn P. Cyclohexenyl nucleic acids: conformationally flexible oligonucleotides. *Nucleic Acids Res* 2005;33:2452–2463.

40. Wang J, Verbeure B, Luyten I, Lescrinier E, Froeyen M, Hendrix C, Rosemeyer H, Seela F, Van Aerschot A, Herdewijn P. Cyclohexen nucleic acids (CeNA): serum stable oligonucleotides that activate RNase H and increase duplex stability with complementary RNA. *J Am Chem Soc* 2000;122:8595–8602.

41. Eschenmoser A. The TNA-family of nucleic acid systems: Properties and prospects. *Orig Life Evol Biosph* 2004;34:277–306.

42. Kempeneers V, Froeyen M, Vastmans K, Herdewijn P. Influence of threose nucleoside units on the catalytic activity of a hammerhead ribozyme. *Chem Biodivers* 2004;1: 112–123.

43. Kempeneers V, Vastmans K, Rozenski J, Herdewijn P. Recognition of threosyl nucleostides by DNA and RNA polymerases. *Nucleic Acid Res* 2003;31:6221–6226.

44. Ichida JK, Zou K, Horhota A, Yu B, McLaughlin LW, Szostak JW. An *in vitro* selection system for TNA. *J Am Chem Soc* 2005;127:2902–2903.

45. Horhota A, Zou K, Ichida JK, Yu B, McLaughlin LW, Szostak JW, Chaput JC. Kinetic analysis of an efficient DNA-dependent TNA polymerase. *J Am Chem Soc* 2005;127: 7427–7434.

46. Ichida JK, Horhota A, Zou K, McLaughlin LW, Szostak JW. High fidelity TNA synthesis by Therminator polymerase. *Nucleic Acids Res* 2005;33:5219–5225.

47. Piccirilli JA, Krauch T, Moroney SE, Benner SA. Enzymatic incorporation of a new base pair into DNA and RNA extends the genetic alphabet. *Nature* 1990;343:33–37.

48. Benner SA. Understanding nucleic acids using synthetic chemistry. *Acc Chem Res* 2004;37:784–797.

49. Yu C, Henry AA, Romesberg FE, Schultz PG. Polymerase recognition of unnatural base pairs. *Angew Chem Int Ed* 2002;41:3841–3844.

50. Atwell S, Meggers E, Spraggon G, Schultz PG. Structure of a copper-mediated base pair in DNA. *J Am Chem Soc* 2001;123:12364–12367.

51. Liu H, Gao J, Lynch SR, Saito YD, Maynard L, Kool ET. A four-base paired genetic helix with expanded size. *Science* 2003;302:868–871.

52. Ohtsuki T, Kimoto M, Ishikawa M, Mitsui T, Hirao I, Yokoyama S. Unnatural base pairs for specific transcription. *Proc Natl Acad Sci USA* 2001;98:4922–4925.

53. Switzer C, Moroney SE, Benner SA. Enzymatic incorporation of a new base pair into DNA and RNA. *J Am Chem Soc* 1989;111:8322–8323.

54. Johnson SC, Marshall DJ, Harms G, Miller CM, Sherrill CB, Beaty EL, Lederer SA, Roesch EB, Madsen G, Hoffman GL, Laessig RH, Kopish GJ, Baker MW, Benner SA, Farrell PM, Prudent JR. Multiplexed genetic analysis using an expanded genetic alphabet. *Clin Chem* 2004;50:2019–2027.

55. Sismour AM, Benner SA. The use of thymidine analogs to improve the replication of an extra DNA base pair: a synthetic biological system. *Nucleic Acids Res* 2005;33:5640–5646.

56. Seela F, Wei C. The base-pairing properties of 7-deaza-2'-deoxyisoguanosine and 2'-deoxyisoguanosine in oligonucleotide duplexes with parallel and antiparallel chain orientation. *Helv Chim Acta* 1999;82:726–745.

57. Benner SA. Redesigning genetics. *Science* 2004;306:625–626.

58. Geyer CR, Battersby TR, Benner SA. Nucleobase pairing in expanded Watson–Crick-like genetic information systems. *Structure* 2003;11:1485–1498.

59. von Krosigk U, Benner SA. Expanding the genetic alphabet: pyrazine nucleosides that support a donor-donor-acceptor hydrogen-bonding pattern. *Helv Chim Acta* 2004;87: 1299–1324.

60. Sismour AM, Lutz S, Park J-H, Lutz MJ, Boyer PL, Hughes SH, Benner SA. PCR amplification of DNA containing non-standard base pairs by variants of reverse transcriptase from human immunodeficiency virus-1. *Nucleic Acids Res* 2004;32:728–735.

61. Liu H, Gao J, Maynard L, Saito YD, Kool ET. Toward a new genetic system with expanded dimensions: size-expanded analogues of deoxyadenosine and thymidine. *J Am Chem Soc* 2004;126:1102–1109.

62. Lu H, He K, Kool ET. yDNA: a new geometry for size-expanded base pairs. *Angew Chem Int Ed* 2004;43:5834–5836.

63. Lee AH, Kool ET. A new four-base genetic helix, yDNA, composed of widened benzopyrimidine-purine pairs. *J Am Chem Soc* 2005;127:3332–3338.

64. Liu H, Lynch SR, Kool ET. Solution structure of xDNA: a paired genetic helix with increased diameter. *J Am Chem Soc* 2004;126:6900–6905.

65. Gao J, Liu H, Kool ET. Expanded-size bases in naturally sized DNA: evaluation of steric effects in Watson–Crick pairing. *J Am Chem Soc* 2004;126:11826–11831.

66. Gao J, Liu H, Kool ET. Assembly of the complete eight-base artificial genetic helix, xDNA, and its interaction with the natural genetic system. *Angew Chem Int Ed* 2005;44: 3118–3122.

67. Liu H, Gao J, Kool ET. Helix-forming properties of size-expanded DNA, an alternative four-base genetic form. *J Am Chem Soc* 2005;127:1396–1402.

68. Henry AA, Olsen AG, Matsuda S, Yu C, Geierstanger BH, Romesberg FE. Efforts to expand the genetic alphabet: identification of a replicable unnatural DNA self-pair. *J Am Chem Soc* 2004;126:6923–6931.

69. Matsuda S, Henry AA, Schultz PG, Romesberg FE. The effects of minor-groove hydrogen-bond acceptors and donors on the stability and replication of four unnatural base pairs. *J Am Chem Soc* 2003;125:6134–6139.

70. Matsuda S, Romesberg FE. Optimization of interstrand hydrophobic packing interactions within unnatural DNA base pairs. *J Am Chem Soc* 2004;126:14419–14427.

71. Henry AA, Yu C, Romesberg FE. Determinants of unnatural nucleobase stability and polymerase recognition. *J Am Chem Soc* 2003;125:9638–9646.

72. Ishikawa M, Hirai I, Yokoyama S. Synthesis of 3-(2-deoxy-β-D-ribofuranosyl)pyridin-2-one and 2-amino-6-(*N,N*-dimethylamino)-9-(2-deoxy-β-D-ribofuranosyl)purine derivatives for an unnatural base pair. *Tetrahedron Lett* 2000;41:3931–3934.

73. Fujiwara T, Kimoto M, Sugiyama H, Hirao I, Yokoyama S. Synthesis of 6-(2-thienyl) purine nucleoside derivatives that form unnatural base pairs with pyridin-2-one nucleosides. *Bioorg Med Chem Lett* 2001;11:2221–2223.

74. Hirao I, Fujiwara T, Kimoto M, Yokoyama S. Unnatural base pairs between 2- and 6-substituted purines and 2-oxo(1*H*)pyridine for expansion of the genetic code. *Bioorg Med Chem Lett* 2004;14:4887–4890.

75. Kimoto M, Yokoyama S, Hirao I. A quantitative, non-radioactive single-nucleotide insertion assay for analysis of DNA replication fidelity by using an automated DNA sequencer. *Biotechnol Lett* 2004;26:999–1005.

76. Hirao I, Harada Y, Kimoto M, Mitsui T, Fujiwara T, Yokoyama S. A two-unnatural-base-pair system toward the expansion of the genetic code. *J Am Chem Soc* 2004;126: 13298–13305.

77. Endo M, Mitsui T, Okuni T, Kimoto M, Hirao I, Yokoyama S. Unnatural base pairs mediate the site-specific incorporation of an unnatural hydrophobic component into RNA transcripts. *Bioorg Med Chem Lett* 2004;14:2593–2596.

78. Kimoto M, Endo M, Mitsui T, Okuni T, Hirao I, Yokoyama S. Site-specific incorporation of a photo-crosslinking component into RNA by T7 transcription mediated by unnatural base pairs. *Chem Biol* 2004;11:47–55.

79. Kawai R, Kimoto M, Ikeda S, Mitsui T, Endo M, Yokoyama S, Hirao I. Site-specific fluorescent labeling of RNA molecules by specific transcription using unnatural base pairs. *J Am Chem Soc* 2005;127:17286–17295.

80. Hirao I, Ohtsuki T, Fujiwara T, Mitsui T, Yokogawa T, Okuni T, Nakayama H, Takio K, Yabuki T, Kigawa T, Kodama K, Yokogawa T, Nishikawa K, Yokoyama S. An unnatural base pair for incorporating amino acid analogs into proteins. *Nature Biotechnol* 2002;20: 177–182.

81. Ferris JP, Ertem G. Montmorillonite catalysis of RNA oligomer formation in aqueous solution. A model for the prebiotic formation of RNA. *J Am Chem Soc* 1993;115: 12270–12275.

82. Kawamura K, Ferris JP. Kinetic and mechanistic analysis of dinucleotide and oligonucleotide formation from the 5′-phosphorimidazolide of adenosine on Na$^+$-montmorillonite. *J Am Chem Soc* 1994;116:7564–7572.

83. Prabahar KJ, Ferris JP. Adenine derivatives as phosphate-activating groups for the regioselective formation of 3′,5′-linked oligoadenylates on montmorillonite: possible phosphate-activating groups for the prebiotic synthesis of RNA. *J Am Chem Soc* 1997;119:4330–4337.

84. Huang W, Ferris JP. Synthesis of 35-40 mers of RNA oligomers from unblocked monomers. A simple approach to the RNA world. *Chem Commun* 20031458–1459.

85. Gao K, Orgel LE. Phosphorylation and non-enzymatic template-directed ligation of oligonucleotides. *Orig Life Evol Biosph* 2000;30:45–51.

86. Joyce GF, Inoue T, Orgel LE. Non-enzymatic template-directed synthesis on RNA random copolymers. *J Mol Biol* 1984;176:279–306.

87. Usher DA. RNA double helix and the evolution of the 3′,5′ linkage. *Nat New Biol* 1972;235: 207–208.

88. Lohrmann R, Orgel LE. Preferential formation of (2′-5′)-linked internucleotide bonds in non-enzymatic reactions. *Tetrahedron* 1978;34:853–855.

89. Zielinski M, Kozlov IA, Orgel LE. A comparison of RNA with DNA in template-directed synthesis. *Helv Chim Acta* 2000;83:1678–1684.

90. Rohatgi R, Bartel DP, Szostak JW. Kinetic and mechanistic analysis of non-enzymatic, template-directed oligoribonucleotide ligation. *J Am Chem Soc* 1996;118: 3332–3339.

91. Rohatgi R, Bartel DP, Szostak JW. Nonenzymatic, template-directed ligation of oligoribonucleotides is highly selective for the formation of 3′-5′ phosphodiester bonds. *J Am Chem Soc* 1996;118:3340–3344.

92. von Kiedrowski G. A self-replicating hexadeoxynucleotide. *Angew Chem Int Ed* 1986;25: 932–935.

93. von Kiedrowski G, Wlotzka B, Helbing J. Sequence dependence of template-directed synthesis of hexadeoxynucleotide derivatives with 3,-5' pyrophosphate linkage. *Angew Chem Int Ed* 1989;28:1235–1237.

94. von Kiedrowski G, Wlotzka B, Helbing J, Matzen M, Jordan S. Parabolic growth of a self-replicating hexadeoxynucleotide bearing a 3'-5'-phosphoramidate linkage. *Angew Chem Int Ed* 1991;30:423–426.

95. Achilles T, von Kiedrowski G. A self-replicating system from three starting materials. *Angew Chem Int Ed* 1993;32:1198–1201.

96. Schöneborn H, Bülle J, von Kiedrowski G. Kinetic monitoring of self-replicating systems through measurement of Fluorescence resonance energy transfer. *ChemBioChem* 2001;922–927.

97. Li T, Nicolaou KC. Chemical self-replication of palindromic duplex DNA. *Nature* 1994;369:218–221.

98. Sievers D, von Kiedrowski G. Self-replication of complementary nucleotide-based oligomers. *Nature* 1994;369:221–224.

99. Sievers D, von Kiedrowski G. Self-replication of hexadeoxynucleotide analogues: autocatalysis versus cross-catalysis. *Chem Eur J* 1998;4:629–641.

100. Luther A, Brandsch R, von Kiedrowski G. Surface-promoted replication and exponential amplification of DNA analogues. *Nature* 1998;396:245–248.

101. Eckardt LH, Naumann K, Pankau WM, Rein M, Schweitzer M, Windhab N, von Kiedrowski G. Chemical copying of connectivity. *Nature* 2002;420:286.

102. Scheffler M, Dorenbeck A, Jordan S, Wüstefeld M, von Kiedrowski G. Self-assembly of trisoligonucleotidyls: the case for nano-acetylene and nano-cyclobutadiene. *Angew Chem Int Ed* 1999;38:3312–3315.

103. Haugen P, Simon DM, Bhattacharya D. The natural history of group I introns. *Trends Genet* 2005;21:111–119.

104. Tucker BJ, Breaker RR. Riboswitches as versatile gene control elements. *Curr Opin Struct Biol* 2005;15:342–348.

105. Noller HF, Hoffarth V, Zimniak L. Unusual resistance of peptidyl transferase to protein extraction procedures. *Science* 1992;256:1416–1419.

106. Nissen P, Hansen J, Ban N, Moore PB, Steitz TA. The structural basis of ribosome activity in peptide bond synthesis. *Science* 2000;289:920–930.

107. Ban N, Nissen P, Hansen J, Moore PB, Steitz TA. The complete atomic structure of the large ribosomal subunit at 2.4 A resolution. *Science* 2000;289:905–920.

108. Joyce GF. Directed evolution of nucleic acid enzymes. *Annu Rev Biochem* 2004;73:791–836.

109. Böhler C, Nielsen PE, Orgel LE. Template switching between PNA and RNA oligonucleotides. *Nature* 1995;376:578–581.

110. Schmidt JG, Christensen L, Nielsen PE, Orgel LE. Information transfer from DNA to peptide nucleic acids by template-directed syntheses. *Nucleic Acids Res* 1997;25:4792–4796.

111. Schmidt JG, Nielsen PE, Orgel LE. Information transfer from peptide nucleic acids to RNA by template-directed synthesis. *Nucleic Acids Res* 1997;25:4797–4802.

112. Chaput JC, Szostak JW. TNA synthesis by DNA polymerases. *J Am Chem Soc* 2003;125:9274–9275.

113. Monnard PA, Kanavarioti A, Deamer DW. Eutectic phase polymerization of activated ribonucleotide mixtures yields quasi-equimolar incorporation of purine and pyrimidine nucleobases. *J Am Chem Soc* 2003;125:13734–13740.

114. Trinks H, Schroder W, Biebricher CK. Ice and the origin of life. *Orig Life Evol Biosph* 2005;35:429–445.

115. Ricardo A, Carrigan MA, Olcott AN, Benner SA. Borate minerals stabilize ribose. *Science* 2004;303:196.

116. Sacerdote MG, Szostak JW. Semipermeable lipid bilayers exhibit diastereoselectivity favoring ribose. *Proc Natl Acad Sci USA* 2005;102:6004–6008.

117. Springsteen G, Joyce GF. Selective derivatization and sequestration of ribose from a prebiotic mix. *J Am Chem Soc* 2004;126:9578–9583.

118. Kozlov IA, Pitsch S, Orgel LE. Oligomerization of activated D- and L-guanosine mononucleotides on templates containing D- and L-deoxycytidylate residues. *Proc Natl Acad Sci USA* 1998;95:13448–13452.

119. Doudna JA, Couture S, Szostak JW. A multisubunit ribozyme that is a catalyst of and template for complementary strand RNA synthesis. *Science* 1991;251: 1605–1608.

120. Paul N, Joyce GF. A self-replicating ligase ribozyme. *Proc Natl Acad Sci USA* 2002;99: 12733–12740.

121. Kim D-E, Joyce GF. Cross-catalytic replication of an RNA ligase ribozyme. *Chem Biol* 2004;11:1505–1512.

122. McGinness KE, Joyce GF. In search of an RNA replicase ribozyme. *Chem Biol* 2003;10:5–14.

123. Johnston WK, Unrau PJ, Lawrence MS, Glasner ME, Bartel DP. RNA-catalyzed RNA polymerization: accurate and general RNA-templated primer extension. *Science* 2001;292:1319–1325.

124. Lawrence MS, Bartel DP. Processivity of ribozyme-catalyzed RNA polymerization. *Biochemistry* 2003;42:8748–8755.

125. Lawrence MS, Bartel DP. New ligase-derived RNA polymerase ribozymes. *RNA* 2005;11:1173–1180.

126. Kun A, Santos M, Szathmary E. Real ribozymes suggest a relaxed error threshold. *Nat Genet* 2005;37:1008–1011.

127. Levy M, Griswold KE, Ellington AD. Direct selection of trans-acting ligase ribozymes by *in vitro* compartmentalization. *RNA* 2005;11:1555–1562.

128. Szabo P, Scheuring I, Czaran T, Szathmary E. *In silico* simulations reveal that replicators with limited dispersal evolve towards higher efficiency and fidelity. *Nature* 2002;420: 340–343.

129. Szostak JW, Bartel DP, Luisi PL. Synthesizing life. *Nature* 2001;409:387–390.

130. Hanczyc MM, Szostak JW. Replicating vesicles as models of primitive cell growth and division. *Curr Opin Chem Biol* 2004;8:660–664.

131. Chakrabarti AC, Breaker RR, Joyce GF, Deamer DW. Production of RNA by a polymerase protein encapsulated within phospholipid vesicles. *J Mol Evol* 1994;39: 555–559.

132. Chen IA, Salehi-Ashtiani K, Szostak JW. RNA catalysis in model protocell vesicles. *J Am Chem Soc* 2005;127:13213–13219.

133. Oberholzer T, Wick R, Luisi PL, Biebricher CK. Enzymatic RNA replication in self-reproducing vesicles: an approach to a minimal cell. *Biochemical and biophysical research communications* 1995;207:250–257.

134. Noireaux V, Libchaber A. A vesicle bioreactor as a step toward an artificial cell assembly. *Proc Natl Acad Sci USA* 2004;101:17669–17674.

135. Atencia J, Beebe DJ. Controlled microfluidic interfaces. *Nature* 2005;437:648–655.

136. Rasmussen S, Chen L, Nilsson M, Abe S. Bridging nonliving and living matter. *Artif Life* 2003;9:269–316.

137. Tawfik DS, Griffiths AD. Man-made cell-like compartments for molecular evolution. *Nat Biotechnol* 1998;16:652–656.

138. Ghadessy FJ, Holliger P. A novel emulsion mixture for *in vitro* compartmentalization of transcription and translation in the rabbit reticulocyte system. *Protein Eng Des Sel* 2004;17:201–204.

139. Ghadessy FJ, Ong JL, Holliger P. Directed evolution of polymerase function by compartmentalized self-replication. *Proc Natl Acad Sci USA* 2001;98:4552–4557.

140. Agresti JJ, Kelly BT, Jaschke A, Griffiths AD. Selection of ribozymes that catalyse multiple-turnover Diels-Alder cycloadditions by using *in vitro* compartmentalization. *Proc Natl Acad Sci USA* 2005;102:16170–16175.

141. Margulies M, Egholm M, Altman WE, Attiya S, Bader JS, Bemben LA, Berka J, Braverman MS, Chen YJ, Chen Z, Dewell SB, Du L, Fierro JM, Gomes XV, Godwin BC, He W, Helgesen S, Ho CH, Irzyk GP, Jando SC, Alenquer ML, Jarvie TP, Jirage KB, Kim JB, Knight JR, Lanza JR, Leamon JH, Lefkowitz SM, Lei M,   Li J, Lohman KL, Lu H, Makhijani VB, McDade KE, McKenna MP, Myers EW, Nickerson E, Nobile JR, Plant R, Puc BP, Ronan MT, Roth GT, Sarkis GJ, Simons JF, Simpson JW, Srinivasan M, Tartaro KR, Tomasz A, Vogt KA, Volkmer GA, Wang SH, Wang Y, Weiner MP, Yu P, Begley RF, Rothberg JM. Genome sequencing in microfabricated high-density picolitre reactors. *Nature* 2005;437:376–380.

142. Bernath K, Magdassi S, Tawfik DS. Directed evolution of protein inhibitors of DNA-nucleases by *in vitro* compartmentalization (IVC) and nano-droplet delivery. *J Mol Biol* 2005;345:1015–1026.

143. Cohen HM, Tawfik DS, Griffiths AD. Altering the sequence specificity of HaeIII methyltransferase by directed evolution using *in vitro* compartmentalization. *Protein Eng Des Select* 2004;17:3–11.

144. Griffiths AD, Tawfik DS. Directed evolution of an extremely fast phosphotriesterase by *in vitro* compartmentalization. *EMBO J* 2003;22:24–35.

145. Ghadessy FJ, Ramsay N, Boudsocq F, Loakes D, Brown A, Iwai S, Vaisman A, Woodgate R, Holliger P. Generic expansion of the substrate spectrum of a DNA polymerase by directed evolution. *Nat Biotechnol* 2004;22:755–759.

146. Spiegelman S. An approach to the experimental analysis of precellular evolution. *Q Rev Biophys* 1971;4:213–253.

147. Breaker RR, Joyce GF. Emergence of a replicating species from an *in vitro* RNA evolution reaction. *Proc Natl Acad Sci USA* 1994;91:6093–6097.

148. Dobson CM, Ellison GB, Tuck AF, Vaida VV. Atmospheric aerosols as prebiotic chemical reactors. *Proc Natl Acad Sci USA* 2000;97:11864–11868.

149. Forterre P. The two ages of the RNA world, and the transition to the DNA world: a story of viruses and cells. *Biochimie* 2005;87:793–803.

150. Seeman NC. DNA in a material world. *Nature* 2003;421:427–431.

151. Mao C, Sun W, Shen Z, Seeman NC. A nanomechanical device based on the B-Z transition of DNA. *Nature* 1999;397:144–146.

152. Yurke B, Turberfield AJ, Mills AP Jr, Simmel FC, Neumann JL. A DNA-fuelled molecular machine made of DNA. *Nature* 2000;406:605–608.

153. Yan H, Zhang X, Shen Z, Seeman NC. A robust DNA mechanical device controlled by hybridization topology. *Nature* 2002;415:62–65.

154. Shin JS, Pierce NA. A synthetic DNA walker for molecular transport. *J Am Chem Soc* 2004;126:10834–10835.

155. Liao S, Seeman NC. Translation of DNA signals into polymer assembly instructions. *Science* 2004;306:2072–2074.

156. Benenson Y, Paz-Elizur T, Adar R, Keinan E, Livneh Z, Shapiro E. Programmable and autonomous computing machine made of biomolecules. *Nature* 2001;414:430–434.

157. Benenson Y, Gil B, Ben-Dor U, Adar R, Shapiro E. An autonomous molecular computer for logical control of gene expression. *Nature* 2004;429:423–429.

158. Shih WM, Quispe JD, Joyce GF. A 1.7-kilobase single-stranded DNA that folds into a nanoscale octahedron. *Nature* 2004;427:618–621.

159. Henry AA, Romesberg FE. The evolution of DNA polymerases with novel activities. *Curr Opin Biotechnol* 2005;16:370–377.

160. Sprinzak D, Elowitz MB. Reconstruction of genetic circuits. *Nature* 2005;438:443–448.

161. Seeman NC. From genes to machines: DNA nanomechanical devices. *Trends Biochem Sci* 2005;30:119–125.

14

THE SYNTHETIC APPROACH FOR REGULATORY AND METABOLIC CIRCUITS

Wilson W. Wong and James C. Liao

Chemical and Biomolecular Engineering Department, University of California, Box 951592, Los Angeles, California 90095

14.1 INTRODUCTION

14.1.1 Motivation for Synthetic Approach to Biology

The study of biology and gene regulation has traditionally been conducted using a reductionist approach, where a complex system of biomolecules is reduced to smaller units and each component is individually investigated. These smaller units, however, are always connected *in vivo* to form a network of interacting molecules, thus, the overall properties of the network are rarely the sum of its individual parts. Network connectivity and topology, as well as biochemical properties of individual components, are therefore required to completely describe the behaviors of an organism.

While the reductionist approach aims to determine the biochemical properties of each individual component, the systems approach focuses on elucidating network connectivity. Although these two complementary approaches hold significant promise for characterizing the behavior of biological systems, they often do not readily yield the design principles behind the complex networks. Due to millions years of evolution, existing intracellular networks are complicated by many auxiliary circuits that may

Systems Biology and Synthetic Biology   Edited by Pengcheng Fu and Sven Panke mask the basic design principle of the system. Therefore, to deduce fundamental principles by illuminating each component in the modern-day cell is as difficult as rediscovering fundamental laws of physics by disassembling an automobile. In an alternative approach, dubbed the synthetic approach, hypothetical operating principles are generated and then tested using artificially synthesized networks. The design approach may avoid second-order functions that are not important for the first principle. Furthermore, synthetic networks are not limited by natural biological systems, providing a wider range of test conditions.

The synthetic approach is initiated by educated creativity, much like the design of any engineering system. At this stage, the principle is inspired by physical and mathematical insights, but constrained by biological and chemical realities. A prototype mathematical model that serves as a conceptual blueprint is useful and often necessary. When such a prototype model is constructed, each component needs to be implemented by biological elements. The proper biological components such as promoters, regulators, enzymes, and metabolites are then identified to fulfill design specifications. Finally, the network is "reconstituted" inside the cell to test the properties of the system.

The creation of artificial systems also allows exploration of potential applications that are not displayed by natural design. An example of such application is the dynamic metabolic feedback loop [1] that addresses the fundamental challenge in metabolic engineering of maintaining a balance of the cell's resources, specifically between cell growth and metabolite production. In this circuit, a synthetic feedback controller was constructed in *Escherichia coli* that allows for gene expression of the key enzyme in lycopene production pathway to be under the control of a metabolite, acetyl-phosphate. When grown in glucose, *E. coli* produces acetate, a metabolic waste, when the tricarboxylic acid cycle (TCA) is no longer able to accommodate the incoming glycolytic flux. Acetate production also serves as an indication that the cells have sufficient energy and material resources and therefore represents a prime opportunity to shift cellular resources from cell growth toward the production of metabolites. When the level of acetate increases, its precursor, acetyl-phosphate, would also increase and activate the production of lycopene (Fig. 14-1).

The network synthesis approach is analogous to *in vitro* protein reconstitution commonly performed in the fields of biochemistry and molecular biology. Instead, the network is reconstituted *in vivo* by the selection of the proper genetic and metabolic components. A design flow diagram is illustrated in Figure 14-2. This approach has also been generalized to cell-free systems [2]. The design principle tested using the synthetic approach may or may not be used in real life. However, these principles provide focal points to search for similar designs in the cell and to examine the gap between theoretical prediction and biological reality.

14.1.2 Challenges in Synthetic Biological Circuit Design and Construction

The design of synthetic biological circuits shares many similarities with engineering constructions, but faces a major complication in the form of biological uncertainties. These uncertainties are manifested at two levels. First, the lack of detailed kinetic parameters for biological elements prevents the precise prediction of network

(a)

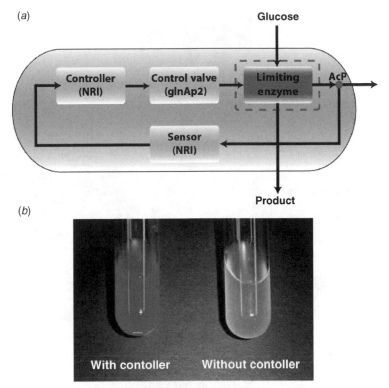

(b)

Figure 14-1 Dynamic metabolic feedback controller. (a) As cells grow on glucose, metabolic waste product, acetate, and it precursor acetyl-phosphate, accumulates. To divert cellular resources toward metabolite production when acetyl-phosphate accumulates, the limiting enzymes of the metabolic pathway are placed under the control of *glnAp2* promoter. The *glnAp2* promoter is activated by the phosphorylated form of NRI and acetyl-phosphate phosphorylates NRI In the absent of NRII [1]. (b) Production of lycopene, a reddish compound, with and without the controller.

behavior. Second, the interaction of these molecules with other cellular components is even less characterized, which causes additional difficulties. Constructions of synthetic circuits are therefore challenging and often iterative. Efforts have been made to expedite the construction process through combinatorial synthesis [3], directed evolution [4], and the creation of standardized biological parts [5]. Nevertheless, several synthetic circuits have already been demonstrated, which provide valuable insights into the design principles of biological networks [3,6–10]. Most of the recent synthetic circuits are reviewed in other chapters. The focus of this review is on oscillation, intercellular communication, and their interaction with metabolism.

14.2 BIOLOGICAL OSCILLATORS

Oscillation is a fascinating and an important phenomenon displayed by biological systems. Biological oscillators are ubiquitous circuits with a wide range of frequencies. They are found in numerous organisms, such as bacteria, plants, insects,

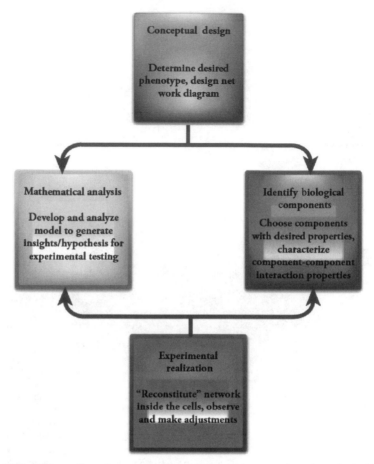

Figure 14-2 Process flow diagram for the design and construction of synthetic biological circuits. As with many engineering projects, the first step is to generate a conceptual design. Biological components are then identified and mathematical models are constructed. After analyzing the feasibility of the concept, the design is implemented inside the cells. Due to uncertainties associated with biological systems, implementation of synthetic circuits tends to be iterative. Efforts are being made to create standardized parts and infrastructures to reduce uncertainties [5].

and mammals. Oscillation also governs numerous vital processes such as global gene expression and cell cycle. Disruptions to circadian rhythm circuits, an oscillator with a 1-day period, have been demonstrated to cause sleep disorders and have also been linked to alcohol [11] and drug abuse [12,13] in mammals. Therefore, understandings of biological oscillators can have far-reaching medical and social implications.

Although many molecular details of biological oscillators have been determined in recent years, the network properties of these oscillators remain elusive. To gain further insight into the design principles of biological oscillators, three synthetic oscillators have been demonstrated so far, each of which is based on a conceptual idea that serves as a design principle. The first two are isolated modules that do not interact with metabolism, but demonstrate the idea of oscillation at the transcriptional level. The third oscillator integrates gene expression with metabolism to drive the oscillatory circuit. Theoretical discussions of biological oscillation have been sufficiently reviewed by others [14,15].

14.2.1 Synthetic Oscillators

14.2.1.1 Three Mutually Repressible Transcription Factors and Promoters Form an Oscillator
The first synthetic biological oscillator to be constructed is the ring oscillator, termed the repressilator, by Elowitz and Leibler [16]. This oscillator involves three mutually repressible promoters that regulate the expression of the three repressors. Repressor 1 inhibits the expression of repressor 2, and similarly repressor 2 inhibits the expression of repressor 3. Finally, repressor 3 inhibits the expression of repressor 1 to complete the circuit (Fig. 14-3a). The final construct used LacI as repressor 1, TetR as repressor 2, and λ cI as repressor 3. As with every oscillator design, not all parameters of the individual components will lead to oscillation. This is where mathematical analysis and intuition can be helpful. Through modeling, Elowitz and Leibler determined that components property such as tightly regulated promoters and shorter protein half-lives can improve the likelihood of oscillation. Hence, promoters were chosen to minimize "leakiness" when fully

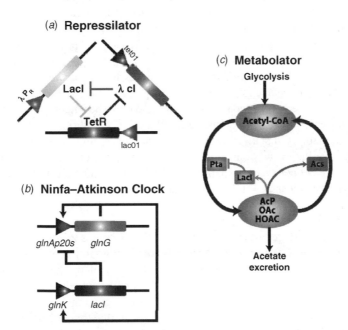

Figure 14-3 Schematic network diagram for (a) repressilator, (b) Ninfa–Atkinson clock, and (c) metabolator. See text for more detailed descriptions.

repressed and degradation peptide sequences were added to each repressor to reduce the half-life of the proteins. To observe the response of the repressilator, green fluorescence protein (*gfp*) was placed under the control of one of the promoters. This system was not designed to be synchronized and therefore the oscillation dynamics can only be observed at the single-cell level. The period of the repressilator is approximately 150 min with 40 percent of the cells exhibiting oscillation.

14.2.1.2 A Predator (lacI)–Prey (glnG) Pair of Regulators Exhibit Oscillation
Atkinson et al. [17] later designed a synthetic oscillator (Fig. 14-3b) with a dramatically longer period of 10–20 h and the oscillation dynamics that can be observed in a continuous culture. The conceptual design is reminiscent of the one proposed by Barkai and Leibler [18], which is inspired from observations of naturally occurring oscillators. In this oscillator design, an activator enhances its own gene expression and the expression of another regulatory protein that inactivates the activator. Through simulation, this design consisting of a positive element coupled with the negative element was determined to be relatively noise resistant. The actual design created by Atkinson et al. is slightly different from the one proposed by Barkai and Leibler. This oscillator involves a positive regulator (NtrC) that activates its own expression and the expression of a repressor (LacI). Instead of antagonizing the activity of NtrC, LacI represses the gene expression of NtrC. The activator, NtrC, can be regarded as a prey that "feeds" into the predator LacI, which decreases the population of the prey. Again, the authors relied on both conceptual reasoning and mathematical modeling in their selection of the appropriate biological components. The repressor, LacI, was placed under the control of *glnK* promoter. The activator, NtrC, which is a part of a two-component system involved in the nitrogen starvation response, was placed under the control of a modified *glnAp2* promoter that contains a LacI-binding site (Fig. 14-3b). Both *glnAp2* and *glnK* promoters are positively activated by NtrC, but the *glnK* promoter requires higher levels of activated NtrC to be fully induced when compared to the *glnAp2* promoter. The design by Atkinson et al. did not involve any degradation sequence and the experiments were performed in a continuous bioreactor under constant cell density condition (turbidostat). This oscillator amazingly displayed oscillation dynamics at the population level, even though the oscillation was dampened. The synchronization is probably due to the exposure of IPTG before the experiment, which sets all the cells to the same states. However, it is unclear how this circuit maintained synchronization throughout the experiment.

14.2.1.3 Two Interconverting Pools of Metabolites with Nonequilibrium Fluxes Display Oscillation
Naturally occurring oscillators found in many organisms rarely operate independently from the rest of the cell's physiology. In fact, the oscillators are usually linked to the global regulation of gene expression and metabolism. As mentioned earlier, circadian rhythms can regulate alcohol and drug intake. Conversely, alcohol intake can also affect the function of the circadian rhythm [19–22]. This intimate relationship between intracellular oscillators and the environment is a critical property of natural oscillators, which allows the circuit to

sense and respond to environmental changes. The work by Fung et al. [23] mimicked this property by integrating genetic oscillators into the central metabolism of *E. coli*.

The integrated gene and metabolic oscillator by Fung et al., termed the metabolator, consists of a flux-carrying network with two interconverting metabolite pools (M1 and M2) catalyzed by two enzymes (E1 and E2), whose expressions are negatively and positively regulated by M2, respectively. In the first stage where the M2 level is low, E1 is expressed, while E2 is not. A high-input metabolic flux converts M1 to M2 rapidly. The accumulation of M2 represses E1 and upregulates E2. When the backward reaction rate exceeds the sum of the forward reaction rate and the output rate, M2 level decreases and M1 level increases. E1 is then expressed again and E2 is degraded, returning to the first stage. On the other hand, if the input flux is low, M2 will not accumulate quickly enough to cause a large swing in gene expression, and thus a stable steady state will be reached. This design allows metabolism to control gene expression cycles, a characteristic commonly seen in circadian regulation.

The experimental design is realized with the promoter *glnAp2* controlling two genes, *lacI* and *acs* (encoding acetyl coenzyme A synthetase) (Fig. 14-3c). Phosphotransacetylase (Pta), a reversible enzyme that catalyzes the conversion of acetyl coenzyme A (AcCoA) to acetyl phosphate (AcP), is placed under the control of the *lacO1* promoter. The *lacO1* promoter is a synthetic promoter designed to reduce the leakiness of gene expression when repressed, while maintaining a large dynamic range of protein expression. To obtain readout of the circuit, a green fluorescence protein is placed under the control of another LacI repressible promoter. All proteins were fused to an *ssrA* degradation peptide to reduce their half-lives. The promoter *glnAp2*, in the absence of NRII (a bifunctional protein kinase/phosphatase regulation involved in the nitrogen starvation response), can be activated by AcP [1]. Here AcCoA corresponds to M1, AcP corresponds to M2, Pta corresponds to E1, and Acs corresponds to E2. When the AcP level is low, LacI and Acs expression levels are also low, and thus derepressing the *lacO1* promoter, which in turn increases the production of Pta. As Pta is being produced, it converts acetyl coenzyme A into AcP, which activates the *glnAp2* promoter and synthesizes Acs and LacI. Increasing the concentration of LacI represses the transcription rate of *pta*, hence lowering the level of AcP. Meanwhile, as the level of Acs increases, it converts more acetate into AcCoA. This removes the downstream product from the AcP degradation pathway and helps lower the level of AcP. One important aspect of this design is the interconversion of two metabolite pools, AcCoA and AcP, through two enzymes, Pta and Acs, which are controlled by the circuit. These two enzymes in turn, indirectly and directly, respond to AcP.

To gain more insight into the properties of the system, nonlinear differential equations and bifurcation analysis were employed to probe the dynamic properties of the system. The analysis predicts that oscillation will be favored when the metabolic influx is high. The metabolic flux-driven dynamics represents a very important feature of this design. An imbalance of metabolic fluxes destabilizes the steady states and leads to oscillation. This allows the metabolator to respond to the glycolytic influx. To test the prediction, the metabolator was cultured in different carbon sources that support different glycolytic rates. When grown in glucose, fructose, and mannose, which support high-glycolytic flux, the metabolator exhibits oscillation. When grown in glycerol however, which yields a low-glycolytic flux, the metabolator did not exhibit oscillation. The experimental results therefore confirmed the mathematical predictions. The construction of the metabolator demonstrates that the two-pool network architecture produces oscillation with metabolic fluxes as the driving force for oscillation.

To successfully design and construct the metabolator, we began with an understanding of the interaction between gene regulation and metabolism. Then a conceptual idea of the network was conceived. With the conceptual framework, biological components were identified and the network was constructed using techniques from microbiology and molecular biology. Mathematical models were constructed with reasonable parameters based on the network connectivity and biological components of choice to identify parameter space that leads to oscillation. Modeling can be very helpful in identifying and exploring parameters that have significant impact on the performance of the system. Many pitfalls, however, can also be associated with modeling, such as choosing inaccurate range of parameters. Artifacts from the simulation can lead to surprising results that defy common sense. Therefore, one cannot blindly trust the results from modeling and the results generated from the model should make intuitive sense.

14.2.2 Circadian Circuits from Cyanobacteria Also Form Two Interconverting Pools

Interestingly, the circadian rhythm network structure found in cynaobacteria *Synechococcus elongatus* also possesses two interconverting pools similar to those described in the metabolator (Fig. 14-4). This circuit from *S. elongatus* is remarkably robust. A study at the single cell level demonstrated that the oscillation is stable at the

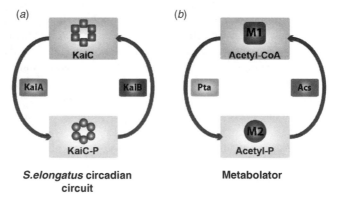

(a)   *(b)*

S.elongatus circadian circuit   **Metabolator**

Figure 14-4 Similarities between the circadian circuit from *S. elongatus* and the metabolator. The overall structure of the metabolator is similar to the Kai system from *S. elongatus*. This structure might represent a fundamental motif for oscillation in bacteria. The major difference between the two systems is that the phosphorylated form of KaiC, KaiC-P, probably does not decrease the activity of KaiA. Moreover, KaiC-P can control gene expression at genomic scale in *S. elongatus*.

individual cell level without the need to synchronize with surrounding cells [24]. The central clock components in *S. elongatus* are coded by the genes *kaiA*, *kaiB*, and *kaiC* ("Kai" is a Japanese word that means cycle). KaiC represses its own expression, as well as KaiB, both of which are in the *kaiBC* operon. KaiA activates KaiB and KaiC expression [25]. The traditional model for generating a circadian rhythm is the transcription–translation oscillator (TTO) model. KaiC represses its own gene expression, and this negative feedback loop is the core oscillator component. KaiA sustains the oscillation by enhancing the expression of KaiC. Recent findings, however, report data that contradicts the TTO model. KaiC phosphorylation, rather than the transcription and translation of KaiC, seems to be the dominating factor in controlling oscillation dynamics [26]. Tomita et al. [27] directly tested the TTO model by growing *S. elongates* under constant darkness. In this condition, transcription terminates in *S. elongates*, but the phosphorylation state of KaiC continues to display circadian rhythmicity. More interestingly, in a subsequent report, KaiC phosphorylation was demonstrated to exhibit oscillation in a mixture that contained only purified KaiA, KaiB, KaiC, and ATP *in vitro* [28]. The current model of the Kai circuit is as follows. When KaiC is not phosphorylated, KaiA repeatedly and rapidly associates with KaiC to enhance the phosphorylation of KaiC. When KaiC phosphorylation reaches a sufficiently high level, its binding with KaiB is promoted, which in turn inactivates KaiA leading to its own dephosphorylation. When KaiC is dephosphorylated, KaiB dissociates from KaiC, which then activates KaiA to repeat the cycle [29].

14.2.3 Metabolism and Circadian Rhythm

Interactions between the circadian rhythm and metabolism in mammals are well documented in literature [30]. In mammals, the main "clock" that controls the rest of the peripheral systems is located in the suprachiasmatic nucleus (SCN) of the hypothalamus. NPAS2 (or its homologue, Clock) and BMAL1 are the major regulators in circadian rhythm. They form a heterodimer (NPAS2:BMAL1 or Clock:BMAL1) in the nucleus. When this complex is activated, it binds to the DNA and expresses the clock genes *Cry* and *Per* and clock output genes such as lactate dehydrogenase *Ldh*. Rutter et al. [31] have shown that reduced nicotinamide adenine dinucleotide (NAD(P)H) can directly activate the NPAS2:BMAL1 complex with near switch-like response *in vitro*.

Aside from NADH, recent work [32,33] suggests that heme negatively regulates the activity of the NPAS2:BMAL1 complex in the presence of carbon monoxide (CO). The complex, in turn, activates the expression of a heme biosynthesis rate-limiting enzyme aminolevulinate synthase 1 (Alas1). As the level of Alas1 increases, the concentration of heme also increases. A sufficiently high level of heme will eventually induce its own degradation enzyme, called heme oxygenase, which generates CO as a final product. CO inhibits the activity of the NPAS2:BMAL1 complex and therefore, the expression of Alas1 as well. Without Alas1, the heme level eventually reaches a sufficiently low level to allow NPAS2:BMAL1 to become active again, thus continues the cycle.

Glucose has also been shown to interact with circadian gene expression in rat fibroblasts [34]. In transgenic mice, the disruption of BMAL1 and *Clock* upsets glucose homeostasis [35]. These mutant mice display altered diurnal variation of plasma glucose and triglycerides, as well as glucose intolerance and insulin resistance with a high-fat diet. In another report, mice with a mutation in the *Clock* gene are obese and develop metabolic syndromes such as hyperleptinemia, hyperlipidemia, hepatic steatosis, hyperglycemia, and hypoinsulinemia [36,37]. These works once again demonstrate the intimate link between biological oscillation and metabolism. They also highlight the importance in elucidating the design principle of oscillation for both fundamental understanding and potential medical applications.

14.2.4 Oscillation Frequency and Responses

As demonstrated in electronic circuits, information can be encoded into the frequency of oscillation. Therefore, an oscillator with tunable frequency can be useful in encoding information, thus allowing more information content to be stored and transmitted with fewer signaling molecules. Examples of the frequency-dependent response in natural system have already been identified in calcium oscillation and NFκB oscillation [38–43]. A successful engineering application of encoding information into the frequency, however, entails another challenge of constructing a biological frequency decoder. Currently, it is not known how such decoding is achieved in biological systems.

14.3 CELL–CELL COMMUNICATION IN BACTERIA

Intercellular communication is of paramount importance to the development of higher organisms. Recently, numerous reports have also demonstrated the importance of cell–cell communication in bacterial physiology. Further insight into biological networks requires a better understanding of molecular details and system level analysis of intercellular communication. Using well-characterized model organisms such as *E. coli* and *S. cerevisiae,* one can construct synthetic intercellular circuits to decipher underlying principles, similar to the approach used in developing synthetic intracellular circuits.

14.3.1 Natural Cell–Cell Communication Systems

Bacteria have long been considered to be unicellular organisms that do not interact with other bacteria. This notion is mainly due to the way in which studies were performed using bacteria. Under most laboratory conditions, bacteria are grown in pure culture. In natural environment, however, bacteria rarely live alone in pure culture planktonic condition. Environmental biologists have long recognized the importance of bacterial communities to biogeochemical cycling that maintain the biosphere [44]. Biofilm is one example of a bacterial community. Biofilm is composed of single specie or multiple species. Such communities can live on biotic and abiotic surfaces and perform diverse metabolic functions.

To coordinate population behavior, intercellular communication is essential. In the 1970s, researchers had identified a communication system in *Vibrio fisheri* with homoserine lactone as the signaling molecule termed autoinducer-1 (AI-1) [45]. *V. fischeri* is a bacterial symbiont that lives in the light-producing organ of the squid *Euprymna scolopes*. At low cell density conditions, *V. fisheri* produces small amounts of AI-1 proportional to growth. Once a threshold concentration of AI-1 is reached, it activates the *lux* operon, which encodes gene products that emits light, while producing more AI-1. Because this gene expression system was found to respond to density, or quorum, this system was termed "quorum" sensing. Since then, numerous examples of quorum sensing have been found in both Gram-positive and Gram-negative bacteria with different types of diffusible molecules and regulation mechanisms. Cellular functions that are under the control of quorum sensing include sporulation, biofilm formation, and virulence factor expression.

In higher organisms, cellular development relies heavily on the signaling cues generated by other cells. Complex spatial patterning has been shown to be the result of multiple intercellular signals coupled with feedbacks. One example is the development of left–right asymmetry in mouse embryos. The heart and other inner organs develop an asymmetrical arrangement during morphogenesis [46]. The expression and relay of TGF-β family signaling molecule, Nodal, has been found to be crucial in the symmetry breaking and differentiation of left–right organs. Nodal couples with extracellular cofactor EGF-CFC to form a positive feedback loop and substantiate its own existence. Moreover, Nodal also induces another intercellular signal called Lefty-2/antivin to form a negative feedback loop. In chicks, Nodal was found to induce a Cre-like molecule, Caronte, to relay the signal to more distal cells [47].

14.3.2 Synthetic Cell–Cell Communication Circuits

Equipping circuits with communication systems will greatly enhance capabilities of circuits, as demonstrated by the marriage of computers and the Internet. Analogously, capabilities of gene circuits can be greatly improved with an intercellular communication network. Engineering a communication system between cells will allow cellular programming at a population level rather than at the single cell level. Basu and Weiss utilized the quorum-sensing system from *V. fisheri* to create a spatiotemporal pulse generator [48] (Fig. 14-5a). This circuit contains sender cells that can produce AI and receiver cells that generate a pulse response. Using a similar concept, but with a different network configuration, Basu et al. created spatial patterns such as a "bulls eye" pattern on solid media (Fig. 14-5b) [49]. These two reports demonstrate that spatial and temporal patterns can be created in single cell organisms, mimicking a powerful capability commonly found in higher organisms.

You et al. [50] developed a circuit in *E. coli* that can sense and control its own cell density. This circuit produces AI continuously and accumulates AI in a cell-density-dependent manner (Fig. 14-6). As the AI level reaches a threshold, it activates a toxic gene, which leads to cell death. This circuit is a negative feedback loop that operates concertedly at the population level. The stability of AI is pH dependent, and therefore the steady-state cell density can be modulated by pH. As with any negative feedback loop, this system can potentially oscillate when operated in the proper parameter space. In this system, the degradation of AI is probably too slow for oscillation to occur.

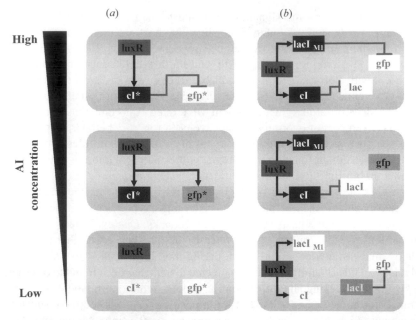

Figure 14-5 Intercellular communication network in *E. coli* based on the lux/AI system from *V. fisheri*. Network configuration for (a) the spatiotemporal pulse generator [48] and (b) the pattern formation [49]. The asterisk denotes that the protein half-life is reduced by the addition of a degradation sequence at the end of the protein. lacI$_{M1}$ is a codon modified mutant of lacI.

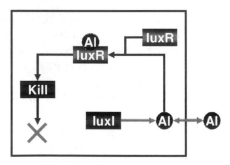

Figure 14-6 Schematic diagram of the population controller. AI is the diffusible autoinducer [50]. "Kill" is a gene that when expressed, can inhibit growth of *E. coli*. The "Kill" gene used in this network is *ccdB*, control cell death B, which interferes with DNA gyrase and causes cell death.

When grown in a microchemostat condition, however, the constant washing of the growth chamber facilitates AI removal, which allows the circuit to display cell-density oscillation [51].

14.3.2.1 *Artificial Cell–Cell Communication System*   With only one channel of communication, the amount of information that can be transmitted is limited.

This limitation constrains the capability of all the networks within the system. To increase the information content, one solution is to generate oscillating signals and to encode the information into frequencies. Another solution is to create additional communication channels. The synthetic oscillators discussed in the earlier sections had the potential to generate oscillating intercellular signals. A separate frequency decoder circuit, however, will be needed to complete the scheme. Bulter et al. [52] addressed the later possibility by demonstrating the design and construction of an artificial communication system in *E. coli* using gene and metabolic network with acetate as the signaling molecule. Chen and Weiss later engineered an artificial communication system in *Saccharomyces cerevisiae* by incorporating *Arabidopsis thaliana* signal synthesis and receptor components into the host [53]. Using the *V. fisheri* AI-1 as the basis, Collins et al. created mutants of the transcription factor LuxR that are sensitive to different autoinducers. The design strategy identified here can also serve as a blueprint for further development of more independent communication channels.

An intercellular communication system can be divided into two modules—signal generation and signal detection. In designing a signal generation system, the signaling molecules must be chosen so that it is diffusible. The production of the signal must also be controlled. In designing the signal-detection module, the response to the signal must be specific and cannot be a general toxic or stress response. The detection system should also be sufficiently sensitive to detect the broad range of signal production and tunable.

14.3.2.2 *Synthetic Communication System in E. coli*

In the artificial communication system in *E. coli*, acetate was chosen as the signaling molecule (Fig. 14-7a). Acetate is mainly produced from the *pta/ackA* pathway in *E. coli* [52]. Acetate is typically considered as a waste product during fermentation. When the TCA cycle cannot oxidize the carbon flux from glycolysis, Accyl-CoA buildup is resulted. Since Acetyl-CoA is the entry point into the TCA cycle, the production of acetate through the *pta/ackA* pathway is intimately linked to the activity of the TCA cycle and the availability of oxygen. When the *pta/ackA* pathway is disrupted, however, a small amount of acetate, about 10 percent of the original level, is still produced through the biosynthesis of arginine and cysteine. This residual production of acetate, however, is no longer sensitive to oxygen levels.

The transport of acetate across the membrane is passive with the permeability of acetic acid across the membrane being three orders of magnitude higher than acetate, the negatively charged conjugate base of acetic acid. The dissociation equilibrium of acetic acid is pH dependent. Since the intracellular pH of *E. coli* is homeostatically regulated near pH 7.6, thus the intracellular acetate concentration depends on the ΔpH, intracellular pH minus extracellular pH, across the membrane. At high extracellular pH, the acid–base equilibrium shifts toward acetate. At low extracellular pH, the reverse is true. Therefore, less acetate is needed to activate the *glnAp2* promoter at low pH and to enhance the sensitivity of the *glnAp2* promoter to acetate. This unique property of weak acids/promoter interaction allows dynamic sensitivity tuning of the system. This tuning property is the opposite of AI. AI stability is lowered when pH decreases, therefore decreasing its sensitivity.

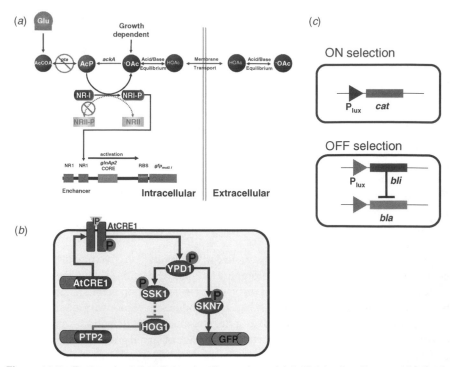

Figure 14-7 *Engineering intercellular-signaling systems.* (a) Artificial cell–cell communication in *E. coli* and (b) *S. cerevisiae*. In (a), the signaling molecule is acetate, a metabolite from the central metabolism. The detection system employs a nitrogen starvation response two-component system, Ntr, to sense acetate. NR-I is responsive to acetate in the absence of it cognate sensor, NR-II [52]. AcCoA, acetyl-CoA; AcP, acetyl-phosphate; OAc, acetate; HOAc, acetic acid. In (b), the signaling molecule and detection module is borrowed from the plant *Arabidopsis thaliana* [53]. For (c), the regulator luxR, which normally is sensitive to 3OC6HSL, is subjected to a dual selection scheme that selects for mutant sensitive to another autoinducer without significant cross talk from the original 3OC6HSL [54]. This selection involves an ON selection that select for luxR with expanded sensitivity to different autoinducer. The ON selection, however, does not eliminate the sensitivity to the original 3OC6HSL. Thus, the OFF selection is employed to select against the sensitivity of 3OC6HSL. *Cat* is a chloramphenicol resistant gene, *bla* is an amplicilin-resistant gene, and *bli* is a gene that inhibits the activity of *bla*.

For the detection module, the *glnAp2* promoter described in the construction of gene-metabolic oscillator was employed. The activity of the promoter was reported by using *gfp*. This circuit can exhibit cell density-dependent gene expression. The sensitivity of this behavior can be modulated by pH and also through promoter engineering. The NRI-P binding sequence was altered to manipulate the enhancer's strength. A stronger and a weaker binding sequence of NRI-P compared to the wild type had been identified and incorporated into the quorum sensing circuit. When performing the quorum sensing experiment with these new enhancer sequences, the strong enhancer requires less cell density to achieve the same level of gene expression as the wild type. Similarly, the wild type required less cell density to achieve the same level of gene expression as the weak enhancer.

Acetate is produced in a wide variety of organisms. Moreover, AcP is also a regulator molecule in many bacteria. Two-component systems such the Che, Pho, and Ntr had been shown to be capable of sensing AcP. Those two-component systems are present in many different bacteria as well. Many other bacteria are known to possess similar two-component systems, which can be modified to respond to AcP. These facts suggest that this cell–cell communication system can be universally adapted to other bacteria. Hence, the artificial system presented here also possesses the possibility for interspecies communication. This system can serve as a model system to understand how cell–cell communication that differs between different species can lead to various behaviors and phenotypes observed in nature.

14.3.2.3 Synthetic Communication System in Saccharomyces cerevisiae

In the yeast communication system, cytokinin isopentenyladenine (IP) from the plant *Arabidopsis thaliana* was chosen [53] as the signaling molecule (Fig. 14-7b). IP can be generated by adenylate isopentenyl-transferases. To sense this signal, the cytokine receptor AtCRE1 from the same plant was chosen. This receptor can interact with yeast's endogenous phosphorylation-signaling pathway YPD1/ SKN7 to activate gene expression. In the wild-type strain, YPD1/SKN7 is part of a phosphorylation pathway with SLN1, a cell–surface osmosensor hisdine kinase, and SSK1, an aspartate response regulator. Under normal condition, SLN1 phosphorylates YPD1, which in turn phosphorylates SSK1 and represses HOG1 activity. HOG1 activity is crucial for survival in high osmolarity conditions. In normal condition, however, HOG1 activity is lethal. The constant phosphorylation of YPD1 by SLN1 is therefore a problem because it will render the system insensitive to IP. Deletion of SLN1, however, results in activation of HOG1. To circumvent this problem, Chen and Weiss removed SLN1, but overexpressed an endogenous HOG1 phosphatase to decrease HOG1 activity. This artificial system exhibits quorum-sensing behavior when both the signal generation module and the receiver module are present in the same cell.

14.3.2.4 Engineering Specificity of Autoinducer Variants in E. coli

Numerous quorum-sensing signals that are based on acyl-homoserine lactones have been identified in nature. These signaling molecules share homoserine lactones as the same core structure, but are differentiated by their unique side chains. The most well-studied acyl-homoserine lactone is 3OC6HSL from *Vibrio fisheri*. 3OC6HSL is synthesized by LuxI. When 3OC6HSL is bound to the transcription LuxR, it activates gene expression from the P_{lux} promoter. LuxR and its homologues had been shown to have crosswalk among different acyl-homoserine lactones. To utilize homoserine lactone variants as signaling molecules, Collins et al. employed a dual selection strategy to evolve and select for LuxR mutants from *V. fisheri* that are only specific to another acyl-homoserine lactone [54]. This dual selection strategy selected for LuxR variants that activate gene expression when different AIs are present, but remain inactive when the original AI is present (Fig. 14-7c).

The above examples represent different strategies for creating artificial communication networks. In the case with acetate, an endogenously produced metabolite is converted to a signaling molecule. In the case for *S. cerevisiae*, however, the communication system was transplanted from another organism.

Moreover, naturally occurring signal sensors can be evolved to enhance specificity to other signaling molecules, thus creating more channels of communication system with less cross talk.

14.3.3 Information and Cell–Cell Communication

In cell–cell communication, gene expression is dependent on the concentration of small molecules, such as acetate, cytokines, or various forms of autoinducers from natural systems. It is important to note the difference between cell–cell communication system and other small molecule inducible promoters, such as the *lac* promoter with IPTG as the inducer. The difference between these systems lies in the information that the signaling molecule carries. Since cells are unable to produce IPTG, the operator of the experiment needs to add this inducer to the culture. The inducer carries only the desire of the operator to activate gene expression and nothing else. The level of the inducer does not reflect any information of the cells. In quorum sensing, the signaling molecules carry cell density information, thus encoding one aspect of the physiological state. Although the artificial cell–cell communication network is incorporated into a quorum sensing system, the use of primary metabolites allows encoding of metabolic states as well. When *pta* is disrupted, the production of acetate is proportional to growth. With intact *pta*, acetate can be used to encode the metabolic state of the central metabolism. The *pta* gene can also be transcriptionally fused to other promoters to transfer information sensed by the promoter to other cells.

14.3.4 Identifying Communication Molecules

Every endogenously produced diffusible small molecule has the potential to be a signaling molecule for cell–cell communication. How to differentiate a *bona fide* communication signal from other metabolites? Winzer et al. [55] contends that the key criterion for signaling molecules is that its ability to respond to the signal should extend beyond the needs to detoxify and metabolize the signals. The autoinducer-1 system seems to satisfy the criteria [55,56]. Whether another autoinducer system, AI-2, satisfies this criteria is less certain. AI-2 is proposed to be a universal signaling molecule because of its primary synthesis gene, *LuxS*, is present in many different organisms. AI-2 is produced from *S*-adenosylmethionine, SAM. SAM is used as a methyl donor to DNA, RNA, and other metabolites, giving *S*-adenosylhomocysteine (SAH). The subsequent step of metabolizing SAH is critical because SAH is a potent inhibitor for SAM-dependent methyltransferases and the cells need to regenerate their building blocks. An AI-2 uptake system, Lsr transporter, has also been discovered. The activation of this transporter is cyclic AMP (cAMP) dependent [57]. Hence, in the presence of glucose, AI-2 can accumulate to a high level whereas, with glycerol, AI-2 will be consumed. The reason why the cells would excrete and later internalize AI-2 remains unclear. Winzer et al. argues that the production of AI-2 is a metabolic side-product used to metabolize SAH, and later reuptake in a controlled catabolite repression-dependent manner. This excretion and reuptake mechanism is very similar to acetate, our artificial signaling molecule. One major difference between the synthetic systems and the AI-2 system is the response to the signal. The acetate system in *E. coli* is purely artificial. The response to AI-2, although unknown, seems to be metabolic. Our work highlights the difficulties in defining a signaling molecule in natural systems. Before the disruption of NRII, acetate is generally considered as a metabolic waste product. After NRII is disrupted, acetate becomes a cell–cell communication signal. One can imagine a scenario where a metabolic waste product in one condition can become a signaling molecule in another condition. Whether a molecule is a signaling molecule can be condition dependent.

14.4 CONCLUSION

Some of the synthetic circuits constructed, especially the ones involve complex dynamics such as oscillation, perform poorly when compared to their naturally evolved counterparts. Such constructions highlight the major challenge in biological network engineering—dealing with biological uncertainties. Therefore, the construction of synthetic circuits can be tedious. Nonetheless, synthetic circuits have served as proof of concept and generated new insights. The shortcomings of these circuits also raise important issues regarding biological network design, such as those related to how fluctuation of the parameters in individual parts can affect the overall system's robustness. Many exciting works have been done recently to quantify the stochasticity, or noise, in gene expression [58–61] and the source of such stochasticity [62–66]. Based on a bioinformatics study, noise in gene expression seems to be minimized for essential genes, suggesting the importance of noise regulation in fitness enhancement [67]. Noise had also been demonstrated to play an important role in *E. coli pap* operon pilli expression [68,69], bacteriophage lytic decision [70], and HIV viral latency decision [71]. Incorporating these findings into the design of synthetic circuits will improve the performance of the system.

The first generation of circuits is designed to operate mainly at the transcription level in simple organisms such as *E. coli* and *S. cerevisiae*. However, life is rarely that one dimensional. Rather, it is organized and regulated at multiple levels. To engineer more complex behaviors, more layers of controls are needed to incorporate into synthetic circuits. Our group is focused on integrating metabolic and transcriptional regulation. Others have engineered control at the translation level by manipulating the three dimensional structure of mRNA and the interaction between ribosomes and mRNA [72–74]. Protein-signaling cascades can also be altered and manipulated by rewiring the input and output domains [75–77]. Borrowing from enzymes' powerful chemical synthetic capabilities, biosynthetic pathways have been engineered and rewired to improve the yield of high-valued compounds and to create completely new compounds [78–87]. Works have also been done to engineer circuits, such as toggle switches, in mammalian cells [88,89]. Coupled with the intercellular communication circuits described earlier, sophisticated networks that operate on a global level and involve multiple species that mimic natural multicellular organisms can be created as a platform for understanding the emergence of complex behavior. More importantly, these systems can also be exploited for biotechnological and medical applications.

REFERENCES

1. Farmer WR, Liao JC. Improving lycopene production in *Escherichia coli* by engineering metabolic control. *Nat Biotechnol* 2000;18:533–537.

2. Noireaux V, Bar-Ziv R, Libchaber A. Principles of cell-free genetic circuit assembly. *Proc Natl Acad Sci USA* 2003;100:12672–12677.

3. Guet C, Elowitz MB, Hsing W, Leibler S. Combinatorial synthesis of genetic networks. *Science* 2002;296:1466–1470.

4. Yokobayashi Y, Weiss R, Arnold FH. Directed evolution of a genetic circuit. *Proc Natl Acad Sci USA* 2002;99:16587–16591.

5. Endy D. Foundations for engineering biology. *Nature* 2005;438:449–453.

6. Alon U. Biological networks: the tinkerer as an engineer. *Science* 2003;301:1866–1867.

7. Becskei A, Serrano L. Engineering stability in gene networks by autoregulation. *Nature* 2000;405:590–593.

8. Becskei A, Seraphin B, Serrano L. Positive feedback in eukaryotic gene networks: cell differentiation by graded to binary response conversion. *EMBO J* 2001;20:2528–2535.

9. Gardner TS, Cantor CR, Collins JJ. Construction of a genetic toggle switch in *Escherichia coli*. *Nature* 2000;403:339–342.

10. Kobayashi H, Kaern M, Araki M, Chung K, Gardner TS, Cantor CR, Collins JJ. Programmable cells: interfacing natural and engineered gene networks. *Proc Natl Acad Sci USA* 2004;101:8414–8419.

11. Spanagel R, Pendyala G, Abarca C, Zghoul T, Sanchis-Segura C, Magnone MC, Lascorz J, Depner M, Holzberg D, Soyka M, et al. The clock gene Per2 influences the glutamatergic system and modulates alcohol consumption. *Nat Med* 2005;11:35–42.

12. McClung CA, Sidiropoulou K, Vitaterna M, Takahashi JS, White FJ, Cooper DC, Nestler EJ. Regulation of dopaminergic transmission and cocaine reward by the Clock gene. *Proc Natl Acad Sci USA* 2005;102:9377–9381.

13. Abarca C, Albrecht U, Spanagel R. Cocaine sensitization and reward are under the influence of circadian genes and rhythm. *Proc Natl Acad Sci USA* 2002;99:9026–9030.

14. Goldbeter A. Biochemical Oscillations and Cellular Rhythms: The Molecular Bases of Periodic and Chaotic Behaviour. New York: Cambridge University Press, 1996.

15. Goldbeter A. Computational approaches to cellular rhythms. *Nature* 2002;420:238–245.

16. Elowitz MB, Leibler S. A synthetic oscillatory network of transcriptional regulators. *Nature* 2000;403:335–338.

17. Atkinson MR, Savageau MA, Myers JT, Ninfa AJ. Development of genetic circuitry exhibiting toggle switch or oscillatory behavior in *Escherichia coli*. *Cell* 2003;113:597–607.

18. Barkai N, Leibler S. Circadian clocks limited by noise. *Nature* 2000;403:267–268.

19. Spanagel R, Rosenwasser AM, Schumann G, Sarkar DK. Alcohol consumption and the body's biological clock. *Alcohol Clin Exp Res* 2005;29:1550–1557.

20. Rosenwasser AM. Alcohol, antidepressants, and circadian rhythms. Human and animal models. *Alcohol Res Health* 2001;25:126–135.

21. Wasielewski JA, Holloway FA. Alcohol's interactions with circadian rhythms. A focus on body temperature. *Alcohol Res Health* 2001;25:94–100.

22. Chen CP, Kuhn P, Advis JP, Sarkar DK. Chronic ethanol consumption impairs the circadian rhythm of pro-opiomelanocortin and period genes mRNA expression in the hypothalamus of the male rat. *J Neurochem* 2004;88:1547–1554.

23. Fung E, Wong WW, Suen JK, Bulter T, Lee SG, Liao JC. A synthetic gene-metabolic oscillator. *Nature* 2005;435:118–122.

24. Mihalcescu I, Hsing W, Leibler S. Resilient circadian oscillator revealed in individual cyanobacteria. *Nature* 2004;430:81–85.

25. Ishiura M, Kutsuna S, Aoki S, Iwasaki H, Andersson CR, Tanabe A, Golden SS, Johnson CH, Kondo T. Expression of a gene cluster kaiABC as a circadian feedback process in cyanobacteria. *Science* 1998;281:1519–1523.

26. Xu Y, Mori T, Johnson CH. Cyanobacterial circadian clockwork: roles of KaiA, KaiB and the kaiBC promoter in regulating KaiC. *EMBO J* 2003;22:2117–2126.

27. Tomita J, Nakajima M, Kondo T, Iwasaki H. No transcription-translation feedback in circadian rhythm of KaiC phosphorylation. *Science* 2005;307:251–254.

28. Nakajima M, Imai K, Ito H, Nishiwaki T, Murayama Y, Iwasaki H, Oyama T, Kondo T. Reconstitution of circadian oscillation of cyanobacterial KaiC phosphorylation *in vitro*. *Science* 2005;308:414–415.

29. Kageyama H, Nishiwaki T, Nakajima M, Iwasaki H, Oyama T, Kondo T. Cyanobacterial circadian pacemaker: Kai protein complex dynamics in the KaiC phosphorylation cycle *in vitro*. *Mol Cell* 2006;23:161–171.

30. Rutter J, Reick M, McKnight SL. Metabolism and the control of circadian rhythms. *Annu Rev Biochem* 2002;71:307–331.

31. Rutter J, Reick M, Wu LC, McKnight SL. Regulation of clock and NPAS2 DNA binding by the redox state of NAD cofactors. *Science* 2001;293:510–514.

32. Kaasik K, Lee CC. Reciprocal regulation of haem biosynthesis and the circadian clock in mammals. *Nature* 2004;430:467–471.

33. Reick M, Garcia JA, Dudley C, McKnight SL. NPAS2: an analog of clock operative in the mammalian forebrain. *Science* 2001;293:506–509.

34. Hirota T, Okano T, Kokame K, Shirotani-Ikejima H, Miyata T, Fukada Y. Glucose down-regulates Per1 and Per2 mRNA levels and induces circadian gene expression in cultured Rat-1 fibroblasts. *J Biol Chem* 2002;277:44244–44251.

35. Rudic RD, McNamara P, Curtis AM, Boston RC, Panda S, Hogenesch JB, Fitzgerald GA. BMAL1 and CLOCK, two essential components of the circadian clock, are involved in glucose homeostasis. *PLoS Biol* 2004;2:e377.

36. Turek FW, Joshu C, Kohsaka A, Lin E, Ivanova G, McDearmon E, Laposky A, Losee-Olson S, Easton A, Jensen DR,et al. AT Obesity and metabolic syndrome in circadian Clock mutant mice. *Science* 2005;308:1043–1045.

37. Staels B. When the Clock stops ticking, metabolic syndrome explodes. (Discussion 55.) *Nat Med* 2006;12:54–55.

38. Dolmetsch RE, Xu K, Lewis RS. Calcium oscillations increase the efficiency and specificity of gene expression. *Nature* 1998;392:933–936.

39. Allen GJ, Chu SP, Harrington CL, Schumacher K, Hoffmann T, Tang YY, Grill E, Schroeder JI. A defined range of guard cell calcium oscillation parameters encodes stomatal movements. *Nature* 2001;411:1053–1057.

40. Nelson DE, Ihekwaba AE, Elliott M, Johnson JR, Gibney CA, Foreman BE, Nelson G, See V, Horton CA, Spiller DG,et al. Oscillations in NF-kappaB signaling control the dynamics of gene expression. *Science* 2004;306:704–708.

41. Hu Q, Deshpande S, Irani K, Ziegelstein RC. [$Ca^{(2+)}$](i) oscillation frequency regulates agonist-stimulated NF-kappaB transcriptional activity. *J Biol Chem* 1999;274: 33995–33998.

42. Lewis RS. Calcium oscillations in T-cells: mechanisms and consequences for gene expression. *Biochem Soc Trans* 2003;31:925–929.

43. Li W, Llopis J, Whitney M, Zlokarnik G, Tsien RY. Cell-permeant caged InsP3 ester shows that Ca^{2+} spike frequency can optimize gene expression. *Nature* 1998;392:936–941.

44. Davey ME, O'Toole GA. Microbial biofilms: from ecology to molecular genetics. *Microbiol Mol Biol Rev* 2000;64:847–867.

45. Waters CM, Bassler BL. Quorum sensing: cell-to-cell communication in bacteria. *Annu Rev Cell Dev Biol* 2005;21:319–346.

46. Gaio U, Schweickert A, Fischer A, Garratt AN, Muller T, Ozcelik C, Lankes W, Strehle M, Britsch S, Blum M,et al. A role of the cryptic gene in the correct establishment of the left–right axis. *Curr Biol* 1999;9:1339–1342.

47. Freeman M. Feedback control of intercellular signalling in development. *Nature* 2000;408:313–319.

48. Basu S, Mehreja R, Thiberge S, Chen MT, Weiss R. Spatiotemporal control of gene expression with pulse-generating networks. *Proc Natl Acad Sci USA* 2004;101:6355–6360.

49. Basu S, Gerchman Y, Collins CH, Arnold FH, Weiss R. A synthetic multicellular system for programmed pattern formation. *Nature* 2005;434:1130–1134.

50. You L, Cox RS, 3rd, Weiss R, Arnold FH. Programmed population control by cell–cell communication and regulated killing. *Nature* 2004;428:868–871.

51. Balagadde FK, You L, Hansen CL, Arnold FH, Quake SR. Long-term monitoring of bacteria undergoing programmed population control in a microchemostat. *Science* 2005;309:137–140.

52. Bulter T, Lee SG, Wong WW, Fung E, Connor MR, Liao JC. Design of artificial cell-cell communication using gene and metabolic networks. *Proc Natl Acad Sci USA* 2004;101:2299–2304.

53. Chen MT, Weiss R. Artificial cell-cell communication in yeast *Saccharomyces cerevisiae* using signaling elements from *Arabidopsis thaliana*. *Nat Biotechnol* 2005;23:1551–1555.

54. Collins CH, Leadbetter JR, Arnold FH. Dual selection enhances the signaling specificity of a variant of the quorum-sensing transcriptional activator LuxR. *Nat Biotechnol* 2006;24:708–712.

55. Winzer K, Hardie KR, Williams P. Bacterial cell-to-cell communication: sorry, can't talk now—gone to lunch! *Curr Opin Microbiol* 2002;5:216–222.

56. Xavier KB, Bassler BL LuxS quorum sensing: more than just a numbers game. *Curr Opin Microbiol* 2003;6:191–197.

57. Xavier KB, Bassler BL. Regulation of uptake and processing of the quorum-sensing autoinducer AI-2 in *Escherichia coli*. *J Bacteriol* 2005;187:238–248.

58. Elowitz MB, Levine AJ, Siggia ED, Swain PS. Stochastic gene expression in a single cell. *Science* 2002;297:1183–1186.

59. Raser JM, O'Shea EK. Control of stochasticity in eukaryotic gene expression. *Science* 2004;304:1811–1814.

60. Ozbudak EM, Thattai M, Kurtser I, Grossman AD, van Oudenaarden A. Regulation of noise in the expression of a single gene. *Nat Genet* 2002;31:69–73.

61. Pedraza JM, van Oudenaarden A. Noise propagation in gene networks. *Science* 2005;307:1965–1969.

62. Becskei A, Kaufmann BB, van Oudenaarden A. Contributions of low molecule number and chromosomal positioning to stochastic gene expression. *Nat Genet* 2005;37:937–944.

63. Rosenfeld N, Young JW, Alon U, Swain PS, Elowitz MB. Gene regulation at the single-cell level. *Science* 2005;307:1962–1965.

64. Volfson D, Marciniak J, Blake WJ, Ostroff N, Tsimring LS, Hasty J. Origins of extrinsic variability in eukaryotic gene expression. *Nature* 2006;439:861–864.

65. Austin DW, Allen MS, McCollum JM, Dar RD, Wilgus JR, Sayler GS, Samatova NF, Cox CD, Simpson ML. Gene network shaping of inherent noise spectra. *Nature* 2006;439:608–611.

66. Colman-Lerner A, Gordon A, Serra E, Chin T, Resnekov O, Endy D, Pesce CG, Brent R. Regulated cell-to-cell variation in a cell-fate decision system. *Nature* 2005;437:699–706.

67. Fraser HB, Hirsh AE, Giaever G, Kumm J, Eisen MB. Noise minimization in eukaryotic gene expression. *PLoS Biol* 2004;2:e137.

68. Zhou B, Beckwith D, Jarboe LR, Liao JC. Markov Chain modeling of pyelonephritis-associated pili expression in uropathogenic *Escherichia coli*. *Biophys J* 2005;88:2541 2553.

69. Jarboe LR, Beckwith D, Liao JC. Stochastic modeling of the phase variable pap operon regulation in uropathogenic *Escherichia coli*. *Biotechnol Bioeng* 2004;88:189–203.

70. Arkin A, Ross J, McAdams HH. Stochastic kinetic analysis of developmental pathway bifurcation in phage lambda-infected *Escherichia coli* cells. *Genetics* 1998;149:1633–1648.

71. Weinberger LS, Burnett JC, Toettcher JE, Arkin AP, Schaffer DV. Stochastic gene expression in a lentiviral positive-feedback loop: HIV-1 Tat fluctuations drive phenotypic diversity. *Cell* 2005;122:169–182.

72. Pfleger BF, Pitera DJ, Smolke CD, Keasling JD. Combinatorial engineering of intergenic regions in operons tunes expression of multiple genes. *Nat Biotechnol* 2006;24:1027–1032.

73. Bayer TS, Smolke CD. Programmable ligand-controlled riboregulators of eukaryotic gene expression. *Nat Biotechnol* 2005;23:337–343.

74. Isaacs FJ, Dwyer DJ, Ding C, Pervouchine DD, Cantor CR, Collins JJ. Engineered riboregulators enable post-transcriptional control of gene expression. *Nat Biotechnol* 2004;22:841–847.

75. Bhattacharyya RP, Remenyi A, Yeh BJ, Lim WA. Domains, motifs, and scaffolds: the role of modular interactions in the evolution and wiring of cell signaling circuits. *Annu Rev Biochem* 2006;75:655–680.

76. Dueber JE, Yeh BJ, Chak K, Lim WA. Reprogramming control of an allosteric signaling switch through modular recombination. *Science* 2003;301:1904–1908.

77. Park SH, Zarrinpar A, Lim WA. Rewiring MAP kinase pathways using alternative scaffold assembly mechanisms. *Science* 2003;299:1061–1064.

78. Ro DK, Paradise EM, Ouellet M, Fisher KJ, Newman KL, Ndungu JM, Ho KA, Eachus RA, Ham TS, Kirby J,et al. Production of the antimalarial drug precursor artemisinic acid in engineered yeast. *Nature* 2006;440:940–943.

79. Yoshikuni Y, Ferrin TE, Keasling JD. Designed divergent evolution of enzyme function. *Nature* 2006;440:1078–1082.

80. Lee TS, Khosla C, Tang Y. Engineered biosynthesis of aklanonic acid analogues. *J Am Chem Soc* 2005;127:12254–12262.

81. Tang Y, Lee TS, Khosla C. Engineered biosynthesis of regioselectively modified aromatic polyketides using bimodular polyketide synthases. *PLoS Biol* 2004;2:E31.

82. Schmidt-Dannert C. Engineering novel carotenoids in microorganisms. *Curr Opin Biotechnol* 2000;11:255–261.

83. Schmidt-Dannert C, Umeno D, Arnold FH. Molecular breeding of carotenoid biosynthetic pathways. *Nat Biotechnol* 2000;18:750–753.

84. Achkar J, Xian M, Zhao H, Frost JW. Biosynthesis of phloroglucinol. *J Am Chem Soc* 2005;127:5332–5333.

85. Guo J, Frost JW. Biosynthesis of 1-deoxy-1-imino-D-erythrose 4-phosphate: a defining metabolite in the aminoshikimate pathway. *J Am Chem Soc* 2002;124:528–529.

86. Wang C, Oh MK, Liao JC. Directed evolution of metabolically engineered *Escherichia coli* for carotenoid production. *Biotechnol Prog* 2000;16:922–926.

87. Wang CW, Oh MK, Liao JC. Engineered isoprenoid pathway enhances astaxanthin production in *Escherichia coli*. *Biotechnol Bioeng* 1999;62:235–241.

88. Kramer BP, Fussenegger M. Hysteresis in a synthetic mammalian gene network. *Proc Natl Acad Sci USA* 2005;102:9517–9522.

89. Kramer BP, Viretta AU, Daoud-El-Baba M, Aubel D, Weber W, Fussenegger M. An engineered epigenetic transgene switch in mammalian cells. *Nat Biotechnol* 2004;22:867–870.

15

SYNTHETIC GENE NETWORKS

David Greber and Martin Fussenegger

Department of Biosystems Science and Engineering, ETH Zurich, Mattenstrasse 26, CH-4058 Basel, Switzerland

15.1 INTRODUCTION

Advances in molecular manipulation techniques, together with an ever-increasing accumulation of genetic information, are progressively opening new possibilities for gene therapy and biomedical engineering. By combining naturally occurring genetic components in unique ways, it has become possible to artificially engineer genetic networks that possess increasingly sophisticated functional capabilities. By analogy to electronic circuit engineering, the desired characteristics of such networks can be rationally designed and tested through predictive modeling. Similarly to electrical networks, genetic networks also possess "input" and "output" functionality such that they are capable of monitoring and responding in highly defined mechanisms. The creation of synthetic networks from well-defined modular components has enabled researchers to investigate and test many network characteristics found in natural genetic networks. It is from an applied perspective, however, that synthetic genetic networks represent a truly exciting innovation. It is not difficult to envisage applications where synthetic networks could be used to manipulate cellular behavior in a highly orchestrated way. While these concepts are still in their infancy, significant progress has been made in the creation of first-generation synthetic networks, which will one day enable the engineered control of cellular function to become a viable reality.

Systems Biology and Synthetic Biology   Edited by Pengcheng Fu and Sven Panke
Copyright © 2009 John Wiley & Sons, Inc.

This chapter begins by describing the modular genetic components that form the building blocks of engineered genetic networks. It then describes the development of both simple and complex networks, many of which were initially developed in prokaryotic systems, but which have been subsequently extended to eukaryotic systems. The focus is upon describing networks that have been experimentally tested and validated. It does not cover the extensive modeling and computational work that has been conducted on either synthetic or natural genetic regulatory networks (readers are referred to Chapter 7). Advances in network functionality have been made on both the input and output dimensions. Examples of output functionality include the generation of stable behavior, such as bistable toggle and hysteric switches, and dynamic behavior such as an oscillatory network. From an input perspective developments include the creation of logical information "gates," where a range of input combinations produce highly defined outputs in a manner directly analogous to electrical circuits; the development of transcriptional cascades, which have enabled the range of inputs to a network to be greatly increased; and the development of novel sensory networks which, for example, can detect inputs within a defined concentration range, or respond precisely to a rising level of an input. The chapter concludes by presenting the initial first steps into the emerging field of semisynthetic networks. These are prosthetic genetic networks that are capable of responding to physiological cues so that they are effectively integrated into the host-cell's biology. Such networks, in response to acute or pathological cues, hold great promise for the controlled manipulation of cellular processes such as protein synthesis, metabolism, cell growth, and differentiation.

15.2   NETWORK BUILDING BLOCKS

While synthetic in the sense that they are artificially designed and created, synthetic genetic networks are actually engineered from naturally occurring genetic components. A discussion of these networks requires a basic understanding of these components and the manner in which they interact. While gene expression can be regulated and artificially manipulated at a number of levels, the networks described below have only utilized a limited number of transcriptional control elements. Hence, this overview is limited to the mechanisms and components that have been used in these systems. A comprehensive overview of other gene control systems and their application can be found in several recent reviews [1–3].

Transcriptional control operates at the level of mRNA synthesis through the use of inducible transcriptional activators and repressors that are capable of binding naturally occurring or specifically engineered promoters. The majority of systems utilize bacterial response regulators or activators that, upon binding to a target promoter, inhibit or activate transcription respectively. Binding of a specific molecule to the response regulator induces an allosteric change leading to disassociation of the regulator from its cognate promoter.

Prokaryotic gene control systems generally use inducible repressors and activators drawn from well-documented genetic operons such as the *lac* operon of *Escherichia coli* [4], the tetracycline-resistance transposon Tn*10* [5], or the λcI repressor of bacteriophage lambda [6]. In each case, the respective response regulator binds to a DNA sequence, typically a short tandem repeat referred to as the "operator," located within or adjacent to a promoter where it either enhances transcription or sterically hinders the initiation of transcription. By substituting operators across different strength promoters it has been possible to generate inducible systems with varied induction characteristics [7].

Bacterial response regulators also form the basis of synthetic eukaryotic gene regulation systems although given transcriptional differences they require adaptation. This has been successfully achieved for many bacterial response regulators by placing the operator for the response regulator adjacent to an eukaryotic compatible promoter [8]. The response regulator thus acts as a heterologous DNA-binding protein (DBP) whose association with the desired promoter can be controlled through addition of an appropriate inducer. If the operator is placed close to an strong constitutive promoter (e.g., P_{CMV}, cytomegalovirus immediate early promoter), DBP binding can sterically prevent the initiation of transcription by RNA polymerase II machinery. Alternatively, transcription can be actively repressed by fusing a eukaryotic transcriptional silencer, such as the Kruppel-associated box protein (KRAB), to the DBP [9]. Such systems are referred to as ON-type systems, as the addition of an inducer leads to derepression of transcription (Fig. 15-1). In an OFF-type configuration, in which addition of inducer leads to transcriptional silencing, a transcriptional activation domain, such as the *Herpes simplex* virus VP16, is fused to the DBP [10]. By placing the corresponding operator site adjacent to a minimal promoter (e.g., $P_{hCMVmin}$, minimal version of the human cytomegalovirus immediate early promoter), DBP binding activates transcription from an otherwise silent minimal promoter. Addition of an inducer results in subsequent deactivation of transcription.

As many prokaryotic antibiotic response regulators have been well described, and given the low interference of many antibiotics with eukaryotic biology, they represent an ideal class of inducible DBPs for eukaryotic gene control. Using the aforementioned configurations, eukaryotic gene control systems responsive to tetracyclines [11], streptogramins [12], and macrolides [13] amongst others have been developed. As these gene control systems do not interfere with each other, they can be readily combined. For this reason, and their nonpleiotrophic effects, they have formed the basis of most eukaryotic synthetic gene networks. A list of the common transcriptional control elements used in the assembly of both prokaryotic and eukaryotic synthetic gene networks is provided in Table 15-1.

15.3 CHARACTERIZATION OF SIMPLE AND COMPLEX NETWORKS

The past decade has seen a progressive increase in the development and application of both prokaryotic and eukaryotic synthetic networks. In some cases, these networks have been relatively simple and have been used to test and investigate naturally occurring phenomena. In other cases, the networks exhibit far greater complexity as they seek to reproduce or create much more sophisticated functionality. When adopting

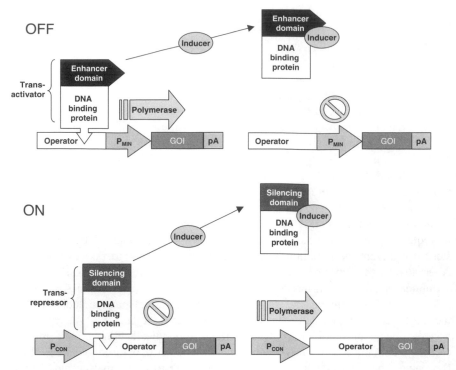

Figure 15-1 Molecular configuration of OFF and ON synthetic eukaryotic gene regulation. In the OFF configuration, a DNA-binding protein—typically a bacterial transcriptional repressor—binds a specific operator site placed adjacent to a minimal promoter (P_{MIN}). An activation domain fused to the DBP activates polymerase-mediated transcription of a gene of interest (GOI). Addition of an inducer specific to the DBP causes an allosteric change resulting in disassociation of the transactivator with subsequent transcriptional arrest. In the ON configuration, the DBP is fused to a repressor domain. Binding of DBP-TR to an operator site placed adjacent to a constitutive promoter (P_{CON}) represses transcription of the GOI. Again, addition of a DBP specific inducer results in transrepressor disassociation although in this configuration, repression is abolished resulting in expression of the GOI [8].

the electrical circuit analogy it is possible to describe synthetic genetic networks in terms of their input functionality—how the network receives and integrates specific signals as well as their output functionality—how the network produces and maintains a specific pattern of expression. Given that much of the pioneering work in synthetic circuits was directed toward producing novel patterns of gene expression, it is expedient to commence with network descriptions of output functionality.

In considering the design of a synthetic genetic network for a biological application it is useful to imagine what kind of functions one might wish to create. Thus, some applications may benefit from a mechanism that ensures a network produces a consistent and stable response even when there are considerable random fluctuations in either network components, inducer concentrations, or cellular components more broadly. For other applications, one may require a system that produces more than one

Table 15-1 Common genetic transcriptional components used in the reaction of synthetic genetic networks

DNA-Binding Protein	System Application	Engineered Regulatory Protein	Inducer	Response to Inducer	References
TetR	Prokaryotic	—	Doxycycline, aTc	Derepression	[5]
LacI	Prokaryotic	—	IPTG	Derepression	[4]
λcI	Prokaryotic	—	Temperature	Derepression	[6]
NRI	Prokaryotic	—	Phosphorylation	Activation	[94]
LuxR	Prokaryotic	—	Acyl-homoserine lactone	Activation	[95]
LacI	Eukaryotic	—	IPTG	Derepression	[96]
TetR	Eukaryotic	TetR-VP_6 (tTA)	Tetracycline,	Deactivation	[11]
rTetR	Eukaryotic	rTetR-VF16	Doxycycline, aTc	Activation	[11]
Pip	Eukaryotic	Pip-KRAB	Streptogramins	Derepression	[12]
Pip	Eukaryotic	Pip-VP16	Streptogramins	Deactivation	[12]
E	Eukaryotic	E-KRAB	Macrolides	Derepression	[13]
E	Eukaryotic	E-VP16	Macrolides	Deactivation	[13]
ScbR	Eukaryotic	ScbR-VP16	Butyrolactones	Deactivation	[97]
Gal4	Eukaryotic	Gal4-VP_6	Mifepristone	Deactivation	[56]
HIF-1α	Eukaryotic	–	Hypoxia	Activation	[98]

aTc, anhydrotetracycline; IPTG, isopropyl-β-D-thiogalactopyranoside KRAB, Kruppel-associated box protein-derived transrepressor domain; VP16, *Herpes simplex* viral protein 16-derived transactivation domain.

discrete expression state. A mechanism that "remembers" what conditions the network has been exposed to may be useful in applications where only a transient pulse of an inducer is required or expected. A mechanism that not only remembers the past but also reacts differently to subsequent changes would also be desirable. Finally, a mechanism that produces continuous oscillations in expression readout may be highly practical where repeated temporal expression is required. All of these mechanisms have their counterpart in natural biological systems where they represent the molecular controls for numerous basic cellular functions ranging from cellular differentiation, cell-cycle control, and circadian rhythms. It is therefore not surprising that genetic engineers have applied considerable effort to synthetically reproduce these mechanisms. Apart from being useful tools, such synthetic networks also shed considerable light on how the equivalent mechanism occurs in a natural system.

15.3.1 Expression Stability

To produce a unified and consistent outcome a biological process, whether it involves metabolic homeostasis or cellular growth and development, must be capable of withstanding a certain degree of variation and difference [14–16]. As cellular biochemical networks are highly interconnected, a perturbation in reaction rates or molecular concentrations may affect multiple cellular processes including transcription, translation, and RNA and protein degradation—all of which impact gene expression. Systems that, despite the influence of considerable variation and random perturbation, are capable of remaining close to a steady state can be characterized as stable (or robust). Existing artificial gene regulation systems are typically highly susceptible to even modest fluctuations in regulatory components, which can significantly affect expression performance. In contrast, many natural gene networks intrinsically exhibit high stability. A natural question, therefore, is which mechanism(s) would enable a network to withstand such variation? A key development in our understanding of how stability is maintained was through the discovery of autoregulatory feedback loops in which proteins, directly or indirectly, influence their own production [17]. An autofeedback mechanism can either be negative, in which a protein inhibits its own production, or positive, in which a protein stimulates its own production.

Although it had been proposed that autoregulatory negative feedback loops provide stability, thereby limiting the range over which the concentrations of network components fluctuate, it was Becskei and Serrano who first demonstrated how a negative feedback mechanism can increase expression stability (Fig. 15-2) [18]. By fusing green fluorescent protein (GFP) to the tetracycline-responsive repressor protein (TetR) they were able to measure variations in TetR expression (measured by coefficient of variation in fluorescence intensity) across a population of *E. coli*. In using an established prokaryotic gene regulation system they created a negatively autoregulated system in which TetR inhibits its own transcription, as well as an unregulated system where TetR has no influence upon its transcription rate. Consistent with predictions from mathematical modeling, the experimental data showed that the autoregulated system exhibited a threefold narrower variation in expression levels

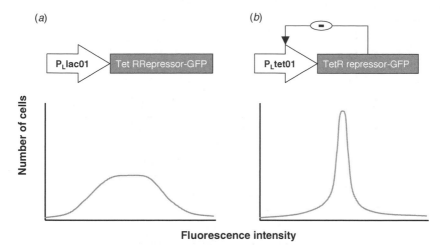

Fluorescence intensity

Figure 15-2 Expression profile of (a) an unregulated genetic system compared to (b) an equivalent system utilizing negative autofeedback. Both systems were based on the same architecture in which a promoter was used to control expression of a fusion protein consisting of the tetracycline repressor (TetR) and GFP in *E. coli*. In the regulated system, the promoter contained two tetracycline repressor operator modules (P_{Ltet01}). Negative feedback occurs as TetR represses transcription from P_{Ltet01}. In the unregulated system, TetR was prevented from interacting with the promoter by substituting the TetR operator with a different (LacR) operator (or by the functionally equivalent step of mutating the TetR-DNA-binding domain). In this way, the feedback mechanism was eliminated without altering other aspects of the genetic system. The resulting distribution of expression states for the unregulated system was wider than the corresponding distribution for the negative feedback system thus demonstrating the higher stability of a genetic system employing autofeedback [18].

than the unregulated system. Furthermore, through the addition of anhydrotetracycline (aTc), which causes TetR to dissociate from its cognate operator thereby reducing feedback repression, it was possible to introduce variation levels into the autoregulated system which approached the variation levels observed in the unregulated system. Hence, in this simple synthetic network negative feedback provides a mechanism for ensuring a more stable expression state. This is consistent with observations of expression stability in natural systems for either prokaryotes or eukaryotes in which transcription factors are known to use both positive and negative autoregulation to control their own production [19,20].

A key requirement for many networks and biological functions is the capacity to produce more than one discrete stable expression state. The creation of binary, or even multiple, expression states raises a number of possibilities for how a network can transition from one state to the other (Fig. 15-3). In a classic graded expression system, an increase in the concentration of an inducer generates a graded (or continuous) transcriptional response that, in a graphical representation, resembles a sigmoid shape. This pattern is due to transcriptional cooperativity in which initial binding of a transcriptional regulator to a promoter enhances subsequent binding of further regulators to the same promoter. This can either be due to cooperative binding or regulator multimerization [21]. Yet, in some systems the switch from one state to

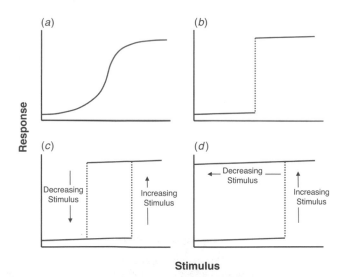

Figure 15-3 Stimulus–response profiles for (a) graded, (b) generic bistable, (c) hysteretic bistable, and (d) self-sustaining bistable genetic networks.(a) In a graded genetic system an increasing stimuli is progressively converted into an increasing response, which often adopts a sigmoidal pattern due to activator or repressor cooperativity. (b) In a generic bistable network the system exhibits quasi-discontinuous behavior whereby it only resides in one of two alternative steady states and not an intermediate state. Through changes in stimuli beyond a threshold point it is possible to switch or "toggle" the system from one state to another. (c) A hysteretic bistable network requires differing threshold stimuli levels to switch between steady states depending upon the starting state of the system. (d) In a self-sustaining bistable network the system remains in one steady state indefinitely even after the stimulus used to create that state has been removed [37,38].

another can be so swift as to almost represent discontinuous behavior. With increasing sophistication such a quasi-discontinuous switch can have different switching dynamics depending upon its starting point (i.e., hysteresis) or may even be self-sustaining (i.e., toggle) and/or irreversible. In addition to providing a means of achieving a single stable expression level, feedback regulation is also an important mechanism for producing a binary, or bistable, expression state in response to different input parameters [22].

15.3.2 Binary Expression

A graded transcriptional response typically results in a unimodal expression pattern where, when viewed across a cell population, there is no evident separation of expression states (Fig. 15-4). This remains so even when an inducer is used to increase expression—the resulting distribution is simply shifted upward reflecting an overall increase in expression across the entire cell population. Using a common tetracycline-responsive transactivator (TetR-VP16) and GFP reporter, it has been demonstrated that a simple autofeedback mechanism can create a binary expression readout in *Saccharomyces cerevisiae* [22]. By introducing positive feedback into the classical

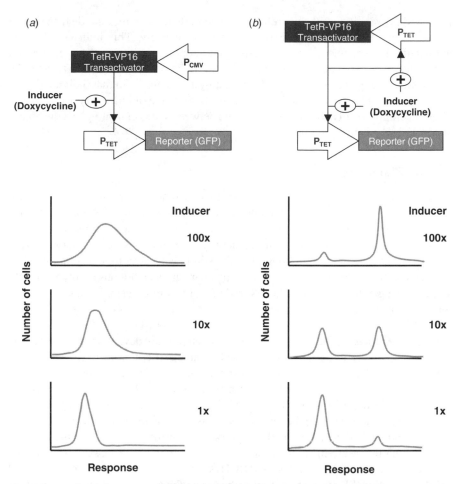

Figure 15-4 Graded response profile of (a) a classic transcription control system relative to (b) a bistable expression profile using a positive autofeedback mechanism. Both the classic and autoregulated systems were based on the tetracycline-dependent transactivator (TetR-VP16) eukaryotic transcriptional system. In the classic graded system, a strong constitutive CMV promoter (P_{CMV} cytomegalovirus immediate early promoter) was used to transcribe TetR. For the autoregulated system, the constitutive CMV promoter was replaced with a TetR-inducible promoter (P_{TET}) thereby creating a positive autofeedback loop. In both cases, a chromosomally integrated TetR-inducible GFP reporter construct was used to assess expression profiles. The classic graded system exhibits a unimodal distribution pattern which, following addition of doxycycline, shifts progressively to the right. The autoregulated system exhibits a bimodal distribution pattern that does not shift upon inducer addition. Rather, doxycycline addition concomitantly alters the proportion of cells residing in either of the two expression states [22,38].

TetR-VP16 transcription system, in which expression of TetR-VP16 positively influences its own production rate, a binary distribution pattern was produced whereby the cell population was clearly divided into discrete pools of ON and OFF cells. Importantly, following progressive administration of increasing inducer levels, the pools did not significantly shift relative to each other, but rather the distribution of cells between the ON and OFF pools changed inversely. This indicates that the autofeedback mechanism prevents cells from adopting an intermediary expression status such that they can only reside in one of the two possible states. Despite, the delineation of expression into one of two states, it was observed that individual cells did not necessarily remain in a fixed state. Across a range of inducer concentrations, a certain proportion of cells randomly flipped between states indicating that the binary states were not entirely stable.

15.3.3 Bistability

A binary expression system that does not exhibit random switching between two expression states is said to be bistable. Bistability is a minimal requirement for a network to possess memory in which the state of the network stores information about its past [23]. In addition to bistability, a network can only possess memory where it remains in an expression state long after the stimulus used to force it into that state has been removed. Such a self-sustaining mechanism is analogous to a typical light switch or toggle. Switching a light ON or OFF only requires a single transient, rather than a persistent, input.

15.3.3.1 Bacterial Toggle   A pioneering step in the development of synthetic networks was the creation of a plasmid-based bistable expression switch in *E. coli* [24]. The switch was constructed from two inducible bacterial repressors, transcribed from two similar strength promoters selected such that each repressor inhibited the promoter of the opposing repressor (Fig. 15-5). By placing a fluorescent reporter gene (GFP) downstream of one of the repressors it was possible to monitor which repressor was currently active, and thereby the expression status of the network. Owing to the mutually inhibitory arrangement of the two repressor genes, the network was capable of one of the two binary states: A HIGH state in which the first repressor and the downstream GFP reporter are transcribed from the second promoter, and a LOW state in which the second repressor is transcribed from the first promoter. In the absence of relevant inducers, the network can initially adopt either state, but once committed remains in the adopted state indefinitely. However, through the addition of a relevant inducer, it was possible to switch the network from one state to the other. The addition of an inducer to the active repressor enables the opposing repressor to be maximally transcribed. Once the opposing repressor has reached a certain level it represses transcription of the initially active repressor. As the prevalence of the opposing repressor over the initially active repressor becomes self-perpetuating, the inducer can be withdrawn and the network continues indefinitely in its altered state. In this manner, the network behaves as a bistable "toggle" switch in which the maintenance of either expression state does not require an ongoing inducer or stimulus. Furthermore, the status of the toggle could be maintained across cell generations indicating that network memory could be passed to progeny cells.

Six different toggle switches, employing different promoter-repressor pairs, were designed and characterized. Together with a mathematical approach it was possible

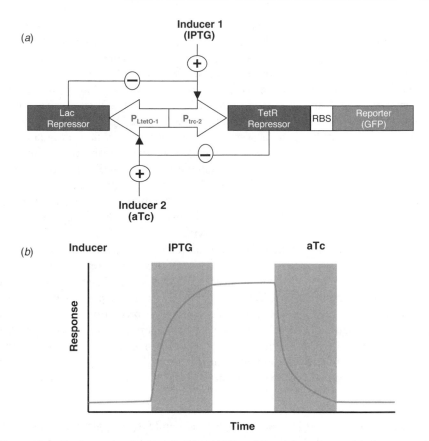

Figure 15-5 Engineered self-sustaining bistable "toggle" switch in *E. coli*. (a) Genetic design and (b) response profile. The genetic toggle switch was constructed from two sets of mutually opposing repressors/promoters. In the depicted configuration, the TetR repressor inhibits transcription of the Lac repressor from the $P_{LtetO-1}$ promoter. The Lac repressor in turn inhibits transcription from a P_{trc-2} promoter of the TetR repressor, a downstream ribosome-binding site (RBS) and a reporter gene (GFP). In the absence of inducers, both repressors mutually inhibit each other resulting in a low expression state. Addition of isopropyl-β-D-thiogalactopyranoside (IPTG) results in derepression of the Lac repressor and subsequent full expression of the TetR repressor and GFP (a HIGH expression state). Conversely, addition of aTc causes deinhibition of the TetR repressor with subsequent full expression of the Lac repressor (a LOW expression state). In both cases, only a transient pulse of inducer is required to enable the opposing repressor to be maximally transcribed until, in a self-perpetuating manner, it stably represses the originally active promoter [24].

to predict and assess many of the properties required for bistable switching. The interaction between toggle components was described using a simple differential equation model based upon rate equations for each repressor's production, repression activity, and degradation/dilution. Importantly, two criteria were found to be critical for robust bistability. First, each repressor had to be capable of cooperative repression at the promoter to which it binds. Mathematical modeling predicted that it is not the strength of the promoters *per se*, but rather the degree of cooperative repression that has a direct impact upon system robustness, defined as the ability to avoid stochastic switching between expression states. Thus, even weak promoters should be capable of bistability as long as cooperative repression is sufficiently high. Second, it was predicted that the rates of synthesis of the two repressors must be evenly balanced. This was empirically confirmed in one set of toggle components that were only capable of a single steady state due to uneven repressor synthesis rates. The different toggles also provided insight into the dynamics of switching time—defined as the time required for the relevant inducer to mediate a sustainable switch—although in this case the primary determinant was surprisingly not the rate of elimination of the initially active repressor protein. In one toggle system, requiring IPTG-induced inhibition of a repressor, the switching time was 6 h. In contrast, when a temperature sensitive repressor was employed, the immediate inhibition of the repressor caused by thermal destabilization resulted in sustainable switching occurring within 35 min.

The construction and characterization of several toggle switches illustrates the increasing utility of synthetic genetic networks. The construction of synthetic networks with varying properties enabled the testing and empirical validation of physical and mathematical approaches to gene regulation. While these approaches have been previously applied it has not been possible to test their predictions. Synthetic gene networks are a useful tool for this purpose and should permit the qualitative behavior of gene regulation to be studied and described in a manner analogous to that already conducted for enzyme regulation. It also highlights the importance of correct component selection and compatibility in creating a network with desired specific behavior [25].

15.3.3.2 *Mammalian Toggle*   A synthetic mammalian toggle switch capable of bistable expression has also been created, employing the same network architecture used in the synthetic *E. coli* toggle switch [26]. In this case, however, two eukaryotic transrepressor control systems were used: the E-KRAB system that is responsive to macrolide antibiotics such as erythromycin (EM) and the Pip-KRAB system responsive to streptogramin antibiotics such as pristinamycin (PI) (Fig. 15-6). A mutually opposing configuration, whereby each system represses expression of the other systems' transrepressor, generated two alternate stable expression states.

In the absence of either inducer molecule the network is balanced so long as both systems exhibit the same (low) expression levels with neither expression system able to prevail over the other. However, this balance can be tipped by addition of either inducer molecule in which case expression from one system is increased while expression from the other system is simultaneously repressed. Depending on the inducer added, the result is one of the two alternate expression states in which one transrepressor is expressed much more highly than the other. By placing a reporter gene (i.e., SEAP) immediately downstream of one of the transrepressors (i.e., Pip-KRAB), it was possible to tie SEAP expression to Pip-KRAB expression thereby obtaining a readout of the network status. A HIGH response, corresponding to high (or derepressed) Pip-KRAB expression, was obtained following induction with erythromycin whereas a LOW response, corresponding to low (increased repression)

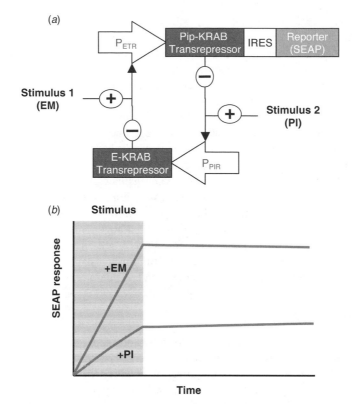

Figure 15-6 (a) Genetic construction and (b) response profile of engineered self-sustaining "toggle" switch. The mammalian toggle switch was assembled using two antibiotic-inducible transrepressor control systems, which were arranged to repress each other's expression. Erythromycin-inducible E-KRAB repressors transcription of Pip-KRAB and the human model reporter protein SEAP (human placental secreted alkaline phosphatase)—whose translation is modulated by an internal ribosome entry site (IRES). Pristinamycin I-inducible Pip-KRAB in turn repressors expression of E-KRAB. In both cases, addition of the respective inducer inhibits the repressive effect of the responsive transrepressor. Transient administration of EM results in P_{ETR}-driven coexpression of Pip-KRAB and SEAP with concomitant repression of E-KRAB (a HIGH response), whereas transient administration of PI results in P_{PIR} driven expression of E-KRAB with concomitant corepression of Pip-KRAB and SEAP (a LOW response). Both responses were maintained in a steady state following removal of relevant inducer molecules (nonshaded region) [26].

Pip-KRAB expression resulted after induction with pristinamycin. Importantly, once the network balance had been tipped toward one state, the change became self-perpetuating and, following removal of the initial inducer, was not lost. This was in contrast to isogenic control experiments using separate Pip-KRAB and E-KRAB systems where expression levels markedly decreased following inducer removal. In addition to self-sustainability, it was also demonstrated that the system was reversible, and that the expression profile could be repeatedly switched between expression states over a two-week period.

These two characteristics, sustained expression stability and reversible switching, are also key requirements for epigenetic imprinting or memory that occurs when differential expression levels are imprinted and passed to subsequent cell generations well after the original signal generating that expression level has been removed. Many natural epigenetic switches have been characterized where their role has been implicated in coordinating diverse processes such as cell fate and memory [27], plant development [28], and lysogeny [29,30]. In this case, the synthetic mammalian switch provides one possible model for how epigenetic imprinting may occur at the transcriptional level in multicellular organisms. Beyond this, however, the toggle switch may also have important therapeutic applications. Classical transcription control systems operate in a dose-dependent manner and therefore require the on-going presence of regulating molecules to sustain transgene expression levels. Prolonged exposure to regulating molecules (e.g., antibiotics) can be associated with clinical ramifications such as the selection of pathogen resistance [31] and the accumulation of antibiotics in bone and teeth [32]. A self-sustaining, yet reversible, genetic network that requires only a transient stimulus to establish a steady state may provide an attractive means of overcoming such considerations.

15.3.4 Hysteresis

In a typical bistable switch movement between expression levels occurs in a quasi-discontinuous manner once a controlling stimulus crosses a specific threshold. This threshold is the same regardless of the direction in which the switch is being moved. A refinement on this switch is where the threshold required to move the switch in one direction is different to the threshold required to move it in the other direction. Thus, the threshold required to flip the switch depends on the starting state of the switch. This phenomenon, which can occur at molecular or macroscopic levels, is known as hysteresis [33]. To use a nonbiological example, traffic jams often exhibit hysteresis because the car density required to alleviate the traffic jam is less than the density that initially caused the jam. In a genetic network, a switch exhibits hysteresis when a different concentration of inducer is required to shift a system from one state to another than is required for the reverse shift [18,22,24,29,34,35]. Hysteretic behavior has been observed in several natural examples including the control of lactose utilization in *E. coli* [33], and ensuring unidirectional cell-cycle progression in eukaryotes [36]. A significant benefit of a hysteretic system is its inherent ability to buffer against modest changes in the inducing molecule. Thus, to switch a system from one state to another and then to back again requires a far greater change in inducer levels than in an equivalent typical bistable switch. Such devices could have broad potential for applications in which the input signal is prone to minor fluctuations but for which a constant all or nothing expression status is required.

Using a positive autofeedback mechanism and competitive transcriptional mechanism, a synthetic hysteretic switch has been constructed in mammalian cells (Fig. 15-7) [37]. The system used a tetracycline-dependent transactivator (TetR-VP16), which induces its own transcription via positive feedback together with a reporter gene (SEAP), as well as a competing erythromycin-dependent

(a)

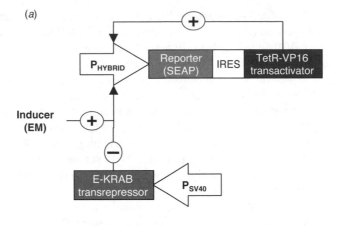

(b)

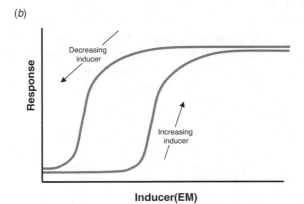

Figure 15-7 (a) Genetic design and (b) response profile of an engineered mammalian hysteretic switch. The hysteretic eukaryotic switch is based upon a chimeric promoter (P_{hybrid}) that drives expression of a SEAP (human placental secreted alkaline phosphatase) reporter gene and, via an IRES, the tetracycline-dependent transactivator (TetR-VP16). P_{hybrid} is responsive to both TetR-VP16, which establishes a positive autofeedback loop, as well as the erythromycin-responsive transrepressor (E-KRAB), which is independently expressed from a separate constitutive promoter (P_{SV40}). E-KRAB inhibits P_{hybrid} in an EM dose-dependent manner whereby a higher concentration of EM is required to switch the system from OFF to ON than is required to return the system from an ON to OFF state. The switching behavior of the network is therefore dependent upon the network's EM cultivation history [37].

transrepressor (E-KRAB), which was capable of inhibiting the TetR-VP16 mediated positive feedback. The hysteretic behavior of the synthetic network results from the competitive interaction of TetR-VP16 and E-KRAB for an engineered hybrid promoter (P_{hybrid}) that was responsive to both TetR-VP16 and E-KRAB. At low-EM concentrations E-KRAB binds P_{hybrid} and inhibits both TetR-VP16 positive feedback and SEAP expression (i.e., an OFF configuration). At high-EM concentrations, disassociation of E-KRAB from P_{hybrid} enables TetR-VP16 mediated transactivation resulting in positive autofeedback and high SEAP expression (i.e., an ON configuration). The observed hysteretic behavior occurs due to the interaction at intermediate EM concentrations where the prevalence of E-KRAB-mediated inhibition versus TetR-VP16-mediated positive feedback depends upon historical EM concentration. A high historical EM concentration means a high level of TetR-VP16 is already present, which therefore requires greater E-KRAB activity, and correspondingly lower EM concentration, to drive the expression state from ON to OFF. The converse applies for low historical EM concentrations where minimal to no TetR-VP16 is present. In this case a significantly higher EM concentration is required before TetR-VP16 autoexpression becomes self-sustaining. For TetR-VP16 to outcompete E-KRAB full derepression of all E-KRAB activity is required which is achieved through a relatively much higher EM concentration. In this process the level of active TetR-VP16, and therefore the extent of positive feedback, acts as a molecular "memory" of the historical EM concentration of the system. If the extent of positive feedback is reduced, for example through tetracycline addition, which reduces the level of active TetR-VP16 in the system, then the EM concentration required to switch the system between ON and OFF configurations begins to resemble a classical graded profile thereby removing the hysteretic effect. While it was not possible to test using the constructed system, it is plausible that if the positive feedback within the system could be rendered sufficiently strong, then even the complete removal of EM would not be sufficient to enable E-KRAB to outcompete TetR-VP16. In such an event the system would exhibit irreversibility.

The importance positive feedback mechanisms has long been recognized as essential for many cellular processes and is increasingly being identified in natural biological systems, including signaling pathways [29]. For example, the maturation of the *Xenopus* oocytes involves the p42 mitogen-activated protein kinase (MAPK) and the cell-division cycle protein kinase Cdc2, which form positive autofeedback loops. Both mediators generate an irreversible switch-like response following transient stimulation with the steroid hormone progesterone. If the feedback loops are selectively disrupted using specific inhibitors, progesterone-induced maturation can still occur, however, the presence of progesterone must be actively maintained. Thus, following disruption of positive feedback the ability of the system to "remember" a transient signal is compromised [35,38]. Using synthetic genetic networks it has now also been possible to empirically demonstrate the role of feedback mechanisms in ensuring expression stability.

The synthetic networks described above show that either a single positive feedback loop or a double negative feedback loop can result in bistability. Future work in designing synthetic systems, as well as the study of naturally occurring networks, may yet identify other mechanisms for switching and the generation of sustainable responses to transient stimuli.

15.3.5 Oscillator

Expression stability is a common element of all the aforementioned networks. Dynamic instability, in which transcriptional components are in a constant state of flux, can result in an equally exciting behavior characterized by periodic, as opposed to stable, expression. Where such periods are of a consistent period and amplitude, and require minimal to no external stimuli, the resulting behavior is oscillatory in character. Such behavior is found in a wide range of natural systems from archaebacteria to eukaryotes with the most well-known example being the circadian rhythm [39]. In humans, processes such as body temperature modulation, endocrine production and release, and immune responses exhibit circadian oscillations [40]. Circadian clocks have been proposed to consist of autoregulatory loops that use transcriptional feedback and high protein decay rates to maintain 24 h periodicity [41–43]. Similarly to the creation of expression stability, several synthetic approaches utilizing transcriptional feedback have successfully resulted in the creation of oscillatory behavior.

15.3.5.1 *Bacterial Oscillator ("Repressilator")*

Elowitz et al. constructed a plasmid-based synthetic oscillator in *E. coli* (termed the "repressilator") from three common bacterial transcriptional repressor systems that are not part of any natural biological clock mechanism [44]. The three repressor systems were interconnected such that they formed a cyclic negative feedback loop or "daisychain" (Fig. 15-8). This configuration produced oscillating levels of each repressor protein. A GFP reporter gene, carried not only on a separate plasmid but also under the control of a promoter induced by one of the repressors, provided a readout of oscillations for that repressor. A mathematical model was again used to predict the parameters required for steady and repeated oscillations. Key requirements included strong promoters with tight induction characteristics and minimal leakiness, cooperative repression, and comparable protein and mRNA decay rates. The first requirement was achieved using engineered *E. coli* promoters that exhibited similar strong induction profiles. To reduce repressor protein half-lives a bacterial destruction tag was fused to the 3'-end of each repressor. The reduction in repressor half-lives approximately from 60 to 4 min ensured that protein decay rates were similar to mRNA decay rates of a 2 min. Finally, to ensure a cyclical readout was technically observable, the GFP reporter gene was also engineered to reduce its effective half-life.

Initial attempts focused on determining whether oscillations could be observed across a population of cells. Using a transient dose of IPTG, an inhibitor of one of the repressors (LacI), an attempt was made to synchronize the population at a common point. While a single damped oscillation was subsequently observed, the lack of any mechanism to ensure the cells remained synchronized, meant that no further oscillations could be discerned at a population level. Although not performed in this case, cell synchronization could potentially be achieved by coupling the oscillating network to a periodic process that is intrinsic to the cell [45], or by using a quorum-sensing mechanism or other intercell signaling to ensure that cells remain synchronized [46–48]. Nonetheless, by following individual cells it was possible to observe repeated oscillations (Fig. 15-8). Despite high variability between cells, which were attributed to random stochastic influences, oscillatory periods of approximately 160 min were observed. Given *E. coli* cell division times of 50–70 min, the almost threefold longer oscillatory periods indicated that the state of the network could be successfully passed to progeny cells.

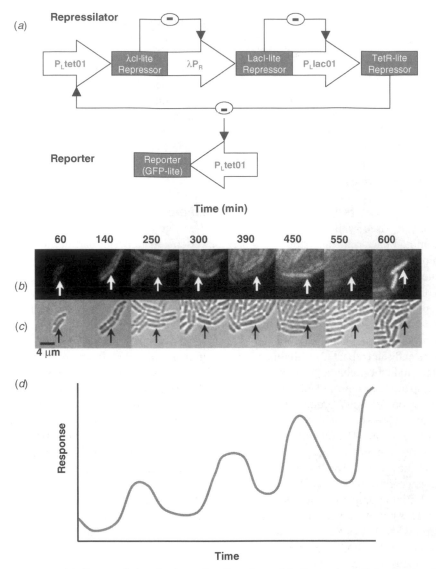

Figure 15-8 (a) Repressilation in bacteria: Genetic architecture of oscillatory network, (b) fluorescent and (c) bright-field snapshots of individual cells, and (d) GFP expression profile. The bacterial oscillatory network was constructed from three bacterial repressor systems arranged in a cyclical negative feedback loop. The first repressor protein, λcl, inhibits transcription of the second repressor protein LacI (from λP_R), which inhibits transcription of the third repressor protein TetR (from $P_L$lac01), which in turn inhibits expression of the first repressor (from $P_L$tet01) thereby completing the feedback loop. A reporter gene (GFP) under the control of a separate Tet-responsive promoter ($P_L$tet01) was used to assess oscillating TetR levels. By engineering short repressor and reporter half-lives (designated as lite) a dynamic unstable state was achieved in which TetR repressor levels cyclically rose and fell as evidenced through direct observation of individual cells and by GFP timecourse [44].

Shortly thereafter, the construction of another bacterial oscillator made from noncircadian components was reported [34]. In this case, it was constructed using a combination of positive (an "activator" module) and negative (a "repressor" module) feedback mechanisms (Fig. 15-9). Critical to producing a dynamic unstable outcome was the use of a hybrid promoter, capable of responding to both an activator and repressor, which effectively integrated the positive and negative feedback modules [49]. The resulting competitive interaction resulted in the "burst" like generation of activator and repressor proteins, which progressively smoothed over time. When coupled to a reporter system capable of measuring repressor levels, the result was a series of oscillations that progressively damped over time. However, in contrast to the bacterial repressilator developed by Elowitz et al., it was possible to synchronize a population of cells, via transient inhibition of the repressor, and observe up to four damped oscillations across the entire population. Oscillatory behavior, exhibiting periods close to 10 h, could be observed in continuous culture for up to 70 h again indicating that the network could be passed to subsequent progeny and that it was much more resistant to intrinsic noise than the "repressilator." Through mathematical modeling it was predicted that the key parameter causing damped, as opposed to sustained, oscillations in the system was the respective differences in half-lives between the activator and repressor proteins. Although not experimentally tested it was predicted that sustained oscillations could be achieved by increasing the half-life of the repressor whilst decreasing that of the activator.

15.3.5.2   A Mammalian Oscillator?   Unlike other expression functions, the development of a synthetic eukaryotic oscillator has not yet mirrored the creation of the bacterial equivalent, although given the pattern for these developments, it will not be surprising to see the emergence of a synthetic eukaryotic network in the near future. However, given the intense interest in understanding the mechanisms responsible for the natural circadian clock, it is also not surprising that attempts have been made to create a synthetic clock from actual clock components.

Using the core set of positive and negative regulatory elements common to all known circadian mechanisms, including the *cryptochrome* genes CRY1 and CRY2, the *period* genes PER1, PER2, and PER3, and the positive transcription factors BMAL1 and CLOCK [50–52], an attempt has been made to artificially engineer an oscillatory clock [53]. Among these components, BMAL1/CLOCK are positive regulators of CRY and PER proteins that, upon accumulation over a specific threshold, translocate to the nucleus where they negatively inhibit not only their own expression but also BMAL1/CLOCK. In this model, BMAL1/CLOCK mediated transcriptional inhibition is eventually relieved by PER and CRY degradation [54]. In the synthetic approach, BMAL1 and CLOCK expressions were placed on a tetracycline inducible "positive" regulation construct while PER, CRY, and a destabilized reporter gene were placed on a "negative" regulation construct in which their expression was under the control of a BMAL1/CLOCK/PER/CRY responsive promoter. Theoretically, turning the system ON by withdrawing tetracycline leads to BMAL1/CLOCK expression that subsequently drives expression from the negative regulation construct. Accumulation of PER/CRY eventually leads to autofeedback inhibition of the negative regulation

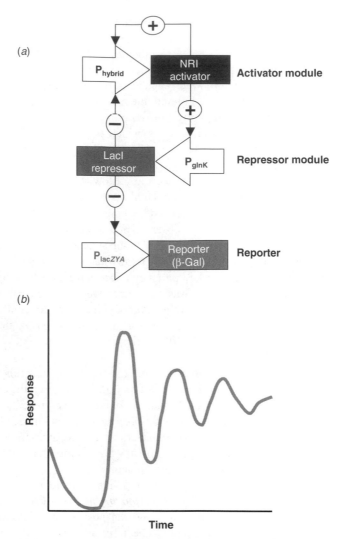

Figure 15-9 (a) Genetic design and (b) response profile of bacterial oscillatory network using positive and negative feedback mechanisms. A bacterial oscillatory network constructed from an activator and repressor module. The activator module forms a positive autofeedback loop in which the NRI transactivator activates its own expression from P_{hybrid}—a modified glnALB promoter (P_{glnk}) engineered to include LacI operator sites in addition to normal NRI operator sites. NRI also activates expression, via P_{glnk}, of the LacI repressor module which in turn repressors expression of NRI via P_{hybrid}. A reporter construct consisting of β-galactosidase and employing a LacI repressible promoter was used to assess oscillating levels of LacI repressor. Following synchronization with a transient pulse of IPTG (an inhibitor of the LacI repressor), up to four damped oscillations were observed across a cell population [34].

construct that, to complete the cycle, is relieved by eventual degradation of PER/CRY. In practice,this system was not capable of producing sustained oscillations, yet it did nevertheless exhibit a single cycle of a clock-like oscillation, which at a minimum establishes the possibility that homologous regulatory components can be used for synthetic constructions and that the design and creation of a successful mammalian clock will necessitate the incorporation of some kind of feedback mechanism. This latter conclusion is supported by recent experimental analysis of the mammalian clock system where directed disruption of CRY-mediated transcriptional autorepression resulted in arrhythmic phenotypes in both single- and multicell populations [51].

The aforementioned networks indicate that transcriptional feedback and feedforward processes are ubiquitous mechanisms for ensuring controlled expression whether that output is stable, binary, toggle, hysteretic, or even for periodic oscillating behavior. The further creation and characterization of synthetic networks will hopefully determine whether feedback is a minimal requirement for all networks or whether any other mechanisms could produce novel functional expression forms.

Alongside developments into unique expression states have been the concomitant development of novel means of integrating signals—so called "input" functionality. This has included the serial linking of transcriptional control systems to form transcriptional cascades, the creation of electronic circuit emulating logic gates, and the development of sophisticated sensors enabling cell-to-cell communication.

15.3.6 Transcriptional Cascades

Initial attempts in constructing regulatory cascades involved the construction of a two-level cascade using the $TetR_{OFF}$ and Lac_{ON} systems in mammalian cells [55]. In this simple system, the TetR-VP16 transactivator was constitutively expressed and, via a TetR-VP16 responsive promoter, drove the expression of a LacI repressor. LacI in turn inhibited expression of a reporter gene, via a LacI-inducible promoter. In this case, reporter gene expression could occur either in the presence of tetracycline, which prevents LacI expression, and/or in the presence of IPTG, which inhibits LacI repression of the reporter gene. This pioneering system established the basis for interconnecting gene control systems and successfully enabled the tight induction characteristics of the Tet_{OFF} system to be used to for an ON-type system (i.e., addition of tetracycline results in reporter gene expression), which in their native form (i.e., TetR-KRAB) do not typically exhibit such tight regulation. However, the high cytotoxicity of IPTG in mammalian cells is likely to prevent any clinical application of this technology.

In a very similar approach, Imhof et al. constructed a regulator network consisting of an engineered tetracycline-dependent transrepressor (TetR-KRAB) that controlled the expression of a Gal4-VP16 transactivator, which in turn controlled its own expression as well as a highly cytotoxic reporter gene (diptheria toxin A) [56]. Gal4-VP16 is an OFF-type system but exhibits typical residual leaky transcriptional control. Tight repression of Gal4-VP16 by TetR-KRAB ensured no reporter gene expression under noninduced conditions, whereas addition of tetracycline resulted in derepression of Gal4-VP16, subsequent autoexpression of further Gal4-VP16, and subsequent strong reporter gene expression. In this manner, a cascade was used together with a regulatory feedback loop to amplify the window of transgene regulation resulting ultimately in extremely tight transcriptional control.

15.3.6.1 Multilevel Gene Control

Most classical transcriptional control systems exhibit sigmoid-shaped dose–response characteristics where the range within the system flips between ON and OFF states is relatively narrow. As previously which mentioned, this is predominantly due to the transcriptional cooperativity inherent to most gene control systems. One consequence is that most current transcription control systems operate in an all or nothing manner (i.e., ON or OFF) and are not reproducibly capable of intermediate levels of adjustment. It is conceivable that future gene therapy applications will require precise dosing of therapeutic genes in much the same way that dosing of pharmaceuticals is critical to their successful application. All or nothing control mechanisms may therefore be of limited use. By combining several typical ON/OFF mechanisms in a network configuration, it has been possible to construct a gene control system where a target gene can be accurately and repeatedly titrated to intermediate levels [57].

Multilevel transgene control was achieved through the cascade arrangement of three heterologous control systems: the tetracycline (Tet_{OFF}), macrolide (E_{OFF}), and streptogramin (PIP_{OFF}) systems (Fig. 15-10). As these systems and their inducers (tetracycline, erythromycin, and pristinamycin, respectively) exhibit minimal to no cross-interference, it was possible to connect them in a linear type fashion whereby each system acts as the activator of the next system. All of the selected systems were OFF-type systems in which transcription is active in the absence of inducer and repressed following addition of inducer. Here, addition of each respective inducer prevents transcription of the next component in the cascade. However, as all of these systems exhibit minimal residual expression following addition of inducer (referred to as "leakiness"), there is nonetheless some activation of lower levels in the cascade. The impact of this leakiness on total expression levels depends upon the point in the cascade at which it occurs. Thus, at "upstream" points within the cascade, transcriptional leakiness is amplified by latter stages thereby limiting the extent of overall OFF switching. For "downstream" interventions within the cascade there is minimal opportunities for transcriptional leakiness to be amplified. The result is that upstream interventions have less impact on overall expression than downstream interventions. Using different inducers it is possible to select the desired intervention point as each inducer affects a different point in the cascade. Thus, expression levels of 100 percent (no cascade intervention), 70 percent (intervention at first level of cascade), 40 percent (intervention at second level), and close to 0 percent (intervention at third and final level) of a target reporter gene were possible. This genetic network demonstrated that the typical ON/OFF switching characteristics of current control systems, together with residual inherent leakiness, could be exploited to produce a system capable of intermediate expression levels in response to up to three different inputs [57].

15.3.6.2 Regulation Sensitivity

In a similar experimental approach, but with a different outcome, up to three bacterial transcriptional repressors were linked in a linear cascade [58]. Unlike the heterologous systems employed above, homologous bacterial

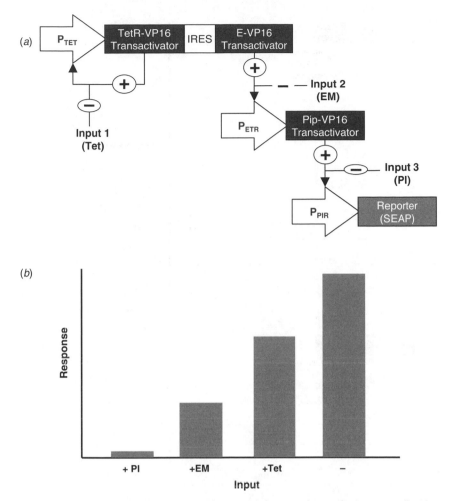

Figure 15-10 Network design (a) and regulation performance of a synthetic mammalian three-level regulatory cascade.The three-level regulatory cascade consists of three heterologous interconnected gene transcription systems. The tetracycline responsive promoter (P_{hCMV^*-1}) drives a dicistronic expression unit encoding the tetracycline-dependent transactivator (TetR-VP16) and, via an IRES, the macrolide-dependent transactivator (E-VP16). E-VP16 subsequently drives expression, via a macrolide-responsive promoter (P_{ETR}), of a streptogramin-responsive transactivator (Pip-VP16). Finally, Pip-VP16 drives expression of the reporter gene human placental secreted alkaline phosphatase (SEAP) from a streptogramin-responsive promoter (P_{PIR}). The linear arrangement ensures that SEAP expression can be controlled from a number of levels. Shutting off expression at the top of the cascade, by inhibiting the autofeedback loop controlling TetR-VP16 expression with tetracycline (Tet), reduces overall expression to approximately 70 percent of maximum noninduced expression. Closing the cascade further downstream, by inhibiting E-VP16 with erythromycin, has a greater impact reducing total expression to approximately 30 percent. Finally, interventions at the bottom level of the cascade, through inhibition of Pip-VP16 with PI, reduces expression within the system to almost baseline levels [57].

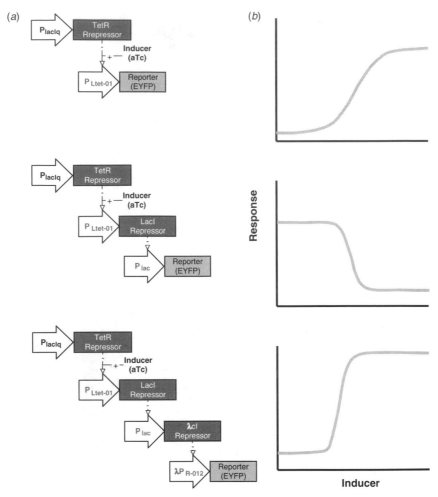

Figure 15-11 (a) Genetic architecture and (b) dose–response curves of single-, two-, and three-level transcriptional cascades. Transcriptional cascades were assembled using up to three bacterial repressors linked to each other in a linear fashion. In each case the tetracyline repressor (TetR) was constitutively expressed, induced with aTc, and system output measured by enhanced yellow fluorescence protein (EYFP) production. In the single-level cascade, TetR bound the $P_{Ltet-01}$ promoter where it directly repressed EYFP production. In the two- and three-level cascades, TetR repressed production of a second repressor (LacI), also from $P_{Ltet-01}$. In the two-level cascade, LacI repressed production of EYFP from P_{lac}. In the three-level cascade, LacI repressed production of yet a third repressor, (λCI) from P_{lac} which in turn repressed production of EYFP from λP_{R-012}. Dose–response curves for the three types of cascades reveal that the inducer range needed to effect a change between ON and OFF states narrows with the length of the cascade thereby increasing sensitivity to the inducer [58].

repressors exhibit much tighter regulation performance with virtually no leakiness. Hence, rather than creating multilevel gene control, the aim was to investigate the impact of multilevel cascades on the regulation performance of a typical bacterial repression system, the TetR system (Fig. 15-11). Three versions were compared; a single-level cascade—where TetR directly represses expression of a reporter gene, a two-level cascade—where TetR represses transcription of a second repressor which in turn controls the reporter gene, and a three-level cascade—where yet another repressor system was introduced between the second repressor and reporter gene.

Dose–response experiments indicated that the number of levels, or depth, of a cascade has a significant impact upon a number of regulation characteristics. First, the sensitivity of the cascade increases with the depth of the cascade. Thus, the system switches between low and high from a smaller range of input values. Second, the extent of noise within the system, as seen by variation in fluorescence across a population, while minimal at input ranges far from the transition region, increases with the length of the cascade. Deeper cascades serve to amplify the noise around the transition point presumably due to the extra number of transition points involved. This may limit the utility of adding even further cascade levels as additional increases in noise amplification around transition points may ultimately offset any further sensitivity gains. Third, the delay in the output response of the system increases commensurately with the depth of the cascade. This is to be expected and is largely the result of protein production and decay rates, and repression thresholds. Interestingly, there is evidence that time delays caused by regulatory cascades may actually be a design parameter required for many natural gene networks [59]. Database analyses of natural networks, which are involved in rapid and reversible gene expression in response to external stimuli (so called "sensory" transcriptional networks), reveal that such networks generally contain short regulatory cascades. Networks involved in slow and irreversible gene expression during development (so called "developmental" transcriptional networks) typically contain longer cascades.

15.3.7 Logic Gates

The expression output of many cell-based regulatory networks is often a logic response generated by one or more input signals. Due to their sigmoid-shaped dose–response curves, most gene control systems can be regarded as the genetic equivalent of an analog-to-digital converter. Their output is either ON or OFF across a wide range of inducer concentrations, except for a small concentration window where transitions between the two states occur. In this regard, the analogy between genetic networks and electronic circuitry is very compelling. This has led to the conceptualization of genetic networks as logic gates with switchboard-type truth-tables and schematic representations that directly mirror electronic circuit diagrams [60–62]. Adapting gene control systems to Boolean language, ON-type gene control systems represent IF type gates in the sense that expression results IF an input is present. Conversely, OFF-type gene control systems represent NOT type gates whereby expression results when an input is NOT present.

By utilizing several compatible heterologous gene control systems responsive to tetracycline, macrolide, streptogramin, and butyrolactone input signals, it has been

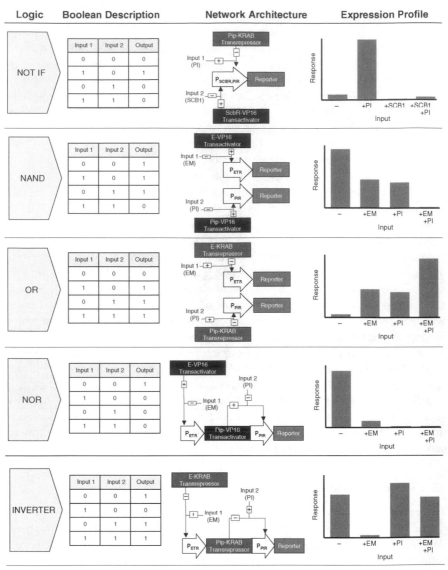

Figure 15-12 Boolean description, network architecture, and expression profile of five mammalian BioLogic Gates. All five mammalian logic gates were constructed from heterologous mammalian transcription systems. In the NOT IF gate, the butyrolactone-responsive transactivator (ScbR-VP16) and the streptogramin-responsive transrepressor (Pip-KRAB) are constitutively expressed and modulate expression of a reporter gene from a chimeric promoter ($P_{SCBR,PIR}$) containing operator sites for both ScbR-VP16 and Pip-KRAB. Input signals, 2-(1'-hydroxy-6-methylheptyl)-3-(hydroxymethyl) butanolide (SCB1) and/or PI result in disassociation of ScbR-VP16 and Pip-KRAB respectively. Expression only occurs when ScbR-VP16 is bound to the chimeric promoter and Pip-KRAB is disassociated therefore requiring the absence of SCB1 and presence of PI. For the NAND gate, both the macrolide-responsive transactivator (E-VP16) and the streptogramin-responsive transactivator (Pip-VP16) are constitutively expressed. Each transactivator binds its cognate promoter (P_{ETR} and P_{PIR}, respectively) which drive separate expression of two copies of

possible to design a range of eukaryotic logic circuits that follow strict Boolean logic in their integration of two input signals (Fig. 15-12) [63]. Hence, in the NOT IF gate, expression of a reporter gene occurs if and only if one specific input is present and the other input is absent. In the NAND gate, expression always occurs unless both inputs are present. The converse, where expression always occurs unless both inputs are absent, is reflected in the OR gate. The inverse, where expression occurs only when both inputs are absent is reflected in the NOR gate. Finally, the INVERTER gate represents the opposite of the NOT IF gate whereby expression always occurs unless one specific input is present and the other input is absent. Analogously to electronic circuit design some of these networks were constructed by linking elements in parallel while others were constructed by combining elements in series through the use of simple transcriptional cascades. These examples demonstrate that a considerable range of logical switches responding in unique ways to the same two input signals can be constructed from modular transcriptional control components. It is imaginable that such networks could be highly useful for gene therapy applications that require a particular response to highly specific inputs, which could vary depending upon the application.

Similar to electronic circuit design, the above switches were based on rational design principles. However, a number of other approaches have also been used to produce electronic-type circuit behavior, which produce a defined output in response to two inputs. Guet et al. used a combinatorial method involving prokaryotic transcriptional control systems that were randomly combined to generate a library of networks with varying connectivity [64]. From this library it was possible to isolate and characterize a range of diverse computational functions that produced unique phenotypes. While such an approach may yield unexpected network architectures for

the same reporter gene. Input signals, EM and/or PI, modulate transactivator activity respectively. Expression occurs when either or both transactivators are bound to their cognate reporter. The presence of both EM and PI are required to disassociate both transactivators to prevent expression. The OR gate is identical in design to the NAND gate but uses the transrepressor versions (i.e., E-KRAB and Pip-KRAB) of the macrolide- and streptogramin-responsive transcription control systems. Again, EM and/or PI modulate transrepressor activity respectively. In this case, expression is blocked only when both transrepressors are operator bound which only occurs when both EM and PI are absent. The NOR gate involves a short linear cascade between a constitutively produced macrolide-responsive transactivator (E-VP16) which drives the expression, via its cognate promoter (P_{ETR}), of the streptogramin-responsive transactivator (Pip-VP16) which in turn drives expression, via its cognate promoter (P_{PIR}), of a reporter gene. Modulation of transactivator activity is achieved through EM and PI, respectively. In this configuration, expression only occurs when E-VP16 is bound to its cognate operator and Pip-VP16 is disassociated from its cognate promoter therefore requiring the absence of both EM and PI. The final gate, the INVERTER, is identical in design to the NOR gate but uses the transrepressor versions (i.e., E-KRAB and Pip-KRAB) of the macrolide and streptogramin responsive transcription control systems. Again, EM and/or PI modulate transrepressor activity respectively. The only conditions under which expression will not occur are when E-KRAB is promoter disassociated and Pip-KRAB is promoter associated which occurs in the presence of EM and absence of PI. For each gate, the input and output characteristics of the Boolean description are reflected in the expression profile of the synthetic system [63].

a given function, the approach is not particularly amenable to forward engineering approaches that seek to design circuits that exhibit specifically required functions. In a related approach, Yokobayashi et al. combined rational design with an evolutionary approach to design specific circuits in *E. coli* [65]. Rational design based upon existing knowledge of well-characterized components was initially used to design a network with a specific function. Given that the synthesized network exhibited sub-optimal behavior, due to unexpected interactions and poor matching of network components, a directed-evolutionary approach was then used to fine-tune (or "debug") the system to obtain the required function. This was achieved through sequential rounds of localized random mutagenesis and recombination followed by phenotype screening. Subsequent sequence analysis of successful networks revealed that many changes, or "solutions", were capable of producing the desired phenotype. This could be manifested in changes which altered either protein-DNA or protein–protein interactions, but which nonetheless enabled superior biochemical matching of genetic components.

15.3.8 Sensory Networks

15.3.8.1 Signal Amplification To extend the electrical circuit analogy further, Karig and Weiss recently developed a highly effective signal-amplifier from prokaryotic bacterial control systems [66]. Their aim was to try and develop a means for detecting weak transcriptional responses that, despite being difficult to detect *in vivo*, are often involved in regulatory functions where only trace amounts of a gene product are required. In typical transcriptional studies aimed at determining the conditions under which a promoter is activated, a reporter gene is placed downstream of the promoter and assayed under varying conditions. However, where the promoter response is weak it is often not possible to discern any kind of activity. By placing a repressor cascade downstream of the promoter it was possible to amplify an otherwise undetectable promoter response. In their system, Karig and Weiss placed the λcI repressor downstream of several Rhl quorum sensing (qsc) promoters from *Pseudomonas aeruginosa*. By coupling the repressor to a fluorescent reporter, under the control of a $\lambda P_{(R-O12)}$ promoter, they were able to monitor the response of selected promoters to acyl-homoserine lactones (AHL). As λcI is a highly efficient repressor, even very low concentrations of λcI can completely repress $\lambda P_{(R-O12)}$ thereby altering the fluorescent reporter readout. The amplifying cascade allowed up to 100-fold differences in fluorescence to be observed, between AHL-induced and -noninduced conditions, for promoters whose responses were otherwise not detectable. Apart from illustrating a biological means by which weak transcriptional responses can be amplified, the amplifying circuit could potentially be useful for a number of applications including the detection of trace toxins or molecules.

15.3.8.2 A Band-Detection Network One can imagine it would be useful for a range of applications to design an input mechanism that can respond to an inducer within a given concentration range, or perhaps one which is capable of a transient response when a progressively increasing inducer reaches a threshold concentration.

In a series of innovative synthetic constructions Basu and colleagues recently created synthetic networks capable of such behavior in *E. coli* [67–69].

The key requirements for band-detection network are the design of modular components that enable the detection of a low-threshold, a high threshold, and a means of integrating the two thresholds. In this case, this was achieved by exploiting differences in repressor activities, and by linking several bacterial repressor systems (Fig. 15-13) [67]. Guided by mathematical analysis the band-detection thresholds were engineered by combining high-detection and low-detection componentry.

For both components the initial input was the same and was represented by the extent of LuxR activity—a bacterial activator that is activated by the inducer compound AHL. In the high detection componentry, the LuxR activator drives expression, via its cognate promoter, of a weakened secondary repressor, $LacI_{M1}$ which if present in sufficiently high quantities prevents expression of a reporter gene from the P_{lac} promoter. Thus, the boundary of the high threshold is determined by the amount of AHL required to produce enough $LacI_{M1}$ to repress the P_{lac} promoter that in turn depends upon the relative activity of the $LacI_{M1}$ repressor. The low detection componentry also relies on the LuxR activator, but to express the strong λcI repressor. This in turn is coupled via a transcriptional cascade to production of wild-type LacI, which, like the $LacI_{M1}$ repressor, also represses expression of the reporter gene. In this case, the boundary of the low threshold is the lowest amount of AHL required to prevent λcI expression thereby enabling the wild-type LacI repressor to be fully expressed resulting in reporter gene repression. It is only between the two thresholds that both the high and low detection componentry fail to repress the reporter gene. Hence, the relative activity of the $LacI_{M1}$ repressor and the AHL concentration that results in λcI expression are the two key components that determine the size and location of the band-detection characteristics. By altering the activity of the $LacI_{M1}$ repressor, Basu et al. were able to create three versions of the band-detection network each with differing upper detection limits.

15.3.8.3 A Pulse-Generating Network

Basu et al. also utilized the above bacterial componentry to develop a network capable of producing a transient pulse when exposed to increasing concentrations of AHL [69]. The pulse-generating network produces output when a threshold concentration of increasing AHL is reached, and then through a feedforward mechanism shuts down reporter expression regardless of whether AHL concentration continues to rise or fall [70]. In this network AHL again activates LuxR, which in this case is constitutively present. Activated LuxR activates both a destabilized λcI repressor as well as directly activating reporter gene expression via a chimeric hybrid promoter responsive to both LuxR and λcI. Hence, increasing levels of AHL initially trigger both reporter and λcI expression. Following a delay, λcI accumulates to a sufficient extent where it eventually shuts down reporter expression. Like the band-detection network the pulse-generating network provides important insights into how pulse-generating behavior could occur in natural systems.

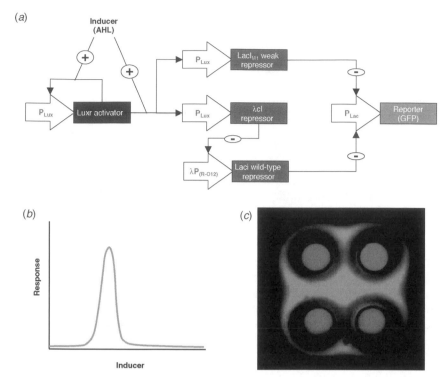

Figure 15-13 (a) Genetic architecture, (b) dose–response profile, and (c) pattern formation of a band-detection network.(a) In the band-detection network acyl-homoserine lactone (AHL) binds LuxR, an AHL-dependent transcriptional activator, which is produced in an autoregulatory manner from a P_{LuxR} promoter. LuxR also drives (from P_{LuxR} promoters) the expression of a λcl repressor and a weakened form of the Lacl repressor (Lacl$_{M1}$). The λcl repressor is coupled to a further regulatory cascade which represses the expression of wild-type Lacl from a $\lambda P_{(R-O12)}$ promoter. Both the weakened and wild-type Lacl repress expression, to a different extent, of a green fluroscence reporter gene (GFP) from a P_{Lac} promoter. At low AHL concentrations, LuxR is not active such that only basal levels of both Lacl$_{M1}$ and λcl repressors are produced. The absence of λcl ensures that wild-type Lacl is fully expressed which consequently represses GFP expression. At high AHL concentrations, the LuxR activator drives both high Lacl$_{M1}$ and λcl expression. The presence of λcl ensures that wild-type Lacl is completely repressed. However, as sufficiently high concentrations of Lacl$_{M1}$ are expressed, GFP expression remains nonetheless repressed. It is only at intermediate concentrations of AHL that a balance is reached between sufficiently low expression of Lacl$_{M1}$ to prevent Lacl$_{M1}$-mediated repression of P_{Lac}, and sufficiently high expression of λcl to prevent Lacl expression and consequent Lacl-mediated repression of P_{Lac}. At this point insufficient repression from either Lacl repressor results in GFP expression. (b) In an AHL dose–response curve, GFP expression is only observed within a band of AHL concentration. (c) If AHL is chemically produced and allowed to diffuse from a defined set of "sender" cells (exhibiting red fluorescence) placed within a lawn of "receiver" cells containing the band-detection network, the resulting AHL gradient produces a distinctive green fluorescence pattern based upon the spatiotemporal location of the receiver cells to the sender cells [67].

15.3.8.4 Cellular Cross Talk and Intercell Communication   The band-detection and pulse-generation networks developed by Basu and colleagues have been successfully used to generate spatiotemporal differentiation patterns that, much like natural pattern formation, rely upon cell-to-cell communication and signal transduction networks [67,69]. In initial work, it was demonstrated that "sender" cells engineered to produce AHL could influence "receiver" cells endowed with synthetic networks capable of responding to AHL, which has diffused from the sender cells [71]. In pattern formation experiments cell-to-cell communication was commenced from "sender" cells, which produced an AHL concentration gradient. In one set of examples, receiver cells containing the band-detection network responded to the chemical gradient and at intermediate distances from the sender cells expressed their reporter gene in accordance with the AHL detection thresholds within their band-detection network (Fig. 15-13). Such cell-to-cell communication or cellular cross talk could be engineered to result in a range of patterns, and by altering the thresholds of the band-detection network and using different fluorescent reporter genes an impressive array of multicolored patterns and shapes could be produced. In addition to representing a sophisticated genetic network, the formation of patterns from a synthetic network together with cellular communication represents a significant step toward reproducing and understanding natural developmental processes. In addition to pattern formation, intercellular communication could also be used to ensure synchronization of cellular populations. Using *E. coli* as a model system it has been demonstrated that synthetic gene networks can be used to engineer an artificial quorum-sensing mechanism that utilizes a common cellular metabolite [72].

15.4 SEMISYNTHETIC NETWORKS

The majority of synthetic genetic networks built and characterized to date have utilized external signals to create a desired function. To reach their therapeutic potential, however, it will be necessary to design networks that are capable of responding not only to external signals but also to endogenous or physiological signals. Hence, one can imagine sophisticated networks that independently provide a therapeutic outcome in response to pathological signals, and can also be overridden or altered through external modulation should the need arise. While still in their infancy, several systems integrating physiological signals—so called "semisynthetic" systems—have already been developed.

In *E. coli*, semisynthetic systems have been designed, which interface various physiological inputs into a bacterial toggle network thereby producing a sustainable switch-like response to a transient physiological input [73]. A DNA damage sensing network was constructed by interfacing the SOS pathway to a bacterial toggle. The SOS pathway detects single-stranded DNA following DNA damage by activating RecA coprotease. Activated RecA subsequently cleaves the λcI repressor in the toggle circuit causing derepression of the λP_{R-O12} promoter, and a sustainable switch to high LacI production (the other repressor in the toggle circuit). If LacI production is linked to a fluorescent output, the system can detect and retain a memory of transient

DNA damage. In an alternate application, if the fluorescent reporter gene is substituted for a biofilm producing gene, transient DNA damage can induce the cells to commence biofilm production. In a separate example, the transgenic AHL quorum sensing pathway was interfaced to the toggle circuit. When AHL reached sufficiently high levels, (e.g., if cell density reaches a critical density), the AHL-dependent activator LuxR repressed the P_{Lac} promoter of the toggle thereby leading to high λcl expression. This semisynthetic system is capable of producing a sustainable output once cell density reaches a critical threshold.

In other prokaryotic systems, Farmer and Liao developed a feedback controller in *E. coli* in which the expressed genes are key enzymes in the lycophene biosynthesis pathway [74]. By engineering the genes to be under the control of a physiological metabolite that is present during periods of high glycolytic flux, it was possible to coordinate lycophene production with the energy status of the cell thereby preventing metabolic imbalance and suboptimal productivity. Using glycolytic flux as a physiological cue Liao and colleagues have also developed an oscillatory network that is coupled to *E. coli* host metabolism (termed the "metabolator") [75]. In this system, a steady state is dependent upon the relative state of two metabolic "pools" which under high glycolytic flux result in instability in the engineered network with consequent oscillations.

Progress has also been made in developing mammalian semisynthetic systems. The mammalian oxygen response system, in which a specific set of endogenous genes is induced in response to low oxygen levels (e.g., VEGF), relies upon the translocation of hypoxia-induced factor 1 alpha (HIF-1α) to the nucleus where through a series of interactions it activates expression from promoters containing hypoxia-response elements (HRE). Under normoxic conditions, HIF-1α is rapidly degraded thereby preventing the low-oxygen response [76,77]. A semisynthetic network has been created by coupling the HIF-1α response system to a mammalian heterologous regulatory cascade resulting in multilevel gene control that can be influenced by endogenous signals (i.e., oxygen levels) as well as external signals (Fig. 15-14) [78]. By combining three inputs, it has been possible to produce six distinct expression states depending upon the combination of signals used.

While representing the first steps toward the therapeutic application of synthetic networks, a major challenge remains to find and/or preferably design transcription control systems that not only detect changes to a specific endogenous inducer but also detect changes within a specified concentration range. The systems constructed to date have largely relied upon serendipity and have sufficed as a proof of concept. Yet to reach their true potential, one will need to find means of detecting and interfacing changes to pathologically relevant molecules.

15.5 THE INFLUENCE OF "NOISE"

A major influence upon the fidelity and function of both synthetic and natural genetic networks is noise. Noise is evidenced by high fluctuation in expression levels, which if sufficiently high enough may produce very different network outcomes both within

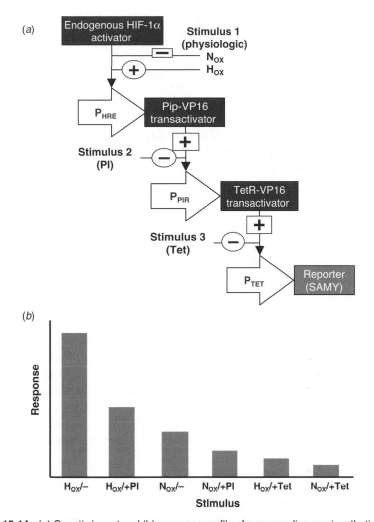

Figure 15-14 (a) Genetic layout and (b) response profile of a mammalian semisynthetic regulatory cascade. The semisynthetic cascade is triggered by endogenous HIF-1α that, under hypoxic conditions (H_{OX}), is mobilized to the nucleus where it binds and activates a synthetic promoter containing hypoxia-response elements (P_{HRE}). Under normoxic conditions (N_{OX}) HIF-1α is rapidly degraded to undetectable levels. Activation of P_{HRE} sets off a transcriptional cascade of two heterologous transcription systems; the streptogramin-responsive transactivator Pip-VP16 which upon expression binds its cognate promoter (P_{PIR}) leading to expression of the tetracycline-responsive transactivator TetR-VP16, which subsequently binds its cognate promoter (P_{TET}) leading to expression of a SAMY *(Bacillus stearothermophilus* derived secreted α-amylase) reporter gene. In addition to sensing physiologic oxygen levels via the HIF-1α activator, the system is also responsive to PI which interrupts the cascade at Pip-VP16, and tetracycline (Tet), which interrupts the cascade at TetR-VP16. Up to six expression levels can be produced by different combinations of the three inputs [78].

and across a cell population. Noise can be generated from a number of sources. In many circuits the genetic components required for gene expression, such as promoter sites, are typically present at very low copy levels. The result is that biochemical rates of transcription and translation are therefore correspondingly low and, compared to other cellular interactions (e.g., protein–protein interactions), occur relatively infrequently. Such infrequency can lead to large fluctuations, which due to their origination within the genetic circuit, are referred to as internal or intrinsic noise. Noise can also be generated externally or extrinsically of the circuit not only through stochastic variation in cellular components required for gene expression (e.g., polymerase, transcription, translation factors, and so on) but also through environmental or global changes that impact all gene activity (e.g., cell division). Given the modular nature of artificial gene networks, and the ability to rationally design them from well-understood components, it has been possible to gain insight into how noise is created, how it is propagated through a network, and finally instances where the existence of noise is in fact crucial to the output function of a network [79]. A significant body of work has focused upon designing synthetic networks to experimentally test predictions relating to noise phenomena. Such insights have also been crucial for the later design of noise-tolerant networks [80–82].

In early work on prokaryotic systems, it was shown that both intrinsic and extrinsic sources contribute significantly to the generation of noise that places certain inherent limits on the precision of gene expression [83,84]. It was also shown that the lower the effective strength of a promoter, whether through reduced gene copy number, repression, or a different cellular environment, the greater the extent of noise [83]. Other work aimed at determining whether transcription or translation is the major point of noise creation suggests that differences in translational efficiency have a greater impact than transcriptional differences [85] and that translational differences can result in variation which persists long after intrinsic noise from transcription has decayed [86].

The design and characterization of toggles, oscillators, and regulatory cascades has led to insights into how noise is propagated through a network [80,82]. As previously described, the presence of a simple autofeedback mechanism can reduce noise in a genetic network [18] and this has certainly been suggested as a major function of both positive and negative feedback mechanisms [70,87]. Failures in such mechanisms have been attributed to certain disease states. For example, the transformed phenotype in tumor formation has in some cases been attributed to instability of autocrine positive feedback loops [88]. Network connectivity can also be a major cause of noise. Using a bacterial regulatory cascade, Pedraza and van Oudenaarden demonstrated that the connectivity of sequential network components can result in an a total variation that is greater than the variation intrinsic to the expression of each component gene [84]. This implies that variation in a cascade can be cumulative. Hooshangi et al. in characterizing differences in their three-level bacterial regulatory cascade (described above) also witnessed greater variability around input transition points [58]. Indeed, the extent of variation can be so significant in a cascade that variations in an upper-level cascade within a cell population can cause the population to display bistable expression states [49,89–91]. That stochastic fluctuation can be crucial for generation of cellular phenotypes is becoming more and more evident. Noise can establish an initial asymmetry that, once propagated and amplified through a network, may result in phenotypic consequences that impact processes such as differentiation and disease [90,92,93].

15.6 CONCLUSION

All of the engineered genetic networks thus described have utilized at least some aspects of rational design to produce a behavior that is based upon the modular interaction of DNA sequences and regulatory proteins. By assembling molecular parts not normally associated with each other into different configurations, it has been possible to produce an already impressive array of robust network behaviors. As the number of available modules increases, and their kinetic parameters become better characterized, and our ability to model and predict their interaction continues to improve, it is inevitable that further novel and increasingly more sophisticated synthetic networks will be created. Existing synthetic networks have already provided important insights and confirmation of hypothesis on a range of natural phenomena such as importance of feedback mechanisms, of balanced genetic componentry, of regulatory cascades, and of noise to name a few. It can also be expected that engineered gene networks will have many important biotechnological and therapeutic applications all of which aim to manipulate cellular processes at the genetic level.

REFERENCES

1. Weber W, Fussenegger M. Inducible gene expression in mammalian cells and mice. *Methods Mol Biol* 2004;267:451–466.

2. Weber W, Fussenegger M. Approaches for trigger inducible viral transgene regulation in gene-based tissue engineering. *Curr Opin Biotechnol* 2004;15:383–391.

3. Weber W, Fussenegger M. Pharmacologic transgene control systems for gene therapy. *J Gene Med* 2006;8:535–556.

4. de Boer HA, Comstock LJ, Vasser M. The tac promoter: a functional hybrid derived from the trp and lac promoters. *Proc Natl Acad Sci USA* 1983;80:21–25.

5. Skerra A. Use of the tetracycline promoter for the tightly regulated production of a murine antibody fragment in *Escherichia coli*. *Gene* 1994;151:131–135.

6. Elvin CM, Thompson PR, Argall ME, Hendry P, Stamford NP, Lilley PE, Dixon NE. Modified bacteriophage lambda promoter vectors for overproduction of proteins in *Escherichia coli*. *Gene* 1990;87:123–126.

7. Lutz R. Bujard H, Independent and tight regulation of transcriptional units in *Escherichia coli* via the LacR/O, the TetR/O and AraC/I1-I2 regulatory elements. *Nucleic Acids Res* 1997;25:1203–1210.

8. Kramer BP, Fussenegger M. Transgene control engineering in mammalian cells. *Methods Mol Biol* 2005;308:123–143.

9. Bellefroid EJ, Poncelet DA, Lecocq PJ, Revelant O, Martial JA. The evolutionarily conserved Kruppel-associated box domain defines a subfamily of eukaryotic multi-fingered proteins. *Proc Natl Acad Sci USA* 1991;88:3608–3612.

10. Triezenberg SJ, Kingsbury RC, McKnight SL. Functional dissection of VP16, the trans-activator of herpes simplex virus immediate early gene expression. *Genes Dev* 1988;2: 718–729.

11. Gossen M., Bujard H. Tight control of gene expression in mammalian cells by tetracycline-responsive promoters. *Proc Natl Acad Sci USA* 1992;89:5547–5551.

12. Fussenegger M., Morris RP, Fux C, Rimann M, von Stockar B, Thompson CJ, Bailey JE. Streptogramin-based gene regulation systems for mammalian cells. *Nat Biotechnol* 2000;18:1203–1208.

13. Weber W, Fux C, Daoud-el Baba M, Keller B., Weber CC, Kramer BP, Heinzen C, Aubel D, Bailey JE, Fussenegger M. Macrolide-based transgene control in mammalian cells and mice. *Nat Biotechnol* 2002;20:901–907.

14. Alon U, Surette MG, Barkai N, Leibler S. Robustness in bacterial chemotaxis. *Nature* 1999;397:168–171.

15. Barkai N, Leibler S. Robustness in simple biochemical networks. *Nature* 1997;387: 913–917.

16. Little JW, Shepley DP, Wert DW. Robustness of a gene regulatory circuit. *EMBO J* 1999;18:4299–4307.

17. Savageau MA. Comparison of classical and autogenous systems of regulation in inducible operons. *Nature* 1974;252:546–549.

18. Becskei A, Serrano L. Engineering stability in gene networks by autoregulation. *Nature* 2000;405:590–593.

19. Thieffry D., Huerta AM, Perez-Rueda E, Collado-Vides J. From specific gene regulation to genomic networks: a global analysis of transcriptional regulation in *Escherichia coli*. *Bioessays* 1998;20:433–440.

20. Bateman E. Autoregulation of eukaryotic transcription factors. *Prog Nucleic Acid Res Mol Biol* 1998;60:133–168.

21. Kringstein AM, Rossi FM, Hofmann A, Blau HM. Graded transcriptional response to different concentrations of a single transactivator. *Proc Natl Acad Sci USA* 1998;95: 13670–13675.

22. Becskei A, Seraphin B, Serrano L. Positive feedback in eukaryotic gene networks: cell differentiation by graded to binary response conversion. *EMBO J* 2001;20:2528–2535.

23. Hasty J, McMillen D, Collins JJ. Engineered gene circuits. *Nature* 2002;420:224–230.

24. Gardner TS, Cantor CR, Collins JJ. Construction of a genetic toggle switch in *Escherichia coli*. *Nature* 2000;403:339–342.

25. Feng XJ, Hooshangi S, Chen D, Li G, Weiss R, Rabitz H. Optimizing genetic circuits by global sensitivity analysis. *Biophys J* 2004;87:2195–2202.

26. Kramer BP, Viretta AU, Daoud-El-Baba M, Aubel D, Weber W, Fussenegger M. An engineered epigenetic transgene switch in mammalian cells. *Nat Biotechnol* 2004;22: 867–870.

27. Orlando V. Polycomb, epigenomes, and control of cell identity. *Cell* 2003;112:599–606.

28. Kohler C, Grossniklaus U. Epigenetic inheritance of expression states in plant development: the role of Polycomb group proteins. *Curr Opin Cell Biol* 2002;14:773–779.

29. Angeli D, Ferrell JE Jr, Sontag ED. Detection of multistability, bifurcations, and hysteresis in a large class of biological positive-feedback systems. *Proc Natl Acad Sci USA* 2004;101:1822–1827.

30. Casadesus J, D'Ari R. Memory in bacteria and phage. *Bioessays* 2002;24:512–518.

31. Wegener HC, Bager F, Aarestrup FM. Surveillance of antimicrobial resistance in humans, food stuffs and livestock in Denmark. *Euro Surveill* 1997;2:17–19.

32. Kapunisk-Uner JE, Sande MA, Chambers HFS. Antimicrobial Agents: Tetracyclines, Chloramphemical, Erythromycin, and miscellaneous Antibacterial Agents. *Goodman and Gilman's The Pharmacological Basis of Therapeutics.* In: Limbierd LE, editors. New York: McGraw-Hill, 1996 Chapter 47: pp 1123–1154.

33. Ozbudak EM, Thattai M, Lim HN, Shraiman BI, Van Oudenaarden A. Multistability in the lactose utilization network of *Escherichia coli*. *Nature* 2004;427:737–740.

34. Atkinson MR, Savageau MA, Myers JT, Ninfa AJ. Development of genetic circuitry exhibiting toggle switch or oscillatory behavior in *Escherichia coli*. *Cell* 2003;113:597–607.

35. Xiong W, Ferrell JE Jr. A positive-feedback-based bistable 'memory module' that governs a cell fate decision. *Nature* 2003;426:460–465.

36. Sha W, Moore J, Chen K, Lassaletta AD, Yi CS, Tyson JJ, Sible JC. Hysteresis drives cell-cycle transitions in *Xenopus laevis* egg extracts. *Proc Natl Acad Sci USA* 2003;100: 975–980.

37. Kramer BP, Fussenegger M. Hysteresis in a synthetic mammalian gene network. *Proc Natl Acad Sci USA* 2005;102:9517–9522.

38. Ferrell JE Jr. Self-perpetuating states in signal transduction: positive feedback, double-negative feedback and bistability. *Curr Opin Cell Biol* 2002;14:140–148.

39. Schibler U, Sassone-Corsi P. A web of circadian pacemakers. *Cell* 2002;111:919–922.

40. Edery I. Circadian rhythms in a nutshell. *Physiol Genom* 2000;3:59–74.

41. Dunlap JC. Molecular bases for circadian clocks. *Cell* 1999;96:271–290.

42. Reppert SM, Weaver DR. Coordination of circadian timing in mammals. *Nature* 2002;418: 935–941.

43. Young MW, Kay SA. Time zones: a comparative genetics of circadian clocks. *Nat Rev Genet* 2001;2:702–715.

44. Elowitz MB, Leibler S. A synthetic oscillatory network of transcriptional regulators. *Nature* 2000;403:335–338.

45. Hasty J, Dolnik M, Rottschafer V, Collins JJ. Synthetic gene network for entraining and amplifying cellular oscillations. *Phys Rev Lett* 2002;88:148101.

46. Garcia-Ojalvo J, Elowitz MB, Strogatz SH. Modeling a synthetic multicellular clock: repressilators coupled by quorum sensing. *Proc Natl Acad Sci USA* 2004;101: 10955–10960.

47. McMillen D, Kopell N, Hasty J, Collins JJ. Synchronizing genetic relaxation oscillators by intercell signaling. *Proc Natl Acad Sci USA* 2002;99:679–684.

48. Wang R, Chen L. Synchronizing genetic oscillators by signaling molecules. *J Biol Rhythms* 2005;20:257–269.

49. Barkai N, Leibler S. Circadian clocks limited by noise. *Nature* 2000;403:267–268.

50. Gekakis N, Staknis D, Nguyen HB, Davis FC, Wilsbacher LD, King DP, Takahashi JS, Weitz CJ. Role of the CLOCK protein in the mammalian circadian mechanism. *Science* 1998;280:1564–1569.

51. Sato TK, Yamada RG, Ukai H, Baggs JE, Miraglia LJ, Kobayashi TJ, Welsh DK, Kay SA, Ueda HR, Hogenesch JB., Feedback repression is required for mammalian circadian clock function. *Nat Genet* 2006;38:312–319.

52. van der Horst GT, Muijtjens M, Kobayashi K, Takano R, Kanno S, Takao M, de Wit J, Verkerk A, Eker AP, van Leenen D, et al. Mammalian Cry1 and Cry2 are essential for maintenance of circadian rhythms. *Nature* 1999;398:627–630.

53. Chilov D, Fussenegger M. Toward construction of a self-sustained clock-like expression system based on the mammalian circadian clock. *Biotechnol Bioeng* 2004;87:234–242.

54. Panda S, Hogenesch JB, Kay SA. Circadian rhythms from flies to human. *Nature* 2002;417:329–335.

55. Aubrecht J, Manivasakam P, Schiestl RH. Controlled gene expression in mammalian cells via a regulatory cascade involving the tetracycline transactivator and lac repressor. *Gene* 1996;172:227–231.

56. Imhof MO, Chatellard P, Mermod N. A regulatory network for the efficient control of transgene expression. *J Gene Med* 2000;2:107–116.

57. Kramer BP, Weber W, Fussenegger M. Artificial regulatory networks and cascades for discrete multilevel transgene control in mammalian cells. *Biotechnol Bioeng* 2003;83: 810–820.

58. Hooshangi S, Thiberge S, Weiss R. Ultrasensitivity and noise propagation in a synthetic transcriptional cascade. *Proc Natl Acad Sci USA* 2005;102:3581–3586.

59. Rosenfeld N, Alon U. Response delays and the structure of transcription networks. *J Mol Biol* 2003;329:645–654.

60. McAdams HH, Shapiro L. Circuit simulation of genetic networks. *Science* 1995;269: 650–656.

61. Weiss R. Cellular computation and communications using engineered genetic regulatory networks. Ph.D. Thesis, Massachusetts Institute of Technology, 2001.

62. The device physics of cellular logic gates. Available at http://www-2.cs.cmu.edu/~phoenix/nsc1/paper/3-2.pdf 2002.

63. Kramer BP, Fischer C, Fussenegger M. BioLogic gates enable logical transcription control in mammalian cells. *Biotechnol Bioeng* 2004;87:478–484.

64. Guet CC, Elowitz MB, Hsing W, Leibler S. Combinatorial synthesis of genetic networks. *Science* 2002;296:1466–1470.

65. Yokobayashi Y, Weiss R, Arnold FH. Directed evolution of a genetic circuit. *Proc Natl Acad Sci USA* 2002;99:16587–16591.

66. Karig D, Weiss R. Signal-amplifying genetic circuit enables *in vivo* observation of weak promoter activation in the Rhl quorum sensing system. *Biotechnol Bioeng* 2005;89: 709–718.

67. Basu S, Gerchman Y, Collins CH, Arnold FH, Weiss R. A synthetic multicellular system for programmed pattern formation. *Nature* 2005;434:1130–1134.

68. Basu S, Karig D, Weiss R. Engineering signal processing in cells: towards molecular concentration band detection. *Natural Comput* 2003;2:463–478.

69. Basu S, Mehreja R, Thiberge S, Chen MT, Weiss R. Spatiotemporal control of gene expression with pulse-generating networks. *Proc Natl Acad Sci USA* 2004;101:6355–6360.

70. Mangan S, Zaslaver A, Alon U. The coherent feedforward loop serves as a sign-sensitive delay element in transcription networks. *J Mol Biol* 2003;334:197–204.

71. You L, Cox RS, 3rd, Weiss R, Arnold FH. Programmed population control by cell–cell communication and regulated killing. *Nature* 2004;428:868–871.

72. Bulter T, Lee SG, Wong WW, Fung E, Connor MR, Liao JC. Design of artificial cell–cell communication using gene and metabolic networks. *Proc Natl Acad Sci USA* 2004;101: 2299–2304.

73. Kobayashi H, Kaern M, Araki M, Chung K, Gardner TS, Cantor CR, Collins JJ. Programmable cells: interfacing natural and engineered gene networks. *Proc Natl Acad Sci USA* 2004;101:8414–8419.

74. Farmer WR, Liao JC. Improving lycopene production in *Escherichia coli* by engineering metabolic control. *Nat Biotechnol* 2000;18:533–537.

75. Fung E, Wong WW, Suen JK, Bulter T, Lee SG, Liao JC. A synthetic gene-metabolic oscillator. *Nature* 2005;435:118–122.

76. Ehrbar M, Djonov VG, Schnell C, Tschanz SA, Martiny-Baron G, Schenk U, Wood J, Burri PH, Hubbell JA, Zisch AH. Cell-demanded liberation of VEGF121 from fibrin implants induces local and controlled blood vessel growth. *Circ Res* 2004;94:1124–1132.

77. Semenza GL. Targeting HIF-1 for cancer therapy. *Nat Rev Cancer* 2003;3:721–732.

78. Kramer BP, Fischer M, Fussenegger M. Semi-synthetic mammalian gene regulatory networks. *Metab Eng* 2005;7:241–250.

79. McAdams HH, Arkin A. Stochastic mechanisms in gene expression. *Proc Natl Acad Sci USA* 1997;94:814–819.

80. Hasty J, Isaacs F, Dolnik M, McMillen D, Collins JJ. Designer gene networks: towards fundamental cellular control. *Chaos* 2001;11:207–220.

81. Hasty J, Pradines J, Dolnik M, Collins JJ. Noise-based switches and amplifiers for gene expression. *Proc Natl Acad Sci USA* 2000;97:2075–2080.

82. Roma DM, O'Flanagan RA, Ruckenstein AE, Sengupta AM, Mukhopadhyay R. Optimal path to epigenetic switching. *Phys Rev E Stat Nonlin Soft Matter Phys* 2005;71.011902.

83. Elowitz MB, Levine AJ, Siggia ED, Swain PS. Stochastic gene expression in a single cell. *Science* 2002;297:1183–1186.

84. Pedraza JM, van Oudenaarden A. Noise propagation in gene networks. *Science* 2005;307: 1965–1969.

85. Ozbudak EM, Thattai M, Kurtser I, Grossman AD, van Oudenaarden A. Regulation of noise in the expression of a single gene. *Nat Genet* 2002;31:69–73.

86. Rosenfeld N, Young JW, Alon U, Swain PS, Elowitz MB. Gene regulation at the single-cell level. *Science* 2005;307:1962–1965.

87. Isaacs FJ, Hasty J, Cantor CR, Collins JJ. Prediction and measurement of an autoregulatory genetic module. *Proc Natl Acad Sci USA* 2003;100:7714–7719.

88. Schulze A, Lehmann K, Jefferies HB, McMahon M, Downward J. Analysis of the transcriptional program induced by Raf in epithelial cells. *Genes Dev* 2001;15:981–994.

89. Blake WJ, M KA, Cantor CR, Collins JJ. Noise in eukaryotic gene expression. *Nature* 2003;422:633–637.

90. Thattai M, van Oudenaarden A. Attenuation of noise in ultrasensitive signaling cascades. *Biophys J* 2002;82:2943–2950.

91. Vilar JM, Kueh HY, Barkai N, Leibler S. Mechanisms of noise-resistance in genetic oscillators. *Proc Natl Acad Sci USA* 2002;99:5988–5992.

92. Arkin A, Ross J, McAdams HH. Stochastic kinetic analysis of developmental pathway bifurcation in phage lambda-infected *Escherichia coli* cells. *Genetics* 1998;149: 1633–1648.

93. Arkin AP. Synthetic cell biology. *Curr Opin Biotechnol* 2001;12:638–644.

94. Ninfa AJ, Magasanik B. Covalent modification of the glnG product, NRI, by the glnL product, NRII, regulates the transcription of the glnALG operon in *Escherichia coli*. *Proc Natl Acad Sci USA* 1986;83:5909–5913.

95. Fuqua WC, Winans SC, Greenberg EP. Quorum sensing in bacteria: the LuxR-LuxI family of cell density-responsive transcriptional regulators. *J Bacteriol* 1994;176:269–275.

96. Hu MC, Davidson N. The inducible lac operator-repressor system is functional in mammalian cells. *Cell* 1987;48:555–566.

97. Weber W, Schoenmakers R, Spielmann M, El-Baba MD, Folcher M, Keller B, Weber CC, Link N, van de Wetering P, Heinzen C, et al. Streptomyces-derived quorum-sensing systems engineered for adjustable transgene expression in mammalian cells and mice. *Nucleic Acids Res* 2003;31:e71.

98. Semenza GL, Wang GL. A nuclear factor induced by hypoxia via de novo protein synthesis binds to the human erythropoietin gene enhancer at a site required for transcriptional activation. *Mol Cell Biol* 1992;12:5447–5454.

16

THE THEORY OF BIOLOGICAL ROBUSTNESS AND ITS IMPLICATION TO CANCER

Hiroaki Kitano

Sony Computer Science Laboratories, Inc., 3-14-13 Higashi-Gotanda, Shinagawa, Tokyo 141-0022, Japan

The Systems Biology Institute, 6-31-15 Jingumae, Shibuya, Tokyo 150-0001, Japan

Department of Cancer Systems Biology, Cancer Institute of Japanese Foundation for Cancer Research, Tokyo, Japan

16.1 INTRODUCTION

Systems biology aims at system-level understanding of biological systems [1,2]. Investigations of biological systems at system level are not a new concept and can be traced back to homeostasis by Walter Cannon, Cybernetics by Norbert Weiner [3], and the general systems theory by von Beltaranffy [4]. Numbers of approaches in physiology have also taken a systemic view of the biological subjects. Systems biology is gaining renewed interest today because of progress in genomics, molecular biology, nonlinear dynamics, computational science, and other related fields.

However, "system-level understanding" is a rather vague notion and is often hard to define. This is due to the fact that the system is not a tangible object. Genes and

Systems Biology and Synthetic Biology   Edited by Pengcheng Fu and Sven Panke
Copyright © 2009 John Wiley & Sons, Inc.

proteins are more tangible because they are identifiable matters. Although the system is composed of these matters and they are components of the system, the system itself cannot be made tangible. Often, diagrams of the gene regulatory networks and the protein interaction networks are shown as representations of systems. It is certainly true that such diagrams capture any one aspect of the structures of the system, but they are still only static slices of the system. The heart of the system lies within the dynamics it creates and the logic behind it. It is science on the dynamical state of affairs.

There are four distinct phases that lead us to system-level understanding at various levels. First, system structure identification enables us to understand the structure of the system. While this may be only a static view of the system, it is an essential first step. The structure shall then be identified, ultimately, in both physical and interaction structures. Interaction structures are represented as gene regulatory networks and biochemical networks that indicate how components interact within and in between cells. Physical details of specific regions of the cell, overall structure of cells, and organisms are also important because such physical structure imposes constraints on possible interactions and the outcome of interactions impacts the formation of physical structures. Nature of interaction could be different if proteins involved in interaction move by simple diffusion or under specific guidance from cytoskeleton.

Second, system dynamics needs to be understood. Understanding the dynamics of the system is an essential aspect of the study in systems biology. This requires integrative efforts of experiments, measurement technology development, computational model development, and theoretical analysis. Various methods, such as bifurcation analysis, have been used, but further investigations are necessary to handle the dynamics of systems with very high dimensional space.

Third, methods to control the system shall be investigated. One of the implications is to find a therapeutic approach based on system-level understanding. Many drugs have been developed through extensive effect-oriented screening. It is only recently that specific molecular targets have been identified and lead compounds are designed accordingly. Success in control methods of cellular dynamics may enable us to exploit intrinsic dynamics of the cell so that its effects can be precisely predicted and controlled.

Finally, designing the system that is to modify and construct biological system with designed features. Bacteria and yeast may be redesigned to yield desired properties for drug production and alcohol production. Artificially created gene regulatory logic could be introduced and linked to innate genetic circuits to attain desired functions [5].

Several different approaches can be taken within systems biology field. One may decide to carry out a large-scale, high-throughput experiment and try to find out the overall picture of the system at coarse-grain resolution [6–9]. Alternatively, working on precise details of specific signal transduction [10,11], cell cycle [12,13], and other biological issues to find out the logic behind them are a viable research approach. Both approaches are essentially complementary, and, together, can reshape our understanding of biological systems.

16.2 ROBUSTNESS IS THE FUNDAMENTAL ORGANIZATIONAL PRINCIPLE OF BIOLOGICAL SYSTEMS

Robustness is a property of the system that maintains a certain function despite external and internal perturbations that are ubiquitously observed in various aspects of biological systems [14]. It is distinctively a system-level property that cannot be observed by just looking at components. Specific aspects of the system, the functions to be maintained, and the types of perturbations that the system is robust against must be well defined to make solid arguments. For example, a modern airplane (system) has a function to maintain its flight path (function) against atmospheric turbulences (perturbations).

Bacteria chemotaxis is one of the most well-documented examples in which chemotaxis is a function maintained against the perturbations that are changes in ligand concentration and rate constants for the interactions involved [15–17]. The network for segmental polarity formation during *Drosophila* embryogenesis robustly produces repetitive stripes of differential gene expressions despite variations in initial concentration of substances involved, as well as kinetic parameters of interactions [18,19]. Various aspects of robustness of biological systems have been studied extensively, but more remains to be explored and formalized to create solid theoretical foundations.

Why is robustness so important? First of all, it is a feature that is observed to be so ubiquitous in biological systems; from such a fundamental process like phage fate decision switch [20] and bacteria chemotaxis [15–17] to developmental plasticity [18] and tumor resistance against therapies [21,22]. This implies that it may be a basis for principles that are universal in biological systems, as well as being opportunistic toward finding cures for cancer and other complicated diseases.

Second, robustness is a system-level property of the system in which interactions of components give rise to this feature. Robustness in this context refers to a feature of the system to maintain its function instead of structures or specific states. Structures or states can be dynamically changed if they lead to maintenance of the function of the system.

Third, robustness against environmental and genetic perturbation is essential for evolvability [23–25]. Evolvability requires generation of variety of nonlethal phenotype and genetic buffering [26,27]. Mechanisms that attain robustness against environmental perturbation may be used also for attaining robustness against mutations, developmental stability, and other features that facilitate evolvability [14,23–25].

Fourth, it is one of the features that distinguish biological systems and man-made engineering systems. Although some man-made systems, such as airplanes, are designed to be robust against the range of perturbations, most man-made systems are not as robust as biological systems. Some engineering systems that are designed to be highly robust entail mechanisms that are also present in life forms, which imply existence of the universal principle.

16.3 UNDERLYING MECHANISMS FOR ROBUSTNESS

16.3.1 System Control

First, extensive systems control is used, mostly saliently negative feedback loops but also feedforward and positive feedback controls, to make a system dynamically stable around the specific state of the system. An integral feedback is used in bacteria with chemotaxis as a typical example [15–17]. Due to integral feedback, bacteria can sense changes of chemoattractant and chemorepellant independent of absolute concentration so that proper chemotaxis behavior is maintained over a wide range of ligand concentration. In addition, the same mechanism makes it insensitive to changes in rate constants involved in the circuit. Positive feedbacks are often used to create bistability in signal transduction and cell cycle, so that the system is tolerant to minor perturbation in the stimuli [10,12,13].

16.3.2 Fault Tolerance (Redundancy and Diversity)

Second, fault tolerance mechanisms increase tolerance against components failure and environmental changes by providing alternative components or methods to ultimately maintain a function of the system. Sometimes there are multiple components that are similar to each other and are redundant. Other cases are different means that they are used to cope with perturbations that cannot be handled by the other means. This is often called phenotypic plasticity [28,29] or diversity. Redundancy and phenotypic plasticity are often considered as opposite things, but it is more consistent to view them as different ways to meet an alternative fail-safe mechanism.

16.3.3 Modularity

Third, modularity provides isolation of perturbation from the rest of the system. The cell is the most significant example. More subtle and less obvious examples are modules of biochemical and gene regulatory networks. Modules also play an important role during developmental processes that buffer perturbations so that proper pattern formation can be accomplished [18,30,31]. The definition of the module and the methods of how to detect such modules are still controversial, but the general consensus is that the module does exist and play an important role [32].

16.3.4 Decoupling (Buffering)

Fourth, decoupling isolates low-level noise and fluctuations from functional-level structures and dynamics. One example here is genetic buffering by Hsp90 in which misfolding of proteins due to environmental stresses is fixed, and thus effects of such perturbations are isolated from the functions of the circuits. This mechanism also applies to genetic variations where genetic changes in coding region that may affect protein structures are masked because protein folding is fixed by Hsp90, unless such masking is removed by extreme stress [24,33,34]. Emergent behaviors of complex networks also exhibit such buffering properties [35]. These effects may constitute canalization proposed by Waddington [36]. A recent discovery by Uri Alon's group on oscillatory expression of p53 upon DNA damage may exemplify decoupling at signal-encoding level [37], because stimuli invoked pulses of p53 activation level, instead of gradual changes, effectively converting analogue into digital signal. Digital pulse encoding may indicate robust information transmission, although further investigations are clearly warranted to draw any conclusion at this moment.

An example of a sophisticated engineering system clearly illustrates how these mechanisms work as a whole system. An airplane is supposed to maintain a flight path following the command of the pilot against atmospheric perturbations and various internal perturbations, including changes in the center of gravity due to fuel consumption and movement of passengers, as well as mechanical inaccuracies. This function is carried out by controlling flight control surfaces (rudder, flaps, elevators, etc.) and a propulsion system (engines) by an automatic flight control system (AFCS). Extensive negative feedback control is used to correct deviations of flight path. The reliability of the AFCS is critically important for stable flight. To increase reliability, the AFCS is composed of three independently implemented modules (a triple redundancy system) all of which meet the same functional specification. Most parts of the AFCS are digitalized, so that low-level noise of voltage fluctuations is effectively decoupled from digital signals that define the function of the system. Due to these mechanisms, modern airplanes are highly robust against various perturbations.

16.4 INTRINSIC FEATURES OF ROBUST SYSTEMS: EVOLVABILITY AND TRADE-OFFS

For the system to be evolvable, it must be able to produce variety of nonlethal phenotypes [27]. At the same time, genetic variations need to be accumulated as a neutral network so that pools of genetic variants are exposed when the environment suddenly changes. Systems that are robust against environmental perturbations entail mechanisms such as system control, alternative, modularity, and decoupling that also support, by congruence, generation of nonlethal phenotype and genetic buffering. In addition, the capability to generate flexible phenotype and robustness requires the emergence of the bow tie structure as an architectural motif [38]. One of the reasons why robustness in biological systems is so ubiquitous is that it facilitates evolution, and evolution tends to select traits that are robust against environmental perturbations. This leads to successive addition of system controls.

Systems that acquire robustness against certain perturbations through design or evolution have intrinsic trade-offs between robustness, fragility, performance, and resource demands. Carlson and Doyle argued, using simple examples from physics and forest fire, that systems that are optimized for specific perturbations are extremely fragile against unexpected perturbations [39,40]. A system that has been designed, or evolved, optimally (either globally optimal or suboptimal) against certain perturbations is called a high optimized tolerance (HOT) system. Ceste and Doyle further argued that robustness is a conserved quantity [41]. This means when robustness is enhanced against a range of perturbations, it must then be paid off by fragility elsewhere as well as compromised performance and increased resource demands.

Robust-yet-fragile trade-offs can be understood intuitively using the airplane example yet again. When comparing modern commercial airplanes with the Wright Flyer, modern commercial airplanes are, by a great magnitude, more robust against atmospheric perturbations than the Wright flyer, and are thus attributed to a sophisticated flight control system. However, such a flight control system fully relies on electricity. In a very unthinkable event of total power failure in which all electricity is lost in the airplane, the airplane cannot be controlled at all. Obviously, airplane manufacturers are well aware of this issue and take all possible counter measures to minimize such a risk. On the other hand, despite its vulnerability against atmospheric perturbations, the Wright flyer will never be affected by the power failure because there is no reliance on electricity. This extreme example illustrates that systems that are optimized for certain perturbations could be extremely fragile against unusual perturbations.

HOT model systems are successively optimized/designed (not necessarily globally optimized, though) against perturbations in contrast to self-organized criticality (SOC) [42] or scale-free networks [43] that are unconstrained stochastic additions of components without design or optimization involved. Such differences actually affect failure patterns of the system, and thus have direct implications on understanding the nature of disease and therapy design.

Unlike scale-free networks, HOT systems are robust against perturbations like removal of hubs as far as systems are optimized against such perturbations. However, systems are generally fragile against "Fail-on" type failure in which components failure results in continuous malfunction, instead of cease to function "Fail-off," so that incorrect signals are kept transmitted. This type of failure is known in the engineering field as the Byzantine Generals Problem [44], named after the problem in the Byzantine army composed of numbers of generals dispersed in the field, some of them traitors who sent incorrect messages to confuse the army.

Disease often reflects the systemic failure of the system triggered by the fragility of the system. Diabetes mellitus is an excellent example of how systems that are optimized for near-starving, intermittent food supply, high energy utilization lifestyle, and highly infectious conditions are fragile against unusual perturbations such as high energy containing foods, and a low energy utilization lifestyle [45]. Due to optimization toward a near-starving condition, the extensive control to maintain a minimum blood glucose level is acquired so that activities of central neural systems and innate immunity are maintained. However, no effective regulatory loop has been developed against excessive energy intake and feedback regulations work to reduce glucose uptake by adipocyte and skeletal muscle cells because it may reduce plasma glucose level below the acceptable level. These mechanisms lead to a state where blood glucose level is chronically maintained higher than the desired level, from the longer time scale that has not been optimized for, further leading to cardiovascular complications. Similar observations have been made for autoimmune disorders where the evolution of robust immunity also entails proinflammatory and hyperactive immune system [46].

16.5 SELF-EXTENDING SYMBIOSIS

So far, robustness and its relationship with evolution have been argued within the framework of Mendel's genetics in a sense that mutation and crossover through mating has been considered as a mechanism for evolutionary innovations. Emergence of specific mechanisms for increasing robustness and enrichment of bow tie structure has been discussed within this paradigm. I have previously proposed that there may be other means of enhancing robustness through evolution, but by extending "self" with foreign biologic substances, a notation that I termed "self-extending symbiosis" [47]. Self-extending symbiosis is a phenomenon where evolvable robust systems continue to extend their system boundary by incorporating foreign biologic forms (genes, microorganisms, etc.) to enhance their adaptive capability against environmental perturbations, hence improving their survivability and reproduction potential. In other words, robust evolvable systems have consistently extended themselves by incorporating nonself into tightly coupled symbiotic states.

Looking at the history of evolutionary innovations, it has become clear that some of the major innovations are the result of acquisition of "nonself" into "self" at various levels. Horizontal gene transfer (HGT) facilitates evolution by exchanging genes of different species that have evolved for different optimization contexts, and was shown to be a frequently observed phenomenon in prokaryotes, archea, and unicellular eukaryotes [48,49]. Microorganisms acquire novel functions, mostly to enhance their robustness against environmental challenges, through horizontal exchange of genes. For example, it has been argued that global emergence of antibiotic-resistant bacteria may be caused by horizontal transfer of antibiotic genes [50–52]. In metazoan species, HGT has not been reported (at best, reported highly controversially) except in some rare instances on insect–bacteria symbiosis between the adzuki bean beetle *Callosobruchus chinensis* and *Wolachia* [53].

The serial endosymbiosis theory by Lynn Margulis [54,55] argues that eukaryotic cells have been created by acquiring bacteria as their organelles. This resulted in greater functionalities of eukaryotic cells, hence more robust against environmental challenges. Here, symbiosis resulted in incorporation of foreign biologic entity into cytoplasm as well as into its own genome.

While HGT and endosymbiosis resulted in incorporation of foreign biologic entity into genome and cellular structure, there are forms of symbiosis that do not directly alter genome but essential to the survival of the species. There are species that allow certain bacteria to be vertically inherited through the host's oocytes as observed in sponges, clams [56], and aphids [57]. Aphids, for example, are infected with the genus *Buchnera*, resulting in an endosymbiotic relationship and acquired dramatically improved energy utilization and terrain exploration capability. It was shown that aphids and *buchnera* undergo parallel evolution where the phylogeny trees of the host (aphids) and symbionts (genus *Buchnera*) are consistent [57]. A case of parallel evolution has also been observed in endosymbiosis of *Psyllid* and *Candidatus* [58].

Apart from such tight coupling of host and symbiont, horizontal (environmental) acquisition of symbionts [59] is yet another approach in extending the self by incorporating a broader range of microbes, thereby allowing the host to be able to adapt to a broader range of environments and nutrients. Commensal bacterial flora are ubiquitously observed in various metazoan species, including termites [60], cockroaches [61], prawns [62], and mammalians, and have established inseparable relationships with the host organisms, and are even considered to have coevolved [63]. In human beings, the commensal bacterial flora in the gut consists of diverse microorganisms up to 500–1000 species, amounting to about 10^{14} bacteria weighing a total of 1.5 kg [64]. The human being as a symbiotic system consists of approximately 90 percent prokaryotes and 10 percent eukaryotes [65], and a random shotgun sequencing of the whole human symbiotic system would result in predominantly bacterial genome readouts of about 2 million genes with sporadic mammalian genes [66]. Such commensal intestinal bacteria play a critical role in various aspects of the host physiology. Mammalian bacterial flora has been considered to constitute an integral part of host protection by mutually beneficial symbiosis with the host immune system.

The line of observations point to the characteristic property of biological systems that the greater levels of robustness and functionalities is gained by incorporating foreign biologic entities into their own system in the form of different degree of symbiosis. HGT and endosymbiosis incorporate foreign entities into genome and cellular structures, where vertical inheritance based endosymbiosis do not directly alter the genome. Bacterial flora simply adds a layer of adaptive system that is symbiotically interacting with mucosal immune system of the host. A general tendency observed here is the continuous addition of external layers by symbiotic incorporation of foreign entities, and increased level of robustness against environmental perturbation is gained in this process.

16.6 CANCER AS A ROBUST SYSTEM

Cancer is a heterogeneous and highly robust disease that represents worse case scenario of system failure; a fail-on fault where malfunction components are protected by mechanisms that support robustness in normal physiology [21,22]. It is a robustness hijack. Survival and proliferation capability of tumor cells are robustly maintained against a range of therapies due to intratumoral genetic diversity, feedback loops for multidrug resistance, tumor–host interactions, and so on.

Intratumoral genetic heterogeneity is a major source of robustness in cancer cells. Chromosome instability facilitates generation of intratumoral genetic heterogeneity through gene amplification, chromosomal translocation, point mutations, aneuploidy, and so on [67–70]. Intratumoral genetic heterogeneity is one of the most important features of cancer that provides alternative, or fail-safe mechanisms for tumor to survive and grow again despite various therapies, because some tumor cells may have genetic profile that are resistant to the therapies carried out. Although there are only a few studies on intratumoral genetic heterogeneity, available observations in certain types of solid tumors indicate that there are multiple subclusters of tumor cells within one tumor cluster in which each subcluster has different chromosomal aberrations [71–75]. This implies that each subcluster is developed as clonal expansion of a single mutant cell, and creation of a new subcluster depends upon the emergence of a new mutant that is viable for clonal expansion. A computational study demonstrates that spatial distribution within a tumor cluster enables the coexistence of multiple subclusters [76].

Multidrug resistance is a cellular-level mechanism that provides robustness of viable tumor cell against toxic anticancer drugs. In general, this mechanism involves overexpression of genes such as MDR1 that encodes ATP-dependent efflux pump, P-glycoprotein (P-gp) that effectively pumps out broad range of cytotoxins [77,78]. Trials to mitigate function of P-gp using verapamil, cyclosporine its derivative PSC833 have been disappointing [79].

Tumor–host interactions play major roles in tumor growth and metastasis [80]. When tumor growth is not balanced by vascular growth, hypoxic condition emerges in a tumor cluster [81]. This triggers HIF-1 upregulation that induces a series of reactions that normally function to maintain normal physiological conditions [82]. Upregulation of HIF-1 induces upregulation of VEGF that facilitates angiogenesis, and uPAR and other genes that enhance cell motility [81]. These responses solve hypoxia of tumor cells either by providing oxygen to tumor cluster or by moving tumor cells to a new environment—resulting in further tumor growth or metastasis. Interestingly, macrophages are found to chemotaxis into tumor cluster. Such macrophages are called tumor-associated macrophage (TAM), and found to over-express HIF-1[83]. This means that the macrophage that is supposed to remove tumor cells may be built-in to feedback loops to facilitate tumor growth and metastasis.

In addition, it can be considered that tumor cells may evolve through self-extending symbiosis. If this is the case, tumor cells shall enhance their robustness against various perturbations through horizontal gene transfer, symbiosis with other cells in the form of cell fusion, and formation of symbiotic relationship with surrounding environments. Interestingly, recent reports indicate that tumor cells may be actively involved in cell fusion and uptake of chromosomes of other cells [84–87]. In addition, artificially produced hybrodimas between antibody-producing plasma cell and tumor cell are used for monoclonal antibody production indicating stable maintenance of cellular function upon hybridization. These series of observations imply that tumor cells may be considered as a group of cells that have become somewhat detached from the host system and have begun evolving independently, so that a wide range of phenomena, such as self-extending symbiosis, also occur on tumor cells and thereby their robustness against perturbation is enhanced (Fig. 16-1).

So far, such phenomena have only been reported independently, and not been placed in the perspective. Reorganizing these findings under the coherent view of cancer robustness will provide us a guideline for further research.

(a)   **Self-extending symbiosis**

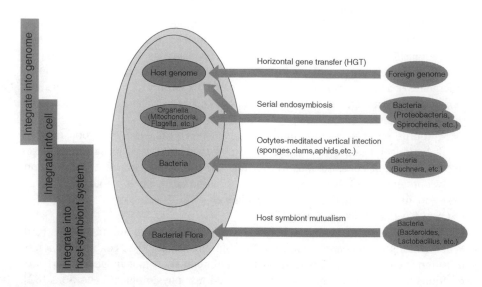

(b)   **Cancer self-extending symbiosis**

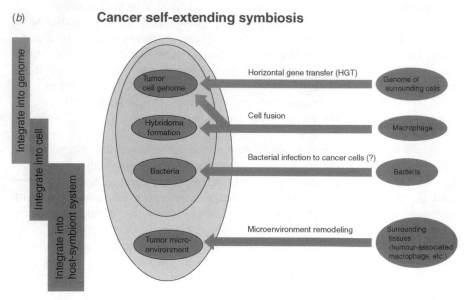

Figure 16-1 Self-extending symbiosis and cancer evolution. Self-extending symbiosis is a path that multicellular organism might have gone through in the course of evolution. Acquisition of nonself into self at various levels of flexibility enhances robustness of organisms against various perturbations. Cancer may also evolve through self-extending symbiosis. Assume cancer as an independent species diverted from somatic cell, it may rapidly evolve through bacteria-like horizontal gene transfer, cell fusion, and microenvironment remodeling to enhance robustness against environmental perturbation. In self-extending symbiosis, there is clear evidence of ootytes-mediated vertical infection. There is no conclusive report if any bacterial infection is observed in any type of cancer that affects robustness of cancer against perturbation. Such phenomena might have been simply unnoticed waiting for future discovery.

16.7 THEORETICALLY MOTIVATED THERAPY STRATEGIES

Given the highly complex control and heterogeneity of tumor, random trial of potential targets is not as effective as one wish it to be. There is a need for theoretically motivated approach that guides us to identify a set of therapies to best counter the disease. The implication of the theory of cancer robustness is that there are specific patterns of behaviors and weakness in robust systems as well as rational way of controlling and fixing the system, and such general principles also apply to cancer. Thus, there must be theoretically motivated approach for the prevention and treatment of cancer. This section discusses therapeutic implications of the theory.

Strategy for cancer therapy may depend upon the level of robustness that the tumor of a specific patient has. When robustness is low, and genetic heterogeneity is low, then there is a good chance that the use of drugs with specific molecular targets may effectively cure cancer by causing the common mode failure: a type of failure in which all redundant subsystems fail for the same reason. An example of CML (chromic myeloid leukemia) therapy by imatinib metylate (Glivec: Novertis) may provide us some insights [88,89]. Dramatic effects of imatinib metylate for early stage CML stem from the fact that it selectively target BCR-ABL protein that is specifically expressed in tumor cells and tumor growth depends on BCR-ABL [90]. Thus, it causes the common mode failure in tumor cells that have similar fragility. However, it is resistant in advanced stage due to heterogeneity of mutations so that the drug cannot inhibit diverse emergent mutant proteins [91]. For this strategy to be effective, there must be proper means to diagnose the degree of intratumoral genetic variations. Then, the most effective molecule as a target needs to be recognized that directs the lead identification and optimization processes.

However, for patients with an advanced stage cancer, intratumoral genetic heterogeneity may be already high and various feedback controls may be significantly upregulated. In these cases, drugs that are effective in the early stage may not work as expected, due to heterogeneous response of tumor cells and feedbacks to compensate for perturbations. For these cases, therapy and drug design need a drastic shift from molecule-oriented approach to a system-oriented approach. Then, the question is which approach shall be taken to target the system, instead of the molecule. I would consider that there are three theoretically motivated countermeasures.

First, robustness/fragility trade-off implies that the cancer cells that have gained increased robustness against various therapies may have a point of extreme fragility. Targeting such a point of fragility may bring dramatic effects for the disease. The major challenge is to find such a point of fragility. Since this trade-off emerged due to successive modifications of the system design to optimally cope with specific perturbations, it is essential to identify the perturbations that the system is optimized against and the underlying mechanisms that enable such optimization. For example, one mechanism for tumor robustness is enhanced genetic heterogeneity that is generated by chromosomal instability, so that some cells may have genetic profile suitable for survival under the specific pressure from the therapy. Then, a method to enhance chromosomal instability selectively in cells that already have unstable chromosome could be one candidate. The point here is whether such effects can be done with sufficient selectivity. Nonselective approach to increase chromosomal instability has been proposed [92], but it may enhance chromosome instability of the cells that are relatively stable now and thus potentially promotes malignancy.

Second, approaches that avoid the increase of robustness constitute the other possibility. Since genetic heterogeneity is enhanced, at least in part, by somatic recombination, selectively inducing cell cycle arrest to tumor cells can effectively control the robustness. There is a theoretical possibility that such subtle control can be done by careful combination of multiple drugs that specifically perturb biochemical interactions. A computational study indicates that the removal or attenuation of specific feedback loops involved in cell cycle reduces the robustness of cell cycle against changes in rate constant [93]. The challenge is to find appropriate combination of drugs that can effectively induce cell cycle arrest only in tumor cells, but not in other cells. Although this approach uses combination of multiple drugs, there is hope to find a set of drugs that can be administered at minimum dosage and toxicity. This approach results in the dormancy of the tumor. Cancer dormancy has already been proposed [94,95] and many report that induced dormancy has been found in mouse [96,97]. However, these studies report cases where tumor cell proliferation is offset by increased apoptosis. Since heterogeneity may increase by cell proliferation, this type of dormancy, which I call "pseudo dormancy" does not prevent increase in heterogeneity, hence robustness is not controlled. Genuine dormancy needs to induce selective cell cycle arrest.

Third, an approach to actively reduce intratumoral genetic heterogeneity followed by a therapy by molecular targeted drugs may be a viable option. If we can design an initial therapy to impose a specific selection pressure on the tumor in which there are only cells with specific genetic variations to survive the therapy, then reduction of genetic heterogeneity may be achieved. Then, if a tumor cell population is sufficiently homogeneous, a drug that specifically targets a certain molecule may have significant impact on the remaining tumor cell population. An important point here is that the drugs used shall not enhance mutation and chromosomal instability. If mutations and chromosomal instability are enhanced, particularly by the initial therapy, heterogeneity may quickly increase so that the second line therapy will be ineffective.

Fourth, one may wish to retake control of the feedback loops that give rise to robustness in an epidemic state. Since the robustness of tumor is often caused by host tumor feedback controls, robustness of tumor can be seriously mitigated if such feedback loops can be controlled. One possible approach is to introduce a decoy that effectively disrupts feedback control or invasive mechanisms of the epidemic. Such an approach is proposed in AIDS therapy where conditionally replicating HIV-1 (crHIV-1) vector that has only *cis* region but no *trans* is introduced [98,99]. This decoy virus dominates the replication machinery, so that HIV-1 virus is pushed into latency, instead of eradication. In solid tumor, an interesting idea has been expressed to use TAM as delivery vehicle of the vector [83,100]. TAM migrates into solid tumor cluster and upregulates HIF-1 that facilitates angiogenesis and metastasis. If TAM can be used to retake a control, robustness may be well controlled and self-extending symbiosis in cancer evolution may be aborted.

Finally, multicomponent drugs may be designed where each component targets molecule in which perturbation differentially affects tumor and normal cells. A certain perturbation affects more the tumor cells but less the normal cells. Even if each of such perturbations does not eliminate tumor cells or is not able to stop their proliferation, there may be specific combination of such drugs that in synergy affects drastically and selectively the tumor cells. One extreme of such approach is the "long-tail drug," recently proposed by the author, that uses large numbers of weakly interacting compounds to affect the tumor cells [101].

16.8 A PROPER INDEX OF TREATMENT EFFICACY

It is important to recognize that, in the light of cancer robustness theory, tumor mass reduction is not an appropriate index for therapy and drug efficacy judgment. As discussed already, reduction of tumor mass does not mean that proliferation potential of tumor has generally decreased. It merely means that subpopulation of tumor cells that are respondent of the therapy were eradicated, or significantly reduced. The problem is that the remaining tumor cells may be more malignant and aggressive, so that therapies for relapsed tumor could be extremely ineffective. This is particularly the case, drugs used to reduce tumor mass are toxic and potentially promote mutations and chromosomal instability in nonspecific ways. It may even enhance malignancy but imposing selective pressures to select resistant phenotype, enhance genetic diversity, as well as providing niche for growth by eradicating fragile subpopulation of tumor cells.

The proper index shall be based on control of robustness: either minimizes the increase of robustness or reduces robustness. This can be achieved by inducing dormancy, actively imposing selective pressure to reduce heterogeneity or exposing fragility that can be the target of therapies to follow, and retaking control of the feedback regulations. The outcome of controlling the robustness may vary from moderate growth of tumor, dormancy that is no tumor mass growth or significant reduction in tumor mass. It should be noted that robustness control does not exclude the possibility of significant tumor mass reduction. If we can target a point of fragility of tumor, it may trigger a common mode failure and may result in significant tumor mass reduction. However, this is a result of controlling robustness, and should not be confused as therapy aimed that tumor mass reduction because robustness has to be controlled to the first to actively exploit a point of fragility. Except for the fragility attack, other options seek for dormancy that results in no tumor growth.

However, this criterion poses a problem for drug design, because current efficacy index of antitumor drugs is measured on the basis of tumor mass reduction. Drugs that induce dormancy will not satisfy this efficacy criterion; thus, they are most likely to be rejected in Phase-II stage. On the other hand, this means that many compounds that have been rejected in Phase-II could be effective from the point of robustness control. Whether such approach can be taken depends on perception change in practitioners, drug industries, and regulatory authorities.

16.9   CONCLUSION

This chapter discussed basic ideas behind the theory of biological robustness and its implications for cancer research and treatment. Biological robustness is one of the essential features of living systems that argued to be tightly coupled with evolution. It may also shape the basic architectural feature of biological systems that are robust and evolvable. One of major consequences is trade-offs between robustness, fragility, resource demands, and performance. Fragility is particularly relevant to diseases. At the same time, cancer established its own robustness. It may be the result of hijacking the robustness intrinsic to the host system. Understanding of this complex nature of biological systems may have profound implications for biomedical research in future.

ACKNOWLEDGMENTS

This study is supported by ERATO-SORST Program (Japan Science and Technology Agency) and the Genome Network Project (Ministry of Education, Culture, Sports, Science, and Technology) of the Systems Biology Institute, the Center of Excellence Program, and the special coordination funds (Ministry of Education, Culture, Sports, Science, and Technology) to Keio University.

REFERENCES

1. Kitano H. Systems biology: a brief overview. *Science* 2002;295(5560):1662–1664.

2. Kitano H. Computational systems biology. *Nature* 2002;420(6912):206–210.

3. Wiener N. *Cybernetics: or Control and Communication in the Animal and the Machine.* Cambridge, MA: The MIT Press, 1948.

4. Bertalanffy, LV. *General System Theory,* New York: George Braziller, 1968.

5. Hasty J, McMillen D, Collins JJ. Engineered gene circuits. *Nature* 2002;420 (6912):224–230.

6. Guelzim N, Bottani S, Bourgine P, Kepes F. Topological and causal structure of the yeast transcriptional regulatory network. *Nat Genet* 2002;31(1):60–63.

7. Ideker T, Ozier O, Schwikowski B, Siegel AF. Discovering regulatory and signalling circuits in molecular interaction networks. *Bioinformatics* 2002;18(Suppl 1): S233–S240.

8. Ideker T, Thorsson V, Ranish JA, Christmas R, Buhler J, Eng JK, Bumgarner R, Goodlett DR, Aebersold R, Hood L. Integrated genomic and proteomic analyses of a systematically perturbed metabolic network. *Science* 2001;292(5518):929–934.

9. Ihmels J, Friedlander G, Bergmann S, Sarig O, Ziv Y, Barkai N. Revealing modular organization in the yeast transcriptional network. *Nat Genet* 2002;31(4):370–377.

10. Ferrell JE Jr. Self-perpetuating states in signal transduction: positive feedback, double-negative feedback and bistability. *Curr Opin Cell Biol* 2002;14(2):140–8.

11. Bhalla US, Iyengar R. Emergent properties of networks of biological signaling pathways. *Science* 1999;283(5400):381–387.

12. Tyson JJ, Chen K, Novak B. Network dynamics and cell physiology. *Nat Rev Mol Cell Biol* 2001;2(12):908–916.

13. Chen KC, Calzone L, Csikasz-Nagy A, Cross FR, Novak B, Tyson JJ. Integrative analysis of cell cycle control in budding yeast. *Mol Biol Cell* 2004;15(8):3841–3862.

14. Kitano H. Biological robustness. *Nat Rev Genet* 2004;5:826–837.

15. Alon U, Surette MG, Barkai N, Leibler S. Robustness in bacterial chemotaxis. *Nature* 1999;397(6715):168–171.

16. Barkai N, Leibler S. Robustness in simple biochemical networks. *Nature* 1997;387 (6636):913–917.

17. Yi TM, Huang Y, Simon MI, Doyle J. Robust perfect adaptation in bacterial chemotaxis through integral feedback control. *Proc Natl Acad Sci USA* 2000;97 (9):4649–4653.

18. von Dassow G, Meir E, Munro EM, Odell GM. The segment polarity network is a robust developmental module. *Nature* 2000;406(6792):188–192.

19. Ingolia NT. Topology and robustness in the *Drosophila* segment polarity network. *PLoS Biol* 2004;2(6):E123.

20. Little JW, Shepley DP, Wert DW. Robustness of a gene regulatory circuit. *Embo J* 1999;18 (15):4299–4307.

21. Kitano H. Cancer as a robust system: implications for anticancer therapy. *Nat Rev Cancer* 2004;4(3):227–235.

22. Kitano H. Cancer robustness: tumour tactics. *Nature* 2003;426(6963):125.

23. Wagner GP, Altenberg L. Complex adaptations and the evolution of evolvability. *Evolution* 1996;50(3):967–976.

24. Rutherford SL. Between genotype and phenotype: protein chaperones and evolvability. *Nat Rev Genet* 2003;4(4):263–274.

25. de Visser J, Hermission J, Wagner GP, Meyers L, Bagheri-Chaichian H, Blanchard J, Chao L, Cheverud J, Elena S, Fontana W, et al. Evolution and detection of genetics robustness. *Evolution* 2003;57(9):1959–1972.

26. Gerhart J, Kirschner M. *Cells, Embryos, and Evolution: Toward a Cellular and Developmental Understanding of Phenotypic Variation and Evolutionary Adaptability.* Malden Massachusetts: Blackwell Science, 1997.

27. Kirschner M, Gerhart J. Evolvability. *Proc Natl Acad Sci USA* 1998;95(15):8420–8427.

28. Agrawal AA. Phenotypic plasticity in the interactions and evolution of species. *Science* 2001;294(5541):321–326.

29. Schlichting C, Pigliucci M. *Phenotypic Evolution: A Reaction Norm Perspective.* Sunderland, MA: Sinauer Associates, Inc., 1998.

30. Eldar A, Dorfman R, Weiss D, Ashe H, Shilo BZ, Barkai N. Robustness of the BMP morphogen gradient in *Drosophila* embryonic patterning. *Nature* 2002;419 (6904):304–308.

31. Meir E, von Dassow G, Munro E, Odell GM. Robustness, flexibility, and the role of lateral inhibition in the neurogenic network. *Curr Biol* 2002;12(10):778–786.

32. Schlosser G, Wagner G, editors. *Modularity in Development and Evolution.* Chicago: The University of Chicago Press, 2004.

33. Rutherford SL, Lindquist S. Hsp90 as a capacitor for morphological evolution. *Nature* 1998;396(6709):336–342.

34. Queitsch C, Sangster TA, Lindquist S. Hsp90 as a capacitor of phenotypic variation. *Nature* 2002;417(6889):618–624.

35. Siegal ML, Bergman A. Waddington's canalization revisited: developmental stability and evolution. *Proc Natl Acad Sci USA* 2002;99(16):10528–10532.

36. Waddington CH. *The Strategy of the Genes: A Discussion of Some Aspects of Theoretical Biology.* New York: Macmillan, 1957.

37. Lahav G, Rosenfeld N, Sigal A, Geva-Zatorsky N, Levine AJ, Elowitz MB, Alon U. Dynamics of the p53-Mdm2 feedback loop in individual cells. *Nat Genet* 2004;36 (2):147–150.

38. Csete ME, Doyle J. Bow ties, metabolism and disease. *Trends Biotechnol* 2004;22 (9):446–450.

39. Carlson JM, Doyle J. Highly optimized tolerance: a mechanism for power laws in designed systems. *Phys Rev E Stat Phys Plasmas Fluids Relat Interdiscip Topics* 1999;60(2 Part A): 1412–1427.

40. Carlson JM, Doyle J. Complexity and robustness. *Proc Natl Acad Sci USA* 2002;99(Suppl 1): 2538–2545.

41. Csete ME, Doyle JC. Reverse engineering of biological complexity. *Science* 2002;295 (5560):1664–1669.

42. Bak P, Tang C, Wiesenfeld K. Self-organized criticality. *Physical Review A* 1988;38 (1):364–374.

43. Barabasi AL, Oltvai ZN. Network biology: understanding the cell's functional organization. *Nat Rev Genet* 2004;5(2):101–113.

44. Lamport L, Shostak R, Pease M. The Byzantine generals problem. *ACM Trans on Program Lang Syst* 1982;4(3):382–401.

45. Kitano H, Kimura T, Oda K, Matsuoka Y, Csete ME, Doyle J, Muramatsu M. Metabolic syndrome and robustness trade-offs. *Diabetes* 2004;53(Suppl 3): S1–S10

46. Kitano H, Oda K. Robustness trade-offs and host? Microbial symbiosis in the immune system. *Mol Syst Biol* 2006;doi:10. 1038/msb4100039.

47. Kitano H, Oda K. Self-extending symbiosis: a mechanism for increasing robustness through evolution. *Biol Theory* 2006;1(1):61–66.

48. Gogarten JP, Townsend JP. Horizontal gene transfer, genome innovation and evolution. *Nat Rev Microbiol* 2005;3(9):679–687.

49. Brown JR. Ancient horizontal gene transfer. *Nat Rev Genet* 2003;4(2):121–132.

50. Monroe S, Polk R. Antimicrobial use and bacterial resistance. *Curr Opin Microbiol* 2000;3(5):496–501.

51. Levy SB, Marshall B. Antibacterial resistance worldwide: causes, challenges and responses. *Nat Med* 2004;10(Suppl 12):S122–S129.

52. Smets BF, Barkay T. Horizontal gene transfer: perspectives at a crossroads of scientific disciplines. *Nat Rev Microbiol* 2005;3(9):675–678.

53. Kondo N, Nikoh N, Ijichi N, Shimada M, Fukatsu T. Genome fragment of Wolbachia endosymbiont transferred to X chromosome of host insect. *Proc Natl Acad Sci USA* 2002;99(22):14280–14285.

54. Margulis L. Symbiosis and evolution. *Sci Am* 1971;225(2):48–57.

55. Margulis L, Bermudes D. Symbiosis as a mechanism of evolution: status of cell symbiosis theory. *Symbiosis* 1985;1:101-24.

56. Cary SC, Giovannoni SJ. Transovarial inheritance of endosymbiotic bacteria in clams inhabiting deep-sea hydrothermal vents and cold seeps. *Proc Natl Acad Sci USA* 1993; 90(12):5695–5699.

57. Baumann P, Baumann L, Lai CY, Rouhbakhsh D, Moran NA, Clark MA. Genetics, physiology, and evolutionary relationships of the genus *Buchnera*: intracellular symbionts of aphids. *Annu Rev Microbiol* 1995;49:55–94.

58. Thao ML, Clark MA, Burckhardt DH, Moran NA, Baumann P. Phylogenetic analysis of vertically transmitted psyllid endosymbionts (*Candidatus Carsonella ruddii*) based on atpAGD and rpoC: comparisons with 16S-23S rDNA-derived phylogeny. *Curr Microbiol* 2001;42(6):419–421.

59. Nyholm SV, McFall-Ngai MJ. The winnowing: establishing the squid–vibrio symbiosis. *Nat Rev Microbiol* 2004;2(8):632–642.

60. Schmitt-Wagner D, Friedrich MW, Wagner B, Brune A. Phylogenetic diversity, abundance, and axial distribution of bacteria in the intestinal tract of two soil-feeding termites (*Cubitermes spp.*). *Appl Environ Microbiol* 2003;69(10):6007–6017.

61. Bracke JW, Cruden DL, Markovetz AJ. Intestinal microbial flora of the American cockroach. *Periplaneta americana L. Appl Environ Microbiol* 1979;38(5):945–955.

62. Oxley AP, Shipton W, Owens L, McKay D. Bacterial flora from the gut of the wild and cultured banana prawn. *Penaeus merguiensis. J Appl Microbiol* 2002;93(2):214–223.

63. Backhed F, Ley R, Sonnenburg J, Peterson D, Gordon JI. Host-bacterial mutualism in the human intestine. *Science* 2005;307:1915–1920.

64. Xu J, Gordon JI. Inaugural article: honor thy symbionts. *Proc Natl Acad Sci USA* 2003; 100(18):10452–10459.

65. Savage DC. Microbial ecology of the gastrointestinal tract. *Annu Rev Microbiol* 1977;31:107–133.

66. Hooper LV, Midtvedt T, Gordon JI. How host–microbial interactions shape the nutrient environment of the mammalian intestine. *Annu Rev Nutr* 2002;22:283 307.

67. Lengauer C, Kinzler KW, Vogelstein B. Genetic instabilities in human cancers. *Nature* 1998;396(6712):643–649.

68. Li R, Sonik A, Stindl R, Rasnick D, Duesberg P. Aneuploidy vs. gene mutation hypothesis of cancer: recent study claims mutation but is found to support aneuploidy. *Proc Natl Acad Sci USA* 2000;97(7):3236–3241.

69. Tischfield JA, Shao C. Somatic recombination redux. *Nat Genet* 2003;33(1):5–6.

70. Rasnick D. Aneuploidy theory explains tumor formation, the absence of immune surveillance, and the failure of chemotherapy. *Cancer Genet Cytogenet* 2002;136(1):66–72.

71. Baisse B, Bouzourene H, Saraga EP, Bosman FT, Benhattar J. Intratumor genetic heterogeneity in advanced human colorectal adenocarcinoma. *Int J Cancer* 2001;93 (3):346–352.

72. Fujii H, Yoshida M, Gong ZX, Matsumoto T, Hamano Y, Fukunaga M, Hruban RH, Gabrielson E, Shirai T. Frequent genetic heterogeneity in the clonal evolution of gynecological carcinosarcoma and its influence on phenotypic diversity. *Cancer Res* 2000;60 (1):114–120.

73. Gorunova L, Hoglund M, Andren-Sandberg A, Dawiskiba S, Jin Y, Mitelman F, Johansson B. Cytogenetic analysis of pancreatic carcinomas: intratumor heterogeneity and nonrandom pattern of chromosome aberrations. *Genes Chromosomes Cancer* 1998;23(2):81–99.

74. Gorunova L, Dawiskiba S, Andren-Sandberg A, Hoglund M, Johansson B. Extensive cytogenetic heterogeneity in a benign retroperitoneal schwannoma. *Cancer Genet Cytogenet* 2001;127(2):148–154.

75. Frigyesi A, Gisselsson D, Mitelman F, Hoglund M. Power law distribution of chromosome aberrations in cancer. *Cancer Res* 2003;63(21):7094–7097.

76. Gonzalez-Garcia I, Sole RV, Costa J. Metapopulation dynamics and spatial heterogeneity in cancer. *Proc Natl Acad Sci USA* 2002;99(20):13085–13089.

77. Juliano RL, Ling V. A surface glycoprotein modulating drug permeability in Chinese hamster ovary cell mutants. *Biochim Biophys Acta* 1976;455(1):152–162.

78. Nooter K, Herweijer H. Multidrug resistance (mdr) genes in human cancer. *Br J Cancer* 1991;63(5):663–669.

79. Tsuruo T, Iida H, Tsukagoshi S, Sakurai Y. Overcoming of vincristine resistance in P388 leukemia *in vivo* and *in vitro* through enhanced cytotoxicity of vincristine and vinblastine by verapamil. *Cancer Res* 1981;41(5):1967–1972.

80. Bissell MJ, Radisky D. Putting tumours in context. *Nat Rev Cancer* 2001;1(1):46–54.

81. Harris AL. Hypoxia: a key regulatory factor in tumour growth. *Nat Rev Cancer* 2002;2 (1):38–47.

82. Sharp F.R, Bernaudin M. HIF1 and oxygen sensing in the brain. *Nat Rev Neurosci* 2004;5 (6):437–448.

83. Bingle L, Brown NJ, Lewis CE. The role of tumour-associated macrophages in tumour progression: implications for new anticancer therapies. *J Pathol* 2002;196(3):254–265.

84. Pawelek JM. Tumour–cell fusion as a source of myeloid traits in cancer. *Lancet Oncol* 2005;6(12):988–993.

85. Pawelek J, Chakraborty A, Lazova R, Yilmaz Y, Cooper D, Brash D, Handerson T. Co-opting macrophage traits in cancer progression: a consequence of tumor cell fusion? *Contrib Microbiol* 2006;13:138–155.

86. Vignery A. Macrophage fusion: Are somatic and cancer cells possible partners? *Trends Cell Biol* 2005;15(4):188–193.

87. Ogle BM, Cascalho M, Platt JL. Biological implications of cell fusion. *Nat Rev Mol Cell Biol*. 2005;6(7):567–575.

88. Hochhaus A. Cytogenetic and molecular mechanisms of resistance to imatinib. *Semin Hematol* 2003;40(2 Suppl 3): 69–79.

89. Hochhaus A, Kreil S, Corbin A, La Rosee P, Lahaye T, Berger U, Cross NC, Linkesch W, Druker BJ, Hehlmann R, et al. Roots of clinical resistance to STI-571 cancer therapy. *Science* 2001;293(5538):2163

90. Druker BJ, Talpaz M, Resta DJ, Peng B, Buchdunger E, Ford JM, Lydon NB, Kantarjian H, Capdeville R, Ohno-Jones S, et al. Efficacy and safety of a specific inhibitor of the BCR-ABL tyrosine kinase in chronic myeloid leukemia. *N Engl J Med* 2001;344 (14):1031–1037.

91. Druker BJ. Overcoming resistance to imatinib by combining targeted agents. *Mol Cancer Ther* 2003;2(3):225–226.

92. Sole RV. Phase transitions in unstable cancer cell populations. *Eur Phys J B* 2003;35:117–123.

93. Morohashi M, Winn AE, Borisuk MT, Bolouri H, Doyle J, Kitano H. Robustness as a measure of plausibility in models of biochemical networks. *J Theor Biol* 2002;216 (1):19–30.

94. Takahashi Y, Nishioka K. Survival without tumor shrinkage: re-evaluation of survival gain by cytostatic effect of chemotherapy. *J Natl Cancer Inst* 1995;87(16):1262–1263.

95. Uhr JW, Scheuermann RH, Street NE, Vitetta ES. Cancer dormancy: opportunities for new therapeutic approaches. *Nat Med* 1997;3(5):505–509.

96. Holmgren L, O'Reilly MS, Folkman J. Dormancy of micrometastases: balanced proliferation and apoptosis in the presence of angiogenesis suppression. *Nat Med* 1995;1(2):149–153.

97. Murray C. Tumour dormancy: not so sleepy after all. *Nat Med* 1995;1(2):117–118.

98. Dropulic B, Hermankova M, Pitha PM. A conditionally replicating HIV-1 vector interferes with wild-type HIV-1 replication and spread. *Proc Natl Acad Sci USA* 1996;93(20):11103–11108.

99. Weinberger LS, Schaffer DV, Arkin AP. Theoretical design of a gene therapy to prevent AIDS but not human immunodeficiency virus type 1 infection. *J Virol* 2003;77(18):10028–10036.

100. Owen MR, Byrne HM, Lewis CE. Mathematical modelling of the use of macrophages as vehicles for drug delivery to hypoxic tumour sites. *J Theor Biol* 2004;226(4):377–391.

101. Kitano H. A robustness-based approach to systems-oriented drug design. *Nat Rev Drug Discov* 2007;6(3):202–210.

17

NUCLEIC ACID ENGINEERING

Wenlong Cheng, Liang Ding, Hisakage Funabashi, Nokyoung Park,
Soong Ho Um, Jianfeng Xu, and Dan Luo

*Department of Biological and Environmental Engineering,
Cornell University, Ithaca, New York 14853*

17.1 INTRODUCTION: NUCLEIC ACIDS

One of the ambitions of synthetic biology is to design and engineer life, a goal regarded by many biological engineers as noble and achievable. After all, life has already been engineered at many different levels and in many different ways, from genetic engineering to tissue engineering to whole animal cloning. In the process, engineering itself can also learn from life, using principles discovered from biology along the way to guide engineering designs and creations. Indeed, integration of engineering with biology is the essence of synthetic biology, and one of its most successful approaches as well.

In particular, the integration of engineering and biology at the molecular level— molecular biological engineering—holds great promise for future systems biology. For example, biological building blocks such as nucleic acids and amino acids can be a novel source for new materials and devices. Indeed, the commonalities between molecular biology and materials engineering are much greater than many of us originally thought—both materials engineering and biology, in one way or the other, build materials from the building blocks that are at the molecular level. Realization of such bioengineering integration needs not only close collaboration between engineers and biologists but also more urgently require a new generation of biological engineers who are well educated in both fields—biology and engineering. Here, we will focus on nucleic acid engineering: the use of deoxyribonucleic acid(DNA) in the engineering of new materials, new devices, and new applications.

This chapter is mainly for engineers; it starts with a brief introduction of DNA, ribonucleic acid (RNA), peptide nucleic acid (PNA), and locked nucleic acid (LNA). A historical review of DNA nanotechnology will be presented followed by a detailed review of nucleic acid engineering and its nonbiological, self-assembled structures. Approaches of manipulation and characterization of oligonucleic acids are discussed and applications of nucleic acid engineering are presented at the end of the chapter.

17.1.1 Nucleic Acids: DNA, RNA, PNA, LNA, GNA, and TNA

Nucleic acids, the most common of which are DNA and RNA, are known to be genetic materials. Artificial nucleic acids resulting from chemical synthesis, including PNA, LNA, glycol nucleic acid (GNA) and threose nucleic acid (TNA), have also been developed for various purposes.

DNA is a genetic information storage biomolecule for almost all life forms, and genetic coding information flows from DNA to RNA to proteins (i.e., the Central Dogma of Molecular Biology). Structurally speaking, DNA is a biopolymer whose monomers (nucleotides) are covalently linked through phosphodiester bonds (Fig. 17-1a). Each nucleotide is composed of three structural constituents: a heterocyclic base, a sugar group (pentose or deoxy-ribose here), and a phosphate group. The heterocyclic bases in DNA are adenine (A), cytosine (C), guanine (G), and thymine (T). Nucleotides are differentiated by the heterocyclic bases they carry. These four different bases recognize each other by forming hydrogen bonds, the so-called Watson–Crick base pairing: A with T and G with C. G–C bonding is stronger than A–T bonding due to its three hydrogen bonds compared with the two hydrogen bonds of an A–T pair. Adjacent heterocyclic bases of the same chain can stack onto each other as a result of the strong π–π hydrophobic interaction between the heterocyclic aromatic rings. Provided with ideal conditions, DNA can exist as a single-stranded DNA (ssDNA) or antiparallel double-stranded DNA(dsDNA) molecule. In most cases, the double-stranded DNA forms a right-handed helical structure with each helical turn of about 10.5 bases. The rise of one turn is about 3.4 nm and the width about 2.0 nm. This most common configuration of DNA is called B-form DNA (B-DNA). DNA can also adopt other forms of helixes (A-DNA or C-DNA) under unusual conditions. In particular, left-handed DNA (Z-DNA) can exist with special sequences and buffer conditions. To make things even more interesting, DNA can also form triple helixes or even quadruplexes—of course, in these cases, special sequences have to be utilized. The following aspects of DNA make it an ideal genetic molecule chosen by nature: DNA is chemically stable; its duplex structure is relatively stable but breakable when desired; Watson–Crick ensures precise processing of genetic information.

RNA is very similar to DNA except for an additional hydroxyl group ($-$OH) at the C2 position of the ribose ring (Fig. 17-1b). This hydroxyl group can participate in hydrolysis of the phosphodiester bond and speed up the reaction; therefore, it is

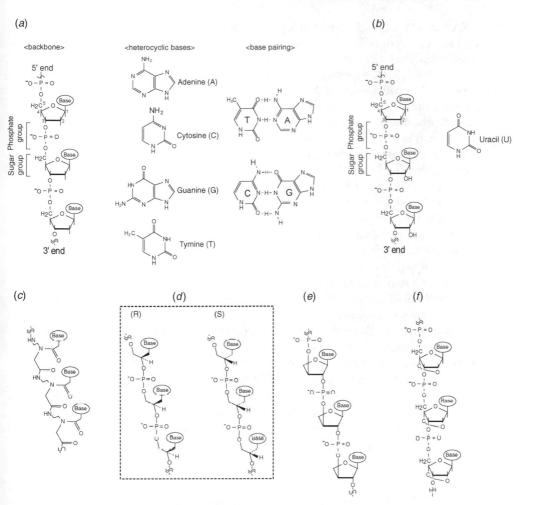

Figure 17-1 Molecular structure of nucleic acids: DNA (a), RNA (b), PNA (c), GNA (d), TNA (e), and LNA (f).

responsible for RNA's lesser stable nature when compared to DNA. RNA usually adopts a single-stranded structure; although it can also form a double-stranded one with RNA itself or with DNA. RNA uses the same bases (A, G, and C) as DNA but substitutes uracil (U) for DNA's thymine (T). There are many types of RNA; they are classified based chiefly on their functions. Messenger RNA (mRNA) serves as a messenger between DNA and proteins; it copies genetic information from DNA in the transcription process and acts as a template for protein synthesis in translation process. Transfer RNA (tRNA) carries a specific amino acid and transfers it to a growing protein chain to which the amino acid should be added. Ribosomal RNA (rRNA) is part of the ribosome where protein synthesis is carried out. Most recently, small inhibitory

RNA (siRNA), a group of small RNA molecules capable of inhibiting specific gene expression, was discovered.

Artificial nucleic acids have been developed for biological research and medical applications. They can recognize and exhibit a much stronger affinity with their corresponding counterparts (DNA or RNA) to form a thermodynamically very stable duplex structure or other types of structures [1], but are resistant to enzymatic degradation, compared to the DNA counterpart. Therefore, they could be used in gene therapy to silence culprit genes. These DNA analogues have chemically modified bases, or backbones, or ribose sugar moiety, compared to DNA. PNA has a neutral pseudo peptide backbone (Fig. 17-1c), thus less soluble than DNA. The lack of charge and electrostatic repulsion confers additional stability on the duplex involving PNA. Both GNA and TNA have a chemically modified backbone, but still negatively charged due to the phosphodiester bonds. The backbone of GNA is composed of glycerol repeats linked by phosphodiester bonds (Fig. 17-1d), while that of TNA consists of threose repeats (Fig. 17-1e). LNA has an extra bridge connecting $2'$ and $4'$ carbons of its ribose (Fig. 17-1f) [2,3]. This structural modification locks ribose in $3'$-endo conformation, which enhances base stacking in duplex and, as a result, thermal stability. Bases of nucleic acids have been modified with novel base pairing patterns by forming *de novo* designed hydrogen bonds, with enhanced π–π hydrophobic interaction that improves thermal stability of duplexes, with fluorescent properties for molecular probing, with additional ligands for forming coordination bonds with metals. A thorough discussion on the subject is beyond the scope of this chapter.

Nucleic acid engineering includes chemical modification of the nucleic acid structures as discussed above, genetic engineering that manipulates genes to express proteins, and nucleic acid structural engineering that uses oligonucleic acids as building blocks of superstructures. The latter has gained more and more attention due to the rapid progress of nanobiotechnology and nanobiomaterials. This is our focus in this chapter.

17.1.2 Self-Assembly

Self-assembly is a process through which molecules recognize each other to form hierarchical orders of complexes and superstructures. The forces employed for molecular recognition are usually weak and noncovalent bonding, as reversibility is desired during self-assembling. They include electrostatic interaction, hydrophobic interaction, Van der Waals forces, and hydrogen bonding. Weak covalent bonds such as disulfide bond and coordination bond can also be used for directing self-assembling. Self-assembly is a fundamental principle for structural organization on all scales from small molecules to macromolecules, to cells, to our ecosystems and space.

When it comes to nanoscience and engineering, two approaches are employed to build nanoscale structures and devices: top-down and bottom-up. In the top-down approach, the desired structure is carved out of a bulk material; as the structure gets smaller and smaller for applications in nanoscience, it becomes harder for this approach to be effective. In the bottom-up approach, designed molecules can recognize each other and position themselves in the final structure through self-assembly.

Self-assembly is more and more recognized and employed in building superstructures, especially at the microscale and nanoscale.

17.1.3 Bottom Up: The First Step Toward a Synthetic World

In his famous presentation in 1959, Nobel laureate, physicist Richard Feynman predicted that "There is plenty of room at the bottom" [4]—a phrase that has become the birth symbol for nanotechnology. His prediction is becoming a reality, and we are indeed moving from the microscale world into the nanoscale world through the top-down and bottom-up approaches. Nanoscience has changed our life fundamentally.

In a broader sense, self-assembly and self-organization can happen at any scale, from atom to molecule to polymer, and, of course, to oligonucleic acids and proteins. Indeed, the entire biological system itself is a result of a complex self-assembling system—from biomolecules including proteins, oligonucleic acids, lipids, carbohydrates, and so on. Naturally, a designer of life could and should learn from the natural self-assemblies of biological molecules in living organisms.

Biological molecules, such as DNA, protein, lipid, and carbohydrates, have nanoscale dimensions and possess molecular recognition capabilities for self-assembly. The processing of genetic information from DNA to mRNA to protein involves self-assembling of different molecules, at various hierarchical levels and with such amazing precision. Therefore, they are especially useful building blocks in the building of a synthetic world. We believe that future intelligent materials and devices will be generated from biomolecules and their derivatives. DNA is probably the first biomolecule to be utilized for that purpose, partially due to DNA's unique properties and advantages including high precision in base pairing, programmability, manipulability, and stability.

17.1.4 DNA: Building Blocks

From a molecular perspective, DNA has many advantages over other synthetic, chemical building blocks.

(1) We understand thoroughly the chemistry, property, and structure of both nucleic acids and duplex structures of DNA. DNA is quite stable. For example, DNA is 1000-fold more stable to hydrolytic destruction than protein and almost 100,000-fold more stable than RNA [5]. DNA can be stored frozen or dried into a powder, stable for thousand of years. Knowing 10.5 nucleotides and 3.4 nm per turn of B-form DNA duplex and its diameter of 2.0 nm, as well as persistence length of 50.0 nm for DNA duplex, we can design DNA sequences to obtain a structure with exact dimensions in our mind. In a word, DNA is programmable and predictable.

(2) DNA with designed sequences can be conveniently obtained by solid-phase synthesis on an automatic DNA synthesizer. In fact, DNA is commercially available. Long and natural DNA strands from PCR and biological sources can also be used to make superstructures.

(3) A variety of enzymes are available for DNA manipulations, see Section 17.15 for details.

(4) Sticky ends can be employed to assemble building blocks into high-order superstructures. Thus, sticky-ended cohesion is a convenient construction tool.

(5) State-of-the-art instrumental techniques can be used for the characterization of DNA superstructures, such as transmission electron microscopy (TEM), atomic force microscopy (AFM).

17.1.5   Enzymes Used in Nucleic Acid Engineering

DNA/RNA are substrates of more than 4000 different enzymes. These enzymes are readily used for the manipulation of DNA/RNA and the analysis of designed DNA/RNA structures. In general, these enzymes are classified into three groups: *Restriction endonucleases*, *Polymerases*, and *DNA/RNA modifying enzymes*. Most of them are usually commercially available from Promega (www.promega.com) and New England Biolab (www.neb.com). *Restriction endonucleases*, also called restriction enzymes, are produced by bacteria that cleave the phosphodiester bond of DNA at specific sites. There are numerous types of restriction enzymes, most of them cleaving double-stranded DNA at specific recognition sequences (usually composed of 4–6 bp and are palindromic), leaving either blunt or sticky ends. However, there are also a few restriction enzymes called nicking endonucleases that can cleave only one strand of DNA on a double-stranded DNA to produce DNA molecules that are "nicked" rather than cleaved. *Polymerases*, including DNA and RNA polymerase, are the enzymes that catalyze the polymerization of new DNA or RNA against an existing DNA or RNA template in the processes of replication and transcription. Of particular importance are groups of DNA polymerases isolated from the thermophilic bacterium such as *Thermus aquaticus* (*Taq*) that is widely used in polymerase chain reaction (PCR), one of the most important tools in molecular biology techniques, *DNA/RNA modifying enzymes*, including ligases, nucleases, kinases, alkaline phosphatases, and others. Of prime importance are ligases such as T4-DNA ligase that catalyzes the joining of two DNA fragments by forming a new phosphodiester bond. Usually, ligases are used together with restriction enzymes in the process of DNA manipulation. Nucleases include deoxyribonucleases (DNase) and ribonucleases (RNase) that nonspecifically cleave DNA and RNA backbones, either from the end of DNA/RNA molecules or from anywhere along the chain. T4-polynucleotide kinases and alkaline phosphatases function oppositely in that the former catalyzes the transfer and exchange of P_i from ATP to the 5′-hydroxyl terminus of polynucleotides (double- and single-stranded DNA and RNA), and the later removes 5′-phosphate groups from DNA and RNA at 5′-hydroxyl terminus.

17.2   DNA NANOTECHNOLOGY

DNA nanotechnology employs DNA as building blocks for construction of DNA superstructures that can act as templates for arrangement of other functional species. Starting from building a stable DNA junction, this field has progressed from building robust and stiff DNA motifs to 2D and 3D DNA structures. Seeman and his colleagues are pioneers in employing DNA to construct 2D and 3D nanostructures, which started as early as 1982 [6]. Seeman pictured DNA as hinges and joints and bolts and braces in his designs, and by careful design of DNA base sequences, they could be programmed to fold and bind to each other. The resulting DNA superstructures can be used for templated assembly of functional species such as metals, proteins, and fullerenes.

17.2.1 DNA Junctions

DNA duplex is a linear molecule, and we cannot do much with a linear structural motif as a material building block. But like RNA, DNA can form "strange" structures besides the usual double helix. These additional structures include hairpins and three- and four-way branch points, which have importance for biological functions. Branched DNA occurs in nature but is not stable due to its sequence symmetry. Such mobile Holliday junction is used for exchange of genetic information by yeasts through a process called homologous recombination. By the *de novo* design of DNA sequences to remove the sequence symmetry in the four-component strands, a stable DNA junction can be formed (Fig. 17-2) [7]. This is regarded as a milestone for DNA nanotechnology. DNA motifs developed later are in essence reminiscent of this DNA junction.

17.2.2 DNA Motifs and DNA Structures

Many DNA junctions [8] and topological DNA objects (DNA-truncated octahedron, DNA cube, Barromean rings) have been reported [9]. However, the DNA junction is not rigid and flat enough for use as a tile in the construction of well-behaved 2D and 3D structures. Therefore, more rigid and flat DNA motifs have been built: double

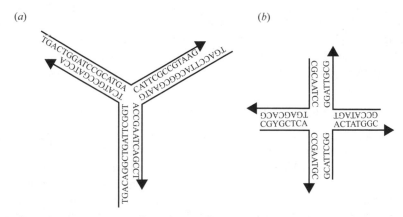

Figure 17-2 DNA junctions: three-arm DNA junction (a) and four-arm DNA junction (b). They are formed by annealing three or four partially complementary DNA strands. (a) reproduced from Ref. 54; (b) reproduced from Ref. 8.

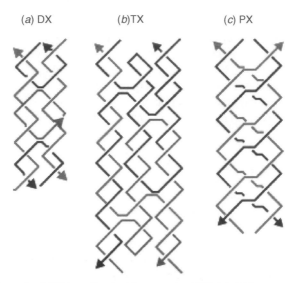

(a) DX   (b)TX   (c) PX

Figure 17-3 Example of DNA motifs: double crossover (DX) (a), triple crossover (TX) (b), and paranemic crossover (PX) (c). Each line represents a DNA strand. (a) reproduced from Ref. 10; (b) reproduced from Ref. 12; (c) reproduced from Ref. 11.

crossover (DX) motif where four strands are bound by two crossover points [10] (Fig. 17-3a); triple crossover (TX) where six strands are bound by three crossover points (Fig. 17-3b); and a paranemic crossover (PX) where strands migrate to the other duplex at every contact point of the two duplexes (Fig. 17-3c) [11,12].

The DX, TX, and PX DNA motifs can be used as tiles in a way similar to floor and wall tiles but they can also self-assemble into DNA complexes via sticky-ended cohesion. Yan et al. created 4 × 4 DNA complexes that can self-assemble into planar square tiles (Fig. 17-4) [13] while Adleman's group assembled hexagonal tiles [14]. Research and interest in DNA building blocks and their self-assembly have started to catch up in the past 10 years, and a variety of crossover DNA along with a variety of different DNA tiles have been added to the DNA construction kit, from which extremely complicated architectures and patterns have been achieved, including 2D lattices, octahedrons, pesudo hexagonal crystals, ribbons, two-dimensional nanogrids, and tubes [12,13,15-18].

17.2.3 Structures from Long Single-Stranded DNA

The work described above uses synthetic DNA as a building material. After assembly, the reaction mixture contains the desired structure, the misfolded assemblies, and the extra short DNA strands. It is difficult, however, if not impossible, to isolate the desired product. The question then arises: can DNA from biological sources or PCR be used for DNA nanotechnology? These are usually long and readily available in bulk. All the sequences necessary for the designed complex can be contained in a single strand, eliminating the need to remove short strands as in the case of synthetic DNA. The successful structure also can be isolated and amplified with PCR or in cells to

(a)   (b)

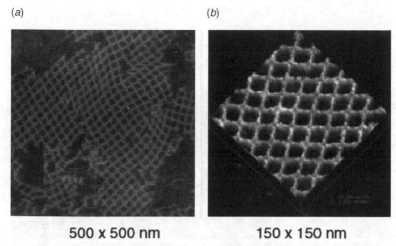

500 x 500 nm   **150 x 150 nm**

Figure 17-4 AFM images of 2D DNA based nanogrids formed from the corrugated design: The right panel measuring 150 × 150 nm (b) is a surface plot of a magnified region from the left panel measuring 500 × 500 nm (a). Reproduced from Ref. 13.

meet with quantity requirements for specific applications. Recent progress has proved that this approach is possible. Shin et al. showed that a single-stranded DNA can fold into an octahedron [19]. The single-stranded DNA is 1.7 kb and contains all the sequences at designed positions of the single strand to form the DX and PX motifs discussed above. The strand is PCR amplifiable and can assemble into an octahedron as it is designed. Rothemund described a smart way to fold plasmid DNA into two-dimensional-shaped DNA origami, including squares, rectangles, five-pointed stars, smiley faces, and so on (Fig. 17-5) [20]. The designed shapes were created by folding the 7 kb single-stranded DNA, similar to paper origami and raster filling, with the help of short "staple strands" that form duplexes with the complementary sequences on the long single strand. In addition to binding and holding a DNA scaffold in shape, staple strands can act as binary pixels to make any desired pattern—words, map, and so on. In addition, the designed 2D structure can act as a nanobreadboard for functional materials.

17.2.4 Applications

One projected application of DNA nanostructures is the templated assembly of proteins, metals, and other materials that have potential in research and other applications. DNA-templated protein arrays and highly conductive nanowires have been reported by Yan et al. as an example [13]. DNA tiles can be treated like Wang tiles, and the self-assembly of such DNA systems can be used to perform algorithmic tasks. The energy involved in folding and unfolding of DNA duplexes and the conformational change associated with the process also can be used in nanomechanical devices [21].

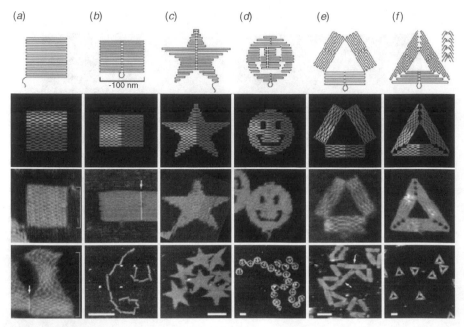

Figure 17-5 DNA origami shapes: top row, folding paths. Square (a); rectangle (b); star (c); disk with three holes (d); triangle with rectangular domains (e); sharp triangle with trapezoidal domains and bridges between them (red lines in inset) (f). Dangling curves and loops represent unfolded sequence. Second row from top, diagrams showing the bend of helices at crossovers (where helices touch) and away from crossovers (where helices bend apart). Grey scale intensity indicates the base-pair index along the folding path. Bottom two rows, AFM images. White lines and arrows indicate blunt-end stacking. White brackets mark the height of an unstretched square and that of a square stretched vertically (by a factor 1.5) into an hourglass. All images and panels without scale bars are the same size, 165 nm × 165 nm. Scale bars for lower AFM images: b, 1 mm; c–f, 100 nm [20].

17.2.4.1 DNA Computing
The algorithmic self-assembly of DNA can be employed to process information. The idea of algorithmic self-assembly arose from three lines of work: DNA computing [22], the theory of tilings [23], and general DNA nanotechnology pioneered by Seeman [24]. In one example, "DNA bricks" were used as molecular Wang tiles [25]. The four arms of the DX molecules (DNA brick or tile) have different sequences corresponding to the labels on the four sides of the Wang tiles. In this way, any chosen Wang tile could be implemented as a DNA molecule. Consequently, such a DNA system becomes programmable. In other words, algorithmic tasks can be carried out with a DNA system such as this. It is envisioned that nanoscale and hierarchically structured materials with properties far beyond the development capability of today's materials engineering technology can be achieved by algorithmically controlled growth processes.

17.2.4.2 DNA and Inorganic Hybrids
There has been significant interest in the rational organization of nanoscale objects, such as C60, metal nanoparticles,

semiconductor nanoparticles, into 1D, 2D, and 3D ordered aggregates that may possess novel electronic and optical properties with potential applications in future nanoelectronics and nanophotonics. Such hybrid systems may also have novel biological applications for synthetic biology. Biomineralization is a natural process used by organisms to produce minerals. Biomolecules such as peptides have been used in the template arrangement of metal crystals. Nucleic acid engineering may assist such a nanoparticle organization—after all, DNA is in a similar length scale with nanoparticles, and it can be self-organized as already reviewed earlier in this chapter. Chemical modification of DNA to carry a sulfhydryl group for metal (Au) attachment can be easily done. Bold DNA strands can even be used to template metal through cation–anion electrostatic interaction. In terms of DNA structure, both linear plasmid DNA and rational designed DNA complexes can be used.

As far as gold nanoparticles are concerned, 1D and 2D assemblies of gold nanoparticles have been achieved using several kinds of DNA nanostructures and various gold–DNA attaching strategies. DNA-templated 1D organization of gold nanoparticles has been obtained through nonspecific electrostatic interactions between positively charged gold nanoparticles and negatively charged DNA [26]. An alternative approach employs DNA sequence complementarity where one strand is a DNA chain and the other is its complementary counterpart attached with gold. Mao's group reported a DNA-encoded self-assembly of AuNPs into 1D micrometers-long gold nanoparticle arrays [27]. An ssDNA template composed of a repetitive unit was generated by rolling circle DNA polymerization. Gold nanoparticles were attached to DNA that is complementary to the sequence on RCA-obtained ssDNA. Aligned gold oligomer (3-, 4-, 5-, and 6-) and extended arrays in the scale of micrometers were obtained. In another report, cysteamine-modified gold nanoparticles with amino moiety were anchored onto cisplatin-functionalized DNA by ligation of Pt atoms [28]. In addition, 2D gold nanoparticle assemblies were reported by using 2D DNA nanogrid templates fabricated from DNA tiles (Fig. 17-6) [29]. Each nanoparticle sit only on one DNA tile within a periodic square lattice; therefore, the nanoparticle–nanoparticle distance was uniformly controlled. As a result, gold nanoparticles ended up in a hexagonal arrangement. In another design, a rigid 120° synthetic vertex and short DNA strands were used to ensure the rigidity of the final assemblies [30].

We noted that there has been growing interest in DNA/carbon nanomaterials, motivated by possible applications in nanoelectronics, drug delivery, and biosensors. Based on the electrostatic interactions between DNA and positively charged fullerene, 1D and 2D fullerene arrays were formed using linear dsDNA [31] and 2D DNA lattices [32]. Interestingly, single-stranded DNA tends to wrap around the exterior surface of a carbon nanotube [33,34]. The wrapping of carbon nanotubes by ssDNA was found to be sequence dependent and the force behind this is the hydrophobic interaction between the bases and the carbon nanotube. DNA has also been demonstrated to insert itself into the interior of carbon nanotubes [35]. In addition, DNA can also be covalently linked to carbon nanotubes via molecular linkers [36,37]. These hybrids may find immediate applications in gene delivery [38].

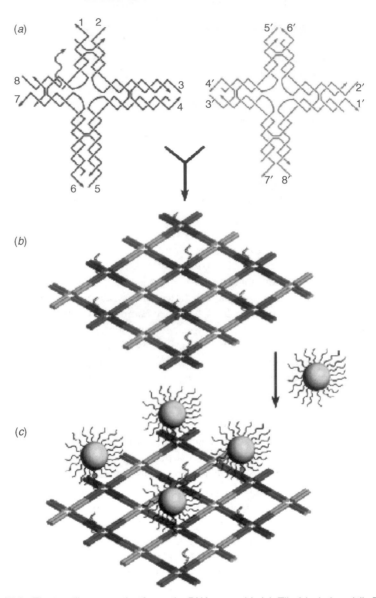

Figure 17-6 The two-tile system that forms the DNA nanogrids (a). Tile A is dark and tile B is grey. The numbers indicate the complementary "sticky" ends that allow the tiles to adhere together, with 1, and so on. The red strand on tile A is A_{15}. The DNA(pairing with 1' nanogrid, showing the A_{15} strand on each A tile (b). Gold nanoparticles on the DNA grids. The zigzagged black lines surrounding the nanoparticles represent T_{15} strands (c) Reproduced from Ref. 29.

17.2.4.3 DNA and Protein Hybrids
Both DNA and protein are functional biomolecules, and combining them will result in multifunctional nanobiosystems with potential applications in the creation of a more functional biological systems.

Hybrid DNA–protein conjugates can be fabricated by either covalent or noncovalent interactions.

The pioneering work on DNA–protein hybrids is the synthesis of the oligonucleotide-staphylococcus nuclease [39]. Oligonucleotide–enzyme conjugates were fabricated from thiol-modified oligonucleotide and an maleimide-modified enzyme (calf intestine alkaline phosphatase, horseradish peroxidase, or β-galactosidase) by coupling thiol group to maleimide [40]. Antibodies have also been coupled with DNA for immunoassay applications. Covalent conjugates of single- and double-stranded DNA fragments and immunoglobulin molecules were used as probes in immuno-PCR in an ultrasensitive antigen detection method [41]. Hundred-fold enhancement in sensitivity was reported compared to a conventional method without the DNA–protein conjugate (Fig. 17-7) [42].

In addition to covalent DNA–protein linkage, a noncovalent approach based on affinity between biotin and homotetrameric streptavidin was reported by Niemeyer' group [43]. By thermal denaturation and rapid cooling, interesting finger ring-like DNA streptavin nanocircles were observed whose dimensions varied from 12 nm to 55 nm [44]. In another example, Tomkins et al. fabricated nanoscale symmetrical and nonsymmetrical dumbbells using DNA as a molecular scaffold for streptavidin attachment [45]. Another interesting noncovalent binding of proteins with DNA is based on DNA aptamer-directed self-assembly [46]. In this system, a DNA docking site containing a DNA aptamer (specific to thrombin) was distributed regularly within ordered DNA lattices. Thrombin proteins bound to these docking sites, formed a DNA–protein hybrid array.

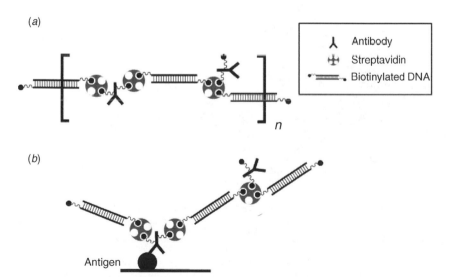

Figure 17-7 The self-assembled DNA-protein hybrid material for immuno-PCR (IPCR). (a) The bis-biotinylated DNA, Streptavidin and biotinylated antibody were mixed to form the hybrid material. (b) This hybrid material can be applied as a regent in IPCR. The antibody part can recognize an antigen (target molecule). The DNA part can be amplified using PCR and detected as an amplified signal. Reproduced from Ref. 43.

17.2.4.4 DNA-Based Mechanical Nanodevices [47,48]

To build a synthetic world, mechanical devices are essential. Although the notion of nanorobotics roaming through the blood stream is quite naïve, the realization of nanoscale mechanical devices, roaming or not, is anything but fantasy. Oligonucleic acids, especially DNA molecules are in nanometer sizes and have already been employed to assemble a variety of nanostructures including 2D arrays and lattices, tubes, scaffolds, and gels. A lot of knowledge has been obtained in controlling the conformation and assembly of DNA molecules. Considering the fact that conformational changes usually accompany movement and considering the fact that DNA's conformations are easily controlled, it seems natural to employ DNA for mechanical nanodevices.

Seeman reported a "nano twister" based on the conformation changes between left-handed and right-handed DNA [48,49]. Later, a DNA-based nanotweezers was reported by Yurke et al. in 2000 (Fig. 17-8a) [50]. (1) This DNA tweezers was composed of several strands of DNA with engineered pairing properties, as illustrated in the scheme. (2) The addition of a fuel strand DNA (F), which hybridized to the open DNA parts, caused the ends to close. (3) The subsequent addition of a complementary strand of DNA (cF) displaced the F strand from the body, forming a very tight F–cF double-stranded DNA. The departing of F strand (i.e., dehybridization) caused the closed ends to open again, returning the structure to the original open state.

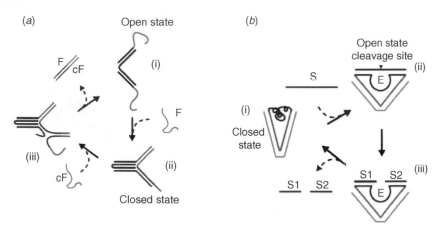

Figure 17-8 (a) Scheme of the DNA tweezers. Addition of a fuel strand DNA (F), which hybridized to the open DNA parts, caused the ends to close (ii). The subsequent addition of a complementary strand of DNA (cF) displaced the F strand from the body, forming a very tight F–cF double-stranded DNA. The departing of F strand caused the closed ends to open again (iii), returning the structure to the original open state (i). Reproduced from Ref. 50. (b) Schematic drawing of the autonomous DNA nanomotor. This DNA nanomotor consists of the deoxyribozyme part that cleaves RNA and the tweezers. The edges of the tweezers that are connected with the compacted deoxyribozyme exist closer (i). The DNA–RNA chimera substrate (S) as a fuel strand hybridizes and stretches the deoxyribozyme, activating its activity (ii). The activated deoxyribozyme cleaves S at the cleavage site (iii) and the substrate fragments (S1 and S2) dissociate from the DNA motor, which returns to the closed state (i). Reproduced from Ref. 51.

The continuous adding of fuel strands (F) and removing of waste (F–cF) cycled this DNA structure from open to closed states. This first DNA mechanical tweezers is not autonomous; fuel and waste need to be added or eliminated by hand. Utilizing the RNA cutting DNAzyme, Chen et al. developed an autonomous DNA nanomotor in 2004 using DNA–RNA chimera as fuel strands (see Fig. 17-8b) [51].

Other mechanical nanodevices have also been reported recently. For example, Yin et al. created an enzyme-assisted DNA walker that moves along a double-stranded DNA track [52]. The enzymes used were T4 ligase and restriction enzymes. The fuels were ATP and were consumed by ligases. Up to now, several versions of DNA walkers have been developed, though none has been sophisticated enough to carry out certain functions.

17.3 DNA MATERIALS

DNA is the nature chosen genetic molecule of life; DNA is also an excellent building block of DNA nanotechnology. From the polymer science perspective, DNA is a biodegradable and biocompatible polymeric material, possibly with many desired functions as well. It is suggested that DNA can also be used in conductor and photonics.

17.3.1 Dendrimer-Like DNA (DL-DNA) [53,54]

Dendrimers are repeatedly branched polymers, characterized by their structural symmetry and polydispersity. As branches of dendrimers can be modified with functional groups or molecules, they have been proposed for applications in drug delivery and for components of advanced materials. Inspired by earlier and impressive achievements from the Seeman group and others, our lab has succeeded in constructing DNA junction molecules. We treat the DNA junction molecule as a repeating unit similar to that of a dendrimer (Fig. 17-9). The junction molecules branch out repeatedly by using sticky-ended cohesion; in this case, each generation of growth entails a specific pair of sticky ends. In addition to sticky-ended cohesion, we use DNA T4 ligase to form covalent bonds. Compared with chemically synthesized dendrimers, dendrimer-like DNA can be assembled in a controlled manner with great precision, high efficiency and anisotropy. We used gel electrophoresis, AFM, and TEM to characterize assembled dendrimer-like DNA.

17.3.2 DNA-Based Fluorescent Nanobarcode [53,55]

The development of techniques for multiplexed, rapid, and sensitive molecular detection has received great attention due to the potential applications in clinical diagnosis, detection systems against terrorists, and environmental analysis. For multiplexed detection, it is a challenge to identify each species with a distinct code, and match each output signal with the species responsible for it.

Fluorescent dyes are very sensitive to their chemical and physical environment, and widely used for probing biomolecular structures, biological processes, and molecular

(a)   (b)

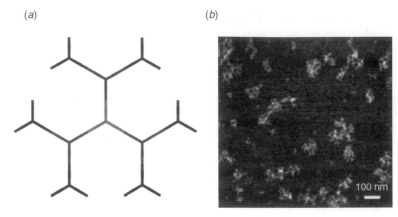

Figure 17-9 Dendrimer-like DNA (DL-DNA). Schematic structure of three-generation DL-DNA (a) and AFM image of DL-DNA (four generations) on the mica surface using a SWNT (single-wall nanotubes) tip (b). Reproduced from Ref. 54.

detection. They can be covalently linked to DNA that is modified to carry a functional group (amine or thiol) reactive to the functional group of the dyes. Our dendrimer-like DNA is multivalent and anisotropic, and can be synthesized in a controlled hierarchical manner. The dye-carrying Y-DNA is a DNA junction molecule composed of three strands: one has a sticky end and the other two are labeled with two different fluorescent dyes. This sticky-ended fluorescent Y-DNA hybridizes with dendrimer-like DNA whose sticky ends are complementary to that of the Y-DNA. This will amplify numbers of dyes on each nanobarcode.

We used our fluorescent nanobarcode to detect multiple DNA pathogens (Fig. 17-10). To be detected by fluorescence microscopy, a strategy has to be employed where the nanobarcodes will be attached, in the presence of its target DNA pathogen, to microbeads whose size is detectable to fluorescence microscopy. We also employed flow cytometry technique to monitor multiplexed detection of DNA pathogen with our DNA-based fluorescent nanobarcode.

17.3.3 DNA Bulk Material: DNA Hydrogel

We recently succeeded in making a DNA hydrogel and showing that it can be used for drug delivery [56]. DNA, generally recognized as the molecule of life, has been employed as a building block of functional materials and self-assembled superstructures that can scaffold or template other functional species. But, no previous work reports 3D nanostructures using DNA as the building block. Neither has a bulk DNA material been constructed before our DNA hydrogel (Fig. 17-11). This is significant to the field of DNA nanotechnology.

DNA is hydrophilic; hydrogel can be formed when network of such polymeric chains takes water as its dispersion medium. Our DNA hydrogel uses the unusual branched DNA motifs as the building unit, like those in our dendrimer-like DNA. Our building units can be designed to have three or four branching points necessary

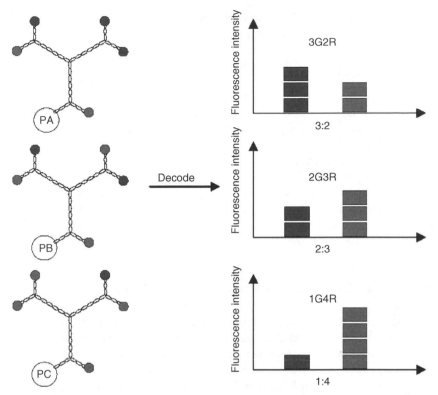

Figure 17-10 Principle of DNA-based fluorescent nanobarcode. Three nanobarcodes 3G2R, 2G3R, and 1G4R, which are decoded based on the ratio of fluorescence density, are synthesized by labeling two types of fluorescent dyes at varying ratios on the dendrimer-like DNA (DL-DNA). A molecular recognition element, a probe (PA, PB, and PC), is also attached to each nanobarcode. With a preassigned code library, these nanobarcodes can be used for molecular detection. Reproduced from Ref. 55.

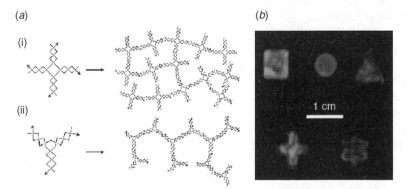

Figure 17-11 Scheme of DNA hydrogel. X- and Y-DNA monomers serve as cross-linkers to form networked gels (a); images of X-DNA hydrogels with different patterns (b): rectangular, circular, triangular, star, and cross (from the top left corner, clockwise). The hydrogel is stained with EtBr (Top left, top right, bottom left) or SYBR I (Top left, top right, bottom left). Reproduced from Ref. 56.

for forming the network of the hydrogel; they assemble into a polymeric network through hybridization of the sticky ends and covalent bonding formed by T4 DNA ligase.

Proteins, small chemical drugs, and even live cells were encapsulated within the gel; making DNA gel a great candidate for use as a drug and cell delivery vector. *In situ* encapsulation efficiency for porcine insulin and camptothecin is close to 100 percent. No burst release is observed. A zero-order release profile is obtained for camptothecin. Up to 60 percent insulin can be released over a period of 12 days.

More recently, by incorporating genes into the DNA gel matrix, we have also created a DNA hydrogel that is able to produce functional proteins without any living organisms (the gel is termed as "P-gel"). The yield and efficiency of the P-gel is many fold higher than current conventional methods. We believe that this is an enabling technology for producing most proteins in high yield and rapidly (usually within 24 h). This *in vitro* protein-producing device should be of significant value in synthetic biology.

17.4 DNA APTAMERS AND NUCLEIC ACID ENZYMES

Aptamers are short oligonucleic acids or peptides that can bind to specific target molecules. DNA and RNA aptamers usually consist of specific sequences and can fold into a specific structural conformation for binding to their target. Thrombin binding DNA aptamers have GC-rich sequences and can form G-quartet structure. RNA aptamers are more versatile in terms of the secondary structures it can form. Although natural aptamers exist in riboswitches, the part of mRNA that can bind to its target molecule; typically, however, they are obtained through an *in vitro* evolution process called SELEX (systematic evolution of ligands by EXponential enrichment). In this method, a library composed of random sequences is generated first. *In vitro* evolution is engineered through a selection process followed by an amplification step (Fig. 17-12). After several rounds, the candidates that have the potential to bind to the target are enriched.

Aptamers have potential applications in medicine and technology. Binding of aptamers to their target molecules can be coupled to signal producing processes in construction of molecular biosensors. When the targets are proteins (antigens) or enzymes, aptamers behave much like antibodies. The activity of proteins or enzymes can be inhibited when they are targeted and bound by aptamers; therefore, it has been proposed that aptamers have the potential to be therapeutic drugs [57]. Since aptamers can be engineered to bind to specific molecules, they also become a promising tool for nucleic acid engineering [58,59].

Some oligonucleic acids have catalytic activities like enzymes; they are called nucleic acid enzymes [60]. There are two types of nucleic acid enzymes: one composed of RNA and the other of DNA. RNA enzymes are called "ribozymes." DNA enzymes are called DNAzymes. While a great number of different activities have been realized with ribozymes (such as strand cutting and ligation), only a few activities have been achieved with DNAzymes. This is likely due to two facts: first, DNA lacks

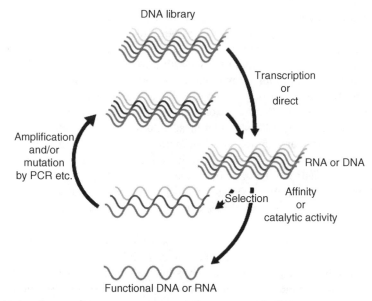

Figure 17-12   Schematic concept of *in vitro* evolution process: A DNA library composed of random sequences is generated first. A corresponding RNA library (optional) can be created through *in vitro* transcription. *In vitro* DNA or RNA evolution is then carried out through selection (for desired properties), followed by an amplification step. This selection and amplification process will be repeated several times until the functional DNA or RNA are obtained.

the active hydroxy groups on its sugar back bone that can participate and catalyze certain reactions; and second, DNA tends to form double-stranded helixes while RNA is able to form various secondary structures such as stem loops and hairpins in addition to duplex structures.

17.5   CHARACTERIZATIONS OF OLIGONUCLEIC ACIDS

Research using DNA as a building block for DNA complexes or for functional materials is at the interface of biology, chemistry, material science, and engineering. Knowledge and experimental techniques from one field is not enough. In the last part of the chapter, we will briefly introduce some experimental techniques. Readers are strongly advised to refer to more specialized discussions of each technique for a better understanding.

17.5.1   UV Spectroscopy [61]

It is important for us to know the concentration of a DNA solution to adjust to a desired stoichiometry. One convenient way is to measure OD260. Since all four nucleotides absorb ultraviolet light at around a wavelength of 260 nm with a similar extinction coefficient (also called molar absorptivity), the concentration of a nucleic acid solution can be easily estimated by measuring its absorbance at 260 nm, according to Beer's law: $A = \varepsilon l c$, where A is the absorbance, ε is the extinction coefficient, l is the cell path length (normally 1 cm), and c is the nucleic acid concentration. As the spectroscopic properties of nucleic acids are quite similar within ssDNA, dsDNA, and RNA, one can estimate their concentrations by reading the absorbance first and then multiplying by their specific unit absorbance. For dsDNA, an absorbance of 1 corresponds to 50 µg/mL. For ssDNA, an absorbance of 1 represents a concentration of 37 µg/mL. For RNA, the value is 40 µg/mL. Note that these values have been calculated using the average ε and molecular weight. If one requires an accurate measurement of a specific oligonucleic acid solution (especially if it is a short chain of nucleotides), the extinction coefficient for the specific sequence should be used in the calculation using Beer's law. It can be calculated from the nucleotide extinction coefficients contained in the sequence. This can be carried out with Web-based oligonucleotide property calculators.

Absorbance ratios between 260 and 280 nm reflect the purity of a nucleic acid solution. In the case of DNA extracted out of cellular mixtures, they reflect the contamination by proteins, as most proteins have a stronger absorbance of 280 nm compared to oligonucleic acids. Thus, the lower ratio of 260 versus 280 nm represents more contamination (and consequently less pure nucleic acid solution). A typical benchmark number used for a pure DNA solution is 1.8 (260 versus 280 nm). A ratio of 2.0 is used for pure RNA solution.

OD260 is also used to characterize secondary structure and affinity of oligos. Bases in duplex adsorb less UV light than in single-stranded state, due to base stacking and electronic interaction in duplex. The difference between the two states is called hypochromicity. OD260 of duplex solution increases gradually as temperature increases. When UV spectroscopy is coupled with a thermal controller, thermal stability of DNA duplexes can be studied. The midpoint of denaturation is called Tm, a parameter used for comparing thermal stability.

17.5.2   Gel Electrophoresis [61]

Electrophoresis is one of the most widely used methods for purification and characterization of oligonucleic acids. The principle of electrophoresis is quite simple: in an electrical field, a molecule carrying a negative charge will migrate toward a positive electrode whereas a molecule with a positive charge will move toward a negative electrode. Since oligonucleic acids (DNA or RNA) are highly negatively charged due to their phosphate backbones, they migrate toward a positive electrode if placed in an electric field. If a media matrix is used in the process to contain the oligonucleic acids (typically an agarose gel or a polyacrylamide gel), the electromobility of oligonucleic acids depends on the general shape of the nucleic acid molecules due to the sieving property of these gels, or more accurately, the mobility is determined by the charge–surface ratio. Generally, smaller fragments move faster than larger fragments. Note that other topological parameters may also influence the electrophoretic mobility of oligonucleic acids. For example, supercoiled DNA moves much faster than linear DNA with the same size. It should be noted that the sieving properties of both agarose and polyacrylamide gels are defined by the effective pore size of the gels. Gels with smaller pores, which correspond to a higher gel percentage or more cross-linking agents used as in the case of PAGE, are more suitable for separation or analysis of smaller oligonucleic acids, and vice versa. Oligonucleic acids can be analyzed under native conditions where formation of duplex is allowed, or under denaturing conditions where formation of duplex is prohibited by thermal conditions (high temperature) and denaturing reagents such as formamide and urea. Denaturing conditions are required for purification of synthetic oligonucleotides.

Agarose is a polysaccharide extracted from seaweed. Typically, it is used for making gels at a concentration of 0.5–3 percent. Typically, DNA fragments from 200 bp to approximately 50 kbp can be separated in an agarose gel at an appropriate concentration. polyacrylamide gel electrophoresis (PAGE) is used to separate relatively small oligonucleic acid molecules (from a few nucleotides to several hundreds), due to much smaller pore sizes within polyacrylamide gels. The gel, usually in the range of 3.5–20 percent, is made by polymerizing acrylamide monomers in the presence of a cross-linking agent: N,N'-methylene bisacrylamide. Ammonium persulfate is typically used as a catalyst, and N,N,N,N'-tetramethylethyldiamine (TEMED) is added as an accelerator.

Besides the consideration of the fragment size each type of gel is best suited for, both gels have their advantages and disadvantages. Agarose gels are easier to cast, nontoxic, and relatively inexpensive. Polyacrylamide gels, on the other hand, have higher resolution in separation. For routine applications, agarose gel electrophoresis is by far the more commonly used method.

After gel electrophoresis, gels can be stained with dyes to locate the position of nucleic acid fragments. All the dyes for gel staining should be able to bind to oligonucleic acids and differentiate them from the background signals. Stains-All, a nonfluorescent dye, and fluorescent dyes such as ethidium bromide or SYBR series are commonly used to stain the gel. The intensity of fluorescence dyes is typically used to measure the quantity of oligonucleic acids. In addition, the bands of nucleic acid fragments can be cut out (not stained by Stains-All) and then extracted for downstream processing.

17.5.3 Imaging

As DNA complexes from building blocks are in the nanoscale or microscale, imaging tools developed for nanoscience have to be used to characterize their structures. Here, we introduce TEM and AFM.

17.5.3.1 Transmission Electron Microscopy
The first TEM was built by Max Knoll and Ernst Ruska in 1931. Since then, much progress has been made in improving instruments and methods for exploring microscale, nanoscale, and atomic scale objects. In nature, TEM is analogue of optical microscopy and it operates based on the same principle as the usual optical microscopy. TEM takes advantage of the much shorter wavelength of electrons than light; therefore, a much higher resolution (10^{-10} m) is achieved. The electrons in TEM are usually generated by a process known

as thermionic discharge or by field emission. They are accelerated by an electric field and focused onto the sample by a lens realized via electrical and magnetic fields. Such a process is possible because electrons have both wave and particle properties. The wave properties make a beam of electrons behave like a beam of radiation, whose wavelength depends on their energy. The wave property of electron beams can be modulated by electrical and magnetic fields, as light is by optics.

By design, TEM is usually composed of an electron gun for generating electron beam, condenser lenses, a sample stage, an objective lens, project lenses, and a fluorescent screen for observing by naked eye (Fig. 17-13). Sometimes, charge-coupled device (CCD) camera is connected to TEM to give real-time digital micrographs. TEM specimens, whether a tissue or a particular material, must be supported on a thin e-beam transparent film to image them. The evaporated carbon film usually acts as a substrate for TEM specimen support.

TEM has been largely used for imaging, atomic structural analysis (diffraction pattern), and chemical elemental analysis (X-ray microanalysis). The phase structure of materials can be known by a combination of imaging and electron diffraction studies. With the help of energy-dispersive X-ray spectroscopy (EDS) and electron

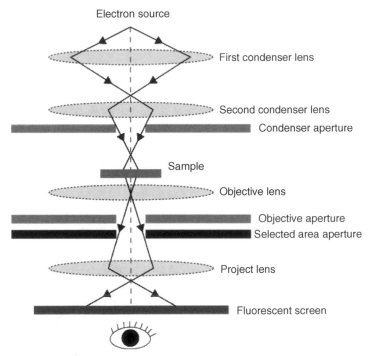

Figure 17-13 Schematic of imaging mechanism of TEM: E-beam was generated by an electron gun, which is accelerated, condensed, and hit by samples. The transmitted portion of E-beam is focused by objective lens into an image, which is passed down the column through project lens to be enlarged. The image strikes the phosphor image screen to emit light, allowing the user to see the image by naked eye or a CCD camera.

energy loss spectroscopy (EELS), TEM has become a comprehensive tool to gather information from a certain material. Since the birth of nanoscience and nanotechnology, TEM has been the most powerful tool for characterizing nanomaterials. For instance, carbon nanotubes were first identified by TEM.

In biology, TEM has made a great contribution to our understanding of the cell, tissues, and even biomacromolecules such as DNA. Unlike metals, biomaterials are usually not electronically dense and are sometimes damaged by electron beams because of electrical charging. A solution is to stain samples with heavy metal salts, which consist of a high atomic number of protons and electrons that can scatter electrons efficiently. Many organic molecules in biomaterials can reduce strongly oxidative heavy metal salts into metals giving improved contrast. These stains include uranyl acetate, lead citrate, osmium tetroxide, and so on. Note that the small size (and width) of the nucleic acid helix makes it impossible to be imaged by conventional electron microscopy directly, and usually needs heavy staining, and sometimes even an additional metallic shadowing.

17.5.3.2 *Atomic Force Microscopy*
Since Binning and Rohrer from IBM invented scanning tunneling microscopy (STM) in 1981, scanning probe microscopy (SPM), which uses an extremely sharp tip (10 nm) as a probe to study local properties of samples by measuring various weak forces, has been developed. Although STM has an atomic level of resolution, it measures the tunneling current and requires the samples to be electron conductive. In 1986, Binning, Quate, and Gerber developed an atomic force microscopy (AFM) that can study the topography of a substrate by measuring the atomic force between the probe tip and the sample. And thus electron conductivity is no longer needed.

Like all other scanning probe microscopes, AFM uses an ultrasharp tip (1–10 nm in radius) as a probe to scan a substrate surface. The tip is on the end of a cantilever that elastically bends in response to the force between the tip and the sample. The bending of the cantilever is detected by laser reflection onto photodiodes. The detected positional signal of a reflected laser beam is then used as a feedback signal to control the piezoelectric nanotranslator to ensure a constant force between the tip and the sample surface (Fig. 17-14). By translating the piezoelectric response during the tip-scanning process, AFM obtains the surface morphology (topography) with an atomic spatial resolution.

There are two working modes for AFM depending on the working distance between the tip and the sample surface: contact mode and noncontact mode. Within the contact mode there is a special mode called the tapping mode.

In the contact mode, the tip and the sample surface make contact during scanning (the distance between the tip and the surface is less than 10 nm). Although the contact mode is most often used due to its higher resolution, the excessive dragging force during a contact scanning can damage the soft sample, such as polymers and biomolecules, and thus limits its usage for fragile samples such as oligonucleic acids. In addition, if the sample surface is covered by layers of water or gas, which is very common for biological substrates, the force can be distorted due to a meniscus on the tip from a contaminated surface, leading to a mistaken interpretation

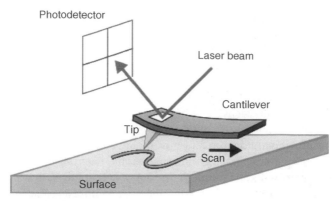

Figure 17-14 Schematic illustration of AFM: The tip mounted to a cantilever is scanned over a surface. The tip–surface interactions cause cantilever deflection, which is monitored by a photodiode.

of the morphology. Contact mode in liquid can sometimes be used to overcome this problem.

In the tapping mode, the cantilever is oscillated at its resonant frequency. When the tip scans above the sample surface, the tip taps the surface for a very small fraction of the oscillating period. Because the contact occurs in an extremely short time, the lateral force is greatly reduced compared to the contact mode. Sample damaging is thus minimized. The tapping mode of AFM is mostly suitable for imaging poorly immobilized samples or soft samples. A more advanced tapping mode, Mac AFM mode, uses a magnetic field to drive a cantilever coated with paramagnetic materials, improving imaging greatly.

For the noncontact mode, the working distance of the tip is about 5–15 nm above the sample surface. The topographic images are constructed by measuring the attractive, very weak forces. Although the noncontact mode does not damage samples, it is a more difficult method due to the possibility of water contamination.

DNA is very thin (2 nm in diameter) and long (from nm to mm) and generally nonconductive. Before AFM was developed, transmission electron microscopy (TEM) or scanning electron microscopy (SEM) had been used for DNA imaging after coating the DNA with a layer of conductive materials. This treatment tended to distort the DNA structure. In the case of AFM, DNA can be viewed directly either in air or in solution. In addition, AFM is much easier to operate, and the cost is much lower than that of TEM or SEM. Because DNA is a very soft material, the tapping mode is preferred in imaging. Also, to strongly fix DNA onto a carrier substrate, positively charged molecules such as aminopropyl-triethoxy silane (APTES) are used to coat the substrate.

17.5.3.3 Fluorescence Microscopy

The easiest way of imaging bulk DNA is to use fluorescence microscope. Note that DNA itself is not fluorescent. However, many dyes can bind to DNA and fluorescence strongly, with many folds of difference compared to nonDNA binding state, such as ethidium bromide, SYBR green, YOYO, and POPO. The fluorescence can be detected by fluorescence microscopy or even by the naked eye. The resolution of fluorescence microcopy is not as high as the imaging tools discussed above, a limitation posed by light diffraction and quality of optics. However, it is improved to some extent in confocal fluorescence microscopy.

REFERENCE

1. Kumar R, et al. The first analogues of LNA (locked nucleic acids): phosphorothioate-LNA and 2'-thio-LNA. *Bioorg Med Chem Lett* 1998;8:2219–2222.

2. Haaima G, Hansen HF, Christensen L, Dahl O, Nielsen PE. Increased DNA binding and sequence discrimination of PNA oligomers containing 2, 6-diaminopurine. *Nucleic Acids Res* 1997;25:4639–4643.

3. Hyrup B, Nielsen PE. Peptide nucleic acids (PNA): synthesis, properties and potential applications. *Bioorg Med Chem* 1996;4:5–23.

4. Feynman RP. There's plenty of room at the bottom. *Eng Sci* 1960;23:22.

5. Breaker RR. Catalytic DNA: in training and seeking employment. *Nat Biotechnol* 1999;17:422–423.

6. Seeman NC. Nucleic acid junctions and lattices. *J Theor Biol* 1982;99:237–247.

7. Kallenbach NR, Ma RI, Seeman NC. An immobile nucleic acid junction constructed from oligonucleotides. *Nature* 1983;305:829–831.

8. Seeman NC, Kallenbach NR. DNA branched junctions. *Annu Rev Biophys Biomol Struct* 1994;23:53–86.

9. Wang YL, Mueller JE, Kemper B, Seeman NC. Assembly and characterization of five-arm and six-arm DNA branched junctions. *Biochemistry* 1991;30:5667–5674.

10. Fu TJ, Seeman NC. DNA double-crossover molecules. *Biochemistry* 1993;32:3211–3220.

11. Shen Z, Yan H, Wang T, Seeman NC. Paranemic crossover DNA: a generalized Holliday structure with applications in nanotechnology. *J Am Chem Soc* 2004;126:1666–1674.

12. Mao CD, LaBean TH, Reif JH, Seeman NC. Logical computation using algorithmic self-assembly of DNA triple-crossover molecules. *Nature* 2000;407:493–496.

13. Yan H, Park SH, Finkelstein G, Reif JH, LaBean TH. DNA-templated self-assembly of protein arrays and highly conductive nanowires. *Science* 2003;301:1882–1884.

14. Chelyapov N, et al. DNA triangles and self-assembled hexagonal tilings. *J Am Chem Soc* 2004;126:13924–13925.

15. He Y, Chen Y, Liu H, Ribbe AE, Mao C. Self-assembly of hexagonal DNA two-dimensional (2D) arrays. *J Am Chem Soc* 2005;127:12202–12203.

16. Rothemund PW, et al. Design and characterization of programmable DNA nanotubes. *J Am Chem Soc* 2004;126:16344–16352.

17. Ding B, Sha R, Seeman NC. Pseudohexagonal 2D DNA crystals from double crossover cohesion. *J Am Chem Soc* 2004;126:10230–10231.

18. Liu D, Park SH, Reif JH, LaBean TH. DNA nanotubes self-assembled from triple-crossover tiles as templates for conductive nanowires. *Proc Natl Acad Sci USA* 2004;101:717–722.

19. Shih WM, Quispe JD, Joyce GF. A 1.7-kilobase single-stranded DNA that folds into a nanoscale octahedron. *Nature* 2004;427:618–621.

20. Rothemund PW. Folding DNA to create nanoscale shapes and patterns. *Nature* 2006;440:297–302.

21. Shen W, Bruist MF, Goodman SD, Seeman NC. A protein-driven DNA device that measures the excess binding energy of proteins that distort DNA. *Angew Chem Int Ed Engl* 2004;43:4750–4752.

22. Adleman LM. Molecular computation of solutions to combinatorial problems. *Science* 1994;266:1021–1024.

23. Grunbaum B, Shephard GC. *Tilings and Patterns*. Freeman, WH & Co. New York 1987.

24. Seeman NC. Biochemistry and structural DNA nanotechnology: an evolving symbiotic relationship. *Biochemistry* 2003;42:7259–7269.

25. Winfree E, Liu F, Wenzler LA, Seeman NC. Design and self-assembly of two-dimensional DNA crystals. *Nature* 1998;394:539–544.

26. Nakao H, et al. Highly ordered assemblies of Au nanoparticles organized on DNA. *Nano Lett* 2003;3:1391–1394.

27. Deng ZX, Tian Y, Lee SH, Ribbe AE, Mao CD. DNA-encoded self-assembly of gold nanoparticles into one-dimensional arrays. *Angew Chem Int Ed* 2005;44:3582–3585.

28. Noyong M, Gloddek K, Simon U. Immobilization of gold nanoparticles on DNA. *Mater Res Soc Symp Proc* 2003;735:153–158.

29. Zhang J, Liu Y, Ke Y, Yan H. Periodic square-like gold nanoparticle arrays templated by self-assembled 2D DNA nanogrids on a surface. *Nano Lett* 2006;6:248–251.

30. Aldaye FA, Sleiman HF. Sequential self-assembly of a DNA hexagon as a template for the organization of gold nanoparticles. *Angew Chem Int Ed* 2006;45:2204–2209.

31. Cassell AM, Scrivens WA, Tour JM. Assembly of DNA/fullerene hybrid materials. *Angew Chem Int Ed* 1998;37:1528–1531.

32. Song C, Chen, YQ, Xiao SJ, Ba, L, Gu Z-Z, Pan Y, You XZ. Assembly of fullerene arrays templated by DNA scaffolds. *Chem Mater* 2005;17:6521.

33. Lu YR, et al. DNA functionalization of carbon nanotubes for ultrathin atomic layer deposition of high kappa dielectrics for nanotube transistors with 60 mV/decade switching. *J Am Chem Soc* 2006;128:3518–3519.

34. Zheng M, et al. Structure-based carbon nanotube sorting by sequence-dependent DNA assembly. *Science* 2003;302:1545–1548.

35. Gao HJ, Kong Y, Cui DX, Ozkan CS. Spontaneous insertion of DNA oligonucleotides into carbon nanotubes. *Nano Lett* 2003;3:471–473.

36. Hazani M, Naaman R, Hennrich F, Kappes MM. Confocal fluorescence imaging of DNA-functionalized carbon nanotubes. *Nano Lett* 2003;3:153–155.

37. McKnight TE, et al. Tracking gene expression after DNA delivery using spatially indexed nanofiber arrays. *Nano Lett* 2004;4:1213–1219.

38. Kam NWS, Liu ZA, Dai HJ. Carbon nanotubes as intracellular transporters for proteins and DNA: an investigation of the uptake mechanism and pathway. *Angew Chem Int Ed* 2006;45:577–581.

39. Corey DR, Schultz PG. Generation of a hybrid sequence-specific single-stranded deoxy-ribonuclease. *Science* 1987;238:1401–1403.

40. Ghosh, SS, Kao, PM, McCue, AW, Chappelle HL. Use of maleimide-thiol coupling chemistry for efficient syntheses of oligonucleotide-enzyme conjugate hybridization probes. *Bioconjugate Chem* 1990;1:71.

41. Hendrickson ER, Truby TM, Joerger RD, Majarian WR, Ebersole RC. High sensitivity multianalyte immunoassay using covalent DNA-labeled antibodies and polymerase chain reaction. *Nucleic Acids Res* 1995;23:522–529.

42. Niemeyer CM, et al. Self-assembly of DNA-streptavidin nanostructures and their use as reagents in immuno-PCR. *Nucleic Acids Res* 1999;27:4553–4561.

43. Niemeyer CM, Sano T, Smith CL, Cantor CR. Oligonucleotide-directed self-assembly of proteins: semisynthetic DNA–streptavidin hybrid molecules as connectors for the generation of macroscopic arrays and the construction of supramolecular bioconjugates. *Nucleic Acids Res* 1994;22:5530–5539.

44. Niemeyer CM, Adler M, Gao S, Chi L. Supramolecular nanocircles consisting of streptavidin and DNA. This work was supported by the Deutsche Forschungsgemeinschaft (SPP 1072), the Fonds der Chemischen Industrie, and the Tonjes-Vagt Stiftung. We thank Prof D. Blohm and Prof H. Fuchs for generous support. *Angew Chem Int Ed Engl* 2000;39:3055–3059.

45. Tomkins JM, et al. Preparation of symmetrical and unsymmetrical DNA–protein conjugates with DNA as a molecular scaffold. *Chembiochem* 2001;2:375–378.

46. Liu Y, Lin C, Li H, Yan H. Aptamer-directed self-assembly of protein arrays on a DNA nanostructure. *Angew Chem Int Ed Engl* 2005;44:4333–4338.

47. Simmel FC, Dittmer WU. DNA nanodevices. *Small* 2005;1:284–299.

48. Seeman NC. From genes to machines: DNA nanomechanical devices. *Trends Biochem Sci* 2005;30:119–125.

49. Mao CD, Sun WQ, Shen ZY, Seeman NC. A nanomechanical device based on the B–Z transition of DNA. *Nature* 1999;397:144–146.

50. Yurke B, Turberfield AJ, Mills AP, Simmel FC, Neumann JL. A DNA-fuelled molecular machine made of DNA. *Nature* 2000;406:605–608.

51. Chen Y, Wang MS, Mao CD. An autonomous DNA nanomotor powered by a DNA enzyme. *Angew Chem Int Ed* 2004;43:3554–3557.

52. Yin P, Yan H, Daniell XG, Turberfield AJ, Reif JH. A unidirectional DNA walker that moves autonomously along a track. *Angew Chem Int Ed* 2004;43:4906–4911.

53. Um S, Lee J, Kwon S, Li Y, Luo D. Dendrimer-like DNA (DL-DNA) based fluorescence nanobarcodes. *Nat Protocols* 2006;1:995–1000.

54. Li Y, et al. Controlled assembly of dendrimer-like DNA. *Nat Mater* 2004;3:38–42.

55. Li Y, Cu YT, Luo D. Multiplexed detection of pathogen DNA with DNA-based fluorescence nanobarcodes. *Nat Biotechnol* 2005;23:885–889.

56. Um SH, et al. Enzyme-catalysed assembly of DNA hydrogel. *Nat Mater* 2006;5:797–801.

57. Blank M, Blind M. Aptamers as tools for target validation. *Curr Opin Chem Biol* 2005;9:336–342.

58. Lin C, Liu Y, Yan H. Self-assembled combinatorial encoding nanoarrays for multiplexed biosensing. *Nano Lett* 2007;7:507–512.

59. Lin C, Katilius E, Liu Y, Zhang J, Yan H. Self-assembled signaling aptamer DNA arrays for protein detection. *Angew Chem Int Ed Engl* 2006;45:5296–5301.

60. Fiammengo R, Jaschke A. Nucleic acid enzymes. *Curr Opin Biotechnol* 2005;16:614–621.

61. Sambrook J, Fritsch, EF, Maniatis T. *Molecular Cloning: A Laboratory Manual*, 2nd ed. New York, NY: Cold Spring Harbor Laboratory Press, 1989.

18

POTENTIAL APPLICATIONS OF SYNTHETIC BIOLOGY IN MARINE MICROBIAL FUNCTIONAL ECOLOGY AND BIOTECHNOLOGY

Guangyi Wang[1] and Juanita Mathews[2]

[1]*Department of Oceanography, University of Hawaii at Manoa, 1000 Pope Road, Honolulu, Hawaii 96822*
[2]*Department of Molecular Biosciences and Bioengineering, University of Hawaii at Manoa, Honolulu, Hawaii 96822*

18.1 INTRODUCTION

The oceans cover 70 percent of the earth's surface and are the most complicated and dynamic of all the earth's ecosystems; they provide the largest inhabitable space for living organisms, particularly microbes [1–3]. Microbes are well known to live in every corner of the oceans. Their habitats are extremely diverse and include, but not limited to, open water, sediment, estuaries, and specialized niches like hydrothermal vents and symbiotic hosts [1,4]. Microbial cells may account for more than 90 percent of the total oceanic biomass [5]. For more than 3 billion years, these microscopic creatures have mediated critical physical, chemical, and biological processes that have shaped the planet's habitability [2,6]. Marine microbes are responsible for about 50 percent of global primary productivity and play a major role in global nutrient cycles, which

Systems Biology and Synthetic Biology   Edited by Pengcheng Fu and Sven Panke
Copyright © 2009 John Wiley & Sons, Inc.

can directly or indirectly impact global climate change [6–9]. Due to their unique living environments in the oceans, marine microbes must adjust metabolically and physically to escape predators and adapt to harsh environments. Accordingly, intensive evolutionary pressures have forced marine microbes to evolve a wide range of metabolic abilities for regulatory function and the production of diverse molecules. Therefore, it comes as no surprise that the microbial diversity of the ocean is vast and a rich source of interesting biological materials for biotechnological applications [3,4,10].

Prior to the 1990s, our understanding of the diversity, ecological function, and biomedical potential of microbial communities was limited by the complexity of marine ecosystems. Recent developments in microbiological oceanography, high throughput screening methods and genomics have revealed new marine microbes and the natural compounds that they produce. However, in marine ecosystems, less than 0.1 percent of the indigenous microorganisms can be readily recovered by standard cultivation techniques. Therefore, our understanding of the ecological function and biotechnological potential of most marine microbes has been greatly limited [11,12]. At present, most studies of marine ecology still focus on mining genetic materials from diverse marine habitats and the understanding of the diversity and structure of marine microbial consortia [2,7,8,11,13–17]. Particularly, molecular approaches have opened the door to the understanding of ecological functions and the discovery of novel metabolic pathways and natural compounds. Because most marine microbes are not amenable to genetic manipulation, little work in synthetic biology has been done using marine microbes.

There are two broad goals for synthetic biology. One is the design and fabrication of biological components and systems using unnatural molecules, and the other is the redesign and fabrication of existing biological systems using interchangeable parts from natural biology [18]. Among the applications of this new field is the creation of bioengineered microbes and possibly other life forms that produce pharmaceuticals, detect toxic chemicals, break down pollutants, repair defective genes, destroy cancer cells, and generate hydrogen for the postpetroleum economy. Synthetic biology is chiefly an engineering discipline, but the ability to design and construct simplified biological systems offers life scientists a useful way to test their understanding of the complex functional networks of genes and biomolecules that mediate life process [19]. In this chapter, we review the use of genetic material from marine microbes to engineer conventional hosts for biotechnological and ecological benefits. The major goal is to illustrate the application of synthetic biology in oceanography and marine biotechnology research.

18.2 MARINE QUORUM SENSING AND SYNTHETIC REGULATORY NETWORK

Many accomplishments have already been made in synthetic biology, including diagnostic tools and diverse regulatory genetic circuits [18,20]. In this section, we only summarize utilization of genetic elements of marine quorum sensing for synthetic cell communication systems.

18.2.1   Quorum Sensing of the Marine Symbiotic Bacterium *Vibrio fisheri*

Quorum sensing is the regulation of gene expression in response to fluctuations in cell population density [21]. Quorum-sensing bacteria synthesize and release chemical signal molecules called autoinducers that increase in concentration as a function of cell density. The detection of a minimal threshold concentration of an autoinducer leads to change in gene expression. Quorum sensing was first described in two symbiotic luminous marine bacteria, *Vibrio fischeri* and *Vibrio harveyi* [22]. In both species, enzymes responsible for light production are encoded by the luciferase structural operon *luxICDABEG*, and light emission only occurs at high cell population density in response to the accumulation of the secreted autoinducer molecules [22–25]. The LuxI/LuxR quorum sensing system of *V. fisheri* is the first and the most thoroughly studied system in quorum sensing. *V. fisheri* is a Gram-negative bacterium that can be free living or can form a symbiotic relationship with a variety of invertebrate and vertebrate marine organisms [25,26]. In these symbiotic associations, the eukaryotic host supplies the bacterium with a nutrient-rich environment so that the bacterial culture can grow to extremely high cell densities, reaching 10^{11} cells/mL and emitting light [26,27]. The quorum sensing of *V. fisheri* depends on the synthesis and recognition of the autoinducer, *N*-(3-oxohexanoyl) homoserine lactone, also called *V. fisheri* autoinducer or VAI. This molecule freely diffuses across the cell membrane, triggering the formation of the enzymes necessary for bioluminescence [28]. The gene product of *luxI*, an acylhomoserine lactone synthase, can use acyl-ACP from the fatty acid metabolic cycle and *S*-adenosylmethionine (SAM) from the methionine pathway to synthesize the autoinducer [24,31].

The quorum sensing mechanism of *V. fisheri* is illustrated in Figure 18-1 and involves several products encoded by *lux* operon and *luxR* gene. The regulatory protein encoded by the *luxR* gene has two binding domains, one that interacts with the autoinducer and the other that binds to the promoter region of the *lux* operon and also to the promoter region of the *luxR* gene itself. The amino terminus contains the binding site for the autoinducer and the carboxyl terminus possesses a helix-turn-helix binding motif, typical of many DNA binding domains. In the absence of the autoinducer, the

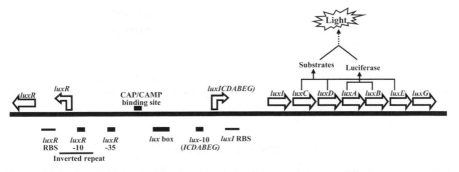

Figure 18-1 Genetic organization of genes and regulatory elements within lux operon on the chromosome of *V. fisheri.*

amino terminus is able to mask the carboxyl terminus, preventing the luxR protein from binding to the *lux* operon promoter region. Once the autoinducer binds to the luxR protein, the newly formed complex binds upstream of the *luxICDABEG*, promoting transcription of all the necessary components of the luciferase system [21]. The complex also acts as a negative autoregulator of the luxR transcriptional unit, by binding near the *luxR* promoter. The genes contained within the *lux* operon encode for several enzymes; *luxAB* encode the subunits of the luciferase enzyme, *luxCDE* encode proteins required for biosynthesis of the aldehyde substrate, and another open reading frame (*luxG*) exists downstream, but its function is still unknown [30]. The two regulatory *lux* genes (*luxR* and *luxI*) exist adjacent to each other, but unlike *luxI*, *luxR* is transcribed divergently from the *lux* operon (Fig. 18-1). At low population density, the *luxICDABEG* operon is transcribed at a basal level. Hence, a low level of autoinducer is constantly produced along with a low level of light [32]. When the autoinducer concentration reaches a threshold level (about 1–10 μg/mL), the cytoplasmic LuxR can detect and bind to it [31]. Interaction of LuxR and the autoinducer unmasks the DNA binding domain of LuxR, allowing LuxR to bind with the *luxICDABEG* promoter and activate its transcription [33]. This reaction causes an exponential increase in both autoinducer production and light emission. In addition, the LuxR and autoinducer complex represses the expression of *luxR*. This negative feedback loop is a compensatory mechanism that decreases *luxICDABEG* expression in response to the positive feedback circuits [21,34]. In the quorum sensing system, the autoinducer functions as a communication signal for the bacteria "inside" the host as opposed to "outside" in the seawater. The quorum sensing system enables *V. fisheri* to produce light only under conditions in which there is a positive selective advantage for the light [21].

The regulatory region of *lux* operon is complicated and contains two divergently transcribed promoters, as illustrated in Figure 18-1. The left promoter P_{luxL} constitutively transcribes the *luxR* gene. This promoter has a standard δ^{70} binding region, consisting of the −10 and −35 sequences, and a CRP/CAMP binding site, which is involved in catabolic repression of LuxR transcription. The right promoter P_{luxR} controls the expression of the *luxICDABEG* transcript [35]. Interestingly, the lux box, a 20 bp inverted palindromic repeat, allows dimeric binding of the LuxR protein in the presence of the autoinducer. This dimeric binding results in a nonlinear concentration response, a transcriptional control behavior of DNA binding proteins that is an essential element of signal restoration and digital control of expression [35,36]. These complicated genetic regulatory elements of quorum sensing allow populations of bacteria to simultaneously regulate gene expression in response to changes in cell density. Quorum sensing has broad biotechnological applications, including pathogen/pest management, recombinant gene expression, food preservation, and drug design [37–39]. Quorum sensing also exists in other bacteria and has been extensively discussed in several reviews [21,29,38–44].

18.2.2   Synthetic Cell Communication Network

Engineering of multicellular systems that utilize cell-to-cell communication to achieve coordinated behavior has been one of the foci for synthetic biology. This type of engineered system can be used to study multicellular phenomena ranging from synchronized gene expression in homogenous populations to spatial patterning in developmental processes [20]. Recently, the genetic elements of *V. fisheri* quorum sensing have been successfully used to engineer several cell–cell communication systems (Fig. 18-2) [45–48].

First, genetic elements of the quorum sensing system are separated into sender and receiver components that are integrated into two different *E. coli* populations (Fig. 18-2a) [20,35]. The sender cells contain the genetic elements responsible for autoinducer production. The receiver cells are engineered with the control element of the *lux* operon, a reporter gene (GFP), and the *luxR* gene. The free diffusion of the autoinducer within the medium and across the cell membranes allows the establishment of chemical gradients and the controlled expression of the reporter gene. For good control, the expression level of the *luxI* gene is placed under the control of the $P_{Ltet0-1}$ promoter, which is upregulated by the *tetR* gene product in the presence of tetracycline [49,50]. The *tetR* gene under the control of the constitutive promoter P_{N25} is chromosomally carried in a special strain of *E. coli*, which harbors the spectinomycin resistance gene. The $P_{Ltet0-1}$ promoter allows the controlled expression of the *luxI* gene using a varying amount of a nongrowth inhibitory version of tetracycline, anhydrotetracycline (aTc). Therefore, the level of the autoinducer in the sender cells can be controlled by varying the aTc concentrations [35]. The autoinducer diffused into the receiver cells can regulate the expression of *luxR* and therefore the reporter gene (GFP). In this engineered system, the levels of fluorescence in the receiver cells are successfully controlled via aTc concentration.

In another synthetic system, positive and negative regulations of gene expression are integrated into multicellular bacterial systems to obtain a transient response in cell-to-cell communication [47] (Fig. 18-2b). Using aTc, the sender cells in the system are induced to produce the autoinducer, which then diffuses to the nearby pulse-generating receiver cells. In response to a long-lasting increase in the autoinducer concentration, the receiver cells are engineered to transiently express a GFP. This is accomplished by a feedforward motif, which is placed in the genetic circuit of the receiver cells and allows them to display an initial excitation followed by a delayed inhibition in the presence of the autoinducer [51]. The feedforward motif is made up of two transcriptional regulators, LuxR and the lambda repressor (CI) that act on the GFP promoter. The LuxR protein when combined with the autoinducer from the sender cells, acts as an activator of CI production, and also acts as an activator of GFP transcription. CI acts as an inhibitor of GFP transcription, but because it has a lower affinity for the promoter than the LuxR and autoinducer complex, it is only able to repress GFP transcription after it has accumulated a threshold concentration. Thus, the receiver circuits can distinguish between various rates of increase in the autoinducer levels and gain ability to generate a spatiotemporal behavior so that the receiver cells only respond transiently to signal from the nearby cells but ignore signal from sender cells, which are farther away.

Using the same cell-to-cell communication mechanism, a "population control" genetic circuit is engineered to program the dynamics of cell population despite variability in the behavior of individual cells by coupling quorum-controlled gene

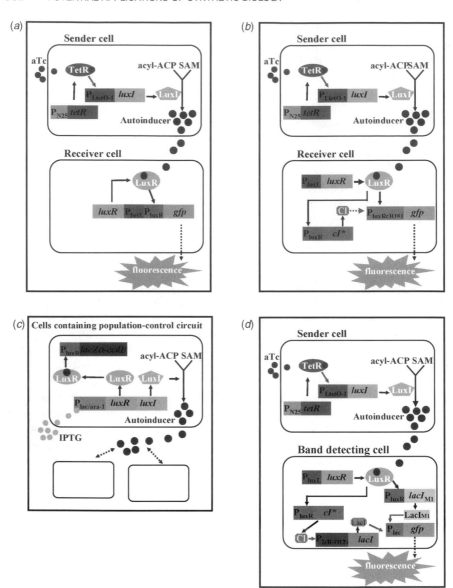

Figure 18-2 Engineered cell-to-cell communication networks using genetic elements of lux operon from *V. fisheri.*(a) diagram of gradient communication system [20,35]; (b) genetic network for pulse signal generation [47]; (c) design of "cell population control"" genetic circuit [48]; (d) design of "band-detect"" gene network [46]. Red arrow means suppression and black arrow induction. See text for abbreviations and details.

expression to cell survival and death [48]. This synthetic circuit can autonomously control the density of an *E. coli* population through a quorum-sensing system and can set a stable steady state in terms of cell density and gene expression that is easily tunable by varying the autoinducer signal. As illustrated in Figure 18-2c, the *luxI* and *luxR* genes are placed under the control of a synthetic promoter Plac/ara^{-1} in the "population control" system [52]. When isopropyl-β-D-thiogalactopyranoside (IPTG) is present, LuxR is produced along with the autoinducer. The activated LuxR transcriptional regulator activates the promoter P_{luxR} from the *lux* operon that controls the expression of the killer gene *lacZα-ccdB*. The killer gene product is a fusion protein containing LacZα and CcdB. The LacZα portion of the fusion protein allows the measurement of fusion protein levels using a LacZ assay. The CcdB portion still has the toxicity of native CcdB, which kills susceptible cells by poisoning the DNA gyrase complex [53]. Therefore, in the presence of IPTG, cells harboring the genetic circuit will produce enough killer protein to maintain a stable cell density.

Another synthetic system utilizes "band-detect" gene networks that are engineered to allow the receiver cells to form diverse patterns around the sender cell colony (Fig. 18-2d) [46]. As in the above two systems, the sender cells produce LuxI protein for the biosynthesis of the autoinducer, which forms a chemical gradient around the sender cell colony. The LuxR protein in the receiver cells activates the expression of lambda repressor (CI) and Lac repressor ($LacI_{M1}$, a product of a codon-modified *lacI*), which are under the control of P_{luxR}. CI then binds to the $P_{λ(R-012)}$ promoter and represses the expression of the wild-type LacI. The GFP reporter gene is under the control of the promoter P_{lac}, which is repressed by $LacI_{M1}$ and LacI. Receiver cells proximal to the senders encounter high concentrations of the autoinducer and produce high levels of CI and $LacI_{M1}$. Hence, receiver cells near the sender cells will not express GFP. The receiver cells that are far from the sender cells will express $LacI_{M1}$ and CI at basal levels. Thus, the wild-type LacI will be expressed and again suppress the expression of GFP. At intermediate distances from the senders, both CI and $LacI_{M1}$ are expressed in moderate levels in the receiver cells. However, due to the different repression efficiency of CI and $LacI_{M1}$, CI effectively represses LacI expression while the $LacI_{M1}$ concentration is below the level required for GFP production. Hence, the GFP is produced. Overall, this feedforward loop, including LuxR, CI, $LacI_{M1}$, LacI, and GFP, attributes the desired nonmonotonic response to the autoinducer concentrations to the genetic circuit [54,55]. By deliberately arranging sender cells on solid-phase media containing a mixture of receiving cells, diverse spatial patterns including bull's-eyes, ellipses, hearts, and clovers can be produced using the system [46].

18.3 RECONSTRUCTING NATURAL SYSTEM OF UNCULTURABLE MARINE MICROBES IN MICROBIAL HOST

One of the potential applications of synthetic biology is testing our understanding of the functions involved in biological systems [56]. Overall, research on this aspect of synthetic biology is rare. In this section, we will briefly discuss the potential application of synthetic biology in a marine functional ecology study. The example described below may be relatively simple, but it illustrates how the concept of synthetic biology can be used to understand the ecological function of unculturable marine microbes.

Major efforts have been made to investigate marine microbial diversity in many different natural habitats [2,6,8,15,16]. However, our current understanding of the ecological function of marine microbes in their natural environments is minimal. The major reason for this lack of understanding can be ascribed to our limited ability to cultivate and to genetically manipulate these marine microbes for physiological and metabolic characterization. The vast majority (>99 percent) of marine microbes are unculturable and therefore their ecological roles in marine natural environments remain largely unknown. Some molecular techniques such as the FISH (fluorescence *in situ* hybridization) have revealed the identity, abundance, and distribution of selected unculturable microbes in natural marine habitats. However, the ecological function of these marine microbes cannot be understood in this way. Genetic materials from unculturable microbes can be recovered using an environmental genomic strategy. Functional biological components or pathways encoded in the genetic materials can then be fabricated using the principles of synthetic biology. Large genomic DNA fragments of uncultured marine microbes are usually recovered from environmental genomic libraries, which are constructed using fosmid or BAC (bacterial artificial chromosome) vectors [57,58]. Recently, a new phototroph in the sea was identified by characterization of new type of rhodopsin from a picoplankton bacterial artificial chromosome library [57]. Analyses of a 130 kb environmental clone revealed a new class of genes for the rhodopsin family (named proteorhodopsin) that has not been observed in bacteria before. Proteorhodopsin (PR) proteins were found to be bacterial retinal-binding membrane pigments that function as light-driven proton pumps in the marine ecosystem [56]. Subsequent investigations indicated that proteorhodopsin occurs in many marine bacteria and evolves for various light wavelengths at different ocean depths [59–66]. However, it is a great challenge to prove bacteria containing proteorhodopsin, are a novel group of marine phototroph. To that end, *E. coli* cells were engineered to use the proteorhodopsin genes. The engineered cells acquired the net-outward transport of protons in the presence of retinal and light [57]. Recently, analysis revealed that PR genes are linked to a carotenoid biosynthesis gene cluster, which encodes proteins responsible for converting geranylgeranly diphosphate to β-carotene [67]. In addition, a gene coding for a homologue of the bacteriorhodopsin-related-protein-like homologue protein (Blh) from the archaeon *Halobacterium* sp. NRC-1 was also found in the marine bacteria BAC clone. Blh has been shown to be involved in the retinal biosynthesis [68]. This indicates that bacteria possessing PR proteins also carry the ability to synthesize the retinal chromophore and to potentially form functional PR holoproteins. Indeed, expression of *blh* in the β-carotene producing *E. coli* cells results in the loss of the yellow color of these cells because β-carotene is converted into a colorless all-*trans* retinal by Blh. When the colorless retinal binds PR protein, the resulting complex becomes red colored and can function as an active proton pump [67]. Thus, proteorhodopsin is proved to have the ability to couple light energy harvesting with carbon cycling through nonchlorophyll-based pathways in the ocean.

In addition, the environmental genomic approach has been used to study methane-oxidizing microbial consortia in deep sea methane seeps [69] and resulted in the identification of the methanogenic pathway of the ANME-1 archeal groups [70].

Unfortunately, no further functional analysis of the pathway was carried out. It is believed that a methanogenic *E. coli* strain could be constructed by using this archeal pathway. Overall, application of synthetic biology in the understanding of marine microbial ecology is still in its infancy. Close collaboration between marine microbial ecologists and synthetic biologists may greatly benefit the development of both fields.

18.4 METABOLIC ENGINEERING FOR THE PRODUCTION OF MARINE NATURAL PRODUCTS

The production of natural compounds through metabolic engineering has been one of the major foci in synthetic biology [18,19]. Tremendous progress has been made in the production of natural compounds of terrestrial origins and is summarized in many excellent reviews [71–76]. In this section, we review the progress of metabolic engineering for the production of marine natural compounds. Production of valuable marine natural products in engineered microbial hosts has been an active research area. Some engineered hosts have shown promise in pharmaceutical and nutraceutical industries. Like terrestrial natural products, marine natural compounds are often produced by enzymes coded in gene clusters. Polyunsaturated fatty acids (PUFAs) are of biotechnological interest for their beneficial properties to human health and their importance in infant development [77]. The most important PUFAs are eicosopentaenic acid (EPA) and docosahexaenic acid (DHA). The 38 kb genomic fragment, which includes all genes responsible for the production of EPA, was recovered from the marine bacterium *Shewanella putrefaciens* strain SCRC-2738 [78]. Engineered *E. coli* cells, with the foreign gene cluster cloned into them, produced EPA in low yield. Also, the same gene cluster was cloned into the marine cyanobacterium *Synechococcus* sp. using a broad host cosmid vector, pJRD215. The engineered cyanobacterial cells produced EPA up to 0.56 mg/g dry cells at 23°C [79]. In addition, the production yield of EPA was further improved by stabilizing the expression and maintenance of the cluster in the host cells [78]. Thus, these studies provide the first examples of EPA production in bioengineered hosts. Also, the increased understanding of PUFA-related genes offers the possibility for the engineering of microbial cell factories suitable for an alternative production of EPA and DHA.

Most microalgae are obligate photoautotroph and their growth strictly depends on the generation of photosynthetically derived energy. *Phaeodactylum tricornutum* is a unicellular nonsilicate diatom and can accumulate EPA up to 30 percent of the total fatty acid content. Furthermore, astaxanthin is an efficient antioxidant and produced by a number of marine bacteria and microalgae. It can be synthesized from β-carotene by the addition of two keto groups to carbons C4 and C4' and two hydroxyl groups to C3 and C3' [81]. The gene *crtO* encoding β-C-4-oxygenase from the green alga *Haematococcus pluvialis* can convert β-carotene to astaxanthin. The cyanobacterium *Synechococcus* PCCC7 can produce astaxanthin as well as other keto-carotenoids [82]. After the introduction of a single gene for glucose transporters *glut1* or *hup1*, the microalga *P. tricornutum* was genetically engineered to thrive on exogenous glucose in the absence of light [80]. The trophic conversion of microalgae has provided an important platform for large-scale production of PUFAs and carotenoids using engineered microalgal cells. Metabolic engineering of conventional noncaroteno-genic bacteria and yeasts using carotenoid metabolic pathways genes (e.g., *crt* genes or IPP synthetic genes) from marine microbes has been intensively studied (for reviews, see Refs [83,84]).

Marine invertebrates such as sponges, ascidians, and bryozoans are well known for their production of bioactive natural products, several of which are currently under-going clinical trials [85,86]. These marine invertebrates also harbor diverse symbiotic microbes [64,87,88]. Because many marine natural products from these marine invertebrates resemble bacterial compounds, some of their natural chemicals have long been proposed to be produced by their bacterial symbionts [88]. Several studies have demonstrated that microbial isolates associated with sponges produced the same compounds formerly isolated from sponges [89–93]. However, these results do not rule out the possibility that substances might be transported between bacterial symbionts and their hosts via export or sequestration mechanisms [94]. Recently, several biosynthetic pathways for anticancer compounds have been isolated from marine invertebrates using molecular approaches [87,95,96]. Particularly, the patellamide A and C biosynthetic pathway was identified from *Prochloron didemni*, a cyanobac-terial symbiont of *Lissoclinum patella*. *E. coli* cells that were engineered to harbor this biosynthetic pathway and its regulatory region produced patellamide A at the level of 20 μg/L. Although the production yield is low, it represents the first successful case of the production of marine natural compounds in a synthetic microbial host.

18.5 CONCLUSION

The world's oceans cover the largest portion of the global surface and contain the most complicated ecosystems. They are home to different biota ranging from tiny plank-tonic organisms that comprise the base of the marine food web (i.e., phytoplankton and zooplankton) to large marine mammals like the whales, manatees, and seals. It has been estimated that the oceans harbor 3.6×10^{29} microbial cells with a total cellular carbon content of about 3×10^{17}g [97]. These microbial cells are responsible for the vast majority of primary production and mediate all biogeochemical cycles in the oceans [5]. Considering the enormous number of microbes, their interaction with other hosts, and their vast metabolic diversity, marine environments can be an enormously rich source for novel molecular regulatory networks and pathways for new natural compounds.

Further environmental genomic investigation of marine microbes will contribute to the development of synthetic biology by providing novel genetic regulatory networks and pathways. On the other hand, synthetic biology can also benefit marine microbial ecology by providing techniques for the functional characterization of unculturable marine microbes. Particularly, the synthetic biology approach can provide a viable solution for the development of interesting marine natural compounds. For example, the valuable and powerful antimalarial drug artemisinin (a sesquiterpene lactone) is isolated from the sweet wormwood, *Artemisia annua*, at very low yield. Recently, its immediate precursors artemisinic acid and amorphadiene have been successfully produced at a significantly high level in engineered *Saccharomyces cerevisiae* and *E. coli*, respectively [98,99]. Thus, metabolically engineered microbial hosts are likely to solve the supply and affordability issues for this effective antimalarial drug. Therefore, many valuable marine terpenoids such as cytotoxic eleutherobin and sarcodictyins could also be produced in engineered microbial hosts using similar strategies because most terpenoids use the same building blocks IPP (isopentenyl diphosphate) and DMAPP (dimethylallyl diphosphate) for their biosynthesis. Unfortunately, most of the key genes responsible for the production and modification of these valuable marine compounds are still not available. At present, the application of synthetic biology to the understanding of marine microbial ecology and marine biotechnology is mainly limited by the availability of the novel genetic materials from the marine environments.

The quorum-sensing system of *V. fisheri* has been successfully used to engineer several cell-to-cell communication systems. It is reasonable to believe that diverse and novel genetic regulatory systems will be found in marine microbial genomes using an environmental genomics approach. Thus, these marine regulatory systems will provide the platform for bioengineers to synthesize novel genetic circuits and cell communication systems for diverse biotechnological applications. Collaborative research of interdisciplinary scientists and researchers from oceanography, microbiology, metabolic engineering, computer science, mathematics, informatics, and marine biology can provide greater progress in understanding marine ecosystems and the discovery of new techniques in synthetic biology.

ACKNOWLEDGMENTS

The authors would like to thank Diane Henderson for her suggestions to greatly improve this manuscript.

REFERENCES

1. Das S, Lyla PS, Khan SA. Marine microbial diversity and ecology: importance and future perspectives. *Curr Sci* 2006;90:1325–1335.
2. DeLong EF. Microbial community genomics in the ocean. *Nat Rev Microbiol* 2005;3: 459–469.
3. Donia M, Hamann MT. Marine natural products and their potential applications as anti-infective agents. *Lancet Infect Dis* 2003;3:338–348.
4. Hunter-Cevera JC, Karl DM, Buckley MR. *Marine Microbial Diversity: The Key to Earth's Habitability.* American Academy of Microbiology, San Francisco, CA: 2005.
5. Sogin ML, Morrison HG, Huber JA, Welch DM, Huse SM, Neal PR, Arrieta JM, Herndl GJ. Microbial diversity in the deep sea and the underexplored rare biosphere. *Proc Natl Acad Sci USA* 2006;103:12115–12120.

6. Giovannoni SJ, Tripp HJ, Givan S, Podar M, Vergin KL, Baptista D, Bibbs L, Eads J, Richardson TH, Noordewier M, Rappe MS, Short JM, Carrington JC, Mathur EJ. Genome streamlining in a cosmopolitan oceanic bacterium. *Science* 2005;309:1242–1245.

7. Arrigo KR. Marine microorganisms and global nutrient cycles. *Nature* 2005;437: 349–355.

8. DeLong EF, Karl DM. Genomic perspectives in microbial oceanography. *Nature* 2005;437:336–342.

9. Rees J. Bio-oceanography. *Nature* 2005;437:335.

10. Haefner B. Drugs from the deep: marine natural products as drug candidates. *Drug Discov Today* 2003;8:536–544.

11. Delong EF. Marine microbial diversity: the tip of the iceberg. *Trends Biotechnol* 1997;15: 203–207.

12. Connon SA, Giovannoni SJ. High-throughput methods for culturing microorganisms in very-low-nutrient media yield diverse new marine isolates. *Appl Environ Microbiol* 2002;68:3878–3885.

13. DeLong EF, Pace NR. Environmental diversity of bacteria and archaea. *Syst Biol* 2001;50: 470–478.

14. DeLong EF. Microbial seascapes revisited. *Curr Opin Microbiol* 2001;4:290–295.

15. DeLong EF. Microbial population genomics and ecology. *Curr Opin. Microbiol* 2002;5: 520–524.

16. Giovannoni SJ, Stingl U. Molecular diversity and ecology of microbial plankton. *Nature* 2005;437:343–348.

17. Rappe MS, Giovannoni SJ. The uncultured microbial majority. *Annu Rev Microbiol* 2003;57:369–394.

18. Benner SA, Sismour AM. Synthetic biology. *Nat Rev Genet* 2005;6:533–543.

19. Tucker JB, Zilinskas RA. The promise and perils of synthetic biology. *New Atlantis* 2006;25–45.

20. McDaniel R, Weiss R. Advances in synthetic biology: on the path from prototypes to applications. *Curr Opin Biotechnol* 2005;16:476–483.

21. Miller MB, Bassler BL. Quorum sensing in bacteria. *Annu Rev Microbiol* 2001;55: 165–199.

22. Nealson KH, Hastings JW. Bacterial bioluminescence: its control and ecological significance. *Microbiol Rev* 1979;43:496–518.

23. Miyamoto CM, Boylan M, Graham AF, Meighen EA. Organization of the Lux structural genes of *Vibrio harveyi* expression under the T7 bacteriophage promoter messenger RNA analysis and nucleotide sequence of the Lux-D gene. *J Biol Chem* 1988;263: 13393–13399.

24. Engebrecht J, Silverman M. Identification of genes and gene products necessary for bacterial bio-luminescence. *Proc Natl Acad Sci USA* 1984;81:4154–4158.

25. Munn CB. *Marine Microbiology.* New York: Garland Science/BIOS Scientific Publishers, 2004.

26. Ruby EG, Nealson KH. Symbiotic association of *Photobacterium fischeri* with the marine fish *Monocentris japonica* model of symbiosis based on bacterial studies. *Bio Bull* 1976;151:574–586.

27. Nyholm S, McFall-Ngai MJ. Sampling the light-organ microenvironment of *Euprymna scolopes*: description of a population of host cells in association with bacterial symbiont *Vibrio fischeri. Bio Bull* 1998;195:89–97.

28. Kaplan HB, Greenberg EP. Diffusion of autoinducer is involved in regulation of the *Vibrio fischeri* luminescence system. *J Bacteriol* 1985;163:1210–1214.

29. Fuqua C, Parsek MR, Greenberg EP. Regulation of gene expression by cell-to-cell communication: Acyl-homoserine lactone quorum sensing. *Annu Rev Genet* 2001;35: 439–468.

30. Meighen EA. Enzymes and genes from the Lux operons of bioluminescent bacteria. *Annu Rev Microbiol* 1988;42:151–176.

31. Cao JG, Meighen EA. Purification and structural identification of an autoinducer for the luminescence system of *Vibrio harveyi*. *J Biol Chem* 1989;264:21670–21676.

32. Atkinson S, Throup JP, Stewart GSAB, Williams P. A hierarchical quorum-sensing system in *Yersinia pseudotuberculosis* is involved in the regulation of motility and clumping. *Mol Microbiol* 1999;33:1267–1277.

33. Hanzelka BL, Greenberg EP. Evidence that the N-terminal region of the *Vibrio fischeri* LuxR protein constitutes an autoinducer-binding domain. *J Bacteriol* 1995;177:815–817.

34. Engebrecht J, Nealson K, Silverman M. Bacterial bioluminescence isolation and genetic analysis of functions from *Vibrio fischeri*. *Cell* 1983;32:773–782.

35. Weiss R, Knight TF. Engineered communications for microbial robotics. In: *DNA6: Sixth International Meeting on DNA-Based Computers, Leiden, The Netherlands*, 2000;1–5.

36. Knight TF, Sussman GJ. Cellular gate technology. *First International Conference on Unconventional Models of Computation, Auckland, NZ*, 1998.

37. Suga H, Smith KM. Molecular mechanisms of bacterial quorum sensing as a new drug target. *Curr Opin Chem Biol* 2003;7:586–591.

38. Smith JL, Fratamico PM, Novak JS. Quorum sensing: a primer for food microbiologists. *J Food Prot* 2004;67:1053–1070.

39. March JC, Bentley WE. Quorum sensing and bacterial cross-talk in biotechnology. *Curr Opin Biotechnol* 2004;15:495–502.

40. Zhang L-H, Dong Y-H. Quorum sensing and signal interference: diverse implications. *Mol Microbiol* 2004;53:1563–1571.

41. Whitehead NA, Barnard AML, Slater H, Simpson NJL, Salmond GPC. Quorum-sensing in gram-negative bacteria. *FEMS Microbiol Rev* 2001;25:365–404.

42. Kuipers OP, De Ruyter PGGA, Kleerebrezem M, De Vos WM. Quorum sensing-controlled gene expression in lactic acid bacteria. *J Biotechnol* 1998;64:15–21.

43. Fuqua WC, Winans SC, Greenberg EP. Quorum sensing in bacteria: the LuxR-LuxI family of cell density-responsive transcriptional regulators. *J Bacteriol* 1994;176: 269–275.

44. Daniels R, Vanderleyden J, Michiels J. Quorum sensing and swarming migration in bacteria. *FEMS Microbiol Rev* 2004;28:261–289.

45. Bassler BL. How bacteria talk to each other: regulation of gene expression by quorum sensing. *Curr Opin Microbiol* 1999;2:582–587.

46. Basu S, Gerchman Y, Collins CH, Arnold FH, Weiss R. A synthetic multicellular system for programmed pattern formation. *Nature* 2005;434:1130–1134.

47. Basu S, Mehreja R, Thiberge S, Chen M-T, Weiss R. Spatiotemporal control of gene expression with pulse-generating networks. *Proc Natl Acad Sci USA* 2004;101: 6355–6360.

48. You L, Cox RS, Weiss R, Arnold FH. Programmed population control by cell-cell communication and regulated killing. *Nature* 2004;428:868–871.

49. Baron U, Schnappinger D, Helbl V, Gossen M, Hillen W, Bujard H. Generation of conditional mutants in higher eukaryotes by switching between the expression of two genes. *Proc Natl Acad Sci USA* 1999;96:1013–1018.

50. Mayford M, Bach ME, Huang Y-Y, Wang L, Hawkins RD, Kandel ER. Control of memory formation through regulated expression of a CaMKII transgene. *Science* 1996;274: 1678–1683.

51. Mangan S, Alon U. Structure and function of the feed-forward loop network motif. *Proc Natl Acad Sci USA* 2003;100:11980–11985.

52. Lutz R, Bujard H. Independent and tight regulation of transcriptional units in *Escherichia coli* via the LacR/O, the TetR/O and AraC/I-1-I-2 regulatory elements. *Nucleic Acids Res* 1997;25:1203–1210.

53. Engelberg-Kulka H, Glaser G. Addiction modules and programmed cell death and antideath in bacterial cultures. *Annu Rev Microbiol* 1999;53:43–70.

54. Milo R, Shen-Orr S, Itzkovitz S, Kashtan N, Chklovskii D, Alon U. Network motifs: simple building blocks of complex networks. *Science* 2002;298:824–827.

55. Shen-Orr SS, Milo R, Mangan S, Alon U. Network motifs in the transcriptional regulation network of *Escherichia coli*. *Nat Genet* 2002;31:64–68.

56. Chin JW, Modular approaches to expanding the functions of living matter. *Nat Chem Biol* 2006;2:304–311.

57. Beja O, Aravind L, Koonin EV, Suzuki MT, Hadd A, Nguyen LP, Jovanovich SB, Gates CM, Feldman RA, Spudich JL, Spudich EN, DeLong EF. Bacterial rhodopsin: evidence for a new type of phototrophy in the sea. *Science* 2000;289:1902–1906.

58. Stein JL, Marsh TL, Wu KY, Shizuya H, Delong EF. Characterization of uncultivated prokaryotes: isolation and analysis of a 40-kilobase-pair genome fragment from a planktonic marine archaeon. *J Bacteriol* 1996;178:591–599.

59. Giovannoni SJ, Bibbs L, Cho J-C, Stapels MD, Desiderio R, Vergin KL, Rappe MS, Laney S, Wilhelm LJ, Tripp HJ, Mathur EJ, Barofsky DF. Proteorhodopsin in the ubiquitous marine bacterium SAR11. *Nature* 2005;438:82–85.

60. Schwalbach MS, Brown M, Fuhrman JA. Impact of light on marine bacterioplankton community structure. *Aquat Microb Ecol* 2005;39:235–245.

61. Bielawski JP, Dunn KA, Sabehi G, Beja O. Darwinian adaptation of proteorhodopsin to different light intensities in the marine environment. *Proc Natl Acad Sci USA* 2004;101: 14824–14829.

62. Kelemen BR, Du M, Jensen RB. Proteorhodopsin in living color: diversity of spectral properties within living bacterial cells. *Biochim Biophys Acta* 2003;1618:25–32.

63. de la Torre JR, Christianson LM, Beja O, Suzuki MT, Karl DM, Heidelberg J, DeLong EF. Proteorhodopsin genes are distributed among divergent marine bacterial taxa. *Proc Natl Acad Sci USA* 2003;100:12830–12835.

64. Hentschel U, Usher KM, Taylor MW. Marine sponges as microbial fermenters. *FEMS Microbiol Ecol* 2006;55:167–177.

65. Friedrich T, Geibel S, Kalmbach R, Chizhov I, Ataka K, Heberle J, Engelhard M, Bamberg E. Proteorhodopsin is a light-driven proton pump with variable vectoriality. *J Mol Biol* 2002;321:821–838.

66. Beja O, Spudich EN, Spudich JL, Leclerc M, DeLong EF. Proteorhodopsin phototrophy in the ocean. *Nature* 2001;411:786–789.

67. Sabehi G, Loy A, Jung K-H, Partha R, Spudich JL, Isaacson T, Hirschberg J, Wagner M, Beja O. New insights into metabolic properties of marine bacteria encoding proteorhodopsins. *PLoS Biol* 2005;3:1409–1417.

68. Peck R.F, Echavarri-Erasun C, Johnson EA, Ng WV, Kennedy SP, Hood L, DasSarma S, Krebs MP. brp and blh are required for synthesis of the retinal cofactor of bacteriorhodopsin in *Halobacterium salinarum. J Biol Chem* 2001;276:5739–5744.

69. Hallam SJ, Putnam N, Preston CM, Detter JC, Rokhsar D, Richardson PM, DeLong EF. Reverse methanogenesis: testing the hypothesis with environmental genomics. *Science* 2004;305:1457–1462.

70. Krueger M, Meyerdierks A, Gloeckner FO, Amann R, Widdel F, Kube M, Reinhardt R, Kahnt J, Boecher R, Thauer RK, Shima S. A conspicuous nickel protein in microbial mats that oxidize methane anaerobically. *Nature* 2003;426:878–881.

71. Barkovich R, Liao JC. Metabolic engineering of isoprenoids. *Metab Eng* 2001;3:27–39.

72. Bongaerts J, Kramer M, Muller U, Raeven L, Wubbolts M. Metabolic engineering for microbial production of aromatic amino acids and derived compounds. *Metab Eng* 2001;3: 289–300.

73. Cameron DC, Chaplen FWR. Developments in metabolic engineering. *Curr Opin Biotechnol* 1997;8:175–180.

74. Chartrain M, Salmon P.M, Robinson DK, Buckland BC. Metabolic engineering and directed evolution for the production of pharmaceuticals. *Curr Opin Biotechnol* 2000;11:209–214.

75. Khosla C, Keasling JD. Timeline: Metabolic engineering for drug discovery and development. *Nat Rev Drug Discov* 2003;2:1019–1025.

76. Stafford DE, Stephanopoulos G. Metabolic engineering as an integrating platform for strain development. *Curr Opin Microbiol* 2001;4:336–340.

77. Gill I, Valivety R. Polyunsaturated fatty acids, part 1: occurrence, biological activities and applications. *Trends Biotechnol* 1997;15:401–409.

78. Yu R, Yamada A, Watanabe K, Yazawa K, Takeyama H, Matsunaga T, Kurane R. Production of eicosapentaenoic acid by a recombinant marine cyanobacterium, *Synechococcus* sp. *Lipids* 2000;35:1061–1064.

79. Takeyama H, Takeda D, Yazawa K, Yamada A, Matsunaga T. Expression of the eicosapentaenoic acid synthesis gene cluster from *Shewanella* sp. in a transgenic marine cyanobacterium, *Synechococcus* sp. *Microbiology* 1997;143:2725–2731.

80. Zaslavskaia LA, Lippmeier JC, Shih C, Ehrhardt D, Grossman AR, Apt KE. Trophic conversion of an obligate photoautotrophic organism through metabolic engineering. *Science* 2001;292:2073–2075.

81. Fan L, Vonshak A, Gabbay R, Hirshberg J, Cohen Z, Boussiba S. The biosynthesis pathway of astaxanthin in a green alga *Haematococcus pluvialis* as indicated by inhibition with diphenylamine. *Plant Cell Physiol* 1995;36:1519–1524.

82. Harker M, Hirschberg J. Biosynthesis of ketocarotenoids in transgenic cyanobacteria expressing the algal gene for beta-C-4-oxygenase, *crtO. FEBS Lett* 1997;404:129–134.

83. Watson SB. Cyanobacterial and eukaryotic algal odour compounds: signals or byproducts? A review of their biological activity. *Phycologia* 2003;42:332–350.

84. Misawa N, Shimada H. Metabolic engineering for the production of carotenoids in non-carotenogenic bacteria and yeasts. *J Biotechnol* 1998;59:169–181.

85. Newman DJ, Hill RT. New drugs from marine microbes: the tide is turning. *J Ind Microbiol Biotechnol* 2006;33:539–544.

86. Newman DJ, Cragg GM. The discovery of anticancer drugs from natural sources. In: Zhang L, Demain AL, editors. *Natural Products: Drug Discovery and Therapeutic Medicine*. Humana Press: Totowa, NJ, 2005,129–168.

87. Schmidt EW. From chemical structure to environmental biosynthetic pathways: navigating marine invertebrate-bacteria associations. *Trends Biotechnol* 2005;23:437–440.

88. Wang G. Diversity and biotechnological potential of the sponge-associated microbial consortia. *J Ind Microbiol Biotechnol* 2006;33:545–551.

89. Bewley CA, Holland ND, Faulkner DJ. Two classes of metabolites from *Theonella swinhoei* are localized in distinct populations of bacterial symbionts. *Experientia* 1996;52:716–722.

90. Flowers AE, Garson MJ, Webb RI, Dumdei EJ, Charan RD. Cellular origin of chlorinated diketopiperazines in the dictyoceratid sponge *Dysidea herbacea* (Keller). *Cell Tissue Res* 1998;292:597–607.

91. Schmidt EW, Nelson JT, Rasko DA, Sudek S, Eisen JA, Haygood MG, Ravel J. Patellamide A and C biosynthesis by a microcin-like pathway in *Prochloron didemni*, the cyanobacterial symbiont of *Lissoclinum patella*. *Proc Natl Acad Sci USA* 2005;102:7315–7320.

92. Stierle AC, Cardellina JH II, Singleton FL. A Marine micrococcus produces metabolites ascribed to the sponge *Tedania ignis*. *Experientia* 1988;44:1021.

93. Thiel V, Imhoff JF. Phylogenetic identification of bacteria with antimicrobial activities isolated from Mediterranean sponges. *Biomol Eng* 2003;20:421–423.

94. Piel J, Butzke D, Fusetani N, Hui D, Platzer M, Wen G, Matsunaga S. Exploring the chemistry of uncultivated bacterial symbionts: antitumor polyketides of the pederin family. *J Nat Prod* 2005;68:472–479.

95. Hildebrand M, Waggoner LE, Liu H, Sudek S, Allen S, Anderson C, Sherman DH, Haygood M. bryA: an unusual modular polyketide synthase gene from the uncultivated bacterial symbiont of the marine bryozoan *Bugula neritina*. *Chem Biol* 2004;11:1543–1552.

96. Piel J, Hui D, Fusetani N, Matsunaga S. Targeting modular polyketide synthases with iteratively acting acyltransferases from metagenomes of uncultured bacterial consortia. *Environ Microbiol* 2004;6:921–927.

97. Whitman WB, Coleman DC, Wiebe WJ. Prokaryotes: the unseen majority. *Proc Natl Acad Sci USA* 1998;95:6578–6583.

98. Martin VJJ, Pitera DJ, Withers ST, Newman JD, Keasling JD. Engineering a mevalonate pathway in *Escherichia coli* for production of terpenoids. *Nat Biotechnol* 2003;21:796–802.

99. Ro D-K, Paradise EM, Ouellet M, Fisher KJ, Newman KL, Ndungu JM, Ho KA, Eachus RA, Ham TS, Kirby J, Chang MCY, Withers ST, Shiba Y, Sarpong R, Keasling JD. Production of the antimalarial drug precursor artemisinic acid in engineered yeast. *Nature* 2006;440:940–943.

19

ON FUNDAMENTAL IMPLICATIONS OF SYSTEMS AND SYNTHETIC BIOLOGY

Cliff Hooker

*Faculty of Education and Arts, School of Humanities and Social Science,
The University of Newcastle, NSW, Australia*

19.1 SETTING SYSTEMS AND SYNTHETIC BIOLOGY IN CONTEXT

19.1.1 Systems and Synthetic Biology in Context

Systems and synthetic biology promise to revolutionize our understanding of biology, blur the boundaries between the living and the engineered in a vital new bioengineering, and transform our daily relationship to the living world. Their emergence thus deserves to be understood in a wider intellectual perspective. Close attention to their relationship to the larger scientific intellectual frameworks within which they function reveals that systems and synthetic biology raise fundamental challenges to scientific orthodoxy, but stand in the vanguard of an emerging new complex dynamical systems paradigm now sweeping across science.

They emerge from a preceding developmental stage of science where, sketching crudely, biology was divided between molecular biology on the one side and, on the other, physiology (functional biology) and, on a larger scale, population genetics (evolutionary biology), and there was relatively little commerce among these approaches. Molecular biology and evolutionary population biology effectively agreed on assuming simple rules for gene expression that had the effect of reducing organism complexity to genetic complexity and so of treating the organism (reduced

Systems Biology and Synthetic Biology   Edited by Pengcheng Fu and Sven Panke
Copyright © 2009 John Wiley & Sons, Inc.

to a phenotype) as if it consisted simply of a bundle of genes. Whence, with genes directly related to produced phenotypes through the simple gene–trait rules, population gene frequencies could be constructed, and the diversity of complex organic processes could be explained in terms of evolutionary natural selection expressed in population frequency shifts. This left molecular biology to focus on the genes, aka DNA, and evolutionary theory to focus on gene population statistics. Caught between them, physiology focused on its own functional descriptions, cast in terms of organism features like energy fluxes and tissue densities, as, in different ways, did its sister domains of embryology and developmental biology.

Though somewhat a caricature, this division of conception and labor leaves the treatment of biosynthetic pathways out of the picture; however, they are essential for biological understanding. For they are the linkages connecting gene activity through intracellular and then intercellular formation and functioning to organism formation and functioning, and on, finally, to an enriched multilayered conception of evolutionary process (see below at footnote 22). It is exactly at this locus that systems and synthetic biology intervene.

These subdisciplines act, severally and together, as an interlevel bridge between molecular biology and physiology, precisely by developing the treatment of biosynthetic pathways, and in this way create a lively, reinvigorating integration to biology. Despite the complexity of biosynthetic pathways, scientists have been able to study them by carrying over into biology certain engineering modeling tools, such as control theory and electrical circuit theory and its generalization to dynamical network theory. With genes, proteins, and metabolites as components and replication, self-assembly, metabolism, repair, growth/death, signaling and regulation as process elements, systems and synthetic biology using these tools to model the complexes of processes that constitute cells, and interacting multicellular bodies like organs, in ways analogous to those in which engineers model and regulate fighter jet acrodynamics and multistage industrial processes.[1]

Of the two, synthetic biology has a wider scope than systems biology since, beyond the actual life forms of systems biology, the domain of synthetic biology also includes novel viable life forms and bioengineering complexes in which specialized organisms and/or biomaterials/processes play important roles. However, the hope underlying work in both studies is that a cell can be adequately modeled as a dynamical pathway network and a multicelled organism can be adequately modeled as a supernetwork of these (and so on up). Adding inanimate engineering network components then suffices to encompass all the wider domain of synthetic biology.

Methodologically, systems biology and synthetic biology are mutually beneficial (symbiotic); systems biology employs to advantage the perturbational and measurement methods developed by synthetic biology, while systems biology provides knowledge of dynamical models of various useful organisms from which synthetic biology may work. The key to the rise of these two interrelated subdisciplines has been the (accelerating) emergence over the past 50 years of high-throughput experimental

---

[1] See, among many recent texts, the nicely diagrammed overview in Ref. [1], Chapter 1.

technologies capable of amplifying trace chemical presences to reliably measurable quantities in practicable times and of doing so simultaneously with increasingly many cellular components. Starting with recombinant DNA techniques for single genes in the 1970s, today the techniques are crossing the threshold of being able to simultaneously monitor all the "omics" for entire, or nearly entire, cellular genomes.[2]

As Palsson says (op cit. footnote 1) the arrival of this data both forces and enables the study of the cell as a system. While the earlier experimental stages were appreciated for their capacity to identify the lists of components involved, once this had been achieved the vast quantities of simultaneous data now available can only be usefully simplified and comprehended in terms the interrelationships they reveal, that is, in terms of a network model.

Method is as yet at a relatively early stage of development compared to engineering theory, confined in many cases to topological considerations backed by stoichiometric considerations like flux measurements.[3] Beyond this "kinetic modeling is still severely hampered by inadequate knowledge of the enzyme–kinetic rate laws and their associated parameter values"[4] and is only recently beginning to enhance stoichiometry with direct dynamical modeling. This is partly because data of the kind and quality required is only recently becoming available,[5] and partly because the dynamical operations of very complex networks are still being only indirectly studied, requiring the development of new data analysis techniques.[6] The methodological challenges in this respect focus around improving the reliable identification of circuit structure, including (1) the discrimination of partial redundancies, (2) the development of recently initiated methods for the treatment of integrated pathways where two or more kinds of links (e.g., metabolic and signaling) are simultaneously partially served by the same chemical elements, (3) better understanding of cross-pathway interaction and whether it should be treated as mere interference or evidence of inappropriate pathway modeling, (4) the resolution of hierarchical functional architectures, and (5) sufficiently increasing the extent and precision of dynamical information required to accomplish all this.

As interlevel bridging theories, the emergence of systems and synthetic biology represents a revolution in scientific biological knowledge. But, as the opening remarks signaled, these developments also have intellectual impacts of a wider and deeper nature that can best be appreciated when set in a wider context. First there is the larger question of the nature of the living domain: against the earlier division between

[2] See Mitsuro Itaya, Chapter 5 herein and, for example, Ref. [2].

[3] Cf. Joyce and Palsson, Chapter 6 herein for deliberate development of this approach as a constraints-based delineation of possibilities.

[4] Ralph Steuer (Humboldt University, Berlin) "From topology to dynamics of metabolitic networks," lecture to the Bio-Modelling Network, Manchester University, UK, August 29, 2007.

[5] For instance, Ref. [3], noting the capacity to directly observe functional units, remarks "By linking genes and proteins to higher level biological functions, the molecular fluxes through metabolic networks (the fluxome) determine the cellular phenotype. Quantitative monitoring of such whole network operations by methods of metabolic flux analysis, thus bridges the gap by providing a global perspective of the integrated regulation at the transcriptional, translational, and metabolic level."

[6] See, for example, Ref. [4] and the discussion of modeling in Section 20.2.6.

a crude mechanism and a mysterious vitalism, systems and synthetic biology hold out the prospect of a reenergized naturalism for biology in which vital characteristics of organisms are captured as natural features of certain kinds of organized chemical systems. But to do so biological theory will have to meet some larger challenges that stem from the nature of complex adaptive systems more generally. For example, we still have no complete and coherent account of organization in complex systems, much less an account that illuminates the nature of life as a particular species of dynamical organization. Second, as this example indicates, there are still larger issues surrounding the introduction of complex dynamical system concepts, principles, tools/ methods, and models into science—where they are now expanding rapidly across most of the sciences.[7] It is to these two larger questions that the remainder of this essay briefly turns—lest, not doing so, they return to confuse us. Only then shall we be able to properly consider the challenges ahead in biology, the topic of the closing chapter.

19.1.2   The Wider Problem of the Life Sciences

During the century bounded by the rise of organized modern public science 1850–1875 and its expansion to the massive institutions of 1950–1975, the intellectual conception of science was dominated by its fundamental and most excitingly progressive discipline: physics. The philosophy of science followed suit, entranced by the prospect of simple universal laws induced from rigorous evidence and with multifarious practical applications as the truest revelation of the creator's rationality, or anyway of the nature of prediction and explanation, theory and justification. This conception encompassed chemistry, if with some difficulty, and also engineering, medicine, and "biophysics," at least while these studies were confined to physics-like objectives such as building houses, simple surgery, and osmotic pressure and all their apparent other complexities were set aside as "merely practical."

However, the hope of a universal "physics vision" would later collapse as more lifelike systems were studied. Indeed, the chief problem with this vision became the lack of any obvious way to incorporate the sciences of living organisms, cellular biology, evolution, and ecology, extending to sociology, economics, and the humanities generally. By the end of the nineteenth century, the prospect of a separate vitalist foundation for these studies, where one looks to principles for living organisms that are fundamentally independent of those for inanimate systems, was successfully exorcised from mainstream science. The vitalist view, the critique of which dates back at least to Robert Boyle, offended against both unity under physics and the practical naturalism—often expressed in terms of materialism, mechanism or both—that has

---

[7] Something of the reach and richness of the complex systems revolution sweeping the sciences will be able to be gleaned from a volume for the first time devoted to this task with 30 plus contributions by researchers across the sciences. For author abstracts see http://www.johnwoods.ca/HPS/#Complexity. Part of a multivolume *Handbook of the Philosophy of Science* now in publication and preparation, the volume's current working details are Cliff Hooker (Ed.) *Philosophy and Foundations of Complex Systems*, Vol. 10 of D Gabbay, Paul Thagard, and John Woods (Eds) *Handbook of the Philosophy of Science*, Amsterdam: Elsevier, 2006–2009.

successfully guided scientific advance for 400 years. So the life sciences were held in abeyance, some day to be somehow subsumed under the great general mechanical laws. Of course, those within the excluded domains felt obliged to declare their difference, perpetuating an often tragic conflict.[8]

Thus providing a more adequate understanding of the nature of the life sciences is an urgent intellectual problem. Indeed, we still see the old opposition in action in the latest volume by Ernst Mayr [5], a book by a prominent research biologist who has been reflecting on the nature of biology for 40 years and through many books. Mayr argues that biology is unique, distinct from physics, chemistry, engineering, and all their applied forms from rockets to robotics. Although biological entities are subject to physical and chemical laws, he says, what makes them unique is essentially that they exhibit a suite of properties not possessed by the inanimate objects of these disciplines, namely metabolism, regeneration, regulation, growth, replication, evolution, and developmental and behavioral teleology.

There is no doubt that Mayr is right that these are significant features of the living world. Thus, it becomes a pressing issue to understand how systems and— especially—synthetic biology are possible, and how they are to be understood. To do that we need to briefly review the historical tradition that culminates in Mayr's contention—for it will also reveal the seeds of the contemporary promise of its resolution through systems and synthetic biology, even while calling attention to outstanding issues.

19.2 FORMATION OF INTELLECTUAL ORTHODOXY FOR THE FIRST SCIENTIFIC–INDUSTRIAL REVOLUTION

19.2.1 Establishment of a Physics-Based Framework for Biology

For roughly 250 years from the publication of Newton's *Principia* to the close of the Second World War in 1945, the defining characteristic of fundamental advance in physics was the understanding of dynamical symmetry and conservation. A symmetry is an invariance under some operation, for example, of spherical shape under rotation. In physics, the relevant symmetries are the invariances, that is, the conservation, of dynamical quantities under various continuous space–time shifts, for example, conservation of linear motion (momentum) under shift in spatial position or of energy under time shift. Noether gave systematic form to this in 1918 and showed that it was the invariance of the form of the dynamical laws themselves that was expressed. Collections of the same space–time shifts form mathematical groups, and the corresponding invariances then form dynamical symmetry groups.

---

[8] This conflict within the research community formed the roots of Snow's two cultures, the gulf between the tools, styles, and goals of the sciences and the humanities. It will take at least another century to tackle this issue properly, but within this and the closing essay the reader will find an array of systems tools, beginning with (but by no means ending with) systems and synthetic biology, that bid fair to resolve the basic root of the problem, if not all of its branches.

For instance, Newton's equations obey the Galilean symmetry group. Symmetry forms the deepest principle for understanding and investigating fundamental dynamical laws.[9]

In addition to their general dynamical symmetries, many states have additional symmetries, for example, the lattice symmetries of a crystal. Within this framework thermodynamics emerged, with thermodynamic equilibrium the only dynamical state condition that could be identified for dealing with complex systems. The advantage of thermodynamic equilibrium states is their greater internal symmetry because all residual motion is random (a gas is stochastically spatially symmetric). When each equilibrium state is invariant with respect to transitory pathways leading to it (the outcome is independent of those initial conditions), so its history can be ignored in studying its dynamics. The dynamics itself can then be developed in a simplified form, namely in terms of local, small and reversible—hence linearizable—departures from stable equilibria, yielding classical thermodynamics.

The study of simple physical systems of a few components and of many component systems at equilibrium supported the idea that the paradigm of scientific understanding was linear causal analysis and reduction to linear causal mechanisms, with the real as what was stable, especially invariant. Paradigm cases were Newton's Laws and two-body solar system dynamics, engineering lever and circuit equations, simple two-component chemical rate equations, crystal lattices, and equilibrium thermodynamics of gases.

The philosophy of science was shaped to suit, focusing on determinism, universal atemporal (hence acontextual) causal laws, analysis into fundamental constituents then yielding bottom-up mechanical synthesis. To this was added a simple deductive model of explanation and prediction—deduction from theory plus initial conditions gives explanation after the event and prediction before it—with reduction to fundamental laws and separate contingent initial conditions becoming the basic explanatory requirement. This supports an ideal of scientific method as logical inference: induction from the data, where the most probable correct theory is logically inferred from the data (cf. statistical inference in bioinformatics), deduction from theory for prediction and explanation, and falsification: deduction from data that conflict with prediction to a failure of the predicting theory (or other assumptions).[10] However, it turns out (interestingly!) that neither the logical nor the

---

[9] For instance, the shift from Newtonian to relativistic dynamics is a shift from Euclidean to Minkowski space–time and a corresponding shift from the Galilean to the Lorentz symmetry group, while the shift to nonrelativistic quantum theory, which exhibits stronger symmetries (expressing indistinguishable states), is a shift to the unitary symmetry group. Currently the as-yet-incomplete development of relativistic quantum theory is explored in terms of the further symmetry groups involved. Further see any of the many textbooks on this subject. If all this seems somewhat impenetrable to a life scientist, it suffices to grasp the idea that symmetry is the central structural feature of dynamics in physics. On the stability–equilibrium framework see, for example, Refs [6–8] and on symmetry disruption by newer systems dynamics ideas see these and, for example, Refs [9,10] and Brading's *Stanford Encyclopedia of Philosophy* entry at http://plato.stanford. edu/entries/symmetry-breaking/.

[10] See classics of the time like Ref. [11] on induction and reduction, and on falsification see Ref. [12]. For a contemporary version in systems and synthetic biology see Breiman in Section 20.2.6.

methodological situation is so simple; both scientific practice and rational method are, and must be, much more complex than this.[11]

The philosophy of science and the scientific paradigm together constituted the intellectual framework of scientific orthodoxy for a century of scientific understanding and the evident fit between philosophy and paradigm supported the conviction that both were right, the logical clarity and elegance of the philosophy reinforcing that conviction. From within this framework, the greatest challenge is that of quantum theory to determinism and simple causality. But while this is a profound problem, the immediate theoretical challenge is also limited since the fundamental dynamical idea of a universal deterministic flow on a manifold characterized by its symmetries remains at the core.[12]

The great formation period of modern biology, characterized by the rise of genetics and its incorporation into evolutionary theory, and the subsequent emergence of elementary molecular genetics in its support, was understood within this orthodox framework. The simple fundamental laws of evolutionary population genetics and of molecular genetics that underlay them were held to provide the universal, unchanging causal reality underlying the apparently bewildering diversity of biological phenomena. The observed diversity was to be seen simply as reflecting a diversity of initial conditions independent of these laws, whether generated as exogenous geoecological events or as endogenous random mutations.

Reduction to molecular genetics thus became a defining issue. Initially this took the form, noted earlier, of treating the phenotype as effectively just a bundle of gene–trait pairs that determined fitness. This simplification sufficed, given their developmental stages, for population genetics and molecular biology, at the time. For the longer term the reductionist paradigm, based on analysis and bottom-up synthesis, assumed that the information gained by unraveling the separate simple mechanisms of all the different molecular components could be used to provide adequate linear assembly models of cellular and multicellular organisms. Functional analysis was based on the similar idea of dissecting a complex system into its functional components, all the way down to its simplest basic functions, then reducing the basic functions to simple mechanisms, and resynthesizing. This research paradigm dominated twentieth century mainstream biology, a time in which enormous progress also took place in accumulating molecular information.

19.2.2   Framework-Induced Dichotomy in the Life Sciences

The consequence of this approach to biology is that either life is radically reduced to simple chemical mechanisms and then to physics, or it has to be taken outside the paradigm altogether and asserted as metaphysically *sui generis*, a realm in itself from which flowed all of the distinctive features Mayr lists (see Section 19.1.2, especially regeneration, replication, and teleology). Both implausible positions had

[11] For overview and discussion of the situation, see, for example, Ref. [13], Chapter 2.

[12] However, as the dispute between Bohr and Einstein suggests, there may be implicit in this challenge more profound issues that do at least call into question the nature of intelligible reality, cf. Ref. [14].

devoted proponents. In particular, expressed as various forms of vitalism, the latter position has had a long history in Western thought; especially as science emerged from a Christian religious framework, it freed up investigation of the physical body while leaving the mind and soul to religious teaching. Descartes, for example, drew a sharp distinction between human body and spirit (other organisms were simply clever automatons), regarding the human organism as a hybrid, hierarchical control system: the spirit carried all the initiative, expressed perhaps through the pineal gland, while the material body was reduced to an automaton responding to control orders, a deterministic machine explained by simple physics. Kant similarly ascribed biological processes to teleology that, while embodied, escaped material scrutiny in themselves.

This situation formed the general approach to the scientific treatment of living entities, whether in psychology, sociology, economics or history, and other cultural studies. In psychology, for example, the corresponding primary choice is that between reductionist materialism and dualism (Cartesianism). Behaviorism was a particularly severe form of reductionist materialism that dominated in the first half of the twentieth century. It was followed by the currently dominant artificial intelligence version, a functionally generalized behaviorism where the mind is modeled as internal deterministic assembly and control programs, ultimately representable as digital software. This respectively parallels the transparent phenotype and molecular mechanism assembly stages of biological theory as successively reductionist input/output black box and then gray box input/output transform models. In economics we similarly begin with Homo economicus, where agents are reduced to sets of preferences, behaviorally revealed (in principle), plus a simple welfare optimization program; only recently are agents beginning to be fleshed out with preference dynamics, decision psychology, and collective interactions (e.g., through multiagent models and evolutionary game theory). The philosophy of these disciplines was shaped to suit in ways analogous to those for biology.

These are undoubtedly the early theory building stages through which any science has to go as it laboriously assembles better understanding. Possibly this was itself intuitively understood by many scientists. Even so, there was enough dogmatic conviction in science, and certainly in philosophy of science, that the results were not pleasant for dissenters who were denied a hearing and research funding and often ostracized. One might recall, as examples of this, the fates of Baldwin and Lamarck and others in biology, and of Piaget in biology and philosophy, all now being at least partially rehabilitated as the old simple dogmas breakdown, not to mention those in entire subdisciplines such as embryology and ecology who were sidelined for many years before they have again returned to the forefront of scientific progress.[13] Yet the problem of reconciling biology and physics was always a dilemma: either the organic

[13] Ultimately, embryology must become a vital application of systems and synthetic biology, since all living systems exhibit complex developmental histories. Similarly, ecological systems theory must ultimately become its sister science focused at the organism and population levels instead of the cellular and cell assembly levels—exhibited, for example, through the network models of Levins [15,16] and the dynamical resilience models of Gunderson, Holling, and others [17,18]. However, these interrelationships are as yet in their early development.

dwelled in a realm *sui generis*, or the reductive paradigm of physics was too restrictive to give a realistic account of any but the simplest of natural systems.

19.3 QUIET PREPARATIONS FOR A REVOLUTION

19.3.1 How the Emergence of the New Orthodoxy-Breaking Concepts is Tied to the Emergence of the Basic System Tools Used by Systems and Synthetic Biology

Yet all the while scientific work itself was quietly and often unintentionally laying the groundwork for superseding these approaches, both scientifically and philosophically. To understand why this might be so one has only to contemplate what the previous paradigm excludes, namely all irreversible, far-from-equilibrium thermodynamic phenomena. This comprises the vast majority of subject matter of interest to science, everything from supergalactic formation in the early cooling of the universe down to planet formation, all or most of our planet's geoclimatic behavior, all phase change behavior, natural to the planet or not, and of course all life forms, since these are irreversible far-from-equilibrium systems. What all of these phenomena exploit is spontaneous instability, specifically nonlocal, irreversible dynamical departure from their present state, whether it be the instability of a gas cloud condensing to a star, or that of a collection of chemicals forming a continuously self-regenerating life form. Moreover, all of these transitions represent the formation of nonequilibrium structures and the formation of increased complexity through symmetry breaking. This is starkly clear for cosmic condensation: the universe begins as a superhot supersymmetric expanding point sphere, but as it expands it cools and differentiates, breaking its natal supersymmetry; the four fundamental forces differentiate out, their nonlinearities amplifying the smallest fluctuational differences into ever-increasing structural features. In sum, all of these vast sweeps of phenomena are characterized by the opposite of the symmetry/equilibrium paradigm.[14]

Thus it is not surprising that from early on, even while the elegantly simple mathematics of the stability–symmetry paradigm were being developed and its striking successes explored, scientists sensed the difficulties of remaining within its constraints, albeit in scattered and hesitant forms. Maxwell, who formulated modern electromagnetic theory in the later nineteenth century and sought to unify physics, drew explicit attention to the challenge posed by instability and failure of

---

[14] An early mathematical classic on nonlinear instabilities referred to the old paradigm as the "stability dogma," see Ref. [19], pp. 256ff. See also the deep discussion of the paradigm by the Nobel prize winning pioneer of irreversibile thermodynamics, Prigogine, in Refs [20–22]. I add the phase-shift cosmogony of Daodejing, Chapter 42, translated by my colleague Dr Yin Gao, because the West has been slow to appreciate the deep dynamical systems orientation of this tradition in Chinese metaphysics, for instance, in medicine [23]:

> The dao (the great void) gives rise to one (singularity)
> Singularity gives rise to two (yin and yang)
> Yin and yang give rise to three (yin, yang, and the harmonizing force)
> Yin, yang, and the harmonizing force give birth to the 10,000 things/creatures.

universality for formulating scientific laws, while his young contemporary Poincaré spearheaded an investigation of both nonlinear differential equations and instability, especially geometric methods for their characterization.[15] By the 1920s static, dynamic, and structural equilibria and instabilities had been distinguished.[16] A static equilibrium requires no irreversible process to maintain it while a dynamic equilibrium does. Living organisms illustrate dynamical equilibria since they only persist if maintained by flows of energy and matter through them. Either equilibrium is unstable if its conditions are sufficiently perturbed. The system is then on a transient trajectory until a new equilibrium is reached. Phase changes illustrate structural instabilities, where the dynamical form itself changes during the transient trajectory. It was discovered in the 1950s and 1960s that simple chemical reaction systems, like that studied by Belousov and Zhabotinskii, show phase changes among dynamical equilibria.

In engineering, nonlinearity and emergent dynamics appeared in an analytically tractable manner with the discovery of feedback and the development of dynamical (as distinct from later programming) control theory. Maxwell in 1868 provided the first rigorous mathematical analysis of a feedback control system (Watt's 1788 steam governor). By the early twentieth century General Systems Theory was developed by von Bertalanffy and others, with notions like feedback/feedforward, homing-in, and homeostasis at their basis, while later Cybernetics (the term coined by Weiner in 1948) emerged from control engineering as its applied counterpart.[17] Classical control theory, which became a disciplinary paradigm by the 1960s, forms the basis of the use of dynamical system models in contemporary systems and synthetic biology.

In 1887 Poincaré had also become the first person to discover a chaotic deterministic system (Newton's three-body system), later introducing ideas that ultimately led to modern chaos theory. Meanwhile Hadamard 1898 studied a system of idealized "billiards" and was able to show that all trajectories diverge exponentially from one another (sensitivity to initial conditions), with a positive Lyapunov exponent. However, it was only with the advent of modern computers in the 1960s that investigation of chaotic dynamics developed, beginning with Lorenz whose model of atmospheric dynamics as a simple convective cell revealed sensitivity to initial

---

[15] This sensitivity was already evident in the 20 years Newton delayed publication of his magisterial *Principia Mathematica*, while he searched for a principled way to encompass the treatment of lunar dynamics within its framework, a classical nonlinear three-body gravitational problem for which his doubts have subsequently been shown amply justified.

[16] Thanks to Birkhoff and Andropov, following Poincaré. Lyapunov's study of the stability of nonlinear differential equations was in 1892, but its significance was not generally realised until the 1960s.

[17] In 1840, Airy developed a feedback device for pointing a telescope, but it was subject to oscillations; he subsequently became the first to discuss the instability of closed-loop systems, and the first to use differential equations in their analysis. Following Maxwell and others, in 1922, Minorsky became the first to use a proportional–integral–derivative (PID) controller (in his case for steering ships), and considered nonlinear effects in the closed-loop system. By 1932, Nyquist derived a mathematical stability criterion for amplifiers related to Maxwell's analysis and in 1934 Házen published the *Theory of Servomechanisms*, establishing the use of mathematical control theory in such problems as orienting devices (e.g., naval guns). Later development of the use of transfer functions, block diagrams, and frequency-domain methods saw the full development of classical control theory.

conditions, which offered a possible explanation of why, even with enormously increased data collection, long-term weather prediction remained elusive. By the mid-1970s chaos had been found in many diverse places, including physics (both empirical and theoretical work on turbulence), chemistry (the Belousov–Zhabotinskii system), and biology (logistic map population dynamics and Lotka–Volterra equations for four or more species), and the mathematical theory behind it was solidly established (Feigenbaum, Mandelbrot, Ruelle, Smale, and others).

This historical account is unavoidably selective and sketchy, but it sufficiently indicates the slow build up of an empirically grounded conceptual break with the simple symmetry/equilibrium orthodoxy. However the new approach still often remained superficial to the cores of the sciences themselves. In physics this is for deep reasons to do with the lack of a way to fully integrate instability processes, especially for structural instabilities, into the fundamental dynamical flow framework (at present they remain interruptions of flows), the lack of integration of irreversibility into fundamental dynamics,[18] and the related difficulty of dealing with global organizational constraints in flow characterization (specifically the difficulty of dealing with the autonomy constraint that characterizes coherent metabolisms for living creatures[19]). For biology all that had really developed was a partial set of mathematical tools applied to a disparate collection of isolated examples that were largely superficial to the then core principles and dynamics of the field.

We now say complexity was discovered, and indeed science came to distinguish a range of new systems from those that could be more thoroughly treated because they were few bodied with simple dynamics or many bodied but either unordered (random) or highly ordered (crystal-like). But there was then no principled framework for understanding systems that were many bodied, nonlinear, sufficiently ordered to be organized, and dynamically labile; indeed, the problem of fully characterizing complexity in a principled manner remains open.[20]

Nonetheless, by the late 1970s it is clear in retrospect that science had begun to pull together many of the major ideas and principles that would undermine the hegemony of the simple symmetry/equilibrium orthodoxy. Instabilities were seen to play crucial roles in many real-life systems—they even conferred sometimes valuable properties on those systems, such as sensitivity to initial conditions and structural lability in response. These instabilities broke symmetries and in doing so produced the only way to achieve more complex dynamical conditions. The phenomenon of deterministic chaos was not only surprising to many, but to some extent it pulled apart determinism from analytic solutions, and so also from prediction, and hence also pulled explanation

---

[18] Prigogine [21] had even proposed to modify the Schrodinger equation to circumvent its entrenched linearity and accommodate irreversible dissipation. However irreversible thermodynamics has subsequently made some internal progress through the work of Morowitz and others. As for quantum theory (about which Einstein had earlier similarly complained) subsequent experience with relativistic quantum theory suggests that the problem has to be tackled at a much deeper level, if it can be tackled at all from within our present flow conception of dynamics.

[19] On this use of autonomy, see, for example, Refs [24–26]. Its incorporation into systems and synthetic biology remains an outstanding theoretical task (see the concluding essay herein).

[20] See further Section 19.7 and the concluding essay herein and Ref. [27].

apart from prediction. It also emphasized a principled, as opposed to a merely pragmatic, role for human finitude in understanding the world.[21] The models of phase change especially, and also those of far-from-equilibrium dynamical stability, created models of emergence with causal power ("top-down" causality), and hence difficulty for any straightforward idea of reduction to components.[22] And, although not appreciated until recently, they created an alternative paradigm for situation-dependent rather than universal, laws.[23] Thus, responses like that of Duhem in *The Aim and Structure of Physical Theory* to retain the simple symmetry/equilibrium orthodoxy despite being aware of the results of Poincaré and Hadamard became less and less reasonable, and a new appreciation for the sciences of complex dynamical systems began to emerge. These are the very ideas that, allied to the development of generalized network analysis emerging from circuit theory, chemical process engineering, and elsewhere would later underlie contemporary systems and synthetic biological modeling.

19.3.2   Preparations for Change in Biology

This period also quietly set the stage for the undoing of geneticism, the simple gene–trait model noted at the outset, and that later paved the way for the more intimate introduction of complex systems methods into the heart of biology. Genetics had of course emphasized the importance of what lay inside the cell but, as noted in Section 19.1.1, geneticism made the phenotype irrelevant to biological theory and explanation. However, in physics Prigogine (following Schrodinger and Turing) worked on irreversible thermodynamics as the foundation for life (Footnotes 14, 18), modeling organisms as far-from-equilibrium systems sustained only by a continuous throughput of matter and energy, thereby importing suitably ordered energy (negative entropy) from their environment in order to create and maintain internal organization and discharging the inevitable less ordered waste products that result. (Biologically, this amounts to food and water intake and excreta output.) This generates a (high level) metabolic picture in which the full internally regulated body is essential to life. In this conception, it is organism activity, metabolic and behavioral, that supports development, regeneration, reproduction, and senescence for individuals and ultimately also for communities and ecosystems. This encompasses all Mayr's distinctive properties (see Sections 19.2 and 19.4.1).[24]

During the immediate postwar period in which Prigogine and others were developing these ideas, cellular biology was revived and underwent a rapid development, partly driven by new, biochemical-based problems (understanding kinds and rates of chemical reactions like electron transport, and so on) and partly by new instrumentation (electron microscope, ultracentrifuge) that allowed much more

---

[21] The point being that any finite creature can only make finitely accurate measurements, independently of any further constraints arising from specific biology or culture; there is always a residual uncertainty, and chaotic dynamics will amplify that uncertainty over time.

[22] For a systems biology illustration and discussion see Refs [28,29].

[23] See further Section 19.7 below and the concluding essay herein; for the basic idea, see Ref. [30].

[24] See also, for example Ref. [31].

detailed examination of intracellular structure and behavior. In consequence, there was an increasing molecular understanding of genetic organization, especially development of RNA roles in relation to DNA, of regulator genes and higher order operon formation and of the roles of intracellular biochemical gradients, intercellular signaling, and the like in cellular specialization and multicellular development. All this prepared the ground for envisioning the cell as a site of many interacting biochemical processes, in which DNA played complex interactive roles as some chemicals among others, rather than the dynamics being viewed as a consequence of a simple deterministic genetic program. Genetics was replaced by "omics" (genomics, proteomics, metabolomics, and so on).[25]

During roughly the same period, 1930–1960, Rashevsky and others pioneered the application of mathematics to biology. With the slogan *mathematical biophysics: biology:mathematical physics:physics*, Rashevsky proposed the creation of a quantitative theoretical biology and was an important figure in the introduction of quantitative dynamical models and methods into biology, ranging from models of fluid flow in plants to various medical applications. That general tradition was continued by his students, among them Rosen, whose edited volumes on mathematical biology of the 1960s and 1970s did much to establish the approach. Indeed, as Rosen remarks, "It is no accident that the initiative for System Theory itself came mostly from Biology; of its founders, only Kenneth Boulding came from another realm, and he told me he was widely accused of 'selling out' to biologists."[26]

In this tradition various physiologists began developing the use of dynamic systems to model various aspects of organism functioning. In 1966, for example, Guyton developed an early computer model that gave the kidney preeminence as the long-term regulator of blood pressure, with other systems only able to regulate pressure in the short term, and went on to develop increasingly sophisticated dynamical network models of this kind. The next generation expanded these models to include intracellular dynamics. Tyson, for example, researched mathematical models of chemical systems like Belousov–Zhabotinskii in the 1970s, passing to cellular aggregation systems like *Dictyostelium* in the 1980s and to intracellular network dynamic models in the 1990s, and this was a common progression.[27] See also the increasingly sophisticated models of timing.[28] In this manner physiology has supported a smooth introduction of increasingly refined dynamical models into biology, providing a direct resource for contemporary systems and synthetic biology.

There has also been a correlative revival of a developmental perspective in biology, in embryology generally and early cellular differentiation in particular. This became

---

[25] See, for example Refs [32,33]

[26] On Rashevsky see, for example Ref. [34], http://www.kli.ac.at/theorylab/AuthPage/R/RashevskyN.html. For many years (1939–1972) he was editor and publisher of the journal *The Bulletin of Mathematical Biophysics*. For Rosen, see http://www.panmere.com/rosen/booklist.htm#bkrosen and the concluding essay. The quote comes from his *Autobiographical Reminiscence* at http://www.rosen-enterprises.com/RobertRosen/rrosenautobio.html.

[27] Among other resources see respectively http://www.umc.edu/guyton/, http://mpf.biol.vt.edu/people/tyson/tyson.html. Compare the work of Hogeweg, for example: http://www.binf.bio.uu.nl/master/.

[28] Refs [35–38].

linked to evolutionary "bottlenecks" and evolutionary dynamics generally to form evo-devo as a research focus. Added to this was work on epigenetics and nonnuclear inheritance, especially maternal inheritance, and early work on enlarging evolutionary dynamics to include roles for communal (selection bias, group selection) and ecological factors, culminating in the holistic "developmental systems" movement.[29]

Ecology too has been studied as a dynamic network (Lotka/Volterra, May, Levins, and others), as an irreversible far-from-equilibrium dissipative flux network (Ulanowicz) or food-web energetics system (Odum), as a spatiotemporally differentiated energy and matter flow pathway network (Pahl–Wostl) self-organizing through interorganism interaction (Holling, Solé/Bascompte) and as an organized complex dynamic system employing threshold (bifurcation) dynamics, spatial organization and exhibiting adaptive resilience (Holling, Walker, and others), responding in complex, often counterintuitive, ways to policy-motivated inputs.[30] All these features are found within cells, albeit more tightly constrained by cellular regenerative coherence, and fruitful cross-fetilization should eventually be expected, perhaps particularly with respect to the recent emphasis in both on understanding the coordination of spatial with functional organization.

All of these scientific developments, still in process, work toward replacing black box geneticism with a larger model of a mutually interacting set of evolutionary/developmental/communal/ecological dynamic processes.[31] Although still a collection of diverse models and methods, dynamical network methods are emerging across these disciplines as a shared methodological toolkit.[32] In combination, these developments present a picture of life as a complex system of dynamic processes running on different groups of timescales at different spatial scales, with longer term, more extended processes setting more local conditions for shorter term, less extended processes, while shorter term, local products accumulate to alter longer term, more extended processes.

This conception now extends into medicine (especially through Chinese medicine), psychology (through mathematical psychology, especially neuropsychological and social interaction dynamical modeling), and economics (through econophysics, evolutionary economics).[33] From there the conception extends still more widely (but more diffusely) through the social sciences and management (dynamical/evolutionary game theory, human–natural interaction dynamical networks), military theory, technology theory, and even the nature of science itself (research resource webs, economic and interactionist dynamics of knowledge).

The earlier physics paradigm of simple universality, symmetry, and (static) equilibrium no longer dominates. The new dynamical ideas are still based in the same fundamental dynamics, but derive from an aspect of them that has hitherto remained hidden, the complex spatiotemporal coordination of nonlinear dynamical

---

[29] See, respectively, for instance, Refs [39–43], the title itself indicating something of the macro intellectual landscape in which the idea emerged, and Ref. [44].

[30] See, among many others, the following works and their references: Refs [16–18,45–53].

[31] Cf., for example, Ref. [54] and references in footnote 29.

[32] For a recent review see Ref. [55].

[33] See, respectively, and among many others, Refs [23,56–59].

interactions to form organized, far-from-equilibrium systems that arise through symmetry-breaking amplification and propagation of asymmetrical variations (however generated). It is this aspect of dynamics that is now coming to dominate research across the sciences. Although biology will come to have an unprecedented centrality for it, this century will not be known as the century of biology (as is sometimes said), but, more fundamentally, as the *century of complex systems dynamics*. And it offers for the first time the genuine prospect of natural, productive integration among a range of scientific disciplines based on interrelating the dynamical models being employed in each.

19.4 TOWARD A NEW COMPLEX SYSTEMS PARADIGM AND PHILOSOPHY

19.4.1 Complexity of Complex Systems and the Uniqueness of Biology

The complex systems that constitute our life world are characterized by deterministic dynamics that manifest the following properties:

(1) Nonlinear interactions; nonadditivity

(2) Irreversibility; nonequilibrium constraints; dynamical stabilities

(3) Amplification; sensitivity to initial conditions, especially to "rare" events

(4) Finite deterministic unpredictability; edge-of-chaos criticality

(5) Symmetry breaking; self-organization; bifurcations; emergence

(6) Enabling and coordinated constraints

(7) Coordinated spatial and temporal differentiation with functional organization

(8) Intrinsically global coherence and organization; modularity; hierarchy

(9) Path dependence and historicity

(10) Constraint duality; supersystem formation

(11) Autonomy; anticipativeness; adaptiveness

(12) Multiscale and multiorder functional organization; learning

(13) Model specificity/model plurality; model centeredness

Roughly, properties lower on the list are increasingly richly possessed by living systems and present increasing contemporary challenges to our dynamical understanding. The diversity and the domain-specificity of these properties explain the diversity of notions of complexity, and the challenges to understanding that they continue to pose undermines hope for any unified account of complexity in the near future.

Many of these terms are well known and have already been explained or illustrated; they will be assumed understood. Some are in common usage and often considered well known but in fact present ongoing challenges to understanding (self-organization, emergence, and organization); these will be assumed here as sufficiently intuited and briefly reconsidered in the closing essay. Finally, some others are likely less well known though straightforward: constraints (enabling, coordinated and dual),[34] path dependence,[35] and model specificity/plurality and centeredness,[36] while autonomy (cf. footnote 20), anticipativeness, and adaptiveness are briefly discussed in the concluding essay.

It should be noted that the shift to complex systems represents enrichment—albeit a massive enrichment—of the classical symmetry–stability–invariance dynamical framework, not its wholesale abrogation. It is just that it took some centuries to understand this. For instance, a dynamical attractor basin specifies a set of dynamical states within which the dynamical laws take on a stable, self-contained form. For a dynamics with several attractor basins and energetically determined transient paths connecting them, this provides an immediate model of sets of local, energetically dependent laws. Parameter-dependent deformations of this basin topology, where some basins may disappear and others arise, then provide cases of the emergence of higher order context (parameter) dependent law domains. More disruptive discontinuous bifurcations, for example, from fluid conduction to convection or blastula formation and internal phase difference, represent a more serious rift in the classical

---

[34] The term "constraint" implies limitation, most generally in the present context it refers to limited access to dynamical states (equivalently limiting dynamical trajectories to subsets of state space); this is the common disabling sense of the term. But it is crucial to appreciate that constraints can at the same time also be enabling, they can provide access to new states. Thus, a skeleton is a disabling constraint, for example limiting the size of hole through which a body can fit; but by providing a jointed frame for muscular attachments it also acts to enable a huge range of articulated motions, transforming an organism's accessible niche, initiating armor and predator/prey races, and so on. This is the general aspect of the duality of constraints, but it has a specific application in the system/supersystem context where system constraints may contribute to enabling supersystem capacities, for example the role of mitochondria in eukaryote energy production, and supersystem constraints may free up system constraints, for example wherever multicellular capacities permit member cells to specialize. In all of these cases there has to be a coordination of component constraints to achieve the final effect: the many component bones of a skeleton have to be quite specifically coordinated so as to achieve an articulation that facilitates fitness-providing behaviors, mitochondrial functioning has to be integrated with the larger cellular processes for its products to innervate the cell.

[35] Path dependence occurs when initially nearby dynamical trajectories subsequently diverge as a function of small differences in their initial conditions. It is brought about by amplification, for example in selection-reinforced amplification of small genetic differences generating diverging developmental or speciation trajectories, where the source of amplification may be bifurcations, feedback, or simply suitable nonlinearities.

[36] Complex systems of the kind described typically require many parameters to adequately characterize, for example specifying the rate and storage characteristics of the many processes they sustain. Model specificity refers to the capacity to select parameter values so as to specialize the model to the characterization of some unique individual and/or situation, while model plurality refers to the converse capacity to capture the characterization of a plurality of individuals/situations within its parameter ranges. These features are the basis for formulating valid generalizations across populations and, conversely, for deducing feature ranges in individuals from more broadly characterized populations. Model centeredness refers to the fact that systems of these kinds typically manifest nonanalytic dynamics (their dynamical equations lack analytical solutions) whence it is necessary to explore their dynamics computationally. This places computational modeling at the center of their scientific investigation in a strong manner and highlights the unique contribution of computers to cognition (all its other uses being pragmatic, if often valuable).

fabric since the idea of an analytic but condition-dependent superdynamics for state space fails, or has so far eluded construction; but even here both the conditions under which they occur and the nature of their outcomes arise from the underlying general dynamics.

Comparing this list of complex system properties with Mayr's earlier list characterizing biological systems (Section 19.2) we can see that, at least in principle, complex systems provide resources for modeling, and hence explaining, each of them: metabolism, regeneration, growth, replication, evolution, regulation, and teleology (both developmental and behavioral). Metabolism, for example, refers to the organized network of biochemical interactions that convert input matter and negentropy (food and water) into usable forms and direct their flows to various parts of the body as required, for example, for cellular respiration. The individual biochemical reactions are largely known. However it remains a challenge to characterize multilevel processes like respiration, comprising processes from intracellular Krebs Cycles to somatic cardiovascular provision of oxygen and removal of carbon dioxide, processes that must be made coherent across the entire body. In this conception, global coherence is a result of internal regulation at various functional levels (intracellular and intercellular, organ and body), and we now have massive information about the individual multifarious feedback and switching processes that contribute to somatic and, neurally, to behavioral regulation. These same capacities, placed in the context of globally organized multiscale functional organization and adaptive retention, in principle also model all the basic properties of agency, including human agency, in particular the teleology distinctive of intentional intelligence.[37] The challenge global coherence poses is to understand how these processes are interrelated so as to produce the regulated dynamical labilities and equilibria that the explanation of organism capacities demands. Here systems and synthetic biology, together with neurobiology, have a central contribution to make.

Pursuit of every scientific framework, that is, of a philosophy and paradigm, is underwritten by a practical act of faith that its cognitive apparatus, including concepts, classes of models and underlying mathematics, and experimental instruments, techniques, and interpretations, is adequate to understand the domain concerned. Here that faith revolves around the adequacy of complex systems concepts, models, and techniques as deployed in roughly the scheme just presented.

19.4.2 Need for a New Scientific Framework

The world has turned. The old orthodox framework for science that sufficed for the study of simpler systems, physical systems of a few components and of many-component equilibrium systems, no longer suffices; science has discovered the power of complex systems. The Bénard cell, Belousov–Zhabotinsky reaction, and *Dictyostelium* aggregation have replaced the slingshot, gas, and crystal as model systems. With them has emerged the general model of a complex organization of dynamic processes running on different groups of timescales at different spatial

---

[37] For an outline, see Refs [25,26] and the concluding essay.

scales, with longer term, more extended processes setting more local conditions for shorter term, less extended processes and shorter term, local products accumulating to alter longer term, more extended processes. Minimally this development calls for the articulation of a new framework adequate for the de facto deployment of complex systems models throughout the sciences, and especially in biology.

The old framework that supported a paradigm of linear causal analysis, reduction, symmetry, static stability (equilibrium), and invariance is being replaced by a new complex systems paradigm focused on multiscale nonlinear networked dynamical interrelationships, symmetry-breaking self-organization of complexity, partial unpredictability, context-dependent laws, complex global organization with partial hierarchical modularity, path dependence and historicity, and the unavailability of closed-formed analytic solutions and consequent model centeredness.

Correlatively, instead of a philosophy of science focusing on determinism, identification of universal atemporal (hence acontextual) causal laws and reduction achieved through analysis followed by linear, bottom-up analysis reduced to closed formed analytic solutions, we now need a philosophy of science focusing on dealing with multiple simultaneous, multiscale interdependencies, validity of top-down as well as bottom-up analysis, the entwinement of emergence and reduction, domain-bound (context-dependent) dynamical laws (causality makes limited sense in these contexts) that accept historically unique individuals as the norm, and limited knowability and controllability. In consequence, the simple induction-based resolution of theory development must be replaced (it never worked anyway), but now with an account rich enough to encompass these complications, and while the bare logical forms of deductive explanation and falsification survive, they too will need correlative enrichment to illuminate realistic scientific method. In short, a substantially revised philosophy of science is required.

These are still new ideas in science, despite their being manifest everywhere, and the new philosophy of science will need to be underpinned by clarified conceptual/theoretical accounts of these new features, especially complexity, self-organization, emergence, order/organization, information, system causality, reduction, and analysis/synthesis. Some progress has been made and will be commented on in the closing essay. Often this will have counterintuitive (really counterclassical) consequences, for example, reduction as naturalization (kind reduction through function-to-dynamics mapping) is entwined with emergence as antireductive top-down constraint formation, each relying on the other and both dictated by nonlinear dynamics. These in turn are needed to rethink explanation, prediction, control, and scientific method, for example, the statistical treatment of data and error identification. A coherent form of all of this is the necessary foundation for continuing to embrace the practical act of faith in the adequacy of the dynamical systems approach (Section 19.7), and its empirical confirmation the necessary ground for affirming that commitment as rational rather than merely faith.

These new ideas and practices also create new, and sometimes unexpected, scientific associations. In particular we note that, whereas biology and engineering were divided literally by the study of the living and the dead under the old paradigm, under the new complex systems paradigm they acquire a mutual affinity. As engineers have increasingly studied systems with similar complexity characteristics to those listed above, whether as sophisticated aeroplane control systems, multisensor, "intelligent" distributed signaling systems, or traffic flow systems, they have been forced to face the same issues over multiscale functional organization as do biologists. They have pursued robotics, their version of organisms, and used such exploratory methods as genetic algorithms, their version of evolution. Bioengineering increasingly integrates organisms into engineering designs, such as using bacteria to process waste water in artificial wetland design or, genetically engineered, to generate energy in industrial photosynthesis, and conversely in the synthesis of artificial life forms using engineering genome models. Thus, contemporary engineers would recognize the complex systems characterizations of Mayr's biologically distinctive features as belonging in principle to their field as well.

Further afield, multiagent adaptive modeling in economics, social organization, intelligent firms, military conflict, and much more now find affinity with the general methods of the complex systems approach. So the developments considered in this book are themselves taking place in the wider context of a systems-led transformation of scientific concepts, principles, and methods that are having an increasingly deep impact.

REFERENCES

1. Palsson B. *Systems Biology: Properties of Reconstructed Networks*. New York: Cambridge University Press, 2006.

2. Borodina I, Nielsen J. From genomes to *in silico* cells via metabolic networks. *Curr Opin Biotechnol* 2005;16(3):350–355.

3. Sauer U. High-throughput phenomics: experimental methods for mapping fluxomes. *Curr Opin Biotechnol* 2004;15(1):58 63.

4. Joyce A, Palsson B. The model organism as a system: integrating 'omics' data sets. *Nat Rev Mol Cell Biol* 2006;7:198–210.

5. Mayr E. *What Makes Biology Unique?* Cambridge, UK: Cambridge University Press, 2004.

6. Brading K, Castellani E, editors. *Symmetries in Physics: Philosophical Reflections*. Cambridge, UK: Cambridge University Press, 2003.

7. Stewart I, Golubitsky M. *Fearful Symmetry. Is God a Geometer?* Oxford: Blackwell, 1992.

8. van Fraassen B. *Laws and Symmetry*. Oxford: Oxford University Press, 1989.

9. Mainzer K. *Symmetry and Complexity: The Spirit and Beauty of Non-Linear Science*. Singapore: World Scientific, 2005.

10. Schmidt J. *Instability in Nature and Science: A Philosophy of Late-Modern Physics*. Berlin: De Gruyter, 2007.

11. Nagel E. *The Structure of Science*. London: Routledge & Kegan Paul, 1961.

12. Popper K. *Conjectures and Refutations*. London: Routledge & Kegan Paul, 1972.

13. Hooker C. *Reason, Regulation and Realism*. Albany, NY: SUNY Press, 1995.

14. Hooker C. Physical intelligibility, projection, objectivity and completeness: the divergent ideals of Bohr and Einstein. *Br J Philos Sci* 1991;42:491–511.

15. Levins R, Lewontin R. *The Dialectical Biologist*. Harvard University Press, 1985.

16. Haila Y, Levins R. *Humanity and Nature: Ecology, Science and Society.* London: Pluto Press, 1992.

17. Gunderson L, Holling C, Pritchard L, Peterson G.Resilience in Ecosystems, Institutions and Societies, Stockholm, Beijer International Institute of Ecological Economics, Beijer Discussion Paper Series, No. 95, 1997.

18. Gunderson L, Holling C, editors. *Panarchy: Understanding Transformations in Human and Natural Systems.* Washington, DC: Island Press, 2002.

19. Guckenheimer J, Holmes P. *Non-Linear Oscillations, Dynamical Systems, and Bifurcations of Vector Fields.* New York: Springer, 1983.

20. Prigogine I. *Thermodynamics of Irreversible Processes*, 3rd ed. Wiley Interscience, 1967.

21. Prigogine I. *From Being to Becoming.* San Francisco: W. H. Freeman, 1980.

22. Prigogine I, Stengers I. *Order Out of Chaos.* Boulder, CO: Shambhala, 1984.

23. Herfel W, Rodrigues D, Gao Y. Chinese medicine and the dynamic conceptions of health and disease. *J Chinese Philos* 2007;3:58–80.

24. Etxeberria A, Moreno A, Umerez J, editors. *Commun Cogn* 2000;17:3–4 (Special ed., *The Contribution of Artificial Life and the Sciences of Complexity to the Understanding of Autonomous Systems*).

25. Christensen W, Hooker C.Self-directed agents. In: MacIntosh J, editor. Contemporary Naturalist Theories of Evolution and Intentionality, Canadian Journal of Philosophy, Special Supplementary Volume, 2002, pp.19–52.

26. Hooker C.Interaction and bio-cognitive order. Synthese, Special Edition on Interactionism, in press.

27. Collier J, Hooker C. Complex organised dynamical systems. *Open Syst Inform Dyn* 1999;6:241–302.

28. Bruggeman FJ, Westerhoff HV, Boogerd FC. BioComplexity: a pluralist research strategy is necessary for a mechanistic explanation of the 'live' state. *Philos Psychol* 2002;15(4): 411–440.

29. Boogerd FC, Bruggeman FJ, Richardson RC, Stephan A, Westerhoff HV. Emergence and its place in nature; a case study of biochemical networks. *Synthese* 2005;145:131–164.

30. Hooker C. Asymptotics, reduction and emergence. *Br J Philos Sci* 2004;55:435–479.

31. Brooks D, Wiley E. *Evolution as Entropy, Toward a Unified Theory of Biology.* Chicago: University of Chicago Press, 1986.

32. Bechtel W. An evolutionary perspective on the re-emergence of cell biology. In: Hahlweg K, Hooker C, editors. *Issues in Evolutionary Epistemology.* Albany, NY: State University of New York Press, 1989.

33. Westerhoff H, Palsson B. The evolution of molecular biology into systems biology. *Nat Biotechnol* 2004;22:1249–1252.

34. Rashevsky N. *Some Medical Aspects of Mathematical Biology,* Springfield, IL: Thomas, 1964.

35. Glass L, Mackey M. *From Clocks to Chaos.* Princeton: Princeton University Press, 1988.

36. Kaplan D, Glass L. *Understanding Nonlinear Dynamics.* New York: Springer-Verlag, 1995.

37. Winfree A. *The Timing of Biological Clocks.* Scientific American Library, New York: W. H. Freeman, 1987.

38. Winfree A. *When Time Breaks Down*. Princeton: Princeton University Press, 1987.

39. Solé R, Salazar I, Garcia-Fernandez J. Common pattern formation, modularity and phase transitions in a gene network model of morphogenesis. *Physica A* 2002;305:640–647.

40. Raff R, Kaufman T. *Embryos, Genes, and Evolution*. New York: Macmillan, 1983.

41. Goodwin B, Saunders P. *Theoretical Biology: Epigenetic and Evolutionary Order from Complex Systems*. Edinburgh: Edinburgh University Press, 1989.

42. Jablonka E, Lamb M. *Epigenetic Inheritance and Evolution: The Lamarckian Dimension*. Oxford: Oxford University Press, 1995.

43. Gray R. Death of the gene: developmental systems strike back. In: Griffiths P, editor. *Trees of Life: Essays in Philosophy of Biology*. Dordrecht: Kluwer, 1992.

44. Oyama S, Griffiths P, Gray R, editors. *Cycles of Contingency: Developmental Systems and Evolution*. Cambridge, MA: MIT Press, 2001.

45. May R, editor. *Theoretical Ecology: Principles and Applications*. Saunders: Philadelphia, 1976.

46. Ulanowicz R. *Ecology: The Ascendant Perspective*. New York: Columbia University Press, 1997.

47. Odum H. *Environment, Power and Society*. New York: Wiley, 1971.

48. Pahl-Wostl C. *The Dynamic Nature of Ecosystem: Chaos and Order Entwined*. New York: John Wiley & Sons, 1995.

49. Holling C. Cross-scale morphology, geometry and dynamics of ecosystems. *Ecol Monogr* 1992;62(4):447–502.

50. Solé R, Bascompte J. *Self-Organization in Complex Ecosystems*. Princeton: Princeton University Press, 2006.

51. Dieckmann U, Law R, Metz J,editors. *The Geometry of Ecological Interactions: Simplifying Spatial Complexity*. Cambridge, UK: Cambridge University Press, 2000.

52. Cash D, editor. Scale and Cross-Scale Dynamics: Governance and Information in a Multilevel World. http://www.ecologyandsociety.org/.

53. Scheffer M, Westley F, Brock W, Holmgren M. Dynamic interaction of societies and ecosystems—linking theories from ecology, economy and sociology. In: Gunderson L, Holling C, editors. *Panarchy: Understanding Transformations in Human and Natural Systems*. Washington, DC: Island Press, 2002.

54. Jablonka E, Lamb M. *Evolution in Four Dimensions: Genetic, Epigenetic, Behavioral, and Symbolic Variation in the History of Life*. Boston: MIT Press, 2005.

55. Strogatz S. Exploring complex networks. *Nature* 2001;410:268–277.

56. Goldberger A. Fractal variability versus pathologic periodicity: complexity loss and stereotypy in disease. *Pers Biol Med* 1997;40(4):543.

57. Anderson R, May R. *Infectious Diseases of Humans: Dynamics and Control*. Oxford: Oxford University Press, 1991.

58. Heath R. *Nonlinear Dynamics: Techniques and Applications in Psychology*. Mahwah, NJ: Lawrence Erlbaum Associates, 2000.

59. Mantegna R, Stanley H. *An Introduction to Econophysics: Correlations and Complexity in Finance*, Cambridge, UK: Cambridge University Press, 1999.

20

OUTSTANDING ISSUES IN SYSTEMS AND SYNTHETIC BIOLOGY

Pengcheng Fu[1] and Cliff Hooker[2]

[1]*Faculty of Chemical Science and Engineering, China University of Petroleum, 18 Fuxue Road, Changping District, Beijing 102249, People's Republic of China*
[2]*Faculty of Education and Arts, School of Humanities and Social Science, The University of Newcastle, NSW, Australia*

In this book, we have discussed a number of applications of systems biology and synthetic biology. In fact, the scope and potential applications of systems biology and synthetic biology are not yet fully defined. As we ponder the future directions in biology research, there remain many open issues, including those that are discussed.

20.1 OUTSTANDING SPECIFIC ISSUES

20.1.1 Systems Biology and Synthetic Biology for the Investigation of Nonprotein-Coding RNAs

The epoch of systems biology and synthetic biology began when whole-genome sequences for various organisms started to accumulate. The amount and precision of this information made it possible to map the coding and noncoding regions and the hierarchy of regulatory mechanisms, relationships among structural and functional assemblies, subcellular organelles and compartments, and interaction with external signals. One of the most important discoveries of the last few years has been the

Systems Biology and Synthetic Biology   Edited by Pengcheng Fu and Sven Panke
Copyright © 2009 John Wiley & Sons, Inc.

identification of small, nonprotein-coding RNAs (ncRNAs) that act as integral regulatory components of cellular networks [1]. ncRNAs serve an astonishing variety of functions and thus play important roles in many intracellular processes, from transcriptional regulation, gene silencing, chromosomal replication, through RNA processing and modification, mRNA stability and translation, to protein degradation and translocation, and so on [2]. The size of ncRNAs range from about 20 nt for the large family of microRNAs (miRNAs) that modulate development in *Caenorhabditis elegans*, Drosophila, and mammals [3–8] to 100–200 nt for small RNAs (sRNAs) commonly found as translational regulators in bacterial cells [9,10] and up to 10,000 nt for RNAs involved in gene silencing in higher eukaryotes [11–13]. There are two approaches to searching for ncRNAs: computation methods that focus on intergenic regions and expression-based methods that examine expression levels of the transcripts [2]. Systematic identification and characterization of ncRNAs in genomes has become one of the most exciting challenges in cellular and development biology.

Among the noncoding RNA genes that produce functional molecules instead of encoding proteins, a large number of newly identified RNAs have been found to function as regulators [1]. These regulatory RNAs (reRNAs) impact all the steps in the genetic information pathways, and may serve as transcriptional regulators, translational regulators, modulators of protein function, or regulators of RNA and protein distribution. Study of many of these RNAs in bacteria and eukaryotes has shown a surprisingly high degree of similarity between regulatory RNAs in all types of organisms [1]. Therefore, insights gained by investigation of the regulatory role of reRNAs using one system are applicable to other systems as well. reRNAs are now recognized to play important roles as regulatory elements, yet little work has been done on a global scale to identify these intracellular regulators. Elucidation of correlations between expression levels of regulatory RNAs and cell metabolism such as photosynthesis and respiration may reveal the occurrence of hitherto unknown regulatory mechanisms. This information may clarify the mechanisms of gene expression and gene regulation. It may also facilitate rational engineering of the signaling, regulatory, and metabolic networks for desirable cellular functions.

Compared to protein-coding RNAs, ncRNAs are relatively small. ncRNAs are hard to find by classical mutational screens because they are inherently immune to frameshift or nonsense mutations [14]. Therefore, limitations exist for both computation-based and expression-based ncRNA detection methods. Recently, research efforts have been made to carry out systematic ncRNA gene-identification screens along three main lines: cDNA cloning and sequencing tailored to find new small non-mRNAs [15]; specially designed cDNA cloning screens for a new regulatory RNA gene family of miRNAs [3–5]; and comparative genome analysis for general ncRNA gene finding [16–18].

Systematic identification and characterization of ncRNAs in bacterial and eukaryotic genomes has become one of the most exciting challenges in cellular and development biology. There exists a controversial "introns-first" theory [19] that states that from the evolutionary point of view, the ncRNA molecules predate the origin of protein translation and therefore predate the exons surrounding them. It is believed that the contemporary introns housing functional RNAs are ancient relics of the RNA world genome organization, and the newer protein regions surrounding them represent sequences that were originally noncoding and from which protein genes were eventually spawned [20]. On the contrary, search for new ncRNAs has resulted in finding many such ncRNAs with apparently well-adapted and specialized biological roles in the cellular transcription machinery [14].

Now can we use systems biology approaches to strive to understand how information flow in cells is adjusted by particular reRNAs, and how expression, function, and turnover of these reRNAs themselves are controlled. Elucidation of correlations between expression levels of reRNAs and metabolic flux distributions under different environmental perturbations may reveal the occurrence of hitherto unknown regulatory mechanisms. This information may clarify the mechanisms of gene expression and gene regulation. The next issue is whether we can facilitate rational engineering of the signaling, regulatory, and metabolic networks containing ncRNAs for desirable cellular functions.

20.1.2 Dimension Reduction in Systems Biology and Synthetic Biology Applications

Systems biology is inherently a universe in which every "ome"—genome, transcriptome, proteome, metabolome, interactome, phenome, and so on, is another dimension. We have to reduce this dimensionality through integration in order to comprehend, evaluate, and make use of the information. Integrating and evaluating the knowledge bases with their highly disparate nomenclature and frames of reference is arguably the greatest methodological challenge in this new discipline. One example is the concisely described work of Toyoda and Wada, who have developed means of defining the dimensions of several data sets in common terms and projecting the intersections of these sets in two dimensions [21]. Their premise is that the intersections have four defining properties: data set, position, dynamics, and probability that the putative relationship actually exists. The implementation of their "genome-phenome superhighway" (GPS) for human, mouse, the worm *Caenorhabditis elegans*, and the mustard plant *Arabidopsis thaliana* may be found at http://omicspace.riken.jp/gps.

Each "omic" domain has its own unique annotation terminology and attributes, which has led to the development of unique "markup languages" compatible with the Internet HTML, Perl, and other computational data-handling conventions. These include "G-language" for the genomics environment [22], CellML, MathML, and SBML (systems biology markup language) [23], to name a few.

Given the huge amount of data produced in array-based studies, how does one (a) assess its reliability, (b) interpret it in a systematic, unbiased way, and (c) determine the completeness of the data set? A substantial literature has been developed just to address each of these questions. Chen et al. [24] provide a brief introduction and guide to the reliability and analysis aspects. Reliability is affected by the physical quality and composition of the array, stringency of the experimental conditions, background gene expression, and similarities among the probes. Nonlinear responses are inherent in biological systems, so appropriate nonlinear multivariate analysis is essential. Further research is needed to enable efficient database search, development of programming language and data fusion for systems level understanding in biology and the integration of well-characterized biological parts into genetic circuits and metabolic networks for desired end products.

20.1.3 The Quest for the "Minimal Organism" and the Creation of Artificial Life Forms

Systems biologists and synthetic biologists are interested in determining the smallest set of genes, molecules, and structures for replication, growth, metabolism, and regulation that comprises life. Study of such a minimal gene set and its features may shed light on the basics of cellular function, help to determine the subset of essential genes in most species, and improve functionality. Theoretical and experimental efforts have been made using comparative genomics and systems analysis to determine the list of essential genes for a suite of minimal functions that many organisms have in common [25]. The smallest possible group of genes from small genomes is presumed necessary and sufficient for sustaining the functional growth of cells in the presence of a full complement of essential nutrients and in the absence of environmental stress [26].

Methods for making estimates of the "minimal gene set" by experimental biology include saturating transposon mutagenesis (gene knockout) [27] and gene silencing with antisense RNA [28], and so on. These genes can also be computationally identified from the well-studied organisms with small genomes by comparison of essential and nonessential proteins across related genera [29], and using a database of essential genes [30]. For example, *Mycoplasma genitalium* contains the smallest genome of any organism, and has a minimal metabolism. Glass et al. [31] have used global transposon mutagenesis to isolate and characterize the gene disruption mutants for 100 different nonessential protein-coding genes. They have identified 382 essential genes from the 482 *M. genitalium* protein-coding genes. Disruption of some genes accelerated the *M. genitalium* growth. The resulting *M. genitalium* mutants represent a close approximation to the minimal set of genes needed to sustain bacterial life, with little genomic redundancy [32]. Another study, analyzing viable gene knockouts in *Bacillus subtilis*, *M. genitalium*, and *Mycoplasma pneumoniae*, has resulted in a similar estimate [33]. It was found that approximately 80 genes out of the 250 in the original minimal gene set are represented by orthologs in all life forms. For ~15 percent of the genes from the minimal number of genes, viable knockouts were obtained in *M. genitalium* [25]. *Escherichia coli* is also used as a model system for gene knockout to create a reduced "clean genome." Fred Blattner's team [34] has removed about 750 "redundant genes" and planned to delete 500–600 more genes to approach the "core genome" that may be common to all organisms. It was claimed that after the gene removal, the constructs were observed to be more genetically stable and to exhibit increased protein synthesis and electroporation efficiency [34].

The quest for the minimal genome will improve our understanding of the workings of bacterial lives at systems level. On the other hand, the ultimate dream of the synthetic biologists is to create novel life forms that do not exist in nature. For this purpose, minimal organisms may be built up by designing a modular system from "ground zero" that can be given functions. It may involve the design and assembly of genetic circuits and metabolic pathways and even whole chromosomes from chemical components of DNA. Researchers in Synthetic Genomics Inc. (http://www.syntheticgenomics.com/index.htm) have achieved this technical feat by chemically making DNA fragments in their laboratory and developing new methods for the assembly and reproduction of the DNA segments.

The goal is to obtain a synthetic chromosome, and eventually a synthetic cell for the construction of "biofactories" for the energy, chemical, and pharmaceutical industries. The synthetic chromosome created in Synthetic Genomics, Inc. was named *Mycoplasma laboratorium*. It can be transplanted into a living cell where it should "take control" of the cellular metabolism. Although *M. laboratorium* was claimed as a man-made bacterium, there exist some questions about it because the partially synthetic life form was composed of building blocks from already existing organisms. Even when the whole chromosome can be synthesized from chemical components, will we consider the engineered cells to be new life forms, or should we also require the synthesis of ribosomes and other components necessary for the expression of genetic information contained in the genome before accepting the result as an "authentic" new cell?

20.1.4 Systems Biology and the Evolution of Organelles

The properties, genomes, and functions of plastids and mitochondria are an obligatory part of systems biology studies of eukaryotes. Genomic and biochemical studies have established that mitochondria most likely evolved from the rickettsial group of α-proteobacteria [35]. The reRNA sequences in the genomes of aerobic mitochondria are most homologous to those of α-proteobacteria, specifically those of *Rhodospirillum*, *Bradyrhizobium*, and *Rickettsia* [36]. Homologues of 18 different rickettsial proteins are encoded in mitochondrial DNA, and in yeast, the nuclear genome encodes more that 150 mitochondrial proteins with homologues in *Rickettsiales* [37]. The rickettsial pathway for ATP production and that of aerobic mitochondria are virtually identical, and the individual enzymes are orthologs. The properties of rickettsia as an obligate intracellular pathogen, its ability to transport molecules in either direction across its cell walls, and other key factors firmly support the concept that aerobic mitochondria evolved from α-proteobacteria.

Anaerobic environments ranging from sea floor sediments to the gastrointestinal tracts of vertebrates and invertebrates are populated by extremely diverse communities of lower single-cell and multicellular eukaryotic life forms. Some eukaryotes have adapted to anaerobic life by using alternate mitochondrial respiratory pathways, such as reduction of fumarate to succinate, using rhodoquinone instead of ubiquinone as electron carrier [38,39]. These organisms retain their mitochondria and have been called "Type I anaerobic eukaryotes." Additional organisms with "anaerobic mitochondria" include the fungus *Fusarium oxysporum* that uses a nitrate respiration pathway, platyhelminthes that utilize fumarate respiration, and trypanosomes that produce succinic acid while making ATP (summarized by Rotte et al. [40]). Perhaps most unusual is the mitochondrion of the anaerobic ciliate protist *Nictotherus ovalis*, which generates ATP with protons as the terminal electron acceptor, thus producing molecular hydrogen [41,42].

A second group of primitive anaerobic eukaryotes, most notably the parasitic trichomonads such as *Tritrichomonas fetus* in cattle, *Trichomonas vaginalis* in humans, some ciliated protozoa, and the cattle rumen chytrid fungi *Neocallimastix* and *Piromyces* have developed hydrogenosomes—an organelle with intriguing similarities and differences compared to anaerobic, as well as aerobic mitochondria [43]. Hydrogenosomes produce ATP as well as hydrogen. Although mitochondria use pyruvate dehydrogenase, the TCA cycle to regenerate CoA~SH, and molecular oxygen as the terminal electron acceptor, hydrogenosomes have pyruvate-ferredoxin oxidoreductase, no TCA cycle, succinate-acetate CoA transferase and succinyl-CoA synthase to regenerate CoA~SH, and protons as the terminal electron acceptor [40]. Hydrogenosomes and mitochondria use the same "transit peptides" for protein importation. Other proteins common to both organelle types include Hsp 10, Hsp 60, and Hsp 70, the succinyl-CoA synthase subunits α and β, and similar variants of ATP-ADP translocase. Early studies found no DNA in hydrogenosomes. In 1998, Akhmanova et al. [42] described genomic DNA in putative "hydrogenosomes" in the anaerobic ciliate *Nictotherus ovalis*. In retrospect, it appears more correct to describe the organelle in this species as an anaerobic mitochondrion.

The discovery of single-celled eukaryotes that had no mitochondria or hydrogenosomes originally suggested that these organisms were ancestors of eukaryotes that had the organelles (presumably endosymbionts of bacterial or archaean origin). These simple eukaryotes of four types—Metamonads, Microsporidia, Parabasalia, and Archamoebae—were grouped as a subkingdom called Archezoa [44] to distinguish them from Mitozoa, the subkingdom of all eukaryotes that contain mitochondria [45]. Ribosomal RNA sequencing indicated that the Archezoa predated the other known eukaryotes, and Archaezoan ribosomes were 70S, corresponding to those of prokaryotes. However, subsequently it was shown that several Archaezoa contained enzyme-coding mitochondrial DNA sequences [46]. Trichomonas, a Parabasalian, were found to contain hydrogenosomes, Microsporidia undergo meiosis and have tubulin genes that relate these taxa to fungi, and at least one type of Metamonad expresses a chaperonin immunochemically homologous to mitochondrial cpn60. In summary, the more recent and definitive research indicates that Archezoa are among the earliest eukaryotes, but they do not predate the endosymbiosis of mitochondria and hydrogenosomes [44]. That various Archezoa lack mitochondria, hydrogenosomes, peroxisomes, or other organelles as presumptive endosymbionts has been called, "...a secondary reduction caused by their parasitic lifestyle." [45]

In summary, a critical mass of genomic, proteomic, and phylogenetic data has finally accrued to support a comprehensive hypothesis for the origin of eukaryotes, consistent with the known properties of anaerobic and aerobic mitochondria, as well as hydrogenosomes. This hypothesis takes into account the metabolic pathways, electron transport chains, protein importation signals [47], gene loss and transfer to the host cell nucleus [48,49], assembly of Fe–S centers and their incorporation into apoproteins [50], chaperonins, enzymes, and lipids of the endoplasmic reticulum and nuclear envelope membranes.

How, and from what progenitors, did the first single-cell and multicelled eukaryotes develop? How did the nucleus, mitochondria, chloroplasts, and other organelles originate? What primal events led to the formation of chromosomes? Or the mechanisms of cell division? Or meiosis and sexual recombination? At the molecular level, the following questions include: How and when did RNA and DNA first develop? What transformed a world based on RNA as a carrier of genetic information and enzymatic activity into one in which the genetic information resided in DNA, and enzymatic catalysis was endowed in proteins?

A great deal of evidence acquired over the past 30 years supports the theory that mitochondria and hydrogenosomes originated as bacteria that developed an endo-symbiotic relationship with eukaryotes. Various hypotheses have been forwarded to reconcile experimental data with how, and at what stage of evolution, the symbiosis occurred.

20.2 OUTSTANDING GENERAL ISSUES

An appropriate framework for systems and synthetic biology requires the construction of a naturalistic paradigm and philosophy of science for biological research. That project remains incomplete. In what follows a number of component issues in that project are discussed. Although progress toward a naturalistic understanding has been made in each case, there are also future unresolved challenges to the task.

20.2.1 Mechanism and Reduction

While the new high-throughput experimental technologies can profile all the chemical components within a cell as whole, and this has an interest in itself, the ultimate goal is to understand cellular physiology, that is, to understand how these components deliver cellular functioning. To do this it is necessary to study the dynamical interrelations among the components. And because the components must clearly interrelate in multiple ways to deliver function, it will be the complex dynamical system they jointly comprise that must be uncovered. It is this object that is the common core of systems and synthetic biology, and it is its representation as a complex dynamical system that forms the basis of their distinctively new and powerful modeling tools—and raises the issues in Section 20.1 (see Section 20.2.6 for a further issue within this claim)

This raises a basic ontological issue (i.e., one concerning what exists): what is the relationship between physiologically described function and biochemically described dynamical states and processes? The obvious response to make is that the two are one and the same; that, for example, aerobic cellular respiration is nothing but ATP synthesis through glycolysis, Krebs cycling, and electron transport. This is reduction by identification. The physiological function of respiration is identically reduced to, is identical to, and so nothing other than the dynamical system process. (All this assuming that the biochemical systems models involved are empirically supported and predictively and explanatorily adequate; an assumption made throughout this discussion.[1])

There is a large philosophical literature on reduction, some of it proclaiming reduction and much arguing against reduction, especially in biology. Yet, from a scientific point of view it would be anomalous to claim anything less than a reduction, for example, to claim instead just a correlation between the occurrence of functional and biochemical systems properties, because this would leave unexplained duplicate realities, one functional and the other dynamical. Against the advice of Occam's razor, it would leave two realms mirroring each other but running in parallel, for no substantive reason. In what follows the state of philosophical debate is briefly summarized, from a commonsense scientist-friendly point of view, in order to focus on the specific issues at stake for systems and synthetic biology.

20.2.1.1 *General Objections* Perhaps surprisingly, one group of philosophical objections to reduction in general argues that correlation must be accepted because identification is impossible. These arguments largely turn on semantic (meaning) considerations: talk of functioning, for example, of respiring, the argument goes, has a very different meaning from talk of biochemical states and state transitions, so the two can never be identified, even if they are 1:1 correlated. The proper response to this kind of objection is to point out that it relies on *a priori* claims about semantics that are very unlikely in the face of what we know scientifically about language: roughly, that its recent evolutionary emergence, rapid dynamical shifts in vocabulary, syntax, and semantics as historical conditions change, and action-centered intentional basis, all suggest that current semantics are better treated as themselves shifting dynamical emergents, not *a priori* constraints.[2] There would need to be better reasons than these to defeat a general identification of the two subject matters described.

Another group of arguments turns on the fact that the mirroring is often not precise, that often there will be particular phenomenological conditions (e.g., "respiration") that do not nicely reduce to exactly corresponding underlying conditions (e.g., "ATP synthesis") of exactly the same scope. This is true, and not only because of the anaerobic organisms and other energy storage molecules, but also because of the complex dynamics. For instance, even Kepler's laws of planetary motion do not reduce exactly to a theorem of Newtonian mechanics because planet–planet interactions produce small deviations from Kepler's generalizations. This will be a common

---

[1] The issue of when and why that assumption is reasonable is just the general issue of the nature of scientific method at large. It turns out that scientific method is much more complex (and interesting!) than the neat logical models to which the philosophers had hoped to reduce it, and must itself be understood in dynamical systems terms, but this is another story—see Ref. [51] and, for example, Ref. [52].

[2] For those interested in the technicalities, a same-dynamical-role criterion of property identity is a useful small first step toward a more plausible alternative semantics and this already suffices to license identification of functions with dynamical processes, should other substantive requirements be met. This is argued in Part II of Ref. [53]. For a bioorganizational approach to the underlying intentionality, see Ref. [54].

situation wherever a more complex dynamics underlies more phenomenological observations. In such cases, surely, so long as the departures from strict correspondence can also be explained by the underlying (reducing) dynamics, the reduction can be considered successful. Call the last the explanatory principle.

This works well for cases where the departures are small. However, there are also large departures, such as in the relationship of phlogiston chemistry to oxygen chemistry, where we deny that phlogiston exists even if its postulation served to codify a number of chemical relationships that survive the replacement. And there are intermediate cases, for example, the imperfections of the thermodynamics–statistical mechanics relation. How are these to be treated? To decide we need to remind ourselves that for science reduction is not only about satisfying metaphysical curiosity, from a methodological point of view, but it is also primarily about extending explanation and evaluating the potential errors involved in using the phenomenological model to explain, in place of the underlying one. (Hence the explanatory principle above.) From this perspective, reduction is ultimately about the capacity to systematically replace one kind of description (the more phenomenological one) with another kind (the more basic, theoretical one) that is equally or more precise and equally or more predictively and explanatorily powerful. This satisfies the key cognitive aims of science. Reduction by identification then forms one extreme of a spectrum, where component ontology as well as relational structure is conserved under the replacement. The other extreme is occupied by cases like phlogiston where significant relational structure, but not ontology, is conserved under replacement.[3] Mismatch along the spectrum means that some nonconservation is melded with identificatory reduction. However, the key point remains that when the explanatory requirement holds, overall reduction is obtained.

20.2.1.2 Geneticism
These general issues aside, an important part of the philosophical objection to specifically biological reduction has really been to geneticism, to the idea that organisms could be reduced to just a collection of genes and gene-determined traits. Modern biology agrees with this objection, DNA is one biochemical component among many—if with a distinguishable role—and it is the dynamical system of all of them that sustains function. But, conversely, the whole biochemical system now becomes the reduction candidate for physiology, so the objection to geneticism does not defeat reduction, but just shifts its focus. Setting aside that literature as well, there remains only those objections that are specific to reduction of functions to systems dynamics.

20.2.1.3 Reduction of Function to Dynamics
Some objections to this have to do with the fact that our commonsense day-to-day function talk is rather

---

[3] Beyond that, sheer discontinuous replacement would occur, but it is hard to think of a substantial case in science. For the replacement view, see Part I of Ref. [53] and, more informally and accessibly, Ref. [55]. P. M. Churchland's elegant overall strategy, more subtle but powerful than it may appear, is itself explained in Ref. [56].

imprecise for marrying up to dynamical systems specifications, while others stem from the related problem that vague function descriptions can seem to cut across what turn out to be the dynamical process distinctions. These can all be resolved through a little careful analysis of language.[4] This is useful to know in a field like biology where talk of functions is ubiquitous, but often more pragmatic than precise, especially considering that only features that have functional consequences are likely to be modeled.

Setting general objections to reduction from function talk aside as well, brings us at last to the substantive conditions for function to system process reduction. First, we specify a function as a map from inputs to outputs. For example, cellular respiration, crudely globally specified, is the function that takes food and water molecules as inputs and outputs carbon dioxide. Note that more specific functional maps capturing the process detail can clearly be constructed as required. Corresponding to this in the molecular description is a dynamical process—that is, a metabolic map carried by (biochemical) dynamical laws—that takes oxygen and glucose as inputs and yields ATP (and perhaps other energy storage) and carbon dioxide as outputs. Then the obvious requirement for identificational reduction is that the respiration functional map be embeddable into the corresponding biochemical process map without distortion (homomorphically embeddable). A further coherence condition is equally obvious: the collection of all such embedded dynamical maps, together with any nonfunctional data concerning the system, should provide a single coherently unified biochemical model of the cell genome that preserves or increases predictive and explanatory power.[5] The embedding criterion essentially captures recent conceptions of a function to mechanism reduction, reducing both the cell and multicellular organisms to complexes of mechanisms.[6]

There is an inherent underdetermination by any function, taken in isolation, of its correct embedding. Although this has sometimes been taken as a fundamental objection to reduction, it ultimately reduces to a pragmatic issue of sufficient data. The problem is nicely illustrated in the case of the output of a network of electrical generators having a frequency variation less than that of any one generator; some kind

---

[4] See Part III of Ref. [53] and, briefly, Ref. [57], Part V, case I and case II end.

[5] See Part III of Ref. [53] and, briefly, Refs [56] and [57]. The basic reduction requirement, that functional maps are mirrored by dynamical maps, is in fact just the application of Nagel's [58] deductive reduction conception, rightly understood. Nagel shows how scientists arrive at reduction of a law L_2 or property P_2 of theory T_2 respectively to a law L_1 or property P_1 of theory T_1 by first showing how to choose conditions (real or idealized) under which it is possible to construct in T_1 a law L_1 or property P_1 that will mirror (be a relevantly isomorphic *dynamical* image of) the dynamical behavior of L_2 or P_2. From that the reduction is shown to be possible through the identification of L_2 or P_2 with the mirroring L_1 or P_1. Indeed, the requisite "bridging" conditions can be deduced from the mirroring condition, and then asserted as identities on the basis that doing so will achieve a reduction, supported in that light by claims of spatiotemporal coincidence or appeal to Occam's razor.

[6] See especially Refs [59, 60]. However, the conception of mechanism here does not yet adequately reflect the importance of process *organization* to cellular function [61], an outstanding issue for future development. Further on organization, see Sections 20.2.3 and 20.2.4.

of feedback governing process is at work, but is it a real governor or simply the functional appearance of one at the network level? The latter is possible because connecting the electrical generators in parallel automatically creates a phase-stabilizing mutual interaction among them without the need for a real governor.[7] This question is resolved by gathering other data about the network–this is the point of the unification criterion above.

Nonetheless serious issues remain with the overall position. Rosen, for example, argued that organisms could not be complexes of mechanisms in any compositional sense and that they were indeed not mechanisms.[8] Disentangling the aspects involved, there remain these systems issues that must be resolved: (1) self-organization and emergence, (2) the nature of the complexity in "complex of mechanisms," and (3) the specific implications of self-regeneration for (1) and (2). Of these (1) and (3) will pose specific challenges for reduction. Conversely, however, a thoroughly dynamical systems approach will allow us to understand the subtle intertwining of reduction and its failure in emergence within a unifying framework, providing a full, naturalist account of reduction and emergence in systems and synthetic biology. These three issues are now discussed separately and in order.

20.2.2 Self-Organization and Emergence

In all systems it is true that the interacting components together create a dynamics that would not otherwise be present. When the outcome is surprising or unexpected or too complex to be readily understood, scientists are apt to talk about self-organized emergent patterns. There are many reasons why leaving things like that is unsatisfactory, among them that (i) no significant feature is addressed, our subjective surprise, and so on, keeps shifting and has no substantive association with reality, and (ii) this criterion is dynamically so weak as to trivialize these ideas. But when it comes to strengthening the requirement, there is currently huge diversity of opinion about both the concepts, self-organization and emergence. Two broad approaches to identifying something more penetrating can be distinguished, one epistemic and the other causal or dynamical.

The epistemic approach tightens up the subjectivity by adding a clause along the lines that self-organization occurs when the resulting system dynamics could not have been predicted from the known interaction rules of the components. Since the dynamics is entirely internal to the system, it is properly referred to a self-organized.

This approach is attractive because there are many complex behavioral patterns that arise from the simplest interaction rules, for example, with social insects (hives of bees and termite mounds), city traffic, and even simple population dynamics as reflected in the logistic equation. However, it still ties the definition of evidently physical

---

[7] For this example see Ref. [62] and further Part III of Ref. [53].

[8] See Refs [63, 64]. Rosen's objections have to do with the role of global organizational constraints on organisms and are discussed under Section 20.2.4 below.

properties to a cognitive test, and anyway proves difficult to formulate satisfactorily.[9] So we pass to the option of a causal/dynamical criterion.

One causal/dynamical distinction stands out, and fixing on this avoids a long detour through a tortuous literature. The distinguished difference is between patterns that dynamically constrain their components—that show "top-down" dynamical constraints—and those that do not. Consider the formation of an iron bar from cooling molten iron. In this phase transition a macroscopic pattern of intermolecular relations is formed, the iron crystal, which does thereafter have the power to constrain the movements of its molecular components through the formation of a new macroscale force constituted in the ionic lattice bonds formed. Its formation alters not only individual component behavior but also the specific dynamics under which they are now able to move: there are lattice vibrations and a Fermi conduction band in place of liquid molecular dynamics, that is, the phase change alters the force form of the dynamical equations that govern component behavior. The new macroscale force is able to retain the constraint relationship invariant under component fluctuations and exogenous perturbations, through lattice dissipation of these perturbing energies as sound and/or heat.[10]

By contrast, from intersecting shallow waves on a gently undulating beach there emerges the most beautiful and intricate patterns, but there is no comparable constraint formed by their interaction; shift the underlying sand structure and the dynamics can shift to entirely other patterns. Similarly, there is no dynamical constraint internal to social insect societies comparable to the ferric crystal force and compelling their insect members to satisfy hive and mound laws, or compelling city drivers to create traffic jams, and so on. All of these patterns are produced by dynamical interactions of components and thus reflect their "bottom-up" dynamical constraints, but only some also express top-down dynamical constraint.

It is natural to choose the formation of a new top-down constraint as a criterion of emergence for just this characterizes the coming into being of a new dynamical existence. The iron top-down constraint formation constitutes the coming into being of

---

[9] As it stands, the text formulation is intolerably vague: Predicted by whom? Knowing what? Using what tools? And it makes an apparently ontological distinction (the existence of emergent behavior) depend on a cognitive condition (human predictive capacity). If, in response, the criterion is instead formulated along the lines of "cannot be derived from the set of interaction rules," then these problems are lessened, but only to be replaced by the problem of what counts as an acceptable derivation. If derivation includes computational modeling of collective dynamics then almost all dynamics counts as derivable and nothing self-organizes. (Perhaps noncomputable dynamics might be considered an exception, but since this occurs in quantum theory and other "wave" dynamics, it seems a peculiar boundary.) If instead derivation is restricted to logical deduction then almost everything self-organizes since the demand for analytic closed-form solutions fails for almost all sets of differential equations. No satisfactory criterion of in-between scope is readily formulable.

[10] The iron bar is a new macroscale level with respect to its molecular constituents because it has its own characteristic dynamical interaction form. All other talk of levels either concerns measurement (liquid level), gravitation (level surface), or is metaphorical (semantic, social, abstraction, theory . . . levels) and can thus be paraphrased away—or is confused. Note that the presence of a top-down constraint does not fully determine the specific dynamical form of the system; both the virtual and real electrical governor arrangements (see footnote 7 and text) exhibit the same phase-stabilizing top-down constraint. Distinguishing between them is the electrical engineering "system identification" problem.

a new, individuated capacity to do work, expressed both endogenously in dissipation of perturbations and exogenously in rigid body action. It is the arrival of a new dynamical individual characterized by a new dynamical form.[11] The character of the new individual is constituted by its capacity to do new work. To broaden the criterion further would be to conflate genuine interactive emergence with the mere emerging in time of a pattern (as "from concealment").

Real emergent dynamical filtering insures that macroscopic properties have the stability we find them to have, making the macroscopic world as viably simple to survive in as it is for macroscopic creatures like us. But it also applies to smaller-scale structures; cellular metabolic regenerative organization and the cellular structures it sustains, for example, are emergent top-down constraints and cellular function would not be stable without them.[12] But by providing higher level structure for lower level processes, all these constraints actually underpin the reduction of the functions served to dynamical processes (and of course the constraints themselves and attendant structures to dynamical compounds of the components whose interactions constitute them).[13] Emergence heralds the presence of an irreducibly new dynamical existent; reduction to the components alone fails. Yet, contrary to the standard view of reduction and emergence as opposed, this discussion shows that emergence and reduction are intricately interwoven and mutually supportive.

There is no physical mystery about this when a dynamical model of emergence is to hand since it is precisely what the filtering consequent upon formation of a new dynamical constraint provides. In this way, we naturalize emergence for science. And it is precisely on that general basis, and only on that basis, that we can track causal paths "up" and "down" through the component/supracomponent levels and thus, come to

---

[11] In a more traditional philosophical language, the iron bar is supervenient on its molecules; nothing about the bar can change without the change being dynamically grounded in appropriate molecular changes. But dynamical analysis provides a much richer language in which to discuss the possibilities. First, it specifies top-down behavioral constraint formation in terms of change in dynamical form, the change in form describing the causal power this novel constraint possesses. (This also distinguishes such effects as nonepiphenomenal.) Second, the dynamics itself shows how the constraint, a (relatively) macrolevel state/property, is determined by the states/properties of its microconstituents and so is supervenient on them, yet can nonetheless also constitute a constraint on them. Here dynamics gives the constraint a subtle status that eludes conventional formal analysis, combining what common philosophical assumption opposes. (See Refs [57] and [65].) Thus, dynamical determination, there being only one dynamical possibility for the collective dynamical state/property, cannot be equated with logical determination—the collective dynamical state/property is logically derivable from but can be expressed as a logical sum of its constituent states/properties. The former is specified as the constituents fixing all space–time trajectories so as to allow only one macropossibility, but these trajectories may be computationally strongly inaccessible, for example, through all critical point phase transitions. The neuroscientist Roger Sperry was among the early adopters of a top-down constraint model of mind emergence, see in later summary [66].

[12] For a systems biology illustration and discussion see Refs [67] and [68].

[13] Metaphysical aside. If there are unique, unchanging, spatiotemporally local, fundamental dynamical entities (e.g., chemical ions as biochemical atoms) then there is no fundamental emergence, only existential emergents having these entities as ultimate components in various dynamical compounds. But top-down constraint formation of itself does not require this. Fundamental nonlinear fields would yield the same emergent result and there are no such local components, while mutant spatiotemporally local fundamental components would issue in fundamental kind emergence.

understand cellular and multicellular organization. However, there remains a challenge for science, though not specifically for biological science, to find a full analytic mathematical treatment of the top-down formation process that permits a more rigorous and general discussion of when, where, and how it occurs, in biological systems in particular. The "how" is the difficult part.

While this seems the proper way to deal with emergence, it might be allowed that self-organization should be more broadly defined to capture simply the central idea that the resulting pattern is brought about through the interactions of the system components. The colloquial term "organize," as in "get organized," encourages this wide connotation. This position is permissible; all that then matters is that the definition of the term is clear, as Alice's Humpty Dumpty allowed. Under the wider usage self-organization is coextensive with organization (widely interpreted) but neither coincides with emergence, while under the narrower constraint-formation usage, self-organization coincides with emergence, but neither coincides with organization. As noted at the outset, there is no worthwhile definition to be had that sits between these two options. In my view, it leads to clearer, stronger, more scientifically useful conceptions of organization, self-organization, and emergence to adopt the latter usage. For instance, the formation of a crystal is a clear case of emergence, but not of any significant organization (see subsequently), yet it is a paradigm self-organizing process in the sense of top-down constraint formation.[14]

An immediate consequence worth noting is that self-*organization* need have little to do with organization proper. This is as it should be. Organization is a relational condition of systems where components play distinct roles but the roles are so interrelated as to produce a coherent global outcome. A simple illustration is found in the way the parts of a car engine are interrelated so as to deliver torque from fuel ignition; a profound example lies in intracellular organization. Self-organization is simply constraint formation and, as the case of crystallization shows, need not involve the emergence of any organization. Crystal formation is, rather, an instance of von Feurster's correctly named principle of order-(not organization)-from-noise (i.e., from random reassortment). von Feurster's own example of shaking coins down through successively smaller size filters orders them by size but does not organize them in any interesting sense. The unfortunately wide colloquial connotation of "organize" conflates order and organization, which are important to distinguish in understanding what is distinctive of living systems (see subsequently).

20.2.3   Organization and Complexity

Complex systems are complex, not only because they have many components, but fundamentally also because they are organized.[15] This raises two complementary

---

[14] For this position see Ref. [69].

[15] All of the further properties they may show—see 19.4.1—are forms of organization.

issues, the nature of organization as a form of complexity and the nature of the constraints that ensures overall functionality. Here we focus on the former issue and address the latter issue in the next section.

We begin with order, the basic relational notion of which organization is a special form. The root notion of complex order is that derived from algorithmic complexity theory: the complexity of the order in a pattern is measured by the length of its shortest, most compressed, complete description. A crystal lattice is simply ordered: it has a short compressed description given by fixing the locations of all ions as multiples of crystal plane distances away from any one reference ion. A gas, by contrast, has a very long minimal description and hence is maximally complexly ordered because its component molecules are all moving at random, so the position of each has to be separately specified. The crystal is highly ordered and the gas highly disordered. By contrast, an organized system is one where a number of distinct kinds of components playing unique roles nonetheless interrelate so that together they support one or more overall, global functions; a car engine and an organism are paradigm cases. The extremes of order are equally inhospitable to organization; a highly ordered system is too uniform, and a highly disordered system is too random, to support the variety of specific interrelationships required for organization. The relation of the piston rod movement in a car engine to that of the behavior of the fuel injector is very different from its relation to the exhaust muffler temperature, yet all combine to produce harmonious functioning. The variety in the relationships explains why too simple or high orderedness restricts organization, while the occurrence of the systematic interrelationships among the components explains why too complex or low orderedness equally restricts organization. Organization occurs in an intermediate "window" of ordered complexity between extremes. Thus, complex organization, as in living cells, is not straightforwardly complex in the sense of algorithmic complexity, but in some other sense.

Subtle, multiple different coordinations—that is, correlations—are required for complex organization if its very different component roles are to jointly serve a function. That is, it involves nested, higher order correlations of correlations. Very complexly organized systems, like cells, multicellular organisms, and cities are characterized by many layers or orders of correlations of correlations. Let us mean by a system's organizational depth roughly the number of nestings of subordering relations within it (cf. cells within organs within bodies within communities). Then the complexity of an organization is better measured by its organizational depth than it is by algorithmic complexity.

But it is still not a very satisfactory measure, primarily because it does not take into account the appearance of top-down constraints within nested systems, that is, it misses regulatory hierarchy and modularity. The class of all merely nested systems includes, but is much wider than, that of the organized systems, since organized systems must also sustain a global function. To achieve a global function an *organized* system exhibits a highest order global correlation expressing a global constraint (to performing its functions), with nested sets of lower order correlations within that, some of them modularized by lower order top-down constraints. We have as yet no way to properly take all these features fully into account, and hence no satisfactory definition of organizational complexity.[16]

More importantly for science, we have as yet neither a real capacity to represent organization mathematically, nor a real capacity to investigate it experimentally. The mathematical framework for dynamical modeling in most of science, including systems and synthetic biology, is that of differential equations (d.e.s) as vector fields on differential manifolds, for example, on system phase space. But these modeling resources, powerful though they are for modeling the energetics of processes, do not explicitly describe the physical organization of the system—a metabolic cycle and a pendulum, for instance, may be modeled as equivalent dynamical oscillators. In a phase space only the global dynamical states and their time evolution are specified, not the organized processes that produce the dynamics; hence, it cannot capture organization. There is at present no obvious resolution to the general theoretical problem of how to incorporate organizational principles into dynamical models in a principled way. Correspondingly, the parameters we can measure are either component features—biochemical concentrations and the like—or higher order regulation parameters, such as respiration rate. There are no experimental techniques for detecting organization directly. Rather, it is reconstructed in retrospect after system relationships, in the genome for instance, have been reconstructed from what we can measure. Thus, a future challenge to systems and synthetic biology is to become more understanding of dynamical organization, both theoretically and experimentally.

20.2.4   Autonomy and Living Organization

The most basic global biological function is the regeneration of the body through metabolism, utilizing intakes of air, water, and food; for without this nothing else is possible. It is clearly a global function because it concerns the regeneration of the whole body. Autonomy, a form of recursive self-maintenance, is the name given to the global organizational constraint that must be met in order to support metabolic function. It is worth explicitly identifying autonomy because of its useful roles. For instance, it uniquely picks out the living systems from within the wider domain of complex, organized, nonlinear, dissipative (entropy increasing) and irreversible, chemical and biological systems, providing an unbiased, operational criterion of life hitherto missing and especially needed in exobiology. It also suffices to provide a naturalistic grounding for agency (see subsequently) and fruitfully frames the evolution of intelligence (see subsequently), thus also providing a framework for (organically) intelligent robotics. Let us explore the idea.

Finite systems sustaining dynamical equilibria far-from-(static)-equilibrium must do so by irreversibly taking in ordered or low entropy energy and material components from their environment and exporting it to material components carrying dissipated,

---

[16] Gell-Mann [70] discusses effective complexity and logical depth (see Ref. [71]) as other possibilities, but neither is satisfactory for various reasons he notices, but fundamentally because these too are general dynamical conceptions and do not directly include top-down constraints (for some further discussion see Ref. [69]).

less ordered, or higher entropy energy. These open systems must be organized: by the Morowitz theorem they must have at least one, and typically have many, closed-loop processes running within them.

For instance, a candle flame creates a thermodynamic asymmetry between itself and its environment, including an organizational asymmetry as it both preheats its own fuel supply (oil or wax) and creates a convection air current that delivers fresh oxygen to the flame. By supporting these two cyclical processes, the candle flame process contributes to the maintenance of the process temperature; in those partial respects, it is self-maintained (including of its self-maintenance capacity). But it has no self-regulatory capacity: should the flame die down, it does not cause more oxygen and wax vapor to flow in to revive it or cause a search to bring about delivery of other means to revive it, in contrast to hungry animals actively searching for food to revive themselves. The locus of regulation of these latter processes, if any, lies outside the flame process.

Living beings from single cells "up" are also among these open, irreversible, partially self-maintenance systems that maintain a state asymmetry with their environment. But unlike the candle they display a self-regulatory capacity that is extensive and active. Internally, as self-regenerating systems their cyclic processes must contribute to re-creating each other, that is, each process must partially regenerate the material constraints for themselves and/or others to work, requiring a highly organized web of cyclic process-constraint interdependencies.[17] Hence there must be strong mutual internal regulation of activity if internal coherence is to be maintained. Externally, organisms actively search for, and intake, requisite ordered energy and materials and excrete wastes, all the while avoiding or ameliorating damage. This requires active regulation of behavior. Even single cells regenerate themselves metabolically and partially regulate their environmental experience. Multicellular animals perform the same overall tasks, only with an expanded range of self-regulatory capacities, for both internal interaction (e.g., the cardiovascular resource delivery and waste removal system) and external interaction (e.g., neurally regulated sensory and neuromuscular motor systems, and so on) to match their expanded regenerative requirements.[18]

There are two broad cyclic processes involved in this activity, internal metabolic interaction and external environmental interaction, and these need to be coordinated: the environmental interaction cycle needs to deliver energy and material components to the organism in a usable form and at times and locations the metabolism requires to complete its regeneration cycles. The presence of these two thus synchronized cyclic processes resulting in system regeneration is the broadest functional sense of what is meant by a system's being autonomous. Though the detail, especially the dynamical boundaries and self-regulatory capacity, vary, this autonomy requirement picks out all and only living individuals—from cells, to multicellular organisms to various multi-organism communities, including many business firms, cities, and nations. In all

---

[17] These are what Kaufman [72] calls work-constraint cycles.

[18] They are models of self-regulation, including active self-maintenance of their self-maintenance capacities. Hence they are recursively self-maintained—see Ref. [73].

autonomous systems, the locus of living process regulation lies more wholly within them than in the environment—hence, the root sense of autonomy in the traditional sense.[19] Birds organize twigs to make nests, but twigs themselves have no tendency to organize nests or birds.

Autonomy is a subtle global constraint on the organization of interaction for whole organisms in their environmental context. In contrast to gases and crystals, dividing a cell in two typically does not produce two new cells because the fundamental global process organization that produces cell-type cohesion has been disrupted. Clearly, autonomy is an emergent property of the cell as a whole. In fact, emergence is a ubiquitous feature of the far-from-equilibrium systems. Comparing living systems to inanimate systems highlights the distinctive character of living interactive organization:

Comparative System Order			
Property	System Kind		
	Gas	Crystal	Cell
Internal bonds	None	Rigid, passive	Adaptive, active
Directive ordering[a]	Very weak, simple	Very strong, simple	Moderate, very complex
Constraints	None	Local	Global
Organization	None	None	Very high

[a]Directive ordering is spatiotemporally selective energy flow.

Entities are properly treated as genuine agents when they have a distinctive wholeness, individuality, and perspective on the world and their activities are self-regulated, normatively self-evaluated, willful, anticipative, and adaptive. Autonomous systems are inherently all of those things:

- *Self-Regulation.* We have already seen that autonomous systems are strongly self-regulated in both their internal and external interaction, making them the distinctive primary locus of their regulation. And because the self-regulation is in service of maintaining an internally coherent whole, they have a distinct, individual reference point for their activity that provides them a distinctive perspective on the world.

- *Normative Self-Evaluation.* Autonomous self-regeneration constitutes the fundamental basis for normative evaluation because it is the sine qua non and reference point for all else. Autonomy is the condition against which the outcomes of system processes are measured for success or failure. In single cells the measurement is simply continued existence or not. Multicellular systems have developed many internal, partial, and indirect surrogate indicators

[19] On autonomy see further Refs [51,54,64,75] and references therein. Self-governance lies at the core of our commonsense conception of autonomy. However, we are most familiar with the idea of autonomy as applied to persons and political governance, but these are sophisticated notions applied to sophisticated systems whose trappings may distract from fundamentals. We need to return to basic principles operating in all living systems to construct a naturalist notion that will "grade up" across the evolutionary sequence to our sophisticated concept.

for autonomy satisfaction and its impending violation, often based around closure conditions for their important subprocesses, for example, hunger (impending violation) and food satiation (satisfaction). It is these specific surrogate signals (cf. also thirst/fluid satiation, pain/pain-freeness) we think of as the basic, primitive norms guiding behavior, but they are literally grounded in turn in the obtaining of autonomy, from which they derive their normative character.

- *Willfulness.* A will is the capacity to do work (i.e., transform energy) in relation to the self whose will it is. The constitution of the autonomy constraint, which focuses directive organization on the generation of behavior to achieve self-regeneration, constitutes just such a distinctive capacity.

- *Anticipation.* To anticipate is to act now in relation to some future state, event, or process. Anticipation is thus an integral feature of autonomous systems because of their need to interact with their environment in ways achieving future closure outcomes that contribute to maintaining autonomy. The interactive relationship between the present action performed and the future, autonomy-evaluated outcome required is the most basic form of anticipation.[20] The willful performance of anticipative interactive activity against a normative evaluation criterion provides a root sense of action.

- *Adaptedness, Adaptiveness.* An organism is adapted when it possesses an autonomy-satisfying set of traits in its life environment. Conversely, an organism's ecological niche comprises the range of life environments for which its traits provide satisfaction of autonomy. An organism's adaptiveness is its capacity to alter its specific traits in mutually coordinated ways so as to adapt to, that is, satisfy autonomy in, a wider range of life environments than its current one.

20.2.4.1 *Intelligence and Intentionality*

Agency of this kind provides an organizational platform for characterizing, and understanding the evolution of, intelligence and intentionality. There are three major aspects determining a system's anticipative capacities: the width of its interactive time window, the degree of articulation of the autonomy-related norms that it can use, and the high-order interactive relationships that it can effectively regulate. Between them, these features characterize the dimensions of intelligent/intentional capacity, and their roughly joint

---

[20] The root notion of anticipative action for Rosen [63] is that of a sequence of subactions that together achieve a closure condition and for which each subaction exists only because it is a member of that closure-achieving sequence. Each element then anticipates the next and the sequence anticipates the closure outcome. While this is too broad to provide any distinctively agency sense of anticipativeness, since any cyclically regenerating system (e.g., an autocatalytic polymer) counts as acting anticipatively, it does capture the central functional character of anticipation. Elementary systems like single cells will only exhibit action sequences where what anchors the repeated activation of the elements is just their belonging to a closure-achieving sequence. A distinctive agency sense of anticipativeness emerges when Rosen's root condition is applied to autonomous systems, since only these define a principled sense of it being the system itself that is anticipatory.

evolution traces the emergence of mind. And because of their preceding properties, autonomous systems can also be provided with action-centered informational and semantic characterizations, to complete the sense of agency. Organism information is modeled as reduction in downstream process regulation uncertainty. ("Shall I do A or B? Given the result of my last interaction, B is the thing to do.") Organism semantics is that of the anticipated norm-referenced, autonomy-satisfaction provided by an action. These conceptions of information and semantics grade back to the actions of single cells, though the stronger the self-directed anticipative organization involved, the richer the semantic and informational structures sustained. In this context intentionality is conceived as a high-order regulatory capacity for fluid, meaningful goal-directed management of interaction. Intelligence and intentionality coevolve making use of a common self-regulatory apparatus. This avoids the common but implausible split between the two, respectively into problem solving and referential capacities.[21]

In sum, autonomy promises to provide the broad organizational framework from within which a fully naturalized conception of organisms can be developed in terms of the naturalistic intertwined emergences and mechanistic reductions that reveal their biochemical organizational depth. Of course, from a scientific point of view, the devil lies in providing the details. And the challenges in doing so are not only to do with coping with complications, but they also run deeper.

20.2.4.2 Challenges Posed by Autonomy

Science, as discussed in Section 20.2.3, has only weak tools for studying organization. It has equally weak tools for studying global constraints, especially spatiotemporally extended global constraints like autonomy. These are at present not representable in the differential equation/phase space formalism. Although autonomy, like any dynamical constraint, must in principle be representable as a limitation on system accessibility to dynamical states (viz., constraint to those satisfying autonomy), there is at present no modeling methodology for constructing its constraint representation. So, while it is always possible to capture the dynamical consequences of internal organization by modeling system plus environment as a system of coupled component subsystems, there is no principled, internally motivated basis for reversing the process to extract organization from the dynamics, that is, for individuating the system in a principled way.[22] This is as much a challenge for theoretical robotics as for theoretical biology.

Dually, the challenge posed to practical construction and regulation/control in biology and robotics is equally deep because, if the account of autonomy (and of autonomy-based cognition) is even roughly correct, it provides a set of organizational requirements for this task that will prove far from simple to meet. For instance, despite using the label "autonomous agent," there are at present no truly autonomous robots in

---

[21] This interaction-centered semantics is very different from, and more powerful than, standard direct referential semantics, for it captures directly the unlimited implicit possibility content in our action-differentiated grasp on reality. Bickhard argues that in this way it resolves the frame problem and is anyway ultimately the only coherent naturalist semantics, see for example Ref. [76]. Further see Refs [54] and [74] and Section 20.2.4.

[22] One thinks instinctively of the coupling of equations as the requisite tool, but so far as I am aware there is as yet no well-defined way to characterize either organization or globalness of constraints in these terms.

this organizational sense. Robotics uses a very limited formal notion of autonomy (something like invariant dynamical form) and limited performance criteria (typically confined to a single task) and an equally limited satisfaction method. This is as yet very far from even incorporating normative signals into the body coherence of robots, let alone the complexity required for self-regeneration and the capacity for fluid management of multidimensional environmental and internal interaction processes in relation to that (cf. Ref. [4], footnote 17). Similar constraints currently apply to our capacity to understand, much less synthesize, real biological systems. Despite calls for the simulation of biological autopoietic cells, we remain far from being capable of doing so.

Robert Rosen argued that living systems were not mechanical, that they could not be reduced to congeries of mechanisms (see Section 20.2.1), not simply because reduction in general failed, but for deeper structural reasons. Yet reduction to mechanisms is evidently what systems and synthetic biology aim to do. The gist of Rosen's objections (their final 1991 version [14] is couched in an arcane modeling language) is that holistic, organizational features like autonomy are central to being alive and these cannot be captured by analysis into mechanisms—indeed, our present general modeling tools must necessarily fail to adequately capture such features. He argued that these limitations, largely unrecognized and unexamined, represented a powerful limitation on the development of biological science. Cloning is hailed still, even while the profession knows that, though a technical feat, it is limited to intercellular nuclear transfer, and the entire cytoplasmic apparatus of the globally coherent regenerative cell is simply ignored.

There is some point to Rosen's line of objection. Metabolic regeneration is central, does exhibit autonomous organization, and currently cannot be adequately modeled dynamically. The emergence of high-order global functional coherence expressed in adaptive intelligence offers another version of this challenge. Rosen argues that this difficulty is made more pointed by the fact that often the components in metabolism are only thus because of the character of the whole (cf. Rosen on anticipation, footnote 20). This seems to make it impossible to understand such systems without postulating the global dynamical organization at the outset, stymieing attempts to synthesize the organization from its components. If, in addition, the components are formed during the self-organization of the whole process, then the argument is reinforced.

Especially these last cases are real challenges to substantive biological theory. However, the scientists involved might argue that new tools to understand them are being developed, albeit slowly, and this shows that they should be recognized as so many methodological challenges rather than overwhelming *a priori* demonstrations of the separation of biology from natural science. Hogeweg, for example, has pioneered the use of computational models to understand the ways in which spatial segregation processes can lead to the survival of entities, whether molecules, viruses, organisms, or even prebiotic entities, where an unsegregated model would predict their extinction, and also illuminate multilevel selection and evolution processes. Hogeweg employs cellular automata (CA) models to capture the spatial organization necessary to explain the outcome. This is possible because CAs are inherently and explicitly relationally organized (though in a generalized, not narrowly spatial, way), even while able to incorporate some aspects of local dynamical interaction. Dynamic networks are similarly inherently and explicitly relationally organized, while currently they are largely used to express functional relationships, they too can be adapted to express spatial relationships. In ecology, for example, there is increasing attention to modeling spatial organization and these kinds of modeling tools can be used to model intracellular spatial relationships.[23] Even so, there is no inherent capacity in any of these tools to represent either organization per se or globalness of constraints. But it may be that in future, as need and capacity to model spatial organization grows, more and powerful such tools will alleviate these problems.

20.2.5 Condition-Dependent Laws and the Unity of Science

Scientists in systems and synthetic biology often regard their approach more as "model building" than as "theory" or "law" centered; this is understandable in a domain where nearly every variation results in differing functional capacities and behavioral patterns. Compared to the grand universal, invariant laws of physics, these local idiosyncratic behavioral patterns do not count as laws; so, especially when most biological systems are yet too complex to predict, it is more useful to simply model each system and try to understand it on that basis. But of course biologists do use laws in constructing their models, the laws of (bio-)chemistry; if these did not operate the same-everywhere biology would be much harder than it already is. Even so, the complication arises from the fact that the operational invariance largely occurs at the ion–ion interaction level. How n-body, k-component ion systems operate is often a strong and sensitive function of the initial and boundary conditions, especially organizational conditions, obtaining and that is why no simple set of laws can be deduced in advance. Indeed, self-organization precisely occurs because of the sensitivity of dynamical form to dynamical initial and boundary conditions (see Section 20.2.2).

But the last equally provides license to extend the notion of law to such cases. For since self-organization involves a new dynamical form, it is reasonable to say that it obeys new dynamical laws characteristic of that form. Moreover, the idea that true laws have to be specified independently of any initial and boundary conditions is a conceit of physics, and perhaps ultimately not true there either considering that even fundamental laws evidently changed form as the big bang cosmos cooled. But once that independence requirement is dropped we are free to see biology as replete with real behavioral laws, it is just that they will be condition-dependent, or "special" (as some philosophers say).[24] For instance, condition—a cooling mould of liquid iron in contact with a heat reservoir of lower temperature, emergent laws—rigid body (not fluid) dynamics, crystalline (not fluid) conduction of electricity, heat, and

---

[23] See, for example, Hogeweg [77–79]. On spatial modeling in ecology see, for example, Refs [80] and [81] and references therein.

[24] See Ref. [82] for an early insight of this kind. We now see that this condition is not unique to life, for instance, it characterizes at least all dynamics that shows self-organization.

sound. Put that way, condition-dependent laws are commonplace even in physics, and certainly throughout all the other sciences. It is just that condition-dependent laws are often on that account hard to predict or use for prediction, but that is a different, epistemic issue.

Why not push condition-dependence further to include every instance of change in initial and/or boundary conditions? For instance, the specific force of gravity changes between the Sun–Saturn and Sun–Earth subsystems because of the changing masses involved. Why not claim all these as equally condition-dependent laws? Well, because it is the same general law that is involved; the diverse cases are unified by a single lawful interaction form. This is not so for the iron bar and other cases involving self-organization. However, surely the self-organization cases are equally a consequence of the underlying universal dynamics, and simply produced under specific initial and/or boundary conditions; if it is just that at present we cannot analytically represent self-organization then that should not stop us from allying them to the previous simpler cases. This is so, but there are two important differences marking off the self-organization cases: (i) it heralds the presence of an irreducibly new dynamical existent, (ii) the dynamical form itself alters accordingly, so there is no common universal law form. Thus, they represent a genuinely interesting set of conditions.

However, there are also interesting mid-way cases. Self-organization through something as radical as phase change is not the only way to induce the formation of a new constraint condition; inducing a Hopf bifurcation of the dynamics (where a smooth parameter change alters the dynamic attractor landscape) is another, as is simply being moved from one local energy well to another in an unchanged dynamical landscape of a system. In each case changed initial and/or boundary conditions lead to changed dynamical laws. Although the well-shift cases may be set aside on the grounds that they too are unified by a common dynamics (represented by the landscape), the former Hopf-bifurcation cases also manifest a changed dynamical form and deserve to belong to the self-organization cases, as do other kinds of dynamical bifurcations. Polanyi once argued, in effect, that what was distinctive of living systems was that their governing laws were so strongly dependent on initial and/or boundary conditions [32]. Polanyi had in mind at least the way that information can alter the basis of behavior in living systems. If the impact of an information-conveying signal on an organism is dynamically equivalent to a Hopf or other bifurcation, then Polanyi's living systems can all be brought under the same dynamical paradigm.

The same considerations apply to dynamical models of the genome. There may be a wide variety of dynamically different forms that a genome can take up as various of its processes alter its own initial and/or boundary conditions so as to induce a dynamical bifurcation, for example, create and insert a new catalyst into the protein dynamics, thus forming special laws for that condition. These effects can propagate historically. The emergence of a new constraint with new dynamics may lead to the subsequent dynamical formation of still further top-down constraints, and so new entities, that would not have been dynamically possible without that preceding formation event. Indeed, something like that must be the overall dynamical form of development. Moreover, this cascade of dynamical consequences is marked by its initiating formation event and thus exhibits dynamical fixation of (these) historical constraints.

Such path-dependent dynamics occur throughout ecology (e.g., somatic and niche symbioses), economics (e.g., in off-highway development and technological learning), and the social sciences (fashion, and so on), but are not unknown in physics, for example, in hysteresis.

Needless to say, biology will not splinter into an unprincipled disunity under these complex dynamics.[25] Once again, these bifurcations will still be dynamically determined by, and identified in terms of, their dynamical constituents and governed by laws that themselves are thus grounded in the underlying universal dynamics. This is precisely what the biochemistry of the dynamical network models is meant to show. It will also encompass the many changes that consist of less profound dynamical transitions falling under the same dynamical form (even shifting between strange and other attractors). And the requirement to "match-up" the dynamics of different spatiotemporal scales and domains provides a further important unifying component. For example, unifying molecular chemistry and cellular biology requires interpreting cellular processes in biochemical terms that immediately generate many penetrating tests because of the requirement to match up the two descriptions—for instance all of the function to process reductions.

All this provides a shared dynamical framework interconnecting emergent variety in intimate ways that make it possible to successfully model complex genome dynamics and even development, navigating through the complex world of emergent but interconnected cellular and intercellular levels and laws. This gives a strong sense in which biological science remains unified even while acknowledging more strongly initial and/or boundary condition-dependent laws than simple physics and chemistry was wont to consider. The challenge to biological science is to recognize explicitly and better understand this plethora of law types and shifts, so as to make explicit their basis and their theoretical and methodological implications.

20.2.6 Limits of Knowability

The advent of complex systems models introduces new considerations concerning the manner and limits of scientific knowability. By this it is not meant the pragmatic fact of vastly more complex systems generating vastly more extensive sets of data than can practically be managed (cf. Section 20.1.2). Rather, the interest here is in principled limits on knowledge. Discussion is limited to knowledge of complex systems and is even so preliminary.[26]

An immediate consideration is the limit on analytic solvability to achieve "closed form" symbolic representation of dynamics, that is, a single formula giving the

---

[25] See, for example, Ref. [83], and for the complex dynamical unity response presented here see Ref. [57], footnote 4, and Section 20.2.5. It should be added that the conception of laws as simple universal generalizations, common among philosophers and scientists alike, is simplistic, science shows a far more complex and rich spectrum of laws—see Ref. [84].

[26] For more fundamental limits on knowability deriving from quantum theory, see for example Ref. [70], and for something of the variety of forms knowledge limits can take, see Ref. [85].

universal solution to a set of dynamical equations.[27] But as we move beyond simple sets of independent linear differential equations toward nonlinear, partial differential equations in interdependent coupled equation sets, we find that the dynamics they represent rapidly becomes very complex and the equation sets lack analytic solutions. Beyond this again lie bifurcations; these have no analytic representation within which their dynamics is exhibited, as the standard dynamical systems do, and cannot have one in standard dynamical terms precisely because they change their dynamical form, that is, change their dynamical representation, and as a function of their own initial/boundary conditions. In all these cases, as noted in Chapter 19 (see footnote 36), it is then necessary to explore their dynamics through numerical approximation and temporal iteration. Their dynamics is exhibited in extended form in space and time, rather than being condensed into a single abstract relation among symbols. This places computational modeling at the center of their scientific investigation in a strong manner and highlights the huge, and unique, contribution of computers to scientific knowledge.

However, it should not be forgotten that in most cases computational modeling provides only a numerical approximation, not exact values. Again, in most cases this is not a problem since the degree of approximation can be increased at will. But mathematical science contains many noncomputable functions,[28] that is, functions where information crucial to identifying it is lost at any level of finite approximation. Many superposition or "wave" phenomena (classical and quantum) are of this kind where wavelet information at indefinitely small scales is important to identifying the whole function. A comparable situation occurs when chaos (a "strange" attractor) is involved. Because nearby chaotic trajectories diverge exponentially from one another (at all points along their trajectories), any approximation will be invalidated by some trajectories within the approximation range—often quickly, one of the earliest discoveries of chaos (by Lorenz, using a coupled triad of partial differential equations) concerned just such divergence brought about from slightly different rounding errors. Thus, though computational numerical approximation represents a huge expansion of our capacity to know complex dynamics, it also represents a selective, but important, diminution in our knowledge capacity.

The exponential divergence of dynamical trajectories characteristic of chaotic attractors manifests sensitivity to initial conditions. Small differences in the conditions determining the initial dynamical state are eventually amplified into large divergences. This can happen with nonlinear dynamics generally, it does not require chaos; bifurcations are examples. In such circumstances, prediction is limited by the accuracy of knowledge of the initial system state, that is, of the initial conditions. This is so even

---

[27] Curiously, this is equivalent to science constructing a compressed symbolic description of reality, in the sense of algorithmic complexity theory. Could the latter's difficulties with defining organization be reflected in some characteristic of the former? And what has this to do with Rosen's [64] more abstract concerns with modeling?

[28] That is, mathematical functions, not biological functions. Mathematical functions are many: one maps from a domain to a range, hence unique on the range. One distinctive merit of the proposal to model biological functions as input/output maps is that this relates them directly to mathematical functions and hence, via modeling, to dynamical maps and so to biochemical processes.

though the system dynamics is deterministic, and so exactly one precise trajectory happens. Since all human knowledge (indeed all creaturely knowledge) has finite resolution, prediction is permanently parted from determinism. Self-organization ensures that this extends to prediction of condition-dependent laws—in particular to predicting self-organized intracellular dynamics.

Limits on predicting the behavior of intelligent agents provides a further class of special cases. Even simple sensory agents can on occasion amplify very small signals (perhaps a few light quanta) into large behavioral differences, so that even smaller uncertainties in those signals or in the internal state created will place limits on predicting behavior. Even where these details are knowable in principle another limit typically bites hard: such complex systems have considerable logical depth and for such systems the time required to make a prediction is in principle large,[29] added to which is the time required to obtain all the relevant state and process information to do so. Often, those times are much longer than the time horizon for relevant prediction and action, whence one's interaction with them is always on the basis of some uncertainty. Slow, long-running "agendas" in human personal development can produce surprising behaviors that defeat even decades of contrary data about a person. Another version of the same limit applies when an agent alters its own environment on too fast a timescale for it to know the consequences of its past actions before it acts again. Humans have always been in this predicament and continue notoriously to be so, as witness climate change, peak oil, nuclear proliferation, stability of financial markets, and so on.

Finally, an important methodological issue has recently opened up concerning the most effective statistical means of extracting knowledge of genome organization and dynamics from the large data sets generated by contemporary high-throughput experimental technologies, data often sparsely distributed in large-dimensional parameter spaces. A classic paper by Breiman [36] opposes two approaches to the data: model learning and machine learning. In model learning, a class of mathematical models specified by parameter values is chosen as a presumed model of the underlying reality from which the data is taken and its parameters are interpreted in terms of the entities and potential dynamical processes thought to constitute the underlying reality. The problem for statistical methodology is then to use the data in an unbiased way to estimate the parameter values and so fix the particular model involved. This model can subsequently be tested by its prediction of new data and the parameter values re-estimated as required.

In machine learning, by contrast, no model is specified, rather the data are used to "tune" a machine-learning process (some one of a large class of convergent mathematical adaptation or self-correction processes, for example, neural nets, run on computers). The tuned machine is then used to predict further data and is tested against it. The tuned machine state may have no obvious understanding in physical model terms; indeed its state dimensions, hence parameters, emerge from the tuning process and may be very large. Nonetheless, in a variety of situations it provides superior predictive performance and, with modeling goodness-of-fit tests often too weak to select among a variety of models, it emphasizes prediction as a self-sufficient goal of

---

[29] For discussion see Refs [69] and [70].

science. This threatens to pull apart prediction from ontological-dynamical under-standing as epistemic goals of science and thus, represents a distinctive constraint on scientific knowledge. Where it provides intelligible insight into underlying processes, it is essentially a form of induction from data, just as the previous model learning approach can be considered a form of hypothetico-deductive falsification, linking these alternatives to a much more general and venerable debate about scientific methodology (empiricism versus Popperianism, and so on). There is a lively version of the debate in systems biology [37,38].

To the philosophically minded the machine-learning language may, however, suggest much more, specifically the prospect of either (1) a new phenomenological empiricism, where data are again uncritically glorified and last century's well-abandoned extreme claims of solely reconstructing theory from it (cf. behaviorism) reemerge, or (2) a kind of postmodern antirealism in which scientific investigations each use their own separate machine-learning states, each state employed while ever it is predictively adequate, with discussion of underlying reality considered a sign of nostalgia for grand schemes, implicit attempts at ideological hegemony or mental confusion. This is not the place to discuss either the foundations of statistical method or the fate of grand conceptions. Rather, abandoning the extremes represented by 1 and 2 above, I want to briefly suggest consideration of a middle-ground approach to method recognizing the utility of both induction and hypothetico-deduction in context.

It is surely a false dichotomy to oppose prediction to understanding, since each is necessary to the other: understanding without prediction is ignorant and uncritical, prediction without understanding is weak and fragmented. The former is obvious for any finite, comprehensively fallible species like us commencing research in ignorance. The outcomes of predictive tests underpin acceptance/rejection of any proposed models and hence of improved understanding. The latter rests on the way a confirmed dynamical model can direct research much more effectively than simply trying to collect more data per se. For instance, such models distinguish law-like relations (as energy transform processes) from mere correlations or noisier relations, and also identify the sources of noise and bias, including in the interaction between system and data-gathering instruments—all of which structure future-testing regimes and assess-ment regimes, including the filtering and correction of data itself. And as before, model matching across scales and domains widens and focuses this role.

In addition, the machine-learning approach still relies on the choice of data categories, experimental setup, and appropriate instruments and probes to generate its data. But all such choices make presumptions about the character and salience of features in the underlying reality. For instance, instruments have systematic errors and limitations and unless we have sound insight into how the instruments work we cannot know what their defects are and hence how to process data. (A striking demonstration of this comes from the quantum theory proof that even core classical measuring devices have inherent error rates, of which science had been entirely unsuspecting.) Instruments themselves are understood through empirically validated theoretical models.[30]

---

[30] Often enough using the very theories they are used to test. But this is largely OK, see Ref. [89]. On data choices, see, for example, Ref. [90].

By comparison, in all these cases machine learning can only combine the data pools without direction, there being no methods within data lists alone for simulating dynamical discrimination and unification, systematic data errors and data limitations. Even identifying random data errors may cause problems here, since these have somehow to be distinguished from inherent dynamical fluctuations in the system, the latter behaving as noise except near bifurcations where their form may be critical to understanding system dynamics. All this leads to insatiable demands for sufficient data, ultimately extending to encompass all science and the entire universe as a block—not a good theory of epistemic (learning) strategy for finite agents beginning in ignorance.

On the other hand, machine learning often finds patterns in high-dimensional data where our knowledge of models is initially poor and complex dynamical process lie behind the data. All of which suggests that a pragmatic mixed strategy is called for, reinforced by the many approaches in use that combined parametric and nonparametric modeling. If you know nothing about the domain but have enough data (data rich, hypothesis poor), then machine learning may be the best approach, while if you know a lot about the domain then, especially if only a small range of data is available (hypothesis rich, data poor), model learning is surely the best bet. And in between, knowledge-wise and data-wise, the features of the best-mixed model will no doubt vary complexly with context.

REFERENCES

1. Storz G, Altuvia S, Wassarmann KM. An abundance of RNA regulators. *Annu Rev Biochem* 2005;74:199–217.

2. Storz G. An expending universe of noncoding RNAs. *Science* 2002;296:1260–1263.

3. Lagos-Quintana M, Rauhut R, Lendeckel W, Tuschl T. Identification of novel genes coding for small expressed RNAs. *Science* 2001;294(5543):853–858.

4. Lau NC, Lim LP, Weinstein EG, Bartel DP. An abundant class of tiny RNAs with probable regulatory roles in *Caenorhabditis elegans*. *Science* 2001;294(5543):858–862.

5. Lee RC, Ambros V. An extensive class of small RNAs in *Caenorhabditis elegans*. *Science* 2001;294:862–864.

6. Mourelatos ZZ, Dostie J, Paushkin S, Sharma M, Charroux B, Abel L, Rappsilber J, Mann M, Dreyfuss G. miRNPs: a novel class of ribonucleoproteins containing numerous microRNAs. *Genes Dev* 2002;16(6):720–728.

7. Ruvkun G. Molecular biology: glimpses of a tiny RNA world. *Science* 2001;294:797–799.

8. Grosshans H, Slack FJ. Micro-RNAs: small is plentiful. *J Cell Biol* 2002;156(1):17–22.

9. Wassarman KM, Zhang A, Storz G. Small RNAs in *Escherichia coli*. *Trends Microbiol* 1999;7(10):37–45.

10. Altuvia S, Wagner EGH, Switching on and off with RNA. *Proc Natl Acad Sci USA* 2000;97:9824–9826.

11. Sleutels F, Zwart R, Barlow DP. The non-coding Air RNA is required for silencing autosomal imprinted genes. *Nature* 2002;415(6873):810–813.

12. Erdmann VA, Szymanski M, Hochberg A, de Groot N, Barciszewski J. Non-coding, mRNA-like RNAs database Y2K. *Nucleic Acids Res* 2000;28(1):197–200.

13. Avner P, Heard E. X-chromosome inactivation: counting, choice and initiation. *Nature Rev Genet* 2001;2:59–67.

14. Eddy SR. Non-coding RNA genes and the modern RNA world. *Nat Rev Gen* 2001;2:919–929.

15. Hüttenhofer A, et al. RNomics: an experimental approach that identifies 201 candidates for novel, small, non-messenger RNAs in mouse. *EMBO J* 2001;20:2943–2953.

16. Argaman L, et al. Novel small RNA-encoding genes in the intergenic regions of *Escherichia coli*. *Curr Biol* 2001;11:941–950.

17. Rivas E, Klein RJ, Jones TA, Eddy SR. Computational identification of noncoding RNAs in *E. coli* by comparative genomics. *Curr Biol* 2001;11:1369–1373.

18. Wassarman KM, Repoila F, Rosenow C, Storz G, Gottesman S. Identification of novel small RNAs using comparative genomics and microarrays. *Genes Dev* 2001;15:1637–1651.

19. Jeffares DC, Poole AM, Penny D, Pre-rRNA processing and the path from the RNA world. *Trends Biochem Sci* 1995;20:298–299.

20. Poole AM, Jeffares DC, Penny D. The path from the RNA world. *J Mol Evol* 1998;46:1–17.

21. Toyoda T, Wada A. Omic space: coordinate-based integration and analysis of genomic phenomic interactions. *Bioinformatics* 2004;vol. 20:1759–1765.

22. Arakawa K, Mori K, Ikeda K, et al. G-language genome analysis environment: a workbench for nucleotide sequence data mining. *Bioinformatics* 2003;19:305–306.

23. Hucka M, Finney A, Sauro HM, et al. The systems biology markup language (SBML): a medium for representation and exchange of biochemical network models. *Bioinformatics* 2003;19:524–531.

24. Chen Y, Bittner ML, Dougherty ER. Issues associated with microarray data analysis and integration (supplementary information to article by Bittner, M, Trent, J, and Meltzer, P.) *Nat Genet* 1999;22:213–215.

25. Luisi PL, Oberholzer T, Lazcano A. The notion of a DNA minimal cell: a general discourse and some guidelines for an experimental approach. *Helv Chim Acta* 2002;85:1759–1777.

26. Koonin EV. How many genes can make a cell: the minimal-gene set concept. *Annu Rev Genomics Hum Genet* 2000;1:99–116.

27. Hutchison CA III, Peterson SN, Gill SR, Cline RT. Global transposon mutagenesis and a minimal mycoplasma genome. *Science* 1999;286:2165–2169.

28. Ji Y, Zhang B, von Horn SF, Warren P, et al. Identification of critical staphylococcal genes using conditional phenotypes generated by antisense RNA. *Science* 2001;293:2266–2269.

29. Chandonia J-M, Konerding DE, Allen DG. Computational structural genomics of a complete minimal organism. *Genome Inform* 2002;13:390–391.

30. Zhang R, Ou H-Y, Zhang CT. DEG: a database of essential genes. *Nucleic Acids Res* 2004;32:D271–D272.

31. Glass JI, et al. Essential genes of a minimal bacterium. *Proc Natl Acad Sci USA* 2006;103 (2):425–430.

32. Mushegian AR, Koonin EV. A minimal gene set for cellular life derived by comparison of complete bacterial genomes. *Proc Natl Acad Sci USA* 1996;93:10268–10273.

33. Arigoni F, Talabot F, Peitsch M, Edgerton MD, Meldrum E. A genome-based approach for the identification of essential bacterial genes. *Nat Biotechnol* 1998;16:851–856.

34. Salisbury MW. Get ready for synthetic biology. *Genome Technol* 2006;26–33.

35. Gray MW. Evolution of organellar genomes. *Curr Opin Genet Dev* 1999;9:678–687.

36. Andersson SGE, Zomorodipour A, Andersson JO, Sicheritz-Pontén T, Alsmark UCM, Podowski RM, Näslund AK, Eriksson A-S, Winkler HH, Kurland CG. The genome sequences of *Rickettsia prowazekii* and the origin of mitochondria. *Nature* 1998;396:133–140.

37. Müller M, Martin W. The genome of *Rickettsia prowazekii* and some thoughts on the origin of mitochondria and hydrogenosomes. *Bioessays* 1999;21(5):377–381.

38. Tielens AGM, van Hellemond JJ. The electron transport chain in anaerobically functioning eukaryotes. *Biochim Biophys Acta* 1998;1365:71–78.

39. Tielens AGM. Energy generation in parasitic Helminths. *Parasitol Today* 1994;10: 346–352.

40. Rotte C, Henze K, Müller M, Martin W. Origins of hydrogenosomes and mitochondria. *Curr Opin Microbiol* 3(5):481–486.

41. Embley TM, Martin W. A hydrogen-producing mitochondrion. *Nature* 1998;398:517–518.

42. Akhmanova A, Voncken F, val Alen T, van Hoek A, Boxma B, Vogels G, Veenhuis M, Hackstein JHP. A hydrogenosome with a genome. *Nature* 1998;396:528–529.

43. Embley TM, van der Giezen M, Horner DS, Dyal PL, Bell S, Foster PG. Hydrogenosomes, mitochondria and early eukaryotic evolution. *IUBMB Life* 2003;55(7):387–395.

44. Keeling PJ. A kingdom's progress: Archezoa and the origin of eukaryotes. *Bioessays* 1998;20:87–95.

45. Cavalier-Smith T. Eukaryotes with no mitochondria. *Nature* 1987;326:332–333.

46. Clark CG, Roger AJ. Direct evidence for secondary loss of mitochondria in *Entamoeba histolytica*. *Proc Natl Acad Sci USA* 1995;92:6518–6521.

47. Pfanner N, Geissler A. Versatility of the mitochondrial protein import machinery. *Nat Rev Mol Cell Biol* 2001;2:339–349.

48. Mourier T, Hansen AJ, Willerslev E, Arctander P. The human genome project reveals a continuous transfer of large mitochondrial fragments to the nucleus. *Mol Biol Evol* 2001;18 (9):1833–1837.

49. Adams KL, Palmer JD. Evolution of mitochondrial gene content: gene loss and transfer to the nucleus. *Mol Phylogenet Evol* 2003;29:380–395.

50. Sutak R, Dolezal P, Fiumera HL, Hrdy I, Dancis A, Delgadillo-Correa M, Johnson PJ, Muller, Miklos T. Mitochondrial-type assembly of FeS centers in the hydrogenosomes of the amitochondriate eukaryote *Trichomonas vaginalis*. *Proc Natl Acad Sci USA* 2004;101 (28):10368–10373.

51. Hooker C, *Reason, Regulation and Realism*. Albany, NY: SUNY Press, 1995.

52. Shi Y. *An Economic Theory of Knowledge*. London: Elgar, 2001.

53. Hooker C, Towards a general theory of reduction. Dialogue XX, Part I: Historical framework, pp. 38–59; Part II: Identity and reduction, pp. 201–236; Part III; Cross-categorical reduction, pp. 496–529; 1981.

54. Christensen W, Hooker C. Self-directed agents. In: MacIntosh J, editor. *Can J Philos* (Naturalism, Evolution and Intentionality, Special Supplementary, Ottawa, Canada) 2002;27: 19–52.

55. Churchland P. *Scientific Realism and the Plasticity of Mind*. London: Cambridge University Press, 1979.

56. Hooker C. Reduction as cognitive strategy. In: Keeley B, editor. *Paul Churchland*. New York: Cambridge University Press, 2005.

57. Hooker C. Asymptotics, reduction and emergence. *Br J Philos Sci* 2004;55:435–479.

58. Nagel E. *The Structure of Science*. London: Routledge & Kegan Paul, 1961.

59. Bechtel W. *Discovering Cell Mechanisms: The Creation of Modern Cell Biology*. Cambridge, UK: Cambridge University Press, 2006.

60. Bechtel W, Abrahamsen A. Explanation: a mechanistic alternative. *Stud Hist Philos Biol Biomed Sci* 2005;36:421–441.

61. Bechtel W. Biological mechanisms: organised to maintain autonomy. In: Boogerd F, Bruggeman F, Hofmeyr J-H, Westerhoff J, editors. *Systems Biology: Philosophical Foundations*. Amsterdam: Elsevier, 2007.

62. Dewan E. Consciousness as an emergent causal agent in the context of control system theory. In: Globus G, Maxwell G, Savodnik I, editors. *Consciousness and the Brain*. New York: Plenum, 1976.

63. Rosen R. *Anticipatory Systems*. New York: Pergamon, 1985.

64. Rosen R. *Life Itself*. New York: Columbia University Press, 1991.

65. Collier J. Supervenience and reduction in biological hierarchies. In: Matthen M, Linsky B, editors. *Can J Philos*, (Philosophy and Biology, Supplementary Volume, Ottawa) 1988;14: 209–234.

66. Sperry R, *Science and Moral Priority: Merging Mind, Brain and Human Values*. New York: Columbia, 1983.

67. Bruggeman FJ, Westerhoff HV, Boogerd FC. BioComplexity: a pluralist research strategy is necessary for a mechanistic explanation of the 'live' state. *Philos Psychol* 2002;15(4): 411–440.

68. Boogerd FC, Bruggeman FJ, Richardson RC, Stephan A, Westerhoff HV. Emergence and its place in nature: a case study of biochemical networks. *Synthese* 2005;145:131–164.

69. Collier J, Hooker C. Complexly organised dynamical systems. *Open Syst Inf Dyn* 1999;6:241–302.

70. Gell-Mann M. *The Quark and the Jaguar: Adventures in the Simple and the Complex*. New York: Henry Holt, 1994.

71. Bennett C. Dissipation, information, computational complexity and the definition of organization. In: Pines D, editor. *Emerging Syntheses in Science, Proceedings of the founding workshops of the Santa Fe Institute. Redwood California. Addison-Wesley*, 1985.

72. Kaufman S. *Investigations*. New York: Oxford University Press, 2000.

73. Bickhard M. Representational content in humans and machines. *J Exp Theor Artif Int* 1993;5:285–333.

74. Christensen W, Hooker C. An interactivist-constructivist approach to intelligence: self-directed anticipative learning. *Philos Psychol* 2000;13:5–45.

75. Moreno A. A systemic approach to the origin of biological organization. In: Boogerd F, Bruggeman F, Hofmeyr J-H, Westerhoff J, editors. *Systems Biology: Philosophical Foundations*. Amsterdam: Elsevier, 2007.

76. Bickhard M, Terveen L. *Foundational Issues in Artificial Intelligence and Cognitive Science—Impasse and Solution*. Amsterdam: Elsevier Scientific, 1995.

77. Hogeweg P. Computing an organism: on the interface between informatic and dynamic processes. *Biosystems* 2002;64:97–109.

78. Hogeweg P. Multilevel processes in evolution and development: computational models and biological insights. In: Lässig M, Valleriani A, editors. *Biological Evolution and Statistical Physics*, Springer lecture notes in physics 585. Berlin: Springer Verlag, 2002;pp. 217–239.

79. Hogeweg P, Takeuchi N. Multilevel selection in models of prebiotic evolution: compartments and spatial self-organization. *Orig Life Evol Biosph* 2003;33:375–403.

80. Dieckmann U, Law R, Metz J, editors. *The Geometry of Ecological Interactions: Simplifying Spatial Complexity*. Cambridge, UK: Cambridge University Press, 2000.

81. Cash D, editor. Scale and Cross-Scale Dynamics: Governance and Information in a Multilevel World. Available at http://www.ecologyandsociety.org/.

82. Polanyi M. Life's irreducible structure. *Science* 1968;160:1308–1312.

83. Dupré J. *The Disorder of Things: Metaphysical Foundations of the Disunity of Science*. Cambridge, MA: Harvard University Press, 1993.

84. Hooker C, Laws, natural. In: Craig E, editor. *Routledge Encyclopedia of Philosophy*. London: Routledge, 1998.

85. McDaniel R, Driebe D, editors. *Uncertainty and Surprise in Complex Systems*. Berlin: Springer, 2005.

86. Breiman L. Statistical modelling: the two cultures. *Stat Sci* 2001;16:199–215.

87. Kell D, Knowles J. The role of modelling in systems biology. In: Szallasi Z, Sterling J, Periwal V, editors. *System Modelling in Cellular Biology*. Cambridge, MA: MIT Press, 2006.

88. Westerhoff H, Kell D. The methodologies of systems biology. In: Boogerd F, Bruggeman F, Hofmeyr J-H, Westerhoff J, editors. *Systems Biology: Philosophical Foundations*. Amsterdam: Elsevier, 2007.

89. Hooker C. Global theories. *Philos Sci* 1975;42:152–179. (Reprinted in Hooker, C., *A Realistic Theory of Science*, Albany, NY, SUNY Press, 1987).

90. Lewontin R, Levins R. Let the numbers speak. *Int J Health Services* 2000;30(4): 873–877.

INDEX

Systems Biology and Synthetic Biology   Edited by Pengcheng Fu and Sven Panke
Copyright © 2009 John Wiley & Sons, Inc.